In Vivo NMR Spectroscopy

In Vivo NMR Spectroscopy

Principles and Techniques

Third Edition

Robin A. de Graaf

Department of Radiology and Biomedical Imaging
Department of Biomedical Engineering
Magnetic Resonance Research Center (MRRC)
Yale University
New Haven, CT, USA

Registered Offices
John Wiley & Sons, Inc., 111 River Street, Hoboken, NJ 07030, USA
John Wiley & Sons Ltd, The Atrium, Southern Gate, Chichester, West Sussex, PO19 8SQ, UK

Editorial Office
The Atrium, Southern Gate, Chichester, West Sussex, PO19 8SQ, UK

For details of our global editorial offices, customer services, and more information about Wiley products visit us at www.wiley.com.

Wiley also publishes its books in a variety of electronic formats and by print-on-demand. Some content that appears in standard print versions of this book may not be available in other formats.

Library of Congress Cataloging-in-Publication Data

Names: De Graaf, Robin A., author.
Title: In vivo NMR spectroscopy : principles and techniques / Robin A. de Graaf, Magnetic Resonance Research Center (MRRC), Yale University, Department of Radiology and Biomedical Imaging, Department of Biomedical Engineering, New Haven, CT, USA.
Description: Third edition. | Hoboken, NJ : Wiley, [2019] | Includes bibliographical references and index. |
Identifiers: LCCN 2018041668 (print) | LCCN 2018044006 (ebook) | ISBN 9781119382577 (Adobe PDF) | ISBN 9781119382515 (ePub) | ISBN 9781119382546 (hardback)
Subjects: LCSH: Nuclear magnetic resonance spectroscopy. | Magnetic resonance imaging. |
BISAC: SCIENCE / Spectroscopy & Spectrum Analysis.
Classification: LCC QP519.9.N83 (ebook) | LCC QP519.9.N83 D4 2018 (print) | DDC 616.07/548—dc23
LC record available at https://lccn.loc.gov/2018041668

Cover design: Wiley
Cover image: Background ©ake1150sb; Other Images: Courtesy of the Author, Robin de Graaf

Set in 10/12pt Warnock by SPi Global, Pondicherry, India

10 9 8 7 6 5 4 3 2 1

In memory of Klaas Nicolay (1951–2017)

Contents

Preface

The main driving force to write a third edition was the inadequate description of several basic NMR phenomena in the earlier editions, as well as in the majority of NMR textbooks. The quantum picture of NMR provides the most general description that is applicable to all NMR experiments. As a result, the quantum description of NMR often takes center stage, but comes at the expense of forfeiting a physically intuitive picture. Inappropriate descriptions of NMR result when the quantum mechanics are incorrectly simplified to a classical picture. However, ever since the very first report on NMR in bulk matter by Felix Bloch, it is known that the NMR phenomenon for many compounds, like water, can be quantitatively described based on classical arguments without the need to invoke quantum mechanics. The current edition adopts this classical description for a very intuitive and straightforward description of NMR. While many aspects of *in vivo* NMR, including MR imaging, magnetization transfer, and diffusion can be successfully described, the classical description does prove inadequate in the presence of scalar coupling. At this point the classical description is replaced with a semiclassical correlated vector model that naturally leads to the quantum-mechanical product operator formalism.

The third edition also takes the opportunity to correct misconceptions about the nature of radiofrequency (RF) pulses and coils, and provides an updated review of novel methods, including hyperpolarized MR, deuterium metabolic imaging (DMI), MR fingerprinting, advanced magnetic field shimming, and chemical exchange saturation transfer (CEST) methods. However, it should be stressed that this book does not set out to present complete, detailed, and in-depth reviews of *in vivo* MRS methods.

The main objective of the book has always been to provide an educational explanation and overview of *in vivo* NMR, without losing the practical aspects appreciated by experimental NMR spectroscopists. This objective has been enhanced in this edition by relegating a significant number of mathematical equations to the exercises in favor of more intuitive, descriptive explanations and graphical depictions of NMR phenomena. The exercises are designed to review, but often also to extend the presented NMR principles and techniques, including a more in-depth exploration of quantitative MR equations. The textual description of RF pulses has been reduced and supplemented with PulseWizard, a Matlab-based RF pulse generation and simulation graphical user interface available for download at the accompanying website (http://booksupport.wiley.com).

Many of the ideas and changes that formed the basis for this third edition came from numerous discussions with colleagues. I would like to thank Henk De Feyter, Chathura Kumaragamage, Terry Nixon, Graeme Mason, Kevin Behar, and Douglas Rothman for many fruitful discussions.

Finally, I would like to acknowledge the contributions of original data from Dan Green and Simon Pittard (Magnex Scientific), Wolfgang Dreher (University of Bremen), Andrew Maudsley

(University of Miami), Yanping Luo and Michael Garwood (University of Minnesota), Bart Steensma, Dennis Klomp, Kees Braun, Jan van Emous, and Cees van Echteld (Utrecht University), and Henk De Feyter, Zachary Corbin, Robert Fulbright, Graeme Mason, Terry Nixon, Laura Sacolick, and Gerald Shulman (Yale University).

May 2018

Robin A. de Graaf
New Haven, CT, USA

Abbreviations

1D	one-dimensional
2D	two-dimensional
2HG	2-hydoxyglutarate
3D	three-dimensional
5-FU	5-fluoruracil
AC	alternating current
Ace	acetate
ADC	analog-to-digital converter
ADC	apparent diffusion coefficient
ADP	adenosine diphosphate
AFP	adiabatic full passage
AHP	adiabatic half passage
Ala	alanine
Asc	ascorbic acid
Asp	aspartate
ATP	adenosine triphosphate
BHB	β-hydroxy-butyrate
BIR	B_1-insensitive rotation
BISEP	B_1-insensitive spectral editing pulse
BOLD	blood oxygen level-dependent
BPP	Bloembergen, Purcell, Pound
BS	Bloch–Siegert
CBF	cerebral blood flow
CBV	cerebral blood volume
CEST	chemical exchange saturation transfer
CHESS	chemical shift selective
Cho	choline-containing compounds
CK	creatine kinase
CMR_{Glc}	cerebral metabolic rate of glucose consumption
CMR_{O_2}	cerebral metabolic rate of oxygen consumption
COSY	correlation spectroscopy
CPMG	Carr–Purcell–Meiboom–Gill
Cr	creatine
CRLB	Cramer–Rao lower bound
Crn	carnitine
CSDA	chemical shift displacement artifact
CSDE	chemical shift displacement error

CSF	cerebrospinal fluid
CW	continuous wave
DANTE	delays alternating with nutation for tailored excitation
dB	decibel
DC	direct current
DEFT	driven equilibrium Fourier transform
DEPT	distortionless enhancement by polarization transfer
DMb	deoxymyoblobin
DMI	deuterium metabolic imaging
DNA	deoxyribonucleic acid
DNP	dynamic nuclear polarization
DQC	double quantum coherence
DSS	2,2-dimethyl-2-silapentane-5-sulfonate
DSV	diameter spherical volume
DTI	diffusion tensor imaging
EA	ethanolamine
EMCL	extramyocellular lipids
EMF	electromotive force
EPI	echo planar imaging
EPSI	echo planar spectroscopic imaging
FDG	2-fluoro-2-deoxy-glucose
FDG-6P	2-fluoro-2-deoxy-glucose-6-phosphate
FFT	fast Fourier transformation
FID	free induction decay
FLASH	fast low-angle shot
fMRI	functional magnetic resonance imaging
FOCI	frequency offset corrected inversion
FOV	field of view
FSW	Fourier series windows
FT	Fourier transformation
FWHM	Frequency width at half maximum
GABA	γ-aminobutyric acid
GE	gradient echo
Glc	glucose
Gln	glutamine
Glu	glutamate
Glx	glutamine and glutamate
Gly	glycine
GOIA	gradient-offset-independent adiabaticity
GPC	glycerophosphorylcholine
GPE	glycerophosphorylethanolamine
GRAPPA	generalized autocalibrating partially parallel acquisitions
GSH	glutathione (reduced form)
HLSVD	Hankel Lanczos singular value decomposition
HMPT	hexamethylphosphorustriamide
HMQC	heteronuclear multiple quantum correlation
HSQC	heteronuclear single quantum correlation
Ile	isoleucine

IMCL	intramyocellular lipids
INEPT	insensitive nuclei enhanced by polarization transfer
IR	inversion recovery
ISIS	image-selected *in vivo* spectroscopy
IT	inversion transfer
IVS	inner volume selection
JR	jump-return
JRES	J-resolved spectroscopy
Lac	lactate
LASER	localization by adiabatic selective refocusing
Leu	leucine
Mb	myoglobin
MC	multi-coil
MEGA	Mescher–Garwood
mI	*myo*-inositol
MLEV	Malcolm Levitt
MM	macromolecules
MQC	multiple quantum coherence
MRF	magnetic resonance fingerprinting
MRI	magnetic resonance imaging
MRS	magnetic resonance spectroscopy
MRSI	magnetic resonance spectroscopic imaging
MT	magnetization transfer
MTC	magnetization transfer contrast
NAA	*N*-acetyl aspartate
NAAG	*N*-acetyl aspartyl glutamate
NAD(H)	nicotinamide adenine dinucleotide oxidized (reduced)
NADP(H)	nicotinamide adenine dinucleotide phosphate oxidized (reduced)
NDP	nucleoside diphosphate
NMR	nuclear magnetic resonance
nOe	nuclear Overhauser effect (or enhancement)
NOESY	nuclear Overhauser effect spectroscopy
NTP	nucleoside triphosphate
OSIRIS	outer volume suppressed image-related *in vivo* spectroscopy
OVS	outer volume suppression
PCA	perchloric acid
PCr	phosphocreatine
PDE	phosphodiesters
PE	phosphorylethanolamine
PET	positron emission tomography
PFC	perfluorocarbons
PHIP	para-hydrogen-induced polarization
P_i	inorganic phosphate
PME	phosphomonoesters
POCE	proton-observed carbon-edited
PPM	parts per million
PRESS	point resolved spectroscopy
PSF	point spread function

QSM	quantitative susceptibility mapping
QUALITY	quantification by converting line shapes to the Lorentzian type
RAHP	time-reversed adiabatic half passage
RARE	rapid acquisition, relaxation enhanced
RF	radiofrequency
RMS	root mean squared
RNA	ribonucleic acid
ROI	region of interest
SABRE	signal amplification by reversible exchange
SAR	specific absorption rate
SE	spin-echo
SENSE	sensitivity encoding
SEOP	spin-exchange optical pumping
SH	spherical harmonics
sI	*scyllo*-inositol
SI	spectroscopic imaging
SLIM	spectral localization by imaging
SLR	Shinnar–Le Roux
S/*N*	signal-to-noise ratio
SNR	signal-to-noise ratio
SPECIAL	spin-echo, full intensity acquired localized
SQC	single quantum coherence
SSAP	solvent suppression adiabatic pulse
SSFP	steady-state free precession
ST	saturation transfer
STE	stimulated echo
STEAM	stimulated echo acquisition mode
SV	single voxel (or volume)
SVD	singular value decomposition
SWAMP	selective water suppression with adiabatic-modulated pulses
Tau	taurine
TCA	tricarboxylic acid
tCho	total choline
tCr	total creatine
TEM	transverse electromagnetic mode
Thr	threonine
TMA	trimethylammonium
TMS	tetramethylsilane
TOCSY	total correlation spectroscopy
TPPI	time proportional phase incrementation
Trp	tryptophan
TSP	3-(trimethylsilyl)-propionate
Tyr	tyrosine
UV	ultraviolet
Val	valine
VAPOR	variable pulse powers and optimized relaxation delays
VARPRO	variable projection
VERSE	variable rate selective excitation

VNA	variable nutation angle
VOI	volume of interest
VSE	volume selective excitation
WALTZ	wideband alternating phase low-power technique for zero residue splitting
WEFT	water eliminated Fourier transform
WET	water suppression enhanced through T_1 effects
ZQC	zero quantum coherence

Symbols

A	absorption frequency domain signal
A_n, B_n	Fourier coefficients
b	b-value (in s/m^2)
b	b-value matrix
B_0	external magnetic field (in T)
B_1	magnetic radiofrequency field of the transmitter (in T)
B_{1max}	maximum amplitude of the irradiating B_1 field (in T)
B_{1rms}	root mean square B_1 amplitude of a RF pulse (in T)
B_{1x}, B_{1y}	real and imaginary components of the irradiating B_1 field (in T)
B_2	magnetic, radiofrequency field of the decoupler (in T)
B_e	effective magnetic field in the laboratory and frequency frames (in T)
B_e'	effective magnetic field in the second rotating frame (in T)
B_{loc}	local magnetic field (in T)
C	capacitance (in F)
C	correction factor for calculating absolute concentrations
D	(apparent) diffusion coefficient (in m^{2}s^{-1})
D	(apparent) diffusion tensor
D	dispersion frequency domain signal
E	energy (in J)
F	Nyquist frequency (in 1 s^{-1})
F	noise figure (in dB)
$f_B(t)$	normalized RF amplitude modulation function
$f_\nu(t)$	normalized RF frequency modulation function
G	magnetic field gradient strength (in T m^{-1})
$G(t)$	correlation function
h	Planck's constant ($6.626\,208 \times 10^{-34}$ Js)
H	Hadamard matrix
I	imaginary time- or frequency-domain signal
I	spin quantum number
I_0	Boltzmann equilibrium magnetization for spin I
I_{nm}	shim current for shim coil of order n and degree m
J	spin–spin or scalar coupling constant (in Hz)
J_0	zero-order Bessel function
$J(\nu)$	spectral density function
k	Boltzmann equilibrium constant ($1.380\,66 \times 10^{-23}$ J K^{-1})
k	k-space variable (in m^{-1})
k_f	k-space variable in frequency-encoding direction (in m^{-1})
k_p	k-space variable in phase-encoding direction (in m^{-1})

k_{AB}, k_{BA}	unidirectional rate constants (in s^{-1})
k_{for}	forward, unidirectional rate constant (in s^{-1})
k_{rev}	reversed, unidirectional rate constant (in s^{-1})
L	inductance (in H)
m	magnetic quantum number
m	mass (in kg)
M	macroscopic magnetization
M	magnitude-mode frequency domain signal
M	mutual inductance (in H)
M_0	macroscopic equilibrium magnetization
M_x, M_y, M_z	orthogonal components of the macroscopic magnetization
N	noise
N	number of phase-encoding increments
N	total number of nuclei or spins in a macroscopic sample
p	order of coherence
Q	quality factor
r	distance (in m)
R	composite pulse (sequence)
R	product of bandwidth and pulse length
R	real time- or frequency-domain signal
R	resistance (in Ω)
R	rotation matrix
R_{1A}, R_{1B}	longitudinal relaxation rate constants for spins A and B in the absence of chemical exchange or cross-relaxation (in s^{-1})
R_2	transverse relaxation rate (in s^{-1})
R_A, R_B	longitudinal relaxation rate constants for spins A and B in the presence of chemical exchange (in s^{-1})
R_{H}	hydrodynamic radius (in m)
S	measured NMR signal
$S(k)$	spatial frequency sampling function
t	time (in s)
t_1	incremented time in 2D NMR experiments (in s)
$t_{1\text{max}}$	maximum t_1 period in constant time 2D NMR experiments (in s)
t_2	detection period in 2D NMR experiments (in s)
t_{diff}	diffusion time (in s)
t_{null}	time of zero-crossing (nulling) during an inversion recovery experiment (in s)
T	absolute temperature (in K)
T	pulse length (in s)
T_1	longitudinal relaxation time constant (in s)
$T_{1,\text{obs}}$	observed, longitudinal relaxation time constant (in s)
T_2	transverse relaxation time constant (in s)
T_2^*	apparent transverse relaxation time constant (in s)
$T_{2,\text{obs}}$	observed, transverse relaxation time constant (in s)
T_{acq}	acquisition time (in s)
TE	echo time (in s)
TE$_{\text{CPMG}}$	echo time in a CPMG experiment (in s)
TI	inversion time (in s)
TI1	first inversion time (in s)
TI2	second inversion time (in s)

TM	delay time between the second and third 90° pulses in STEAM (in s)
TR	repetition time (in s)
v	velocity (in m s^{-1})
W	transition probability (in 1 s^{-1})
W_{nm}	angular function of spherical polar coordinates
$W(k)$	spatial frequency weighting function
x	molar fraction
X_C	capacitive reactance (in Ω)
X_L	inductive reactance (in Ω)
Z	impedance (in Ω)
α	nutation angle (in rad)
β	precession angle of magnetization perpendicular to the effective magnetic field B_e (in rad)
γ	gyromagnetic ratio (in rad T^{-1} s^{-1})
δ	chemical shift (in ppm)
δ	gradient duration (in s)
Δ	separation between a pair of gradients (in s)
ΔB_0	magnetic field shift (in T)
$\Delta\nu$	frequency offset (in Hz)
$\Delta\nu_{1/2}$	full width at half maximum of an absorption line (in Hz)
$\Delta\nu_{max}$	maximum frequency modulation of an adiabatic RF pulse (in Hz)
ε	gradient rise time for a trapezoidal magnetic field gradient (in s)
η	nuclear Overhauser enhancement
η	viscosity (in Ns m^{-2})
θ	nutation angle (in rad)
μ	magnetic moment (in A·m^2)
μ_0	permeability constant in vacuum ($4\pi\cdot10^{-7}$ kg·m·s^{-2}·A^{-2})
μ_e	electronic magnetic moment (in A·m^2)
ν_0	Larmor frequency (in Hz)
ν_A	frequency of a non-protonated compound A (in Hz)
ν_{HA}	frequency of a protonated compound HA (in Hz)
ν_{ref}	reference frequency (in Hz)
ξ	electromotive force (in V)
σ	density matrix
τ_c	rotation correlation time (in s)
τ_m	mixing time in 2D NMR experiments (in s)
ϕ	phase (in rad)
ϕ_0	zero-order (constant) phase (in rad)
ϕ_1	first-order (linear) phase (in rad)
ϕ_c	phase correction (in rad)
χ	magnetic susceptibility
ω_0	Larmor frequency (in rad s^{-1})
[]	concentration (in M)

Supplementary Material

To access supplementary materials for this book please use the download links shown below. There you will find valuable material designed to enhance your learning, including:

- Solutions to the exercises in the book
- Download option for PulseWizard
- Short video
- PPTs of all the figures

This book is accompanied by a companion website:

http://booksupport.wiley.com

Please enter the book title, author name or ISBN to access this material.

1

Basic Principles

1.1 Introduction

Spectroscopy is the study of the interaction between matter and electromagnetic radiation. Atoms and molecules have a range of discrete energy levels corresponding to different, quantized electronic, vibrational, or rotational states. The interaction between atoms and electromagnetic radiation is characterized by the absorption and emission of photons with an energy that exactly matches the energy level difference between two states. Since the energy of a photon is proportional to the frequency, the different forms of spectroscopy are often distinguished on the basis of the frequencies involved. For instance, absorption and emission between the electronic states of the outer electrons typically require frequencies in the ultraviolet (UV) range, hence giving rise to UV spectroscopy. Molecular vibrational modes are characterized by frequencies just below visible red light and are thus studied with infrared (IR) spectroscopy. Nuclear magnetic resonance (NMR) spectroscopy uses radiofrequencies, which are typically in the range of 10–1000 MHz.

NMR is the study of the magnetic properties and related energies of nuclei. The absorption of radiofrequency energy can be observed when the nuclei are placed in a (strong) external magnetic field. Purcell et al. [1] at MIT, Cambridge and Bloch et al. [2–4] at Stanford simultaneously, but independently discovered NMR in 1945. In 1952, Bloch and Purcell shared the Nobel Prize in Physics in recognition of their pioneering achievements [5, 6]. At this stage, NMR was purely an experiment for physicists to determine the nuclear magnetic moments of nuclei. NMR could only develop into one of the most versatile forms of spectroscopy after the discovery that nuclei within the same molecule absorb energy at different resonance frequencies. These so-called chemical shift effects, which are directly related to the chemical environment of the nuclei, were first observed in 1949 by Proctor and Yu [7], and independently by Dickinson [8]. The ability of NMR to provide detailed chemical information on compounds was firmly established when Arnold et al. [9] in 1951 published a high-resolution ^{1}H NMR spectrum of ethanol in which separate signals from methyl, methylene, and hydroxyl protons could be clearly recognized.

In the first two decades, NMR spectra were recorded in a continuous wave mode in which the magnetic field strength or the radio frequency was swept through the spectral area of interest, while keeping the other fixed. In 1966, NMR was revolutionized by Ernst and Anderson [10] who introduced pulsed NMR in combination with Fourier transformation. Pulsed or Fourier transform NMR is at the heart of all modern NMR experiments.

The induced energy level difference of nuclei in an external magnetic field is very small when compared to the thermal energy at room temperature, making it that the energy levels

In Vivo *NMR Spectroscopy: Principles and Techniques*, Third Edition. Robin A. de Graaf.
© 2019 John Wiley & Sons Ltd. Published 2019 by John Wiley & Sons Ltd.

are almost equally populated. As a result the absorption of photons is very low, making NMR a very insensitive technique when compared to the other forms of spectroscopy. However, the low-energy absorption makes NMR also a noninvasive and nondestructive technique, ideally suited for *in vivo* measurements. It is believed that, by observing the water signal from his own finger, Bloch was the first to perform an *in vivo* NMR experiment. Over the following decades, NMR studies were carried out on various biological samples like vegetables and mammalian tissue preparations. Continued interest in defining and explaining the properties of water in biological tissues led to the promising report of Damadian in 1971 [11] that NMR properties (relaxation times) of malignant tumorous tissues significantly differs from normal tissue, suggesting that proton NMR may have diagnostic value. In the early 1970s, the first experiments of NMR spectroscopy on intact living tissues were reported. Moon and Richards [12] used ^{31}P NMR on intact red blood cells and showed how the intracellular pH can be determined from chemical shift differences. In 1974, Hoult et al. [13] reported the first study of ^{31}P NMR to study intact, excised rat hind leg. Acquisition of the first ^{1}H NMR spectra was delayed by almost a decade due to technical difficulties related to spatial localization, and water and lipid suppression. Behar et al. [14] and Bottomley et al. [15] reported the first ^{1}H NMR spectra from rat and human brain, respectively. Since the humble beginnings, *in vivo* MR spectroscopy (MRS) has grown as an important technique to study static and dynamic aspects of metabolism in disease and in health.

In parallel with the onset of *in vivo* MRS, the world of high-resolution, liquid-state NMR was revolutionized by the introduction of 2D NMR by Ernst and coworkers [16] based on the concept proposed by Jeener in 1971 [17]. The development of hundreds of 2D methods in the following decades firmly established NMR as a leading analytical tool in the identification and structure determination of low-molecular weight chemicals. Richard Ernst was awarded the 1991 Nobel Prize in Chemistry for his contributions to the methodological development of NMR [18]. The application of multidimensional NMR to the study of biological macromolecules allowed determination of the 3D structure of proteins in an aqueous environment, providing an alternative to X-ray crystallography. Kurt Wuthrich was awarded the 2002 Nobel Prize in Chemistry for his contributions to the development of protein NMR and 3D protein structure determination [19].

Around the same time reports on *in vivo* MRS appeared, Lauterbur [20] and Mansfield and Grannell [21] described the first reports on a major constituent of modern NMR, namely *in vivo* NMR imaging or magnetic resonance imaging (MRI). By applying position-dependent magnetic fields in addition to the static magnetic field, they were able to reconstruct the spatial distribution of nuclear spins in the form of an image. Lauterbur and Mansfield shared the 2003 Nobel Prize in Medicine [22, 23]. Since its inception, MRI has flourished to become the leading method for structural and functional imaging with methods like diffusion tensor imaging (DTI) and blood oxygenation level-dependent (BOLD) functional MRI.

As a leading clinical and research imaging modality, the theoretical and practical aspects of MRI are covered in a wide range of excellent textbooks [24–26]. While MRS is based on the same fundamental principles as MRI, the practical considerations for high-quality MRS are very different. This book is dedicated to providing a robust description of current *in vivo* MRS methods, with an emphasis on practical challenges and considerations. This chapter covers the principles of NMR that are common to both MRI and MRS. Starting with classical arguments, the concepts of precession, coherence, resonance, excitation, induction, and relaxation are explained. The quantum mechanical view of NMR is briefly reviewed after which the phenomena of chemical shift and scalar coupling will be described, as well as some elementary processing of the NMR signal.

1.2   Classical Magnetic Moments

The discovery of NMR by Bloch and Purcell in 1945 was not a serendipitous event, but was based on the work by Rabi [27, 28] in the previous decade on magnetic resonance of individual particles in a molecular beam for which he received the 1944 Nobel Prize in Physics. While both groups reported the detection of signal associated with proton magnetic moments, the experimental setups as well as the conceptualization of the NMR phenomenon were very different.

Bloch approached NMR from a classical point of view in which the orientation of magnetic moments is gradually changed by an oscillating magnetic field. This would ultimately lead to the detection of NMR signal from water protons through electromagnetic induction in a nearby receiver coil. Purcell viewed the NMR phenomenon based on quantum mechanics, in close analogy to other spectroscopic methods in which transitions are induced between energy levels by quanta of energy provided by radiofrequency (RF) waves. Purcell described the absorption of energy provided by an oscillating RF field by the protons in solid paraffin. A wonderful overview of the two discoveries of NMR is given by Rigden [29] and Becker et al. [30] as well as by the Nobel lectures of Bloch [5] and Purcell [6].

The spectroscopic or quantum mechanical view often takes center stage in the introduction of many text books, including the previous editions of this book. The main reason for this approach is that a full quantum mechanical description of NMR can account for all observed phenomena, including those that have no classical analog, like scalar or J-coupling. However, as the quantum description of NMR does not deal directly with observable magnetization, but rather with the energetic state of the system, it does not provide an intuitive, physical picture. In the classical view of NMR, the magnetic moments of the individual nuclear spins are summed up to form a macroscopic magnetization vector that can be followed over time using classical electromagnetism concepts. This provides a familiar picture that can be used to follow the fate of magnetization under a wide range of experimental conditions. The classical picture is advocated here, starting with a magnetized needle as found in a compass.

As with all magnets, the compass needle is characterized by a magnetic north and south pole from which the magnetic field lines exit and enter the needle, respectively (Figure 1.1A). The magnetic field lines shown in Figure 1.1A can be summarized by a magnetic moment, μ, describing both the amplitude and direction. In the absence of an external magnetic field the compass needle has no preference in spatial orientation and can therefore point in any direction.

When placed in an external magnetic field, such as the Earth's magnetic field, the compass needle experiences a torque (or rotational force) that rotates the magnetic moment towards a parallel orientation with the external field (Figure 1.1B). As the magnetic moment "overshoots" the parallel orientation, the torque is reversed and the needle will settle into an oscillation or frequency that depends on the strengths of the external magnetic field and the magnetic moment. Due to friction between the needle and the mounting point, the amplitude of the oscillation is dampened and will ultimately result in the stabile, parallel orientation of the needle with respect to the external field (Figure 1.1C) representing the lowest magnetic energy state (the antiparallel orientation represents the highest magnetic energy state).

The equilibrium situation (Figure 1.1C) can, besides mechanical means, be perturbed by additional magnetic fields as shown in Figure 1.1D. When a bar magnet is moved towards the compass, the needle experiences a torque and is pushed away from the parallel orientation. When the bar magnet is removed, the needle oscillates as shown in Figure 1.1B before returning to the equilibrium situation (Figure 1.1C). However, if the bar magnet is moved back and

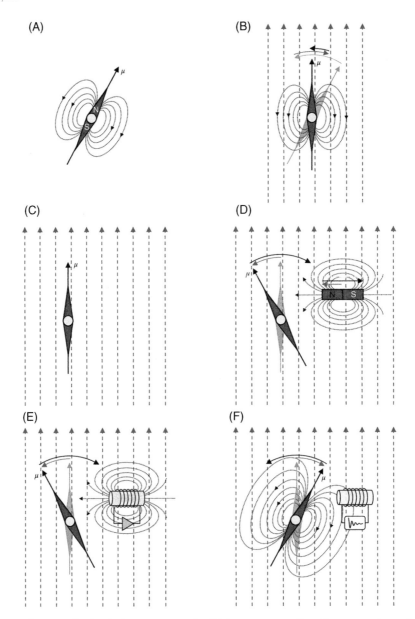

Figure 1.1 **Oscillations of a classical compass needle.** (A) A compass needle with a magnetic north and south pole creates a dipolar magnetic field distribution of which the amplitude and direction are characterized by the magnetic moment μ. (B) When placed in an external magnetic field the magnetic moment oscillates a number of times before (C) settling in a parallel orientation with the external magnetic field. Note that in Earth's magnetic field the compass needle points to the magnetic south, which happens to be close to geographical north. (D) The needle can be perturbed with a bar magnet, whereby the perturbation reaches maximum effect when the bar movement matches the natural frequency of the needle. (E) The bar magnet can be replaced by an alternating current in a coil. (F) The same coil can also be used to detect the oscillating magnetic moment of the needle through electromagnetic induction.

forth relative to the compass, the needle can be made to oscillate continuously. When the movement frequency of the bar magnet is very different from the natural frequency of the needle (Figure 1.1B), the effect of the bar magnet is not constructive and the needle never deviates far from the parallel orientation. However, when the frequency of the bar magnet movement matches the natural frequency of the needle, the repeated push from the bar magnet on the needle is constructive and the needle will deviate increasingly further from the parallel orientation. When the bar magnet has a maximum effect on the needle, the system is in resonance and the oscillation is referred to as the resonance frequency. A similar situation arises when pushing a child in a swing; only when the child is pushed in synchrony with the natural or resonance frequency of the swing set does the amplitude get larger.

The bar magnet can be replaced with an alternating current in a copper coil as shown in Figure 1.1E. The alternating current generates a time-varying magnetic field that can perturb the compass needle. When the frequency of the alternating current matches the natural frequency of the needle, the system is in resonance and large deviations of the needle can be observed with modest, but constructive "pushes" from the magnetic field produced by the coil.

The compass needle continues to oscillate at the natural frequency for some time following the termination of the alternating current (Figure 1.1F). The compass needle creates a time-varying magnetic field that can be detected through Faraday electromagnetic induction in the same coil previously used to perturb the needle. The induced voltage, referred to as the Free Induction Decay (FID), will oscillate at the natural frequency and will gradually reduce in amplitude as the compass needle settles into the parallel orientation.

Figure 1.1 shows that the MR part of NMR can be completely described by classical means. It is therefore also not surprising that Bloch titled his seminal paper "Nuclear induction" [2, 4] as the electromagnetic induction is an essential part of MR detection. The magnetic effects summarized in Figure 1.1 are readily reproduced "on the bench" and provide an excellent means of experimentally demonstrating some of the concepts of MR [31].

1.3   Nuclear Magnetization

Any rotating object is characterized by angular momentum, describing the tendency of the object to continue spinning. Subatomic particles like electrons, neutrons, and protons have an *intrinsic* angular moment, or spin that is there even though the particle is not actually spinning. Electron spin results from relativistic quantum mechanics as described by Dirac in 1928 [32] and has no classical analog. For the purpose of this book the existence of spin is simply taken as a feature of nature. Particles with spin always have an intrinsic magnetic moment. This can be conceptualized as a magnetic field generated by rotating currents within the spinning particle. This should, however, not be taken too literal as the particle is not actually rotating. Note that in the NMR literature, spin and magnetic moments are used interchangeably.

Protons are abundantly present in most tissues in the form of water or lipids. In the human brain, a small cubic volume of $1 \times 1 \times 1$ mm contains about 6×10^{19} proton spins (Figure 1.2A and B). In the absence of an external magnetic field, the spin orientation has no preference and the spins are randomly oriented throughout the sample (Figure 1.2B). For a large number of spins this can also be visualized by a "spin-orientation sphere" (Figure 1.2C) in which each spin has been placed in the center of a Cartesian grid. Summation over all orientations leads to a (near) perfect cancelation of the magnetic moments and hence to the absence of a macroscopic magnetization vector. It should be noted that the concept of a spin-orientation sphere has been used throughout the NMR literature [33–35], albeit sporadically. The description of the NMR phenomenon based on a spin-orientation sphere will be advocated here as a classical, intuitive alternative to the quantum-mechanical view.

Up to this point the nuclear magnetic moments behave similarly to the magnetic moments associated with classical compass needles. However, unlike compass needles nuclear magnetic moments have intrinsic angular momentum or spin which can be visualized as a nucleus spinning around its own axis (Figure 1.2D). When a nuclear spin is placed in an external

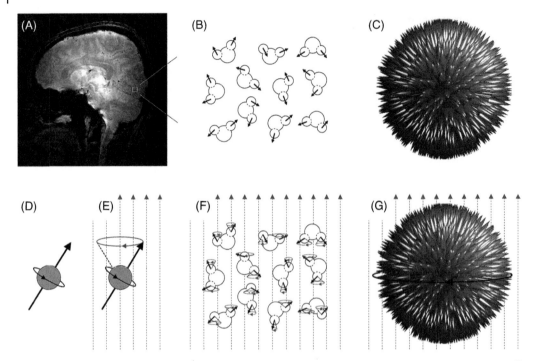

Figure 1.2 Precession of nuclear spins. (A, B) A small 1 μl volume from the human brain contains about 6×10^{19} protons, primarily located in water molecules. (B, C) In the absence of a magnetic field the proton spins have no orientational preference, leading to a randomly distributed "spin-orientation sphere." (D) Unlike compass needles, nuclear magnetic moments have intrinsic angular momentum or spin that leads to (E) a precessional motion when placed in a magnetic field. (F) All spins attain Larmor precession, but retain their random orientation to a good first approximation. (G) While the spin-orientation sphere also remains random when placed in a magnetic field, the entire sphere will attain Larmor precession.

magnetic field (Figure 1.2E) the presence of angular momentum makes the magnetic moment precess around the external magnetic field (Figure 1.2E). This effect is referred to as Larmor precession and the corresponding Larmor frequency ν_0 (in MHz) is given by

$$\nu_0 = \frac{\omega_0}{2\pi} = \frac{\gamma}{2\pi} B_0 \tag{1.1}$$

where γ is the gyromagnetic (or magnetogyric) ratio (in rad·MHz T^{-1}) and B_0 is the magnetic field strength (in T). The gyromagnetic ratio, which is constant for a given nucleus, is tabulated in Table 1.1. For protons at 7.0 T the Larmor frequency is 298 MHz. It should be noted that Larmor precession occurs for any spinning magnetic moment in a magnetic field, including classical objects and that it was described decades before the discovery of NMR [36].

When the protons depicted in Figure 1.2B are subjected to an external magnetic field, every spin starts to precess around the magnetic field with the same Larmor frequency. The Larmor frequency is independent of the angle between the external magnetic field and an individual spin. As the orientation of the magnetic moment with respect to the main magnetic field does (initially) not change, the spin-orientation sphere representation of Figure 1.2C remains unchanged with the exception that the entire sphere is rotating around the magnetic field at the Larmor frequency. If Larmor precession would be the only effect induced by the external magnetic field, then NMR would never have developed into the versatile technique as we know it today.

Table 1.1 NMR properties of biologically relevant nuclei encountered in *in vivo* NMR.

Isotope	Spin	Gyromagnetic ratio (rad·MHzT^{-1})	NMR frequency ratio (% of ^{1}H)	Natural abundance (%)
^{1}H	1/2	267.522	100.000	99.985
^{2}H	1	41.066	15.351	0.015
^{3}He	1/2	−203.802	76.179	0.000 14
^{7}Li	3/2	103.977	38.864	92.58
^{13}C	1/2	67.283	25.145	1.108
^{14}N	1	19.338	7.226	99.630
^{15}N	1/2	−27.126	10.137	0.370
^{17}O	5/2	−36.281	13.556	0.037
^{19}F	1/2	251.815	94.094	100.000
^{23}Na	3/2	70.808	26.452	100.000
^{29}Si	1/2	−53.190	19.867	4.7
^{31}P	1/2	108.394	40.481	100.000
^{33}S	3/2	20.557	7.676	0.76
^{35}Cl	3/2	26.242	9.798	75.53
^{37}Cl	3/2	21.844	8.156	24.47
^{39}K	3/2	12.501	4.667	93.100
^{129}Xe	1/2	−74.521	27.810	26.44

Fortunately, there is a second, more subtle effect that ultimately leads to a net, macroscopic magnetization vector that can be detected. The water molecules in Figure 1.2B are in the liquid state and therefore undergo molecular tumbling with a range of rotations, translations, and collisions. As a result, the amplitude and orientation of the magnetic field generated by one proton at the position of another proton changes over time (Figure 1.3A). When the local field fluctuation matches the Larmor frequency, it can perturb the spin orientation. These perturbations are largely, but not completely, random. The presence of a strong external magnetic field slightly favors the parallel spin orientation. As a result, over time the completely random spin orientation distribution (Figure 1.3B) changes into a distribution that is slightly biased towards a parallel spin orientation (Figure 1.3C). Visually, the spin distributions in the absence (Figure 1.3B) and presence (Figure 1.3C) of an external magnetic field look similar because the net number of spins that are biased towards the parallel orientation is very small, on the order of one in a million. The situation becomes visually clearer when the spin distribution is separated into spins that have a random orientation distribution (Figure 1.3D) and spins that are slightly biased towards a parallel orientation (Figure 1.3E). Adding the magnetic moments of Figure 1.3D does not lead to macroscopic magnetization similar to the situation in Figure 1.2G. However, adding the magnetic moments of Figure 1.3E leads to a macroscopic magnetization vector parallel to the external magnetic field. As the external magnetic field only biases the spin distribution along its direction, the spin distribution in the two orthogonal, transverse directions is still random.

The microscopic processes detailed in Figure 1.3A–E can be summarized at a macroscopic level as shown in Figure 1.3F. In the absence of a magnetic field ($t < 0$) the sample does not produce macroscopic magnetization. When an external magnetic field is instantaneously turned on ($t = 0$), the macroscopic magnetization exponentially grows over time where it plateaus at

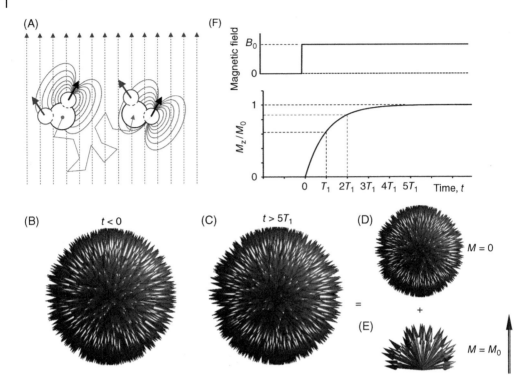

Figure 1.3 Appearance of macroscopic magnetization through T_1 relaxation. (A) Molecular tumbling and Brownian motion causes spin 1 (gray) to experience a wide range of magnetic field fluctuations originating from spin 2 (black) and other spins outside the water molecule. Magnetic field fluctuations of the proper frequency can change the spin orientation. While the perturbations are largely random, there is a very slight bias towards a parallel orientation with the external magnetic field. Over time the almost random perturbations transform a completely random spin-orientation sphere (B) into one that has a small polarization M_0 (C). The small polarization M_0 can be visualized better when the spins with a random orientation (D) are separated from the spins that have attained a slight bias (E). (F) Macroscopically the small polarization M_0 appears exponentially over time with a characteristic T_1 relaxation time constant according to Eq. (1.2).

a value corresponding to the thermal equilibrium magnetization, M_0. The appearance of macroscopic magnetization can be described by

$$M_z\left(t\right) = M_0 - \left(M_0 - M_z\left(0\right)\right)e^{-t/T_1} \tag{1.2}$$

where T_1 is the longitudinal relaxation time constant and $M_z(0)$ is the longitudinal magnetization at time zero. In the case of Figure 1.3, the initial longitudinal magnetization is zero, i.e. $M_z(0) = 0$. At the time of the first NMR studies, little was known about T_1 relaxation times in bulk matter. Both originators of NMR, Bloch and Purcell, were acutely aware that a very long T_1 relaxation time constant could seriously complicate the detection of nuclear magnetism. As a precaution, Purcell used an exceedingly small RF field such as not to saturate the sample [1], whereas it is rumored that Bloch left his sample in the magnet to reach thermal equilibrium while on a skiing trip [29]. Following the initial experiments it became clear that T_1 relaxation time constants can range from milliseconds to minutes, with water establishing thermal equilibrium in seconds. Extraordinarily long T_1 relaxation times may, however, have been the main reason for earlier, negative reports by Gorter [37, 38] on the detection of NMR in bulk matter.

The longitudinal magnetization vector represents the signal that will be detected in an NMR experiment. However, the static, longitudinal magnetization is never detected directly as its small contribution would be overwhelmed by much larger contributions from magnetization associated with electron currents within atoms and molecules. Instead, the longitudinal magnetization is brought into the transverse plane where the precessing magnetization can induce signal in a receiver coil at the very specific Larmor frequency.

The size of the longitudinal equilibrium magnetization and thereby the strength of the induced NMR signal is proportional to the number of spins that are biased towards a parallel orientation with the main magnetic field. The distribution of spin orientations and hence the bias in it can be calculated through the Boltzmann distribution which provides the probability P that a spin is in a certain orientation with an associated energy E according to

$$P(E) \propto e^{-E/kT} \tag{1.3}$$

where k is the Boltzmann constant (1.38066×10^{-23} J K^{-1}) and T is the absolute temperature in Kelvin. Equation (1.3) expresses the chance P of finding a particle with energy E in that state rather than in a random state determined by the available thermal energy of the environment (kT). For nuclear spins the energy limits are $\pm\mu B$, where μ is the magnetic moment and B is the external magnetic field. The lower and higher energies correspond to nuclear spins parallel and antiparallel to the external magnetic field, respectively. Using either a continuous distribution of spin orientations as shown in Figure 1.3C or a quantized distribution (see Section 1.11) the longitudinal equilibrium magnetization can be calculated from the Boltzmann distribution and is given by

$$M_0 = \frac{N\gamma^2 h^2 B_0}{16\pi^2 kT} \tag{1.4}$$

Equation (1.4) reveals several important features of the signals detected by NMR. Firstly, the thermal equilibrium magnetization M_0 is directly proportional to the number of spins N in the sample. This feature makes NMR a quantitative method in which the detected signals are, in principle, proportional to the concentration. The quadratic dependence of M_0 on the gyromagnetic ratio γ implies that nuclei resonating at high frequency (see Eq. (1.1)) generate the strongest NMR signals. Hydrogen has the highest γ of the commonly encountered nuclei, and has therefore the highest relative sensitivity. The linear dependence of M_0 on the magnetic field strength B_0 implies that higher magnetic fields improve the sensitivity. In fact this argument (and the related increase in chemical shift dispersion) has caused a steady drive towards higher magnetic field strength which now typically range from 1.5 T to circa 24 T (or up to circa 11.7 T for human applications). Finally, the inverse proportionality of M_0 to the temperature T indicates that sensitivity can be enhanced at lower sample temperatures. While the latter option is unrealistic for direct *in vivo* applications, it is being utilized to increase M_0 by orders of magnitude in hyperpolarized MR (see Chapter 3). The actual experimental sensitivity is determined by many additional factors, like RF coil characteristics, pulse sequence details, sample volume, natural abundance of the nucleus studied, sample noise, relaxation parameters, and spectral resolution. Although some factors can be predicted by Eq. (1.4), others need a more detailed treatment and can be found throughout the book.

1.4 Nuclear Induction

The orientation of a compass needle in Figure 1.1 could be changed with an additional magnetic field perpendicular to the magnetic moment. Similarly, a second magnetic field, B_1, is used to change the orientation of the longitudinal magnetization, M_0. Figure 1.4A shows the spin-orientation sphere at thermal equilibrium as being composed of randomly oriented

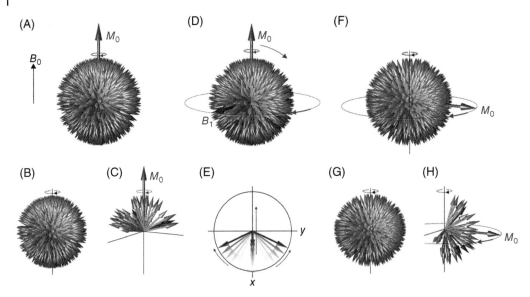

Figure 1.4 Excitation of nuclear spins. (A) A collection of nuclear spins give rise to macroscopic magnetization (red vector) parallel to B_0 at thermal equilibrium. (B) The nuclear spin orientations are largely random, with (C) only a minor amount contributing to the macroscopic magnetization. (D) A secondary magnetic field, B_1, perpendicular to B_0 and oscillating at the Larmor frequency has a constant phase relation with the spin magnetic moments. As a result, the B_1 magnetic field applies a constant torque during each Larmor revolution, leading to a rotation of the entire spin-orientation sphere. (E) The rotating B_1 field shown in (D) is typically achieved by a co-sinusoidally varying magnetic field (red arrow) which is equivalent to the sum of clockwise and anticlockwise vectors (blue arrows). Whereas the clockwise component achieves excitation as shown in (D), the anticlockwise component can for most practical purposes be ignored. (F–H) When the length and amplitude of the B_1 magnetic field is adjusted to provide a 90° rotation, the RF pulse has achieved excitation of the nuclear spins. Note that the entire spin-orientation sphere has been rotated by 90°, whereby (G) the randomly-oriented spins are still random and (H) the biased spins now provide a macroscopic magnetization vector in the transverse plane.

spins (Figure 1.4B) and a small number of spins with a net bias towards a parallel orientation (Figure 1.4C). Each of the spins undergoes Larmor precession around the magnetic field on its own cone, dictated by the initial orientation of the magnetic moment. When a stationary magnetic field B_1 is applied, the spins would not be perturbed in any significant manner since any rotation achieved during the first half of a Larmor precession cycle $1/(2\nu_0)$ (=1.68 ns for ^{1}H at 7 T) would be undone during the second half. In other words, a stationary magnetic field B_1 is not in resonance with the spins and its effect is therefore negligible. However, as shown for the compass needle in Figure 1.1, the second magnetic field can have a large effect when its frequency matches the natural frequency of the compass needle. Figure 1.4D shows the same spin-orientation sphere in the presence of a rotating magnetic field B_1 whose frequency equals the Larmor frequency. Since the angle between the magnetic moments and the magnetic field B_1 is constant, the entire spin-orientation sphere and hence the net macroscopic magnetization experiences a coherent rotation towards the transverse plane. When the amplitude and duration of the B_1 field is adjusted as to achieve a 90° rotation of the magnetization from the longitudinal axis into transverse plane (Figure 1.4F–H), the spins have undergone an excitation and the B_1 field is referred to as an excitation pulse. When the amplitude or duration of the B_1 field is doubled, the initial thermal equilibrium magnetization undergoes a 180° rotation, referred to as an inversion which is achieved by an inversion pulse. Following excitation, the B_1 field is removed leaving the magnetization in the transverse plane to be detected through electromagnetic induction in a nearby receiver coil. It should be noted that the magnitude of the rotating B_1 magnetic field is typically five to six orders of magnitude smaller than the static B_0 magnetic field. In addition, it should be realized that NMR uses *magnetic* fields rotating at the RF frequency, not electromagnetic RF waves [39, 40]. The electric component of electromagnetic RF waves is not relevant for NMR signal excitation or detection and proper RF coil design is aimed at minimizing its contribution. Unfortunately, electric fields cannot be eliminated entirely and are responsible for RF-induced sample heating (see Chapter 10).

At this point it should be mentioned that the secondary magnetic field B_1 is typically not applied as a rotating magnetic field, but rather as a cosine-modulated magnetic field traversing between the $+x$ and $-x$ axes (Figure 1.4E). However, a linear cosine-modulated magnetic field can be seen as the vector sum between clockwise and counterclockwise rotating components (Figure 1.4E). The counterclockwise component influences the spins to the order $(B_1/2B_0)^2$ and is known as the Bloch–Siegert shift [41]. Under most experimental conditions the counterclockwise rotating component can be ignored as it is not in resonance with the spins, leaving just the clockwise-rotating component to act on the nuclear magnetic moments.

1.5 Rotating Frame of Reference

The rotation of the magnetization vector during an RF pulse is fairly complex in a standard nonrotating frame-of-reference xyz, also known as the laboratory frame. In the laboratory frame the magnetization follows a complex path (Figure 1.5A) rotating around B_0 at the Larmor frequency while simultaneously also rotating around the secondary magnetic field B_1. For most applications, the Larmor precession is a constant motion that complicates visualization of rotations induced by the RF pulse. The Larmor precession can be visually removed by switching from a nonrotating laboratory frame to a frame-of-reference $x'y'z'$ that is rotating at the RF pulse frequency (Figure 1.5B). In the rotating frame the B_1 magnetic field is always along the x' axis, whereas the magnetization rotates in the $y'z'$ plane. With the observer still in the stationary laboratory frame, the motion of the magnetization is still as complicated as in Figure 1.5B. However, when the observer is placed within the rotating $x'y'z'$ frame, the observer also rotates at the RF pulse frequency making the associated motion invisible. The only motion that remains in the rotating frame $x'y'z'$ is the simple rotation of the magnetization around the B_1 magnetic field from the z' axis towards the $x'y'$ plane (Figure 1.5C). As is the case for the majority of NMR literature, all subsequent discussions will take place in the rotating frame of reference. Switching from a laboratory frame xyz to a rotating frame $x'y'z'$ is mathematically equivalent to reducing the very large B_0 (in T) or $(\gamma/2\pi)B_0$ (in Hz) magnetic field vector along the z/z' axis to zero. When the RF pulse and Larmor frequencies are equal, the RF pulse is said to be on-resonance and the $(\gamma/2\pi)B_0$ vector is effectively absent in the rotating frame $x'y'z'$. Any transverse magnetization will not show precession, but appears static along a fixed axis. However, when the RF pulse frequency ν is not equal to the Larmor frequency ν_0, the RF pulse is said to be off-resonance by $\Delta\nu = \nu_0 - \nu$ Hz. In that case a small $\Delta\nu$ vector appears along the z' axis of the rotating frame and any transverse magnetization will rotate around the z' axis at a frequency $\Delta\nu$ (Figure 1.5D). A quantitative treatment of rotating frames can be found in Section 1.7 during description of the Bloch equations.

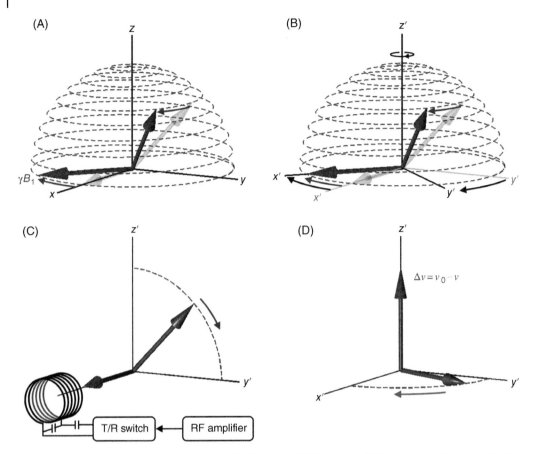

Figure 1.5 Laboratory and rotating frames of references. (A) Rotations of the macroscopic magnetization vector (red) in the nonrotating laboratory frame *xyz*. The magnetization precesses under the influence of the main magnetic field B_0 along the z axis and the oscillating magnetic field B_1 in the *xy* plane. Note that the relative oscillations are not drawn to scale. During the time over which the B_1 magnetic field rotates the spins by circa 45°, the spins precess millions of revolutions around the B_0 magnetic field. (B) In a rotating frame *x'y'z'* that rotates at the RF frequency, the B_1 magnetic field is always aligned along the *x'* axis. As a nonrotating observer in the laboratory frame, the oscillations of the magnetization have not changed. (C) However, as a rotating observer in the rotating frame, the rotations around the B_0 magnetic field are no longer visible, leaving only the simple rotation of the magnetization around the B_1 magnetic field. For an RF pulse of amplitude B_1 (in T) and duration T (in s), the nutation or flip angle θ (in rad) of the magnetization is given by $\theta = \gamma B_1 T$. RF pulses with nutation angles of 90° and 180° are typically referred to as excitation and inversion pulses. The power for an RF pulse is generated by an RF amplifier and is, via a transmit/receive (T/R) switch and an RF coil, delivered to the sample. Details on RF coils and related hardware can be found in Chapter 10. (D) On-resonance spins, for which the RF pulse frequency v equals the Larmor frequency v_0, remain along the *y'* axis following excitation since the effective magnetic field along the *z'* axis has been reduced to zero. Off-resonance spins, for which $v \neq v_0$, experience a small residual magnetic field along the *z'* axis of $(2\pi\Delta v/\gamma)$ which leads to precession at an effective off-resonance frequency Δv. The fate of off-resonance spins *during* an RF pulse is discussed in Chapter 5.

1.6 Transverse T_2 and T_2^* Relaxation

Figure 1.6B shows the situation right after excitation of the longitudinal magnetization into the transverse plane. The spins with a slight bias towards a parallel orientation with the magnetic field before the RF pulse now display a bias towards the *y'* axis. Addition of all individual magnetic moments leads to a macroscopic magnetization vector along the *y'* axis (Figure 1.6B,

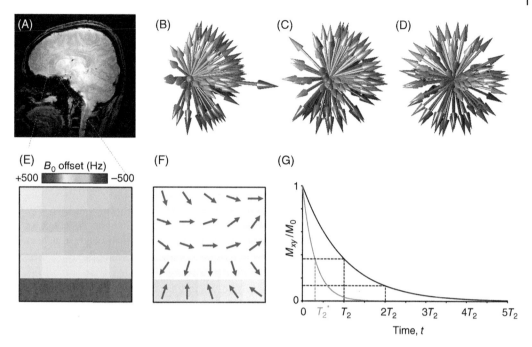

Figure 1.6 **Disappearance of macroscopic magnetization through T_2 and T_2^* relaxation.** (A) Sagittal MR image through the human head. (B) Collection of nuclear spins that give rise to macroscopic transverse magnetization (red vector) immediately following excitation. (C) Over time random molecular tumblings, similar to those depicted in Figure 1.3A, lead to a gradual loss in phase coherence. (D) After a sufficient amount of time has passed, all phase coherence has been lost and no macroscopic signal can be observed. The irreversible loss of phase coherence due to random molecular processes is an exponential process characterized by a T_2 relaxation time constant (G, black line). Under many circumstances the loss of detectable signal is much faster than that governed by T_2 relaxation and is often due to local magnetic field inhomogeneity. (E) Magnetic field B_0 map from a small part of the human brain depicted in (A). Variation of up to a few ppm in the magnetic field B_0 across the small, but macroscopic volume elements leads to a range of Larmor frequencies according to Eq. (1.1). (F) After some time following excitation, spins with different Larmor frequencies will have attained different phases. As the detected signal originates from all volume elements combined, a variation in phase leads to phase cancelation and thus signal loss. The loss of phase coherence due to magnetic field inhomogeneity in addition to T_2 relaxation is typically described as an exponential process with a T_2^* relaxation time constant (G, gray line). Since the T_2^* relaxation contains contributions from magnetic field inhomogeneity and T_2 relaxation, it can be stated that the T_2^* relaxation time constant is always smaller or equal to the T_2 relaxation time constant. The effects of magnetic field inhomogeneity can be completely removed for a single time point following excitation with a spin-echo pulse sequence (see Chapter 3). The effects of macroscopic magnetic field inhomogeneity can be reduced through shimming, as will be discussed in Chapter 10.

red). Note that the vectors are on-resonance and shown in the rotating frame $x'y'z'$ so that the macroscopic magnetization as well as the individual spins do not display Larmor precession. The situation in Figure 1.6B is a nonequilibrium condition and the spins want to return to thermal equilibrium where a longitudinal magnetization vector exists without any macroscopic transverse component. The reappearance of the longitudinal magnetization is governed by T_1 relaxation and at a time $>5T_1$ the thermal equilibrium magnetization will have been reestablished (see also Figure 1.3). The disappearance of macroscopic transverse magnetization is governed by T_2 relaxation. Right after excitation (Figure 1.6B) the individual spins precess with the same Larmor frequency. However, molecular tumbling and Brownian motion lead to random local field fluctuations as shown in Figure 1.3, such that the Larmor frequencies of different spins start to run out of sync over time (Figure 1.6C). As a result, the total sum of all magnetic moments is reduced leading to a smaller, macroscopic transverse magnetization vector. After a sufficient amount of time has passed, the Larmor frequencies are completely out of sync leading to the disappearance of macroscopic transverse magnetization. The ability to attain phase coherence in MR is remarkable; for many compounds the spins can undergo millions of Larmor revolutions before any noticeable loss of phase coherence can be detected. The long lifetime of transverse magnetization is one of the most important features of NMR and has allowed the development of a rich variety of pulse sequences interspersing RF pulses with delays. The disappearance of macroscopic magnetization from the transverse plane can be described by

$$M_{xy}(t) = M_{xy}(0)e^{-t/T_2} \tag{1.5}$$

where T_2 is the transverse relaxation time constant. For ^{1}H NMR in biological tissues the T_2 relaxation time constants (10–300 ms) are typically much shorter than the T_1 relaxation time constants (500–3000 ms). Chapter 3 will discuss T_1 and T_2 relaxation in greater detail. From Figures 1.4 and 1.6 it is clear that the presence of detectable, macroscopic magnetization in the transverse plane relies on phase coherence among the spins that were biased along the z' axis before excitation. As a result, macroscopic transverse magnetization is sometimes referred to as coherence. As will be seen in later chapters, coherence is a broader term that also describes unobservable and correlated transverse magnetization among scalar-coupled spins.

For most MR applications the signal decay observed during the recording of an FID is much faster than governed by T_2 relaxation according to Eq. (1.5). This is due to macroscopic magnetic field inhomogeneity and is illustrated in Figure 1.6E and F. As will be discussed in Chapter 10, differences in magnetic susceptibility between water and air leads to magnetic field inhomogeneity in the proximity to their boundary. A prime example is the transition between air in the nasal cavities and water in the brain, leading to an inhomogeneous magnetic field in the frontal cortex. Figure 1.6E shows a magnetic field map from a small section of the human frontal cortex (Figure 1.6A). While the magnetic field at each location is about 4 T, or 170 MHz, there is a small variation of circa 500 Hz across the spatial locations. When the transverse magnetization is initially formed following excitation, the signal at each location is along the y' axis of the rotating frame. After time t, the spins at position r that are off-resonance by an amount $\Delta\nu = (\gamma/2\pi)\Delta B$ acquire a phase difference given by

$$\Delta\phi(r,t) = 2\pi\Delta\nu(r)t \tag{1.6}$$

leading to the situation in Figure 1.6F where the transverse magnetization in different sub-volumes has acquired various amounts of phase. Summation over the entire macroscopic volume will lead to phase cancelation and thus a smaller macroscopic magnetization vector. The process can be written as the combination of Eqs. (1.5) and (1.6) according to

$$M_{xy}(t) = M_{xy}(0)e^{-t/T_2}\int_r e^{i\Delta\phi(r,t)}dr = M_{xy}(0)e^{-t/T_2^*} \tag{1.7}$$

assuming spatially homogeneous M_{xy} and T_2 relaxation distributions. It follows that in the presence of magnetic field inhomogeneity the disappearance of transverse magnetization is governed by T_2^* relaxation. While T_2^* relaxation is often modeled as an exponential function, the signal decay is governed by the complex integral in Eq. (1.7) which can take on a wide range of (non-exponential) shapes. Whereas both T_2 and T_2^* relaxation are caused by magnetic field

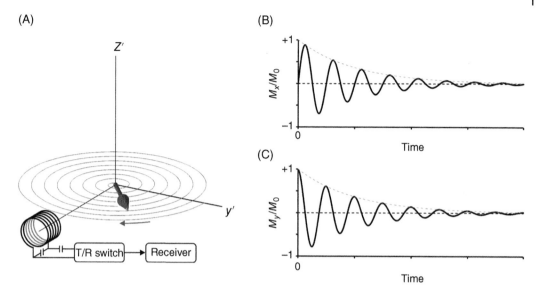

Figure 1.7 Nuclear induction. (A) A rotating, macroscopic magnetization vector induces a voltage in a nearby receive RF coil through electromagnetic induction. Note that the same RF coil that was used to transmit the RF pulse (Figure 1.5C) can be used to receive the FID. The transmit/receive (T/R) switch ensures that the small NMR signal enters the receive chain during reception, while preventing high-powered RF pulses to damage the receive chain during transmission. (B, C) The FID signal is typically sampled along a single axis (x' in (A)), giving rise to M_x in (B)) after which the second, orthogonal component is created during the phase-sensitive, quadrature detection step (see Chapter 10).

inhomogeneity leading to a decrease in phase coherence and an overall signal loss, there is an important difference. T_2 relaxation is caused by the inherently random process of fluctuating magnetic fields among the individual spins leading to irreversible signal loss. The additional dephasing associated with T_2^* relaxation is caused by static magnetic field inhomogeneity. Since the cause of the additional dephasing is constant over time, its effects can be reversed with a spin-echo sequence as will be discussed in Chapter 3. In addition, improvements in the magnetic field homogeneity obtained through shimming (Chapter 10) directly lead to longer T_2^* relaxation times and a smaller difference between T_2 and T_2^*.

The rotating, transverse magnetization (Figure 1.7A) gives rise to the NMR FID signal through electromagnetic induction into a nearby receiver coil. Often the same RF coil that was used for RF transmission (or B_1 magnetic field generation, Figure 1.5C) is also used for NMR signal reception (Figure 1.7A). The transmit/receive (T/R) switch ensures that the sensitive receiver electronics are protected when the high-powered RF pulses are transmitted. More details on RF transmit and receive chain elements can be found in Chapter 10. The x and y components of the FID are shown in Figure 1.7B and C. The signal in Figure 1.7 presents the fundamental NMR signal that, despite the extensive discussions up to this point, can be generated with only three elements. Firstly, a strong external magnetic field is required to create the net thermal equilibrium magnetization. Secondly, a time-varying magnetic field rotates the magnetization into the transverse plane where, lastly the precessing magnetization induces the FID signal in a receiver coil. In the following paragraphs and chapters the information content of the FID signal is greatly enhanced through detection of chemical shifts and scalar couplings and through spatial encoding. However, all NMR experiments are fundamentally based on the three operations described in Figures 1.3, 1.4, and 1.7.

1.7 Bloch Equations

Felix Bloch described the motion of the macroscopic magnetization by three phenomenological differential equations that are commonly referred to as the Bloch equations [4]. In the absence of relaxation the rotations of the macroscopic magnetization vector $\boldsymbol{M} = (M_x, M_y, M_z)$ in the laboratory frame is described by

$$\frac{\mathrm{d}M(t)}{\mathrm{d}t} = M(t) \times \gamma B(t) \tag{1.8}$$

where $\boldsymbol{B} = (B_x, B_y, B_z)$ represents the three orthogonal magnetic fields. B_x and B_y are part of the oscillatory magnetic field of the RF pulse and B_z represents the static magnetic field B_0. Expanding the cross product in Eq. (1.8) yields three coupled differential equations (see Exercises for more detail) for the three components of $\boldsymbol{M}$ that are given by

$$\frac{\mathrm{d}M_x(t)}{\mathrm{d}t} = \gamma M_y(t) B_0 - \gamma M_z(t) B_{1y} \tag{1.9}$$

$$\frac{\mathrm{d}M_y(t)}{\mathrm{d}t} = \gamma M_z(t) B_{1x} - \gamma M_x(t) B_0 \tag{1.10}$$

$$\frac{\mathrm{d}M_z(t)}{\mathrm{d}t} = \gamma M_x(t) B_{1y} - \gamma M_y(t) B_{1x} \tag{1.11}$$

Equations (1.9)–(1.11) (or the compact form of Eq. (1.8)) describe Larmor precession of the nuclear magnetization $\boldsymbol{M}$ in an external magnetic field B_0, as well as the rotation of $\boldsymbol{M}$ by a time-varying RF field with components B_{1x} and B_{1y}. The presence of relaxation requires an additional term to Eqs. (1.8)–(1.11) that describes the disappearance of magnetization from the transverse plane due to T_2 relaxation and the reappearance of the thermal equilibrium magnetization M_0 due to T_1 relaxation.

$$\frac{\mathrm{d}M_x(t)}{\mathrm{d}t} = -\frac{M_x(t)}{T_2} \tag{1.12}$$

$$\frac{\mathrm{d}M_y(t)}{\mathrm{d}t} = -\frac{M_y(t)}{T_2} \tag{1.13}$$

$$\frac{\mathrm{d}M_z(t)}{\mathrm{d}t} = -\frac{M_z(t) - M_0}{T_1} \tag{1.14}$$

Adding Eqs. (1.9)–(1.11) and (1.12)–(1.14) provides the complete Bloch equations in the nonrotating, laboratory frame. Simple solutions are readily obtained for the transverse magnetization following excitation and T_1 and T_2 relaxation in the absence of an RF field (see Exercises). A general solution typically requires numerical integration, although analytical solutions have been obtained for specific RF driving functions. In Chapter 5 it will be shown that in the absence of relaxation, the Bloch equations can be reduced to 3×3 rotation matrices. Under many circumstances NMR experiments are more conveniently described in a Cartesian frame that is rotating around the main magnetic field B_0 at the frequency ν of the B_1 field. Transformation of the Bloch equations from the laboratory frame to the rotating frame (see Exercises) yields

$$\frac{\mathrm{d}M'_x(t)}{\mathrm{d}t} = (\omega_0 - \omega)M'_y(t) - \frac{M'_x(t)}{T_2} \tag{1.15}$$

$$\frac{\mathrm{d}M'_y(t)}{\mathrm{d}t} = -(\omega_0 - \omega)M'_x(t) + \gamma B_1 M'_z(t) - \frac{M'_y(t)}{T_2} \tag{1.16}$$

$$\frac{\mathrm{d}M'_z(t)}{\mathrm{d}t} = -\gamma B_1 M'_y(t) - \frac{M'_z(t) - M_0}{T_1} \tag{1.17}$$

Equations (1.15)–(1.17) quantitatively describe the features that were qualitatively described with the introduction of rotating frames (Figure 1.5). In the rotating frame the RF field B_1 is along the x' axis and thus does not affect M'_x. In addition, the large B_0 magnetic field (or large $\omega_0 = \gamma B_0$ angular frequency offset) is reduced to $(\omega_0 - \omega)$ or even zero when the RF frequency ω and Larmor frequency ω_0 are exactly matched. In the on-resonance condition ($\omega = \omega_0$) the transverse magnetization does not precess over time and in the absence of an RF pulse only decays due to T_2 relaxation. The Bloch equations provide a quantitative description of any NMR experiment that involves RF pulses and relaxation on a sample of noninteracting spins (i.e. without scalar coupling). Modifications of the Bloch equations to include the effects of diffusion and chemical exchange give rise to the Bloch–Torrey [42] and Bloch–McConnell [43] equations, respectively.

1.8 Fourier Transform NMR

The FID of a sample with a single Larmor frequency following excitation into the transverse plane of the laboratory frame (Figure 1.7) can be obtained by solving the Bloch equations of Eqs. (1.9)–(1.14). The resulting FID signal can be described with a complex function $f(t) = R(t) + iI(t)$, where the real $R(t)$ and imaginary $I(t)$ components are given by (see Exercises)

$$R(t) = M_x(t) = M_0 \cos(2\pi\nu_0 t + \phi)e^{-t/T_2^*} \tag{1.18}$$

$$I(t) = M_y(t) = -M_0 \sin(2\pi\nu_0 t + \phi)e^{-t/T_2^*} \tag{1.19}$$

where ϕ represents the phase of the transverse magnetization relative to the x axis immediately following excitation. For an excitation pulse along the x axis, the transverse magnetization is excited along the y axis, making the phase ϕ equal to $-90°$. It follows that the FID is described by four independent parameters, M_0, ν_0, T_2^*, and ϕ, whereby it is assumed that the sample was in a state of thermal equilibrium before excitation. All four parameters describing the FID can be easily recognized in the case of a single Larmor frequency (Figure 1.8A). Especially, the M_0 parameter is of great importance in many NMR applications as it is directly related to the concentration (Eq. (1.4)). For a single frequency signal, M_0 is proportional to the intensity of FID immediately following excitation.

As will be discussed in Section 1.9, one of the strengths of NMR is that different compounds and different chemical groups within a compound can be identified based on differences in their Larmor frequencies. In the case of multiple frequencies, the deduction of the four parameters for each signal will prove difficult, if not impossible, by visual inspection of a time-domain FID signal (Figure 1.8G). Fortunately, a standard processing tool exists that allows the extraction and separation of different frequencies from a time-domain FID and presents it as signals

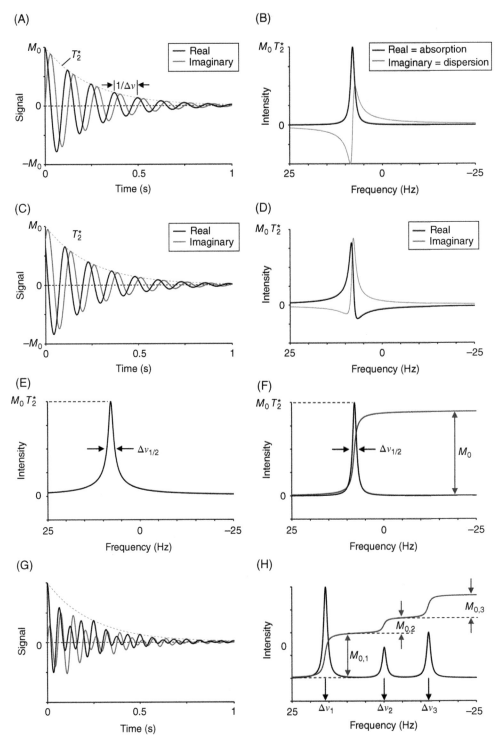

Figure 1.8 **Fourier transformation of time and frequency-domain signals.** (A) Complex, exponentially-decaying, single-frequency FID as described by four parameters M_0, $\Delta\nu$, T_2^*, and ϕ. The phase ϕ is defined as the angle of the transverse magnetization at the start of signal acquisition. (B) FT of the time-domain FID gives a complex, frequency-domain spectrum. By convention only the real part of the spectrum (black line) is shown. When $\phi = 0$, the real spectrum equals the absorption spectrum. (C, D) When $\phi \neq 0$, the real (and imaginary) spectrum represents a mixture of absorptive and dispersive lines according to Eqs. (1.21) and (1.22). The dispersive contribution is undesirable due to the long "tails" and can be eliminated through zero-order phase correction (see text for details). (E) A magnitude representation of the frequency-domain spectrum (Eq. (1.26)) can be used to eliminate all phase information, but comes at the expense of an increase line width $\Delta\nu_{1/2}$. (F) The peak height and integral of the absorption spectrum are proportional to $M_0 T_2^*$ and M_0, respectively, whereby the line width $\Delta\nu_{1/2}$ equals $1/(\pi T_2^*)$. (G, H) Fourier transformation of a multifrequency time-domain signal (G) readily reveals the frequency components and amplitude in the frequency-domain (H).

in a frequency-domain spectrum. The Fourier transformation calculates a frequency-domain spectrum $F(v)$ from a time-domain signal $f(t)$ according to

$$F(v) = \int_{-\infty}^{+\infty} f(t) e^{-2\pi i v t} dt \tag{1.20}$$

whereby the time-domain FID, $f(t) = R(t) + iI(t)$ is given by Eqs. (1.18) and (1.19). For an FID signal the integration boundaries run from 0 to $+\infty$. The Fourier transformation is a reversible operation, such that the inverse Fourier transformation of the spectrum $F(v)$ yields the original time-domain function $f(t)$. Since the integral of Eq. (1.20) involves a complex exponential, the frequency-domain spectrum $F(v) = R(v) + iI(v)$ will also be complex. The real and imaginary components are given by (see Exercises)

$$R(v) = A(v)\cos\phi - D(v)\sin\phi \tag{1.21}$$

$$I(v) = D(v)\cos\phi + A(v)\sin\phi \tag{1.22}$$

whereby the absorption $A(v)$ and dispersion $D(v)$ spectral components are given by

$$A(v) = \frac{M_0 T_2^*}{1 + 4\pi^2 (v_0 - v)^2 T_2^{*2}} \tag{1.23}$$

$$D(v) = \frac{2\pi M_0 (v_0 - v) T_2^{*2}}{1 + 4\pi^2 (v_0 - v)^2 T_2^{*2}} \tag{1.24}$$

The curves described by Eqs. (1.23) and (1.24) are collectively known as a Lorentzian line shape. Figure 1.8B shows the real and imaginary spectra obtained following Fourier transformation of the FID signal shown in Figure 1.8A. When the phase ϕ equals zero, the real spectrum reduces to the absorption spectrum ($R(v) = A(v)$) and the imaginary spectrum reduces to the dispersion spectrum ($I(v) = D(v)$). However, in general, the phase is not zero and the spectrum will be a mixture of absorption and dispersion spectra (Figure 1.8C and D). From Figure 1.8B and D it is clear that the dispersion spectrum is not desirable as it has broad "tails" that span over a much wider spectral range than the absorption spectrum. In case of multiple signals the dispersive part of the spectrum will display a much larger amount of overlap between signals than the absorptive part. Zero-order phase correction is a standard routine available on any spectrometer that aims at minimizing the dispersive component in the real spectrum such that the real spectrum equals the absorption spectrum. Zero-order phase correction is most easily performed in the time domain, by recognizing that the time-domain FID given by Eqs. (1.18) and (1.19) can be written as a complex exponential using Euler's formula (see Appendix A.2)

$$f(t) = M_0 e^{-2\pi i v_0 t} e^{-t/T_2^*} e^{i\phi} \tag{1.25}$$

By multiplying the FID signal with a zero-order phase correction $e^{-i\phi_c}$, the phase can be removed from the FID and the resulting spectrum when $\phi = \phi_c$. Under certain conditions, when zero-order phase correction is not available or viable, the phase of the spectrum can be removed by calculating the absolute-valued or magnitude spectrum (Figure 1.8E) according to

$$|F(v)| = \sqrt{R(v)^2 + I(v)^2} \tag{1.26}$$

Since the magnitude spectrum holds both real and imaginary components, it is always broader than the corresponding, phased absorption spectrum (Figure 1.8F).

The four parameters describing the time-domain FID can also be recognized in the absorption spectrum. The first point of the FID, M_0, equals the area under the curve of the absorption spectrum and can be obtained by numerical integration (summation) of the spectrum (Figure 1.8F). Note that the peak height of a Lorentzian line is equal to $M_0 T_2^*$. While it is attractive to use peak height as a substitute for relative M_0 between various signals, it has to be used with great caution as a change in T_2^* can change the relative peak heights without a true change in M_0. The oscillation or frequency ν_0 of the signal can be recognized as the horizontal position on the frequency axis. The T_2^* relaxation time can be obtained from the frequency width at half the peak height, $\Delta\nu_{1/2}$ according to

$$\Delta\nu_{1/2} = \frac{1}{\pi T_2^*} \tag{1.27}$$

Whereas Figure 1.8A–F dealt with single-frequency signals that could have been analyzed in the time domain, the power of Fourier transformation becomes indispensable when multifrequency signals are acquired. Figure 1.8G shows an FID containing three different frequencies with different amplitudes M_0. While it is apparent that more than one frequency is involved, it is not obvious how to extract the different parameters for each signal in the time domain. Following Fourier transformation, the spectrum in Figure 1.8H immediately reveals the three distinct frequencies, whereas spectral integration gives the relative M_0 amplitudes.

1.9 Chemical Shift

In the years following the discovery of NMR, the technique was primarily used by the physics community. As the Larmor frequency of a given nucleus only depended on the external magnetic field strength, according to Eq. (1.1), it provided physicists the opportunity to measure the magnetic moments of different nuclei with unprecedented precision. At this point in time NMR was of little interest to chemists. That all changed by a number of papers published in 1950 and 1951 that demonstrated different resonance frequencies for various chemical groups within the same molecule. Proctor and Yu [7] described two distinct ^{14}N resonances in an ammonium nitrate (NH_4NO_3) solution, whereas Dickinson [8] described the dependence of ^{19}F resonance frequencies on chemical composition. The ability of NMR to provide detailed chemical information on compounds was firmly established when Arnold et al. [9] in 1951 published a high-resolution ^1H NMR spectrum of ethanol in which separate signals from methyl, methylene, and hydroxyl protons could be clearly recognized. This effect, in which the local chemical environment surrounding a nucleus influences the resonance frequency, is known as the chemical shift.

The phenomenon of chemical shift is caused by a shielding of the magnetic field at the nucleus by the surrounding electron cloud. Figure 1.9A shows a hypothetical, single proton in an external magnetic field B_0. The magnetic field experienced by the nucleus, B_n, is identical to the external magnetic field B_0 and the ^1H Larmor frequency is given by Eq. (1.1). The hypothetical, single proton would normally be surrounded by an electron cloud (Figure 1.9C). Classically, the electron cloud can be seen as a current rotating around the nucleus, which then generates a magnetic field B_e at the nucleus that opposes the external magnetic field B_0. The magnetic field B_n at the nucleus, being the sum of B_0 and B_e, is thus reduced relative to Figure 1.9A leading to a lower Larmor resonance frequency (Figure 1.9D). In general, the proton in Figure 1.9C is part of a larger molecule and is bound directly to a carbon or other

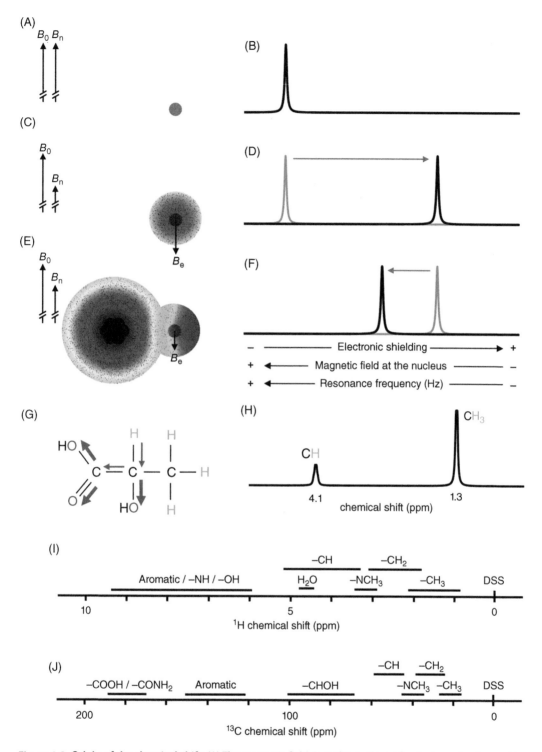

Figure 1.9 **Origin of the chemical shift.** (A) The magnetic field B_n at the nucleus of a single proton is equal to the external magnetic field B_0 leading to (B) a spectrum with a signal at Larmor frequency $\nu_0 = (\gamma/2\pi)B_0 = (\gamma/2\pi)B_n$. (C) The electron density in a hydrogen atom can be regarded as small currents that generate a magnetic field B_e that opposes B_0. The magnetic field at the nucleus is therefore reduced, (D) leading to a lower Larmor frequency. (E, F) The electron density of hydrogen in larger molecules depends on the electronegativity of nearby atoms, such that the exact Larmor frequency becomes very sensitive to the chemical environment. (G, H) The electronegative oxygen atoms in lactate shift the electron density away from the protons, leading to reduced electronic shielding. Since the single methine proton is closer to electronegative oxygen atoms than the three methyl protons, its Larmor frequency and chemical shift are higher. (I, J) Chemical shift ranges from various chemical groups as detected with (I) ^{1}H NMR and (J) ^{13}C NMR. The methyl protons of DSS (2,2-dimethyl-2-silapentane-5-sulfonate) are assigned a chemical shift $\delta = 0$.

atom (Figure 1.9E). Most atoms in organic compounds, like carbon, nitrogen, and oxygen, have a stronger electronegativity than protons meaning that they tend to pull electron cloud density away from protons. As the electron cloud density surrounding the proton decreases, so does the electron-induced magnetic field B_e. The total magnetic field at the nucleus B_n will therefore be larger than in Figure 1.9C leading to a higher proton resonance frequency (Figure 1.9F). The effect of chemical shift can be included in the Larmor Eq. (1.1) by understanding that B_0 means the magnetic field *at* the nucleus B_n, which includes besides the main magnetic field also shielding effects.

Using the arguments laid out for Figure 1.9A–F, the ^{1}H NMR spectrum of lactic acid (Figure 1.9G) can be qualitatively understood. Lactic acid has four types of protons in the methyl, methine, hydroxyl, and carbonyl groups. When lactic acid is dissolved in water, the hydroxyl and carbonyl protons undergo fast exchange with the water protons making them effectively invisible in a normal ^{1}H NMR spectrum (see Chapter 3 for more details on chemical exchange). The methyl and methine protons give rise to resonance lines at lower and higher Larmor resonance frequencies, respectively. The methyl protons have a relatively high electron cloud density due to the absence of atoms with a high electronegativity, like oxygen, in the immediate vicinity. The high electronic shielding leads to a lower magnetic field at the nucleus and hence a lower Larmor resonance frequency. The electron density around the methine proton is much lower due to the presence of three oxygen atoms within two to three chemical bonds. As a result, the magnetic field at the nucleus and thus the Larmor resonance frequency is higher. This assignment can be confirmed by noting that the relative signal areas of the high- and low-frequency signals are 1 and 3, corresponding to the single methine and three methyl protons.

The Larmor resonance frequencies have so far been expressed in units of Hertz (Hz), based on the Larmor Eq. (1.1). However, for chemical identification and comparisons between laboratories this proves to be an inconvenient choice as the Larmor frequency (in Hz) is dependent on the external magnetic field B_0 (in T). The Larmor resonance frequency can become independent of the magnetic field strength when expressed in parts-per-million (ppm) relative to a reference compound according to

$$\delta = \frac{\nu - \nu_{ref}}{\nu_{ref}} \times 10^6 \tag{1.28}$$

where δ is referred to as the chemical shift (in ppm) and ν and ν_{ref} are the Larmor frequencies (in Hz) of the compound under investigation and of a reference compound, respectively.

The reference compound should ideally be chemically inert and its chemical shift should be independent of external variables (temperature, ionic strength, shift reagents) and produce a strong (singlet) resonance signal well separated from all other resonances. A widely accepted reference compound for ^{1}H and ^{13}C NMR is tetramethylsilane (TMS) to which $\delta = 0$ has been assigned. However, the use of TMS is restricted to NMR on compounds in organic solvents. For aqueous solutions, 3-(trimethylsilyl)propionate (TSP) or 2,2-dimethyl-2-silapentane-5-sulfonate (DSS) is typically used, of which DSS is more desirable as the chemical shift is temperature and pH independent [44]. Unfortunately, none of these compounds are found in *in vivo* systems and can therefore never be used as internal references. TSP and DSS can in principle be used as an external reference compound, being placed adjacent to the object under investigation. However, under these circumstances the observed chemical shift needs to be corrected for bulk magnetic susceptibility effects as well as macroscopic inhomogeneity of the main magnetic field (see Chapter 10). Differences in susceptibility and local magnetic field strength between the object under investigation and the adjacent external reference make the use of external chemical shift referencing undesirable. For *in vivo* applications other resonances have been used as an internal reference. Commonly used internal references are the methyl resonance of *N*-acetyl aspartate (2.01 ppm) for ^{1}H MRS of the brain and the phosphocreatine resonance (0.00 ppm) for ^{31}P MRS of brain and muscle. The International Union of Pure and Applied Chemistry (IUPAC) recommends the use of universal referencing for *any* NMR nucleus based on the proton Larmor frequency of DSS [45]. This excellent recommendation is currently only viable under the highly controlled conditions encountered with high-resolution, liquid-state NMR.

Figure 1.9I and J provide an overview of the chemical shifts observed for common chemical groups with ^{1}H and ^{13}C NMR. The overall pattern can be qualitatively explained by shielding effects due to neighboring electronegative atoms. In many NMR publications the terms "downfield" and "upfield" are encountered when the chemical shift of one signal is expressed relative to another signal. The origin of these terms can be found in the early days of NMR in which a spectrum was obtained by varying the magnetic field through a fixed frequency RF source. Signals with a high Larmor frequency and hence a high chemical shift required a lower magnetic field to be in resonance with the fixed frequency RF source. In the NMR spectrum, in which the horizontal axis was expressed in Tesla, signals with a high Larmor frequency would appear "downfield" from signals with a lower Larmor frequency. In modern NMR experiments where the magnetic field is constant the terms "downfield" and "upfield" are confusing. However, since the terms see continued usage in modern NMR literature, one could reinterpret them as reduced and increased electronic shielding fields B_e at the nucleus for "downfield" and "upfield," respectively. In Figure 1.9I and J the resonances in or adjacent to electronegative moieties, such as carbonyl and hydroxyl groups, appear in the downfield region of the spectrum, whereby highly shielded methyl groups appear in the upfield region.

1.10 Digital NMR

The NMR signal as presented up to this point and as detected by Bloch in 1946 is a continuous, analog signal. While acceptable for scanning 1D MR spectra, analog acquisition and processing of NMR signals are incompatible with modern multidimensional NMR and MRI experiments. Since the revolutions in NMR (FT NMR, 2D NMR, MRI) ran parallel to the development of modern computing, NMR could take full advantage of all the benefits provided by digital acquisition and processing.

1.10.1 Analog-to-Digital Conversion

The first step in bringing NMR into the digital domain is the conversion of the analog FID signal induced in a receiver coil into a digital signal. This is achieved with an analog-to-digital converter (ADC), which measures the instantaneous value of the FID at equal time intervals (Figure 1.10). The speed of the analog-to-digital conversion is prescribed by the sampling theory [46, 47]. This theory states that any sinusoidal signal of frequency F can be accurately described when it is sampled at least twice per cycle. This minimum sampling rate is called the Nyquist frequency F_N. The spectral bandwidth SW equals $2F_N$, since frequencies between $-F_N$ and $+F_N$ are accurately sampled. The time between the data points, Δt (Figure 1.10B), is known as the dwell time and equals 1/SW. If a signal is present with a frequency greater than the Nyquist frequency, then this signal will still be digitized, but at an incorrect frequency (Figure 1.10E). A resonance with frequency $\nu_{real} = F_N + \Delta\nu$ relative to the center of the spectral bandwidth will appear after Fourier transformation at an apparent position with frequency $\nu_{app} = -F_N + \Delta\nu$.

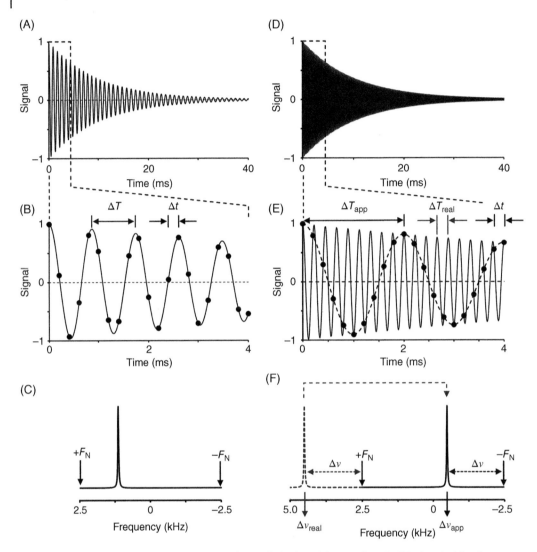

Figure 1.10 Analog-to-digital conversion and signal aliasing. (A) Time-domain FID signal with a frequency $\Delta\nu$ of 1.15 kHz acquired over a spectral bandwidth (SW) of 5.0 kHz, making the Nyquist sampling frequency $F_N = SW/2$. (B) The analog-to-digital converter (ADC) samples the signal at each dwell-time $\Delta t = 1/SW = 0.2$ ms (black dots). Since each period of the signal $\Delta T = 1/\Delta\nu$ is sampled multiple times by the ADC (i.e. $\Delta t < \Delta T$) the Nyquist sampling condition is satisfied, giving (C) a spectrum with the proper frequency. (D) Time-domain FID signal with a frequency $\Delta\nu_{real} = \Delta\nu + F_N$ of 4.5 kHz. (E) Each period ΔT_{real} of the signal is sampled less than twice (i.e. $\Delta t > \Delta T_{real}$) by the ADC such that the Nyquist sampling condition is not satisfied. The sampled data points do still represent a sinusoidal signal, but are sampled at an apparent, lower frequency with period ΔT_{app}. (F) Following Fourier transformation the signal is aliased, appearing at an apparent frequency $\Delta\nu_{app} = \Delta\nu - F_N = -0.5$ kHz.

This so-called aliasing of resonances can lead to erroneous identification of signals. Aliasing can easily be avoided by increasing the spectral bandwidth (i.e. decreasing the dwell time Δt), after which the minimum spectral bandwidth needed to unambiguously observe all the resonances can be determined. Aliasing of signal seems at first sight a large problem in FT NMR, since noise from outside the spectral region would be folded back into the spectrum thereby dramatically decreasing the obtainable S/N ratio. However, high-frequency noise components can easily be filtered out before the ADC sampling by bandpass filters (see Chapter 10). A cut-off filter, such as a Butterworth filter does not affect signals within the spectral range, while suppressing (i.e. multiplying by zero) all signals (i.e. noise) outside the spectral range. In this case, real NMR signals outside the spectral bandwidth will also be suppressed and will therefore be absent in the final NMR spectrum. Aliasing is readily observed in applications where frequencies are sampled indirectly, such as the phase-encoding directions in MR imaging and spectroscopic imaging (see Chapters 4 and 7) and the second dimension in 2D NMR (see Chapter 8).

1.10.2 Signal Averaging

Once the NMR signal is digitized, it can be stored in memory for processing or signal averaging. In signal averaging the signal-to-noise ratio (SNR) of the FID is improved by adding or averaging the FID signals of several consecutive, identical experiments. Averaging of n FID signals leads to an improvement in SNR of a factor $\sqrt{n}$ [48, 49]. This is because the voltage of the signal S increases linearly with n, while for the random processes of noise N the power increases linearly. Since power is proportional to the square of voltage, the noise voltage increases as $\sqrt{n}$, leading to an overall improvement of the SNR of $n/\sqrt{n} = \sqrt{n}$. In practice, the improvement in SNR for *in vivo* NMR by time-averaging is limited, since an improvement of a factor 10 requires a prolongation of measurement time by a factor $10^2 = 100$. Typically, an *in vivo* NMR experiment is a compromise between sufficient SNR and the allowable duration of the experiment.

1.10.3 Digital Fourier Transformation

The stored, digital FID signal can be processed by Fourier transformation to produce a digital MR spectrum. The analog Fourier transformation (Eq. (1.20)) which entails a continuous integration is then replaced with a digital or discrete Fourier transformation (DFT) composed of a discrete summation. However, direct computation of a discrete summation is very inefficient, requiring N^2 calculations for a signal containing N data points. For 1D MR spectra direct summation may be acceptable, but for larger 2D and 3D MRI and MRSI datasets, direct summation would lead to unacceptable calculation times. By using symmetry and recursive characteristics of DFT, Cooley and Tukey [50] proposed in 1965 a new algorithm for the special case of N being a power of 2. This fast Fourier transform (FFT) algorithm requires only $N\log_2 N$ calculations, leading to a 32-fold reduction in calculation time for a dataset with $N = 256$. The improvement increases to 1024 for a MR image with a standard 256×256 matrix. Whereas modern FFT algorithms now also give substantial improvements for powers other than 2, MR images often continue to be acquired on a "power of 2" matrix size.

1.10.4 Zero Filling

In general, the FID of a spectrum with spectral width SW = $2F_N$ is sampled by the ADC over N points spaced by the dwell time $\Delta t = 1/SW$. The total duration of signal acquisition, T_{acq}, is therefore $N \times \Delta t$. Following Fourier transformation of the FID, the NMR spectrum will also contain N points spread out over the spectral width SW. The spectral digital resolution is therefore SW/N, which is equivalent to the reciprocal of the total acquisition, i.e. $1/T_{acq}$. For an experiment with 256 points sampled and $T_{acq} = 64\,ms$ (Figure 1.11A), leading to SW = $2F_N = 4.0\,kHz$, the spectral resolution is 15.63 Hz per point. When the digital Fourier transform spectrum (Figure 1.11B) is compared to the continuous Fourier transform spectrum it is clear that the digital resolution is

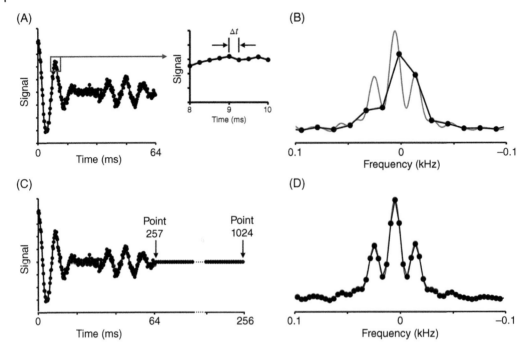

Figure 1.11 **Zero filling and spectral resolution.** (A) Time-domain FID signal acquired and digitized as 256 points over a spectral width of 4.0 kHz (i.e. $\Delta t = 0.25$ ms). (B) Fourier transform spectrum of the digitized time-domain FID signal (black) compared to an analog spectrum (gray). The limited digital resolution of 15.63 Hz per point prevents the visualization of all spectral features. (C) Adding an additional 768 points of zero amplitude to the experimental data, known as zero filling, leads (D) to an increased spectral resolution of 3.91 Hz per point upon Fourier transformation. Zero filling leads to an improved visualization of the spectral features that were already present in the digitized data of (A).

insufficient to properly visualize all spectral features. The digital resolution can be improved by increasing the total acquisition time T_{acq}. This can be achieved by decreasing the spectral width, SW, or increasing the number of acquisition points, N. A decrease in the spectral bandwidth is often limited by the minimum width required to avoid aliasing. Increasing the number of acquisition points comes at the cost of a decreased SNR as more noise and less signal are acquired. The process of extending the acquisition time can be simulated during post-acquisition processing by extending the acquired FID (which has decayed to zero amplitude) artificially by adding a string of points with zero amplitude to the FID prior to (discrete) Fourier transformation (Figure 1.11C). This process is known as zero filling and while it does not increase the information content of data, it can greatly improve the visual appearance of spectra (Figure 1.11D).

1.10.5 Apodization

The appearance of an NMR spectrum can often be improved when the Fourier transformation is combined with additional processing steps. Zero filling was shown to be an effective tool to enhance the digital resolution and visualize all features of NMR spectra (Figure 1.11). Apodization (or time-domain filtering) is another tool used prior to Fourier transformation to enhance the SNR and/or spectral resolution of spectra. During apodization the time-domain FID signal, $f(t)$, is multiplied with a filter function, $f_{\text{filter}}(t)$. Common filter functions are exponential weighting

$$f_{\text{filter}}(t) = e^{-t/T_L} \tag{1.29}$$

or a Lorentzian-to-Gaussian transformation

$$f_{\text{filter}}(t) = e^{+t/T_L} e^{-t^2/T_G^2} \tag{1.30}$$

where T_L and T_G are user-adjustable parameters. A decreasing mono-exponential filter function according to Eq. (1.29) with $T_L > 0$ improves the SNR of the spectrum since the data points at the end of the FID containing primarily noise are attenuated (Figure 1.12B). An unavoidable side effect is that the filter function causes the FID signal to decay faster than the natural T_2^*, leading to a faster, apparent $T_{2,\text{app}}^*$ relaxation time constant $((T_{2,\text{app}}^*)^{-1} = (T_2^*)^{-1} + (T_L)^{-1})$ and hence broader resonance lines (Eq. (1.29), Figure 1.12B). As a result, time-domain apodization with a decreasing exponential function is colloquially referred to as "line broadening the spectrum."

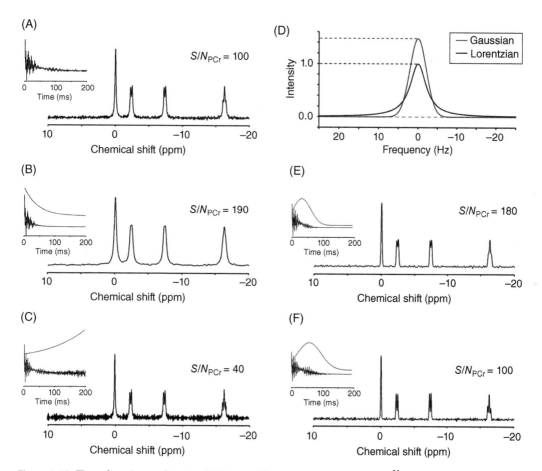

Figure 1.12 **Time-domain apodization.** (A) Time and frequency domain signal of ^{31}P-containing phantom (PCr and ATP) processed without apodization. (B) Time-domain multiplication with a decaying exponential function leads to increased SNR at the expense of spectral resolution. (C) Apodization with an increasing exponential improves the spectral resolution but comes with an SNR penalty. (D) Gaussian and Lorentzian line shapes with identical line widths (at half maximum) and integrated intensities. (E, F) Lorentzian-to-Gaussian transformation with T_L and T_G parameters optimized to provide (E) improved SNR at identical spectral resolution compared to (A) and (F) improved spectral resolution while maintaining the same SNR.

When sufficient data has been recorded to minimize truncation artifacts ($T_{acq} > 3T_2^*$), then optimal sensitivity is obtained by using a so-called matched filter in which $T_L = T_2^*$. The improved SNR comes at the expense of a doubling of the spectral line widths, i.e. spectral resolution has been traded for sensitivity. Besides improving the SNR, this apodization can also be used on FIDs where the last data points have been truncated resulting in oscillatory artifacts in the frequency domain. For $T_L < 0$ the apodization according to Eq. (1.29) leads to a resolution enhancement, since the apparent T_2^* becomes longer, resulting in line narrowing (Figure 1.12C). However, the SNR is decreased since the data points at the end of the FID with a relatively high noise contribution are emphasized.

The Lorentzian-to-Gaussian filtering function (Eq. (1.30)) converts a Lorentzian line shape to a Gaussian line shape. A Gaussian line shape decays to the baseline in a narrower frequency range as would a Lorentzian line shape with the same line width at half height. In other words, a Lorentzian line shape produces longer "tails" that are a disadvantage when accurate determination (by integration) of overlapping resonance lines is required (Figure 1.12D). The principle of the Lorentzian-to-Gaussian transformation is to cancel (or decrease) the Lorentzian part of the FID (by multiplying with $\exp(+t/T_L)$, where $T_L = T_2^*$, such that $\exp(+t/T_L) \cdot \exp(-t/T_2^*) = 1$) while increasing the Gaussian character of the FID (by multiplying with $\exp(-t^2/T_G^2)$). Using a sufficiently long T_G value, significant resolution enhancement can be achieved. Figure 1.12E shows the process of time-domain filtering on a ^{31}P FID. Even though an increasing exponential filter (Figure 1.12C) and a Lorentzian-to-Gaussian transformation can achieve the same resolution enhancement, the former is accompanied by a significant decrease in sensitivity, which can be minimized with a Lorentzian-to-Gaussian transformation. In fact, the two adjustable parameters in Eq. (1.30) can be used to improve the SNR without a significant decrease in spectral resolution (Figure 1.12E) or to improve the spectral resolution without a significant decrease in sensitivity (Figure 1.12F). Besides the mentioned, most commonly used apodization functions, a wide range of other functions are available each with specific characteristics regarding sensitivity and resolution [51].

1.11 Quantum Description of NMR

The description of NMR presented so far has largely followed the classical treatment laid out by Bloch [2–4]. Besides the fact that nuclear spin is an intrinsic quantum property, the classical vector model did not rely on quantum mechanics and was able to provide a physically intuitive description of NMR. However, NMR can also be approached from a quantum mechanical point of view as was done by Purcell and colleagues [1]. One of the fundamentals of quantum mechanics is that energy, momentum, and other properties are quantized, meaning that they are restricted to specific, discrete values. Transitions between discrete energy levels can be investigated with spectroscopy, which in general terms is the study of the interaction between matter and electromagnetic radiation. For example, IR spectroscopy investigates energy-level transitions associated with different vibrational modes (Figure 1.13A). As the name implies, IR spectroscopy uses electromagnetic waves in the IR part of the electromagnetic spectrum (10^{12}–10^{14} Hz). Different vibrational modes are characterized by quantized energy levels given by $E = (n + \frac{1}{2})h\upsilon$, where n is an integer quantum number ($n = 0, 1, 2, ...$) corresponding to the separate levels. There are two primary modes to "sample" the energy-level differences, by absorption or by emission. Molecules that vibrate at a lower energy level with quantum number n can move to a higher energy level with quantum number ($n + 1$) by *absorption* of electromagnetic energy with a frequency υ that exactly matches the energy-level difference ΔE, according to $\Delta E = h\upsilon$. Conversely, a molecule with high vibrational energy can drop down to a lower vibrational

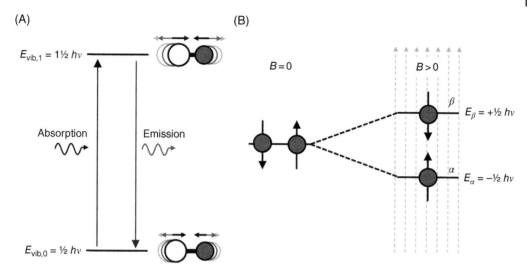

Figure 1.13 **Quantum-mechanical view of spectroscopy.** (A) A diatomic molecule can vibrate at different, but quantized frequencies. The energy levels associated with quantized vibrational modes can be investigated through the absorption or emission of electromagnetic radiation with a frequency that exactly matches the energy-level difference (i.e. $\nu = \Delta E/h$). Transitions in vibrational modes require frequencies in the infrared (IR) region of the electromagnetic spectrum (10^{12}–10^{14} Hz), hence giving rise to IR spectroscopy. (B) Quantum-mechanical view of NMR. In the absence of an external magnetic field ($B = 0$), the spin states are degenerate and no energy-level difference exists. In the presence of an external magnetic field ($B > 0$), the antiparallel (β) spin state is quantized at a higher energy level than the parallel (α) spin state. In analogy with IR spectroscopy in (A), the energy-level difference ΔE can be studied with electromagnetic waves of specific frequencies that match ΔE. For NMR, the frequencies are in the radiofrequency (RF) part of the electromagnetic spectrum (10^6–10^9 Hz).

energy level through the *emission* of electromagnetic waves of specific frequency ν. The IR spectrum of a compound contains unique frequencies related to the various vibrational modes allowed within its structure. IR spectroscopy is routinely used as an analytical tool for the identification of compounds or groups within compounds.

Intrinsic angular momentum, or spin, is also quantized both in amplitude and orientation. In the quantum mechanical picture of NMR spins are either in the parallel (α) or antiparallel (β) orientation with respect to the external magnetic field (Figure 1.13B). In the absence of an external magnetic field, the energy levels of the α and β spin-state are equal or degenerate. Application of an external magnetic field breaks the degeneracy, leading to different energy levels for the two spin orientations. This is referred to as the Zeeman effect, after Pieter Zeeman who in 1897 described the splitting of spectral lines of light in the presence of a magnetic field [52]. As shown in Figure 1.13B, the discrete energy levels associated with NMR are similar to that of any other quantum system (e.g. Figure 1.13A) and can thus be studied by spectroscopy. The spin states are characterized by energy $E = -mh\nu$ with m representing the spin quantum number associated with α ($m = \frac{1}{2}$) and β ($m = -\frac{1}{2}$) spin orientations. Similar to IR spectroscopy (Figure 1.13A), NMR transitions are only allowed between energy levels for which the spin quantum number m changes by ± 1 (Figure 1.13B). Emission is highly unlikely in NMR, as the energy levels are very close making the chance for spontaneous emission of a photon essentially zero. In the first observation of NMR by Purcell [1], the system was sampled by absorption of radiofrequency (RF) energy by protons in a strong magnetic field. Whereas both Bloch and Purcell detected the NMR phenomenon, their methodology and conceptualization were so different that it was not immediately obvious that they were observing the same effect.

In the quantum mechanical picture it may appear that all spins are either in the α or β spin state and that no other orientations are allowed. However, from earlier sections using classical arguments it is known that in a typical macroscopic sample most of the spins are in random orientations. The quantum mechanical view and the classical view can be reconciled to a certain degree by knowing that spins can, besides the pure α and β quantum states, also be in so-called superposition states. A superposition state is a spin state in which the pure or eigen α and β spin states are mixed and in which a spin can therefore attain any orientation. When an individual spin would be observed, as in the classical Stern–Gerlach experiment [53], the mixed or superposition state "collapses" into one of the pure α or β spin states. This forms one of the pillars of quantum mechanics. However, since NMR does not observe single spins, the collapse into pure spin states never happens and the spins within the sample remain in a superposition state throughout the experiment [35]. Since mixed states are simply a combination of the eigen or pure states, a quantum description of NMR typically uses the pure or eigen states to simplify the calculation.

In the case of noninteracting spins, like the protons in water, the quantum mechanical description of NMR presented in Figure 1.13 does not provide any additional insights. In fact it often causes confusion in the reconciliation between the spin-orientation sphere of Figures 1.2–1.4 and the pure α and β spin states of Figure 1.13. In the case of noninteracting spins it is therefore strongly advised to use the intuitive, classical description of NMR. However, in the case of interacting spins, as described in the next section, the classical picture will prove to be incomplete and the quantum description is required for a complete quantitative description.

1.12   Scalar Coupling

The NMR resonance frequencies, or chemical shifts, give direct information about the chemical environment of nuclei, thereby greatly aiding in the unambiguous detection and assignment of compounds. The integrated resonance area is, in principle (see Figure 1.8), directly proportional to the concentration of the compounds, thereby making NMR a quantitative technique. An additional feature that can be observed in NMR spectra is the splitting of resonances into several smaller lines, a phenomenon often referred to as scalar coupling, J coupling, or spin–spin coupling. Scalar coupling was first observed by Proctor and Yu [54] with ^{121}Sb/^{123}Sb NMR on an aqueous solution of $NaSbF_6$ and by Gutowsky and McCall [55] with ^{19}F and ^{31}P NMR on liquid phosphorus halides. Using the previously developed spin-echo method [56], Hahn and Maxwell [57] described scalar coupling as an additional modulation on the spin-echo envelope. Ramsey and Purcell [58] and Gutowsky et al. [59] subsequently provided a firm theoretical description.

Scalar coupling originates from the fact that nuclei with magnetic moments can influence each other, besides directly through space (dipolar coupling) also through electrons in chemical bonds (scalar coupling). Even though dipolar interactions are the main mechanism for relaxation in a liquid, there is no net interaction between nuclei since rapid molecular tumbling averages the dipolar interactions to zero. However, interactions through chemical bonds do not average to zero and give rise to the phenomenon of scalar coupling. Scalar coupling is a quantum effect that cannot be understood with the classical vector model alone. In Chapter 8 the classical description will be extended into a correlated vector model that can describe scalar coupling in a physically intuitive framework. For the remainder of this chapter scalar coupling will be explored within the quantum description of NMR.

Consider an isolated proton and an isolated carbon-13 atom as depicted in Figure 1.14A. Electrons in orbitals with an s character have a finite probability of being *inside* the nucleus, an example of a quantum effect without an exact classical analog. The Fermi contact governs the interaction between the nuclear and electron spins within the nucleus and energetically

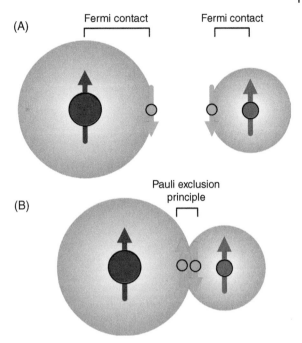

Figure 1.14 Nuclear and electron spin interactions. (A) In isolated atoms, the Fermi contact energetically favors an antiparallel orientation between nuclear and electronic spins. Proton, carbon-13 and electron spins are shown in blue, red and green respectively. (B) In chemical bonds, the Pauli exclusion principle demands that the electron spins are in an antiparallel orientation thereby potentially forcing nuclear and electron spins in an energetically higher parallel orientation, depending on the nuclear spin state.

favors an antiparallel over a parallel arrangement. In terms of energy-level diagrams, the ^{1}H and ^{13}C spins can be drawn as two separate, two-level diagrams (Figure 1.15A) or be combined into a single, four-level diagram (Figure 1.15B) with four energy levels, corresponding to the four nuclear spin combinations. In both cases, the energy level transitions in which the proton spin-orientation changes correspond to a resonance line at the proton frequency, whereas a spin flip for the ^{13}C nucleus corresponds to a resonance line in the ^{13}C NMR spectrum. The four allowed energy-level transitions (for which the spin quantum number m changes by ±1) give rise to two resonance frequencies, ν_H and ν_C, in agreement with the equivalent diagrams shown in Figure 1.15A.

Now consider the situation where the proton and carbon-13 nuclei are covalently bound, as for example in [1-^{13}C]-glucose (Figure 1.14B). The interaction between the two electrons inside a chemical bond is governed by the Pauli exclusion principle, which demands that the electron spins have an antiparallel orientation. When both nuclear spins are antiparallel to the external magnetic field B_0, i.e. the high-energy $\beta\beta$ state, the two bonding electrons cannot both be antiparallel to the nuclear spins, leading to an energetically less favorable state (Figure 1.15C). The $\beta\beta$ energy level increases by an amount proportional to $^1J_{CH}/4$ where $^1J_{CH}$ is the one-bond, heteronuclear scalar coupling constant expressed in Hz. Similar arguments can be used to describe the energy increase for the $\alpha\alpha$ state. However, for the $\alpha\beta$ and $\beta\alpha$ states both electron spins can be antiparallel to the nuclear spins leading to an energetically more favorable situation and a drop in energy level by an amount $^1J_{HC}/4$. In other words, an antiparallel orientation of the two nuclear spins is preferred for scalar coupling over a single bond. The energy level diagram for a scalar coupled two-spin-system still only allows four transitions, but they now correspond to four different frequencies at $\nu_H + ^1J_{CH}/2$ and $\nu_H - ^1J_{CH}/2$ on the proton channel and at $\nu_C + ^1J_{CH}/2$ and $\nu_C - ^1J_{CH}/2$ on the carbon-13 channel. Each of the resonances has been divided into two new resonances of equal intensity separated by $^1J_{CH}$, giving rise to the NMR spectrum shown in Figure 1.16.

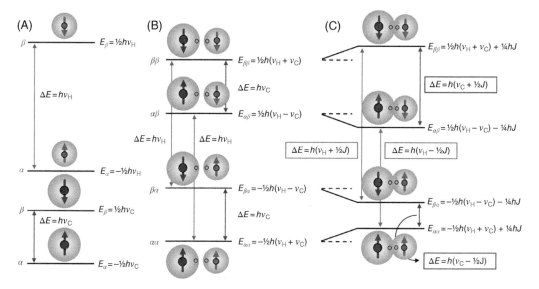

Figure 1.15 The effect of scalar coupling on the energy levels of a two-spin-system. (A) Energy-level diagram for isolated carbon-13 (red) and hydrogen (blue) nuclear spins drawn (A) separately or (B) combined. In the combined mode the energy levels simply represent the sum of the respective proton and carbon-13 energy levels. In both cases the blue and red energy-level differences correspond to proton and carbon-13 transitions of frequencies ν_H and ν_C, respectively. (C) Energy-level diagram for hydrogen and carbon-13 atoms attached via a chemical bond. The $\beta\beta$ spin state (i.e. the nuclear spin for both ^{13}C and ^{1}H is in the β state) becomes energetically less favorable as one of the two nuclear–electronic spin orientations is forced to be parallel (electron spins are shown in green). The same is true for the $\alpha\alpha$ spin-state, whereas in the $\alpha\beta$ and $\beta\alpha$ spin-states all spin orientations can be antiparallel. The shift in energy levels results in different frequencies for the proton ($\nu_H \pm J/2$) and carbon-13 ($\nu_C \pm J/2$) transitions. In other words, the single frequency ^{1}H and ^{13}C resonances in the absence of scalar coupling (B) are split into two resonances of half intensity, separated by the J coupling constant in the presence of scalar coupling (C) (see also Figure 1.16).

Similar arguments can be used to explain scalar coupling over two or three chemical bonds. In that case Hund's rule comes into play which states that the orbitals of a subshell are occupied with single electrons of parallel orientation before double occupation occurs. In the case of a sp^3 hybridization of the orbitals around a carbon nucleus, the electrons occupying each of the four orbital lobes are all in a parallel configuration. This fact, together with the Fermi contact interaction and the Pauli exclusion principle, make it that the $\alpha\alpha$ and $\beta\beta$ spin states between two protons attached to the carbon nucleus become the energetically favorable condition, moving down by an amount $^1J_{CH}/4$ (which is equivalent to moving up by an amount $-^1J_{CH}/4$). In other words, the sign of the scalar coupling constant becomes negative for scalar coupling over two bonds during which the parallel orientation between the nuclear spins becomes preferable. Using similar arguments it can be shown that the scalar coupling constant over three bonds becomes positive, whereby an antiparallel orientation between the two nuclear spins takes preference. The scalar coupling constant rapidly decreases with increasing number of chemical bonds and can typically be ignored for four or more bonds. The scalar coupling constant is independent of the applied external magnetic field, since it is based on the fundamental principle of spin–spin pairing and is therefore expressed in Hertz. Typical magnitudes of scalar coupling constants are, ^{1}H–^{1}H (1–15 Hz), ^{1}H–^{13}C (100–200 Hz), ^{1}H–^{15}N (70–110 Hz), ^{1}H–^{31}P (10–20 Hz), ^{13}C–^{13}C (30–80 Hz), and ^{31}P–O–^{31}P (15–20 Hz). All scalar coupling constants are for one chemical bond, except for ^{1}H–^{1}H and ^{31}P–O–^{31}P interactions which stretch over three and two bonds, respectively.

Figure 1.16 Resonance splitting due to scalar coupling. Scalar coupling between carbon-13 and hydrogen nuclei leads to a splitting of the singlet resonances into so-called doublet resonances. The resonances at the lower and higher frequencies are associated with energy-level transitions in which the nuclear spin of the scalar-coupling partner is in the α and β spin-state, respectively.

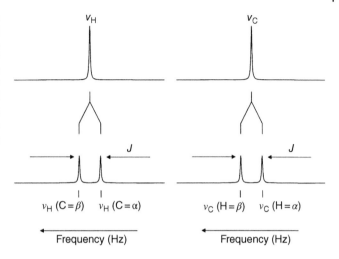

Figure 1.16 Resonance splitting due to scalar coupling. Scalar coupling between carbon-13 and hydrogen nuclei leads to a splitting of the singlet resonances into so-called doublet resonances. The resonances at the lower and higher frequencies are associated with energy-level transitions in which the nuclear spin of the scalar-coupling partner is in the α and β spin-state, respectively.

The spectrum shown in Figure 1.16 is only obtained when the frequency difference between the two scalar-coupled spins is much larger than the scalar coupling constant between them. For a heteronuclear interaction as shown in Figure 1.14 this requirement is certainly valid, as the frequency difference at magnetic fields commonly used for NMR is typically several tens of MHz, while the heteronuclear scalar coupling constant is less than 200 Hz. When the absolute frequency difference between two spins, $|\nu_1 - \nu_2|$, is much larger than their scalar coupling constant, J_{12}, the two spins form a weakly-coupled spin-system and the corresponding NMR spectrum is often referred to as a first-order spectrum. However, for many homonuclear interactions the frequency difference $|\nu_1 - \nu_2|$ is on the same order of magnitude as the homonuclear scalar coupling constant J_{12}, giving rise to so-called strongly-coupled spin-systems. In a strongly-coupled two-spin-system, the $\alpha\beta$ and $\beta\alpha$ spin-states become mixed such that a transition does no longer correspond to either spin 1 or 2 (as is the case in Figure 1.16), but contains contributions from both spins. As a result of this mixing of spin-states, the simple four-resonance-line spectrum (Figure 1.16) becomes more complicated as shown in Figure 1.17A. Strongly-coupled spin-systems produce so-called second-order spectra that are characterized by features not present in first-order spectra. Most noticeably from Figure 1.17A is the so-called "roof effect" in which a line from the outer to the inner resonances forms an imaginary roof. This effect is another feature of NMR spectra that indicates that two multiplets belong to the same molecule and can therefore aid in the identification of compounds. In the case of a two-spin-system, strong coupling only leads to a distortion of the resonance intensities and frequencies. However, for more complicated spin-systems strong coupling can lead to additional resonances (Figure 1.17B). Prime examples of strongly-coupled spin-systems *in vivo* are glutamate and glutamine.

It should be realized that while the behavior and spectral appearance of strongly-coupled spin-systems is no longer intuitive, it can still be quantitatively calculated through the use of the density matrix formalism, as will be discussed in Chapter 8. Note that the "roof effect" is also visible for more complicated spin-systems, as evident in Figure 1.17B.

1.13 Chemical and Magnetic Equivalence

The complexity of an NMR spectrum depends on the number of spins in different chemical environments within a chemical compound. Molecular symmetry can result in two or more spins having the exact same environment and hence having the same chemical shift. These

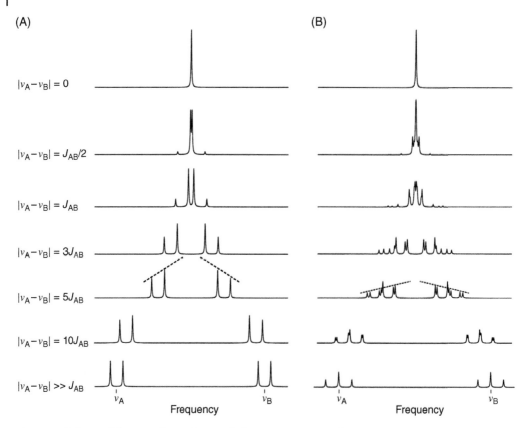

Figure 1.17 Spectral features for scalar-coupled spin-systems. Simulated (A) *AB* and (B) A_2B_2 NMR spectra showing the effects of varying the ratio of the scalar coupling constant to the frequency difference between the *A* and *B* resonances. The lower NMR spectra are indicative of weakly-coupled *AX* and A_2X_2 spin-systems, producing a first-order NMR spectrum, while the higher spectra are indicative of strongly-coupled *AB* and A_2B_2 spin-systems, displaying strong second-order effects, like the appearance of additional resonances, as well as the so-called "roof effect" (dotted lines). See text for details on spin-system nomenclature.

nuclei are said to be chemically equivalent. All physical and chemical properties, like reaction rates, exchange processes, and NMR chemical shift, are the same for chemically equivalent nuclei. Figure 1.18A and B shows an example of how a small chemical change can lead to a reduction in molecular symmetry and hence an increase in spectral complexity. *Scyllo*-inositol (Figure 1.18A) is characterized by six protons attached to a six-member carbon ring in which each proton is in a *trans* configuration relative to the adjacent protons. As a result, *scyllo*-inositol has a high degree of molecular symmetry making it that all six protons have identical chemical environments, leading to a single resonance line. *Myo*-inositol is an isomer of *scyllo*-inositol in which the proton on carbon position 2 is in a *cis* configuration with respect to its neighbors. This seemingly small change reduces the molecular symmetry of *myo*-inositol to a single plane. The remaining symmetry gives rise to four types of protons in four distinct chemical environments resulting in four NMR resonances (ignoring scalar coupling for the moment).

In order to understand the splitting pattern due to scalar coupling it is important to discriminate between magnetically equivalent and nonequivalent nuclei. Magnetic equivalence is a stronger version of chemical equivalence whereby nuclei have to satisfy an additional criterion; Nuclei are magnetically equivalent if they have identical scalar couplings with all other

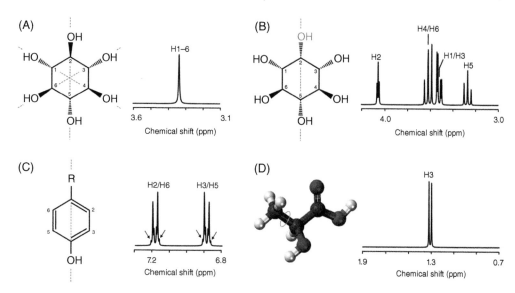

Figure 1.18 Chemical and magnetic equivalence. (A) Structure and ^{1}H NMR spectra of *scyllo*-inositol. The high degree of molecular symmetry makes all six protons chemically and magnetically equivalent, leading to the absence of scalar coupling. (B) Structure and ^{1}H NMR spectra of *myo*-inositol. The reduced molecular symmetry compared to *scyllo*-inositol leads to limited chemical equivalence and no magnetic equivalence. (C) Structure and ^{1}H NMR spectra of tyrosine, whereby R represents an alanine group. Protons 2/6 and 3/5 are chemical equivalent, but since the scalar coupling with other protons is not identical, they are not magnetically equivalent. Chemical equivalence without magnetic equivalence leads to a second-order ^{1}H NMR spectrum (arrows). (D) Structure and ^{1}H NMR spectra of lactate. Rapid rotation around the C2—C3 bond leads to chemical and magnetic equivalence of the three methyl protons, leading to a first-order ^{1}H NMR spectrum.

nonequivalent nuclei in the molecule. Magnetically equivalent nuclei do not produce an observable splitting among themselves, whereas magnetically nonequivalent (but chemically equivalent) nuclei always produce a second-order, strongly-coupled spectrum.

In *scyllo*-inositol (Figure 1.18A) all six protons are chemically equivalent and since there are no other nonequivalent nuclei in the molecule the six protons are also magnetically equivalent. This leads to a single resonance line without splitting due to scalar coupling. In *myo*-inositol (Fig. 1.18B), protons 1 and 3 are chemical equivalent. However, they are not magnetically equivalent since the scalar coupling with a nucleus outside the chemically equivalent group, e.g. proton 4, is not equal for protons 1 and 3 (i.e. $J_{14} = 0\,Hz$ and $J_{34} = 10\,Hz$, making $J_{14} \neq J_{34}$). As a result, protons 1 and 3 *do* split among themselves due to scalar coupling. However, scalar coupling over four bonds in a non-aromatic ring is typically negligible and the splitting between protons 1 and 3 is not observed.

A similar situation arises in the aromatic ring of tyrosine (Figure 1.18C). Due to molecular symmetry protons 2 and 6 and protons 3 and 5 are chemically equivalent. However, since the scalar coupling between protons 2 and 3 and protons 6 and 3 are different, protons 2 and 6 are not magnetically equivalent. Since a four-bond coupling in an aromatic ring can often be readily detected, the NMR signal from protons 2 and 6 displays splitting due to scalar coupling which is automatically second-order since the chemical shifts are identical.

Chemical equivalence can also be achieved in the presence of rapid internal mobility. Consider the methyl protons of lactic acid (Figure 1.18D). The lack of molecular symmetry would suggest the absence of chemical equivalence. However, as the methyl group undergoes rapid rotation around the C2—C3 bond, each proton experiences a chemical environment that is averaged over all orientations. As the average chemical environment is identical for the three protons, the protons in a methyl group are chemically equivalent. Since the coupling with the methine proton outside the methyl group is identical for all three protons, the methyl group is also magnetically equivalent.

A simple method of describing the effects of chemical and magnetic equivalence for any compound is provided by the Pople notation [60], after Nobel laureate John Pople. In the Pople notation, nuclei are indicated by letters, whereby the difference in chemical shift relative to the J-coupling constant is mirrored by the separation of the letters in the alphabet. Chemically equivalent nuclei are indicated by the same letter, whereby magnetically inequivalent nuclei are distinguished with a prime (e.g. A and A'). For the example shown in Figure 1.18, the Pople notations would be A_6 for *scyllo*-inositol and $AA'BB'MX$ for *myo*-inositol. Tyrosine is indicated as $AA'BB'$ and lactate as AX_3.

In the case of multiple, weakly-coupled spins the scalar coupling between each spin pair can be applied consecutively to arrive at the final splitting pattern. For example, in an AMX spin-system ($J_{AX} = 0$) the M spin couples with the A spin forming a doublet with scalar coupling J_{AM}. The M spin also couples with the X spin, such that each of the two M doublet lines splits again in two lines separated by J_{MX}, forming a final four-resonance, doublet-of-doublets signal. The presence of magnetically equivalent nuclei can lead to substantial simplification of the resulting NMR spectrum. If the aforementioned AMX spin-system is transformed to a MX_2 spin-system with two magnetically equivalent X spins, the final splitting pattern for the M spin would be composed of three signals; a doublet between M and X starts off the splitting pattern, after which each M signal is split again due to interaction with the second X spin. Since the two

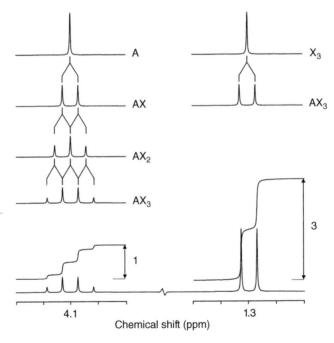

Figure 1.19 **^{1}H NMR spectrum of lactate.** The method of successive splitting for an AX_3 spin system (e.g. lactate, with a methyl group resonating at 1.31 ppm and a methine proton resonating at 4.10 ppm). The A resonance splits successively in a doublet, a triplet, and a quartet, while the X resonance splits into a doublet. Note that the binomial distribution (e.g. 1 : 3 : 3 : 1 for a quartet) only arises for magnetically equivalent nuclei which have an identical scalar coupling constant with a common coupling partner. Since the multiplet resonances at 4.10 and 1.31 ppm originate from one and three protons, respectively, the relative integrated areas of the two multiplets are therefore also one and three.

J-coupling constants are identical, the center peak gains double intensity resulting in a three-resonance, triplet signal with relative intensities of $1:2:1$ among its resonances. The splitting pattern of spin *A* in an AX_n spin system (where *n* is the number of magnetically equivalent spins) is simply given by a binomial distribution in which the lines are separated by the scalar coupling constant (e.g. $n = 3$ results in a four-resonance quartet signal with amplitudes in a $1:3:3:1$ ratio among its resonances).

Using these rules the appearance of all first-order spectra can be predicted. For example, Figure 1.19 shows the ^{1}H spectrum of lactic acid (lactate). Lactate can be seen as an AX_3 spin system, i.e. three magnetically equivalent methyl protons coupled to a nonequivalent single methine proton. Generally, the carbonyl and hydroxyl protons are invisible due to rapid exchange with water protons. Since the magnetically equivalent methyl protons do not produce any splitting among themselves, they only "feel" the methine proton, resulting in a doublet signal. The methine proton experiences three spins with an identical scalar coupling constant, resulting in four lines (a quartet) with a $1:3:3:1$ binomial signal distribution. All the signals in the doublet and in the quartet are separated by the same scalar coupling constant. The relative integrated amplitude of the peaks at 1.31 and 4.10 ppm is $3:1$, respectively, since there are three methyl protons versus one methine proton.

Exercises

1.1 A 2-l water-filled sphere ($T = 298.15$ K) is placed on a bench outside a 3.0 T MR magnet.

    A Calculate the Larmor frequency of the proton spins outside the MR magnet (hint: Earth's magnetic field = 40 μT = 0.4 G).

    B Calculate the Larmor frequency of the proton spins after the sphere is placed inside the MR magnet.

    C For spins with a T_1 relaxation time constant of 4 s, calculate how much time is needed to reach 99% of the thermal equilibrium magnetization M_0 following placement of the water-filled sphere inside the MR magnet.

1.2 Derive the Bloch equations in the laboratory frame in the absence of relaxation (Eqs. (1.9)–(1.11)) from Eq. (1.8).

1.3 Show that free precession of the transverse magnetization according to

$$M_x(t) = M_x(0)\cos \omega t + M_y(0)\sin \omega t$$

and

$$M_y(t) = M_y(0)\cos \omega t - M_x(0)\sin \omega t$$

is a solution of the Bloch equations in the laboratory frame (Eqs. (1.9)–(1.11)) in the absence of a perturbing magnetic RF field.

1.4 **A** Derive the Bloch equations in the rotating frame (Eqs. (1.15)–(1.17)) from the Bloch equations in the laboratory frame (Eqs. (1.9)–(1.14)).

    B Show that Eqs. (1.18) and (1.19) are solutions of the Bloch equations in the rotating frame (assume that $T_2 = T_2^*$).

1.5 **A** Derive the expression for the full line width at half maximum (FWHM) for the absorption component of a Lorentzian line (e.g. Eq. (1.23)).
B Derive the expression for the FWHM for the magnitude component of a Lorentzian line.
C Derive the expression for the absorption and dispersion parts of a resonance line originating from a full spin-echo (as opposed to an FID).
D Determine the peak heights and integrals of (the absorption component of) Lorentzian lines originating from an FID and a full spin-echo.

1.6 Longitudinal magnetization can be excited into the transverse plane by a 90° (or $\pi/2$) pulse.
A Starting from the Bloch equations in the rotating frame, derive an expression for the conversion of longitudinal magnetization M_z into transverse magnetization M_y by a RF pulse of length T and amplitude B_1 applied on-resonance along the x' axis. Show how the nutation angle depends on the pulse amplitude and length.
B If the pulse length of the 90° pulse is 1.0 ms, what is the required B_1 magnitude in μT to achieve excitation?
C How many Larmor precession cycles will occur in the laboratory frame at $B_0 = 3.0$ T during the 90° excitation pulse?

1.7 On a high-resolution NMR magnet (500 MHz for protons) signal is acquired as 16 384 points over a spectral width of 10 kHz. The sample consists of creatine in 99% D_2O and is acquired without water suppression. The magnetic field homogeneity is high, providing a spectral line width of 0.7 Hz for the creatine methyl signal. The relative peak heights of the creatine methylene (CH_2) and methyl (CH_3) signals are 1.2 and 3.0. Provide a likely explanation for the deviation from the expected 2 : 3 peak height ratio.

1.8 In a properly executed spin-echo sequence, the resonances of all (uncoupled) spins appear with the same relative phase. The absolute phase of all resonances can be made zero by a simple zero-order phase correction.
A Calculate the phase difference between the creatine methyl (3.03 ppm) and NAA methyl (2.01 ppm) proton resonances at 7.05 T in the presence of a 500 μs timing error.
B In a proton spectrum acquired at 3.0 T (water is on-resonance), the choline methyl (3.22 ppm) and NAA methyl (2.01 ppm) resonances appear with relative phases of 30° and 210°, respectively. Calculate the required zero- and first-order phase corrections to properly phase the spectrum to pure absorption lines.

1.9 Given the Gaussian line shape:

$$F_G(\omega) = \sqrt{\frac{\pi}{4}} M_0 T_{2G} e^{-(\omega_0 - \omega)^2 T_{2G}^2 / 4}$$

A Find the expression for the full width at half maximum.
B For single Lorentzian and Gaussian resonance lines of equal line width and area, calculate the signal height-to-noise advantage of a Gaussian line (assuming equal noise levels).
C For single Lorentzian and Gaussian resonance lines of equal line width and area, calculate the line width advantage of a Gaussian line at 10% of the respective peak heights.

1.10 Consider a (hypothetical) ^{1}H NMR spectrum with the following five resonance:

Resonance 1: Triplet resonance ($^3J_{HH}$ = 7 Hz) at 1.1 ppm with relative intensity (as determined by numerical integration) of 307.

Resonance 2: Quartet resonance ($^3J_{HH} = 7\,$Hz) at 3.9 ppm with relative intensity 198.
Resonance 3: Doublet-of-doublets ($^3J_{HH} = 11$ and 8 Hz) at 7.2 ppm with relative intensity 102.
Resonance 4: Doublet resonance ($^3J_{HH} = 8\,$Hz) at 8.5 ppm with relative intensity 105.
Resonance 5: Doublet resonance ($^3J_{HH} = 11\,$Hz) at 10.0 ppm with relative intensity 96.

With the knowledge that the ^{1}H NMR spectrum originates from an organic molecule $C_5H_8O_2$, determine the complete chemical structure of the compound.

1.11 Consider a weakly-coupled four-spin-system AMX_2 with chemical shift positions given by $\delta_A = 5.0\,$ppm, $\delta_M = 1.5\,$ppm, and $\delta_X = 3.0\,$ppm.
 A Sketch the NMR spectrum for this compound when $J_{AM} = 20\,$Hz, $J_{MX} = 10\,$Hz, and $J_{AX} = 0\,$Hz. Assume equal T_1 and T_2 characteristics for all resonances.
 B Sketch the NMR spectrum for this compound when $J_{AM} = 20\,$Hz, $J_{MX} = 10\,$Hz, and $J_{AX} = 5\,$Hz. Assume equal T_1 and T_2 characteristics for all resonances.
 C When the NMR spectrum is acquired with a pulse-acquire sequence (90° nutation angle, 500 ms repetition time TR, number of averages 8192) sketch the NMR spectrum for this compound when $J_{AM} = 0\,$Hz, $J_{MX} = 10\,$Hz, and $J_{AX} = 0\,$Hz and $T_{1A} = 5.0\,$s, $T_{1M} = 1.0\,$s, and $T_{1X} = 2.0\,$s. Assume equal T_2 characteristics for all resonances.

1.12 A proton NMR signal is acquired as 512 complex points during an acquisition time of 102.4 ms.
 A Determine the spectral width of the experiment.
 B Determine the apparent spectral frequency position of a signal with a frequency of +3800 Hz.
 C Determine the apparent spectral frequency position of a signal with a frequency of −16 000 Hz.

1.13 In a proton NMR spectrum the resonance from DSS is detected at 170.345 213 MHz. Two other resonances occur at 170.345 453 and 170.354 668 MHz, respectively.
 A Calculate the chemical shifts of the two resonances in ppm.
 B Calculate the frequencies of the two compounds at 7.05 T when DSS appears at a frequency of 300.176 544 MHz.

1.14 The time-domain data from a sample consists of three sinusoidal functions ($M_0 = 150$, 300, 200) oscillating at different frequencies (250, 300, and 500 Hz) and decaying at different rates ($T_2 = 50$, 50, and 100 ms).
 A Sketch the Fourier transform spectrum acquired from the sample when the initial phase is zero for all resonances. Indicate line widths (in Hz) and relative peaks heights.
 B Sketch the Fourier transform spectrum acquired from the sample when the initial phases are 0°, 90°, and 135°, respectively.

1.15 Show that the real and imaginary frequency domain signals given by Eqs. (1.21) and (1.22) reduce to pure absorption and dispersion signals following a zero-order phase correction with $\phi_c = \phi$.

1.16 Hund's rule states that if two or more empty orbitals are available, electrons occupy each with spins parallel until all orbitals have one electron. When chemical bonds in sp^3 hybridized structures are considered, describe the signs of $^2J_{HH}$ and $^3J_{HH}$ relative to $^1J_{CH}$ using similar arguments as used for Figure 1.15.

References

1 Purcell, E.M., Torrey, H.C., and Pound, R.V. (1946). Resonance absorption by nuclear magnetic moments in a solid. *Phys. Rev.* 69: 37–38.

2 Bloch, F., Hansen, W.W., and Packard, M.E. (1946). Nuclear induction. *Phys. Rev.* 69: 127.

3 Bloch, F., Hansen, W.W., and Packard, M.E. (1946). The nuclear induction experiment. *Phys. Rev.* 70: 474–485.

4 Bloch, F. (1946). Nuclear induction. *Phys Rev* 70: 460–473.

5 Bloch, F. (1964). The principle of nuclear induction. In: *Nobel Lecturers in Physics 1942–1962*, 203–216. New York: Elsevier.

6 Purcell, E.M. (1964). Research in nuclear magnetism. In: *Nobel Lectures in Physics 1942–1962*, 219–231. New York: Elsevier.

7 Proctor, W.G. and Yu, F.C. (1950). The dependence of a nuclear magnetic resonance frequency upon chemical compound. *Phys. Rev.* 77: 717.

8 Dickinson, W.C. (1950). Dependence of the F19 nuclear resonance position on chemical compound. *Phys. Rev.* 77: 736.

9 Arnold, J.T., Dharmatti, S.S., and Packard, M.E. (1951). Chemical effects on nuclear induction signals from organic compounds. *J. Chem. Phys.* 19: 507.

10 Ernst, R.R. and Anderson, W.A. (1966). Applications of Fourier transform spectroscopy to magnetic resonance. *Rev. Sci. Instrum.* 37: 93–102.

11 Damadian, R. (1971). Tumor detection by nuclear magnetic resonance. *Science* 171: 1151–1153.

12 Moon, R.B. and Richards, J.H. (1973). Determination of intracellular pH by ^{31}P magnetic resonance. *J. Biol. Chem.* 248: 7276–7278.

13 Hoult, D.I., Busby, S.J., Gadian, D.G. et al. (1974). Observation of tissue metabolites using ^{31}P nuclear magnetic resonance. *Nature* 252: 285–287.

14 Behar, K.L., den Hollander, J.A., Stromski, M.E. et al. (1983). High-resolution ^{1}H nuclear magnetic resonance study of cerebral hypoxia *in vivo*. *Proc. Natl. Acad. Sci. U. S. A.* 80: 4945–4948.

15 Bottomley, P.A., Edelstein, W.A., Foster, T.H., and Adams, W.A. (1985). *In vivo* solvent-suppressed localized hydrogen nuclear magnetic resonance spectroscopy: a window to metabolism? *Proc. Natl. Acad. Sci. U. S. A.* 82: 2148–2152.

16 Aue, W.P., Bartholdi, E., and Ernst, R.R. (1976). Two-dimensional spectroscopy. Application to nuclear magnetic resonance. *J. Chem. Phys.* 64: 2229–2246.

17 Jeener, J. (1971). Ampere Summer School. Basko Polje, Yugoslavia.

18 Ernst, R.R. (1997). Nuclear magnetic resonance fourier transform spectroscopy. In: *Nobel Lectures, Chemistry 1991–1995*, 12–57. Singapore: World Scientific Publishing.

19 Wuthrich, K. (2003). NMR studies of structure and function of biological macromolecules. In: *Les Prix Nobel 2002*, 235–267. Sweden: Almquist and Wiksell International.

20 Lauterbur, P.C. (1973). Image formation by induced local interactions: examples employing nuclear magnetic resonance. *Nature* 242: 190–191.

21 Mansfield, P. and Grannell, P.K. (1973). NMR "diffraction" in solids? *J. Phys. C: Solid State Phys.* 6: L422–L427.

22 Lauterbur, P.C. (2004). All science is interdisciplinary – from magnetic moments to molecules to men. In: *The Nobel Prizes 2003*, 245–251. Stockholm: Nobel Foundation.

23 Mansfield, P. (2004). Snap-shot MRI. In: *The Nobel Prizes 2003*, 266–283. Stockholm: Nobel Foundation.

24 Haacke, E.M., Brown, R.W., Thompson, M.R., and Venkatesan, R. (1999). *Magnetic Resonance Imaging. Physical Principles and Sequence Design*. New York: Wiley-Liss.

25 Stark, D.D. and Bradley, W.G. (1999). *Magnetic Resonance Imaging*. New York: C. V. Mosby.
26 Bernstein, M.A., King, K.F., and Zhou, X.J. (2004). *Handbook of MRI Pulse Sequences*. Amsterdam: Academic Press.
27 Rabi, I.I. (1937). Space quantization in a gyrating magnetic field. *Phys. Rev.* 51: 652–654.
28 Rabi, I.I., Zacharias, J.R., Millman, S., and Kusch, P. (1938). A new method of measuring nuclear magnetic moment. *Phys. Rev.* 63: 318.
29 Rigden, J.S. (1986). Quantum states and precession: the two discoveries of NMR. *Rev. Mod. Phys.* 58: 433–448.
30 Becker, E.D., Fisk, C.L., and Khetrapal, C.L. (2007). Development of NMR: from the early beginnings to the early 1990s. *eMagRes* doi: 10.1002/9780470034590.emrhp0001.
31 Olson, J.A., Nordell, K.J., Chesnik, M.A. et al. (2000). Simple and inexpensive classroom demonstrations of nuclear magnetic resonance and magnetic resonance imaging. *J. Chem. Ed.* 77: 882–889.
32 Dirac, P.A.M. (1928). The quantum theory of the electron. *Proc. Roy. Soc. A* 117: 610–624.
33 Lowry, D.F. (1994). Correlated vector model of multiple-spin systems. *Concepts Magn. Reson.* 6: 25–39.
34 Levitt, M.H. (2005). *Spin Dynamics. Basics of Nuclear Magnetic Resonance*. New York: Wiley.
35 Hanson, L.G. (2008). Is quantum mechanics necessary for understanding magnetic resonance? *Concepts Magn. Reson.* 32: 329–340.
36 Larmor, J. (1897). On the theory of the magnetic influence on spectra; and on the radiation from moving ions. *Philos. Mag.* 44: 503–512.
37 Gorter, C.J. (1936). Negative result of an attempt to detect nuclear magnetic spins. *Physica* 3: 995–998.
38 Gorter, C.J. and Broers, L.J.F. (1942). Negative result of an attempt to observe nuclear magnetic resonance in solids. *Physica* 9: 591–596.
39 Hoult, D.I. (1989). The magnetic resonance myth of radio waves. *Concepts Magn. Reson.* 1: 1–5.
40 Hoult, D.I. (2009). The origin and present status of the radio wave controversy in NMR. *Concepts Magn. Reson.* 34: 193–216.
41 Bloch, F. and Siegert, A. (1940). Magnetic resonance for nonrotating fields. *Phys. Rev.* 57: 522–527.
42 Torrey, H.C. (1956). Bloch equations with diffusion terms. *Phys. Rev.* 104: 563–565.
43 McConnell, H.M. (1958). Reaction rates by nuclear magnetic resonance. *J. Chem. Phys.* 28: 430–431.
44 Wishart, D.S., Bigam, C.G., Yao, J. et al. (1995). ^{1}H, ^{13}C and ^{15}N chemical shift referencing in biomolecular NMR. *J. Biol. NMR* 6: 135–140.
45 Harris, R.K., Becker, E.D., Cabral de Menezes, S.M. et al. (2002). NMR nomenclature: nuclear spin properties and conventions for chemical shifts. IUPAC recommendations 2001. *Magn. Reson. Chem.* 40: 489–505.
46 Nyquist, H. (1928). Certain topics in telegraph transmission theory. *Trans. IEEE* 47: 617–644.
47 Bracewell, R.M. (1965). *The Fourier Transform and Its Applications*. New York: McGraw-Hill.
48 Ernst, R.R. (1966). *Sensitivity Enhancement in Magnetic Resonance*, Advances in Magnetic Resonance, vol. 2. New York: Academic Press.
49 Traficante, D.D. (1991). Time averaging, does the noise really average towards zero? *Concepts Magn. Reson.* 3: 83–87.
50 Cooley, J.W. and Tukey, J.W. (1965). An algorithm for machine calculation of complex Fourier series. *Math. Comput.* 19: 297–301.
51 Ernst, R.R., Bodenhausen, G., and Wokaun, A. (1987). *Principles of Nuclear Magnetic Resonance in One and Two Dimensions*. Oxford: Clarendon Press.

52 Zeeman, P. (1897). The effect of magnetisation on the nature of light emitted by a substance. *Nature* 55: 347.

53 Gerlach, W. and Stern, O. (1922). Der experimentelle nachweis des magnetischen moments des silberatoms. *Z. Phys.* 8: 110–111.

54 Proctor, W.G. and Yu, F.C. (1951). On the nuclear magnetic moments of several stable isotopes. *Phys. Rev.* 81: 20–30.

55 Gutowsky, H.S. and McCall, D.W. (1951). Nuclear magnetic resonance fine structure in liquids. *Phys. Rev.* 82: 748–749.

56 Hahn, E.L. (1950). Spin echoes. *Phys. Rev.* 80: 580–594.

57 Hahn, E.L. and Maxwell, D.E. (1952). Spin echo measurements of nuclear spin coupling in molecules. *Phys. Rev.* 88: 1070–1086.

58 Ramsey, N.F. and Purcell, E.M. (1952). Interactions between nuclear spins in molecules. *Phys. Rev.* 85: 143–144.

59 Gutowsky, H.S., McCall, D.W., and Slichter, C.P. (1953). Nuclear magnetic resonance multiplets in liquids. *J. Chem. Phys.* 21: 279–292.

60 Bernstein, H.J., Pople, J.A., and Schneider, W.G. (1957). The analysis of nuclear magnetic resonance spectra. *Can. J. Chem.* 35: 65–81.

2

In Vivo NMR Spectroscopy – Static Aspects

2.1 Introduction

Magnetic resonance spectroscopy (MRS) is feasible on any nucleus possessing nuclear spin. For *in vivo* MRS applications the most commonly observed nuclei which have this property are proton (^{1}H), carbon-13 (^{13}C), phosphorus (^{31}P), and sodium (^{23}Na). Even though the number of relevant nuclei is limited, each nucleus provides a wealth of information, since a large number of metabolites can be detected simultaneously. ^{1}H MRS allows the detection of a number of important amino acids, such as glutamate, GABA, and aspartate and related compounds, like glutamine, as well as the end product of glycolysis, lactate. ^{13}C MRS offers the possibility to study noninvasively the fluxes through important metabolic pathways, like the tricarboxylic acid (TCA) cycle, *in vivo*. ^{31}P MRS provides information about energetically important metabolites, intracellular pH, magnesium concentration, and reaction fluxes. Besides the mentioned nuclei, *in vivo* MRS is in special cases also relevant on other nuclei, like deuterium (^{2}H), helium (^{3}He), lithium (^{7}Li), nitrogen (^{15}N), oxygen (^{17}O), fluorine (^{19}F), silicon (^{29}Si), potassium (^{39}K), and xenon (^{129}Xe).

This chapter will review the static aspects of MRS, like chemical shifts, scalar coupling constants, and concentrations. The next chapter will focus more on the dynamic processes underlying MRS, like T_1 and T_2 relaxation, diffusion, and chemical exchange.

2.2 Proton NMR Spectroscopy

The proton nucleus is, besides the low abundance, radioactive hydrogen isotope tritium, the most sensitive nucleus for NMR both in terms of intrinsic NMR sensitivity (high gyromagnetic ratio) and high natural abundance (>99.9%). Since nearly all metabolites contain protons, *in vivo* ^{1}H NMR spectroscopy is, in principle, a powerful technique to observe, identify, and quantify a large number of biologically important metabolites in intact tissue.

However, the application of ^{1}H MRS to intact tissues *in vivo* is challenging for a number of reasons. Firstly, the water resonance is several orders of magnitude larger than the signals originating from low-concentration metabolites. Secondly, other large signals like extracranial lipids can also overwhelm the small metabolite signals and thirdly, heterogeneous magnetic field distributions significantly decrease the spectral resolution. Therefore, water suppression and spatial localization are prerequisites for meaningful *in vivo* ^{1}H MRS studies and will be discussed in detail in Chapter 6. Chapter 10 will deal with methods to optimize the magnetic field homogeneity. An inherent limitation of ^{1}H MRS is the narrow chemical shift range of circa

In Vivo NMR Spectroscopy: Principles and Techniques, Third Edition. Robin A. de Graaf.
© 2019 John Wiley & Sons Ltd. Published 2019 by John Wiley & Sons Ltd.

5 ppm for non-exchangeable protons. This causes a large number of metabolite resonances to overlap, making their detection and quantification difficult. Separation of metabolites by spectral editing is discussed in Chapter 8, while the quantification of spectra is detailed in Chapter 9. Finally, even though the proton nucleus is the most sensitive for NMR studies, NMR is in general a very insensitive technique, making the detection of low-concentration metabolites a compromise between time resolution and signal-to-noise ratio.

Despite the challenges, *in vivo* ^{1}H MRS is a powerful technique to observe a large number of biologically relevant metabolites. On a state-of-the-art MR system, up to circa 15–20 different metabolites can be extracted simultaneously from short-TE ^{1}H NMR spectra of normal brain [1]. While the next chapters deal with the acquisition and processing of NMR spectra, this chapter will discuss the information content of ^{1}H NMR spectra. In particular, the structure, NMR characteristics (scalar coupling constants/patterns and chemical shifts), and metabolic functions of the major resonances appearing in *in vivo* ^{1}H NMR spectra will be discussed.

Figure 2.1 shows a typical short-TE ^{1}H NMR spectrum acquired from rat brain *in vivo* at 11.7 T. The identification of the various metabolites is based on a variety of methods, including prior biochemical knowledge, systemic perturbations, physiological alterations (i.e. diseased states), and information obtained from *in vitro* biophysical techniques like gas-chromatography mass spectrometry, high-resolution liquid-state NMR, and biochemical assays. In the next sections the NMR characteristics of the most commonly encountered metabolites will be discussed.

All chemical shifts and scalar coupling constants mentioned throughout this chapter are taken from the literature [2, 3] or are measured on 25 mM solutions at 500 MHz, 310 K, and pH = 7.0. The chemical shifts are referenced against nine equivalent protons of 2,2-dimethyl-2-silapentane-5-sulfonate (DSS) at 0.00 ppm [4]. All simulated ^{1}H NMR spectra are calculated as pulse-acquire spectra at 300 MHz with a line width of 1 Hz. All chemical structures are drawn in neutral acid form, although most compounds will occur in a charged or zwitter ion form at neutral pH. Chemical names provided are the common name and the official name according to guidelines established by the International Union of Pure and Applied Chemistry (IUPAC, [5]). Many biological compounds have chiral centers that lead to two enantiomers (or optical isomers)

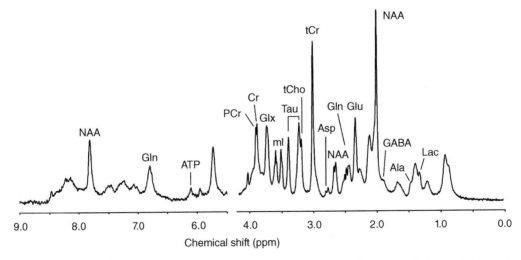

Figure 2.1 **Short-TE ^{1}H MRS of brain.** LASER-localized ^{1}H NMR spectrum from rat brain *in vivo* (11.74 T, TR/TE = 4000/12 ms, 100 μl) acquired without direct water suppression (see Chapter 6). The excellent spectral resolution obtainable at high magnetic field strengths allows the visual separation of >15 metabolites.

that are related by reflection. Enantiomers will be denoted by the older, but commonly used *D/L* system, as well as by the more modern and preferred *R/S* system. An educational overview of enantiomer nomenclature is given by Slocum et al. [6]. The labeling of carbon positions within a chemical structure follows the functional group priorities as defined by the IUPAC [5]. A noticeable exception is that separate numbering is used for subparts of larger structures. For example, the six carbon positions of *N*-acetyl aspartate (NAA) are labeled from 1 to 4 for the aspartate moiety and 1 to 2 for the acetate moiety (Table 2.1).

Table 2.1 ^{1}H chemical shifts and scalar coupling constants for metabolites commonly detected in ^{1}H NMR spectra of normal or diseased cerebral tissues *in vivo* or *in vitro*.

Compound	Chemical group	Chemical shift (ppm)^a	Multiplicity^b	Interaction	Scalar coupling (Hz)^c
Acetate	$^2CH_3$	1.904	s	—	—
Acetoacetate	$^2CH_2$	3.432	s	—	—
	$^4CH_3$	2.270	s	—	—
Acetone	$^2CH_3$	2.222	s	—	—
N-acetyl aspartate (NAA)					
Acetyl moiety	$^2CH_3$	2.008	s	—	—
Aspartate moiety	^2CH	4.382	dd	2–3	3.86
	$^3CH_2$	2.673	dd	2–3′	9.82
		2.486	dd	3–3′	−15.59
	NH	7.820	d	2–NH	7.90
N-acetyl aspartyl glutamate (NAAG)					
Acetyl moiety	$^2CH_3$	2.042	s	—	—
Aspartyl moiety	^2CH	4.607	dd	2–3	4.41
	$^3CH_2$	2.721	dd	2–3′	9.52
		2.519	dd	3–3′	−15.91
	NH	8.260		2–NH	7.32
Glutamate moiety	^2CH	4.128	dd	2–3 / 2–3′	4.61 / 8.42
	$^3CH_2$	1.881	m	3–3′	−14.28
		2.049		3–4 / 3–4′	10.56 / 6.09
	$^4CH_2$	2.190	m	3′–4 / 3′–4′	4.90 / 11.11
		2.180		4–4′	−15.28
	NH	7.950		2–NH	7.46
Adenosine triphosphate (ATP)					
Ribose moiety	^1CH	6.127	d	1–2	5.70
	^2CH	4.796	dd	2–3	5.30
	^3CH	4.616	dd	3–4	3.80
	^4CH	4.396	dd	4–5 / 4–5′	3.00 / 3.10

(Continued)

Table 2.1 (Continued)

Compound	Chemical group	Chemical shift (ppm)^a	Multiplicity^b	Interaction	Scalar coupling (Hz)^c
	$^5CH_2$	4.295	m	5–5′	−11.80
		4.206	m	4–P	1.90
				5–P / 5′–P	6.50 / 4.90
Adenine moiety	^2CH	8.224	s	—	—
	^8CH	8.514	s	—	—
	NH	6.755	s	—	—
Alanine	^2CH	3.775	q	2–3	7.23
	$^3CH_3$	1.467	d		
γ-Aminobutyric acid (GABA)	$^2CH_2$	2.287	t	2–3 / 2–3′	7.66 / 7.10
		2.287		2′–3 / 2′–3′	7.06 / 7.69
	$^3CH_2$	1.892	m	3–3′	−10.82
		1.895		3–4 / 3–4′	5.70 / 8.02
	$^4CH_2$	3.003	t	3′–4 / 3′–4′	10.03 / 6.53
		3.005		4–4′	−9.93
Ascorbic acid (vitamin C)	^4CH	4.492	d	4–5	2.07
	^5CH	4.002	m	5–6	6.00
	$^6CH_2$	3.743	m	5–6′	7.60
		3.716		6–6′	−11.50
Aspartate	^2CH	3.891	dd	2–3	3.65
	$^3CH_2$	2.804	dd	2–3′	9.11
		2.670	dd	3–3′	−17.43
Choline	$(CH_3)_3$	3.185	s	—	—
	$^1CH_2$	4.054	m	1–2 / 1–2′	3.14 / 6.98
				1′–2 / 1′–2′	7.01 / 3.17
	$^2CH_2$	3.501	m	1–1′	−14.1
				2–2′	−14.1
Creatine	CH_3	3.027	s	—	—
	CH_2	3.913	s	—	—
	NH	6.65	s	—	—
Ethanol	$^1CH_2$	3.645	q	1–2	7.08
	$^2CH_3$	1.177	t		
Ethanolamine	$^1CH_2$	3.818	m	1–2 / 1′–2′	3.85
				1′–2 / 1–2′	6.75
	$^2CH_2$	3.147	m	1–1′ / 2–2′	−10.2
Glucose, α-anomer	^1CH	5.216	d	1–2	3.80
	^2CH	3.519	dd	2–3	9.60
	^3CH	3.698	dd	3–4	9.40
	^4CH	3.395	dd	4–5	9.90

Table 2.1 (Continued)

Compound	Chemical group	Chemical shift (ppm)^a	Multiplicity^b	Interaction	Scalar coupling (Hz)^c
	^{5}CH	3.822	m	5–6	1.50
	^{6}CH	3.826	dd	5–6′	6.0
	$^{6′}$CH	3.749	dd	6–6′	−12.1
Glucose, β-anomer	^{1}CH	4.630	d	1–2	8.00
	^{2}CH	3.230	dd	2–3	9.10
	^{3}CH	3.473	dd	3–4	9.40
	^{4}CH	3.387	dd	4–5	8.90
	^{5}CH	3.450	m	5–6	1.60
	^{6}CH	3.882	dd	5–6′	5.40
	$^{6′}$CH	3.707	dd	6–6′	−12.3
Glutamate	^{2}CH	3.748	dd	2–3 / 2–3′	7.33 / 4.65
	^{3}CH$_2$	2.046	m	3–3′	−14.76
		2.120		3–4 / 3–4′	6.28 / 8.70
	^{4}CH$_2$	2.334	m	3′–4 / 3′–4′	8.77 / 6.73
		2.352		4–4′	−16.03
Glutamine	^{2}CH	3.767	dd	2–3 / 2–3′	6.71 / 5.91
	^{3}CH$_2$	2.121	m	3–3′	−14.45
		2.137		3–4 / 3–4′	6.18 / 9.40
	^{4}CH$_2$	2.431	m	3′–4 / 3′–4′	9.34 / 6.32
		2.458		4–4′	−15.61
Glutathione					
Glycine moiety	^{2}CH$_2$	3.769	s	—	—
	NH	7.154	s	—	—
Cysteine moiety	^{2}CH	4.561	dd	2–3	7.09
	^{3}CH$_2$	2.926	dd	2–3′	4.71
		2.975		3–3′	−14.06
	NH	8.177	s	—	—
Glutamate moiety	^{2}CH	3.769	dd	2–3 / 2–3′	6.34 / 6.36
	^{3}CH$_2$	2.146	m	3–3′	−15.48
		2.159		3–4 / 3–4′	6.70 / 7.60
	^{4}CH$_2$	2.510	m	3′–4 / 3′–4′	7.60 / 6.70
		2.560		4–4′	−15.92
Glycerol	^{1}CH$_2$	3.552	dd	1–2 / 2–3	4.43
		3.640	dd	1′–2 / 2–3′	6.49
	^{2}CH	3.770	m	1–1′ / 3–3′	−11.72
	^{3}CH$_2$	3.552	dd		
		3.640	dd		

(*Continued*)

Table 2.1 (Continued)

Compound	Chemical group	Chemical shift (ppm)[a]	Multiplicity[b]	Interaction	Scalar coupling (Hz)[c]
Glycerophosphocholine					
Glycerol moiety	$^1CH_2$	3.605	dd	1–2 / 1'–2	5.77 / 4.53
		3.672	dd	1–1'	−14.78
	^2CH	3.903	m	2–3 / 2–3'	5.77 / 4.53
	$^3CH_2$	3.871	m	3–3'	−14.78
		3.946	m	3, 3'–P	6.03
Phosphocholine moiety	$(CH_3)_3$	3.212	s	—	—
	$^1CH_2$	4.312	m	1–2 / 1–2'	3.10 / 5.90
				1'–2 / 1'–2'	5.90 / 3.10
	$^2CH_2$	3.659	m	1–1' / 2–2'	−9.32
				1, 1'–P	6.03
Glycine	$^2CH_2$	3.547	s	—	—
Histidine					
Alanine moiety	^2CH	3.975	dd	2–3	7.92
	$^3CH_2$	3.120	dd	2–3'	4.81
		3.211	dd	3–3'	−15.5
Imidazole moiety	^2CH	7.79[d]	s	—	—
	^4CH	7.06[d]	s	—	—
Homocarnosine					
GABA moiety	$^2CH_2$	2.944	m	2–2'	−12.5
		2.969	m	2–3 / 2–3'	8.0
	$^3CH_2$	1.881	m	2'–3 / 2'–3'	7.5
		1.896	m	3–3'	−13.9
	$^4CH_2$	2.348	m	3–4 / 3–4'	7.5
		2.378	m	3'–4/3'–4'	7.5
				4–4'	−15.2
Alanine moiety	^2CH	4.467	dd	2–3	8.64
	$^3CH_2$	3.013	dd	2–3'	5.02
		3.191	dd	3–3'	−15.3
Imidazole moiety	^2CH	8.08[d]	s	—	—
	^4CH	7.08[d]	s	—	—
β-Hydroxy-butyrate (BHB)	$^2CH_2$	2.294	dd	2–2'	−14.5
		2.388	dd	2–3	6.30
	^3CH	4.133	m	2'–3	7.30
	$^4CH_3$	1.186	d	3–4	6.28
2-Hydroxyglutarate (2HG)	^2CH	4.008	dd	2–3 / 2–3'	7.65 / 4.04
	$^3CH_2$	1.822	m	3–3'	−14.00
		1.980	m	3–4 / 3–4'	5.25 / 10.66

Table 2.1 (Continued)

Compound	Chemical group	Chemical shift (ppm)[a]	Multiplicity[b]	Interaction	Scalar coupling (Hz)[c]
	$^4CH_2$	2.221	m	3′–4 / 3′–4′	10.84 / 5.86
		2.270	m	4–4′	−14.91
Myo-inositol	^1CH	3.522	dd	1–2	2.89
	^2CH	4.054	dd	2–3	3.01
	^3CH	3.522	dd	3–4	10.00
	^4CH	3.614	dd	4–5	9.49
	^5CH	3.269	dd	5–6	9.48
	^6CH	3.614	dd	1–6	10.00
Scyllo-inositol	^{1-6}CH	3.340	s	—	—
Isoleucine	^2CH	3.663	d	2–3	3.97
	^3CH	1.970	m	3–4 / 3–4′	9.34 / 4.78
	$^4CH_2$	1.250	m	3–6	7.00
		1.458		4–4′	−13.5
	$^5CH_3$	0.928	t	4, 4′–5	7.41
	$^6CH_3$	0.999	d		
Lactate	^2CH	4.097	q	2–3	6.93
	$^3CH_3$	1.313	d		
Leucine	^2CH	3.724	dd	2–3 / 2–3′	n.d.
	$^3CH_2$	1.669	m	3–3′	n.d.
		1.736	m	3–4 / 3′–4	n.d.
	^4CH	1.7	m	4–5 / 4–5′	6.48
	$^5CH_3$	0.943	d		
	$^{5'}CH_3$	0.955	d		
Methanol	$^1CH_3$	3.341	s	—	—
Nicotinamide adenine dinucleotide (NAD$^+$)					
Adenine moiety	^2CH	8.184	s	—	—
	^8CH	8.415	s	—	—
Nicotinamide moiety	^2CH	9.334	m	2–4 / 2–5	1.3 / 0.7
	^4CH	8.849	m	2–6 / 2–N	0.5 / 1.8
	^5CH	8.210	m	4–5 / 4–6	8.1 / 1.8
	^6CH	9.158	m	5–6 / 6–N	6.3 / 1.1
Ribose (adenine)	$^{1'}CH$	6.040	d	1′–2′ / 2′–3′	5.9 / 5.2
	$^{2'}CH$	4.749	m	3′–4′ / 4′–5′	3.6 / 2.7
	$^{3'}CH$	4.498	m	4′–5″ / 4′–P	3.4 / 2.2
	$^{4'}CH$	4.364	m	5′–5″	−11.7
	$^{5'}CH_2$	4.242	m	5′–P / 5″–P	4.8 / 5.3
		4.196	m		

(*Continued*)

Table 2.1 (Continued)

Compound	Chemical group	Chemical shift (ppm)[a]	Multiplicity[b]	Interaction	Scalar coupling (Hz)[c]
Ribose (nicotinamide)	^1CH	**6.091**	d	1–2 / 2–3	5.5 / 5.1
	^2CH	**4.494**	m	3–4 / 4–5	2.9 / 2.4
	^3CH	**4.432**	m	4–5' / 4–P	2.5 / 2.9
	^4CH	**4.543**	m	5–5'	−11.9
	$^5CH_2$	**4.353**	m	5–P/5'–P	4.3 / 5.5
		4.228	m		
Phenylalanine					
Alanine moiety	^2CH	3.975	dd	2–3 / 2–3'	8.01 / 5.21
	$^3CH_2$	3.105	dd	3–3'	−14.57
		3.273	dd		
Phenyl moiety	^1CH	**7.369**	m	1–2 / 1–2'	7.2 / 7.5
	^2CH	**7.420**	m	1–3 / 1–3'	1.6 / 1.0
	^3CH	**7.322**	m	2–3 / 2–3'	7.9 / 0.5
				2'–3 / 2'–3'	0.5 / 7.4
				3–3'	1.4
Phosphocreatine	CH_3	3.029	s	—	—
	CH_2	3.930	s	—	—
	NH	**6.58 / 7.30**	s	—	—
Phosphocholine	$(CH_3)_3$	3.209	s	—	—
	$^1CH_2$	**4.282**	m	1–2 / 1–2'	2.28 / 7.23
			m	1'–2 / 1'–2'	7.33 / 2.24
	$^2CH_2$	3.643	m	1–1' / 2–2'	−14.9 / −14.2
			m	1, 1'–P	6.27
Phosphoethanolamine	$^1CH_2$	3.977	m	1–2 / 1–2'	3.18 / 6.72
			m	1'–2 / 1'–2'	7.20 / 2.98
	$^2CH_2$	3.216	m	1–1' / 2–2'	−14.6 / −14.7
			m	1, 1'–P	7.18
Pyruvate	$^3CH_3$	2.358	s	—	—
Serine	^2CH	3.835	dd	2–3 / 2–3'	5.98
	$^3CH_2$	3.937	dd	3–3'	3.56
		3.976	dd		−12.3
Succinate	$^2CH_2$	2.394	s	—	—
Taurine	$^1CH_2$	3.420	dd	1–2 / 1–2'	6.74 / 6.46
	$^2CH_2$	3.246	dd	1'–2 / 1'–2'	6.40 / 6.79
				1–1' / 2–2'	−12.4 / −12.9
Threonine	^2CH	3.578	d	2–3	4.92
	^3CH	**4.246**	m	3–4	6.35
	$^4CH_3$	**1.316**	d		

Table 2.1 (Continued)

Compound	Chemical group	Chemical shift (ppm)^a	Multiplicity^b	Interaction	Scalar coupling (Hz)^c
Tryptophan					
Alanine moiety	^{2}CH	4.047	dd	2–3 / 2–3′	8.15 / 4.85
	^{3}CH$_2$	3.290	dd	3–3′	–15.37
		3.475	dd		
Indole moiety	^{2}CH	7.312	s		
	^{4}CH	7.726	m	4–5 / 4–6	7.60 / 1.00
	^{5}CH	7.278	m	4–7	0.95
	^{6}CH	7.197	m	5–6 / 5–7	7.51 / 1.20
	^{7}CH	7.536	m	6–7	7.68
Tyrosine					
Alanine moiety	^{2}CH	3.928	dd	2–3 / 2–3′	7.88 / 5.15
	^{3}CH$_2$	3.037	dd	3–3′	–14.73
		3.192	dd		
Phenol moiety	^{2}CH	6.890	m	2–3 / 2–3′	7.98 / 0.46
	^{3}CH	7.186	m	2′–3 / 2′–3′	0.31 / 8.65
				2–2′ / 3–3′	2.45 / 2.54
Valine	^{2}CH	3.595	d	2–3	4.41
	^{3}CH	2.259	m	3–4	6.97
	^{4}CH$_3$	1.028	d	3–4′	7.07
	$^{4'}$CH$_3$	0.977	d		

^{a}Chemical shifts are color-coded according to the chemical shift ranges 0–2 ppm (red), 2–3 ppm (orange), 3–4 ppm (green), 4–5 ppm (blue), and downfield from water (purple).
^{b}Multiplicities are defined as: singlet (s), doublet (d), triplet (t), quartet (q), multiplet (m), and double doublet (dd).
^{c}Heteronuclear ^{1}H–^{14}N couplings are mentioned in the text when relevant.
^{d}Sensitive to pH in the physiological range.
n.d., not determined.

The World Wide Web provides a wealth of resources regarding NMR spectra from pure compounds, chemical naming conventions, and general information on chemicals. Among others, useful websites for compound identification and assignment by NMR are the Human Metabolome Database (HMDB, www.hmdb.ca), the Spectral Database for Organic Compounds (SDBS, https://sdbs.db.aist.go.jp/sdbs/cgi-bin/cre_index.cgi), the Biological Magnetic Resonance Data Bank (BMRB, www.bmrb.wisc.edu), and NMRShiftDB (https://nmrshiftdb.nmr.uni-koeln.de). In addition, a number of websites offer extended NMR databases for purchase.

Other useful websites on small chemical structure and function are the Chemical Entities of Biological Interest (ChEBI, www.ebi.ac.uk/chebi), PubChem (pubchem.ncbi.nlm.nih.gov), and ChemSpider (www.chemspider.com).

2.2.1 Acetate (Ace)

Acetic acid (Ace; $C_2H_4O_2$; molar mass, 60.05 g mol^{-1}; pK_a, 4.76) or its anionic form, acetate, is a small molecule containing a single methyl group that provides a singlet resonance at 1.90 ppm, directly overlapping with a multiplet of GABA-H3 at 1.89 ppm. Acetate is under normal conditions not observed in *in vivo* ^{1}H NMR spectra of brain. However, acetate is readily taken up by the brain and observed by ^{1}H NMR when the plasma acetate levels are raised by intravenous infusion. This is of particular importance for studies on energy and neurotransmitter metabolism by ^{13}C NMR, since the absence of neuronal acetate transporters leads to a selective brain uptake by the astroglia [7]. This has been used to confirm and extend earlier [1-^{13}C]-glucose studies on brain energy metabolism and allows more sensitive detection of astroglial metabolism [8, 9]. More discussion on acetate metabolism can be found in Chapter 3.

The presence of acetate in ^{1}H NMR spectra of tissue extracts can be indicative of postmortem breakdown of NAA and other compounds containing an acetyl group. When accompanied by typical markers of ongoing postmortem metabolism (high level of lactate, absence of phosphocreatine), the acetate signal should be interpreted with caution. If lyophilization is one of the steps in making the tissue extract, acetate may be lost due to its high volatility.

2.2.2 *N*-Acetyl Aspartate (NAA)

N-Acetyl aspartic acid (NAA, $C_6H_9NO_5$; molar mass, 175.14 g mol^{-1}; IUPAC name, (2*S*)-2-Acetamidobutanedioic acid; pK_a, 3.14) or the anionic form of NAA is the second most concentrated metabolite in the brain after the amino acid glutamate. The *L* or (*S*) isomer of NAA is the biologically prevalent form. In ^{1}H NMR spectra of normal brain tissue the most prominent resonance originates from the methyl group of NAA at 2.01 ppm (Figure 2.2A). Smaller resonances appear as doublet-of-doublets at 2.49, 2.67, and 4.38 ppm corresponding to the protons

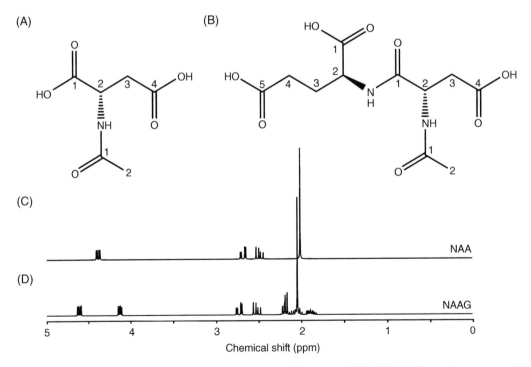

Figure 2.2 **N-acetyl aspartate (NAA) and N-acetyl aspartyl glutamate (NAAG).** (A, B) Chemical structure and (C, D) simulated ^{1}H NMR spectrum of (A, C) NAA and (B, D) NAAG. The exchangeable amide protons give broad resonances around 8 ppm downfield from water (not shown).

of the aspartate CH_2 and CH groups. The amide NH proton is exchangeable with water and gives a broad, temperature-sensitive resonance at 7.82 ppm.

NAA is exclusively localized in the central and peripheral nervous systems. Its concentration varies in different parts of the brain [10–14] and undergoes large developmental changes [15–17], increasing in rat brain from <1 mM at birth to >5 mM in adulthood [18, 19]. Even though there is a substantial amount of literature on the synthesis, distribution, and possible function [20–23], the exact function of NAA remains largely unknown. Suggested functions for NAA include osmoregulation [24] and a precursor of the neurotransmitter NAAG [25]. Others have suggested the involvement of NAA in fatty acid and myelin synthesis [26, 27] by acting as a storage form for acetyl groups. ^{13}C NMR spectroscopy offers the possibility of measuring NAA turnover directly in human [28] and rat [29] brain *in vivo*. Using [1-^{13}C]-glucose as the substrate, NAA turnover is extremely slow with a time constant of circa 14 hours [29], thereby supporting the notion that NAA is not significantly involved in glucose brain energy metabolism at rest.

For *in vivo* ^1H MRS applications, NAA has played two roles. Firstly, the NAA resonance has been used as a marker of neuronal density. This is supported by the observation of a decrease in NAA intensity in disorders which are accompanied by neuronal loss, such as the chronic stages of stroke [30, 31], tumors [32–35], and multiple sclerosis [36, 37]. However, care is required when using NAA as a neuronal density marker since NAA concentrations differ among neuron types [38] and it has also been found in other cells, like immature oligodendrocytes [39]. Dynamic changes of neuronal NAA concentrations have been observed, indicating that NAA levels may reflect neuronal dysfunction rather than neuronal loss. This is substantiated by a recovery of NAA levels during incomplete reversible ischemia [40] and brain injury [41]. Furthermore, reduced NAA levels have been observed in multiple sclerosis in the absence of neuronal loss [23]. Canavan's syndrome [42, 43] and possibly sickle cell disease [44, 45] in newborn infants and young children are the only pathologies known to date that are accompanied by a permanent increase in NAA intensity.

Secondly, NAA has been used as a concentration marker, since its concentration is relatively immune to acute metabolic disturbances such as ischemia or hypoxia. A disadvantage of using NAA as an internal reference is that the concentration is not uniform over the entire brain, with higher concentrations in gray matter (~8–11 mM) as compared to white matter (~6–9 mM) [10–14]. Furthermore, at shorter echo-times NAA is overlapping with glutamate and macromolecules. In addition, resonances from other *N*-acetyl containing metabolites, such as *N*-acetyl aspartyl glutamate (NAAG), will essentially coincide with the NAA methyl signal, thereby further complicating the NAA quantification [46]. Figure 2.3 gives the cerebral concentration of NAA, together with those of other metabolites. The values tabulated in Figure 2.3 represent the concentration ranges found in the literature and should be used with caution, as they include a range of different techniques, conditions, and brain areas (e.g. gray and white matter concentrations are not separated) performed by researchers in different laboratories. As such, Figure 2.3 should be used as a rough guideline for the expected metabolite concentrations, rather than indicating the exact values.

2.2.3 *N*-Acetyl Aspartyl Glutamate (NAAG)

N-Acetyl aspartyl glutamic acid (NAAG; $C_{11}H_{16}N_2O_8$; molar mass, 304.26 g mol^{-1}; IUPAC name, 2-(3-carboxy-2-acetamidopropanamido)-pentanedioic acid) is a dipeptide of *N*-substituted L-aspartate and L-glutamate with a nonuniform distribution within the brain across a concentration range of 0.6–3.0 mM [47–51]. NAAG is suggested to be involved in excitatory neurotransmission as well as a source of glutamate. However, its exact role remains unclear.

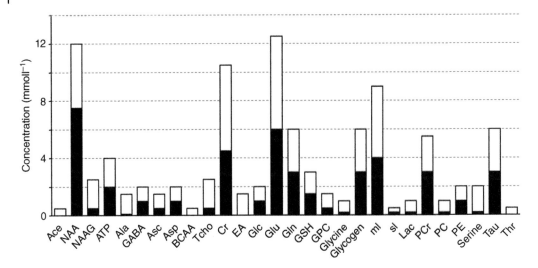

Figure 2.3 Cerebral metabolite concentration ranges. Concentration ranges indicate literature values for normal adult human and rat brain. Solid black and open white bars represent the approximate minimum and maximum reported concentration values in the literature. Note that the wide ranges are not indicative of the metabolite detection accuracy, but are rather a reflection of the range of methods, tissues types, and laboratories.

NMR detection of NAAG was first reported in the human brain by Frahm et al. [46] and confirmed on rat brain extracts by Holowenko et al. [52], after which regional variations across the human brain were established [51]. The largest resonance of NAAG at 2.04 ppm resonates very close to the larger methyl resonance from NAA at 2.01 ppm (Figure 2.2). This requires excellent magnetic field homogeneity to unambiguously detect NAAG separately from NAA. In cases where this separation cannot be achieved, the sum of NAA and NAAG provides a reliable estimate of NAA-containing molecules. Besides the three methyl protons, NAAG has eight other non-exchangeable protons, as well as three water-exchangeable protons. However, these other resonances are typically not observed directly due to the lower intensity and spectral overlap with more concentrated metabolites.

2.2.4 Adenosine Triphosphate (ATP)

Adenosine triphosphate (ATP; $C_{10}H_{16}N_5O_{13}P_3$; molar mass, 507.18 g mol^{-1}) is a nucleotide consisting of an adenine group, a ribose ring, and a triphosphate unit (Figure 2.4). Together with phosphocreatine (PCr), ATP is the principal donor of free energy in biological systems. When one phosphate group is donated, for example, to creatine in the creatine–kinase-catalyzed reaction to form phosphocreatine, ATP is converted to adenosine diphosphate (ADP). While the normal ATP concentration in the human brain is circa 3 mM [53], the ADP concentration is typically well below 100 μM. In contrast to the phosphocreatine concentration which is relatively stable across the brain at circa 3.5 mM, the ATP concentration shows significant variation between gray matter (2 mM) and white matter (3.5 mM) [54].

ATP is normally detected with ^{31}P NMR spectroscopy, where it produces three well-separated resonances. ATP detection by ^{1}H NMR spectroscopy is more difficult because (i) the coupling patterns for most protons are complex and (ii) most resonances appear close to the water resonance or are otherwise in exchange with the water. In the ATP adenine group, the ^{2}CH and ^{8}CH protons resonate as singlets at 8.22 and 8.51 ppm. However, observation of these

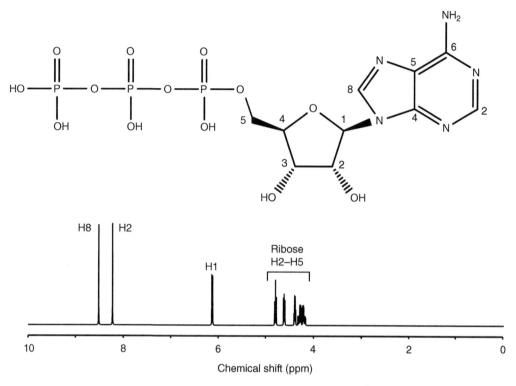

Figure 2.4 **Adenosine triphosphate (ATP).** Chemical structure and simulated ^{1}H NMR spectrum of adenosine triphosphate (ATP). While observable in ^{1}H NMR spectra *in vivo*, ATP detection and quantification is typically performed with ^{31}P NMR.

(pH-dependent) resonances *in vivo* is difficult since the lines are broadened due to slow exchange with water and possibly cross-relaxation with exchangeable protons in adjacent amide groups. While two separate resonances can be seen *in vitro*, only one broadened resonance is observed at 8.22 ppm *in vivo*. The water-exchangeable NH$_2$ protons give rise to a broad resonance at 6.75 ppm while the ribose-1′CH proton resonates as a doublet at 6.13 ppm. While relatively easy to detect (Figure 2.1), the ribose-1′CH resonance is not very specific and may contain contributions from ADP, NAD$^+$, NADH, and other ribose-containing compounds. The other ATP ribose-ring protons resonate as multiplets between 4.2 and 4.8 ppm. The structure and ^{1}H NMR spectrum of ATP are shown in Figure 2.4.

2.2.5 Alanine (Ala)

Alanine (Ala; C$_3$H$_7$NO$_2$; molar mass, 89.09 g mol^{-1}; IUPAC name, (2S)-2-aminopropanoic acid; pK_a, 2.34 and 9.69) is a nonessential amino acid that is present in mammalian brain at a concentration below 0.5 mM. *L* or (*S*)-alanine is the biologically prevalent isomer used in metabolism and protein synthesis, where it is the second-most commonly occurring amino acid after leucine.

Alanine can be manufactured in the body from pyruvate and branched chain amino acids such as valine, leucine, and isoleucine as well as from the breakdown of proteins and the dipeptides carnosine and anserine. Alanine is commonly synthesized through the reductive amination of pyruvate, thereby closely linking alanine to metabolic pathways such as glycolysis,

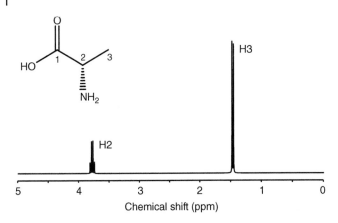

Figure 2.5 **Alanine.** Chemical structure and simulated ^{1}H NMR spectrum of alanine.

gluconeogenesis, and the TCA cycle. The biosynthesis of alanine can be observed with ^{13}C labeling studies using ^{13}C-glucose [55] or ^{13}C-pyruvate [56, 57] as the substrate (see Chapter 3). Increased alanine has been observed in meningiomas [58] and following ischemia [40].

The three methyl protons couple to a single methine proton to form a weakly-coupled AX_3 spin-system, closely resembling lactate, with a doublet resonance at 1.47 ppm and a quartet at 3.77 ppm (Figure 2.5). In high-field NMR spectra from rat brain, as well as in high-resolution spectra from brain slices, the doublet of alanine is often observable as a shoulder on the macromolecular resonance at 1.4 ppm. In lower-field ^{1}H NMR spectra and in the presence of spectral overlap from lipids, alanine can be observable by spectral editing [59] or at longer echo-times. Figure 2.5 shows the structure and ^{1}H NMR spectrum of alanine.

2.2.6   γ-Aminobutyric Acid (GABA)

GABA ($C_4H_9NO_2$; molar mass, 103.12 g mol^{-1}; IUPAC name, 4-aminobutanoic acid; pK_a, 4.23 and 10.43) is an inhibitory neurotransmitter that has a brain concentration of circa 1 mM, although altered concentrations are associated with the menstrual cycle [60], acute deafferentation [61], visual light–dark adaptation [62], alcohol and substance-abuse [63, 64] as well as several neurological and psychiatric disorders [65], including epilepsy [66], depression [67, 68], and panic disorder [69]. Several antiepileptic drugs, like vigabatrin, have been designed to raise cerebral GABA levels, which have been extensively studied with (edited) ^{1}H MRS [70, 71]. More recently, the clinical viability of cerebral GABA detection has been demonstrated on a wide range of clinical MR scanners [72].

As a gamma amino acid, in which the amino group is attached to the gamma carbon, GABA is not used in the biosynthesis of proteins, i.e. GABA is a non-proteinogenic amino acid. GABA is synthesized from glutamate using the enzyme glutamate decarboxylase (GAD) with pyridoxal phosphate as a cofactor. GABA is broken down by GABA transaminase (GABA-T) which catalyzes the conversion of GABA and 2-oxoglutarate into succinic semialdehyde (SSA) and glutamate. SSA can be oxidized to succinic acid by SSA dehydrogenase (SSADH) which can enter the TCA cycle. Reentry of the GABA carbon skeleton into the TCA cycle via GABA-T and SSADH is referred to as the GABA shunt.

GABA has six NMR observable protons (Figure 2.6) in three methylene groups, forming what is historically described as an $A_2M_2X_2$ spin system. However, recent studies have demonstrated that GABA is more accurately described as a strongly-coupled $A_2MM'XX'$ spin system [73, 74]. The triplet resonance for GABA-H2 appears at 2.28 ppm in close proximity of the intense glutamate-H4 (2.34 ppm) and macromolecular M6 (2.29 ppm) resonances.

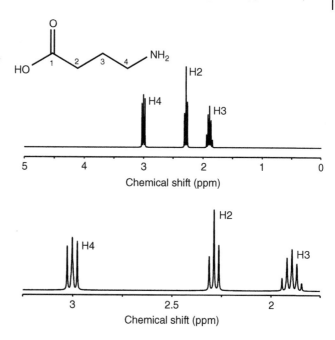

Figure 2.6 **γ-Amino butyric acid (GABA).** Chemical structure and simulated ^{1}H NMR spectrum of GABA. At most magnetic fields, all three resonances of GABA are overlapping with more intense resonances, necessitating the use of spectral editing methods for unambiguous GABA detection (see Chapter 8). Note that GABA is an *A2MM'XX'* spin-system with strong coupling effect on the H3 and H4 multiplets.

The approximate triplet signal for GABA-H4 appears at 3.00 ppm, overlapping with intense signals from creatine and phosphocreatine and smaller signals from macromolecules and glutathione. The approximate GABA-H3 quintet is centered at 1.89 ppm. Since all three resonances of GABA overlap with other, more intense resonances, GABA detection typically relies on spectral editing methods (see Chapter 8).

2.2.7 Ascorbic Acid (Asc)

Ascorbic acid (Asc; $C_6H_8O_6$; molar mass, 176.12 g mol^{-1}; IUPAC name, (5R)-5-[(1S)-1,2-dihydroxy-ethyl]-3,4-dihydroxy-2,5-dihydrofuran-2-one; pK_a, 4.10 and 11.79), commonly known as vitamin C, occurs physiologically as the ascorbate anion. Ascorbate is a water-soluble antioxidant that is found throughout the body with the highest concentration and retention capacities in the brain, spinal cord, and adrenal glands [75]. The average brain concentration of ascorbate is around 1.0 mM with local neuronal and astroglial concentrations of 10 and 1 mM. Ascorbate is heterogeneously distributed throughout the brain with higher concentrations in the cortex and hippocampus as compared to brain stem and spinal cord [75]. Ascorbate acts as a reducing agent by donating electrons to various enzymatic and nonenzymatic reactions. The oxidized form of vitamin C, dehydroascorbic acid, can be reduced in the body by glutathione and NADPH-dependent enzymatic mechanisms. The glutathione–ascorbate cycle is aimed at the removal of reactive oxygen species, such as hydrogen peroxide, while keeping ascorbate in a reduced state. In addition to its antioxidant and neuroprotective properties, ascorbate is also involved as an electron donor in collagen hydroxylation, biosynthesis of carnitine and norepinephrine, tyrosine metabolism, and amidation of peptide hormones. Humans have lost the ability to synthesize ascorbic acid and must obtain it from their food. Ascorbic acid and dehydroascorbic acid are transported into cells by glucose transporters. The fact that dehydroascrobic acid is absorbed at a higher rate than ascorbate has made it a target for hyperpolarized ^{13}C applications to image the redox status *in vivo* [76, 77]. Ascorbate is composed of four NMR-observable protons (Figure 2.7) distributed over one methylene and two methine groups.

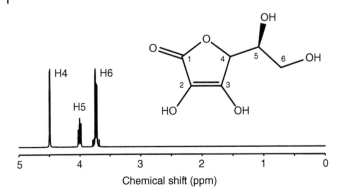

The ^{4}CH proton gives rise to a doublet at 4.49 ppm, while the ^{5}CH and ^{6}CH$_2$ protons give multiplet resonances at 4.00 and 3.73 ppm, respectively. Terpstra and Gruetter [78, 79] were able to observe ascorbate noninvasively in human brain by edited ^{1}H MRS. By selectively perturbing ascorbate-H5, the protons of ascorbate-H6 could be selectively observed and were quantified at a concentration of $1.3 \pm 0.3\,\mu mol\,g^{-1}$. In rat brain, the ascorbic acid concentration shows a strong age-dependence with levels dropping from $4.0\,\mu mol\,g^{-1}$ at P7 to $2.0\,\mu mol\,g^{-1}$ at adulthood [80].

2.2.8 Aspartic Acid (Asp)

Aspartic acid (Asp; C$_4$H$_7$NO$_4$; molar mass, $133.10\,g\,mol^{-1}$; IUPAC name, (2S)-2-aminobutan-edioic acid; pK_a, 1.88, 3.65, and 9.60) or the anionic form of aspartate is a nonessential alpha amino acid that acts as an excitatory neurotransmitter. The L or S-isomer is the biologically prevalent form and is used in the biosynthesis of proteins. It does not cross the blood–brain barrier, but is instead synthesized from glucose and possibly other precursors. The formation of ^{13}C-labeled aspartate from ^{13}C-labeled glucose is readily observed in ^{13}C and ^{1}H-[^{13}C]-NMR experiments, as will be discussed in Chapter 3.

Aspartate is a metabolite in the urea cycle and participates in gluconeogenesis. It is an essential part of the malate–aspartate shuttle, a biochemical system for transporting electrons (reducing equivalents) produced during glycolysis across the mitochondrial inner membrane for use in oxidative phosphorylation.

Aspartate has an approximate brain concentration of 1–2 mM [1, 81]. The ^{3}CH$_2$ and ^{2}CH groups give a typical *ABX* spectral pattern consisting of a doublet-of-doublets for the CH group at 3.89 ppm and a pair of doublet-of-doublets from the methylene protons at 2.67 and 2.80 ppm. The structure and ^{1}H NMR spectrum of L-aspartic acid are shown in Figure 2.8.

2.2.9 Branched-chain Amino Acids (Isoleucine, Leucine, and Valine)

Branched-chain amino acids (BCAAs) are amino acids with a branched, aliphatic side-chain. The proteinogenic BCAAs are isoleucine, leucine, and valine. BCAAs are involved in stress, energy, and muscle metabolism. Whereas many amino acids are catabolized in the liver, BCAAs are primarily oxidized in skeletal muscle. Many types of inborn errors of BCAA metabolism exist, of which maple syrup urine disease is the most common. Increased BCAAs have been detected with ^{1}H MRS in the brain of young children with maple syrup urine disease [82, 83]. All three BCAAs are essential and have recommended daily intake recommendations. The L or S-isomer is the prevalent isoform for all three compounds.

Figure 2.8 Aspartate. Chemical structure and simulated ^{1}H NMR spectrum of aspartate. Note that aspartate forms a classical *ABX* spin system, displaying two doublet-of-doublets around 2.7 ppm with a strong second-order "roof" effect.

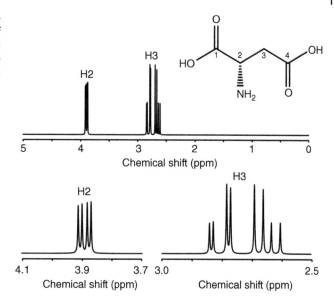

Isoleucine (Ile; $C_6H_{13}NO_2$; molar mass, 131.18 g mol^{-1}; IUPAC name, (2S,3S)-2-amino-3-methylpentanoic acid) and leucine (Leu; $C_6H_{13}NO_2$; molar mass, 131.18 g mol^{-1}; IUPAC name, (2S)-2-amino-4-methylpentanoic acid) are structural isomers. The ^{1}H NMR spectrum for both compounds (Figure 2.9) is dominated by strong signals in the 0.9–1.0 ppm range originating from the two methyl groups in each structure. Smaller resonances originate from the methylene and methine protons. Under normal conditions the brain concentration for leucine and isoleucine is below 0.1 mM. All three BCAAs are readily detectable in blood plasma [84, 85].

Valine (Val; $C_5H_{11}NO_2$; molar mass, 117.15 g mol^{-1}; IUPAC name, (2S)-2-amino-3-methylbutanoic acid; pK_a, 2.32 and 9.62) has a normal brain concentration on the order of 0.1 mM. Hypervalinemia, branched-chain ketonuria, and brain abscesses are three of several diseases in which valine concentrations are increased. In sickle-cell disease, it substitutes for the hydrophilic amino acid glutamic acid in hemoglobin leading to an incorrect folding of the hemoglobin molecule.

The NMR spectrum of valine (Figure 2.9F) can be traced back to the six protons in two methyl groups resonating at 0.98 and 1.03 ppm and two protons in two methine groups resonating at 2.26 and 3.60 ppm. The *in vivo* detection of valine, as well as isoleucine and leucine, is not trivial as the most intense resonances (methyl protons) are broad multiplets overlapping with broad resonances from macromolecules (see also Figure 2.26).

2.2.10 Choline-containing Compounds (tCho)

Besides resonances from NAA and total creatine (tCr), the most prominent resonance in ^{1}H NMR spectra from brain arises from the methyl protons of choline-containing compounds at circa 3.2 ppm (Figure 2.10). Since the resonance contains contributions from free choline (Cho), glycerophoshorylcholine (GPC), and phosphorylcholine (PC), it is often referred to as "total choline" (tCho). At short echo-times overlapping resonances from ethanolamine, *myo*-inositol, glucose, and especially taurine should be taken into account. In the liver and kidney the resonance at 3.26 ppm is almost entirely composed of betaine. The tCho concentration in human brain is approximately 1–2 mM with a nonuniform distribution [13, 14, 86]. Cho is a minor contributor to the tCho resonance, since the concentration is well below the NMR

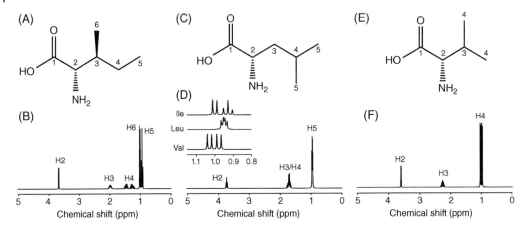

Figure 2.9 Branched-chain amino acids (BCAA). Chemical structure of (A) isoleucine, (C) leucine, and (E) valine together with (B, D, F) their simulated ^{1}H NMR spectrum at 7T. The inset in (D) shows the spectral region of the methyl protons. At 500 MHz the strongly-coupled doublet-of-doublets for leucine (Leu) will appear at an apparent triplet resonance.

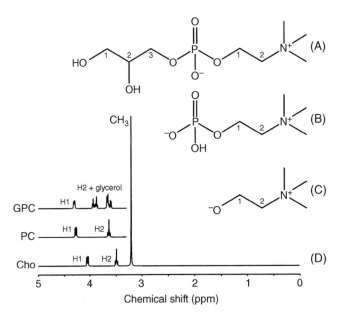

Figure 2.10 **Choline-containing compounds.** (A–C) Chemical structures and (D) simulated ^{1}H NMR spectra of choline-containing compounds. Structures shown are for (A) glycerophosphorylcholine (GPC), (B) phosphorylcholine (PC), and (C) choline. All NMR spectra exhibit a strong singlet resonance at 3.18–3.22 ppm originating from nine equivalent protons, whereby small differences can be observed between 3.5 and 4.5 ppm.

detection limit. The two main contributors, PC and GPC, cannot be separated based on the methyl protons due to the small chemical shift difference relative to the spectroscopic line widths. However, separation can, in principle, be achieved through the smaller resonances from methylene protons [87]. Furthermore, with ^{31}P MRS the relative contributions of PC and GPC can be readily established [88].

Glycerophosphorylcholine (GPC; $C_8H_{20}NO_6P$; molar mass, 257.22 g mol^{-1}) has a total of 18 non-exchangeable protons from the glycerol (5 protons) and choline moieties (13 protons of which 9 protons in three magnetically equivalent methyl groups). The trimethyl protons resonate at 3.21 ppm as a singlet. The ^{1}CH$_2$ and ^{2}CH$_2$ protons of the choline-moiety resonate at 4.31 and 3.66 ppm. The five glycerol protons resonate as a strongly-coupled spin-system between

3.6 and 4.0 ppm. Besides the homonuclear scalar coupling interactions, GPC also displays a heteronuclear interaction between the $^1CH_2$ protons and the adjacent nitrogen-14 head group ($^1J_{HN} = 2.7$ Hz), as well as an interaction between the $^2CH_2$ protons and the nearby ^{31}P nucleus ($^2J_{HP} = 6.0$ Hz). Choline (Cho; $C_5H_{14}NO$; molar mass, 104.17 g mol^{-1}; IUPAC name, (2-hydroxy-ethyl)trimethylazanium) and phosphorylcholine (PC; $C_5H_{15}NO_4P$; molar mass, 184.07 g mol^{-1}; IUPAC name, [2-(trimethylazaniumyl)ethoxy]phosphonic acid) both give rise to 13 non-exchangeable protons. The trimethyl protons appear as a singlet at 3.21 ppm, whereas the remaining $^1CH_2$ and $^2CH_2$ protons appear as multiplets between 3.6 and 4.3 ppm.

Choline-containing compounds are involved in pathways of phospholipid synthesis and degradation, thereby reflecting membrane turnover. Increased choline signal is observed in cancer [89], Alzheimer's disease [90], and multiple sclerosis [91, 92], while decreased choline levels are associated with liver disease and stroke [93]. However, the exact interpretation of changes in tCho signal is complicated due to the multiple contributions to the observed tCho resonance. In breast and other lipid-rich tissues an additional contribution can be observed due to artifacts arising from vibration-induced sidebands of large signals (e.g. lipids), thereby further complicating interpretation if the data acquisition is not properly modified [94]. In tissues outside the central nervous system, the resonance at circa 3.2 ppm contains a dominant contribution from betaine (trimethylglycine).

2.2.11 Creatine (Cr) and Phosphocreatine (PCr)

Creatine (Cr; $C_4H_9N_3O_2$; molar mass, 131.14 g mol^{-1}; IUPAC name, 2-[carbamimidoyl(methyl)amino]acetic acid) and phosphocreatine (PCr; $C_4H_{10}N_3O_5P$; molar mass, 211.11 g mol^{-1}; IUPAC name, 2-(3-methyl-1-phosphonocarbamimidamido)acetic acid) play a central role in cellular energy metabolism. In ^1H NMR spectra of normal tissue which contains creatine kinase, the singlet resonances at 3.03 and 3.93 ppm arise from the methyl and methylene protons of creatine and phosphorylated creatine, i.e. phosphocreatine. Together they are often referred to as "total creatine". In the brain, creatine and phosphocreatine are present in both neuronal and glial cells. Creatine and phosphocreatine, together with ATP play a crucial role in the energy metabolism of tissues [95]. Although some controversy remains about the exact role, it has been suggested that tCr (in combination with creatine kinase) serves as (i) an energy buffer, retaining constant ATP levels through the creatine kinase reaction and (ii) an energy shuttle, diffusing from energy producing (i.e. mitochondria) to energy utilizing sites (i.e. myofibrils in muscle or nerve terminals in brain). The concentrations in human brain for phosphocreatine and creatine are circa 4.0–5.5 mM and 4.5–6.0 mM, respectively, with higher levels in gray matter (6.4–9.7 mM) than in white matter (5.2–5.7 mM). The concentration of tCr is relatively constant, with no changes reported with age [96] or a variety of diseases. As such, the resonance is frequently used as an internal concentration reference. While convenient and under most conditions reliable, the use of any internal concentration reference should be used with caution. Decreased creatine levels have observed in the chronic phases of many pathologies, including tumors and stroke. Furthermore, it has been shown that part of the tCr resonance is NMR invisible due to restricted rotational mobility [97, 98]. Creatine is not an essential nutrient as it is produced from the amino acids arginine and glycine, primarily in the liver. The majority of creatine and phosphocreatine stores are found in skeletal muscle. The liver is characterized by a complete absence of phosphocreatine.

The ^1H NMR spectra of creatine and phosphocreatine are very similar. The chemical shift difference between the creatine (3.027 ppm) and phosphocreatine (3.029 ppm) methyl resonances is too small to allow a reliable separation *in vivo*. However, the chemical shift difference between the creatine (3.913 ppm) and phosphocreatine (3.930 ppm) methylene resonance is large enough to separate the two compounds at magnetic fields of circa 7 T and higher. The structures and ^{1}H NMR spectra of creatine and phosphocreatine are shown in Figure 2.11.

2.2.12 Ethanol

Ethanol (C_2H_6O; molar mass, 46.07 g mol^{-1}) is the principal alcohol found in alcoholic beverages. Within the human body ethanol is converted into acetaldehyde by alcohol dehydrogenase and then into acetic acid by acetaldehyde dehydrogenase. The product of the first step of this breakdown, acetaldehyde, is more toxic than ethanol and has been linked to most of the clinical effects of alcohol. Ethanol is not detectable by ^{1}H MRS in normal brain tissues, but is readily observed after alcohol consumption [99, 100]. The reduced NMR visibility of ethanol in the brain has been a long-standing issue which may be explained by pulse sequence parameters, reduced relaxation constants, and magnetization transfer effects between mobile and immobile ethanol fractions [101, 102].

The ^{1}H NMR spectrum of ethanol (Figure 2.12) is characterized by a triplet resonance at 1.18 ppm and a quartet resonance at 3.65 ppm. Ethanol, together with methanol (singlet, 3.34 ppm), is a common impurity observed in ^{1}H NMR spectra of tissue extracts or body fluids.

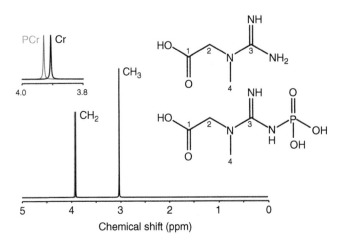

Figure 2.11 **Creatine and phosphocreatine.** Chemical structures of creatine (Cr, top) and phosphocreatine (PCr, bottom) and the simulated ^{1}H NMR spectra at 7T. The methyl resonances are indistinguishable *in vivo* at any magnetic field, but the methylene resonances are separated enough (~0.02 ppm) to allow routine detection of magnetic fields higher than circa 7T. Both resonances have broad resonances downfield from water that are typically not observed directly.

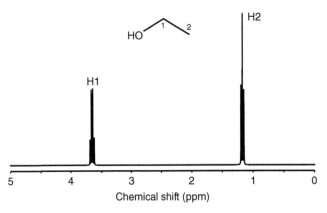

Figure 2.12 **Ethanol.** Chemical structure and simulated ^{1}H NMR spectrum of ethanol.

2.2.13 Ethanolamine (EA) and Phosphorylethanolamine (PE)

Ethanolamine (EA; C_2H_7NO; molar mass, $61.08 \, g \, mol^{-1}$; IUPAC name, 2-aminoethan-1-ol) and phosphorylethanolamine (PE; $C_2H_8NO_4P$; molar mass, $141.06 \, g \, mol^{-1}$; 2-aminoethyl dihydrogen phosphate) are common moieties of phospholipids and sphingomyelin, whereby EA is a precursor of PE. EA is synthesized through the decarboxylation of serine.

Figure 2.13 shows their chemical structures and ^1H NMR spectra. Both compounds have four NMR-observable protons in two methylene groups which give rise to two multiplet resonances around 3.1–3.2 ppm and 3.8–4.0 ppm. EA has been detected in high-resolution of brain extracts [103] with an estimated concentration of several millimoles. However, the *in vivo* detection of EA is not straightforward, as the two resonances overlap with peaks from more concentrated compounds. PE can be detected with ^{31}P NMR spectroscopy and was measured at a concentration of 0.5–1.0 mM in human brain [88]. Increased levels of EA and PE have been observed during ischemia [103] and seizures [104], respectively. PE can be observed with enhanced sensitivity through polarization transfer [105–107] utilizing the heteronuclear ^1H-^{31}P scalar coupling between the ^{31}P nucleus and the $^1CH_2$ protons ($^3J_{1HP} = 7.1 \, Hz$ and $^3J_{1'HP} = 7.3 \, Hz$).

2.2.14 Glucose (Glc)

D-Glucose (Glc; $C_6H_{12}O_6$, molar mass, $180.16 \, g \, mol^{-1}$; IUPAC name, (2R,3S,4R,5R)-2,3,4,5,6-pentahydroxyhexanal) is one of the 16 aldohexose stereoisomers. D-glucose, also known as dextrose, is the ubiquitous source of energy from bacteria to humans. In mammals, glucose is broken down to CO_2 and water in the citric acid or TCA cycle to provide energy in the form of ATP. Glucose can exist in an open-chain aldohexose form in which the C1 carbon is part of an aldehyde group. However, when dissolved in water the dominant form is a six-membered pyranose ring, formed when the aldehyde group at C1 and the hydroxyl group C5 undergo a nucleophilic addition reaction. In addition to the chiral centers at C2, C3, C4, and C5, the C1 position also becomes chiral in the closed ring form thereby giving rise to two configurations, in which the ^1CH proton exists in an axial or equatorial orientation relative to the ring. The so-called α and β anomers coexist in aqueous solutions, with an equilibrium concentration of 36% for the α anomer and 64% for the β anomer. Note that when dissolving glucose crystals of a single anomer, it takes up to several hours for the anomeric equilibrium to be established. This can become important when acquiring "pure solution spectra" for spectral fitting (see Chapter 8).

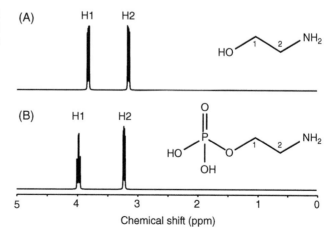

Figure 2.13 **Ethanolamine (EA) and phoshorylethanolamine (PE).** Chemical structures and simulated ^1H NMR spectra of (A) EA and (B) PE.

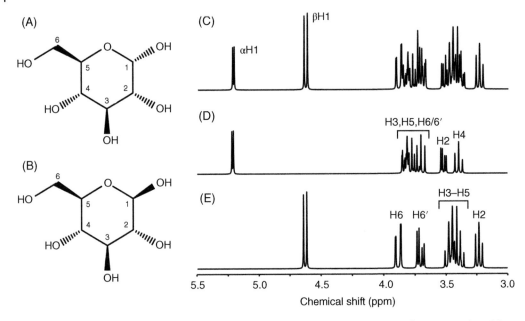

Figure 2.14 Glucose. (A, B) Chemical structure of (A) α-glucose and (B) β-glucose drawn as a closed-form Haworth projection. Note that the anomeric proton on the C1 position is the only difference, with the proton being above and below the plane for α and β-glucose, respectively. (C) ¹H NMR spectra of glucose at anomeric equilibrium with 36% α-glucose and 64% β-glucose and ¹H NMR spectra of (D) pure α-glucose and (E) pure β-glucose. While most glucose signals overlap with resonances from other metabolites, the αH1-glucose signal is readily observed.

The resting cerebral glucose concentration is circa 1.0 mM but can be increased to >5.0 mM by intravenous glucose infusion. More details on the measurement of glucose transport kinetics as well as the role of glucose in energy and neurotransmitter metabolism will be given in Chapter 3.

Glucose contains seven non-exchangeable protons, as well as five water-exchangeable protons in hydroxyl groups. Figure 2.14 shows the closed ring structure and ¹H NMR spectrum of glucose. The β-1CH proton resonance appears at 4.63 ppm and is typically eliminated during water suppression. The other six non-exchangeable protons resonate as a strongly-coupled spin-system between 3.2 and 3.9 ppm, which somewhat simplifies *in vivo* to two broad resonances at 3.43 and 3.8 ppm due to the larger line widths. Glucose was initially detected in rat brain through the inner resonances at 3.43 and 3.80 ppm [108], although severe spectral overlap limited the detection reliability. Gruetter et al. [109] were able to detect changes in glucose levels and study cerebral glucose transport kinetics by performing difference spectroscopy following the intravenous infusion of glucose. With the increased availability of higher magnetic fields, glucose is readily detected from the α-1CH resonance at 5.22 ppm [110, 111]. Alternatively, glucose can be unambiguously detected through spectral editing methods [112–114].

2.2.15 Glutamate (Glu)

Glutamate or glutamic acid (Glu; $C_5H_9NO_4$; molar mass, 147.13 g mol⁻¹; IUPAC name, (2S)-2-aminopentanedioic acid; pK_a, 2.19, 4.25, and 9.67) is a nonessential, proteinogenic alpha amino acid. The *L* or *S* isomer is the one most widely occurring in nature, but the *D* or *R* isomer occurs in some special contexts, such as the cell walls of *Escherichia coli* and the liver of mammals. Glutamate is the major excitatory neurotransmitter in mammalian brain [115]. It is also the

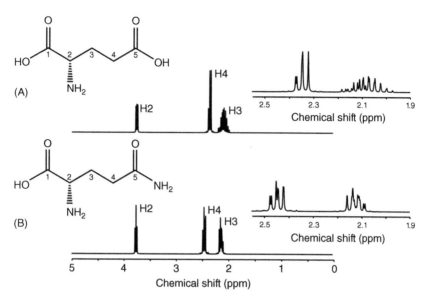

Figure 2.15 **Glutamate and glutamine.** Chemical structure and ^{1}H NMR spectra of (A) glutamate and (B) glutamine. The exchangeable amide protons of glutamine appearing downfield from water are not shown.

direct precursor for the major inhibitory neurotransmitter, GABA. In addition, glutamate is also an important component in the synthesis of other small metabolites (e.g. glutathione), as well as larger peptides and proteins [115]. Glutamate is present at an average concentration of 8–12 mM with significant differences between gray and white matter [116–120]. Glutamate is present in all cell types with the largest pool in glutamatergic neurons and smaller pools in GABAergic neurons and astroglia. The central role of glutamate in the glutamate–glutamine neurotransmitter cycle will be discussed in more detail in Chapter 3.

Glutamate has two methylene groups and a methine group that are strongly coupled, forming an *AMNPQ* spin system (Figure 2.15A). As a result of the extensive and strong scalar coupling interactions glutamate has a complex NMR spectrum with signal spread out over many low-intensity resonances. Signal from the single [2]CH methine proton appears as a doublet-of-doublets at 3.75 ppm, while the resonances from the other four protons appear as multiplets between 2.04 and 2.35 ppm. Glutamate and glutamine are virtually indistinguishable at lower magnetic fields, whereas the glutamate-H4 and glutamine-H4 protons become separate resonances at a magnetic field strength of 7 T and higher [120] (Figure 2.16).

2.2.16   Glutamine (Gln)

Glutamine (Gln; $C_5H_{10}N_2O_3$; molar mass, 146.15 g mol^{-1}; IUPAC name, (2S)-2-amino-4-carbamoylbutanoic acid; pK_a, 2.17 and 9.13) is one of the 20 amino acids encoded by the standard genetic code, whereby the L or S isomer is used in metabolism and biosynthesis of proteins. Glutamine is an important component of intermediary metabolism and is primarily located in astroglia at a concentration of 2–4 mM [1, 14, 81]. Glutamine is synthesized from glutamate by glutamine synthethase in the astroglia and it is broken down to glutamate by phosphate-activated glutaminase in neurons. Glutamine is one of the few amino acids that can cross the blood–brain barrier. Besides its role in the glutamate–glutamine neurotransmitter cycle (see Chapter 3), a main function of glutamine is in ammonia detoxification. Glutamine markedly increases during hyperammonemia and under those conditions brain glutamine is a good

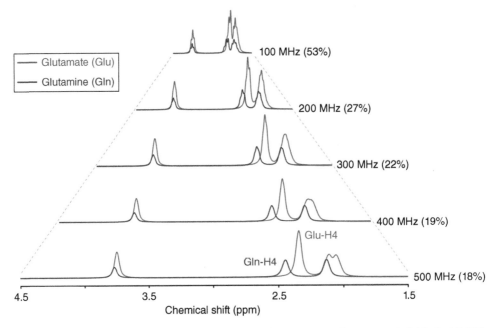

Figure 2.16 Magnetic field dependence for glutamate and glutamine. At low magnetic fields the H3/H3′ and H4/H4′ protons from glutamate and glutamine are strongly coupled leading to complicated multiplets with strong spectral overlap. However, with increasing magnetic field strength the spectral patterns simplify and shift apart, thereby allowing a reliable separation of glutamate-H4 and glutamine-H4. The percentages indicate the spectral overlap between the Glu-H4 and Gln-H4 multiplets.

indicator of the liver disease [121, 122]. Glutamine has also been implicated as a fuel source for a number of cancers [123].

Glutamine is structurally similar to glutamate with a single methine group and two methylene groups (Figure 2.15B). As a result, the chemical shifts and scalar coupling interactions are also similar. The ^{2}CH methine proton resonates as a doublet-of-doublets at 3.77 ppm, while the multiplets of the four methylene protons are closely grouped between 2.12 and 2.46 ppm. In addition, glutamine has two NMR detectable amide protons at 6.8 and 7.5 ppm. The relative intensity of the 6.8 ppm downfield resonance is typically much higher than the 7.5 ppm resonance because the water–amide exchange rate is different.

In order to understand the important roles of glutamate and glutamine in intermediary metabolism, the separate detection of these two compounds is essential. Unfortunately, the similar chemical structures lead to very similar ^{1}H NMR spectra. The increase in magnetic field strength has greatly benefited the separation between the H4-resonances of glutamate and glutamine (Figure 2.16). At magnetic fields of 7 T or higher, the glutamate and glutamine H4 resonances are visually separated, leading to greatly enhanced quantification accuracy. At lower fields, the separation between glutamate and glutamine becomes unreliable, although the sum (often referred to as Glx) can be quantified with high accuracy.

2.2.17 Glutathione (GSH)

Glutathione (GSH; $C_{10}H_{17}N_3O_6S$; molar mass, 307.32 g mol^{-1}; IUPAC name, (2S)-2-amino-4-[[(1R)-1-[(carboxymethyl)carbamoyl]-2-sulfanylethyl]carbamoyl]butanoic acid; pK_a = 2.12 (COOH), 3.59 (COOH), 8.75 (NH$_2$), and 9.65 (SH)) is a tripeptide consisting of glycine, cysteine, and glutamate (Figure 2.17). It can exist in reduced (GSH) or oxidized (GSSG) forms, although

Figure 2.17 Glutathione. Chemical structure and simulated ^{1}H NMR spectrum of the reduced form of glutathione (GSH). The oxidized form of glutathione (GSSG) is a dimer of GSH forming a disulfide bridge on the cysteine moiety. GSSG is present at negligible levels under normal physiological conditions.

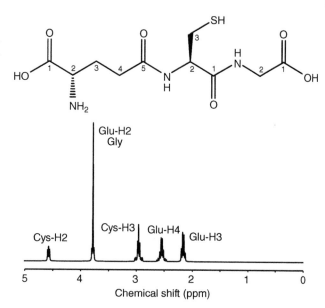

it is present in the living brain almost entirely as GSH at concentrations of 1–3 mM [124–126], primarily located in astrocytes [126]. Once oxidized, glutathione can be reduced back by glutathione reductase using NADPH as an electron donor. The ratio of GSH to GSSG within cells is often used as a measure of cellular oxidative stress. GSH is an antioxidant, essential for maintaining normal red-blood-cell structure and keeping hemoglobin in the ferrous state. GSH participates directly in the neutralization of free radicals and reactive oxygen species, as well as maintaining exogenous antioxidants such as vitamins C and E in their reduced (active) forms. Other functions include that of amino acid transport system, as well as a storage form of cysteine [126, 127]. Altered GSH levels have been reported in Parkinson's disease and other neurodegenerative diseases affecting the basal ganglia [128].

The methylene protons of the glycine moiety resonate at 3.77 ppm, overlapping with the methine proton of the glutamate moiety which appears as a doublet-of-doublets at the same chemical shift position. The glycine moiety is best described as a broadened doublet signal due to scalar coupling with the amide proton [129]. The four glutamate methylene protons give rise to two separate multiplets at circa 2.15 and 2.55 ppm. The ^{2}CH and ^{3}CH$_2$ protons of the cysteine moiety form an *ABX* spin system with three doublet-of-doublets at 2.93, 2.98, and 4.56 ppm. All resonances of GSH are overlapping with more intense signals from other metabolites, most noticeably glutamate, glutamine, creatine, and NAA. *In vivo* detection of GSH has therefore mainly been achieved through spectral editing methods [124, 130, 131], although it has been shown that GSH can be detected directly when combining short-TE spectra with spectral fitting at high magnetic fields [1, 125]. In ^{1}H NMR spectra of tissue extracts both GSH and GSSG can be detected, since GSH is nonenzymatically converted to GSSG.

2.2.18 Glycerol

Glycerol (C$_3$H$_8$O$_3$; molar mass, 92.09 g mol^{-1}; IUPAC name, propane-1,2,3-triol) is a major constituent of phospholipids and free glycerol is imbedded into membrane phospholipids (Figure 2.18). However, NMR resonances from glycerol are not observed in ^{1}H NMR spectra from normal brain, most likely due to line broadening caused by limited rotational mobility.

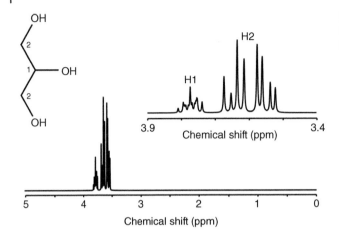

Figure 2.18 **Glycerol.** Chemical structure and simulated ^{1}H NMR spectrum of glycerol.

Free glycerol obtained as an end product of membrane phospholipids breakdown may become observable. NMR observation of glycerol has been described for ^{1}H NMR spectra of human brain homogenates following autopsy, where the glycerol concentration increases with postmortem sample preparation time [132]. The ^{1}H NMR spectrum originates from five spins in two equivalent methylene groups and a single CH-proton. At the magnetic field strengths employed for *in vivo* NMR, glycerol is characterized as a strongly-coupled spin system giving resonances centered at 3.55 and 3.64 ppm (methylene protons) and 3.77 ppm (CH proton multiplet). Unambiguous detection of glycerol is not straightforward due to significant spectral overlap with *myo*-inositol in the 3.5–3.65 ppm range. In adipose tissue and skeletal muscle the glycerol moiety of triglycerides can be observed as a broad signal around 4.0 ppm [133, 134] (Figure 2.35).

2.2.19 Glycine

Glycine ($C_2H_5NO_2$; molar mass, 75.07 g mol^{-1}; IUPAC name, 2-aminoacetic acid; pK_a, 2.34 and 9.60) is the simplest proteinogenic amino acid, having only a single proton in its side chain. Glycine is an inhibitory neurotransmitter and is distributed throughout the central nervous system with higher levels in the brain stem and spin cord. Glycine is primarily being synthesized from glucose through serine with a concentration of approximately 1 mM in human brain. Glycine together with arginine can be converted, via guanidinoacetate, into creatine. Glycine forms a significant fraction of collagen.

Glycine has two methylene protons that co-resonate as a singlet peak at 3.55 ppm. For *in vivo* NMR measurement, the glycine resonance overlaps with those of *myo*-inositol, making unambiguous observation of glycine difficult in short echo-time ^{1}H NMR spectra. As a compound without ^{1}H-^{1}H scalar coupling, improved detection of glycine with regular spectral editing methods is not an option. However, judicious choice of a longer echo-time or echo-time averaging that minimizes the *myo*-inositol signal has been shown to reliably detect glycine in the human brain [135, 136]. The concentration of glycine in gray matter is several times higher than in white matter [137] and increased in infants with hyperglycinemia [138] and patients with brain tumors [139, 140].

2.2.20 Glycogen

Carbohydrate reserves are mainly stored as glycogen in animals and humans. It is particularly abundant in muscle and liver, reaching concentrations up to 30–100 mmol kg^{-1} and

$100–500\,mmol\,kg^{-1}$, respectively (see Ref. [141] for a review). Glycogen is also present in the brain, residing in astroglia at a concentration of circa $5\,mmol\,kg^{-1}$ [142–145]. The regulation of glycogen synthesis and breakdown plays an important role in systemic glucose metabolism and is crucial in the understanding of diseases such as diabetes mellitus.

The primary structure of glycogen consists of α-[1,4]-linked glucose chains containing 12–13 glucose residues. The chains are linked together through α-[1,6] branch points to form larger, spherical units referred to as β-particles. The β-particles can be further arranged into larger rosette-type structures termed α-particles. Despite its high molecular weight ($10^7–10^9$ Da), glycogen gives rise to relatively narrow ^{1}H and ^{13}C NMR resonances both *in vitro* and *in vivo*, indicating a high degree of internal mobility which was confirmed by Shulman and coworkers [146–148]. Natural abundance ^{13}C NMR glycogen detection is now a routine method to observe glycogen *in vivo* [149–151]. Combined with the infusion of [1-^{13}C]-glucose, the formation of [1-^{13}C]-glycogen can be detected in muscle [152], liver [153], and brain [142–145]. While ^{13}C NMR detection of glycogen is relatively straightforward, ^{1}H detection of glycogen remains largely elusive. Only a single report [154] described the proton detection of glycogen in the liver. The ^{1}H detection of glycogen is likely to be complicated by cross-relaxation and exchange pathways between glycogen and water. Glycogen has been detected indirectly through a CEST effect with water [155] (see Chapter 3 for more details).

2.2.21 Histidine

Histidine ($C_6H_9N_3O_2$; molar mass, $155.16\,g\,mol^{-1}$; IUPAC name, (2S)-2-amino-3-(1H-imidazol-5-yl)propanoic acid; pK_a, 1.82, 9.17, and 6.12) is an essential amino acid of which the biologically active *L* or *S* isomer is used for protein synthesis. Histidine is a precursor for histamine and carnosine biosynthesis. While normal brain concentrations are circa 0.1 mM, histidine freely passes the blood–brain barrier such that its concentration can be increased by increasing plasma levels [156]. Increased histidine and histamine have been observed in hepatic encephalopathy [157, 158] and histidinemia [159], a defect of amino acid metabolism. Even under elevated conditions, the absolute concentration of histamine likely remains below the *in vivo* NMR detection limit [156].

Histidine has five non-exchangeable protons of which two reside on the imidazole ring and three in the aliphatic side chain (Figure 2.19). The two imidazole protons at circa 7.1 and 7.8 ppm are sensitive to pH in the physiological range and have been used with *in vivo* ^{1}H NMR to establish intracellular pH [156].

Figure 2.19 **Histidine.** Chemical structure and simulated ^{1}H NMR spectrum of histidine. The proton resonances downfield from water are pH-sensitive in the physiological pH range (see Figure 2.21).

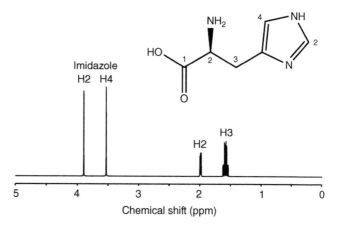

2.2.22 Homocarnosine

Homocarnosine ($C_{10}H_{16}N_4O_3$; molar mass, $240.26\,g\,mol^{-1}$; IUPAC name, (2*S*)-2-(4-aminobu-tanamido)-3-(1H-imidazol-4-yl)propanoic acid; pK_a, 6.86) is a dipeptide consisting of histidine and GABA. In the human brain it has a concentration range of 0.3–0.6 mM [160, 161], while the concentration in rodent brain is negligible. Homocarnosine has been associated with epileptic seizure control, showing a better correlation with the frequency of seizures than the inhibitory neurotransmitter GABA [66]. Homocarnosine levels can be raised by antiepileptic drugs, like gabapentin [162]. Elevated brain and CSF levels are typically associated with homocarnosinosis, a metabolic disorder caused by a deficiency in the homocarnosine-metabolizing enzyme, homocarnosinase [160, 161].

Homocarnosine has 11 non-exchangeable protons of which 2 reside in the histidine–imidazole ring, 3 in the histidine side chain, and 6 in the GABA moiety (Figure 2.20). The two imidazole protons resonate at circa 7.1 and 8.1 ppm and are sensitive to pH in the physiological range [163]. The nonaromatic histidine protons give rise to multiplets between 3.0 and 4.5 ppm, whereas the resonance positions of the GABA-moiety protons are very similar to those of free GABA. As a result, the GABA-H4 resonance at 3.01 ppm as observed in spectral editing experiments should typically be referred to as "total GABA" as it represents the sum of free GABA and homocarnosine. The homocarnosine fraction can be estimated from the downfield imidazole proton resonances which can be observed with short-TE ^{1}H MRS [163].

Figure 2.21 shows the pH dependence of the ^{1}H chemical shift for histidine and homocarnosine. In the physiological pH range the chemical shift of both compounds show a strong pH dependence, making them a good choice for *in vivo* pH monitoring. Imidazole (pK_a, 7.28) and 4-nitrophenol (4NP; pK_a, 7.07) are structurally simpler compounds that can function as *in vitro* pH sensors, for example, in ^{1}H NMR of brain extracts or body fluids.

2.2.23 β-Hydroxybutyrate (BHB)

β-Hydroxybutyric acid (BHB, $C_4H_8O_3$; molar mass, $104.11\,g\,mol^{-1}$; IUPAC name, (3*R*)-3-hydroxybutanoic acid) or its anionic form β-hydroxybutyrate is, together with acetone and acetoacetate, referred to as a ketone body. A ketone body can function as an alternative sub-strate for metabolism, typically under conditions of long fasting or high fat diets [164]. Under

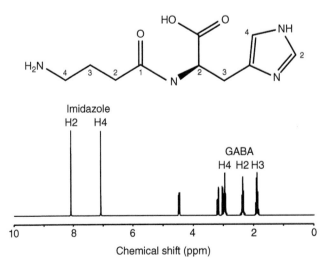

Figure 2.20 Homocarnosine. Chemical structure and simulated ^{1}H NMR spectrum of homocarnosine. The chemical shift of the two downfield resonances of homocarnosine are sensitive to pH in the physiological range (see Figure 2.21).

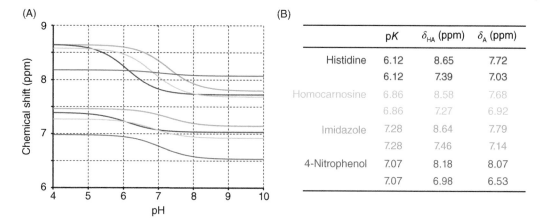

	pK	δ_{HA} (ppm)	δ_A (ppm)
Histidine	6.12	8.65	7.72
	6.12	7.39	7.03
Homocarnosine	6.86	8.58	7.68
	6.86	7.27	6.92
Imidazole	7.28	8.64	7.79
	7.28	7.46	7.14
4-Nitrophenol	7.07	8.18	8.07
	7.07	6.98	6.53

Figure 2.21 **pH dependence of *in vivo* and *in vitro* pH sensors.** (A) ^{1}H chemical shift as a function of pH for histidine (red), homocarnosine (orange), imidazole (green), and 4-nitrophenol (4NP, blue). Note that each compound has two pH-sensitive resonances. (B) pK_a and chemical shifts of the protonated (δ_{HA}) and non-protonated (δ_A) forms of the four pH-sensitive compounds. Imidazole and 4-nitrophenol are not present in biological tissues, preventing their use as an *in vivo* pH sensor. However, their simple NMR spectra make them ideal candidates for pH monitoring on *in vitro* preparations such as brain and tissue extracts, (filtered) blood plasma, urine, and CSF.

normal (fed) conditions the plasma ketone bodies are below 0.1 mM but can rise to 2–3 mM after 72 hours of fasting [164]. ^{1}H MRS has shown that brain BHB levels increase to >1.5 mM under those conditions [165]. ^{13}C NMR studies with ^{13}C-labeled BHB have shown that BHB is readily consumed by the brain [166]. Increased plasma ketone levels, achieved through a ketogenic diet, have shown great potential for seizure control in childhood epilepsy (e.g. Ref. [167]). Besides being an important source of energy during periods of low glucose, BHB and ketone bodies are also used to maintain acetoacetyl-CoA and acetyl-CoA for the synthesis of cholesterol, fatty acids, and complex lipids.

BHB has two enantiomers due to the chiral center on the third carbon, of which the D or (R)-isomer is biologically active. BHB has six non-exchangeable protons of which the three methyl protons resonate as a doublet at 1.19 ppm and are the common target for NMR observation. The methine proton resonates as a complex pattern at 4.13 ppm, while the two nonequivalent methylene protons resonate as doublet-of-doublets at 2.29 and 2.39 ppm. Figure 2.22A shows the structure and ^{1}H NMR spectrum of BHB. The detection of BHB in short-TE ^{1}H MR spectra is complicated by spectral overlap with macromolecules. However, spectral editing of lactic acid provides 100% coediting and an unambiguous detection of BHB [165].

2.2.24 2-Hydroxyglutarate (2HG)

2-Hydoxyglutaric acid (2HG; $C_5H_8O_5$; molar mass, 148.11 g mol^{-1}; IUPAC name, 2-hydroxypentanedioic acid) or 2-hydroxy glutarate exists in D and L isomers, both of which are biologically active. The concentration of 2HG in normal brain tissue is well below the NMR detection limit. However, in brain tumors bearing a mutation in isocitrate dehydrogenase 1 and 2 (IDH1 and 2) D-2HG is overproduced and can reach concentrations of up to 10 mM [168].

2HG has five non-exchangeable protons in three groups that generate complex multiplets at 4.01, 1.90, and 2.25 ppm (Figure 2.23). Direct detection of 2HG in short-TE MR spectra is challenging due to overlap with resonances from NAA, Glu, Gln, GABA, and macromolecules.

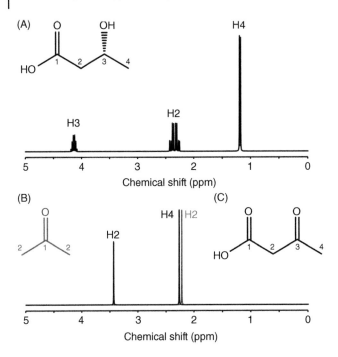

Figure 2.22 **Ketone bodies.** Chemical structure and simulated ^{1}H NMR spectra of (A) β-hydroxybutyrate, (B) acetone, and (C) acetoacetate.

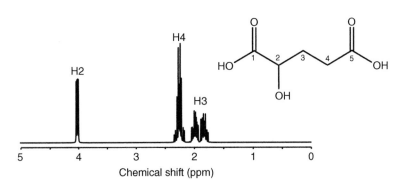

Figure 2.23 **2-Hydroxyglutarate (2HG).** Chemical structure and simulated ^{1}H NMR spectrum of 2-hydroxyglutarate.

Detection of 2HG with optimized echo-time selection [169], spectral editing [169, 170], and 2D NMR [170] has been described.

2.2.25   *myo*-Inositol (mI) and *scyllo*-Inositol (sI)

myo-Inositol (mI; IUPAC name, (1*R*,2*S*,3*R*,4*R*,5*S*,6*S*)-cyclohexane-1,2,3,4,5,6-hexol) and scyllo-inositol (sI; IUPAC name, (1*R*,2*R*,3*R*,4*R*,5*R*,6*R*)-cyclohexane-1,2,3,4,5,6-hexol) are two of the nine possible stereoisomers of inositol. They have the same chemical formula ($C_6H_{12}O_6$) and molar mass (180.16 g mol^{-1}) and only differ in the orientation of proton H2.

  myo-Inositol is a cyclic sugar alcohol and represents the most abundant isomer form found in tissues. *myo*-Inositol has six NMR detectable protons that give rise to four groups of resonances. A doublet-of-doublets centered at 3.52 ppm originates from the equivalent ^{1}CH and ^{3}CH protons, while the equivalent ^{4}CH and ^{6}CH protons give rise to a triplet at 3.61 ppm.

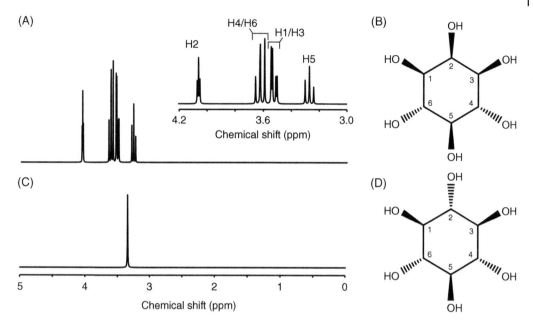

Figure 2.24 Myo- and *scyllo*-inositol. Chemical structures and simulated ^{1}H NMR spectra of (A, B) *myo*-inositol and (C, D) *scyllo*-inositol. The high degree of molecular symmetry leads to six magnetically equivalent protons and a singlet resonance for *scyllo*-inositol.

A smaller triplet from the ^{5}CH protons resides at 3.27 ppm and is typically obscured by the larger tCho resonances. Finally, the ^{2}CH proton is observed as a triplet at 4.05 ppm. The structure and ^{1}H NMR spectrum of *myo*-inositol are shown in Figure 2.24A and B.

myo-Inositol is readily detected in short-TE ^{1}H NMR spectra of the brain due to its high concentration of 4–8 mM [14, 171]. The exact function of *myo*-inositol is not known [122], although is plays an important function in osmotic regulation in the kidney [172] and has a biochemical relationship to messenger-inositol polyphosphates [173]. It has been proposed as a glial marker [174], although that characteristic has been challenged by the detection of *myo*-inositol in various neuronal cell types [175, 176]. Altered levels have been encountered in patients with mild cognitive impairment and Alzheimer's disease [177–180] and brain injury [181].

Similar to *myo*-inositol, *scyllo*-inositol is a cyclic sugar alcohol with six NMR detectable protons (Figure 2.24C and D). However, due to molecular symmetry all six protons are magnetically equivalent giving rise to a singlet resonance at 3.34 ppm. *scyllo*-Inositol is the second-most abundant isomer of inositol and has been detected in a wide range of species. *scyllo*-Inositol levels have been detected by ^{1}H MRS in normal brain [182, 183] and elevated levels during chronic alcoholism [184]. During the identification of *scyllo*-inositol in perchloric acid extracts, care should be taken to avoid misidentification due to methanol, which may be a contaminant of the extraction procedure [185].

2.2.26 Lactate (Lac)

Lactic acid (Lac; $C_3H_6O_3$; molar mass, 90.08 g mol^{-1}; IUPAC name, (2S)-2-hydroxypropanoic acid; pK_a, 3.86 and 15.1) or the anionic form of lactate is the end product of anaerobic glycolysis. The *L* or (*S*)-isomer is the biologically active form of lactate. In normal brain, lactate is present

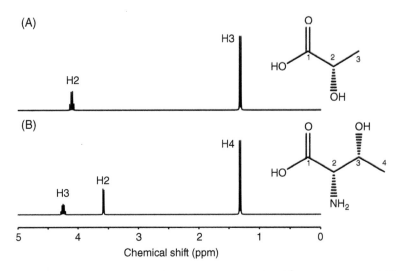

Figure 2.25 **Lactate and threonine.** Chemical structure and simulated ^{1}H NMR spectra of (A) lactate and (B) threonine.

at a low concentration of circa 0.5 mM. The lack of spectral overlap makes it readily observable in short-TE ^{1}H NMR spectra (see Figure 2.1) due to three equivalent protons forming at doublet signal at 1.31 ppm. Increased lactate concentrations have been observed under a wide variety of conditions in which blood flow (and hence oxygen supply) is restricted such as ischemic stroke [30, 186], hypoxia [187, 188], and tumors [32, 33]. Transient increases in lactate levels have also been observed in human brain during and following functional activation [189–192] and hyperventilation [193, 194]. It has been proposed that lactate links astroglial glucose uptake and metabolism to neuronal neurotransmitter cycling and metabolism in the so-called astroglial-neuronal lactate shuttle (ANLS) hypothesis [195, 196]. However, this hypothesis is not uniformly accepted and the definite experimental proof remains to be established [197].

Lactate has four non-exchangeable protons in methine and methyl groups (Figure 2.25A). The three equivalent methyl protons give rise to a doublet resonance at 1.31 ppm, while the single methine proton resonates as a quartet at 4.10 ppm. In normal brain tissue, lactate only overlaps with threonine and resonances from macromolecules. However, in tumors, stroke, or with inadequate localization, lactate can be overlapping with large lipid resonances. In these cases, lactate is best observed with spectral editing techniques, as will be discussed in Chapter 8. During ^{1}H MRS on tissue extracts the threonine signal can be shifted upfield from lactate through the use of a higher pH buffer (pH ~ 10). Lactate is one of the few spin-systems that can be considered weakly-coupled, even at low magnetic fields.

2.2.27 Macromolecules

In short-echo-time ^{1}H NMR spectra a significant fraction of the observed signal must be attributed to macromolecular resonances underlying those of the metabolites. Behar et al. [198–200] and Kauppinen et al. [201, 202] have extensively studied macromolecular resonances in human and animal brain. Based on dialysis and fractionation studies and *in vivo* detection a minimum of 10 macromolecular resonances can be observed (Figure 2.26B) at 0.93 ppm (M1), 1.24 ppm (M2), 1.43 ppm (M3), 1.72 ppm (M4), 2.05 ppm (M5), 2.29 ppm (M6), 3.00 ppm (M7), 3.20 (M8), 3.8–4.0 (M9), and 4.3 ppm (M10). While assignment to specific proteins is virtually impossible, the individual resonances can be tentatively assigned [199, 200] to methyl and

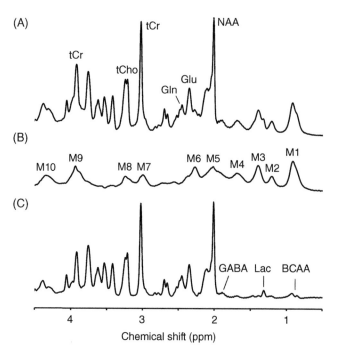

Figure 2.26 Macromolecule detection and suppression in ^{1}H NMR spectra from brain *in vivo*. (A) ^{1}H NMR spectrum acquired with LASER localization from rat brain at 9.4 T (TR/TE = 4000/12 ms, 180 μl). The sharp resonances from metabolites are superimposed on a baseline of broader resonances from macromolecules that are only visible between 0.5 and 2.0 ppm. (B) Same experiment as (A) with the extension of two nonselective inversion pulses (TR/TI1/TI2 = 3250/2100/630 ms). The double inversion methods "nulls" signals over a wide range of T_1 relaxation times (1100–1900 ms), while retaining signal from compounds with shorter T_1 relaxation times, like macromolecules. Ten upfield macromolecular resonances (M1–M10) are consistently detected from normal brain. (C) Subtracting (B) from (A) results in a macromolecule-free spectrum, allowing direct observation of lactate, alanine, and possibly GABA and branched-chain amino acids. The spectrum in (C) can also be obtained directly by suppression of metabolite signals with a single inversion-recovery sequence (TI = 200 ms, see Chapter 7).

methylene resonances of protein amino acids, such as leucine, isoleucine, and valine (M1), threonine and alanine (M2 and M3), lysine and arginine (M4 and M7), glutamate and glutamine (M5 and M6), and the less well-defined band of resonance between 3.0 and 4.5 ppm (M8–M10) to α-methine protons. The macromolecular resonances are further characterized by extensive scalar coupling patterns among the resonances, in particular between J_{M1-M5} = 7.3 Hz (doublet) and J_{M4-M7} = 7.8 Hz (triplet). The scalar coupling between M4 and M7 is of importance for the detection of γ-amino-butyric acid (GABA), as detailed in Chapter 8. The T_1 and T_2 relaxation times of macromolecular resonances are significantly shorter than those of the metabolites (see Chapter 3). The short T_2 relaxation time constants effectively eliminate macromolecular resonances from long-echo-time ^{1}H NMR spectra, leading to significant spectral simplification. The large difference between macromolecular and metabolite T_1s offers an opportunity for the selective detection of metabolites or macromolecules [1, 200, 203]. Figure 2.26 shows the selective detection of macromolecules in rat brain at 9.4 T. Using a double-inversion-recovery sequence (90° – TR – 180° – TI1 – 180° – TI2 – 90° – acquisition) with TR = 3250 ms, TI1 = 2100 ms, and TI2 = 630 ms [204], the longitudinal magnetization of the metabolites is close to zero, while the shorter T_1 of macromolecules leads to greater signal recovery. Therefore, this method "nulls" the metabolite signals, allowing selective observation of the macromolecular resonances. While a similar effect can be achieved with a single-inversion-recovery method, the suppression of metabolites would be incomplete due to significant variation in metabolite T_1s. A double-inversion-recovery method is relatively insensitive to metabolite T_1s. Macromolecular ^{1}H NMR spectra as shown in Figure 2.26B can be used as prior knowledge in spectral fitting algorithms to increase the accuracy and reliability of spectral quantification (see Chapter 9). Alterations in the macromolecular spectrum have been observed in stroke [205–207], tumors [208, 209], multiple sclerosis [210], and with aging [203, 211]. Relatively small differences in the macromolecular profile were found between human brain gray and white matter [212, 213], justifying the use of a single, average macromolecule spectrum for quantification purposes. The difference in T_1 relaxation between metabolites and macromolecules can also be used to eliminate the macromolecular resonances through judicious choice of repetition and inversion times [214]. This approach is typically used to reduce the contribution from extracranial lipid signal in MR spectroscopic imaging and will be discussed in more detail in Chapter 7.

2.2.28 Nicotinamide Adenine Dinucleotide (NAD$^+$)

Nicotinamide adenine dinucleotide (NAD$^+$, $C_{21}H_{27}N_7O_{14}P_2$; molar mass, 664.43 g mol^{-1}) is a coenzyme found in all tissues and plays a central role in cellular reduction–oxidation (redox) reactions. NAD$^+$, the oxidized form, accepts electrons from other molecules during which it is transformed into NADH, the reduced form. The NAD$^+$/NADH ratio is referred to as the redox state of a cell and is an important indicator of the activity and health of cells. In addition to its well-known role as a coenzyme for electron-transfer enzymes, NAD$^+$ is also a substrate for ADP-ribose transferases, poly(ADP-ribose) polymerases, cADP-ribose synthases, and sirtuins [215–217]. NAD$^+$ is therefore involved in gene expression and repair, calcium mobilization, metabolism and aging [218, 219], cancer and cell death [220, 221], and the timing of metabolism via the circadian rhythm [222, 223]. Cellular NAD$^+$ levels are maintained via active biosynthesis through salvage or *de novo* pathways involving tryptophan or niacin.

The *in vivo* detection of both NAD$^+$ and NADH has become possible with ^{31}P MR spectroscopy at high magnetic fields (see Section 2.3). In the normal brain NAD$^+$ and NADH concentration are on the order of 0.2–0.4 mM and 0.05–0.10 mM, respectively, with an age-dependent decrease in NAD$^+$ and an almost proportional increase in NADH [224]. *In vivo* ^{1}H MRS does not allow the detection of NADH due to extensive overlap with more concentrated compounds. However, the *in vivo* detection of NAD$^+$ has been demonstrated on rat [225] and human [226] brain via nonoverlapping resonances around 9 ppm originating from protons in the nicotinamide group. NAD$^+$ consists of two nucleotides that are connected via the phosphate group (Figure 2.27A). The ribose moieties and the adenine base give resonances in the 4.0–6.0 ppm and 8.0–8.5 ppm chemical shift ranges, respectively (Figure 2.27B and C). Significant spectral overlap makes detection of these resonances difficult. The nicotinamide moiety gives unique resonances at 9.33 ppm (H2), 8.85 ppm (H4), and 9.16 ppm (H6) that can be detected without spectral overlap (Figure 2.26B and C). However, the *in vivo* detection of NAD$^+$ with ^{1}H MRS requires the avoidance of water suppression in order to minimize signal loss due to exchange and nuclear Overhauser effects [226].

2.2.29 Phenylalanine

Phenylalanine ($C_9H_{11}NO_2$; molar mass, 165.19 g mol^{-1}; IUPAC name, (2*S*)-2-amino-3-phenyl-propanoic acid; pK_a, 1.83 and 9.13) is an essential aromatic amino acid composed of a phenyl aromatic ring connected to the methyl group of alanine. It is a precursor for the amino acid

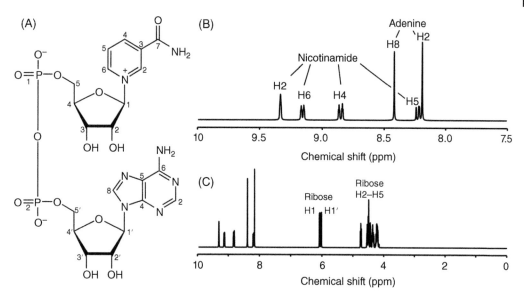

Figure 2.27 **Nicotinamide adenine dinucleotide (NAD⁺).** (A) Chemical structure and (B) upfield and (C) downfield parts of the simulated ¹H NMR spectrum of NAD⁺. The H2, H4, and H6 protons in the nicotinamide group provide three unique resonances around 9.0 ppm.

tyrosine which in turn is used for catecholamine (dopamine, epinephrine, and norepinephrine) synthesis. The *L* or (*S*)-isomer of phenylalanine is normally present in the human brain at circa 0.2 mM, following an age-dependent decrease. ¹H MRS has been used to detect elevated cerebral phenylalanine levels in human [227] and animal [228] brain. At lower magnetic field, the vascular contribution of phenyl alanine is NMR visible and should be accounted for during signal quantification [229]. The concentration is elevated up to 5.0 mM in phenylketonuria (PKU), an inborn error of phenylalanine metabolism. In PKU patients, the hydroxylation of phenylalanine to tyrosine is disturbed due to a deficiency of the liver enzyme phenylalanine hydroxylase or its cofactor. Proton ¹H MRS has been used to detect and quantify phenylalanine in PKU patients [230–232]. PKU is treatable by restricted dietary phenylalanine intake.

Phenylalanine has eight NMR observable protons, five of which reside in the phenyl ring and three in the aliphatic side chain. The five nonequivalent phenyl ring protons give a complex multiplet between 7.30 and 7.45 ppm and represent the typically observed resonance *in vivo*. The ²CH proton resonates as a doublet-of-doublets at 3.98 ppm, while the ³CH₂ protons give rise to two doublet-of-doublets centered at 3.11 and 3.27 ppm. Figure 2.28A shows the structure and ¹H NMR spectrum of L-phenylalanine.

2.2.30 Pyruvate

Pyruvic acid ($C_3H_4O_3$; molar mass, 88.06 g mol⁻¹; IUPAC name, 2-oxopropanoic acid) or its conjugate base, pyruvate, is the simplest alpha-keto acid and plays an important intermediate role in glucose metabolism. One molecule of glucose breaks down in two molecules of pyruvate during glycolysis. Provided that sufficient oxygen is available, pyruvate is converted into acetyl-coenzyme A, which is the main input for a series of reactions known as the TCA cycle, also known as the Krebs cycle or the citric acid cycle. Pyruvate is also broken down to oxaloacetate by a so-called anaplerotic reaction with the purpose of replenishing TCA cycle intermediates. If insufficient oxygen is available, the acid is broken down anaerobically by

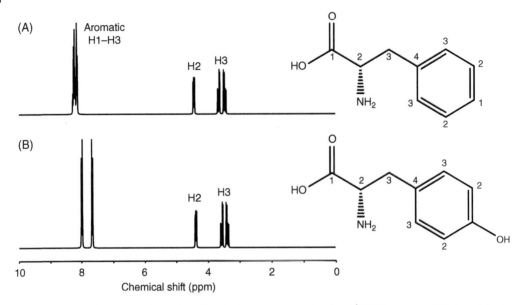

Figure 2.28 **Phenylalanine and tyrosine.** Chemical structures and simulated ^{1}H NMR spectra of (A) phenylalanine and (B) tyrosine.

the enzyme lactate dehydrogenase (LDH) and the coenzyme NADH, creating lactic acid (lactate). Pyruvate is readily transported from the blood and may have neuroprotective properties in stroke [233].

The singlet resonance of pyruvate at 2.36 ppm is readily observable in high-resolution ^{1}H NMR spectra of brain extracts. However, the typical *in vivo* concentration is below 0.2 mM, such that detection has been limited to pathologies, like cystic lesions [234] and neonatal pyruvate dehydrogenase deficiency [235].

Pyruvate is at the center of the hyperpolarized ^{13}C MR field in that it is currently the only FDA approved compound to be used in humans. Intravenous administration of hyperpolarized [1-^{13}C]-pyruvate leads to rapid conversion to [1-^{13}C]-lactate, which provides information on LDH activity at high spatial and temporal resolution. More details on hyperpolarized MRS can be found in Chapter 3.

2.2.31 Serine

Serine ($C_3H_7NO_3$; molar mass, 105.09 g mol^{-1}; IUPAC name, (2S)-2-amino-3-hydroxypropanoic acid; pK_a, 2.21 and 9.15) is a nonessential amino acid of which the *L* or (*S*)-isomer is commonly found in mammalian proteins. Serine can be synthesized in the body from other metabolites, including glycine and 3-phosphoglycerate. Serine participates in the biosynthesis of purines and pyrimidines, cysteine, and a large number of other metabolites. The ^{1}H NMR spectrum (Figure 2.29) is characterized by a cluster of resonances between 3.8 and 4.0 ppm originating from three protons in methine and methylene groups that form an *ABX* spin-system. Serine is present throughout the brain at a concentration of 0.5–1.5 mM. However, since serine resonates in a complicated region of the ^{1}H NMR spectrum with significant spectral overlap, the detection of serine is difficult, even at high magnetic fields [1]. Serine has been detected in the frontal cortex at 7 T through the use of spectral editing [236].

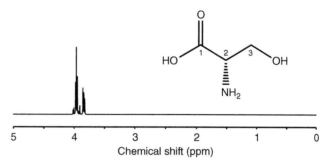

Figure 2.29 **Serine.** Chemical structure and simulated ^{1}H NMR spectrum of serine.

2.2.32   Succinate

Succinic acid ($C_4H_6O_4$; molar mass, 118.09 g mol^{-1}; IUPAC name, butanedioic acid, pK_a, 4.2 and 5.6) or its conjugate base, succinate, is part of the TCA cycle, where it is formed from succinyl CoA by succinyl CoA synthetase. Alternatively, succinate can be formed from succinate semi-aldehyde by succinate semialdehyde dehydrogenase (SSADH) as part of the GABA shunt. Succinate is converted to the TCA cycle intermediate fumarate by succinate dehydrogenase, whereby FAD is converted to FADH2. Succinate is typically present at concentrations well below 0.5 mM.

The four protons in succinate are magnetically equivalent, giving rise to a singlet resonance at 2.39 ppm. While this peak is readily observed in high-resolution ^{1}H NMR spectra of extracts, the *in vivo* observation is complicated by overlapping resonances from glutamate, glutamine, and pyruvate [237]. Increased succinate levels have been reported in brain abcesses [238] and leukoencephalopathy [239, 240].

2.2.33   Taurine (Tau)

Taurine (Tau; $C_2H_7NO_3S$; molar mass, 125.14 g mol^{-1}; IUPAC name, 2-aminoethane-1-sulfonic acid) has two methylene groups with nonequivalent protons, thereby forming an *AA'XX'* spin system (Figure 2.30). At higher magnetic fields taurine can be considered as an A_2X_2 spin system giving two triplets centered at 3.25 and 3.42 ppm. When quantifying the tCho methyl resonance at 3.21 ppm, the upfield taurine multiplet should be taken into account as it forms a significant fraction of the observed "choline" resonance. Especially at lower magnetic fields, taurine is essentially overlapping with *myo*-inositol and choline, such that unambiguous detection requires the use of spectral editing methods [241–243].

The exact function of taurine is not known, but it has been proposed as an osmoregulator and a modulator of neurotransmitter actions. Taurine is present in all cells of the CNS, but is spatially heterogeneous with higher levels in the olfactory bulb, retina, and cerebellum [244].

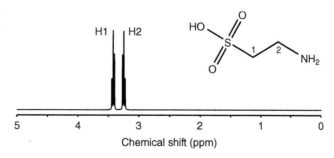

Figure 2.30 **Taurine.** Chemical structure and simulated ^{1}H NMR spectrum of taurine.

The concentration of taurine has a strong age dependence, decreasing from circa 12 mM at birth to 6–8 mM in adult rat brain [81]. The concentration in human brain is significantly lower at circa 1.5 mM [2]. Taurine is largely obtained through food, but it is a nonessential compound as it can be synthesized from other sulfur-containing amino acids, like cysteine.

2.2.34 Threonine (Thr)

Threonine (Thr; $C_4H_9NO_3$; molar mass, 119.12 g mol^{-1}; IUPAC name, (2S,3R)-2-amino-3-hydroxybutanoic acid, pK$_a$, 2.09 and 9.10) is obtained from food and is therefore an essential amino acid. While mammals cannot synthesize threonine, plants and microorganisms are capable of synthesizing threonine from aspartate. Threonine is an important amino acid for the nervous system, with relatively high levels of circa 0.3 mM. It has been used as a supplement to help alleviate anxiety and some cases of depression. Threonine is one of two proteinogenic amino acids with two chiral centers, the other being isoleucine. Threonine can therefore exist in four possible stereoisomers whereby the name L-threonine is used for one single diastereomer, (2S,3R)-2-amino 3-hydroxybutanoic acid. Threonine is a precursor of glycine and can be degraded into α-ketobutyrate.

Threonine has five non-exchangeable protons in a methyl and two methine groups (Figure 2.25B). The ^{2}CH and ^{3}CH protons resonate as a doublet and multiplet at 3.58 and 4.25 ppm, respectively. The most intense resonance of the three methyl protons appears as a doublet at 1.32 ppm and is therefore essentially overlapping with the methyl protons of lactate. Since the scalar coupling patterns of threonine and lactate are very similar, special care should be taken when trying to separate them by spectral editing [245]. When separation is not possible, it is advised to refer to the 1.3 ppm resonance as total lactate, being the sum of lactate and threonine.

2.2.35 Tryptophan (Trp)

Tryptophan (Trp; $C_{11}H_{12}N_2O_2$; molar mass, 204.23 g mol^{-1}; IUPAC name, (2S)-2-amino-3-(1H-indol-3-yl)propanoic acid; pK$_a$, 2.38 and 9.39) is an essential, proteinogenic amino acid that is necessary for the production of the neurotransmitter serotonin, the neurohormone melatonin, vitamin B$_3$ (niancin), and NAD$^+$. Normally present at low concentrations of circa 0.03 mM, the brain concentration can be increased by tryptophan consumption. Higher tryptophan levels lead to a twofold increase in serotonin synthesis, which has led to investigations for use in the treatment of mild insomnia as well as antidepressant. Increased brain tryptophan levels are also associated with hepatic encephalopathy [246].

Tryptophan has eight NMR detectable protons. The ^{2}CH proton gives a singlet at 7.31 ppm, whereas the four phenyl ring protons give two multiplets centered at 7.20 and 7.28 ppm. The three aliphatic side chain protons give three doublet-of-doublets 3.29, 3.48, and 4.05 ppm. Although tryptophan is normally present at low concentrations, the indole ring resonances may combine sufficiently to give an identifiable signal even at low magnetic fields. Figure 2.31 shows the structure and ^{1}H NMR spectrum of tryptophan.

2.2.36 Tyrosine (Tyr)

Tyrosine (Tyr; $C_9H_{11}NO_3$; molar mass, 181.19 g mol^{-1}; IUPAC name, (2S)-2-amino-3-(4-hydroxyphenyl)propanoic acid; pK$_a$, 2.20 and 9.11) is a nonessential amino acid that can be synthesized from phenylalanine. It is a precursor of the neurotransmitters epinephrine, norepinephrine, and dopamine and of the thyroid hormones thyroxine and triiodothyronine.

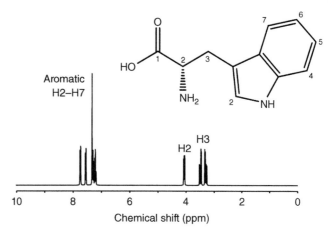

Figure 2.31 **Tryptophan.** Chemical structure and simulated ^{1}H NMR spectrum of tryptophan.

Tyrosine is converted to DOPA by tyrosine dehydroxylase. It plays a key role in signal trans-duction, since it can be phosphorylated by protein kinases to alter the functionality and activity of certain enzymes. (In its phosphorylated state, it is sometimes referred to as phosphotyrosine.) Normal brain concentrations decrease from circa 0.25 mM after birth to 0.05 mM in adulthood, although it can be increased following oral consumption or hepatic encephalopathy.

Tyrosine has seven NMR observable protons, distributed over a phenyl ring and an aliphatic side chain. The four phenyl protons give a multiplet between 6.89 and 7.19 ppm. The CH and CH_2 protons in the aliphatic side chain form an *ABX* spin-system, giving rise to three doublet-of-doublets centered at 3.04, 3.19, and 3.93 ppm. Figure 2.28B shows the ^{1}H NMR spectrum and structure of tyrosine.

2.2.37 Water

Water (H_2O; molar mass, $18.02 \, g \, mol^{-1}$) is the most abundant compound in the human and animal body with water contents ranging from 73% for cerebral white matter, 82% for cerebral gray matter, >95% for cerebrospinal fluid (CSF), and 78% for skeletal muscle [247, 248]. Cerebral water content has a strong age-dependence with 92% in 20–22 week fetuses, 90% in newborns, and 77% (average) in adults. Since the water content only changes moderately among different pathologies, water is a reasonable candidate for internal concentration referencing (see Chapter 9). The T_1 and T_2 relaxation times, as well as magnetization transfer characteristics of water also vary greatly among tissues, leading to excellent image contrast in MRI applications (see Chapters 3 and 4).

Besides many other unique applications that will be discussed in forthcoming chapters, the water chemical shift can be used to detect temperature changes noninvasively *in vivo*. The temperature dependence of the proton resonance frequency was first described by Hindman [249]. In pure water the proton resonance frequency changes by approximately −0.01 ppm per °C and is primarily caused by changes in the electronic shielding of the proton nucleus, with smaller contributions from changes in bulk magnetic susceptibility. Quantitative descriptions of this phenomenon are given by Hindman [249] and de Poorter [250]. A qualitative, general understanding of this effect can be gained by noting that the shielding of the proton nuclei by the surrounding electrons in an isolated water molecule is better than for a water molecule in solution. This is because in solution, adjacent water molecules form hydrogen bonds/bridges between hydrogen and oxygen nuclei. The electronegative oxygen atom partially "pulls" the electron cloud from the hydrogen nucleus, leading to a reduced electronic shielding and hence a higher Larmor resonance frequency. When the temperature increases, the hydrogen bonds stretch, bend, and break more frequently such that the hydrogen nuclei spend less time on average in a hydrogen bonded state to another water molecule. As a consequence, as the temperature rises the electronic shielding of the proton nucleus increases, leading to a lower local magnetic field and hence a lower Larmor resonance frequency. Change in the bulk or volume magnetic susceptibility is a relatively minor contribution to the temperature dependence of water. Pure water is diamagnetic and has a bulk magnetic susceptibility of -9.05 ppm at room temperature. Water becomes less diamagnetic with increasing temperature by approximately $+0.001$ ppm/°C mainly due to a decrease in water density [251].

Temperature changes can, in principle, be obtained by monitoring the Larmor frequency of the water resonance. However, many other effects, like drift of the magnetic field and motion, also lead to a shift in the water frequency. For NMR spectroscopy applications, the method can be made more reliable by measuring the chemical shift difference between water and a reference compound that does not have a temperature-dependent shift. The methyl group of NAA has been used for this purpose [252–254]. For MRI applications, the shift in water resonance frequency is typically detected as a phase change [255, 256]. Temperature mapping using the water resonance has a typical accuracy of circa 0.1 °C, which is mainly due to the relatively small temperature dependence upon the chemical shift. Other compounds have been proposed and used that can have frequency shifts of >1.0 ppm per °C with the possible option of simultaneous pH determination [257–259]. However, the use of exogenous compounds makes the measurement invasive, while the lower concentration may not necessarily improve the accuracy of the temperature measurement.

2.2.38 Non-cerebral Metabolites

The metabolites discussed so far are primarily located in the central and peripheral nervous systems. These metabolites can be detected with relative ease due to the absence of mobile lipids within the healthy brain. Tissue extract studies and tissue magic angle spinning studies have shown that the metabolic content of non-cerebral tissues can be similar to that of brain tissue. Many metabolites including alanine, ATP, choline, creatine, glucose, glutamate, glutamine, lactate, NAD$^+$, pyruvate, and succinate are ubiquitous throughout the human and animal body. However, in many non-cerebral tissues the MR resonances from metabolites are often overwhelmed by the presence of dominant signals from triglycerides. In the next sections a number of metabolites are discussed that can be uniquely detected in non-cerebral tissues despite the presence of strong lipid signals. The chemical shifts and scalar coupling constants of non-cerebral metabolites are summarized in Table 2.2.

2.2.39 Carnitine and Acetyl-carnitine

Carnitine (Crn, $C_7H_{15}NO_3$; molar mass, 161.20 g mol^{-1}; IUPAC name, (3R)-3-hydroxy-4-(trimethylazaniumyl)butanoate) and acetylcarnitine ($C_9H_{17}NO_4$; molar mass, 203.24 g mol^{-1}; IUPAC name, 3-(acetyloxy)-4-(trimethylazaniumyl)butanoate) are involved in transporting long-chain fatty acids across the mitochondrial membrane for β-oxidation. Carnitine may also play a regulatory role in substrate switching and glucose homeostasis. Acetylcarnitine is formed under conditions in which acetyl-CoA formation, either as an end product of glycolysis or β-oxidation, exceeds the amount required by the TCA cycle. Free carnitine can act as a sink for excess acetyl groups in a reversible reaction catalyzed by the enzyme carnitine acetyltransferase (CRAT). Acetylcarnitine, like other acylcarnitine esters, can readily be exported out of the mitochondria. Elevated acetylcarnitine levels have been observed following exercise

Table 2.2 ^{1}H chemical shifts and scalar coupling constants for metabolites commonly detected in ^{1}H NMR spectra of non-cerebral tissues *in vivo* or *in vitro*.

Compound	Chemical group	Chemical shift (ppm)	Multiplicity^a	Interaction	Scalar coupling (Hz)
Acetyl carnitine			s	—	—
Acetyl moiety	$^2CH_3$	2.131	s	—	—
Carnitine moiety	$(CH_3)_3$	3.183	s	—	—
	$^2CH_2$	2.498	dd	2–2′	−15.48
		2.629	dd	2–3 / 2′–3	7.71 / 5.56
	^3CH	5.593	m	3–4 / 3–4′	1.31 / 8.77
	$^4CH_2$	3.598	dd	4–4′	−14.49
		3.842	dd		
Carnitine	$(CH_3)_3$	3.212	s	—	—
	$^2CH_2$	2.411	dd	2–2′	−15.36
		2.443	dd	2–3 / 2′–3	6.17 / 7.18
	^3CH	4.558	m	3–4 / 3–4′	2.09 / 9.26
	$^4CH_2$	3.409	dd	4–4′	−13.94
		3.421	dd		
Carnosine	^2CH	8.05^b	s	—	—
	^4CH	7.04^b	s	—	—
Citrate	$^2CH_2$	2.53^b	d	2–2′	−15.3^b
		2.66^b	d		
Deoxymyoglobin	NH	79	s	—	—
Putrescine (polyamine)	$^1CH_2$	3.041	t	n.d.	n.d.
	$^2CH_2$	1.754	m	n.d.	n.d.
Spermidine (polyamine)	$^1CH_2$	2.579	t	n.d.	n.d.
	$^2CH_2$	1.473	m	n.d.	n.d.
	$^3CH_2$	1.461	t	n.d.	n.d.
	$^4CH_2$	2.574	m	n.d.	n.d.
	$^5CH_2$	2.623	t	n.d.	n.d.
	$^6CH_2$	1.612	m	n.d.	n.d.
	$^7CH_2$	2.636	t	n.d.	n.d.
Spermine (polyamine)	$^1CH_2$	2.698	t	n.d.	n.d.
	$^2CH_2$	1.674	m	n.d.	n.d.
	$^3CH_2$	2.698	t	n.d.	n.d.
	$^4CH_2$	2.665	t	n.d.	n.d.
	$^5CH_2$	1.533	m	n.d.	n.d.

^{a}Multiplicities are defined as: singlet (s), doublet (d), triplet (t), quartet (q), multiplet (m), and double doublet (dd).
^{b}Sensitive to pH in the physiological range.
n.d., not determined.

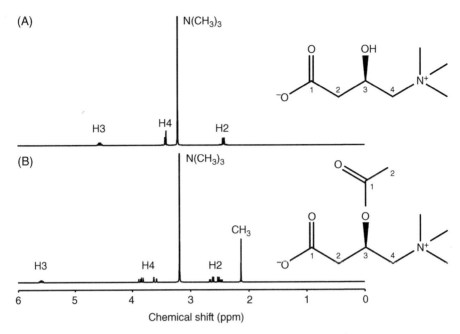

(A)

(B)

H3   H4   H2

N(CH₃)₃

N(CH₃)₃

CH₃

6   5   4   3   2   1   0

Chemical shift (ppm)

Figure 2.32 **Carnitine and acetyl-carnitine.** Chemical structures and simulated ^{1}H NMR spectra of (A) carnitine and (B) acetyl-carnitine.

[260, 261], whereby lower levels in type 2 diabetes patients showed a strong correlation with decreased insulin sensitivity [262]. The majority of carnitine comes from dietary sources such as red meat and dairy products. However, carnitine can be synthesized in the liver, kidney, and brain from the amino acids lysine and methionine.

Carnitine produces a prominent singlet signal at 3.21 ppm originating from the nine equivalent protons in the trimethyl ammonium (TMA) group (Figure 2.32). However, since this resonance overlaps with signal from other TMA-containing molecules (choline, betaine) it is not possible to disentangle the carnitine contribution. Acetyl-carnitine produces a second singlet peak at 2.13 ppm originating from the three acetyl protons. This signal has been detected in ^{1}H MRS spectra of skeletal muscle at short [260] and long [262] echo-times. At high magnetic fields (7 T) the chemical shift difference in the TMA group of carnitine and acetylcarnitine is sufficiently large to provide a visual separation between the compounds [261].

2.2.40   Carnosine

Carnosine, also known as β-alanyl-L-histidine ($C_9H_{14}N_4O_3$; molar mass, 226.24 g mol^{-1}; IUPAC name, (2S)-2-(3-aminopropanamido)-3-(1H-imidazol-5-yl)propanoic acid; pK_a, 6.83 (imidazole)) is a dipeptide between alanine and histidine and part of a series of compounds referred to as aminoacyl-histidine dipeptides. Other members include homocarnosine (γ-aminobutyryl-L-histidine) and anserine (β-alanyl-L-1-methyl-histidine). This group of naturally-occurring histidine-containing molecules is particularly abundant in excitable tissues, such as muscle and nervous tissue. The biological functions of the aminoacyl-histidine dipeptides remain enigmatic, although the roles of pH buffer and antioxidant have been proposed. Carnosine was first observed over a century ago and has subsequently been found by a wide range of techniques, including ^{1}H NMR spectroscopy (see Figure 2.33). The pK of the C-2 and C-4 protons

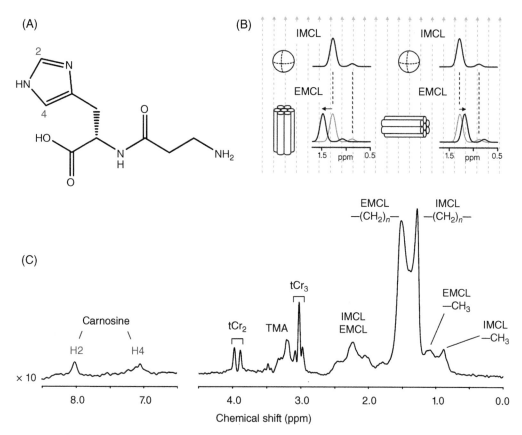

Figure 2.33 **^{1}H MRS on human skeletal muscle.** (A) Chemical structure of carnosine. (B) Resonance shifts due to different magnetic susceptibility distributions. IMCL resides in small spherical droplets that do not show a shift in resonance frequency with respect to the magnetic field direction. EMCL reside in bundles or sheets that show a downfield shift when the bundles are parallel to the main magnetic field. EMCL resonance shows an upfield shift when oriented perpendicular to the main magnetic field. (C) ^{1}H NMR spectrum acquired from human skeletal muscle (TR/TE = 4000/15 ms, 6 ml). Residual dipolar interactions lead to a splitting of the creatine methyl and methylene groups. IMCL and EMCL resonances are well separated due to the approximately parallel orientation of the muscle fibers to the main magnetic field. The two downfield resonances of carnosine are sensitive to pH in the physiological range.

on the imidazole ring of histidine are in the physiological pH range ($pK_a = 6.7$), providing a noninvasive method to measure intracellular pH with ^{1}H NMR spectroscopy *in vivo*. Yoshizaki et al. [263] initially used carnosine to determine pH in excised frog muscle, which has subsequently been followed by studies *in vivo*, including human muscle. Several studies [264, 265] have shown an excellent correlation between the determination of intracellular pH by ^{1}H (carnosine) and ^{31}P (inorganic phosphate) NMR.

Using carnosine for intracellular pH determination has several advantages over the more traditional methods using the chemical shift of inorganic phosphate in ^{31}P NMR spectra. First, the sensitivity of carnosine detection is very high due to the inherently high sensitivity of ^{1}H NMR, the high concentration of carnosine in skeletal muscle (up to 20 mM for human muscle), and the relatively short T_1 relaxation times. This is especially important in studies in which the concentration of inorganic phosphate significantly decreases (e.g. the recovery stage in muscle exercise studies). The high sensitivity of carnosine detection would allow a substantially improved time resolution in exercise studies. Second, the measurement of intracellular pH using carnosine is relatively insensitive to the presence of divalent cations, like free magnesium (Mg^{2+}). This is especially important in exercise studies where intracellular free Mg^{2+} concentrations increase.

2.2.41 Citric Acid

Citric acid ($C_6H_8O_7$; molar mass, 192.12 g mol^{-1}; IUPAC name, 2-hydroxypropane-1,2,3-tricarboxylic acid; pK_a, 3.13, 4.76, and 6.39) or the anionic form of citrate is best known as an intermediate of the Krebs or citric acid cycle, where it is formed when acetyl-CoA donates a carbonyl group to oxaloacetate. In brain tissue the concentration of citrate as well as that of all other intermediates of the citric acid cycle is below the NMR detection limit. However, it is well known [266] that healthy prostate epithelial cells synthesize and secrete citrate in relatively large amounts (~60 mM), due to the fact that high levels of zinc inhibit the oxidation of citrate in the Krebs cycle. In cancerous epithelial cells, zinc levels are drastically reduced, resulting in greatly reduced citrate levels. As a result, monitoring citrate levels in the prostate has been used to obtain direct biochemical information to aid in the diagnosis of malignant adenocarcinoma and benign prostatic hyperplasia [267–271].

Citrate was first detected in human prostate *in vivo* by natural abundance ^{13}C NMR spectroscopy as described by Sillerud et al. [272]. Current detection methods of prostate citrate rely almost exclusively on ^{1}H NMR spectroscopy. The proton NMR spectrum of citrate arises from two magnetically equivalent CH_2 moieties (see Figure 2.34). While the methylene groups are magnetically equivalent due to the molecular symmetry, the two protons in each methylene group are magnetically inequivalent. As a result, citrate is a classic example of a strongly coupled *AB* spin system. The exact chemical shift and scalar coupling constants are dependent on the pH [273], as well as the cation concentration [274]. Average chemical shift and scalar coupling values for the physiological range are 2.57 ppm, 2.72 ppm, and 15.5 Hz. The exact appearance of the four citrate resonances is also a function of magnetic field strength, echo-time, as well as other timing parameters (like TM in STEAM) and RF pulses, all of which require careful optimization to achieve high sensitivity citrate detection [275–277]. Citrate detection is often achieved through direct ^{1}H MRS, due to the absence of significant spectral overlap. However, some authors have argued that detection of reduced citrate levels in cancerous prostate can benefit from spectral editing methods which remove contributions of partially overlapping resonances from tCr and glutamine [278]. Increased citrate concentrations have been reported in brain tumors [279–281].

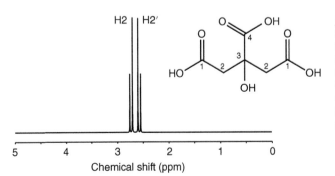

Figure 2.34 **Citrate.** Chemical structure and simulated ^{1}H NMR spectrum of citrate. While the two methylene groups are equivalent due to molecular symmetry, the H2 and H2′ protons within one methylene group at not equivalent leading to the characteristic NMR spectrum of a strongly coupled *AB* spin-system (compare with Figure 1.17A).

2.2.42 Deoxymyoglobin (DMb)

Myoglobin (Mb), a 16.7 kD protein, plays an important role in muscle physiology as an oxygen storage compound and a facilitator of oxygen diffusion. It has been shown by several groups that oxygen saturation in human skeletal and cardiac muscle can be determined by detecting the deoxymyoglobin (DMb) signal by ^{1}H NMR spectroscopy at ~79 ppm [282–285]. Despite its low concentration (~300 μM during muscle ischemia) the detection of DMb by ^{1}H NMR spectroscopy is possible because the resonance position of the N-δ proton in the proximal F8 histidine of DMb is sufficiently shifted downfield, away from the more intense resonances of water and lipids. The short T_1 relaxation time of ~10 ms allows substantial signal averaging and hence an improved sensitivity. The oxygenated form of Mb does not have any resonances with paramagnetic shifts and therefore is unobservable. Under normoxic conditions Mb is completely oxygenated, such that no signal from DMb can be detected at rest. However, under ischemic conditions, as achieved by using a pressure cuff or during heavy exercise, a large DMb signal can be readily observed [286]. Erythrocyte deoxyhemoglobin in the vascular compartment also gives a NMR signal at a similar chemical shift. However, the contribution of the deoxyhemoglobin signal can often be considered negligible due to the small vascular volume.

2.2.43 Lipids

The ^{1}H signal from triglycerides is only second to that of water and most non-cerebral tissues display prominent lipid signals. The high abundance of lipids is one of the main reasons that ^{1}H MRS outside the brain has seen limited applications. However, the characterization of tissue lipid content and composition is in itself a valuable goal for which ^{1}H MRS is well suited. Figure 2.35 holds a high-resolution ^{1}H NMR spectrum of pure vegetable oil showing up to 10 resonances originating from distinct chemical groups within the triglyceride structure. The terminal methyl group (A, red) gives a triplet resonance at circa 0.9 ppm. The bulk of the methylene proton signal resides in a complex multiplet at circa 1.3 ppm. However, several specific methylene groups provide signal at unique chemical shift positions. Methylene protons next to a single double bond (C, green) give a signal at circa 2.0 ppm, whereas methylene protons next to two double bonds (E, blue) resonate at circa 2.8 ppm. Finally, the methylene protons closest to the carbonyl group (F, G) have unique chemical shifts at circa 1.7 and 2.3 ppm. The methine protons that are part of the double bond (D, purple) can be observed around 5.3 ppm. Signal E is unique for poly-unsaturated lipids, whereas signals C and D represent all unsaturated lipids. Together with resonances that are common to all lipids (e.g. F or G) the ^{1}H NMR spectrum can be used to determine the relative amounts of saturated, mono-unsaturated, and poly-unsaturated triglycerides [133, 287]. Figure 2.35C shows a ^{1}H NMR spectrum from human adipose tissue. Even though the *in vivo* line widths are much wider, the information content is similar to that obtained under high-resolution conditions.

The characterization of lipids in skeletal muscle is a special case of that shown in Figure 2.35 in that lipids can reside in different compartments [288–291]. The lipids detected in typical ^{1}H NMR spectra of skeletal muscle are located in either an intracellular or an extracellular compartment. Lipids in the intracellular compartment are generally shaped as spherical droplets and in close proximity to the mitochondria. These lipids are referred to as intramyocellular lipids (IMCL) and represent a pool that displays active turnover and metabolism, for example, during exercise. In the extracellular environment, lipids are located in interstitial adipocytes clustered in relatively long strands between the muscle fibers. These lipids are referred to as

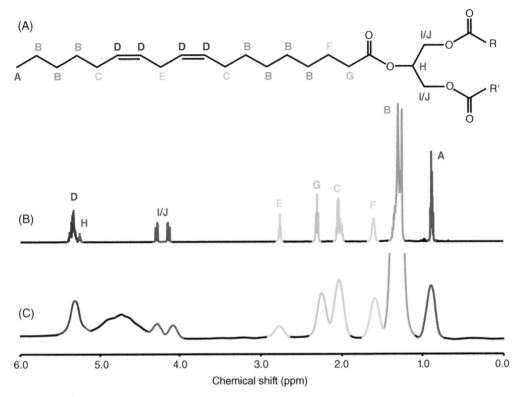

Figure 2.35 **¹H NMR of triglycerides.** (A) Structure of a triglyceride with the top and bottom fatty acids undefined (groups *R* and *R'*), and the middle fatty acid chosen as linoleic acid, a polyunsaturated omega-6 fatty acid. (B) 500 MHz ¹H NMR spectrum of vegetable oil. Ten distinct resonances can be observed that correspond to the various protons in (A). (C) STEAM-localized ¹H NMR spectra of human leg adipose tissue. Chemical shifts of triglyceride resonances are shown in Table 2.3.

extramyocellular lipids (EMCL) and represent a more inert lipid pool. The IMCL and EMCL lipid resonances can be distinguished due to a magnetic susceptibility induced shift in resonance frequency that depends on the lipid compartment shape and orientation [292]. The magnetic susceptibility difference between water ($\chi = -9.05$ ppm) and lipids ($\chi = -8.44$ ppm) leads to a small magnetic field difference (see Chapter 10 for more details on magnetic susceptibility effects) between the water and lipid compartments. The magnetic field difference is dependent on the shape and orientation of the lipid compartment. For the spherical IMCL compartment, the chemical shift of the bulk methylene groups is circa 1.28 ppm and independent of the orientation (Figure 2.33B). For the cylindrical EMCL compartment, the magnetic field and thus the chemical shift becomes dependent on the angle α of the cylinder relative to the magnetic field B_0 proportional to $\Delta\chi(3\cos^2\alpha - 1)$. For EMCL cylinders parallel to B_0, the methylene protons resonate around 1.48 ppm, whereas cylinders perpendicular to B_0 will lead to resonances around 1.18 ppm. The ability to separate IMCL from EMCL therefore depends on the muscle fiber orientation, which in turn depends on the muscle type. In addition, some muscles have a wider dispersion of fiber orientations, leading to significant line broadening for EMCL. The ability to reliably detect IMCL signals also rests on excluding large patches of EMCL or adipose lipids that can quickly overwhelm the smaller IMCL signals. Figure 2.33C shows a typical ¹H MR spectrum from human muscle, displaying a clear differentiation between EMCL and IMCL.

Table 2.3 ^{1}H and ^{13}C chemical shifts and heteronuclear scalar coupling constants of triglycerides.

Carbon position	Label[a]	Chemical shift (ppm)[b]		Scalar coupling $^1J_{CH}$ (Hz)[c]	Comments
		^{1}H	^{13}C		
CH_3-CH_2-	A	0.89	14.16	126.6	
$CH_3-CH_2-CH=$	B_{Ln}	2.04	20.73	132.8	Linolenic ω3 acid
$CH_3-CH_2-CH_2-$	B	1.31	22.94	130.0	
$-CH_2-CH_2-COOR$	F	1.58	25.04	136.1	
$=CH-CH_2-CH=$	E	2.76	25.82	127.4	
$=CH-CH_2-CH_2-$	C	2.03	27.41	129.4	
$-(CH_2)_n-$	B	1.33	29.31	133.8	
$-(CH_2)_n-$	B	1.31	29.63	128.6	
$-(CH_2)_n-$	B	1.32	30.09	132.9	
$CH_3-CH_2-CH_2-CH_2-$	B_L	1.29	31.88	127.8	Linoleic ω6 acid
$CH_3-CH_2-CH_2-CH_2-$	B	1.27	32.24	125.8	
$-CH_2-CH_2-COOR\ (\alpha)^d$	G	2.25	33.85	131.7	
$-CH_2-CH_2-COOR\ (\beta)$	G	2.25	34.12	132.9	
H_2COOR	I	4.30	62.05	161.3	Glycerol H1
H_2COOR	J	4.07	62.06	159.3	Glycerol H1′
$HCOOR$	H	5.22	69.21	155.4	Glycerol H2
$-CH=CH-CH_2-CH=CH-$	D_i	5.31	128.16	159.8	
$-CH=CH-CH_2-CH=CH-$	D_o	5.33	129.68	157.5	
$-CH_2-CH=CH-CH_2-$	D	5.33	129.91	156.4	
$-CH_2-COOR\ (\alpha)$	K	—	—	—	
$-CH_2-COOR\ (\beta)$	K	—	—	—	

[a]Proton and carbon labels and colors indicated in Figure 2.35 and Figure 2.40.
[b]Errors in proton and carbon chemical shifts typically less than 0.01 and 0.03 ppm, respectively.
[c]Errors in heteronuclear scalar couplings less than 2.5 Hz.
[d]Chemical shift reference for ^{1}H (2.25 ppm) and ^{13}C (33.85 ppm). α and β refer to the two outer and single inner fatty acid chains in a triglyceride molecule, respectively.

2.2.44 Spermine and Polyamines

Spermine ($C_{10}H_{26}N_4$; molar mass, 202.35 g mol^{-1}) belongs to the group of polyamines which have two or more primary amino groups. Other prominent polyamines are spermidine ($C_7H_{19}N_3$; molar mass, 145.25 g mol^{-1}) and putrescine ($C_4H_{12}N_2$; molar mass, 88.15 g mol^{-1}). In animals the polyamines are formed from the amino acid L-ornithine. In subsequent steps L-ornithine is converted to putrescine, spermidine, and finally spermine. Polyamines are involved in a wide range of cellular processes, including DNA structure maintenance, RNA processing and translation, protein activation, regulation of ion channels, and providing protection from oxidative damage.

While present in many organs at low levels, polyamines are robustly detected in ^{1}H NMR spectra from prostate. Healthy prostate epithelial cells contain high concentrations of polyamines,

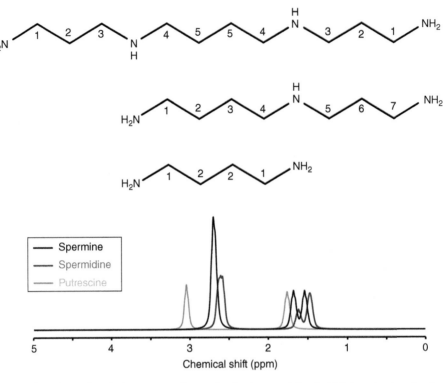

Figure 2.36 **Polyamines.** Chemical structures of spermine (top), spermidine (middle), and putrescine (bottom) and simulated ^{1}H NMR spectra.

particularly spermine. Polyamines are, in addition to citrate, greatly reduced in prostate cancer [293]. Polyamines produce broad, strongly-coupled signals at 2.6–2.7 ppm and 1.3–1.7 ppm (Figure 2.36), often complicating the separation between choline and tCr.

2.3 Phosphorus-31 NMR Spectroscopy

The success of *in vivo* proton NMR spectroscopy in routine (clinical) MR is only matched by phosphorus NMR. The relatively high sensitivity of phosphorus NMR (circa 7% of protons), together with a 100% natural abundance, allows the acquisition of high-quality spectra within minutes. Furthermore, the chemical shift dispersion of the phosphates found *in vivo* is relatively large (~30 ppm), resulting in excellent spectral resolution even at low (clinical) magnetic field strengths. Phosphorus NMR is very useful because with simple NMR methods it is capable of detecting all metabolites that play key roles in tissue energy metabolism. Furthermore, biologically relevant parameters such as intracellular pH may be indirectly deduced.

2.3.1 Chemical Shifts

Phosphorus NMR spectra from tissues *in vivo* typically hold a limited number of resonances (Figure 2.37). The exact chemical shift position of many resonances is sensitive to physiological parameters like intracellular pH and ionic (e.g. magnesium) strength. By convention, the phosphocreatine resonance is used as an internal chemical shift reference and has been assigned a

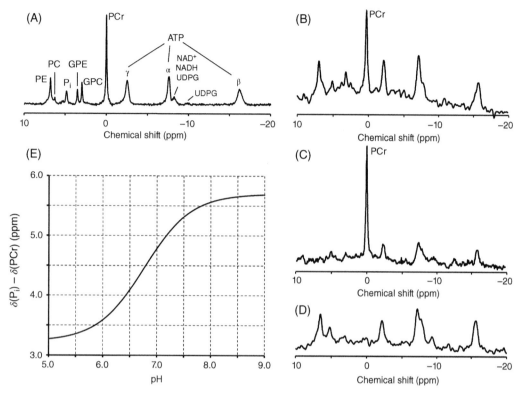

Figure 2.37 **^{31}P MRS *in vivo*.** (A) Unlocalized and (B–D) 3D localized ^{31}P NMR spectra obtained from (A) human brain *in vivo* at 7 T and (B) rat brain, (C) skeletal muscle, and (D) liver *in vivo* at 4.7 T. Note the complete absence of phosphocreatine in liver. (E) pH calibration curve for the P$_i$-PCr system as used for the determination of intracellular pH with *in vivo* ^{31}P MRS. The curve is described by Eq. (2.1) with pK_a =6.77, δ_{HA} =3.23 ppm, and δ_A =5.70 ppm.

chemical shift of 0.0 ppm. At a pH of 7.2, with full magnesium complexation, the resonances of ATP appear at −7.5 ppm (α), −16.3 (β), and −2.4 ppm (γ). The resonance of inorganic phosphate appears around 5.0 ppm. Under favorable, high-sensitivity conditions phosphorus NMR spectra can also hold resonances from phosphomonoesters (PME) and diesters. Table 2.4 summarizes the chemical shift of the most commonly observed ^{31}P-containing metabolites. Note that phosphocreatine is completely absent in ^{31}P NMR spectra from liver. The appearance of phosphocreatine in ^{31}P NMR spectra from liver is a good indication of signal contamination from surrounding muscle tissue.

Besides the chemical shifts, a ^{31}P NMR spectrum is further characterized by homonuclear scalar coupling for ATP and heteronuclear (^{31}P-^{1}H) scalar coupling for the PME, PE and phosphorylcholine, and the phosphodiesters (PDE), glycerol 3-phosphorylethanolamine and glycerol 3-phosphorylcholine. The two-bond homonuclear scalar couplings for ATP have been determined as $^{2}J\alpha\beta$ = 16.3 Hz and $^{2}J\beta\gamma$ = 16.1 Hz in brain, $^{2}J\alpha\beta$ = 16.0 Hz and $^{2}J\beta\gamma$ = 17.2 Hz in skeletal muscle, and $^{2}J\alpha\beta$ = 15.8 Hz and $^{2}J\beta\gamma$ = 16.0 Hz in myocardium [294]. Similar to most ^{31}P chemical shifts, the scalar coupling constants are sensitive to the pH and the magnesium (Mg^{2+}) concentration. The three-bond heteronuclear scalar couplings for the PME and PDE resonances are in the 6–7 Hz range and generally lead to undesirable line-broadening in that crowded spectral region. Several authors have shown that heteronuclear decoupling (see also Chapter 8) can greatly enhance the spectral resolution for the PME and PDE resonances [295, 296].

Table 2.4 Chemical shifts of biologically relevant ^{31}P containing metabolites[a].

Compound	Chemical shift (ppm)
Adenosine monophosphate (AMP)	6.3
Adenosine diphosphate (ADP)	−7.1 (α)
	−3.1 (β)
Adenosine triphosphate (ATP)	−7.5 (α)
	−16.3 (β)
	−2.5 (γ)
Fructose-1-phosphate	7.1
Fructose-6-diphosphate	6.2
Fructose-1,6-diphosphate	7.2
Glucose-1-phosphate	5.2
Glucose-6-phosphate	7.1
Glycerol phosphorylcholine (GPC)	2.8
Glycerol phosphorylethanolamine (GPE)	3.2
Inorganic phosphate (P_i)	5.0
Phosphocreatine (PCr)	0.0
Phosphorylcholine (PC)	5.9
Phosphorylethanolamine (PE)	6.8
Nicotinamide adenine dinucleotide, oxidized (NAD$^+$)	−8.2
	−8.5
Nicotinamide adenine dinucleotide, reduced (NADH)	−8.1
NADP/NADPH[b]	6.7
UDG-glucose	−8.2
	−9.8

[a]The chemical shift of many ^{31}P compounds change as a function of pH and ionic content. The reported values should therefore be considered approximate and used with caution.
[b]The structure and ^{31}P NMR spectrum of NAD$^+$/NADH and NADP$^+$/NADPH are identical with the exception of an additional phosphorus atom on the 2 position of the adenosine ribose ring.

Furthermore, PME and PDE resonances can be observed with enhanced sensitivity by utilizing the heteronuclear scalar coupling in polarization transfer experiments [105–107].

2.3.2 Intracellular pH

The chemical shift of many phosphorus-containing compounds is dependent on a number of physiological parameters, in particular intracellular pH and magnesium concentration. The cause of this phenomenon can be found in the fact that protonation (or complexation with magnesium) of a compound changes the chemical environment of nearby nuclei and hence changes the chemical shift of those nuclei. When the chemical exchange between the protonated and unprotonated forms is slow, the two forms will have two separate resonance frequencies with the resonance amplitudes indicating the relative amounts of the two forms.

However, for most compounds observed with phosphorus NMR, the chemical exchange rate is fast relative to the reciprocal of their frequency difference such that only a single, average resonance is observed. The resonance frequency is now indicative of the relative amounts of protonated and unprotonated form, and hence the pH can be described by a modified Henderson–Hasselbalch relationship according to

$$pH = pK_A + {}^{10}\log\left(\frac{\delta - \delta_{HA}}{\delta_A - \delta}\right) \tag{2.1}$$

where δ is the observed chemical shift, δ_A and δ_{HA} the chemical shifts of the unprotonated and protonated forms of compound A, and pK_A the logarithm of the equilibrium constant for the acid–base equilibrium between HA and A.

Even though most resonances in ^{31}P NMR spectra have a pH dependence in the physiological pH range, the resonance of inorganic phosphate relative to phosphocreatine is most commonly used for several reasons. The pK_a of inorganic phosphate is in the physiological range ($pK_a = 6.77$), it is readily observed in most tissues (with muscle being a possible exception) and it has a large dependence on pH. The chemical shift of creatine can be assumed constant in the physiological pH range ($pK_a = 4.30$). Figure 2.37 shows a pH curve for the P_i-PCr system with $pK = 6.77$, $\delta_{HA} = 3.23$ ppm, and $\delta_A = 5.70$ ppm, which represent average literature values [297–300]. It follows that the greatest sensitivity towards pH changes is achieved when pH $\sim pK$. While the P_i-PCr system is most widely used, it is not applicable under all conditions. For instance, PCr is not detectable in liver and kidney or under several pathological conditions like ischemia, hypoxia, anoxia, and in tumors. Under normal conditions, the inorganic phosphate resonance may be very low or is overlapping with other, more intense resonances. The latter can be the case for *in vivo* ^{31}P NMR spectra of the heart, which are usually contaminated by 2,3-diphosphoglycerate signals from the blood. If the P_i-PCr system can be used, then the accuracy of pH determination is typically 0.05 pH units. Following similar arguments as outlined for intracellular pH, the free magnesium concentration can be deduced from the chemical shifts of ATP [301–306].

2.4   Carbon-13 NMR Spectroscopy

Phosphorus MRS and proton MRS have successfully been employed in a wide range of *in vivo* NMR studies. Nevertheless, their application faces several limitations. The *in vivo* ^{31}P MRS signal usually originates from a limited number of low molecular weight compounds, while ^{1}H MRS is hindered by the small chemical shift range in which the many detectable compounds resonate.

Carbon-13 NMR can offer complementary information to that obtained with ^{1}H and/or ^{31}P MRS. Since most metabolically relevant compounds contain carbon, ^{13}C MRS is, in principle, able to detect many metabolites. While carbon-13 NMR is often used in combination with ^{13}C-labeled substrate infusion, as will be described in Chapter 3, natural abundance ^{13}C NMR spectra also hold valuable information.

2.4.1   Chemical Shifts

Carbon-13 (^{13}C) NMR spectroscopy is generally characterized by a large spectral range of >200 ppm, narrow line widths, and a relatively low sensitivity, due to the 1.108% natural abundance and the low gyromagnetic ratio (γ_{13C}/γ_{1H}) = 0.251. However, when the low sensitivity can

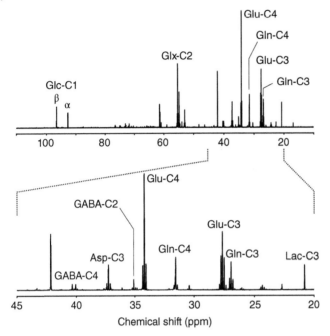

Figure 2.38 **^{13}C MRS *in vitro*.** Pulse-acquire ^{13}C NMR spectrum with heteronuclear decoupling of rat brain extract. The extract was made 2 h following the intravenous infusion of [1-^{13}C]-glucose and allows the detection of a wide range of ^{13}C-labeled metabolites. In particular, the neurotransmitters glutamate and GABA and the related compound glutamine are readily detected. See Chapter 3 for more details on dynamic ^{13}C NMR studies to determine metabolic fluxes *in vivo*. See Table 2.5 for chemical shift assignments.

be overcome (for example, by averaging, polarization transfer, larger volumes, and decoupling) ^{13}C MRS allows the detection of a large number of metabolite resonances with excellent spectral resolution (Figure 2.38). Table 2.5 lists the carbon-13 chemical shifts of the most commonly encountered brain metabolites. As a rule of thumb, a ^{13}C NMR spectrum can be divided in several spectral ranges that are indicative for carbon atoms in particular chemical groups (see also Figure 1.9J). ^{13}C chemical shifts above 150 ppm are typically indicative of carbon atoms in carbonyl groups. Carbon atoms adjacent to hydroxyl groups, like those in glucose, glycogen, and *myo*-inositol, typically resonate in the 60–100 ppm range. Carbon atoms in CH, CH$_2$, and CH$_3$ groups resonate in the 45–60 ppm, 25–45 ppm, and <25 ppm chemical shift ranges. ^{13}C NMR spectra acquired from organs other than brain are typically dominated by dominant lipid resonances in the 20–50 ppm and >120 ppm ranges. Furthermore, two distinct resonances from the glycerol backbone appear at 63 and 73 ppm. The presence of inadequate spatial localization during ^{13}C MRS studies of the brain is readily recognized by small, broader resonances with the most intense one typically appearing around 31 ppm. While natural abundance ^{13}C NMR is useful in many applications, like the identification of unknown compounds, the low sensitivity and the fact that most information can also be obtained from ^{1}H NMR spectra have prevented widespread application of natural abundance ^{13}C MRS. However, there are several areas where ^{13}C NMR can provide unique information and relate to the detection of metabolic fluxes from ^{13}C-labeled precursor (see Chapter 3) and the detection of glycogen and triglycerides.

Carbohydrate reserves are mainly stored as glycogen in animals and humans. It is particularly abundant in muscle and liver, reaching concentrations up to 30–100 mmol kg^{-1} and 100–500 mmol kg^{-1}, respectively (see Ref. [141] for a review). Glycogen is also present in the brain, residing in astroglia at a concentration of circa 5 mmol kg^{-1} [142–145]. The regulation of glycogen synthesis and breakdown plays an important role in systemic glucose metabolism and is crucial in the understanding of diseases such as diabetes mellitus. Despite its high molecular weight (10^{+7}–10^{+9} Da), glycogen gives rise to relatively narrow ^{13}C NMR resonances both *in vitro* and *in vivo*, indicating a high degree of internal mobility (see Figure 2.39). Natural

Table 2.5 Chemical shifts of biologically relevant ^{13}C containing metabolites^{a}.

Compound	Carbon position					
	C1	C2	C3	C4	C5	C6
Acetone	216.2	31.0				
Acetate	182.6	24.5				
Acetoacetate	175.5	54.1	211.0	30.3		
Alanine	176.6	51.5	17.1			
Aspartate	175.1	53.2	37.4			
Bicarbonate	161.0					
Creatine	175.4	37.8	158.0	54.7		
Ethanol	58.2	17.5				
GABA	182.3	35.2	24.6	40.2		
BHB	181.2	47.6	66.8	22.9		
2HG	182.0	72.8	31.7	34.2	183.6	
Glucose, α	92.7	72.1	73.5	70.4	72.1	61.4
Glucose, β	96.6	79.9	76.5	70.4	76.5	61.4
Glutamate	175.3	55.6	27.8	34.2	182.0	
Glutamine	174.8	55.1	27.1	31.7	178.5	
GSH, Cys	172.4	56.4	26.3			
GSH, Glu	174.7	55.0	26.9	32.1	175.7	
GSH, Gly	176.9	44.1				
Glycogen	100.5	72.3	74.0	78.1	72.1	61.4
Myo-inositol	73.3	73.1	73.3	71.9	75.1	71.9
Lactate	183.3	69.3	21.0			
Malate	182.1	71.7	43.9	180.9		
Pyruvate	207.9	172.9	29.1			
NAA, Ace	174.3	22.8				
NAA, Asp	179.7	54.0	40.3	179.7		
Succinate	183.4	35.3				
Taurine	48.4	36.2				

^{a} ^{13}C chemical shifts are referenced against the four methyl carbons in TMS. ^{13}C chemical shift referenced against DSS can be found by adding 1.7 ppm [4] to the reported values.

abundance ^{13}C NMR detection of glycogen is typically done via the glycogen-C1 resonance at 100.5 ppm. Sillerud and Shulman [307] reported in 1983 the surprising result that the [1-^{13}C]-glycogen resonance at 100.5 ppm is ~100% visible, a finding which has subsequently been confirmed [150, 308–310], but also challenged by others [311].

Natural abundance ^{13}C MRS is also very suitable for the detection and characterization of triglycerides (Figure 2.40). The wide spectral dispersion of ^{13}C MRS allows the detection of triglyceride signals that are not detectable by ^{1}H MRS, such as the carboxylic carbon positions (Figure 2.40C). In addition, unique spectral signatures can be observed for linoleic acid

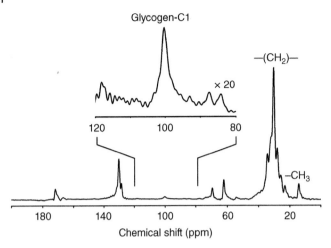

Figure 2.39 **Natural abundance** [13]**C MRS of glycogen** *in vivo.* Glycogen detection in human skeletal muscle with direct, pulse-acquire ^{13}C NMR spectroscopy (TR = 1000 ms, NA = 3000) in the presence of heteronuclear decoupling. Besides the singlet resonance of glycogen-C1 at 100.5 ppm, the ^{13}C NMR spectrum is dominated by resonances from lipids. *Source:* Data from G. I. Shulman.

(an omega-3 fatty acid) and linolenic acid (an omega-6 fatty acid) (Figure 2.40B). The low intrinsic sensitivity of ^{13}C MRS is typically not an obstacle due to the high lipid content of many tissues, thereby allowing high-quality ^{13}C MRS of triglycerides *in vivo* (Figure 2.40F). Table 2.3 summarizes ^{13}C chemical shifts of triglycerides and their relation with the corresponding protons.

The appearance of ^{13}C NMR spectra is, besides the wide chemical shift dispersion, dominated by heteronuclear scalar coupling since most carbon nuclei are directly bonded to one, two, or three protons. As a result the acquisition of *in vivo* ^{13}C NMR spectra is, with the exception of more recent reports at high magnetic fields [312, 313], always performed in the presence of heteronuclear broadband decoupling in order to increase the sensitivity and simplify the spectral appearance. Table 2.6 summarizes the heteronuclear scalar couplings for the most commonly observed metabolites, including the neurotransmitters glutamate and GABA. Homonuclear [13]C–[13]C scalar couplings do not have to be considered in natural abundance ^{13}C NMR as the probability that two adjacent carbon nuclei are both of the ^{13}C isotope is <0.015%. However, during the infusion of ^{13}C-enriched substrates, the probability of ^{13}C isotopes in adjacent positions can increase to >20%. The splitting of resonances due to homonuclear [13]C–[13]C scalar coupling gives rise to so-called isotopomer resonances and isotopomer analysis can provide additional information on the underlying metabolic pathways (see also Chapter 3). Table 2.6 summarizes the homo- and heteronuclear scalar coupling for the most commonly observed metabolites.

2.5 Sodium-23 NMR Spectroscopy

Sodium ([23]Na) is a quadrupolar nucleus with spin 3/2 that provides 9.2% of the proton sensitivity. However, the achievable SNR for sodium NMR *in vivo* is much lower due to low *in vivo* sodium concentration (30–100 mM) compared to protons (40–50 M). As a result, sodium MR images are typically acquired with a spatial resolution of several millimeters in 10–20 minutes. Interest in sodium MRI has recently sparked due to significant gains in SNR achievable at high magnetic fields [314–316].

In the spectroscopic or quantum mechanical view of NMR, a nucleus with spin I will have $2I + 1$ energy levels characterized by spin quantum number $m = -I, -I + 1, ..., +I$. For sodium

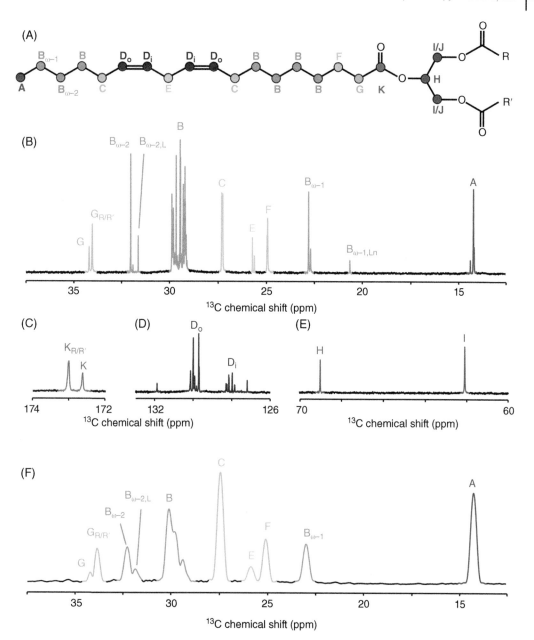

Figure 2.40 **^{13}C NMR of triglycerides.** (A) Structure of a triglyceride with the top and bottom fatty acids undefined (groups *R* and *R'*), and the middle fatty acid chosen as linoleic acid, a polyunsaturated omega-6 fatty acid. The color coding is identical to that used for ^{1}H NMR assignment of triglycerides (Figure 2.35) with the extension of the carbonyl carbons (gray, K). (B–E) 500 MHz ^{13}C NMR spectrum of vegetable oil. The ^{13}C NMR information content is higher than for ^{1}H NMR with unique resonances for inner and outer double-bonded carbons (D_o and D_i), middle and top/bottom chain carbonyl carbons (K and $K_{R/R'}$), and several methylene carbons (B_{Ln}, $B\omega_{-1}$, $B\omega_{-2}$). (F) Nonlocalized ^{13}C NMR spectra of human leg adipose tissue. Chemical shifts of triglyceride resonances are shown in Table 2.3.

Table 2.6 Homonuclear $^{13}C-^{13}C$ and heteronuclear $^{1}H-^{13}C$ scalar coupling constants over one ($^{1}J_{CC}$, $^{1}J_{HC}$) chemical bond of biologically relevant ^{13}C containing metabolites[a].

Compound	Interaction	Scalar coupling constant (Hz)		Compound	Interaction	Scalar coupling constant (Hz)	
		$^{1}J_{CC}$	$^{1}J_{HC}$			$^{1}J_{CC}$	$^{1}J_{HC}$
Acetate	C1–C2	56.7	—	Glutamate	C1–C2	53.4	—
	C2–H	—	127.1		C2–C3	34.8	—
Alanine	C1–C2	54.2	—		C3–C4	34.5	—
	C2–C3	35.4	—		C4–C5	51.3	—
	C2–H	—	145.1		C2–H	—	145.2
	C3–H	—	129.7		C3–H	—	131.4
Aspartate	C1–C2	53.8	—		C4–H	—	126.7
	C2–C3	36.4	—	Glutamine	C1–C2	53.4	—
	C3–C4	50.8	—		C2–C3	34.9	—
	C2–H	—	144.3		C3–C4	34.9	—
	C3–H	—	129.3		C4–C5	48.4	—
GABA	C1–C2	51.0	—		C2–H	—	145.4
	C2–C3	35.1	—		C3–H	—	131.4
	C3–C4	35.1	—		C4–H	—	128.3
	C2–H	—	127.2	Lactate	C1–C2	55.0	—
	C3–H	—	129.4		C2–C3	36.4	—
	C4–H	—	143.1		C2–H	—	145.6
Glucose α	C1–C2	44.8	—		C3–H	—	127.8
	C1–H	—	169.8				
Glucose β	C1–C2	43.9	—				
	C1–H	—	161.3				

[a]Heteronuclear scalar couplings over two and three chemical bonds ($^{2}J_{HC}$ and $^{3}J_{HC}$) are between 3.0 and 5.5 Hz.

($I = 3/2$) this translates into four energy levels that allows three "allowed" or single-quantum transitions for with $\Delta m = \pm 1$ (Figure 2.41B). Up to this point the discussion was limited to spin 1/2 nuclei such as ^{1}H, ^{13}C, and ^{31}P for which the charge distribution within the nucleus is symmetric. The interaction of the nucleus and the electric field gradients originating from the symmetric charge distribution is independent of the orientation and can be ignored for practical MR applications. However, spins with a spin quantum number >½ have an asymmetric charge distribution, giving these nuclei a quadrupole moment. The interaction between the nucleus and the asymmetric electric field gradients becomes orientation dependent and generally leads to enhanced relaxation. For more details on quadrupolar nuclei in NMR, the reader is referred to the literature [317, 318].

In aqueous environments where thermal energy allows rapid molecular tumbling, magnetic and electric fields associated with the quadrupole moment average out, leading to stable, average energy levels (Figure 2.41A and B). The allowable transitions all occur at the same Larmor frequency and rate constants, such that the detected ^{23}Na spectrum is characterized by a single resonance line with single-exponential longitudinal T_1 and transverse T_2 relaxation.

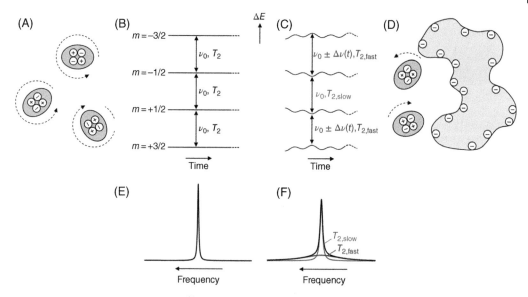

Figure 2.41 **Spectroscopic view of ^{23}Na NMR.** Cartoons of sodium ions in (A) water and (D) tissue-like environments and (B, C) the corresponding nuclear spin energy levels. (A) Rapid tumbling leads to an averaged effect of the quadrupolar moment resulting in constant, time-independent energy levels (B). (C, D) The interaction of sodium and macromolecular electric fields leads to hindered, slower tumbling. As a result, the effects of the sodium quadrupolar moment are not completely averaged out, making the energy levels time-dependent. (E) ^{23}Na NMR spectrum for sodium in water is characterized by a single Larmor frequency and single-exponential T_1 and T_2 relaxation. (F) The time-dependence of the energy levels in (C) lead to bi-exponential relaxation and a "super Lorentzian" spectrum with the two satellite and single central transitions having short and long T_2 relaxation times, respectively.

This situation is referred to as isotropic motion with motional narrowing and is encountered in a simple solution of sodium in water such as cerebrospinal fluid.

In tissues, sodium encounters a wide range of mostly negatively charged proteins and other macromolecules. The nonspherical distribution of electric charge within the sodium nucleus interacts with the charged groups within the macromolecules, therefore slowing down the molecular tumbling. When the molecular tumbling frequency is on the same order of the Larmor frequency, the electric and magnetic fields associated with the quadrupole moment are not completely averaged out. In this case the four ^{23}Na energy levels are perturbed in a time-dependent manner (Figure 2.41C), leading to different Larmor frequencies and relaxation rates for the three allowable transitions. The two outer, satellite transitions will exhibit faster relaxation times $T_{1,fast}$ and $T_{2,fast}$, whereas the inner, central transition is less affected with corresponding slower relaxation times ($T_{1,slow}$ and $T_{2,slow}$). The situation in Figure 2.41C is referred to as isotropic motion without motional narrowing and leads to bi-exponential T_1 and T_2 relaxation. The ^{23}Na NMR spectrum is described as a "super Lorentzian," being composed of a narrow Lorentzian line from the central transition accounting for 40% of the signal and a broad Lorentzian line from the satellite transitions accounting for 60% of the signal. For *in vivo* tissues the range of T_2 relaxation times for the fast and slow components is 0.5–2.5 ms and 5–30 ms, respectively. Additional motional regimes of sodium can be found under highly structured and oriented conditions and in solids. However, these scenarios are typically not encountered *in vivo* and will not be discussed. More extensive descriptions on motional narrowing in various regimes can be found in Refs. [317, 318].

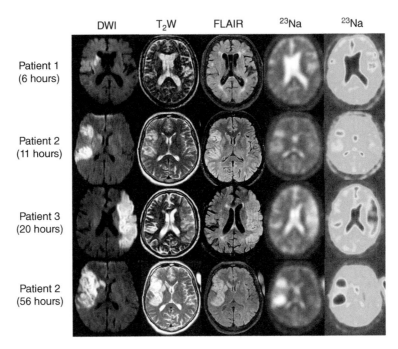

Figure 2.42 **Sodium MRI.** Diffusion-weighted (DWI), T_2-weighted (T_2W), and fluid-attenuation inversion recovery (FLAIR) proton images acquired at 3 T together with sodium images in gray and color scales for three patients acquired between 6 and 56 h following stroke onset. Note that the second and fourth rows are from the same patient at different time points. *Source:* Neumaier-Probst et al. [320]. Reproduced with permission of John Wiley & Sons.

As the sensitivity of ^{23}Na NMR increases due to higher magnetic fields, improved coil designs, and optimized pulse sequences, ^{23}Na MRI is gradually gaining momentum. The bi-exponential and short T_2 relaxation times of sodium require special MRI pulse sequences in order to capture both components. In general, variations of ultra-short TE (UTE) 3D MRI sequence have been employed [319] at the expense of time resolution. Nevertheless, sodium MRI is starting to reach a level that is comparable to early-day proton MRI at 1.5 T. Figure 2.42 shows an example of state-of-the-art ^{23}Na MRI to follow stroke progression at 3 T [320].

Sodium is inhomogeneously distributed in many tissues, including the brain. Normal cells maintain a concentration gradient across the cell membrane of intracellular sodium (~10 mM) against extracellular sodium (~150 mM). The electrochemical gradient of ions across the plasma membrane plays an important role in a variety of cellular processes, such as the generation of action potentials for the transmission of nervous impulses and the regulation of cell volume. The maintenance of the concentration gradients of ions requires metabolic energy and is the result of the combined action of several ion transporters. The ion gradients will be transiently or chronically disrupted in damaged cells as encountered in ischemia, during cortical spreading depressions, in certain tumor cells and in sickle cell anemia. Therefore, detection of abnormal intracellular sodium *in vivo* may have significant diagnostic potential. There are three practical methods by which the intra- and extracellular contributions of sodium can be separated: (i) the use of so-called shift reagents [321–325], (ii) T_1-based inversion recovery [319], and (iii) the utilization of relaxation-allowed multiple quantum coherence filtering [319, 326–330]. Shift reagents are in general anionic chelate complexes of paramagnetic lanthanides (Dy^{3+}, Tm^{3+}, and Tb^{3+}). Popular shift reagents are dysprosium bis(tripolyphosphate), $Dy(PPPi)_2^{7-}$,

Figure 2.43 **Shift reagents in ^{23}Na NMR.** (A) ^{23}Na NMR spectrum of a Langendorff perfused heart shows a single resonance representing the sum of intracellular (Na_i) and extracellular (Na_e) sodium. The smaller signal, Na_{ref}, is an external reference compound (250 mM Na^+ in solution with 5 mM TmDOTP^{5-} shift reagent) used for signal quantification. (B) Following addition of 3.5 mM TmDOTP^{5-} shift reagent to the perfusion medium, the Na_e resonance shifts to a higher frequency such that the (unaffected) Na_i signal becomes visible. *Source:* Data from J. van Emous and C. J. A. van Echteld.

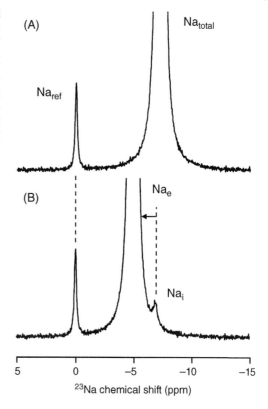

which induces very large shifts but can be toxic after decomposition, and dysprosium triethylene tetraamine hexaacetate, DyTTHA^{3-}, which induces smaller shifts but is much less poisonous. Since the negatively charged shift reagents cannot cross the plasma membrane, they can only interact with extracellular sodium. Through dynamic exchange with the shift reagent, the Larmor frequency of extracellular sodium is changed. Since the Larmor frequency of intracellular sodium remains unaffected, this approach can achieve complete separation of intracellular from extracellular sodium (Figure 2.43). Shift reagents are mainly used on cell systems and perfused organs, partly due to their toxicity at the required concentrations and their inability to cross the blood–brain barrier. The inversion recovery method utilizes the fact that the T_1 relaxation time constant of extracellular sodium can be significantly longer than for intracellular sodium. Similar to the ^{1}H NMR applications of inversion-recovery-based signal "nulling" the suppression of the undesired component is typically incomplete due to variations in T_1 and comes at the expense of signal loss for the desired component. A noninvasive alternative for shift reagents is to utilize differences in rotational motion of intra- and extracellular sodium for multiple quantum coherence filtering. The principles underlying this discrimination rely on the fact that bi-exponential relaxation of spins with $I \geq 1$ induces violations of the coherence transfer selection rules (e.g. $\Delta m = \pm 1$). By utilizing a multiple quantum filter (i.e. a pulse sequence which preserves coherence of a particular order and eliminates all other coherence orders by phase cycling and/or magnetic field gradients, see Chapter 8 for more details), it is possible to discriminate sodium on the basis of its correlation time (and related bi-exponential relaxation). Under the assumption that extracellular sodium (free and mobile) satisfies the motional narrowing condition and intracellular sodium (being partly bound) does not, a double quantum filter should be able to separate intra- and extracellular sodium. The method of relaxation-allowed multiple quantum filtering is completely noninvasive and therefore very suitable for repeated *in vivo* measurements. The main problem of the technique is that if the extracellular sodium does not completely satisfy the motional narrowing condition (e.g. when it interacts with binding sites on the cell surface), the intracellular sodium resonance is rapidly overwhelmed by the extracellular sodium [331]. This can be alleviated to a large extent by selectively quenching the extracellular double-quantum sodium with a relaxation agent (e.g. a Gd-chelate) at a low, non-perturbing, and nontoxic concentration [329, 330]. Since the reagent is exclusively localized in the extracellular space, the intracellular sodium remains unaffected. At present the use of relaxation agents has not found widespread applications in *in vivo* cation NMR.

Sodium has been imaged *in vivo* for over 30 years [332, 333]. With advances in magnetic field strength and RF coil design, the quality of ^{23}Na MRI is now sufficiently high to provide clinically relevant images (Figure 2.42). ^{23}Na MRI focuses on unique aspects of tissue electrolyte homoeostasis that has shown abnormalities in stroke [320], tumors [334, 335], and multiple sclerosis [336, 337] that are not detectable with standard proton MRI. However, like all other nonproton MR methods, ^{23}Na MRI has to demonstrate an undeniable "added value" to routine ^{1}H MRI in order to justify the added costs involved with a second non-proton channel.

2.6 Fluorine-19 NMR Spectroscopy

As a 100% naturally abundance spin 1/2 nucleus, fluorine-19 has an NMR sensitivity of 83% relative to protons. The chemical shift range of ^{19}F NMR is more than four times that of phosphorus, while line widths are comparable. Since there are no endogenous fluorine-19-containing compounds in biological tissues, ^{19}F MRS is not hampered by interference from background signals (such as water in *in vivo* ^{1}H MRS).

^{19}F MRS can be used *in vivo* to monitor the uptake and metabolism of drugs [338–343], to determine cerebral blood flow [344–347], study the metabolism of anesthetics [348–351], and with the aid of specific probe molecules, to measure a variety of parameters such as pH, oxygen levels, and metal ion concentration [352–358]. More recently, ^{19}F MRI has been used in molecular imaging applications to follow the fate of ^{19}F-labeled probes and cells [353, 359–362]. In the next paragraphs, a short description of the potential of ^{19}F MRS will be given. More detailed reviews of applications of *in vivo* ^{19}F MRS can be found in the literature [361, 363–365].

Fluorine-19 has a very large chemical shift range, stretching out over 300 ppm for organic compounds. The chemical shifts of fluorine-containing compounds relevant for *in vivo* NMR spectroscopy can be found in Refs. [338, 339, 341–343, 363, 364, 366–368]. Trifluoroacetate is often used as a chemical shift reference ($\delta = 0.00$), but ^{19}F NMR does not have a single, generally accepted reference compound. For *in vivo* applications, the reference is inevitably an external reference compound (since no endogenous fluorine compounds exist) with the associated problems of magnetic susceptibility differences between the sample and reference. This can lead to errors of several ppm, making *in vivo* assignments of resonances difficult. Although the total chemical shift range spans more than 300 ppm, spectral overlap is a significant problem as many relevant compounds have almost identical chemical environments and hence similar chemical shifts. Furthermore, many fluorinated compounds also contain protons which lead to heteronuclear scalar couplings and a further complication of ^{19}F NMR spectra. It has been shown that the *in vivo* spectral resolution can be greatly improved by proton decoupling [369].

^{19}F MRS allows the direct detection of fluorinated drugs, metabolites, and anesthetics. One of the earlier examples is in pharmacokinetic studies of the anticancer agent, 5-fluorouracil (5-FU). ^{19}F MRS studies on 5-FU metabolism has facilitated the understanding of the metabolic and cytotoxic aspects of 5-FU. Since novel drugs often contain fluorine atoms, ^{19}F MRS continues to be used in their characterization [370].

Several inhalation anesthetics which are commonly used in experimental and clinical settings are fluorinated and can be monitored by ^{19}F MRS [348–351, 371]. These include halothane, isoflurane, and enflurane of which the chemical structures are shown in Figure 2.44A–C. As an example, Figure 2.44D and F shows the clearance and metabolism of halothane and enflurane from rat brain as measured by unlocalized ^{19}F NMR spectroscopy. Many functional MRI studies on animals start with a preparation period under halothane or enflurane anesthesia during which surgery can be performed. However, leading back to the 2-deoxyglucose autoradiography experiments of Ueki et al. [372], it was shown that halothane did not preserve the neuronal–hemodynamic coupling leading to a very small functional response during somatosensory stimulation. α-Chloralose on the other hand showed robust and large functional responses. Because of this result [372], which was confirmed by others [373], the standard

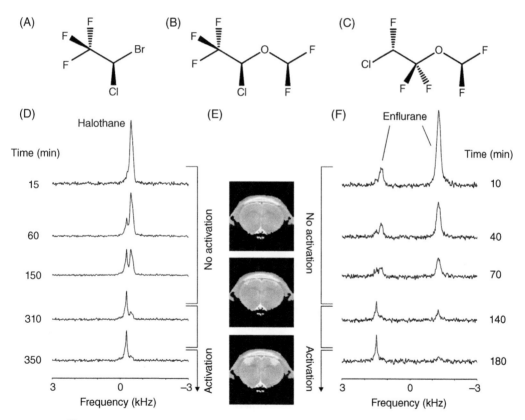

Figure 2.44 **^{19}F MRS on fluorinated anesthetics** *in vivo.* (A–C) Chemical structure of commonly used fluorinated anesthetics, namely (A) halothane, (B) isoflurane, and (C) enflurane. (D) Halothane and (F) enflurane elimination from rat brain following the administration of 1% of the corresponding anesthetic for 2 h. At zero time the fluorinated inhalation anesthetic was replaced with the intraperitoneally infused α-chloralose anesthetic. Note that both anesthetics break down over time, presumably into trifluoroacetate and difluoromethoxy-2-difluoroacetic acid for halothane and enflurane, respectively. Furthermore, note that the washout of enflurane ($t_{1/2} = 42$ min) is considerably faster than for halothane ($t_{1/2} = 123$ min). (E) BOLD fMRI activation maps (yellow) overlaying an anatomical MR image through the somatosensory cortex during double forepaw stimulation. The appearance of BOLD activation is correlated with the washout of the fluorinated anesthetics.

protocol for fMRI experiments on animals involves a discontinuation of the gaseous anesthesia following the preparation period and a switch to α-chloralose infusion. However, while only sporadically or anecdotally reported (e.g. Ref. [374]), it is well known that strong and robust fMRI activation is not obtained until several hours following the switch to α-chloralose. Figure 2.44E shows BOLD fMRI images of double forepaw stimulation in relation to the time of halothane or enflurane discontinuation. No significant BOLD signal is obtained during the first 4 and 1.5 hours following the discontinuation of halothane and enflurane, respectively. The BOLD fMRI signal becomes robust and reliable after circa 6 and 2.5 hours for halothane and enflurane, respectively. These time points correspond roughly with the complete washout of both gaseous anesthetics from the brain. While the results in Figure 2.44 are not conclusive and require further investigations, they do show the utility of ^{19}F NMR spectroscopy to study the dynamics and metabolism of fluorinated anesthetics *in vivo*.

Positron emission tomography (PET) of radioactive ^{18}F-labeled 2-fluoro-2-deoxy-glucose (FDG) is a standard clinical method to map glucose uptake and metabolism [375]. Since the radiation detected by PET cannot distinguish between different metabolites, FDG is designed to become metabolically trapped as 2-fluoro-2-deoxy-glucose-6-phosphate (FDG-6P) following the hexokinase step in glycolysis. Under these conditions, the detected PET signal is directly proportional to glucose metabolism. The assumption of a metabolically inert FDG-6P pool has been challenged by *in vivo* ^{19}F MRS studies that could detect FDG-6P metabolites based on their chemical shifts [366–368, 376, 377].

There are a number of fluorinated compounds ("probes") whose chemical shift is sensitive to physiological variables as pH [357], oxygen tension [353, 378], and the free concentration of intracellular metal ions, such as calcium [352, 356, 379]. Some probes can be made to target specific locations, for instance to discriminate between normal and tumor tissue. Although at this time only used on perfused organs, cell systems, and animals, fluorinated and nontoxic pH indicators may be of great value *in vivo*, especially when other NMR methods are not an option.

More recently, molecular imaging applications have used ^{19}F MRI to map the delivery, migration, and interactions of a wide range of ^{19}F-labeled probes and cells [359–362]. The typical ^{19}F label employed comes from perfluorocarbons (PFC), which are hydrocarbons in which all protons have been replaced by ^{19}F atoms. An ideal PFC will have many magnetically equivalent fluorine positions, leading to an intense, single resonance. Fluorinated crown ethers satisfy this requirement and are a popular choice. The fate of the fluorinated probe within the body can then be followed with ^{19}F MRI, whereby conventional ^{1}H MRI provides the anatomical landmarks. The field of ^{19}F-based molecular imaging is rapidly expanding and the reader is referred to the literature for up-to-date reviews [359–362].

2.7   *In vivo* NMR on Other Non-proton Nuclei

The five nuclei discussed in earlier sections, namely proton (^{1}H), carbon (^{13}C), fluorine (^{19}F), sodium (^{23}Na), and phosphorus (^{31}P), represent the overwhelming majority of *in vivo* NMR studies. However, since almost every element in the periodic table has an isotope with a nuclear spin, NMR can provide information on a wide range of biological processes. For NMR in general, but *in vivo* NMR in particular, a number of prerequisites must be satisfied in order to allow robust signal detection. The native or increased concentration of the nuclear spin must be high enough to provide adequate sensitivity (a rule-of-thumb concentration of 100 μM or higher can be used as an initial estimate of viability). The T_1 relaxation time constant must be short enough to allow rapid pulsing and signal averaging, while the T_2^* relaxation time constant must be long enough to provide a relatively narrow resonance line. Finally, radioactive isotopes are generally not compatible with noninvasive *in vivo* NMR studies. Even with these stringent limitations, *in vivo* NMR studies on a wide range of nuclei have been reported.

Helium (^{3}He) and xenon (^{129}Xe) NMR have been used primarily for the visualization of lungs and related airways. The low proton density in lungs and the poor magnetic field homogeneity have largely prevented the use of standard ^{1}H MRI. Noble gas MRI can overcome the sensitivity limitations through hyperpolarization. The spin polarization of helium and xenon can be increased several orders of magnitude by creating a nonequilibrium condition through spin exchange optical pumping [380–382] (see Chapter 3). Through inhalation the noble gas images provide superb detail on lung structure in health and disease, as well as transport from the lung to the blood. Noble gas MRI was initially demonstrated in 1994 with ^{129}Xe [383], soon followed by first results in humans [384]. Early limitations on the available ^{129}Xe polarization sparked the use of ^{3}He MRI for which a polarization of 30% is common. Almost all human MRI results have been obtained using hyperpolarized ^{3}He [385, 386]. However, the future of ^{3}He MRI is uncertain due to the limited supply of ^{3}He worldwide, which is primarily produced through the decay of tritium (^{3}H) in nuclear warheads. In addition, the limited supply is further restricted by homeland security applications in which ^{3}He is used to detect emitted neutrons from plutonium. This shortage has led to unsustainable pricing of ^{3}He gas for medical imaging.

Fortunately, recent advances in ^{129}Xe polarization has greatly increased the viability of ^{129}Xe-based lung MRI. Besides anatomical images, noble gas MRI often also generates images based on the apparent diffusion coefficient (ADC). The value of ADC mapping rests on the fact that gas diffusion is highly constrained by the architecture of the normal lung. However, in diseases such as emphysema, where airspaces become significantly enlarged, the gases have greater freedom to diffuse further leading to a larger ADC or a greater signal drop in diffusion-weighted MRI. A unique feature of ^{129}Xe (in contrast to ^{3}He) is its moderate solubility in water, which allows the study of lung-to-blood gas exchange. When ^{129}Xe diffuses into alveolar capillary membrane, its chemical shift changes by 198 ppm. Further transport of ^{129}Xe into the capillary blood stream leads to transient binding with hemoglobin, which produces a chemical shift of 217 ppm. The three components of ^{129}Xe in the gas phase, interstitial space, and blood compartment are readily detectable by MR spectroscopy [387]. Since ^{129}Xe follows the same transport pathways as oxygen, the spectroscopic measurement of the three components provides a powerful method to assess pulmonary gas exchange.

Lithium (^{7}Li) NMR has been performed since the early days of *in vivo* NMR [388, 389]. ^{7}Li is relatively easy to observe due to its high NMR sensitivity (27.2% of proton) and high natural abundance of 92.58%. The long T_1 relaxation times and narrow spectral dispersion are complicating factors. While the lithium ion does not play a role in normal metabolism, it becomes important as a treatment for acute mania and bipolar disorder [390]. ^{7}Li NMR has been used to map the spatial distribution across the human brain and establish a relation between brain and blood lithium levels.

Nitrogen NMR is a mainstay in high-resolution NMR studies of proteins, DNA, and other biomolecules, whereby a combination of ^{1}H, ^{13}C, and ^{15}N NMR acquired as a 3D or 4D dataset can provide the entire protein backbone sequence. Nitrogen NMR applications *in vivo* are scarcer and have largely focused on glutamine detection [391–393]. Nitrogen has two isotopes with nuclear spin, namely ^{14}N (spin 1, 99.63% abundance) and ^{15}N (spin 1/2, 0.37% abundance). ^{14}N NMR has primarily been used for natural abundance feasibility studies [394], whereas ^{15}N NMR employs ^{15}N-enriched substrates, such as ^{15}N-ammonia [391–393] and ^{15}N-glycine [395]. Since nitrogen positions typically involved exchangeable protons, the spectral line widths for many compounds can be broad and pH-dependent [396].

The abundant isotope of oxygen (^{16}O) does not have a nuclear spin. The low 0.037% abundant oxygen-17 (^{17}O) isotope has a nuclear spin 5/2 and has been observed *in vivo* [397]. Due to its quadrupolar moment, the T_1 and T_2 relaxation time constants of ^{17}O are short (3–5 ms, [398]) and this allows rapid pulsing and signal averaging. At natural abundance, ^{17}O NMR provide a wide spectral dispersion of >200 ppm [397], albeit at a very low NMR sensitivity. The main application of ^{17}O NMR is in metabolic imaging of the oxygen consumption rate in heart and brain [399–403]. Inhalation of ^{17}O-enriched oxygen gas leads to the formation of ^{17}O-labeled water as the oxygen is consumed as part of oxidative metabolism. The measurement of ^{17}O-labeled water accumulation is directly proportional to the oxygen consumption rate, provided that the ^{17}O-labeled gas inhalation and ^{17}O NMR measurement is short enough that the contribution of water from other organs is minimized. ^{17}O-labeled water can be measured directly with ^{17}O MRS/MRI or indirectly with ^{1}H MRS/MRI through an effect on the ^{1}H relaxation caused by the presence of ^{17}O.

Chlorine has two stable isotopes with a nuclear spin 3/2. Chlorine-35 (^{35}Cl) is the common choice due to the higher 75.53% natural abundance and higher NMR sensitivity. While ^{37}Cl has a lower sensitivity due to its 24.47% natural abundance, it may become a valid choice at higher magnetic fields since its smaller quadrupole moment can lead to more favorable relaxation times. ^{35}Cl MRI has been described in animals [404] and humans [405] and greatly benefits from the higher 7 T magnetic field strength now available for human studies [405, 406]. The combination of ^{23}Na and ^{35}Cl MRI can provide detailed spatial maps of ion homeostasis in health and disease.

Other nuclei that have sporadically been studied are boron (^{10}B/^{11}B), silicon (^{29}Si), sulfur (^{33}S) [407], and potassium (^{39}K and ^{41}K) [408, 409]. Boron NMR has shown applicability in studying boron neutron capture therapy during tumor treatment [410]. The collection of reports, rebuttals, and critical review [411] on ^{29}Si NMR to study silicon degradation of breast implants present an interesting case of the dangers of improper scientific conduct in general and improper use of NMR spectroscopy in particular. The considerations for potassium NMR are similar to those for sodium, albeit at an even lower sensitivity [409].

Exercises

2.1 In a proton spectrum acquired from rat brain at 7.05 T, the water resonance appears on-resonance while the NAA methyl resonance appears −801 Hz off-resonance. On a phantom the relation between the temperature T (in Kelvin) and the chemical shift difference (in ppm) between water and NAA, $\delta_{water-NAA}$, was established as:

$$T = -95.24\delta_{water-NAA} + 564.15$$

A Calculate the brain temperature (in Kelvin) using the phantom calibration data. Following a period of ischemia a proton spectrum is acquired in which the water appears +11 Hz off-resonance. The NAA methyl resonance now appears at −796 Hz off-resonance.

B Calculate the brain temperature (in Kelvin).

2.2 The proton NMR spectrum from rat brain shows an unknown resonance at 3.40 ppm. Name at least five methods by which a full or partial assignment can be made.

2.3 Choline, phosphorylcholine, and glycerophosphoryl choline all predominantly resonate at 3.22 ppm. Name at least three methods by which the three compounds can be detected separately.

2.4 The chemical shift difference between P_i and PCr equals 4.62 ppm. Calculate the intra-cellular pH with the parameters used in Figure 2.37.

2.5 **A** Using the chemical shift and scalar coupling information provided in Tables 2.5 and 2.6 and the structure shown in Figure 2.8, sketch the high-field pulse-acquire ^{13}C NMR spectrum of a mixture of 0.4 mM $[2-^{13}C]$-aspartate, 0.6 mM $[3-^{13}C]$-aspartate, and 0.6 mM $[2,3-^{13}C_2]$-aspartate in the presence of heteronuclear broad-band decoupling during acquisition. Assume equal T_1 and T_2 relaxation parameters for all resonances and ignore the contributions of natural abundance ^{13}C NMR resonances.
B Sketch the ^{13}C NMR spectrum in the absence of heteronuclear decoupling.
C Sketch the ^1H NMR spectra in the presence and absence of heteronuclear decoupling. Assume weak scalar-coupling for all resonances and ignore heteronuclear scalar coupling over more than one chemical bond.

2.6 The macromolecular resonances are characterized by T_1 relaxation time constants in the range of 300–500 ms. Selective observation of macromolecular resonances can be achieved with a double inversion recovery method with TR = 3250 ms, $TI_1 = 2100$ ms, and $TI_2 = 630$ ms. Calculate the signal recovery for metabolites with $T_1 = 1250$, 1500, and 1750 ms and for macromolecules with $T_1 = 300$, 400, and 500 ms. Comment on the results regarding macromolecule detection.

2.7 **A** In a proton spectrum from human brain the tCho/tCr ratio, as determined by numerical integration, is 2.0 when measured at TE = 10 ms, while it is 1.5 when measured at TE = 75 ms. Given that the T_2 relaxation time for tCho and tCr are 200 and 100 ms, respectively, give a likely explanation for the observed ratios.
B Discuss the required steps to obtain identical tCho/tCr ratios at all echo-times TE.

2.8 Draw the theoretical, high-resolution, pulse-acquire ^{19}F NMR spectra of halothane, iso-flurane, and enflurane (see Figure 2.44) obtained with and without proton decoupling. Assume weak coupling for all interactions, $^2J_{HF} = 40$ Hz, $^3J_{HF} = 8$ Hz, $^2J_{FF} = 25$ Hz, and magnetically equivalent nuclei within methyl and methylene groups. Only consider protons and fluorine atoms.

2.9 Sketch all nine stereoisomers of inositol and indicate the magnetically equivalent nuclei.

2.10 **A** A given glycogen pool is characterized by a T_2 relaxation time constant of 10 ms. When the residual magnetic field inhomogeneity is 5 Hz and the observed glycogen peak can be described by a pure Lorentzian line shape, calculate the integration boundaries for which the resulting integral represents 90 and 95% of the glycogen resonance.
B Calculate the integral as a percentage of the maximum integral when the integration boundaries are limited to +/−2 FWHM.

2.11 Name at least five factors that can affect the chemical shift of a metabolite.

2.12 Describe at least two *in vivo* NMR methods by which the fractions of saturated, mono-unsaturated, and poly-unsaturated triglycerides can be determined. Provide a quantitative description and assume the absence of more than two double bonds per fatty acid.

References

1 Pfeuffer, J., Tkac, I., Provencher, S.W., and Gruetter, R. (1999). Toward an *in vivo* neurochemical profile: quantification of 18 metabolites in short-echo-time ^{1}H NMR spectra of the rat brain. *J. Magn. Reson.* 141: 104–120.

2 Govindaraju, V., Young, K., and Maudsley, A.A. (2000). Proton NMR chemical shifts and coupling constants for brain metabolites. *NMR Biomed.* 13: 129–153.

3 Govind, V. (2016). ^{1}H-NMR chemical shifts and coupling constants for brain metabolites. *eMagRes* 5: 1347–1362.

4 Wishart, D.S., Bigam, C.G., Yao, J. et al. (1995). ^{1}H, ^{13}C and ^{15}N chemical shift referencing in biomolecular NMR. *J. Biol. NMR* 6: 135–140.

5 Favre, H.A. and Powell, W.H. (2013). *Nomenclature of Organic Chemistry. IUPAC Recommendations and Preferred Names*. Cambridge, UK: The Royal Society of Chemistry.

6 Slocum, D.W., Sugarman, D., and Tucker, S.P. (1971). The two faces of D and L nomenclature. *J. Chem. Ed.* 48: 597–600.

7 Waniewski, R.A. and Martin, D.L. (1998). Preferential utilization of acetate by astrocytes is attributable to transport. *J. Neurosci.* 18: 5225–5233.

8 Lebon, V., Petersen, K.F., Cline, G.W. et al. (2002). Astroglial contribution to brain energy metabolism in humans revealed by ^{13}C nuclear magnetic resonance spectroscopy: elucidation of the dominant pathway for neurotransmitter glutamate repletion and measurement of astrocytic oxidative metabolism. *J. Neurosci.* 22: 1523–1531.

9 Bluml, S., Moreno-Torres, A., Shic, F. et al. (2002). Tricarboxylic acid cycle of glia in the *in vivo* human brain. *NMR Biomed.* 15: 1–5.

10 Petroff, O.A., Ogino, T., and Alger, J.R. (1988). High-resolution proton magnetic resonance spectroscopy of rabbit brain: regional metabolite levels and postmortem changes. *J. Neurochem.* 51: 163–171.

11 Frahm, J., Bruhn, H., Gyngell, M.L. et al. (1989). Localized proton NMR spectroscopy in different regions of the human brain *in vivo*. Relaxation times and concentrations of cerebral metabolites. *Magn. Reson. Med.* 11: 47–63.

12 Hetherington, H.P., Mason, G.F., Pan, J.W. et al. (1994). Evaluation of cerebral gray and white matter metabolite differences by spectroscopic imaging at 4.1 T. *Magn. Reson. Med.* 32: 565–571.

13 Wang, Y. and Li, S.-J. (1998). Differentiation of metabolic concentrations between gray matter and white matter of human brain by *in vivo* ^{1}H magnetic resonance spectroscopy. *Magn. Reson. Med.* 39: 28–33.

14 Pouwels, P.J. and Frahm, J. (1998). Regional metabolite concentrations in human brain as determined by quantitative localized proton MRS. *Magn. Reson. Med.* 39: 53–60.

15 van der Knaap, M.S., van der Grond, J., van Rijen, P.C. et al. (1990). Age-dependent changes in localized proton and phosphorus MR spectroscopy of the brain. *Radiology* 176: 509–515.

16 Huppi, P.S., Posse, S., Lazeyras, F. et al. (1991). Magnetic resonance in preterm and term newborns: ^{1}H-spectroscopy in developing human brain. *Pediatr. Res.* 30: 574–578.

17 Hashimoto, T., Tayama, M., Miyazaki, M. et al. (1995). Developmental brain changes investigated with proton magnetic resonance spectroscopy. *Dev. Med. Child Neurol.* 37: 398–405.

18 Tallan, H.H. (1957). Studies on the distribution of *N*-acetyl-L-aspartic acid in brain. *J. Biol. Chem.* 224: 41–45.

19 Bates, T.E., Williams, S.R., Gadian, D.G. et al. (1989). ^{1}H NMR study of cerebral development in the rat. *NMR Biomed.* 2: 225–229.

20 Birken, D.L. and Oldendorf, W.H. (1989). *N*-acetyl-L-aspartic acid: a literature review of a compound prominent in ¹H-NMR spectroscopic studies of brain. *Neurosci. Biobehav. Rev.* 13: 23–31.

21 Clark, J.B. (1998). *N*-acetyl aspartate: a marker for neuronal loss or mitochondrial dysfunction. *Dev. Neurosci.* 20: 271–276.

22 Baslow, M.H. (2003). *N*-Acetylaspartate in the vertebrate brain: metabolism and function. *Neurochem. Res.* 28: 941–953.

23 Edden, R.A. and Harris, A.D. (2016). *N*-Acetyl aspartate. *eMagRes* 5: 1131–1138.

24 Taylor, D.L., Davies, S.E., Obrenovitch, T.P. et al. (1995). Investigation into the role of *N*-acetylaspartate in cerebral osmoregulation. *J. Neurochem.* 65: 275–281.

25 Blakely, R.D. and Coyle, J.T. (1988). The neurobiology of *N*-acetylaspartylglutamate. *Int. Rev. Neurobiol.* 30: 39–100.

26 D'Adamo, A.F. Jr. and Yatsu, F.M. (1966). Acetate metabolism in the nervous system. *N*-Acetyl-L-aspartic acid and the biosynthesis of brain lipids. *J. Neurochem.* 13: 961–965.

27 D'Adamo, A.F. Jr., Gidez, L.I., and Yatsu, F.M. (1968). Acetyl transport mechanisms. Involvement of *N*-acetyl aspartic acid in de novo fatty acid biosynthesis in the developing rat brain. *Exp. Brain Res.* 5: 267–273.

28 Moreno, A., Ross, B.D., and Bluml, S. (2001). Direct determination of the *N*-acetyl-L-aspartate synthesis rate in the human brain by ¹³C MRS and [1-¹³C]glucose infusion. *J. Neurochem.* 77: 347–350.

29 Choi, I.Y. and Gruetter, R. (2004). Dynamic or inert metabolism? Turnover of *N*-acetyl aspartate and glutathione from D-[1-¹³C]glucose in the rat brain *in vivo*. *J. Neurochem.* 91: 778–787.

30 Bruhn, H., Frahm, J., Gyngell, M.L. et al. (1989). Cerebral metabolism in man after acute stroke: new observations using localized proton NMR spectroscopy. *Magn. Reson. Med.* 9: 126–131.

31 Gideon, P., Henriksen, O., Sperling, B. et al. (1992). Early time course of *N*-acetylaspartate, creatine and phosphocreatine, and compounds containing choline in the brain after acute stroke. A proton magnetic resonance spectroscopy study. *Stroke* 23: 1566–1572.

32 Bruhn, H., Frahm, J., Gyngell, M.L. et al. (1989). Noninvasive differentiation of tumors with use of localized H-1 MR spectroscopy *in vivo*: initial experience in patients with cerebral tumors. *Radiology* 172: 541–548.

33 Frahm, J., Bruhn, H., Hanicke, W. et al. (1991). Localized proton NMR spectroscopy of brain tumors using short-echo time STEAM sequences. *J. Comput. Assist. Tomogr.* 15: 915–922.

34 Preul, M.C., Caramanos, Z., Collins, D.L. et al. (1996). Accurate, noninvasive diagnosis of human brain tumors by using proton magnetic resonance spectroscopy. *Nat. Med.* 2: 323–325.

35 Majos, C., Julia-Sape, M., Alonso, J. et al. (2004). Brain tumor classification by proton MR spectroscopy: comparison of diagnostic accuracy at short and long TE. *AJNR Am. J. Neuroradiol.* 25: 1696–1704.

36 Arnold, D.L., Matthews, P.M., Francis, G., and Antel, J. (1990). Proton magnetic resonance spectroscopy of human brain *in vivo* in the evaluation of multiple sclerosis: assessment of the load of disease. *Magn. Reson. Med.* 14: 154–159.

37 Larsson, H.B., Christiansen, P., Jensen, M. et al. (1991). Localized *in vivo* proton spectroscopy in the brain of patients with multiple sclerosis. *Magn. Reson. Med.* 22: 23–31.

38 Simmons, M.L., Frondoza, C.G., and Coyle, J.T. (1991). Immunocytochemical localization of *N*-acetyl-aspartate with monoclonal antibodies. *Neuroscience* 45: 37–45.

39 Urenjak, J., Williams, S.R., Gadian, D.G., and Noble, M. (1992). Specific expression of *N*-acetylaspartate in neurons, oligodendrocyte-type-2 astrocyte progenitors, and immature oligodendrocytes *in vitro*. *J. Neurochem.* 59: 55–61.

40 Brulatout, S., Meric, P., Loubinoux, I. et al. (1996). A one-dimensional (proton and phosphorus) and two-dimensional (proton) *in vivo* NMR spectroscopic study of reversible global cerebral ischemia. *J. Neurochem.* 66: 2491–2499.

41 De Stefano, N., Matthews, P.M., and Arnold, D.L. (1995). Reversible decreases in N-acetylaspartate after acute brain injury. *Magn. Reson. Med.* 34 (5): 721–727.

42 Grodd, W., Krageloh-Mann, I., Petersen, D. et al. (1990). *In vivo* assessment of N-acetylaspartate in brain in spongy degeneration (Canavan's disease) by proton spectroscopy. *Lancet (London, England)* 336 (8712): 437–438.

43 Austin, S.J., Connelly, A., Gadian, D.G. et al. (1991). Localized ^{1}H NMR spectroscopy in Canavan's disease: a report of two cases. *Magn. Reson. Med.* 19 (2): 439–445.

44 Steen, R.G. and Ogg, R.J. (2005). Abnormally high levels of brain N-acetylaspartate in children with sickle cell disease. *AJNR Am. J. Neuroradiol.* 26: 463–468.

45 Emam, A.T., Ali, A.M., and Babikr, M.A. (2009). Magnetic resonance spectroscopy of the brain in children with sickle cell disease. *Neurosciences (Riyadh, Saudi Arabia)* 14: 364–367.

46 Frahm, J., Michaelis, T., Merboldt, K.D. et al. (1991). On the N-acetyl methyl resonance in localized ^{1}H NMR spectra of human brain *in vivo*. *NMR Biomed.* 4: 201–204.

47 Curatolo, A., D Arcangelo, P., Lino, A., and Brancati, A. (1965). Distribution of N-acetyl-aspartic and N-acetyl-aspartyl-glutamic acids in nervous tissue. *J. Neurochem.* 12: 339–342.

48 Miyake, M., Kakimoto, Y., and Sorimachi, M. (1981). A gas chromatographic method for the determination of N-acetyl-L-aspartic acid, N-acetyl-alpha-aspartylglutamic acid and beta-citryl-L-glutamic acid and their distributions in the brain and other organs of various species of animals. *J. Neurochem.* 36 (3): 804–810.

49 Koller, K.J., Zaczek, R., and Coyle, J.T. (1984). N-Acetyl-aspartyl-glutamate: regional levels in rat brain and the effects of brain lesions as determined by a new HPLC method. *J. Neurochem.* 43 (4): 1136–1142.

50 Fuhrman, S., Palkovits, M., Cassidy, M., and Neale, J.H. (1994). The regional distribution of N-acetylaspartylglutamate (NAAG) and peptidase activity against NAAG in the rat nervous system. *J. Neurochem.* 62 (1): 275–281.

51 Pouwels, P.J. and Frahm, J. (1997). Differential distribution of NAA and NAAG in human brain as determined by quantitative localized proton MRS. *NMR Biomed.* 10: 73–78.

52 Holowenko, D., Peeling, J., and Sutherland, G. (1992). ^{1}H NMR properties of N-acetylaspartylglutamate in extracts of nervous tissue of the rat. *NMR Biomed.* 5: 43–47.

53 Erecinska, M. and Silver, I.A. (1989). ATP and brain function. *J. Cereb. Blood Flow Metab.* 9 (1): 2–19.

54 Hetherington, H.P., Spencer, D.D., Vaughan, J.T., and Pan, J.W. (2001). Quantitative ^{31}P spectroscopic imaging of human brain at 4 Tesla: assessment of gray and white matter differences of phosphocreatine and ATP. *Magn. Reson. Med.* 45: 46–52.

55 Pfeuffer, J., Tkac, I., Choi, I.Y. et al. (1999). Localized *in vivo* ^{1}H NMR detection of neurotransmitter labeling in rat brain during infusion of [1-^{13}C] D-glucose. *Magn. Reson. Med.* 41: 1077–1083.

56 Kohler, S.J., Yen, Y., Wolber, J. et al. (2007). *In vivo* 13 carbon metabolic imaging at 3T with hyperpolarized ^{13}C-1-pyruvate. *Magn. Reson. Med.* 58: 65–69.

57 Cunningham, C.H., Chen, A.P., Albers, M.J. et al. (2007). Double spin-echo sequence for rapid spectroscopic imaging of hyperpolarized ^{13}C. *J. Magn. Reson.* 187: 357–362.

58 Poptani, H., Gupta, R.K., Roy, R. et al. (1995). Characterization of intracranial mass lesions with *in vivo* proton MR spectroscopy. *AJNR Am. J. Neuroradiol.* 16 (8): 1593–1603.

59 Rothman, D.L., Behar, K.L., Hetherington, H.P., and Shulman, R.G. (1984). Homonuclear ^{1}H double-resonance difference spectroscopy of the rat brain *in vivo*. *Proc. Natl. Acad. Sci. U. S. A.* 81: 6330–6334.

60 Epperson, C.N., Haga, K., Mason, G.F. et al. (2002). Cortical gamma-aminobutyric acid levels across the menstrual cycle in healthy women and those with premenstrual dysphoric disorder: a proton magnetic resonance spectroscopy study. *Arch. Gen. Psychiatry* 59 (9): 851–858.

61 Levy, L.M., Ziemann, U., Chen, R., and Cohen, L.G. (2002). Rapid modulation of GABA in sensorimotor cortex induced by acute deafferentation. *Ann. Neurol.* 52 (6): 755–761.

62 Babak, B., Rothman, D.L., Petroff, O.A.C. et al. (2001). Involvement of GABA in rapid experience-dependent plasticity in the human visual cortex. *Biol. Psychiatry* 49: 148S.

63 Behar, K.L., Rothman, D.L., Petersen, K.F. et al. (1999). Preliminary evidence of low cortical GABA levels in localized ^{1}H-MR spectra of alcohol-dependent and hepatic encephalopathy patients. *Am. J. Psychiatry* 156: 952–954.

64 Ke, Y., Streeter, C.C., Nassar, L.E. et al. (2004). Frontal lobe GABA levels in cocaine dependence: a two-dimensional, J-resolved magnetic resonance spectroscopy study. *Psychiatry Res.* 130: 283–293.

65 Brambilla, P., Perez, J., Barale, F. et al. (2003). GABAergic dysfunction in mood disorders. *Mol. Psychiatry* 8 (8): 721–737, 715.

66 Petroff, O.A., Hyder, F., Rothman, D.L., and Mattson, R.H. (2001). Homocarnosine and seizure control in juvenile myoclonic epilepsy and complex partial seizures. *Neurology* 56: 709–715.

67 Sanacora, G., Mason, G.F., Rothman, D.L. et al. (1999). Reduced cortical gamma-aminobutyric acid levels in depressed patients determined by proton magnetic resonance spectroscopy. *Arch. Gen. Psychiatry* 56: 1043–1047.

68 Sanacora, G., Gueorguieva, R., Epperson, C.N. et al. (2004). Subtype-specific alterations of gamma-aminobutyric acid and glutamate in patients with major depression. *Arch. Gen. Psychiatry* 61: 705–713.

69 Goddard, A.W., Mason, G.F., Almai, A. et al. (2001). Reductions in occipital cortex GABA levels in panic disorder detected with ^{1}H-magnetic resonance spectroscopy. *Arch. Gen. Psychiatry* 58: 556–561.

70 Rothman, D.L., Petroff, O.A., Behar, K.L., and Mattson, R.H. (1993). Localized ^{1}H NMR measurements of gamma-aminobutyric acid in human brain *in vivo*. *Proc. Natl. Acad. Sci. U. S. A.* 90: 5662–5666.

71 Petroff, O.A. and Rothman, D.L. (1998). Measuring human brain GABA *in vivo*: effects of GABA-transaminase inhibition with vigabatrin. *Mol. Neurobiol.* 16 (1): 97–121.

72 Mikkelsen, M., Barker, P.B., Bhattacharyya, P.K. et al. (2017). Big GABA: edited MR spectroscopy at 24 research sites. *Neuroimage* 159: 32–45.

73 de Graaf, R.A., Chowdhury, G.M., and Behar, K.L. (2011). Quantification of high-resolution ^{1}H NMR spectra from rat brain extracts. *Anal. Chem.* 83: 216–224.

74 Kreis, R. and Bolliger, C.S. (2012). The need for updates of spin system parameters, illustrated for the case of gamma-aminobutyric acid. *NMR Biomed.* 25: 1401–1403.

75 Rice, M.E. (2000). Ascorbate regulation and its neuroprotective role in the brain. *Trends Neurosci.* 23 (5): 209–216.

76 Bohndiek, S.E., Kettunen, M.I., Hu, D.E. et al. (2011). Hyperpolarized [1-^{13}C]-ascorbic and dehydroascorbic acid: vitamin C as a probe for imaging redox status *in vivo*. *J. Am. Chem. Soc.* 133 (30): 11795–11801.

77 Keshari, K.R., Kurhanewicz, J., Bok, R. et al. (2011). Hyperpolarized ^{13}C dehydroascorbate as an endogenous redox sensor for *in vivo* metabolic imaging. *Proc. Natl. Acad. Sci. U. S. A.* 108: 18606–18611.

78 Terpstra, M. and Gruetter, R. (2004). ^{1}H NMR detection of vitamin C in human brain *in vivo*. *Magn. Reson. Med.* 51 (2): 225–229.

79 Terpstra, M., Marjanska, M., Henry, P.G. et al. (2006). Detection of an antioxidant profile in the human brain *in vivo* via double editing with MEGA-PRESS. *Magn. Reson. Med.* 56: 1192–1199.

80 Terpstra, M., Tkac, I., Rao, R., and Gruetter, R. (2006). Quantification of vitamin C in the rat brain *in vivo* using short echo-time ^{1}H MRS. *Magn. Reson. Med.* 55: 979–983.

81 Tkac, I., Rao, R., Georgieff, M.K., and Gruetter, R. (2003). Developmental and regional changes in the neurochemical profile of the rat brain determined by *in vivo* ^{1}H NMR spectroscopy. *Magn. Reson. Med.* 50: 24–32.

82 Felber, S.R., Sperl, W., Chemelli, A. et al. (1993). Maple syrup urine disease: metabolic decompensation monitored by proton magnetic resonance imaging and spectroscopy. *Ann. Neurol.* 33: 396–401.

83 Heindel, W., Kugel, H., Wendel, U. et al. (1995). Proton magnetic resonance spectroscopy reflects metabolic decompensation in maple syrup urine disease. *Pediatr. Radiol.* 25: 296–299.

84 de Graaf, R.A. and Behar, K.L. (2003). Quantitative ^{1}H NMR spectroscopy of blood plasma metabolites. *Anal. Chem.* 75: 2100–2104.

85 de Graaf, R.A., Prinsen, H., Giannini, C. et al. (2015). Quantification of ^{1}H NMR spectra from human plasma. *Metabolomics* 11: 1702–1707.

86 Tan, J., Bluml, S., Hoang, T. et al. (1998). Lack of effect of oral choline supplement on the concentrations of choline metabolites in human brain. *Magn. Reson. Med.* 39 (6): 1005–1010.

87 Wijnen, J.P., Klomp, D.W., Nabuurs, C.I. et al. (2016). Proton observed phosphorus editing (POPE) for *in vivo* detection of phospholipid metabolites. *NMR Biomed* 29: 1222–1230.

88 Bluml, S., Seymour, K.J., and Ross, B.D. (1999). Developmental changes in choline- and ethanolamine-containing compounds measured with proton-decoupled ^{31}P MRS in *in vivo* human brain. *Magn. Reson. Med.* 42: 643–654.

89 Gillies, R.J. and Morse, D.L. (2005). *In vivo* magnetic resonance spectroscopy in cancer. *Annu. Rev. Biomed. Eng.* 7: 287–326.

90 Firbank, M.J., Harrison, R.M., and O'Brien, J.T. (2002). A comprehensive review of proton magnetic resonance spectroscopy studies in dementia and Parkinson's disease. *Dement. Geriatr. Cogn. Disord.* 14 (2): 64–76.

91 Narayana, P.A. (2005). Magnetic resonance spectroscopy in the monitoring of multiple sclerosis. *J. Neuroimaging* 15 (4 Suppl): 46S–57S.

92 Gonzalez-Toledo, E., Kelley, R.E., and Minagar, A. (2006). Role of magnetic resonance spectroscopy in diagnosis and management of multiple sclerosis. *Neurol. Res.* 28 (3): 280–283.

93 Malisza, K.L., Kozlowski, P., and Peeling, J. (1998). A review of *in vivo* [1]H magnetic resonance spectroscopy of cerebral ischemia in rats. *Biochem. Cell Biol.* 76: 487–496.

94 Bolan, P.J., DelaBarre, L., Baker, E.H. et al. (2002). Eliminating spurious lipid sidebands in ^{1}H MRS of breast lesions. *Magn. Reson. Med.* 48: 215–222.

95 Wallimann, T., Wyss, M., Brdiczka, D. et al. (1992). Intracellular compartmentation, structure and function of creatine kinase isoenzymes in tissues with high and fluctuating energy demands: the "phosphocreatine circuit" for cellular energy homeostasis. *Biochem. J.* 281: 21–40.

96 Saunders, D.E., Howe, F.A., van den Boogaart, A. et al. (1999). Aging of the adult human brain: *in vivo* quantitation of metabolite content with proton magnetic resonance spectroscopy. *J. Magn. Reson. Imaging* 9 (5): 711–716.

97 Dreher, W., Norris, D.G., and Leibfritz, D. (1994). Magnetization transfer affects the proton creatine/phosphocreatine signal intensity: *in vivo* demonstration in the rat brain. *Magn. Reson. Med.* 31: 81–84.

98 de Graaf, R.A., van Kranenburg, A., and Nicolay, K. (1999). Off-resonance metabolite magnetization transfer measurements on rat brain *in situ. Magn. Reson. Med.* 41: 1136–1144.

99 Hanstock, C.C., Rothman, D.L., Shulman, R.G. et al. (1990). Measurement of ethanol in the human brain using NMR spectroscopy. *J. Stud. Alcohol* 51: 104–107.

100 Mendelson, J.H., Woods, B.T., Chiu, T.M. et al. (1990). *In vivo* proton magnetic resonance spectroscopy of alcohol in human brain. *Alcohol (Fayetteville, NY)* 7: 443–447.

101 Meyerhoff, D.J., Rooney, W.D., Tokumitsu, T., and Weiner, M.W. (1996). Evidence of multiple ethanol pools in the brain: an *in vivo* proton magnetization transfer study. *Alcohol. Clin. Exp. Res.* 20: 1283–1288.

102 Hetherington, H.P., Telang, F., Pan, J.W. et al. (1999). Spectroscopic imaging of the uptake kinetics of human brain ethanol. *Magn. Reson. Med.* 42: 1019–1026.

103 Smart, S.C., Fox, G.B., Allen, K.L. et al. (1994). Identification of ethanolamine in rat and gerbil brain tissue extracts by NMR spectroscopy. *NMR Biomed.* 7: 356–365.

104 Lehmann, A., Hagberg, H., Jacobson, I., and Hamberger, A. (1985). Effects of status epilepticus on extracellular amino acids in the hippocampus. *Brain Res.* 359: 147–151.

105 Gonen, O., Mohebbi, A., Stoyanova, R., and Brown, T.R. (1997). *In vivo* phosphorus polarization transfer and decoupling from protons in three-dimensional localized nuclear magnetic resonance spectroscopy of human brain. *Magn. Reson. Med.* 37: 301–306.

106 Payne, G.S. and Leach, M.O. (2000). Surface-coil polarization transfer for monitoring tissue metabolism *in vivo*. *Magn. Reson. Med.* 43: 510–516.

107 Weber-Fahr, W., Bachert, P., Henn, F.A. et al. (2003). Signal enhancement through heteronuclear polarisation transfer in *in vivo* ^{31}P MR spectroscopy of the human brain. *MAGMA* 16: 68–76.

108 Gyngell, M.L., Michaelis, T., Horstermann, D. et al. (1991). Cerebral glucose is detectable by localized proton NMR spectroscopy in normal rat brain *in vivo*. *Magn. Reson. Med.* 19 (2): 489–495.

109 Gruetter, R., Rothman, D.L., Novotny, E.J. et al. (1992). Detection and assignment of the glucose signal in ^{1}H NMR difference spectra of the human brain. *Magn. Reson. Med.* 27: 183–188.

110 Gruetter, R., Garwood, M., Ugurbil, K., and Seaquist, E.R. (1996). Observation of resolved glucose signals in ^{1}H NMR spectra of the human brain at 4 Tesla. *Magn. Reson. Med.* 36: 1–6.

111 de Graaf, R.A., Pan, J.W., Telang, F. et al. (2001). Differentiation of glucose transport in human brain gray and white matter. *J. Cereb. Blood Flow Metab.* 21: 483–492.

112 Keltner, J.R., Wald, L.L., Ledden, P.J. et al. (1998). A localized double-quantum filter for the *in vivo* detection of brain glucose. *Magn. Reson. Med.* 39 (4): 651–656.

113 de Graaf, R.A., Dijkhuizen, R.M., Biessels, G.J. et al. (2000). *In vivo* glucose detection by homonuclear spectral editing. *Magn. Reson. Med.* 43: 621–626.

114 Kaiser, L.G., Hirokazu, K., Fukunaga, M., and B Matson, G. (2016). Detection of glucose in the human brain with ^{1}H MRS at 7 Tesla. *Magn. Reson. Med.* 76: 1653–1660.

115 Erecinska, M. and Silver, I.A. (1990). Metabolism and role of glutamate in mammalian brain. *Prog. Neurobiol.* 35: 245–296.

116 Pan, J.W., Mason, G.F., Pohost, G.M., and Hetherington, H.P. (1996). Spectroscopic imaging of human brain glutamate by water-suppressed J-refocused coherence transfer at 4.1 T. *Magn. Reson. Med.* 36: 7–12.

117 Srinivasan, R., Cunningham, C., Chen, A. et al. (2006). TE-averaged two-dimensional proton spectroscopic imaging of glutamate at 3 T. *Neuroimage* 30 (4): 1171–1178.

118 Pan, J.W., Venkatraman, T., Vives, K., and Spencer, D.D. (2006). Quantitative glutamate spectroscopic imaging of the human hippocampus. *NMR Biomed.* 19 (2): 209–216.

119 Choi, C., Coupland, N.J., Bhardwaj, P.P. et al. (2006). T_2 measurement and quantification of glutamate in human brain *in vivo*. *Magn. Reson. Med.* 56: 971–977.

120 Tkac, I., Andersen, P., Adriany, G. et al. (2001). *In vivo* ^{1}H NMR spectroscopy of the human brain at 7 T. *Magn. Reson. Med.* 46: 451–456.

121 Kreis, R., Farrow, N., and Ross, B.D. (1991). Localized ^{1}H NMR spectroscopy in patients with chronic hepatic encephalopathy. Analysis of changes in cerebral glutamine, choline and inositols. *NMR Biomed.* 4: 109–116.

122 Ross, B.D. (1991). Biochemical considerations in ^{1}H spectroscopy. Glutamate and glutamine; myo-inositol and related metabolites. *NMR Biomed.* 4: 59–63.

123 Medina, M.A., Sanchez-Jimenez, F., Marquez, J. et al. (1992). Relevance of glutamine metabolism to tumor cell growth. *Mol. Cell. Biochem.* 113: 1–15.

124 Trabesinger, A.H., Weber, O.M., Duc, C.O., and Boesiger, P. (1999). Detection of glutathione in the human brain *in vivo* by means of double quantum coherence filtering. *Magn. Reson. Med.* 42 (2): 283–289.

125 Terpstra, M., Vaughan, T.J., Ugurbil, K. et al. (2005). Validation of glutathione quantitation from STEAM spectra against edited ^{1}H NMR spectroscopy at 4T: application to schizophrenia. *MAGMA* 18: 276–282.

126 Cooper, A.J. and Kristal, B.S. (1997). Multiple roles of glutathione in the central nervous system. *Biol. Chem.* 378 (8): 793–802.

127 Meister, A. and Anderson, M.E. (1983). Glutathione. *Annu. Rev. Biochem.* 52: 711–760.

128 Sian, J., Dexter, D.T., Lees, A.J. et al. (1994). Alterations in glutathione levels in Parkinson's disease and other neurodegenerative disorders affecting basal ganglia. *Ann. Neurol.* 36 (3): 348–355.

129 Kaiser, L.G., Marjanska, M., Matson, G.B. et al. (2010). ^{1}H MRS detection of glycine residue of reduced glutathione *in vivo*. *J. Magn. Reson.* 202: 259–266.

130 Trabesinger, A.H. and Boesiger, P. (2001). Improved selectivity of double quantum coherence filtering for the detection of glutathione in the human brain *in vivo*. *Magn. Reson. Med.* 45 (4): 708–710.

131 Terpstra, M., Henry, P.G., and Gruetter, R. (2003). Measurement of reduced glutathione (GSH) in human brain using LCModel analysis of difference-edited spectra. *Magn. Reson. Med.* 50 (1): 19–23.

132 Michaelis, T., Helms, G., and Frahm, J. (1996). Metabolic alterations in brain autopsies: proton NMR identification of free glycerol. *NMR Biomed.* 9 (3): 121–124.

133 Ren, J., Dimitrov, I., Sherry, A.D., and Malloy, C.R. (2008). Composition of adipose tissue and marrow fat in humans by ^{1}H NMR at 7 Tesla. *J. Lipid Res.* 49: 2055–2062.

134 Lindeboom, L. and de Graaf, R.A. (2018). Measurement of lipid composition in human skeletal muscle and adipose tissue with ^{1}H-MRS homonuclear spectral editing. *Magn. Reson. Med.* 79: 619–627.

135 Prescot, A.P., de B Frederick, B., Wang, L. et al. (2006). *In vivo* detection of brain glycine with echo-time-averaged ^{1}H magnetic resonance spectroscopy at 4.0 T. *Magn. Reson. Med.* 55: 681–686.

136 Choi, C., Bhardwaj, P.P., Seres, P. et al. (2008). Measurement of glycine in human brain by triple refocusing ^{1}H-MRS *in vivo* at 3.0T. *Magn. Reson. Med.* 59: 59–64.

137 Banerjee, A., Ganji, S., Hulsey, K. et al. (2012). Measurement of glycine in gray and white matter in the human brain *in vivo* by ^{1}H MRS at 7.0 T. *Magn. Reson. Med.* 68: 325–331.

138 Heindel, W., Kugel, H., and Roth, B. (1993). Noninvasive detection of increased glycine content by proton MR spectroscopy in the brains of two infants with nonketotic hyperglycinemia. *AJNR Am. J. Neuroradiol.* 14: 629–635.

139 Ganji, S.K., Maher, E.A., and Choi, C. (2016). *In vivo* ^{1}H MRSI of glycine in brain tumors at 3T. *Magn. Reson. Med.* 75: 52–62.

140 Tiwari, V., An, Z., Ganji, S.K. et al. (2017). Measurement of glycine in healthy and tumorous brain by triple-refocusing MRS at 3 T *in vivo*. *NMR Biomed.* 30: doi: 10.1002/nbm.3747.

141 Shulman, R.G. and Rothman, D.L. (2005). *Metabolomics by In Vivo NMR*. Hoboken: Wiley.

142 Choi, I.Y., Tkac, I., Ugurbil, K., and Gruetter, R. (1999). Noninvasive measurements of [1-^{13}C]glycogen concentrations and metabolism in rat brain *in vivo*. *J. Neurochem.* 73: 1300–1308.

143 Choi, I.Y. and Gruetter, R. (2003). *In vivo* ^{13}C NMR assessment of brain glycogen concentration and turnover in the awake rat. *Neurochem. Int.* 43: 317–322.

144 Oz, G., Henry, P.G., Seaquist, E.R., and Gruetter, R. (2003). Direct, noninvasive measurement of brain glycogen metabolism in humans. *Neurochem. Int.* 43: 323–329.

145 Oz, G., Seaquist, E.R., Kumar, A. et al. (2007). Human brain glycogen content and metabolism: implications on its role in brain energy metabolism. *Am. J. Physiol. Endocrinol. Metab.* 292: E946–E951.

146 Zang, L.H., Rothman, D.L., and Shulman, R.G. (1990). ^{1}H NMR visibility of mammalian glycogen in solution. *Proc. Natl. Acad. Sci. U. S. A.* 87 (5): 1678–1680.

147 Zang, L.H., Laughlin, M.R., Rothman, D.L., and Shulman, R.G. (1990). ^{13}C NMR relaxation times of hepatic glycogen *in vitro* and *in vivo*. *Biochemistry* 29: 6815–6820.

148 Chen, W., Zhu, X.H., Avison, M.J., and Shulman, R.G. (1993). Nuclear magnetic resonance relaxation of glycogen H1 in solution. *Biochemistry* 32: 9417–9422.

149 Avison, M.J., Rothman, D.L., Nadel, E., and Shulman, R.G. (1988). Detection of human muscle glycogen by natural abundance ^{13}C NMR. *Proc. Natl. Acad. Sci. U. S. A.* 85: 1634–1636.

150 Taylor, R., Price, T.B., Rothman, D.L. et al. (1992). Validation of ^{13}C NMR measurement of human skeletal muscle glycogen by direct biochemical assay of needle biopsy samples. *Magn. Reson. Med.* 27: 13–20.

151 Buehler, T., Bally, L., Dokumaci, A.S. et al. (2016). Methodological and physiological test-retest reliability of ^{13}C-MRS glycogen measurements in liver and in skeletal muscle of patients with type 1 diabetes and matched healthy controls. *NMR Biomed.* 29 (6): 796–805.

152 Shulman, G.I., Rothman, D.L., Jue, T. et al. (1990). Quantitation of muscle glycogen synthesis in normal subjects and subjects with non-insulin-dependent diabetes by ^{13}C nuclear magnetic resonance spectroscopy. *N. Engl. J. Med.* 322: 223–228.

153 Shulman, G.I., Rothman, D.L., Chung, Y. et al. (1988). ^{13}C NMR studies of glycogen turnover in the perfused rat liver. *J. Biol. Chem.* 263 (11): 5027–5029.

154 Chen, W., Avison, M.J., Bloch, G. et al. (1994). Proton NMR observation of glycogen *in vivo*. *Magn. Reson. Med.* 31: 576–579.

155 van Zijl, P.C., Jones, C.K., Ren, J. et al. (2007). MRI detection of glycogen *in vivo* by using chemical exchange saturation transfer imaging (glycoCEST). *Proc. Natl. Acad. Sci. U. S. A.* 104: 4359–4364.

156 Vermathen, P., Capizzano, A.A., and Maudsley, A.A. (2000). Administration and ^{1}H MRS detection of histidine in human brain: application to *in vivo* pH measurement. *Magn. Reson. Med.* 43: 665–675.

157 Fogel, W.A., Andrzejewski, W., and Maslinski, C. (1990). Neurotransmitters in hepatic encephalopathy. *Acta Neurobiol. Exp. (Wars)* 50 (4–5): 281–293.

158 Fogel, W.A., Andrzejewski, W., and Maslinski, C. (1991). Brain histamine in rats with hepatic encephalopathy. *J. Neurochem.* 56 (1): 38–43.

159 Gadian, D.G., Proctor, E., Williams, S.R. et al. (1986). Neurometabolic effects of an inborn error of amino acid metabolism demonstrated *in vivo* by ^{1}H NMR. *Magn. Reson. Med.* 3: 150–156.

160 Kish, S.J., Perry, T.L., and Hansen, S. (1979). Regional distribution of homocarnosine, homocarnosine-carnosine synthetase and homocarnosinase in human brain. *J. Neurochem.* 32 (6): 1629–1636.

161 Perry, T.L., Kish, S.J., Sjaastad, O. et al. (1979). Homocarnosinosis: increased content of homocarnosine and deficiency of homocarnosinase in brain. *J. Neurochem.* 32 (6): 1637–1640.

162 Petroff, O.A., Hyder, F., Rothman, D.L., and Mattson, R.H. (2000). Effects of gabapentin on brain GABA, homocarnosine, and pyrrolidinone in epilepsy patients. *Epilepsia* 41 (6): 675–680.

163 Rothman, D.L., Behar, K.L., Prichard, J.W., and Petroff, O.A. (1997). Homocarnosine and the measurement of neuronal pH in patients with epilepsy. *Magn. Reson. Med.* 38 (6): 924–929.

164 Robinson, A.M. and Williamson, D.H. (1980). Physiological roles of ketone bodies as substrates and signals in mammalian tissues. *Physiol. Rev.* 60 (1): 143–187.

165 Pan, J.W., Rothman, T.L., Behar, K.L. et al. (2000). Human brain beta-hydroxybutyrate and lactate increase in fasting-induced ketosis. *J. Cereb. Blood Flow Metab.* 20: 1502–1507.

166 Pan, J.W., Telang, F.W., Lee, J.H. et al. (2001). Measurement of beta-hydroxybutyrate in acute hyperketonemia in human brain. *J. Neurochem.* 79: 539–544.

167 Yudkoff, M., Daikhin, Y., Nissim, I. et al. (2001). Ketogenic diet, amino acid metabolism, and seizure control. *J. Neurosci. Res.* 66: 931–940.

168 Dang, L., White, D.W., Gross, S. et al. (2009). Cancer-associated IDH1 mutations produce 2-hydroxyglutarate. *Nature* 462: 739–744.

169 Choi, C., Ganji, S.K., DeBerardinis, R.J. et al. (2012). 2-Hydroxyglutarate detection by magnetic resonance spectroscopy in IDH-mutated patients with gliomas. *Nat. Med.* 18: 624–629.

170 Andronesi, O.C., Kim, G.S., Gerstner, E. et al. (2012). Detection of 2-hydroxyglutarate in IDH-mutated glioma patients by *in vivo* spectral-editing and 2D correlation magnetic resonance spectroscopy. *Sci. Transl. Med.* 4 (116): 116ra114.

171 Kreis, R. (1997). Quantitative localized ^{1}H MR spectroscopy for clinical use. *Prog. Nucl. Magn. Reson. Spectrosc.* 31: 155–195.

172 Nakanishi, T., Turner, R.J., and Burg, M.B. (1989). Osmoregulatory changes in *myo*-inositol transport by renal cells. *Proc. Natl. Acad. Sci. U. S. A.* 86: 6002–6006.

173 Berridge, M.J. and Irvine, R.F. (1984). Inositol trisphosphate, a novel second messenger in cellular signal transduction. *Nature* 312 (5992): 315–321.

174 Brand, A., Richter-Landsberg, C., and Leibfritz, D. (1993). Multinuclear NMR studies on the energy metabolism of glial and neuronal cells. *Dev. Neurosci.* 15 (3–5): 289–298.

175 Sherman, W.R., Packman, P.M., Laird, M.H., and Boshans, R.L. (1977). Measurement of *myo*-inositol in single cells and defined areas of the nervous system by selected ion monitoring. *Anal. Biochem.* 78: 119–131.

176 Godfrey, D.A., Hallcher, L.M., Laird, M.H. et al. (1982). Distribution of *myo*-inositol in the cat cochlear nucleus. *J. Neurochem.* 38: 939–947.

177 Miller, B.L., Moats, R.A., Shonk, T. et al. (1993). Alzheimer disease: depiction of increased cerebral *myo*-inositol with proton MR spectroscopy. *Radiology* 187: 433–437.

178 Shonk, T.K., Moats, R.A., Gifford, P. et al. (1995). Probable Alzheimer disease: diagnosis with proton MR spectroscopy. *Radiology* 195 (1): 65–72.

179 Kantarci, K., Jack, C.R. Jr., Xu, Y.C. et al. (2000). Regional metabolic patterns in mild cognitive impairment and Alzheimer's disease: A ^{1}H MRS study. *Neurology* 55: 210–217.

180 Catani, M., Cherubini, A., Howard, R. et al. (2001). ^{1}H-MR spectroscopy differentiates mild cognitive impairment from normal brain aging. *Neuroreport* 12: 2315–2317.

181 Ross, B.D., Ernst, T., Kreis, R. et al. (1998). ^{1}H MRS in acute traumatic brain injury. *J. Magn. Reson. Imaging* 8: 829–840.

182 Michaelis, T., Helms, G., Merboldt, K.D. et al. (1993). Identification of scyllo-inositol in proton NMR spectra of human brain *in vivo*. *NMR Biomed.* 6 (1): 105–109.

183 Seaquist, E.R. and Gruetter, R. (1998). Identification of a high concentration of scyllo-inositol in the brain of a healthy human subject using ^{1}H- and ^{13}C-NMR. *Magn. Reson. Med.* 39 (2): 313–316.

184 Viola, A., Nicoli, F., Denis, B. et al. (2004). High cerebral scyllo-inositol: a new marker of brain metabolism disturbances induced by chronic alcoholism. *MAGMA* 17 (1): 47–61.

185 Michaelis, T. and Frahm, J. (1995). On the 3.35 ppm singlet resonance in proton NMR spectra of brain tissue: scyllo-inositol or methanol contamination? *Magn. Reson. Med.* 34 (5): 775–776.

186 Berkelbach van der Sprenkel, J.W., Luyten, P.R., van Rijen, P.C. et al. (1988). Cerebral lactate detected by regional proton magnetic resonance spectroscopy in a patient with cerebral infarction. *Stroke* 19: 1556–1560.

187 Behar, K.L., den Hollander, J.A., Stromski, M.E. et al. (1983). High-resolution ^{1}H nuclear magnetic resonance study of cerebral hypoxia *in vivo*. *Proc. Natl. Acad. Sci. U. S. A.* 80: 4945–4948.

188 Behar, K.L., Rothman, D.L., Shulman, R.G. et al. (1984). Detection of cerebral lactate *in vivo* during hypoxemia by ^{1}H NMR at relatively low field strengths (1.9 T). *Proc. Natl. Acad. Sci. U. S. A.* 81: 2517–2519.

189 Prichard, J., Rothman, D., Novotny, E. et al. (1991). Lactate rise detected by ^{1}H NMR in human visual cortex during physiologic stimulation. *Proc. Natl. Acad. Sci. U. S. A.* 88: 5829–5831.

190 Sappey-Marinier, D., Calabrese, G., Fein, G. et al. (1992). Effect of photic stimulation on human visual cortex lactate and phosphates using ^{1}H and ^{31}P magnetic resonance spectroscopy. *J. Cereb. Blood Flow Metab.* 12: 584–592.

191 Frahm, J., Kruger, G., Merboldt, K.D., and Kleinschmidt, A. (1996). Dynamic uncoupling and recoupling of perfusion and oxidative metabolism during focal brain activation in man. *Magn. Reson. Med.* 35: 143–148.

192 Mangia, S., Garreffa, G., Bianciardi, M. et al. (2003). The aerobic brain: lactate decrease at the onset of neural activity. *Neuroscience* 118: 7–10.

193 van Rijen, P.C., Luyten, P.R., van der Sprenkel, J.W. et al. (1989). ^{1}H and ^{31}P NMR measurement of cerebral lactate, high-energy phosphate levels, and pH in humans during voluntary hyperventilation: associated EEG, capnographic, and Doppler findings. *Magn. Reson. Med.* 10: 182–193.

194 Posse, S., Dager, S.R., Richards, T.L. et al. (1997). *In vivo* measurement of regional brain metabolic response to hyperventilation using magnetic resonance: proton echo planar spectroscopic imaging (PEPSI). *Magn. Reson. Med.* 37: 858–865.

195 Pellerin, L. and Magistretti, P.J. (1994). Glutamate uptake into astrocytes stimulates aerobic glycolysis: a mechanism coupling neuronal activity to glucose utilization. *Proc. Natl. Acad. Sci. U. S. A.* 91 (22): 10625–10629.

196 Pellerin, L., Pellegri, G., Bittar, P.G. et al. (1998). Evidence supporting the existence of an activity-dependent astrocyte-neuron lactate shuttle. *Dev. Neurosci.* 20: 291–299.

197 Chih, C.P. and Roberts, E.L. (2003). Energy substrates for neurons during neural activity: a critical review of the astrocyte-neuron lactate shuttle hypothesis. *J. Cereb. Blood Flow Metab.* 23: 1263–1281.

198 Behar, K.L. and Ogino, T. (1991). Assignment of resonance in the ^{1}H spectrum of rat brain by two-dimensional shift correlated and J-resolved NMR spectroscopy. *Magn. Reson. Med.* 17: 285–303.

199 Behar, K.L. and Ogino, T. (1993). Characterization of macromolecule resonances in the ^{1}H NMR spectrum of rat brain. *Magn. Reson. Med.* 30: 38–44.

200 Behar, K.L., Rothman, D.L., Spencer, D.D., and Petroff, O.A. (1994). Analysis of macromolecule resonances in ^{1}H NMR spectra of human brain. *Magn. Reson. Med.* 32: 294–302.

201 Kauppinen, R.A., Kokko, H., and Williams, S.R. (1992). Detection of mobile proteins by proton nuclear magnetic resonance spectroscopy in the guinea pig brain ex vivo and their partial purification. *J. Neurochem.* 58 (3): 967–974.

202 Kauppinen, R.A., Niskanen, T., Hakumaki, J., and Williams, S.R. (1993). Quantitative analysis of ^{1}H NMR detected proteins in the rat cerebral cortex *in vivo* and *in vitro*. *NMR Biomed.* 6 (4): 242–247.

203 Hofmann, L., Slotboom, J., Boesch, C., and Kreis, R. (2001). Characterization of the macromolecule baseline in localized [1]H-MR spectra of human brain. *Magn. Reson. Med.* 46: 855–863.

204 de Graaf, R.A., Brown, P.B., McIntyre, S. et al. (2006). High magnetic field water and metabolite proton T_1 and T_2 relaxation in rat brain *in vivo*. *Magn. Reson. Med.* 56: 386–394.

205 Hwang, J.H., Graham, G.D., Behar, K.L. et al. (1996). Short echo time proton magnetic resonance spectroscopic imaging of macromolecule and metabolite signal intensities in the human brain. *Magn. Reson. Med.* 35: 633–639.

206 Saunders, D.E., Howe, F.A., van den Boogaart, A. et al. (1997). Discrimination of metabolite from lipid and macromolecule resonances in cerebral infarction in humans using short echo proton spectroscopy. *J. Magn. Reson. Imaging* 7: 1116–1121.

207 Graham, G.D., Hwang, J.H., Rothman, D.L., and Prichard, J.W. (2001). Spectroscopic assessment of alterations in macromolecule and small-molecule metabolites in human brain after stroke. *Stroke* 32: 2797–2802.

208 Howe, F.A., Barton, S.J., Cudlip, S.A. et al. (2003). Metabolic profiles of human brain tumors using quantitative *in vivo* [1]H magnetic resonance spectroscopy. *Magn. Reson. Med.* 49: 223–232.

209 Opstad, K.S., Murphy, M.M., Wilkins, P.R. et al. (2004). Differentiation of metastases from high-grade gliomas using short echo time ^{1}H spectroscopy. *J. Magn. Reson. Imaging* 20: 187–192.

210 Mader, I., Seeger, U., Weissert, R. et al. (2001). Proton MR spectroscopy with metabolite-nulling reveals elevated macromolecules in acute multiple sclerosis. *Brain* 124: 953–961.

211 Marjanska, M., Deelchand, D.K., Hodges, J.S. et al. (2018). Altered macromolecular pattern and content in the aging human brain. *NMR Biomed.* 31.

212 Schaller, B., Xin, L., and Gruetter, R. (2014). Is the macromolecule signal tissue-specific in healthy human brain? A ^{1}H MRS study at 7 Tesla in the occipital lobe. *Magn. Reson. Med.* 72: 934–940.

213 Snoussi, K., Gillen, J.S., Horska, A. et al. (2015). Comparison of brain gray and white matter macromolecule resonances at 3 and 7 Tesla. *Magn. Reson. Med.* 74: 607–613.

214 Knight-Scott, J. (1999). Application of multiple inversion recovery for suppression of macromolecule resonances in short echo time ^{1}H NMR spectroscopy of human brain. *J. Magn. Reson.* 140: 228–234.

215 Houtkooper, R.H., Canto, C., Wanders, R.J., and Auwerx, J. (2010). The secret life of NAD$^+$: an old metabolite controlling new metabolic signaling pathways. *Endocr. Rev.* 31: 194–223.

216 Stein, L.R. and Imai, S. (2012). The dynamic regulation of NAD metabolism in mitochondria. *Trends Endocrinol. Metab.* 23: 420–428.

217 Bai, P. and Canto, C. (2012). The role of PARP-1 and PARP-2 enzymes in metabolic regulation and disease. *Cell Metab.* 16: 290–295.

218 Imai, S. (2009). The NAD World: a new systemic regulatory network for metabolism and aging – Sirt1, systemic NAD biosynthesis, and their importance. *Cell Biochem. Biophys.* 53: 65–74.

219 Imai, S. and Yoshino, J. (2013). The importance of NAMPT/NAD/SIRT1 in the systemic regulation of metabolism and ageing. *Diabetes Obes. Metab.* 15: 26–33.

220 Liu, T.F. and McCall, C.E. (2013). Deacetylation by SIRT1 reprograms inflammation and cancer. *Genes Cancer* 4: 135–147.

221 Chiarugi, A., Dolle, C., Felici, R., and Ziegler, M. (2012). The NAD metabolome – a key determinant of cancer cell biology. *Nat. Rev. Cancer* 12: 741–752.

222 Nakahata, Y., Shara, S., Astarita, G. et al. (2009). Circadian control of the NAD$^+$ salvage pathway by CLOCK-SIRT1. *Science* 324: 654–657.

223 Peek, C.B., Affinati, A.H., Ramsey, K.M. et al. (2013). Circadian clock NAD$^+$ cycle drives mitochondrial oxidative metabolism in mice. *Science* 342: 1243417.

224 Zhu, X.H., Lu, M., Lee, B.Y. et al. (2015). *In vivo* NAD assay reveals the intracellular NAD contents and redox state in healthy human brain and their age dependences. *Proc. Natl. Acad. Sci. U. S. A.* 112: 2876–2881.

225 de Graaf, R.A. and Behar, K.L. (2014). Detection of cerebral NAD$^+$ by *in vivo* ^{1}H NMR spectroscopy. *NMR Biomed.* 27: 802–809.

226 de Graaf, R.A., De Feyter, H.M., Brown, P.B. et al. (2017). Detection of cerebral NAD$^+$ in humans at 7T. *Magn. Reson. Med.* 78: 828–835.

227 Pietz, J., Lutz, T., Zwygart, K. et al. (2003). Phenylalanine can be detected in brain tissue of healthy subjects by ^{1}H magnetic resonance spectroscopy. *J. Inherit. Metab. Dis.* 26 (7): 683–692.

228 Avison, M.J., Herschkowitz, N., Novotny, E.J. et al. (1990). Proton NMR observation of phenylalanine and an aromatic metabolite in the rabbit brain *in vivo*. *Pediatr. Res.* 27 (6): 566–570.

229 Kreis, R., Salvisberg, C., Lutz, T. et al. (2005). Visibility of vascular phenylalanine in dynamic uptake studies in humans using magnetic resonance spectroscopy. *Magn. Reson. Med.* 54 (2): 435–438.

230 Novotny, E.J. Jr., Avison, M.J., Herschkowitz, N. et al. (1995). *In vivo* measurement of phenylalanine in human brain by proton nuclear magnetic resonance spectroscopy. *Pediatr. Res.* 37: 244–249.

231 Kreis, R., Pietz, J., Penzien, J. et al. (1995). Identification and quantitation of phenylalanine in the brain of patients with phenylketonuria by means of localized *in vivo* ^{1}H magnetic-resonance spectroscopy. *J. Magn. Reson. B* 107: 242–251.

232 Leuzzi, V., Bianchi, M.C., Tosetti, M. et al. (2000). Clinical significance of brain phenylalanine concentration assessed by *in vivo* proton magnetic resonance spectroscopy in phenylketonuria. *J. Inherit. Metab. Dis.* 23 (6): 563–570.

233 Gonzalez, S.V., Nguyen, N.H., Rise, F., and Hassel, B. (2005). Brain metabolism of exogenous pyruvate. *J. Neurochem.* 95: 284–293.

234 Kohli, A., Gupta, R.K., Poptani, H., and Roy, R. (1995). *In vivo* proton magnetic resonance spectroscopy in a case of intracranial hydatid cyst. *Neurology* 45: 562–564.

235 Zand, D.J., Simon, E.M., Pulitzer, S.B. et al. (2003). *In vivo* pyruvate detected by MR spectroscopy in neonatal pyruvate dehydrogenase deficiency. *AJNR Am. J. Neuroradiol.* 24 (7): 1471–1474.

236 Choi, C., Dimitrov, I., Douglas, D. et al. (2009). *In vivo* detection of serine in the human brain by proton magnetic resonance spectroscopy (^{1}H-MRS) at 7 Tesla. *Magn. Reson. Med.* 62: 1042–1046.

237 Chawla, S., Kumar, S., and Gupta, R.K. (2004). Marker of parasitic cysts on *in vivo* proton magnetic resonance spectroscopy: is it succinate or pyruvate? *J. Magn. Reson. Imaging* 20: 1052–1053.

238 Kim, S.H., Chang, K.H., Song, I.C. et al. (1997). Brain abscess and brain tumor: discrimination with *in vivo* H-1 MR spectroscopy. *Radiology* 204: 239–245.

239 Brockmann, K., Bjornstad, A., Dechent, P. et al. (2002). Succinate in dystrophic white matter: a proton magnetic resonance spectroscopy finding characteristic for complex II deficiency. *Ann. Neurol.* 52: 38–46.

240 Helman, G., Caldovic, L., Whitehead, M.T. et al. (2016). Magnetic resonance imaging spectrum of succinate dehydrogenase-related infantile leukoencephalopathy. *Ann. Neurol.* 79: 379–386.

241 Hardy, D.L. and Norwood, T.J. (1998). Spectral editing technique for the *in vitro* and *in vivo* detection of taurine. *J. Magn. Reson.* 133 (1): 70–78.

242 Lei, H. and Peeling, J. (1999). A localized double-quantum filter for *in vivo* detection of taurine. *Magn. Reson. Med.* 42 (3): 454–460.

243 Lei, H. and Peeling, J. (2000). Simultaneous spectral editing for gamma-aminobutyric acid and taurine using double quantum coherence transfer. *J. Magn. Reson.* 143 (1): 95–100.

244 Huxtable, R.J. (1989). Taurine in the central nervous system and the mammalian actions of taurine. *Prog. Neurobiol.* 32 (6): 471–533.

245 Choi, C., Coupland, N.J., Kalra, S. et al. (2006). Proton spectral editing for discrimination of lactate and threonine 1.31 ppm resonances in human brain *in vivo*. *Magn. Reson. Med.* 56: 660–665.

246 Herneth, A.M., Steindl, P., Ferenci, P. et al. (1998). Role of tryptophan in the elevated serotonin-turnover in hepatic encephalopathy. *J. Neural Transm.* 105: 975–986.

247 Kreis, R., Ernst, T., and Ross, B.D. (1993). Absolute quantification of water and metabolites in the human brain: I. Compartments and water. *J. Magn. Reson. B* 102: 1–8.

248 Kreis, R., Ernst, T., and Ross, B.D. (1993). Development of the human brain: *in vivo* quantification of metabolite and water content with proton magnetic resonance spectroscopy. *Magn. Reson. Med.* 30: 424–437.

249 Hindman, J.C. (1966). Proton resonance shift of water in gas and liquid states. *J. Chem. Phys.* 44: 4582–4592.

250 De Poorter, J. (1995). Noninvasive MRI thermometry with the proton resonance frequency method: study of susceptibility effects. *Magn. Reson. Med.* 34 (3): 359–367.

251 Philo, J.S. and Fairbank, W.M. (1980). Temperature dependence of the diamagnetism of water. *J. Chem. Phys.* 72: 4429–4433.

252 Arus, C., Chang, Y.-C., and Barany, M. (1985). *N*-acetylaspartate as an intrinsic thermometer for ^{1}H NMR in brain slices. *J. Magn. Reson.* 63: 376–379.

253 Corbett, R.J., Laptook, A.R., Tollefsbol, G., and Kim, B. (1995). Validation of a noninvasive method to measure brain temperature *in vivo* using ^{1}H NMR spectroscopy. *J. Neurochem.* 64: 1224–1230.

254 Cady, E.B., D'Souza, P.C., Penrice, J., and Lorek, A. (1995). The estimation of local brain temperature by *in vivo* [1]H magnetic resonance spectroscopy. *Magn. Reson. Med.* 33: 862–867.

255 Ishihara, Y., Calderon, A., Watanabe, H. et al. (1995). A precise and fast temperature mapping using water proton chemical shift. *Magn. Reson. Med.* 34: 814–823.

256 De Poorter, J., De Wagter, C., De Deene, Y. et al. (1995). Noninvasive MRI thermometry with the proton resonance frequency (PRF) method: *in vivo* results in human muscle. *Magn. Reson. Med.* 33: 74–81.

257 Zuo, C.S., Metz, K.R., Sun, Y., and Sherry, A.D. (1998). NMR temperature measurements using a paramagnetic lanthanide complex. *J. Magn. Reson.* 133: 53–60.

258 Sun, Y., Sugawara, M., Mulkern, R.V. et al. (2000). Simultaneous measurements of temperature and pH *in vivo* using NMR in conjunction with TmDOTP5. *NMR Biomed.* 13: 460–466.

259 Trubel, H.K., Maciejewski, P.K., Farber, J.H., and Hyder, F. (2003). Brain temperature measured by ^{1}H-NMR in conjunction with a lanthanide complex. *J. Appl. Physiol.* 94: 1641–1649.

260 Kreis, R., Jung, B., Rotman, S. et al. (1999). Non-invasive observation of acetyl-group buffering by [1]H-MR spectroscopy in exercising human muscle. *NMR Biomed.* 12: 471–476.

261 Ren, J., Lakoski, S., Haller, R.G. et al. (2013). Dynamic monitoring of carnitine and acetylcarnitine in the trimethylamine signal after exercise in human skeletal muscle by 7T ^{1}H-MRS. *Magn. Reson. Med.* 69: 7–17.

262 Lindeboom, L., Nabuurs, C.I., Hoeks, J. et al. (2014). Long-echo time MR spectroscopy for skeletal muscle acetylcarnitine detection. *J. Clin. Invest.* 124: 4915–4925.

263 Yoshizaki, K., Seo, Y., and Nishikawa, H. (1981). High-resolution proton magnetic resonance spectra of muscle. *Biochim. Biophys. Acta* 678: 283–291.

264 Pan, J.W., Hamm, J.R., Rothman, D.L., and Shulman, R.G. (1988). Intracellular pH in human skeletal muscle by ^{1}H NMR. *Proc. Natl. Acad. Sci. U. S. A.* 85: 7836–7839.

265 Damon, B.M., Hsu, A.C., Stark, H.J., and Dawson, M.J. (2003). The carnosine C-2 proton's chemical shift reports intracellular pH in oxidative and glycolytic muscle fibers. *Magn. Reson. Med.* 49: 233–240.

266 Kurhanewicz, J., Swanson, M.G., Nelson, S.J., and Vigneron, D.B. (2002). Combined magnetic resonance imaging and spectroscopic imaging approach to molecular imaging of prostate cancer. *J. Magn. Reson. Imaging* 16: 451–463.

267 Yacoe, M.E., Sommer, G., and Peehl, D. (1991). *In vitro* proton spectroscopy of normal and abnormal prostate. *Magn. Reson. Med.* 19: 429–438.

268 Schiebler, M.L., Miyamoto, K.K., White, M. et al. (1993). *In vitro* high resolution ^{1}H-spectroscopy of the human prostate: benign prostatic hyperplasia, normal peripheral zone and adenocarcinoma. *Magn. Reson. Med.* 29: 285–291.

269 Kurhanewicz, J., Vigneron, D.B., Nelson, S.J. et al. (1995). Citrate as an *in vivo* marker to discriminate prostate cancer from benign prostatic hyperplasia and normal prostate peripheral zone: detection via localized proton spectroscopy. *Urology* 45: 459–466.

270 Kim, J.K., Kim, D.Y., Lee, Y.H. et al. (1998). *In vivo* differential diagnosis of prostate cancer and benign prostatic hyperplasia: localized proton magnetic resonance spectroscopy using external-body surface coil. *Magn. Reson. Imaging* 16: 1281–1288.

271 Garcia-Segura, J.M., Sanchez-Chapado, M., Ibarburen, C. et al. (1999). *In vivo* proton magnetic resonance spectroscopy of diseased prostate: spectroscopic features of malignant versus benign pathology. *Magn. Reson. Imaging* 17: 755–765.

272 Sillerud, L.O., Halliday, K.R., Griffey, R.H. et al. (1988). *In vivo* ^{13}C NMR spectroscopy of the human prostate. *Magn. Reson. Med.* 8: 224–230.

273 Moore, G.J. and Sillerud, L.O. (1994). The pH dependence of chemical shift and spin-spin coupling for citrate. *J. Magn. Reson. B* 103: 87–88.

274 van der Graaf, M. and Heerschap, A. (1996). Effect of cation binding on the proton chemical shifts and the spin-spin coupling constant of citrate. *J. Magn. Reson. B* 112: 58–62.

275 Wilman, A.H. and Allen, P.S. (1995). The response of the strongly coupled AB system of citrate to typical ^{1}H MRS localization sequences. *J. Magn. Reson. B* 107: 25–33.

276 Mulkern, R.V., Bowers, J.L., Peled, S., and Williamson, D.S. (1996). Density-matrix calculations of the 1.5 T citrate signal acquired with volume-localized STEAM sequences. *J. Magn. Reson. B* 110: 255–266.

277 Trabesinger, A.H., Meier, D., Dydak, U. et al. (2005). Optimizing PRESS localized citrate detection at 3 Tesla. *Magn. Reson. Med.* 54: 51–58.

278 Gambarota, G., van der Graaf, M., Klomp, D. et al. (2005). Echo-time independent signal modulations using PRESS sequences: a new approach to spectral editing of strongly coupled AB spin systems. *J. Magn. Reson.* 177: 299–306.

279 Seymour, Z.A., Panigrahy, A., Finlay, J.L. et al. (2008). Citrate in pediatric CNS tumors? *AJNR Am. J. Neuroradiol.* 29: 1006–1011.

280 Bluml, S., Panigrahy, A., Laskov, M. et al. (2011). Elevated citrate in pediatric astrocytomas with malignant progression. *Neuro Oncol.* 13: 1107–1117.

281 Choi, C., Ganji, S.K., Madan, A. et al. (2014). *In vivo* detection of citrate in brain tumors by ^{1}H magnetic resonance spectroscopy at 3T. *Magn. Reson. Med.* 72: 316–323.

282 Wang, Z.Y., Noyszewski, E.A., and Leigh, J.S. Jr. (1990). *In vivo* MRS measurement of deoxymyoglobin in human forearms. *Magn. Reson. Med.* 14: 562–567.

283 Wang, Z., Wang, D.J., Noyszewski, E.A. et al. (1992). Sensitivity of *in vivo* MRS of the N-delta proton in proximal histidine of deoxymyoglobin. *Magn. Reson. Med.* 27: 362–367.

284 Kreutzer, U. and Jue, T. (1991). [1]H-nuclear magnetic resonance deoxymyoglobin signal as indicator of intracellular oxygenation in myocardium. *Am. J. Physiol.* 261: H2091–H2097.

285 Chen, W., Cho, Y., Merkle, H. et al. (1999). *In vitro* and *in vivo* studies of ^{1}H NMR visibility to detect deoxyhemoglobin and deoxymyoglobin signals in myocardium. *Magn. Reson. Med.* 42: 1–5.

286 Kreis, R., Bruegger, K., Skjelsvik, C. et al. (2001). Quantitative ^{1}H magnetic resonance spectroscopy of myoglobin de- and reoxygenation in skeletal muscle: reproducibility and effects of location and disease. *Magn. Reson. Med.* 46: 240–248.

287 Dimitrov, I., Ren, J., Douglas, D. et al. (2010). Composition of fatty acids in adipose tissue by *in vivo* ^{13}C MRS at 7T. *Proc. Int. Soc. Magn. Reson. Med.* 18: 320.

288 Schick, F., Eismann, B., Jung, W.I. et al. (1993). Comparison of localized proton NMR signals of skeletal muscle and fat tissue *in vivo*: two lipid compartments in muscle tissue. *Magn. Reson. Med.* 29: 158–167.

289 Boesch, C., Decombaz, J., Slotboom, J., and Kreis, R. (1999). Observation of intramyocellular lipids by means of ^{1}H magnetic resonance spectroscopy. *Proc. Nutr. Soc.* 58: 841–850.

290 Boesch, C. and Kreis, R. (2000). Observation of intramyocellular lipids by [1]H-magnetic resonance spectroscopy. *Ann. N. Y. Acad. Sci.* 904: 25–31.

291 Boesch, C. and Kreis, R. (2016). Muscle studies by ^{1}H MRS. *eMagRes* 5: 1097–1108.

292 Chu, S.C., Xu, Y., Balschi, J.A., and Springer, C.S. (1990). Bulk magnetic susceptibility shifts in NMR studies of compartmentalized samples: use of paramagnetic reagents. *Magn. Reson. Med.* 13: 239–262.

293 van der Graaf, M., Schipper, R.G., Oosterhof, G.O. et al. (2000). Proton MR spectroscopy of prostatic tissue focused on the detection of spermine, a possible biomarker of malignant behavior in prostate cancer. *MAGMA* 10: 153–159.

294 Jung, W.I., Widmaier, S., Seeger, U. et al. (1996). Phosphorus J coupling constants of ATP in human myocardium and calf muscle. *J. Magn. Reson. B* 110: 39–46.

295 Luyten, P.R., Bruntink, G., Sloff, F.M. et al. (1989). Broadband proton decoupling in human ^{31}P NMR spectroscopy. *NMR Biomed.* 1: 177–183.

296 Freeman, D.M. and Hurd, R. (1997). Decoupling: theory and practice. II. State of the art: *in vivo* applications of decoupling. *NMR Biomed.* 10: 381–393.

297 Madden, A., Leach, M.O., Sharp, J.C. et al. (1991). A quantitative analysis of the accuracy of *in vivo* pH measurements with ^{31}P NMR spectroscopy: assessment of pH measurement methodology. *NMR Biomed.* 4: 1–11.

298 Petroff, O.A., Prichard, J.W., Behar, K.L. et al. (1985). Cerebral intracellular pH by ^{31}P nuclear magnetic resonance spectroscopy. *Neurology* 35: 781–788.

299 Pettegrew, J.W., Withers, G., Panchalingam, K., and Post, J.F. (1988). Considerations for brain pH assessment by ^{31}P NMR. *Magn. Reson. Imaging* 6: 135–142.

300 Roberts, J.K., Wade-Jardetzky, N., and Jardetzky, O. (1981). Intracellular pH measurements by ^{31}P nuclear magnetic resonance. Influence of factors other than pH on ^{31}P chemical shifts. *Biochemistry* 20: 5389–5394.

301 Gupta, R.K., Benovic, J.L., and Rose, Z.B. (1978). The determination of the free magnesium level in the human red blood cell by ^{31}P NMR. *J. Biol. Chem.* 253: 6172–6176.

302 Gupta, R.K. and Moore, R.D. (1980). ^{31}P NMR studies of intracellular free Mg^{2+} in intact frog skeletal muscle. *J. Biol. Chem.* 255: 3987–3993.

303 Gupta, R.K., Gupta, P., Yushok, W.D., and Rose, Z.B. (1983). On the noninvasive measurement of intracellular free magnesium by ^{31}P NMR spectroscopy. *Physiol. Chem. Phys. Med. NMR* 15: 265–280.

304 Gupta, R.K., Gupta, P., and Moore, R.D. (1984). NMR studies of intracellular metal ions in intact cells and tissues. *Annu. Rev. Biophys. Bioeng.* 13: 221–246.

305 Halvorson, H.R., Vande Linde, A.M., Helpern, J.A., and Welch, K.M. (1992). Assessment of magnesium concentrations by ^{31}P NMR *in vivo. NMR Biomed.* 5: 53–58.

306 Mosher, T.J., Williams, G.D., Doumen, C. et al. (1992). Error in the calibration of the MgATP chemical-shift limit: effects on the determination of free magnesium by ^{31}P NMR spectroscopy. *Magn. Reson. Med.* 24: 163–169.

307 Sillerud, L.O. and Shulman, R.G. (1983). Structure and metabolism of mammalian liver glycogen monitored by carbon-13 nuclear magnetic resonance. *Biochemistry* 22: 1087–1094.

308 Gruetter, R., Prolla, T.A., and Shulman, R.G. (1991). ^{13}C NMR visibility of rabbit muscle glycogen *in vivo. Magn. Reson. Med.* 20: 327–332.

309 Shalwitz, R.A., Reo, N.V., Becker, N.N., and Ackerman, J.J. (1987). Visibility of mammalian hepatic glycogen to the NMR experiment, *in vivo. Magn. Reson. Med.* 5: 462–465.

310 Rothman, D.L., Magnusson, I., Katz, L.D. et al. (1991). Quantitation of hepatic glycogenolysis and gluconeogenesis in fasting humans with ^{13}C NMR. *Science* 254: 573–576.

311 Murphy, E. and Hellerstein, M. (2000). Is *in vivo* nuclear magnetic resonance spectroscopy currently a quantitative method for whole-body carbohydrate metabolism? *Nutr. Rev.* 58: 304–314.

312 de Graaf, R.A., De Feyter, H.M., and Rothman, D.L. (2015). High-sensitivity, broadband-decoupled ^{13}C MR spectroscopy in humans at 7T using two-dimensional heteronuclear single-quantum coherence. *Magn. Reson. Med.* 74: 903–914.

313 Cheshkov, S., Dimitrov, I.E., Jakkamsetti, V. et al. (2017). Oxidation of [U-^{13}C]glucose in the human brain at 7T under steady state conditions. *Magn. Reson. Med.* 78: 2065–2071.

314 Qian, Y., Zhao, T., Wiggins, G.C. et al. (2012). Sodium imaging of human brain at 7 T with 15-channel array coil. *Magn. Reson. Med.* 68: 1807–1814.

315 Mirkes, C.C., Hoffmann, J., Shajan, G. et al. (2015). High-resolution quantitative sodium imaging at 9.4 Tesla. *Magn. Reson. Med.* 73: 342–351.

316 Thulborn, K., Lui, E., Guntin, J. et al. (2016). Quantitative sodium MRI of the human brain at 9.4 T provides assessment of tissue sodium concentration and cell volume fraction during normal aging. *NMR Biomed.* 29: 137–143.

317 Springer, C.S. Jr. (1987). Measurement of metal cation compartmentalization in tissue by high-resolution metal cation NMR. *Annu. Rev. Biophys. Biophys. Chem.* 16: 375–399.

318 Rooney, W.D. and Springer, C.S. Jr. (1991). A comprehensive approach to the analysis and interpretation of the resonances of spins 3/2 from living systems. *NMR Biomed.* 4: 209–226.

319 Madelin, G. and Regatte, R.R. (2013). Biomedical applications of sodium MRI *in vivo. J. Magn. Reson. Imaging* 38: 511–529.

320 Neumaier-Probst, E., Konstandin, S., Ssozi, J. et al. (2015). A double-tuned ^{1}H/^{23}Na resonator allows ^{1}H-guided ^{23}Na-MRI in ischemic stroke patients in one session. *Int. J. Stroke* 10 (Suppl A100): 56–61.

321 Gupta, R.K. and Gupta, P. (1982). Direct observation of resolved resonances from intra and extracellular sodium-23 ions in NMR studies of intact cells and tissues using dysprosium(III)-tripolyphosphate as paramagnetic shift reagent. *J. Magn. Reson.* 47: 344–350.

322 Pike, M.M., Fossel, E.T., Smith, T.W., and Springer, C.S. Jr. (1984). High-resolution ^{23}Na-NMR studies of human erythrocytes: use of aqueous shift reagents. *Am. J. Physiol.* 246: C528–C536.

323 Pike, M.M., Frazer, J.C., Dedrick, D.F. et al. (1985). ^{23}Na and ^{39}K nuclear magnetic resonance studies of perfused rat hearts. Discrimination of intra- and extracellular ions using a shift reagent. *Biophys. J.* 48: 159–173.

324 Balschi, J.A., Kohler, S.J., Bittl, J.A. et al. (1989). Magnetic field dependence of Na-23 NMR spectra of rat skeletal muscle infused with shift reagents *in vivo*. *J. Magn. Reson.* 83: 138–145.

325 Balschi, J.A., Bittl, J.A., Springer, C.S. Jr., and Ingwall, J.S. (1990). ^{31}P and ^{23}Na NMR spectroscopy of normal and ischemic rat skeletal muscle. Use of a shift reagent *in vivo*. *NMR Biomed.* 3: 47–58.

326 Pekar, J. and Leigh, J.S. (1986). Detection of biexponential relaxation in sodium-23 facilitated by double-quantum filtering. *J. Magn. Reson.* 69: 582–584.

327 Pekar, J., Renshaw, P.F., and Leigh, J.S. (1987). Selective detection of intracellular sodium by coherence-transfer NMR. *J. Magn. Reson.* 72: 159–161.

328 Rooney, W.D., Barbara, T.M., and Springer, C.S. (1988). Two-dimensional double quantum NMR spectroscopy of isolated spin 3/2 systems: ^{23}Na examples. *J. Am. Chem. Soc.* 110: 674–681.

329 Jelicks, L.A. and Gupta, R.K. (1989). Observation of intracellular sodium ions by double-quantum-filtered ^{23}Na NMR with paramagnetic quenching of extracellular coherence by gadolinium tripolyphosphate. *J. Magn. Reson.* 83: 146–151.

330 Jelicks, L.A. and Gupta, R.K. (1989). Double-quantum NMR of sodium ions in cells and tissues. Paramagnetic quenching of extracellular coherence. *J. Magn. Reson.* 81: 586–592.

331 Hutchison, R.B., Malhotra, D., Hendrick, R.E. et al. (1990). Evaluation of the double-quantum filter for the measurement of intracellular sodium concentration. *J. Biol. Chem.* 265: 15506–15510.

332 Hilal, S.K., Maudsley, A.A., Simon, H.E. et al. (1983). *In vivo* NMR imaging of tissue sodium in the intact cat before and after acute cerebral stroke. *AJNR Am. J. Neuroradiol.* 4: 245–249.

333 Hilal, S.K., Maudsley, A.A., Ra, J.B. et al. (1985). *In vivo* NMR imaging of sodium-23 in the human head. *J. Comput. Assist. Tomogr.* 9: 1–7.

334 Thulborn, K.R., Davis, D., Adams, H. et al. (1999). Quantitative tissue sodium concentration mapping of the growth of focal cerebral tumors with sodium magnetic resonance imaging. *Magn. Reson. Med.* 41: 351–359.

335 Ouwerkerk, R., Bleich, K.B., Gillen, J.S. et al. (2003). Tissue sodium concentration in human brain tumors as measured with ^{23}Na MR imaging. *Radiology* 227: 529–537.

336 Zaaraoui, W., Konstandin, S., Audoin, B. et al. (2012). Distribution of brain sodium accumulation correlates with disability in multiple sclerosis: a cross-sectional ^{23}Na MR imaging study. *Radiology* 264: 859–867.

337 Eisele, P., Konstandin, S., Griebe, M. et al. (2016). Heterogeneity of acute multiple sclerosis lesions on sodium ^{23}Na MRI. *Mult. Scler.* 22: 1040–1047.

338 Stevens, A.N., Morris, P.G., Iles, R.A. et al. (1984). 5-Fluorouracil metabolism monitored *in vivo* by ^{19}F NMR. *Br. J. Cancer* 50: 113–117.

339 Wolf, W., Albright, M.J., Silver, M.S. et al. (1987). Fluorine-19 NMR spectroscopic studies of the metabolism of 5-fluorouracil in the liver of patients undergoing chemotherapy. *Magn. Reson. Imaging* 5: 165–169.

340 McSheehy, P.M. and Griffiths, J.R. (1989). ^{19}F MRS studies of fluoropyrimidine chemotherapy. A review. *NMR Biomed.* 2: 133–141.

341 Semmler, W., Bachert-Baumann, P., Guckel, F. et al. (1990). Real-time follow-up of 5-fluorouracil metabolism in the liver of tumor patients by means of F-19 MR spectroscopy. *Radiology* 174: 141–145.

342 Schlemmer, H.P., Bachert, P., Semmler, W. et al. (1994). Drug monitoring of 5-fluorouracil: *in vivo* ^{19}F NMR study during 5-FU chemotherapy in patients with metastases of colorectal adenocarcinoma. *Magn. Reson. Imaging* 12: 497–511.

343 Kamm, Y.J., Heerschap, A., van den Bergh, E.J., and Wagener, D.J. (2004). ^{19}F-magnetic resonance spectroscopy in patients with liver metastases of colorectal cancer treated with 5-fluorouracil. *Anticancer Drugs* 15: 229–233.

344 Eleff, S.M., Schnall, M.D., Ligetti, L. et al. (1988). Concurrent measurements of cerebral blood flow, sodium, lactate, and high-energy phosphate metabolism using ^{19}F, ^{23}Na, ^{1}H, and ^{31}P nuclear magnetic resonance spectroscopy. *Magn. Reson. Med.* 7: 412–424.

345 Barranco, D., Sutton, L.N., Florin, S. et al. (1989). Use of ^{19}F NMR spectroscopy for measurement of cerebral blood flow: a comparative study using microspheres. *J. Cereb. Blood Flow Metab.* 9: 886–891.

346 Detre, J.A., Eskey, C.J., and Koretsky, A.P. (1990). Measurement of cerebral blood flow in rat brain by ^{19}F-NMR detection of trifluoromethane washout. *Magn. Reson. Med.* 15: 45–57.

347 Pekar, J., Ligeti, L., Sinnwell, T. et al. (1994). ^{19}F magnetic resonance imaging of cerebral blood flow with 0.4-cc resolution. *J. Cereb. Blood Flow Metab.* 14: 656–663.

348 Mills, P., Sessler, D.I., Moseley, M. et al. (1987). An *in vivo* ^{19}F nuclear magnetic resonance study of isoflurane elimination from the rabbit brain. *Anesthesiology* 67: 169–173.

349 Wyrwicz, A.M., Conboy, C.B., Ryback, K.R. et al. (1987). *In vivo* ^{19}F-NMR study of isoflurane elimination from brain. *Biochim. Biophys. Acta* 927: 86–91.

350 Wyrwicz, A.M., Conboy, C.B., Nichols, B.G. et al. (1987). *In vivo* ^{19}F-NMR study of halothane distribution in brain. *Biochim. Biophys. Acta* 929: 271–277.

351 Lockhart, S.H., Cohen, Y., Yasuda, N. et al. (1991). Cerebral uptake and elimination of desflurane, isoflurane, and halothane from rabbit brain: an *in vivo* NMR study. *Anesthesiology* 74: 575–580.

352 Smith, G.A., Hesketh, R.T., Metcalfe, J.C. et al. (1983). Intracellular calcium measurements by ^{19}F NMR of fluorine-labeled chelators. *Proc. Natl. Acad. Sci. U. S. A.* 80: 7178–7182.

353 Clark, L.C. Jr., Ackerman, J.L., Thomas, S.R. et al. (1984). Perfluorinated organic liquids and emulsions as biocompatible NMR imaging agents for ^{19}F and dissolved oxygen. *Adv. Exp. Med. Biol.* 180: 835–845.

354 Metcalfe, J.C., Hesketh, T.R., and Smith, G.A. (1985). Free cytosolic Ca^{2+} measurements with fluorine labelled indicators using ^{19}F NMR. *Cell Calcium* 6: 183–195.

355 Smith, G.A., Morris, P.G., Hesketh, T.R., and Metcalfe, J.C. (1986). Design of an indicator of intracellular free Na^+ concentration using ^{19}F NMR. *Biochem. Biophys. Acta* 889: 72–83.

356 Bachelard, H.S., Badar-Goffer, R.S., Brooks, K.J. et al. (1988). Measurement of free intracellular calcium in the brain by ^{19}F-nuclear magnetic resonance spectroscopy. *J. Neurochem.* 51: 1311–1313.

357 Deutsch, J.C. and Taylor, J.S. (1989). New class of ^{19}F pH indicators: fluoroanilines. *Biophys. J.* 55: 799–804.

358 Levy, L.A., Murphy, E., Raju, B., and London, R.E. (1988). Measurement of cytosolic free magnesium concentration by ^{19}F NMR. *Biochemistry* 27: 4041–4048.

359 Lanza, G.M., Winter, P.M., Neubauer, A.M. et al. (2005). ^{1}H/^{19}F magnetic resonance molecular imaging with perfluorocarbon nanoparticles. *Curr. Top. Dev. Biol.* 70: 57–76.

360 Srinivas, M., Heerschap, A., Ahrens, E.T. et al. (2010). ^{19}F MRI for quantitative *in vivo* cell tracking. *Trends Biotechnol.* 28: 363–370.

361 Ruiz-Cabello, J., Barnett, B.P., Bottomley, P.A., and Bulte, J.W. (2011). Fluorine (^{19}F) MRS and MRI in biomedicine. *NMR Biomed.* 24: 114–129.

362 Ahrens, E.T. and Zhong, J. (2013). *In vivo* MRI cell tracking using perfluorocarbon probes and fluorine-19 detection. *NMR Biomed.* 26: 860–871.

363 Prior, M.J.W., Maxwell, R.J., and Griffiths, J.R. (1992). Fluorine ^{19}F NMR spectroscopy and imaging *in vivo*. In: *NMR Basic Principles and Progress*, vol. 28 (ed. P. Diehl, E. Fluck, H. Gunther, et al.), 101–130. Berlin: Springer-Verlag.

364 Selinsky, B.S. and Burt, C.T. (1992). *In vivo* ^{19}F NMR. In: *Biological Magnetic Resonance*, vol. 11 (ed. L.J. Berliner and J. Reuben), 241–276. New York: Plenum Press.

365 Heerschap, A. (2016). *In vivo* ^{19}F magnetic resonance spectroscopy. *eMagRes* 5: 1283–1290.

366 Nakada, T., Kwee, I.L., and Conboy, C.B. (1986). Noninvasive *in vivo* demonstration of 2-fluoro-2-deoxy-D-glucose metabolism beyond the hexokinase reaction in rat brain by ^{19}F nuclear magnetic resonance. *J. Neurochem.* 46: 198–201.

367 Kwee, I.L., Nakada, T., and Card, P.J. (1987). Noninvasive demonstration of *in vivo* 3-fluoro-3-deoxy-D-glucose metabolism in rat brain by ^{19}F nuclear magnetic resonance spectroscopy: suitable probe for monitoring cerebral aldose reductase activities. *J. Neurochem.* 49: 428–433.

368 Nakada, T., Kwee, I.L., Card, P.J. et al. (1988). Fluorine-19 NMR imaging of glucose metabolism. *Magn. Reson. Med.* 6: 307–313.

369 Berkowitz, B.A. and Ackerman, J.J. (1987). Proton decoupled fluorine nuclear magnetic resonance spectroscopy *in situ*. *Biophys. J.* 51: 681–685.

370 Wang, J., Sanchez-Rosello, M., Acena, J.L. et al. (2014). Fluorine in pharmaceutical industry: fluorine-containing drugs introduced to the market in the last decade (2001–2011). *Chem. Rev.* 114: 2432–2506.

371 Wyrwicz, A.M., Pszenny, M.H., Schofield, J.C. et al. (1983). Noninvasive observations of fluorinated anesthetics in rabbit brain by fluorine-19 nuclear magnetic resonance. *Science* 222: 428–430.

372 Ueki, M., Mies, G., and Hossmann, K.A. (1992). Effect of alpha-chloralose, halothane, pentobarbital and nitrous oxide anesthesia on metabolic coupling in somatosensory cortex of rat. *Acta Anaesthesiol. Scand.* 36: 318–322.

373 Austin, V.C., Blamire, A.M., Allers, K.A. et al. (2005). Confounding effects of anesthesia on functional activation in rodent brain: a study of halothane and alpha-chloralose anesthesia. *Neuroimage* 24: 92–100.

374 Ogawa, S., Lee, T.M., Stepnoski, R. et al. (2000). An approach to probe some neural systems interaction by functional MRI at neural time scale down to milliseconds. *Proc. Natl. Acad. Sci. U. S. A.* 97: 11026–11031.

375 Reivich, M., Kuhl, D., Wolf, A. et al. (1979). The [18F]fluorodeoxyglucose method for the measurement of local cerebral glucose utilization in man. *Circ. Res.* 44: 127–137.

376 Pouremad, R. and Wyrwicz, A.M. (1991). Cerebral metabolism of fluorodeoxyglucose measured with ^{19}F NMR spectroscopy. *NMR Biomed.* 4: 161–166.

377 Southworth, R., Parry, C.R., Parkes, H.G. et al. (2003). Tissue-specific differences in 2-fluoro-2-deoxyglucose metabolism beyond FDG-6-P: a ^{19}F NMR spectroscopy study in the rat. *NMR Biomed.* 16: 494–502.

378 Hamza, M.A., Serratrice, G., Stebe, M.-J., and Delpuech, J.-J. (1981). Fluorocarbons as oxygen carriers. II. An NMR study of partially or totally fluorinated alkanes and alkenes. *J. Magn. Reson.* 42: 227–241.

379 Badar-Goffer, R.S., Ben-Yoseph, O., Dolin, S.J. et al. (1990). Use of 1,2-bis-(2-amino-5-fluorophenoxyl)-ethane-*N,N,N′,N′*-tetraacetic acid (5FBAPTA) in the measurement of free intracellular calcium in the brain by 19F nuclear magnetic resonance spectroscopy. *J. Neurochem.* 55: 878–884.

380 Mugler, J.P. 3rd and Altes, T.A. (2013). Hyperpolarized ^{129}Xe MRI of the human lung. *J. Magn. Reson. Imaging* 37: 313–331.

381 Roos, J.E., McAdams, H.P., Kaushik, S.S., and Driehuys, B. (2015). Hyperpolarized gas MR imaging: technique and applications. *Magn. Reson. Imaging Clin. N. Am.* 23: 217–229.

382 Goodson, B.M., Whiting, N., Coffey, A.M. et al. (2015). Hyperpolarization methods for MRS. *eMagRes* 4: 797–810.

383 Albert, M.S., Cates, G.D., Driehuys, B. et al. (1994). Biological magnetic resonance imaging using laser-polarized ^{129}Xe. *Nature* 370: 199–201.

384 Mugler, J.P. 3rd, Driehuys, B., Brookeman, J.R. et al. (1997). MR imaging and spectroscopy using hyperpolarized ^{129}Xe gas: preliminary human results. *Magn. Reson. Med.* 37: 809–815.

385 Ebert, M., Grossmann, T., Heil, W. et al. (1996). Nuclear magnetic resonance imaging with hyperpolarised helium-3. *Lancet* 347: 1297–1299.

386 MacFall, J.R., Charles, H.C., Black, R.D. et al. (1996). Human lung air spaces: potential for MR imaging with hyperpolarized He-3. *Radiology* 200: 553–558.

387 Wagshul, M.E., Button, T.M., Li, H.F. et al. (1996). *In vivo* MR imaging and spectroscopy using hyperpolarized ^{129}Xe. *Magn. Reson. Med.* 36: 183–191.

388 Renshaw, P.F., Haselgrove, J.C., Leigh, J.S., and Chance, B. (1985). *In vivo* nuclear magnetic resonance imaging of lithium. *Magn. Reson. Med.* 2: 512–516.

389 Komoroski, R.A., Newton, J.E., Walker, E. et al. (1990). *In vivo* NMR spectroscopy of lithium-7 in humans. *Magn. Reson. Med.* 15: 347–356.

390 Komoroski, R.A. (2000). Applications of ^{7}Li NMR in biomedicine. *Magn. Reson. Imaging* 18: 103–116.

391 Kanamori, K., Parivar, F., and Ross, B.D. (1993). A ^{15}N NMR study of *in vivo* cerebral glutamine synthesis in hyperammonemic rats. *NMR Biomed.* 6: 21–26.

392 Kanamori, K. and Ross, B.D. (1993). ^{15}N NMR measurement of the *in vivo* rate of glutamine synthesis and utilization at steady state in the brain of the hyperammonaemic rat. *Biochem. J.* 293: 461–468.

393 Shen, J., Sibson, N.R., Cline, G. et al. (1998). ^{15}N-NMR spectroscopy studies of ammonia transport and glutamine synthesis in the hyperammonemic rat brain. *Dev. Neurosci.* 20: 434–443.

394 Balaban, R.S. and Knepper, M.A. (1983). Nitrogen-14 nuclear magnetic resonance spectroscopy of mammalian tissues. *Am. J. Physiol.* 245: C439–C444.

395 Grunder, W., Krumbiegel, P., Buchali, K., and Blesin, H.J. (1989). Nitrogen-15 NMR studies of rat liver *in vitro* and *in vivo*. *Phys. Med. Biol.* 34: 457–463.

396 Preece, N.E. and Cerdan, S. (1993). Determining ^{15}N to ^{14}N ratios in biofluids by single-pulse ^{1}H nuclear magnetic resonance. *Anal. Biochem.* 215: 180–183.

397 de Graaf, R.A., Brown, P.B., Rothman, D.L., and Behar, K.L. (2008). Natural abundance ^{17}O NMR spectroscopy of rat brain *in vivo*. *J. Magn. Reson.* 193: 63–67.

398 Zhu, X., Merkle, H., Kwag, J. et al. (2001). ^{17}O relaxation time and NMR sensitivity of cerebral water and their field dependence. *Magn. Reson. Med.* 45: 543–549.

399 Pekar, J., Ligeti, L., Ruttner, Z. et al. (1991). *In vivo* measurement of cerebral oxygen consumption and blood flow using ^{17}O magnetic resonance imaging. *Magn. Reson. Med.* 21: 313–319.

400 Fiat, D., Dolinsek, J., Hankiewicz, J. et al. (1993). Determination of regional cerebral oxygen consumption in the human: ^{17}O natural abundance cerebral magnetic resonance imaging and spectroscopy in a whole body system. *Neurol. Res.* 15: 237–248.

401 Zhu, X.H., Zhang, Y., Tian, R.X. et al. (2002). Development of ^{17}O NMR approach for fast imaging of cerebral metabolic rate of oxygen in rat brain at high field. *Proc. Natl. Acad. Sci. U. S. A.* 99: 13194–13199.

402 Atkinson, I.C. and Thulborn, K.R. (2010). Feasibility of mapping the tissue mass corrected bioscale of cerebral metabolic rate of oxygen consumption using 17-oxygen and 23-sodium MR imaging in a human brain at 9.4 T. *Neuroimage* 51: 723–733.

403 Lu, M., Atthe, B., Mateescu, G.D. et al. (2012). Assessing mitochondrial respiration in isolated hearts using ^{17}O MRS. *NMR Biomed.* 25: 883–889.

404 Kirsch, S., Augath, M., Seiffge, D. et al. (2010). *In vivo* chlorine-35, sodium-23 and proton magnetic resonance imaging of the rat brain. *NMR Biomed.* 23: 592–600.

405 Nagel, A.M., Lehmann-Horn, F., Weber, M.A. et al. (2014). *In vivo* ^{35}Cl MR imaging in humans: a feasibility study. *Radiology* 271: 585–595.

406 Weber, M.A., Nagel, A.M., Marschar, A.M. et al. (2016). 7-T ^{35}Cl and ^{23}Na MR imaging for detection of mutation-dependent alterations in muscular edema and fat fraction with sodium and chloride concentrations in muscular periodic paralyses. *Radiology* 280: 848–859.

407 Musio, R. and Sciacovelli, O. (2001). Detection of taurine in biological tissues by ^{33}S NMR spectroscopy. *J. Magn. Reson.* 153: 259–261.

408 Augath, M., Heiler, P., Kirsch, S., and Schad, L.R. (2009). *In vivo* ^{39}K, ^{23}Na and ^{1}H MR imaging using a triple resonant RF coil setup. *J. Magn. Reson.* 200: 134–136.

409 Umathum, R., Rosler, M.B., and Nagel, A.M. (2013). *In vivo* ^{39}K MR imaging of human muscle and brain. *Radiology* 269: 569–576.

410 Bendel, P. (2005). Biomedical applications of ^{10}B and ^{11}B NMR. *NMR Biomed.* 18: 74–82.

411 Hull, W.E. (1999). A critical review of MR studies concerning silicone breast implants. *Magn. Reson. Med.* 42: 984–995.

3

In Vivo NMR Spectroscopy – Dynamic Aspects

3.1 Introduction

The chemical specificity of NMR spectroscopy, observed through the chemical shift and scalar coupling patterns, allows the detection of more than 15 metabolites *in vivo* and more than 50 chemical compounds in bodily fluids *in vitro* [1–3]. The detection and quantification of a wide range of metabolites has led to the characterization of disease progression, allows the study of intervention by medication or surgery, and allows identification or categorization of diseases by observing specific metabolic markers. For example, using pattern recognition algorithms on ^{1}H NMR spectra, a wide range of tumor types can be reliably identified [4]. In epilepsy research, *N*-acetyl aspartate has been identified as an important marker for the detection of the foci of epileptic seizures [5], while the cerebral levels of GABA and homocarnosine have shown to be correlated with the frequency of epileptic seizures [6].

However, despite its great importance, detection of static metabolite concentrations alone provides only a partial description of metabolism. *In vivo* metabolism is largely characterized by dynamic processes, like enzyme-catalyzed chemical exchange, transfer of chemical groups through metabolic pathways and, specific for NMR, relaxation processes. This chapter is dedicated to the description of dynamic processes *in vivo* that can be measured by NMR spectroscopy. The dynamics of T_1 and T_2 relaxation are discussed in Section 3.2. Using appropriate experimental techniques, NMR can be sensitized to a variety of dynamic processes, of which chemical exchange and diffusion are discussed in Sections 3.3 and 3.4, respectively. When combined with the infusion of exogenous compounds, NMR allows the detection of metabolic fluxes noninvasively *in vivo*, and will be the subject of Section 3.5. Deuterium metabolic imaging (DMI) and advances in the application of hyperpolarization methods will also be discussed in Section 3.5.

3.2 Relaxation

3.2.1 General Principles of Dipolar Relaxation

In Chapter 1, relaxation was qualitatively described as the process by which the spins return to the thermal equilibrium state following a perturbation. The restoration of the longitudinal equilibrium magnetization is characterized by the longitudinal or spin–lattice relaxation time constant T_1, while the disappearance of transverse magnetization is described by the transverse or spin–spin relaxation time constant T_2. In this section the nature of the processes responsible for relaxation will be considered in a more quantitative manner.

In Vivo *NMR Spectroscopy: Principles and Techniques*, Third Edition. Robin A. de Graaf.
© 2019 John Wiley & Sons Ltd. Published 2019 by John Wiley & Sons Ltd.

It was shown that the orientation of nuclear spins can be changed by a magnetic field in the transverse plane that is rotating at or near the Larmor frequency of the spins. A coherent perturbation of the spins can be achieved with an external coherent magnetic field B_I^+, whereas incoherent, random magnetic fields internal to the sample lead to incoherent, random perturbations of the nuclear spin orientation.

For spins in solution the randomly fluctuating magnetic fields are predominantly caused by the magnetic moment of other, nearby spins (Figure 3.1). The magnetic moment of one spin affects the local field of another spin in a random manner (both in amplitude and orientation), due to Brownian motion and molecular tumbling. Figure 3.2A shows a typical distribution of the local magnetic field at a nucleus as a function of time. Since relaxation will be induced when the frequency of the local magnetic field is close to the Larmor frequency, a quantitative description of relaxation will require the characterization of the frequency components in the local magnetic field over time. For this purpose the autocorrelation function $G(\tau)$ is introduced which is defined as

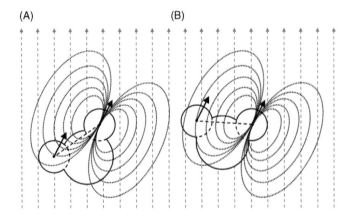

Figure 3.1 Dipole–dipole interactions for water. (A) The magnetic moment of one proton in an external magnetic field B_0 (parallel gray lines) perturbs the magnetic field at the position of another proton, typically within the same water molecule. (B) As the water rotates and translates randomly, the direction and magnitude of the dipole–dipole interaction changes, giving rise to the time-varying magnetic fields required for relaxation.

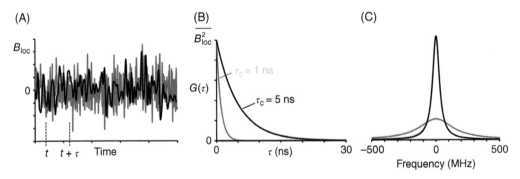

Figure 3.2 Random magnetic field fluctuations. (A) Rapid (gray) and slow (black) magnetic field fluctuations encountered by two different spins. (B) Quantitative description of the magnetic field fluctuations through the use of autocorrelation functions (Eq. (3.2)). The rapid and slow fluctuations are characterized by rotation correlation times τ_c of 1 and 5 ns, respectively. (C) Spectral density functions corresponding to the rapid and slow rotations. Rapid fluctuations contain higher frequency contributions. Note that the total area under both curves is identical.

$$G(\tau) = \overline{B_{\text{loc}}(t) B_{\text{loc}}(t + \tau)} \qquad (3.1)$$

where the bar indicates that the "ensemble average" (i.e. an average over all spins in a macroscopic part of the sample) needs to be taken. The correlation function $G(\tau)$, which is independent of t, is a measure for the correlation between the local magnetic fields as time τ progresses. For a short delay τ, the orientation and amplitude of the local magnetic field will not have changed much and the ensemble average would be large (i.e. high correlation). After longer τ delays the local magnetic field is drastically altered due to Brownian motion and molecular tumbling, leading to a low correlation. Clearly, the correlation function $G(\tau)$ is a decaying function and is usually taken as a decaying exponential, i.e.

$$G(\tau) = \overline{B_{\text{loc}}^2} e^{-|\tau|/\tau_c} \qquad (3.2)$$

where τ_c is the rotation correlation time. For random molecular tumbling, τ_c roughly corresponds to the average time for a molecule to rotate over one radian. Note that for mobile spins in solution τ_c is short (10^{-12}–10^{-10} s), while for immobile spins in solids or high-viscosity liquids, τ_c is longer (10^{-8}–10^{-6} s). Fourier transformation of the correlation function yields the spectral density function $J(\nu)$:

$$J(\nu) = 2\overline{B_{\text{loc}}^2} \frac{\tau_c}{1 + 4\pi^2 \nu^2 \tau_c^2} \qquad (3.3)$$

The spectral density function expresses the motional characteristics (described by τ_c) in terms of the power at frequency ν. Figure 3.2C shows the spectral density function for spins exhibiting slower rotations ($\tau_c = 5\,\text{ns}$) and for spins displaying faster rotations ($\tau_c = 1\,\text{ns}$). As expected, slow rotational motions emphasize the low-frequency components, while fast rotational motion also has significant power at high frequencies.

The remaining point in describing dipolar relaxation is to establish relations between the relaxation time constants T_1 and T_2 and the frequency components given by the spectral density function. In the classical view the local magnetic fields perturb the orientation of the nuclear spins. In the quantum-mechanical or spectroscopic view this is equivalent in stating that the local magnetic fields induce transitions between spin states. The original publication on dipolar relaxation [7] used the quantum mechanical view to link the local magnetic field behavior to transition probabilities between spin states (Figure 3.3). For a dipolar-coupled homonuclear two-spin-system there are three transition types, each with a transition

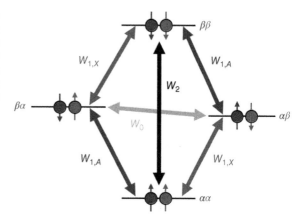

Figure 3.3 **Energy-level diagram for a dipolar coupled two-spin system.** W_A and W_X represent the single-quantum transition probabilities that spin A (red) or X (blue) change energy level, respectively. W_0 and W_2 correspond to the probability of a zero- or double quantum transition during which both spins change at the same time.

probability W that depends on the frequency components of the local magnetic field. A single-quantum transition with probability W_1 occurs when one of the two spins flips while leaving the other unperturbed. A zero-quantum transition with probability W_0 occurs when both spins flip, albeit in opposite direction (i.e. $\alpha \rightarrow \beta$ for spin 1 and $\beta \rightarrow \alpha$ for spin 2). Finally, a double-quantum transition with probability W_2 occurs when both spins flip in the same direction. Detailed calculations outside the scope of this book provide a link between the transition probabilities and the spectral density $J(\nu)$ according to

$$W_0 = \frac{1}{10}C_1^2 J(0), \quad W_1 = \frac{3}{20}C_1^2 J(\nu), \quad \text{and } W_2 = \frac{3}{5}C_1^2 J(2\nu) \quad \text{with } C_1 = \frac{\mu_0}{8\pi^2}\frac{h\gamma^2}{r^3} \quad (3.4)$$

Note that zero-quantum transitions require spectral density at frequencies close to zero, whereas double-quantum transitions require spectral density at frequencies close to twice the Larmor frequency. The Solomon differential equations [8] describe how the transition probabilities W_0, W_1, and W_2 are linked to T_1 and T_2 relaxation and give

$$\frac{1}{T_1} = \frac{3}{10}C_1^2\left[J(\nu) + 4J(2\nu)\right] \quad (3.5)$$

$$\frac{1}{T_2} = \frac{3}{20}C_1^2\left[3J(0) + 5J(\nu) + 2J(2\nu)\right] \quad (3.6)$$

Equations (3.5) and (3.6) show that T_1 and T_2 relaxation are both affected by frequency components at the Larmor frequency and at twice the Larmor frequency. T_2 relaxation is in addition also affected by low-frequency components. The implication of Eqs. (3.5) and (3.6) is that in the presence of slow molecular tumbling (i.e. water bound to macromolecular structures or water in ice) T_2 becomes very small. The T_1 relaxation constant is not affected by low-frequency magnetic field fluctuations.

Figure 3.4 gives a graphical depiction of Eqs. (3.5) and (3.6) for Larmor frequencies of 128 and 298 MHz, corresponding to protons at 3T and 7T, respectively. In the "extreme narrowing" limit ($\nu^2\tau_c^2 \ll 1$) the T_1 and T_2 relaxation times are equal. As the rotation correlation time τ_c increases, the T_1 relaxation time constant hits a minimum before increasing. The T_2 relaxation

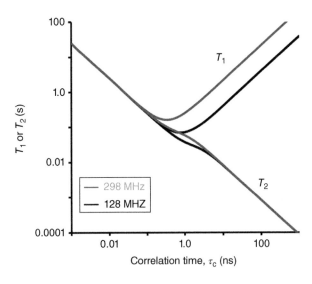

Figure 3.4 **Dipolar relaxation as a function of the correlation time τ_c.** The longitudinal and transverse relaxation times T_1 and T_2 are shown for protons at 3T (128 MHz, black) and 7T (298 MHz, gray). The upper left corner where $T_1 = T_2$ corresponds to the extreme motional narrowing limit and described dipolar relaxation for gases and small molecules in water. The lower right corner where $T_1 \gg T_2$ describes the situation for large molecules, viscous liquids and solids.

time constant continues to decrease with increasing τ_c. While the qualitative features of Figure 3.4 do not change with magnetic field, the T_1 relaxation times tend to increase with increasing magnetic field strength. Based on dipolar relaxation alone, the T_2 relaxation is virtually independent of the magnetic field strength.

Besides the dipole–dipole interactions between two identical nuclei, the required fluctuating dipolar fields can also arise from unpaired electrons, if present. Since the electron magnetic moment is >650 times that of a proton, such relaxation can be extremely effective (i.e. resulting in very short T_1 and T_2 relaxation times), provided that the distance between electron and nucleus is small enough. Because of the high efficiency of paramagnetic centers, only trace amounts are necessary for a significant increase in relaxation rates. Endogenous paramagnetic compounds are molecular oxygen and deoxyhemoglobin. Exogenous paramagnetic compounds, as used for T_1 or T_2^* image contrast, are often chelates of lanthanides (e.g. gadolinium-DTPA), iron oxide particles, or divalent ions like Mn^{2+}. Water is especially sensitive to paramagnetic centers, since it can come in close proximity to the unpaired electron.

3.2.2   Nuclear Overhauser Effect

A phenomenon closely related to magnetic dipolar interactions is the nuclear Overhauser effect (nOe) [9, 10]. Originally, the nOe was defined as a change in the integrated NMR absorption intensity of a nuclear spin when the NMR absorption of another, dipolar-coupled spin is saturated. Since the nOe arises from dipolar relaxation, the formalism of dipole–dipole relaxation can be used to describe the phenomenon. Figure 3.5A shows the energy-level diagram for a heteronuclear, two-spin-system HX, whereby H represents a proton and for *in vivo* MR applications X most commonly describes carbon-13 or phosphorus-31. At thermal equilibrium each energy level contains N_H proton spins and N_X X spins for a total of $4N_H$ and $4N_X$ spins. The presence of a strong external magnetic field increases the number of spins in the α state by Δ_H or Δ_X, with an equivalent decrease of spins in the β state. Detection of the thermal equilibrium system following an excitation pulse would give an X spin resonance of intensity $2\Delta_X$. Saturation of the proton transitions $W_{1,H}$ results in a nonequilibrium situation in which all proton energy levels are equally populated (Figure 3.5B). The X spin populations are not directly affected by the proton saturation. With continuing saturation, the protons will establish a new equilibrium situation through the cross-relaxation pathways W_0 and W_2. When the double-quantum W_2 pathway is dominant, the proton and X spins flip simultaneously from the β to the α orientation and a new equilibrium situation has been established in which the intensity of the X resonance has been enhanced to $2\Delta_X + \delta_1$, whereby δ_1 represents a positive nuclear Overhauser enhancement (nOe). When the zero-quantum W_0 pathway is dominant, the proton and X spins flip orientation in opposite directions leading to a new equilibrium situation in which the intensity of the X resonance has been changed to $2\Delta_X - \delta_2$, whereby δ_2 represents a negative nOe. In reality, the W_0 and W_2 relaxation pathways are both active and the final nOe will have an intermediate value given by

$$nOe = 1 + \frac{\gamma_H}{\gamma_X}\left(\frac{W_2 - W_0}{W_0 + 2W_{1,X} + W_2}\right) \tag{3.7}$$

where W_0, W_1, and W_2 are given by Eqs. (3.4). The maximum nOe of $1 + (\gamma_H/2\gamma_X)$ is achieved when the W_2 pathway is dominant, as is the case in the extreme narrowing regime. Figure 3.6A shows the nOe as a function of rotation correlation time τ_c according to Eq. (3.7) for ^{13}C and ^{31}P NMR at 3 and 7 T. The maximal nOe for ^{13}C-[^1H] (i.e. the nOe on ^{13}C is generated by dipolar interactions with ^1H) and ^{31}P-[^1H] are 2.988 and 2.235, respectively. For slower rotating

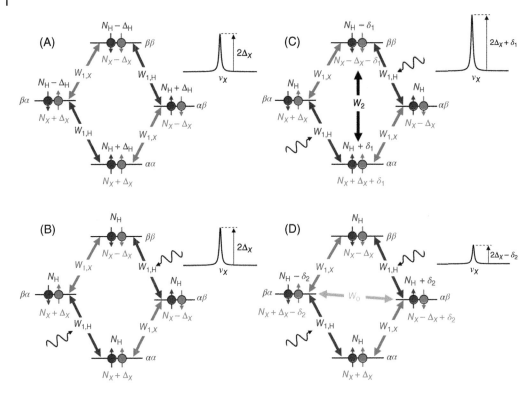

Figure 3.5 Principle of nuclear Overhauser enhancement (nOe). (A) Energy-level diagram of a dipolar-coupled HX ($X =$ ^{13}C, ^{15}N, ^{31}P) spin-system at thermal equilibrium. Spin populations in the α and β states have a surplus and shortage of Δ spins compared to the total spins N, respectively. The intensity of the X spin resonance is therefore proportional to $(N_x + \Delta_x) - (N_x - \Delta_x) = 2\Delta_x$. (B) Saturation of the protons leads to equally populated proton energy levels, but has no direct effect on the X spin populations or X spin resonance intensity. (C, D) The proton spin populations will attain a new equilibrium state, mediated through (C) double and (D) zero-quantum cross-relaxation pathways. (C) During the double-quantum W_2 pathway, every proton flip is accompanied by an X spin flip, leading to an increased X spin population of $N_x + \Delta_x + \delta_1$ for the $\alpha\alpha$ energy level (and an equivalent decrease for the $\beta\beta$ energy level). As a result, the X resonance intensity will *increase* to $2\Delta_x + \delta_1$. (D) During the zero-quantum W_0 pathway, every proton flip is accompanied by an *opposite* X spin flip, leading to an increased X spin population of $N_x - \Delta_x + \delta_2$ for the $\alpha\beta$ energy level (and an equivalent decrease for the $\beta\alpha$ energy level). As a result, the X resonance intensity will *decrease* to $2\Delta_x - \delta_2$. In reality, the W_0 and W_2 relaxation pathways are competing, such that the theoretical maximum nOe is given by Eq. (3.7).

molecules in the presence of alternative relaxation pathways, typical *in vivo* nOe values are 1.3–2.9 and 1.4–1.8 for ^{13}C-[^{1}H] [12–16] and ^{31}P-[^{1}H] [11, 15, 17, 18], respectively. For most practical purposes the nOe factor should be considered an unknown that requires an empirical determination. Figure 3.6B–D shows a typical *in vivo* ^{31}P NMR example of nOe in human calf muscle. The nOe factor of, for example, γ-ATP can be determined from the peak height (or integral) of spectra acquired with (Figure 3.6D, height $h2$) and without (Figure 3.6C, height $h1$) proton saturation. This approach assumes that proton decoupling has a negligible effect on nOe generation.

3.2.3 Alternative Relaxation Mechanisms

The relaxation theory outlined above is generally referred to as the Bloembergen–Purcell–Pound (BPP) theory [7]. After its publication in 1948 the theory was an immediate success, since it could accurately describe dipole–dipole relaxation in liquids such as water and glycerol,

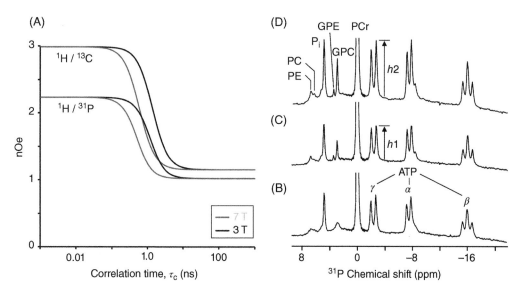

Figure 3.6 **Nuclear Overhauser enhancement (nOe) for ^{13}C and ^{31}P NMR.** (A) Theoretical nOe for ^{1}H/^{13}C and ^{1}H/^{31}P dipolar-coupled spin-systems as a function of the rotation correlation time τ_c at 3 and 7 T. The maximum nOe for ^{1}H/^{13}C and ^{1}H/^{31}P is 2.988 and 2.235, respectively. (B–D) Nonlocalized ^{31}P NMR spectra from human calf muscle (TR = 20 s, 32 averages, 1.5 T) acquired (B) without decoupling or nOe, (C) with proton decoupling, and (D) with proton decoupling and nOe. Note that proton decoupling enhances the spectral resolution, while nOe generation increases the signal intensity. The experimental nOe can be determined as h2/h1. *Source:* Brown et al. [11]. Reproduced with permission of John Wiley & Sons.

as well as in solids. However, apart from dipole–dipole interactions a variety of other relaxation mechanisms may be operative. In principle, any interaction that causes fluctuating magnetic fields can induce relaxation. The interactions which can generate the appropriate conditions for relaxation are (i) magnetic dipole–dipole interactions, (ii) electric quadrupole interactions, (iii) chemical shift anisotropy, (iv) spin rotation interactions, and (v) scalar coupling interactions.

A nucleus with spin $I > 1/2$ possesses an electric quadrupole moment since the charge distribution is no longer spherical (as is the case for spin $I = 1/2$). The electric quadrupole moment interacts with local electric field gradients, whereby fluctuations in the strength of this interaction, as caused by molecular tumbling, will induce relaxation. For *in vivo* NMR, prime example of quadrupolar relaxation can be found for ^{17}O, ^{23}Na, and ^{39}K with T_1 and T_2 relaxation time constants in the order of milliseconds. The quadrupolar interaction does not contribute to relaxation when the quadrupole coupling constant is zero due to molecular symmetry. For example, ammonium (^{14}NH$_4^+$) and sulfate (SO$_4^{2-}$) give narrow lines in ^{14}N and ^{33}S NMR spectra due to their highly symmetrical structures, whereas creatine and glutathione are essentially unobservable due to the broadened lines. The small quadrupole moment of deuterium (^{2}H) is a noticeable exception among the quadrupolar nuclei and can provide high-resolution MR spectra for a wide range of molecules (see also Section 3.5.5).

In Chapter 1, the chemical shift was presented as a single number, proportional to the effective magnetic field at the nucleus, which included the effects of an external magnetic field, as well as electronic shielding of the nucleus. However, because the chemical shift of a nucleus depends upon the orientation of the molecule relative to the main magnetic field direction, a proper representation of the chemical shift is a 3×3 chemical shift tensor (see Appendix A.1). Rapid molecular motions in the liquid state result in an averaging of all possible orientations and chemical shifts, resulting in an average chemical shift represented by the trace of the tensor. Even though the observed chemical shift may not be affected by chemical shift anisotropy (as is the case in the extreme narrowing limit), the nucleus will nevertheless experience fluctuations in the local magnetic field as the molecule tumbles. These fluctuations provide yet another mechanism for relaxation. In contrast to the situation under exclusive dipole–dipole relaxation, T_1 does not equal T_2 under extreme narrowing conditions when relaxation via chemical shift anisotropy is involved. Note that the relaxation rates through chemical shift anisotropy are proportional to the square of the main magnetic field. It can therefore be expected that this mechanism becomes increasingly important as the magnetic field strength B_0 increases. Unlike dipole–dipole relaxation, the relaxation rates under exclusive chemical shift anisotropy will increase with increasing B_0, such that the exact B_0 dependence of T_1 and T_2 relaxation will strongly depend on the relative contributions of the different mechanisms. For ^{31}P MRS, dipolar relaxation and chemical shift anisotropy are the two major, competing relaxation mechanisms [19, 20]. Experiments on model solutions indicate that chemical shift anisotropy is the dominant mechanism for relaxation of ^{31}P nuclei in ATP at high magnetic fields, whereas dipolar interactions dominate the T_1 relaxation rate at lower fields and for the monophosphate groups in phosphocreatine and inorganic phosphate. An advantage of the reduced T_1 relaxation times for ^{31}P NMR at higher magnetic fields is that shorter repetition times can be employed, such that SNR per unit of time can be improved. However, chemical shift anisotropy also reduces the T_2 relaxation time, thereby significantly compromising the enhancement in spectral resolution that may be expected upon increasing the magnetic field strength. If chemical shift anisotropy is a dominant mechanism, the line widths increase linearly with B_0^2, whereas the frequency range only increases according to B_0. It follows that in this case the best spectral resolution is not necessarily obtained at the highest magnetic field.

Reports on the measurement of ^{31}P longitudinal and transverse relaxation times *in vivo* [11, 21–35] support the prediction that ^{31}P T_1 relaxation reduces at higher magnetic fields. T_2 relaxation for ATP is noticeably faster compared to many of the other phosphorus containing metabolites. It has been suggested that the short transverse relaxation time of ATP is partly due to exchange between free and bound states of ATP. Interactions with enzymes like creatine kinase and ATPases, for example, which effectively act as solid matrices may result in long T_1 and short T_2 relaxation times. Furthermore, the strong dipole–dipole interaction between ATP and the unpaired electron of a complexed paramagnetic ion could also play a role. Note that some of the reported T_2 values of ATP in the literature represent an underestimation of the true T_2 relaxation time due to interference of homonuclear scalar coupling. Especially, in the early stages of *in vivo* ^{31}P MRS the homonuclear scalar coupling was ignored, leading to a sinusoidally modulated T_2 relaxation curve. The introduction of selective refocusing pulses and/or homonuclear decoupling [28–31] refocuses homonuclear scalar coupling evolution, thereby allowing the accurate determination of T_2.

Spin rotation relaxation arises from magnetic fields at the nucleus generated by coherent rotational motion of the entire molecule, which can couple with the nuclear spin. Interruption of this coupling (e.g. by collisions) provides a relaxation mechanism. This effect is most significant for small, symmetric molecules with short correlation times or for similar parts of molecules, like methyl groups. This mechanism is of little importance for most molecules observed with *in vivo* NMR.

As described in Section 1.10, apart from direct, through-space dipolar interaction two nuclear spins can also experience indirect coupling through the electrons in a chemical bond. This is referred to as scalar coupling, the strength of which is independent of the orientation of the molecules within the applied field. A scalar interaction, which involves a magnetic field produced by spin 1 acting (indirectly) on spin 2 (and vice versa) can lead to relaxation of spin

2, if a time-dependence in the scalar coupling occurs. This can happen if the scalar coupling constant becomes time-dependent due to exchange processes or if the energy-level populations of spin 1 become time-dependent as a result of relaxation of spin 1. These two possible causes for scalar relaxation are known as scalar relaxation of the first and second kind, respectively. Scalar relaxation is most commonly observed when spin 1 is a quadrupole nucleus with short relaxation times. Most often scalar relaxation only has a pronounced effect on T_2 relaxation, leading, for instance, to line broadening in the ^{1}H spectrum of nuclei coupled to ^{14}N.

3.2.4 Effects of T_1 Relaxation

NMR as we know it today would not be possible without T_1 relaxation to establish the thermal equilibrium magnetization. As an essential phenomenon, T_1 relaxation effects almost all MR studies, thereby making a proper understanding critical for experimental design and interpretation. For a simple pulse-acquire experiment, the effects of T_1 relaxation can be completely eliminated when the repetition time TR is longer than five times the longest T_1 relaxation time in the sample. While this approach is often advocated for quantitative MR spectroscopy, it is not optimal in terms of signal-to-noise per unit measurement time as the majority of time is spent waiting for the thermal equilibrium magnetization to be reestablished. An improved sensitivity can be achieved when TR and possibly also the nutation angle α are reduced, as shown in Figure 3.7. For the three [α, TR] combinations shown, the [α, TR] = [30°, 500 ms] combination provides the highest steady-state M_z. However, the highest SNR per unit time is obtained (through extended signal averaging) when α and TR equal 30° and 250 ms, respectively. It should be noted that the steady-state condition for M_z is not immediately achieved. Initially the reduction in M_z due to excitation is larger than the increase in M_z due to T_1 relaxation. After a number of excitation and relaxation cycles, the longitudinal magnetization M_z reaches a steady-state in which the reduction in M_z due to excitation equals the increase in M_z due to T_1 relaxation. For most studies, the signal variations before steady-state achieved are undesirable and are typically eliminated by running a number of "dummy scans" in which the pulse sequence is executed without acquiring the signal. Note that the number of required dummy scans depend on α, T_1, and TR and can add up to tens or even hundreds.

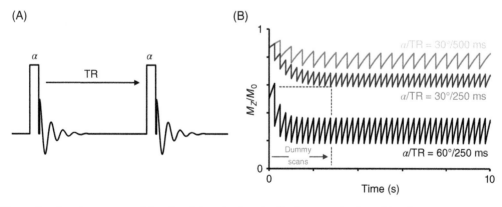

Figure 3.7 Steady-state condition involving the longitudinal magnetization. (A) Pulse-acquire sequence executed with excitation angle α and repeated with repetition time TR. The transverse magnetization prior to the excitation pulse is assumed zero. (B) Temporal dynamics of the longitudinal magnetization ($T_1 = 1000$ ms) during repeated application of the pulse-acquire sequence for three combinations of α and TR. After a number of "dummy scans" the longitudinal magnetization reaches a steady-state condition in which the decrease in M_z due to excitation equals the increase in M_z due to T_1 relaxation.

The steady-state longitudinal magnetization can be calculated through the use of Eq. (1.14) and is given by (see Exercises)

$$M_z\left(\alpha, \text{TR}\right) = M_0 \frac{\left(1 - e^{-\text{TR}/T_1}\right)}{\left(1 - \cos\alpha\, e^{-\text{TR}/T_1}\right)} \tag{3.8}$$

The detectable, steady-state transverse magnetization $M_{xy}(\alpha, \text{TR})$ is given by $M_z(\alpha, \text{TR})\sin(\alpha)$. Figure 3.8A shows the steady-state, transverse magnetization M_{xy} as a function of nutation angle α and TR/T_1 ratio. Each TR/T_1 curve displays a maximum where the corresponding nutation angle provides the highest steady-state M_{xy}. This nutation angle is often referred to as the Ernst angle and can be calculated by solving $dM_{xy}(\alpha, \text{TR})/d\alpha = 0$. The Ernst angle is given by

$$\alpha_{\text{opt}} = \arccos\left(e^{-\text{TR}/T_1}\right) \tag{3.9}$$

and is graphically shown in Figure 3.8B. It follows that a nutation angle of 90° is only optimal when the repetition time TR is much larger than the T_1 relaxation time constant. For shorter TR, the optimal nutation angle is always lower than 90° and is given by Eq. (3.9). The relative SNR per unit measurement time is shown in Figure 3.8C. It follows that for a given T_1 relaxation time constant the relative SNR per unit measurement time is highest for the shortest repetition time TR (and corresponding smallest Ernst angle α_{opt}). The gain in SNR for shorter TR is offset against the signal variation introduced for different T_1 relaxation times. Figure 3.9A shows the detected signal M_{xy} as a function of TR for three different T_1 relaxation times, 500, 1000, and 2000 ms. The optimal nutation angle for each TR is calculated for $T_1 = 1000$ ms. Figure 3.9B shows that the shortest TR achieves the highest SNR at the expense of variation in SNR for different T_1 relaxation times. Only when the TR is increased to at least five times the longest T_1 do the curves converge, making the signal intensity and SNR independent of T_1. For MRS application the use of a long repetition time TR is often advocated to eliminate T_1 relaxation effects. MRI applications often select a shorter repetition time as to enhance the contrast between tissues with different T_1 relaxation time constants (see Chapter 4).

3.2.5 Effects of T_2 Relaxation

The effects of T_1 relaxation on MR signal detection can be minimized or even eliminated by choosing a sufficiently long repetition time. The effects of T_2 relaxation are unavoidable and will always influence MR signal detection. Signal obtained with MRS pulse sequences is especially susceptible since T_2 and T_2^* relaxation are present during the entire acquisition time window. In a perfectly homogenous magnetic field signal would decrease due to intrinsic T_2 relaxation according to Eq. (1.5). However, a combination of microscopic and macroscopic inhomogeneity leads to additional signal dephasing and a more rapid signal decrease, which is characterized by T_2^* relaxation (Eq. (1.7)). Under most *in vivo* conditions the inhomogeneity contribution to T_2^* relaxation is dominant, such that resonance line widths are approximately equal despite differences in T_2 relaxation.

As briefly mentioned in Chapter 1, the contribution of macroscopic magnetic field inhomogeneity to the observed T_2^* relaxation can be eliminated with a spin-echo (or stimulated echo) pulse sequence as shown in Figure 3.10A. Following signal excitation the MR signal is dephased by T_2^* relaxation, similar to that observed during a pulse-acquire method. After a delay TE/2 following the initial excitation pulse, a refocusing pulse rotates the transverse

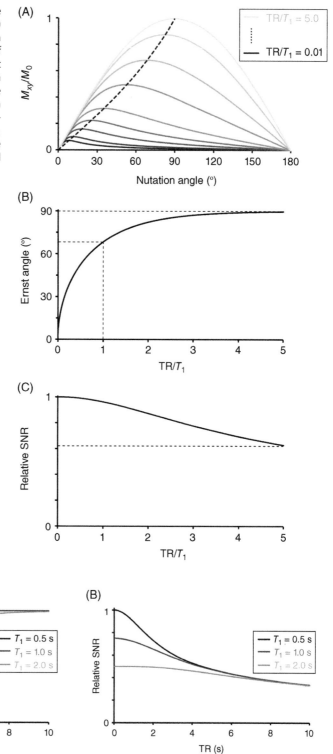

Figure 3.8 T_1, **TR, and α dependence of the steady-state magnetization.** (A) Transverse magnetization as a function of the nutation angle α for a range of TR/T_1 values. The nutation angle that provides the highest signal for a given TR/T_1 is referred to as the Ernst angle (dotted line) and is displayed in (B) as a function of TR/T_1. (C) Relative signal-to-noise ratio (SNR) as a function of TR/T_1 for a fixed total measurement time. The SNR for each TR/T_1 ratio was calculated with the optimal Ernst angle.

Figure 3.9 TR-dependent signal saturation. (A) Signal recovery curves (M_{xy}/M_0) for three T_1 species as a function of the repetition time TR. All time points are calculated with an optimal nutation angle ("Ernst angle") for $T_1 = 1000\,\text{ms}$. (B) SNR per unit measurement time for three T_1 species. For a T_1 of 500 ms the SNR at TR = 10 000 ms is 3× lower than the highest SNR possible at short TR, whereas for a T_1 of 2000 ms the SNR is only 1.5× lower than the highest SNR. The SNR penalty at long TRs is offset by attaining the quantitative aspects of NMR independent of T_1.

plane by 180°. In other words, the phase $\Delta\phi$ acquired by individual spins before the 180° pulse due to magnetic field inhomogeneity has been inverted to a phase $-\Delta\phi$ after the refocusing pulse. During the second TE/2 period, the spins continue to accumulate phase due to the macroscopic magnetic field inhomogeneity that is part of T_2^* relaxation. However, since the acquired phase was inverted by the 180° pulse, the phase accumulation during the second TE/2 period leads to signal rephasing and to spin-echo formation at the echo-time TE. A spin-echo method can therefore eliminate the effects of T_2^* relaxation, albeit only at the time of echo formation. The random magnetic field variations underlying T_2 relaxation cannot be refocused by a spin-echo, leading to a reduced signal intensity at the echo-time TE. Note that a spin-echo sequence also refocuses phase accumulation due to differences in chemical shift. In Chapter 8, it will be shown that a spin-echo does not refocus evolution due to homonuclear scalar coupling, thereby providing the basis for spectral editing.

From Figure 3.10 it is clear that the time-domain signal intensity detected with a spin-echo sequence is affected by T_2 relaxation, whereas T_2^* relaxation is present during the following acquisition time window. Especially for longer echo-times, differences in T_2 relaxation can manifest themselves in the resulting spectrum and obscure the actual spin densities of various compounds. In many publications on long-echo-time MRS it is assumed that all detected chemicals have equal T_2 relaxation time constants, such that the relative peak intensities are still proportional to the concentration. However, there are many conditions in which this assumption is not valid. For example, protons in methylene CH_2 groups have generally a shorter T_2 than protons in methyl CH_3 groups. Under pathological conditions such as stroke or tumors, various metabolites may be present in different cellular environments with different relaxation characteristics. In these cases, the T_2 relaxation time constants should be measured as discussed next.

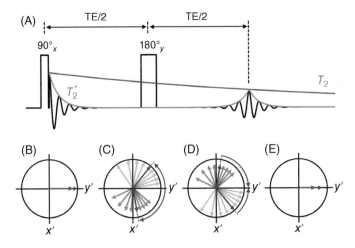

Figure 3.10 T_2 and T_2^* **relaxation during a Hahn spin-echo sequence.** (A) Hahn spin-echo pulse sequence showing the rapid signal decay due to T_2^* relaxation and echo formation at echo-time TE, whereby the echo intensity is reduced by T_2 relaxation. (B–E) Transverse magnetization for two compounds (red and blue) (B) immediately after excitation, (C) before and (D) after the 180° refocusing pulse, and (E) at the top of the echo. Individual spins at different spatial positions evolve under the effects of chemical shift and magnetic field inhomogeneity, leading to phase dispersal across the sample (C). The 180° refocusing pulse inverts the acquired phases, such that the phase acquired during the second TE/2 period exactly cancels that acquired during the first TE/2 period. Since T_2 relaxation is intrinsically random, it cannot be refocused such that it reduces the echo intensity according to Eq. (1.5).

3.2.6 Measurement of T_1 and T_2 Relaxation

The importance of T_1 and T_2 relaxation in NMR cannot be overstated. In MRI the majority of methods rely on image contrast based on differences in T_1 and/or T_2 relaxation, whereas in MRS the relaxation parameters can have a profound effect on metabolite quantification. Both for MRI and MRS it is often necessary to quantify the absolute T_1 and T_2 relaxation time constants. For example, clinical MRI often relies on T_1- or T_2-weighted images, in which the majority of image contrast is provided by differences in T_1 or T_2 relaxation, respectively. However, other parameters like proton density also contribute to the image contrast. This can lead to ambiguous interpretation when one or several parameters change due to disease progression. The ambiguity can be eliminated by generating quantitative T_1, T_2, and proton density maps. Similar arguments can be used for MRS, where a signal intensity can have contributions from T_1, T_2, and T_2^* relaxation, as well as metabolite concentration. The measurement of water T_1 and T_2 relaxation by MRI methods is discussed in Chapter 4. Here, the principles of T_1 and T_2 relaxation time measurements are provided, with an emphasis on MRS.

3.2.6.1 T_1 Relaxation

The measurement of T_1 relaxation is traditionally performed with an inversion recovery method. However, since the majority of the time is spent waiting for the reestablishment of the thermal equilibrium magnetization, several more time-efficient methods are also discussed. These include saturation recovery, variable nutation angle (VNA), and MR fingerprinting (MRF).

3.2.6.2 Inversion Recovery

The inversion recovery method is the most robust and accurate method to measure T_1 relaxation, but comes at the cost of long acquisition times. The method starts with a recovery period TR that is long enough to establish the thermal equilibrium magnetization M_0 for all nuclear spins (Figure 3.11A). This means that TR must be set to at least five times the longest T_1 relaxation time constant in the sample. Next, the magnetization is inverted by a 180° pulse followed by an inversion time TI during which the inverted magnetization partially recovers back to M_0. Finally, the longitudinal magnetization is excited to the transverse plane and detected. T_1 relaxation is discretely sampled by increasing the inversion time TI from the minimum allowable delay to a delay equal to five times the longest T_1 in the sample. The longitudinal magnetization recovers according to Eq. (1.2) with $M_z(0) = -M_0$, corresponding to a perfect inversion. Since T_1 relaxation occurs exponentially over time (Figure 3.11B), the inversion times TI are typically chosen on a logarithmic scale (see Exercises to calculate delays corresponding to $M_z/M_0 = -0.9 + 0.1n$). A two-parameter fit of the observed T_1 recovery data points provides estimates of T_1 and M_0. However, in reality the inversion pulse is not perfect, making a three-parameter fit of (T_1, M_0, $M_z(0)$) much more robust. Furthermore, a three-parameter fit makes the estimated T_1 relaxation times independent of the recovery delay TR and independent of constant, but unknown, delays to the inversion delays TI. A rough estimate of the T_1 relaxation time constant can be obtained from the "zero-crossing" of the detected signal. When t_{null} represents the delay during which the signal crosses zero, the T_1 can be estimated as $T_1 = t_{null}/\ln(2)$. While inversion recovery provides robust estimates of T_1 relaxation, more than half of the total scan duration is spent waiting for the thermal equilibrium magnetization to be reestablished. This has sparked the development of more time-efficient T_1 relaxation measurement methods.

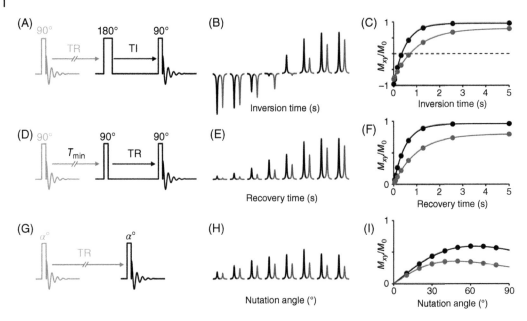

Figure 3.11 **Measurement of T_1 relaxation.** Commonly used methods for the *in vivo* measurement of T_1 relaxation include (A–C) inversion recovery, (D–F) saturation recovery, and (G–I) variable nutation angles. (B, C) Inversion recovery and (E, F) saturation transfer data can be modeled by Eq. (1.2) to obtain quantitative T_1 relaxation time constants. For inversion recovery data a three-parameter fit (M_0, $M_z(0)$, and T_1) is strongly recommended, whereby saturation recovery data can be reliably quantified with a two-parameter fit (M_0 and T_1). (H, I) Variable nutation angle data can be modeled with Eq. (3.8), but requires accurate knowledge on the nutation angle. For all three methods the spectra and recovery curves are simulated for $T_1 = 500\,ms$ (black) and $T_1 = 1000\,ms$ (gray).

3.2.6.3 Saturation Recovery

Saturation recovery is a faster alternative to inversion recovery in which enhanced acquisition speed is traded off against reduced dynamic range. Figure 3.11D shows the basic pulse sequence for saturation recovery, where the repetition time TR is varied. The primary difference with inversion recovery is that saturation recovery does not need a long delay to establish M_0 by setting the longitudinal magnetization to zero immediately following signal acquisition. The longitudinal magnetization still recovers according to Eq. (1.2), but now with $M_z(0) = 0$. The dynamic range of saturation recovery $[0 \dots M_0]$ is half of the inversion recovery range $[-M_0 \dots M_0]$, but the total scan duration is often several times shorter.

3.2.6.4 Variable Nutation Angle

The VNA method is another strategy to increase the time efficiency of T_1 relaxation measurements [36, 37]. Instead of varying the repetition time, as done for saturation recovery, the VNA method varies the excitation angle while keeping the repetition time fixed. The longitudinal magnetization evolves according to Eq. (3.8), leading to sinusoidal curves as shown in Figure 3.11H and I. With repetition times on the order of 0.2–1.0 times T_1, the VNA method is typically faster than saturation recovery. However, the VNA method does require accurate knowledge of the nutation angle. In the presence of RF magnetic field inhomogeneity, the method should be combined with measurement of the nutation angle which may be an obstacle for non-proton nuclei.

3.2.6.5 MR Fingerprinting

MRF is a more recent development to simultaneously measure T_1, T_2, M_0, and other parameters that affect the experiment [38]. The conventional T_1 and T_2 measurement methods described earlier rely on a consistent starting point (e.g. $M_z(0) = 0$) and/or the creation of a steady-state condition. MRF purposely avoids the creation of a steady-state by continuously changing the excitation angle and repetition time as shown in Figure 3.12A and B. The signal acquired after each excitation is simultaneously encoded by different amounts of T_1 and T_2 relaxation (and any other parameters that the sequence has sensitivity towards such as frequency offsets or RF magnetic field inhomogeneity) leading to unique signal intensity patterns for different T_1, T_2, and frequency offset combinations (Figure 3.12C). While the signal variations appear complicated they can be quantitatively calculated using the Bloch equations (Eqs. (1.15)–(1.17)). For a given pulse sequence, the signal patterns for a wide range of T_1, T_2, and frequency offset can be calculated and stored as a "dictionary." An experimental signal acquired with the same pulse sequence (Figure 3.12D, black line) can then be compared with the entries in the dictionary until the best match is found (Figure 3.12D, red line) thereby providing simultaneous estimates of T_1, T_2, and frequency offset. M_0 is obtained as the scaling between the experimental signal and the dictionary match. The procedure is referred to as MRF since matching of the experimental MR signal against a dictionary is reminiscent of matching fingerprints against a database.

MRF was initially described for MRI applications and has seen continued development by a large number of MR groups. Reports of MRF in MRS are scarce, although feasibility has been shown for quantitative creatine kinase reaction rate mapping in ^{31}P MRS [39]. Publications that use a range of repetition and echo times together with multi-parametric fitting [40] can be seen as a special case of MRF in which coherences between excitations are purposely destroyed. True MRF for ^{1}H MRS may be difficult due to the rapid expansion and proliferation of higher-order coherences for scalar-coupled spins.

3.2.6.6 T_2 Relaxation

Transverse T_2 relaxation is traditionally measured with the Hahn spin-echo method (Figure 3.13A). The signals acquired at different echo-times decay exponentially according to Eq. (1.5) (Figure 3.13B and C). Deviations from the simple, exponential decay curve can be seen for water (Figure 9.15) and scalar-coupled spins (Figure 3.13D and E). Water in different compartments

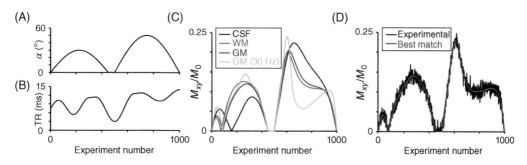

Figure 3.12 MR fingerprinting (MRF). (A) Nutation angle and (B) repetition time variation used during a steady-state free precession sequence provides (C) dynamic signal intensity profiles or "fingerprints" for tissues with different T_1 or T_2 relaxation characteristics or B_0 magnetic field offsets. (D) The experimental MRF data (black line) can be compared to a precompiled dictionary of signal intensity "fingerprints" whereby the best match (red line) provides estimates of T_1 and T_2 relaxation time constants. MRF data can be sensitized for additional parameters such as the B_1^+ transmit magnetic field.

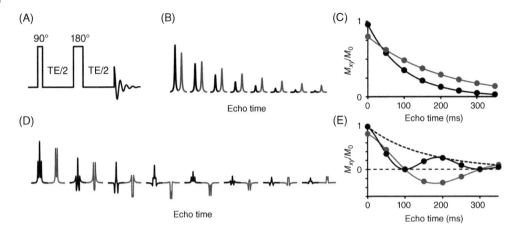

Figure 3.13 Measurement of T_2 relaxation. (A) Hahn spin-echo sequence. (B) MR spectra obtained with increasing echo time TE. (C) Integrated signal intensities as function of echo time TE. The spin-echo T_2 data can be analyzed with a two-parameter fit (M_0, T_2) according to Eq. (1.5). (D) MR spectra obtained with increasing echo time TE for triplet (black) and doublet (gray) signals ($J = 10$ Hz). The scalar coupling evolution complicates the simple mono-exponential decay observed for non-coupled spins. (E) Inclusion of the appropriate scalar coupling modulation allows quantitative modeling of the observed signal to reveal the T_2 relaxation time constants. Note that the maximum of the signal modulation directly reveals signal decay due to T_2 relaxation (dotted line). All spectra and recovery curves are simulated for $T_2 = 100$ ms (black) and $T_2 = 200$ ms (gray).

(gray matter, white matter, interstitial space, cerebrospinal fluid [CSF]) is characterized by different T_2 relaxation times, leading to multi-exponential relaxation. Analysis of a water T_2 decay curve with a bi-exponential T_2 relaxation model allows the separation between brain water ($T_2 < 200$ ms) and CSF water ($T_2 > 200$ ms) (Figure 9.15). For small molecules, like water, care should be taken to minimize the amount of diffusion weighting at longer echo-times in order to avoid an underestimation of the actual T_2 relaxation time constant (see also Section 3.4). In addition to the exponential T_2 relaxation decay, scalar-coupled spins exhibit an echo-time-dependent signal modulation (Figure 3.13D and E). This is because a regular, nonselective spin-echo method does not refocus scalar coupling evolution. For small, weakly-coupled spin-systems the echo-time-dependent modulation can be described by an analytical, trigonometric function (e.g. $M_{xy}(t) = M_{xy}(0) \cdot \cos(\pi J \text{TE}) \cdot \exp(-\text{TE}/T_2)$ for a two-spin-system). Larger or strongly-coupled spin-systems have very complicated modulation functions that can only be obtained through phantom measurements or quantum-mechanical simulations (see Chapter 8). In all cases, the modulation due to scalar coupling needs to be quantitatively known in order to extract the T_2 relaxation time constant. For weakly-coupled spins the TE modulation can be eliminated through inhibition of scalar coupling evolution with frequency-selective spin echo methods (see Chapter 8). Many *in vivo* MRS localization methods (PRESS, LASER) are based on spin-echoes, thereby providing a straightforward means for T_2 relaxation measurement without sequence modification. For MRI applications, the spin-echo method can be extended with repeated refocusing pulses according to the Carr–Purcell–Meiboom–Gill (CPMG) principle for an efficient, single-shot T_2 measurement. The CPMG method is not commonly used for MRS since the typical acquisition time of 100–200 ms makes the inter-echo spacing prohibitively long.

3.2.7 *In Vivo* Relaxation

From the preceding discussions it is clear that relaxation in general is a complex process involving several competing mechanisms. Relaxation in biological tissues is complicated even further due to compartmentalization. In contrast to metabolites, the relaxation characteristics of water

Figure 3.14 Two-compartment model for water in biological systems. Tissue water can be divided in bulk or free water and protein-associated water. Free water is rotationally mobile with a short rotation correlation time τ_c, which according to BPP theory (Figure 3.4) leads to a long T_2 relaxation time constant. Water can also form a hydration layer (of approximately three water molecules) around proteins or other hydrophilic structures. This structured protein-associated water is characterized by restricted mobility, a long rotation correlation time τ_c, and consequently a significantly reduced T_2 relaxation time. The structured and bulk water pools are in fast exchange such that the relaxation characteristics of the immobile pools can significantly shorten the observed relaxation times. The two-compartment model can be extended to more compartments, for instance, by the discrimination between dipolar and ionic bound water.

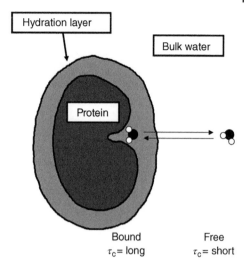

have been extensively studied and it has been established that dipole–dipole interactions give rise to the dominant relaxation pathway. However, applying the BPP theory as outlined earlier to water in biological tissues gives incorrect results. This is mainly because BPP theory assumes isotropic molecular motion, while in tissues the exchange of molecules between different molecular environments plays an important role. The relaxation of water in tissues can largely be explained by hydration-induced changes in rotational motion of the water in the vicinity of macromolecular surfaces [41–44]. Figure 3.14 shows a schematic drawing of the water in the vicinity of a protein. Besides the "bulk" fraction of the water which can freely rotate (and hence has a short $\tau_c \sim 0.1$ ns), there is also a fraction of the water which forms a hydration layer surrounding proteins. Normally this hydration layer has a thickness of several water molecules. The molecules in a hydration layer are more structured as a result of hydrogen or ionic bonds to hydrophilic or ionic sites in the protein. The binding reduces the rotational mobility of the water in the hydration layer, resulting in longer correlation times τ_c (τ_c can be as long as $1-10$ μs for ionic bonds). Although Figure 3.14 shows a hydration layer surrounding a protein, these layers can form along any hydrophilic surface, including membranes and DNA. When a fast exchange between the bulk water and the rotationally restricted water is assumed, the observed relaxation rate can simply be calculated as the weighted sum of free and restricted water fractions. This two-compartment model can be further refined to discriminate between ionic bound water (which behaves as if it were a solid) and hydrogen bound water (which behaves as a viscous liquid with $\tau_c \sim 1$ ns). Using this two-compartment, fast exchange model or other more extensive multi-compartment models, a semiquantitative description of water relaxation in biological tissues has been established. Figure 3.15 provides an overview of T_1 and T_2 relaxation times for water in the human brain as a function of magnetic field strengths [41, 44–47]. Despite the large inter-study variation, a steady increase in longitudinal relaxation time T_1 with increasing magnetic field strength is apparent, indicating that dipole–dipole interactions are a dominant relaxation mechanism. On average the spin–spin T_2 relaxation is relatively constant with a tendency towards a decrease with increasing magnetic fields.

Reports on the proton relaxation times of cerebral metabolites in human [48–57] and animal [58–60] brain are scarcer, which, in combination with the greatly decreased sensitivity of metabolite detection, leads to large inter-study variations. As a result, Figure 3.16 provides the trend established by multiple studies at a range of magnetic fields [48–60]. It follows that the relaxation pattern of metabolites tracks that of water, with increasing T_1 and decreasing T_2

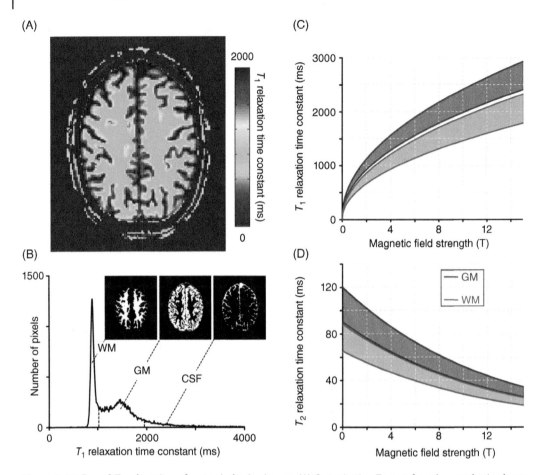

Figure 3.15 T_1 **and** T_2 **relaxation of water in brain tissues.** (A) Quantitative T_1 map from human brain shows high visual contrast between gray matter (GM), white matter (WM), and cerebrospinal fluid (CSF). (B) Histogram of T_1 relaxation time constants is readily segmented into WM, GM, and CSF maps using T_1 value thresholds. More sophisticated methods could be used to segment pixels with multiple contributions based on weighted T_1 values. (C, D) Literature ranges for (C) T_1 and (D) T_2 relaxation time constants of GM and WM in the human brain as a function of the magnetic field strength. T_1 and T_2 relaxation time constants generally increase and decrease with increasing magnetic fields, respectively.

relaxation times at higher magnetic fields. For macromolecular resonances the T_1 relaxation rapidly increases with magnetic field strength, whereas the T_2 relaxation is virtually field independent. T_1 and T_2 relaxation times of scalar-coupled spins have not been reported frequently, but largely follow the trends shown in Figure 3.16.

The decrease of water and metabolite T_2 relaxation times with increasing magnetic field strength is in apparent contradiction with BPP dipolar relaxation theory, which predicts field-independent T_2 relaxation for a wide range of rotation correlation times (see Figure 3.4). Michaeli et al. [61] explained the field-dependent T_2s as the result of increased dynamic dephasing due to increased local (microscopic) susceptibility gradients. It is well known that macroscopic, as well as microscopic magnetic field inhomogeneity due to susceptibility differences between tissues (and air) increases linearly with magnetic field strength (see Chapter 10). As molecules diffuse through these microscopic field gradients they lose phase coherence, resulting in a shorter apparent T_2 relaxation time constant, $T_2^{\dagger}$. Since the dynamic dephasing

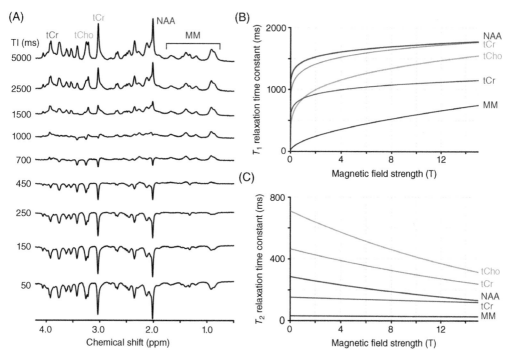

Figure 3.16 **Proton T_1 and T_2 relaxation of cerebral metabolites.** (A) Experimental inversion recovery T_1 relaxation measurement of metabolites in rat brain *in vivo* at 11.7 T. (B, C) Averaged literature values for metabolite (B) T_1 and (C) T_2 relaxation time constants in brain tissue *in vivo*. Macromolecular (MM) resonances exhibit shorter T_1 and T_2 relaxation time constants compared to metabolites.

process is dependent on diffusion, the decrease in observed T_2 relaxation times is expected to be greatest for water ($D \sim 0.7 \ \mu m^2 \ (ms)^{-1}$) and smallest for large macromolecules ($D < 0.02 \ \mu m^2 \ (ms)^{-1}$). Metabolites, with intermediate diffusion coefficients ($D \sim 0.2 \ \mu m^2 \ (ms)^{-1}$), are expected to display a moderate decrease in $T_2^{\dagger}$ with increasing magnetic field strength.

3.3 Magnetization Transfer

NMR is a powerful technique for structure elucidation due to the presence of chemical shifts and scalar couplings among nuclear spins in different chemical environments. Another strength of NMR is that the detected signal can be made exquisitely sensitive to dynamic processes spanning many orders in time. T_1 and T_2 relaxation are caused by dynamic processes around the Larmor frequency or in the nanosecond range. Scalar couplings are present in the Hz to kHz range such that their temporal evolution corresponds to time scales between milliseconds and seconds. Diffusion-sensitized NMR (Section 3.4) is sensitive to processes in the millisecond to second range. When combined with the administration of isotope-enriched substrates, NMR can detect chemical reactions over time scales spanning minutes to hours and even days. Magnetization transfer (MT) methods make NMR sensitive to (much) faster chemical reactions of which the most important categories are depicted in Figure 3.17. The "classical" MT studies focus on enzyme-catalyzed chemical reactions that occur on the T_1 relaxation rate time scale. The reactions most often studied with "classical" MT methods *in vivo* are the creatine kinase-catalyzed transfer of a high-energy phosphate group between phosphocreatine/ADP

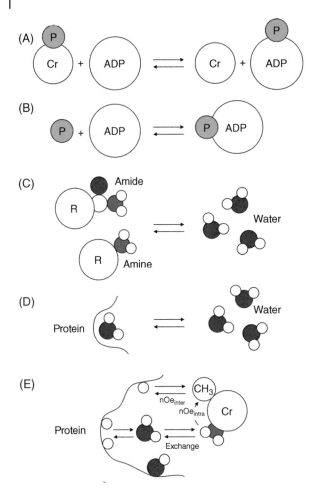

Figure 3.17 Chemical exchange reactions studied with magnetization transfer. ^{31}P NMR can be used to study the (A) exchange reaction between PCr/ADP and Cr/ATP catalyzed by creatine kinase and the (B) exchange reaction between P$_i$/ADP and ATP catalyzed by various ATPases. (C) Chemical exchange between amide, amine, and hydroxyl groups and water/metabolites can be studied with ^{1}H MRS or chemical exchange saturation transfer (CEST) MRI. (D) Magnetization transfer effects between protein-immobilized, bound water/metabolites, and free, mobile water/metabolites can be observed through magnetization transfer contrast (MTC) imaging/spectroscopy. (E) Magnetization transfer between exchangeable and non-exchangeable protons is indicative of nuclear Overhauser enhancement effects and can be observed by MRI or MRS.

and creatine/ATP (Figure 3.17A) and synthesis (or degradation) of ATP from ADP and inorganic phosphate by a range of ATPases (Figure 3.17B). More recently, MT is used for Chemical Exchange Saturation Transfer (CEST) studies which are sensitive to the exchange between compounds containing hydroxyl, amide, or amine groups and the large water pool (Figure 3.17C). Whereas the goal of "classical" MT studies is the determination of absolute reaction rates, the goal of CEST studies is to generate novel MR image contrast or enhance the detection of low-concentration metabolites through the large water signal. During MRS studies the exchange reactions of Figure 3.17C can lead to an underestimation of the metabolite concentrations when the water spins are saturated during water suppression. Another type of exchange is depicted in Figure 3.17D and involves water (or metabolites) in a bound or immobile state and "bulk" water in a mobile state. In a clinical MRI setting, this type of exchange gives rise to MT contrast (MTC) and provides a powerful contrast mechanism to study white matter disease. The exchange processes shown in Figure 3.17A–D all involve exchange of a physical chemical group, being phosphate (Figure 3.17A and B) or protons (Figure 3.17C and D). However, in NMR, magnetization can also be transferred from one spin to another spin through dipolar interactions (Figure 3.17E). These nOe effects play a crucial role in the determination of the 3D structure of proteins and other macromolecules. However, they can also be detected *in vivo*, thereby providing another route to generate image contrast. Even though the goals of the various MT studies are often different, they all use the same MR methods and can be described in the same framework. In the next section a general description of MT is provided, after which the details of "classical" MT, MTC, and CEST in MRI and MRS applications are discussed.

3.3.1   Principles of MT

The exchange reactions shown in Figure 3.17 can all be described by the same two-compartment model composed of an equilibrium between reactants A and B. The pool sizes are indicated by the longitudinal magnetizations $M_{0,A}$ and $M_{0,B}$, whereby the spins are characterized by $T_{1,A}$ and $T_{1,B}$ relaxation parameters. For simplicity the effects of T_2 are ignored. The two pools are in exchange with unidirectional rate constants k_{AB} and k_{BA} for the forward ($A \rightarrow B$) and reverse ($B \rightarrow A$) reactions, respectively. The reaction flux is defined as $k_{AB}M_{0,A}$ and $k_{BA}M_{0,B}$, whereby under steady-state conditions the reaction fluxes are equal, i.e. $k_{AB}M_{0,A} = k_{BA}M_{0,B}$. For this system, the Bloch–McConnell equations that incorporate chemical exchange [62–64] provide a complete description according to

$$\frac{\mathrm{d}M_{zA}(t)}{\mathrm{d}t} = \frac{\left[M_{zA}^0 - M_{zA}(t)\right]}{T_{1A}} - k_{AB}M_{zA}(t) + k_{BA}M_{zB}(t) \tag{3.10}$$

$$\frac{\mathrm{d}M_{zB}(t)}{\mathrm{d}t} = \frac{\left[M_{zB}^0 - M_{zB}(t)\right]}{T_{1B}} + k_{AB}M_{zA}(t) - k_{BA}M_{zB}(t) \tag{3.11}$$

These differential equations can be solved analytically to provide the time-dependence of longitudinal magnetization in the presence of chemical exchange

$$M_{zA}(t) = M_{zA}^0 + c_1 e^{a_1 t} + c_2 e^{a_2 t} \tag{3.12}$$

$$M_{zB}(t) = M_{zB}^0 + c_3 e^{a_1 t} + c_4 e^{a_2 t} \tag{3.13}$$

where

$$a_{1,2} = \frac{1}{2}\left\{-\left(R_A + R_B\right) \pm \left[\left(R_A - R_B\right)^2 + 4k_{AB}k_{BA}\right]^{1/2}\right\} \tag{3.14}$$

$$c_2 = \frac{\left\{\left[M_{zA}(0) - M_{zA}^0\right]\left(a_1 + R_A\right) + k_{BA}\left[M_{zB}^0 - M_{zB}(0)\right]\right\}}{\left(a_1 - a_2\right)} \tag{3.15}$$

$$c_1 = M_{zA}(0) - M_{zA}^0 - c_2 \tag{3.16}$$

$$c_3 = c_1 \frac{\left(a_1 + R_A\right)}{k_{BA}} \tag{3.17}$$

$$c_4 = c_2 \frac{\left(a_2 + R_A\right)}{k_{BA}} \tag{3.18}$$

R_A and R_B are the longitudinal relaxation rate constants for spins A and B, respectively, in the presence of chemical exchange, i.e. R_A and R_B are given by

$$R_A = \frac{1}{T_{1A}} + k_{AB} \tag{3.19}$$

$$R_B = \frac{1}{T_{1B}} + k_{BA} \tag{3.20}$$

where T_{1A} and T_{1B} are the longitudinal relaxation times in the absence of chemical exchange.

3.3.2 MT Methods

Figure 3.18 shows the MR methods to achieve MT. All methods are based on a selective perturbation of one of the exchange partners (e.g. PCr) and observing the effect on the other (e.g. γ-ATP). Selective perturbation can be achieved through frequency-selective inversion (Figure 3.18A), low-power, on- or off-resonance saturation (Figure 3.18B), or more sophisticated pulse sequences composed of RF pulses and delays (Figure 3.18C). Classical MR studies (Figure 3.17A and B) most commonly use inversion transfer (Figure 3.18A) or saturation transfer (ST, Figure 3.18B) sequences, whereas CEST imaging (Figure 3.18C) can benefit from more elaborate schemes to separate out various contributions (see Section 3.3.5).

Figure 3.19 shows experimental results for inversion transfer and ST studies of the PCr-ATP system. During an inversion transfer experiment [65–67], the magnetization of one of the reactants is inverted with a selective inversion pulse such that, for instance, $M_{zB}(0) = -M_{zB}^0$ and $M_{zA}(0) = M_{zA}^0$. These initial conditions do not simplify the evolution curves given by Eqs. (3.12)–(3.18). Figure 3.19A shows *in vivo* ^{31}P NMR spectra from rat skeletal muscle obtained with an inversion transfer sequence in which the γ-ATP resonance was selectively inverted. Due to the short T_2 relaxation time of γ-ATP, relaxation during the long selective inversion pulse prevents a complete inversion of the γ-ATP resonance. However, this can be accounted for by using the appropriate $M_{zB}(0)$ in Eqs. (3.12) and (3.13). It can be seen from Figure 3.19A that while the γ-ATP resonance gradually recovers, the intensity of the PCr resonance initially decreases due to chemical exchange, after which it returns to its equilibrium value through T_1 relaxation. Figure 3.19B shows the theoretical inversion transfer curves for typical values for the creatine kinase equilibrium in brain *in vivo*. Inversion transfer has been successfully used to study the creatine kinase reaction *in vivo* [66–69].

ST (Figure 3.18B) is the most commonly used MT method used *in vivo* and involves a long, low-power saturation pulse that continuously destroys the longitudinal magnetization of one of the two exchanging pools. The magnetization of the other pool is reduced indirectly through

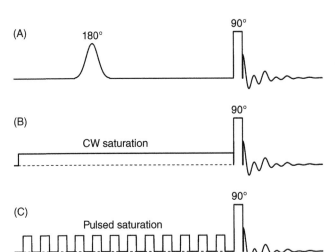

Figure 3.18 Magnetization transfer MR pulse sequences. All methods achieve selective perturbation of one of the exchange partners, but differ in the type of perturbation. (A) Frequency-selective inversion recovery, or inversion transfer (IT), (B) continuous wave (CW) saturation or saturation transfer (ST), and (C) pulsed, gapped saturation transfer. Saturation pulses in (B) and (C) can be applied on- or off-resonance.

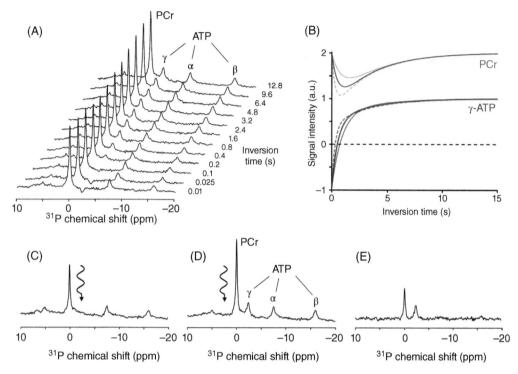

Figure 3.19 **Magnetization transfer on the creatine-kinase-catalyzed PCr/ATP system.** (A) Experimental inversion transfer (IT) ^{31}P NMR dataset with selective inversion of γ-ATP. The time-dependent PCr signal dynamics are clearly visible. (B) Theoretical inversion transfer curves are calculated with Eqs. (3.12) and (3.13) for [PCr] = 5 mM, [ATP] = 2.5 mM, $T_{1PCr} = 5$ s, and $T_{1ATP} = 2$ s. The middle curve is calculated for $k_{PCr→ATP} = 0.5$ s^{-1} and $k_{ATP→PCr} = 1.0$ s^{-1}. The dotted and transparent curves are calculated for ($k_{PCr→ATP} = 1.0$ s^{-1}, $k_{ATP→PCr} = 2.0$ s^{-1}) and ($k_{PCr→ATP} = 0.25$ s^{-1}, $k_{ATP→PCr} = 0.50$ s^{-1}), respectively. (C–E) Experimental steady-state saturation transfer (ST) ^{31}P NMR dataset on rat skeletal muscle with (C) selective saturation of γ-ATP and a (D) control position mirrored relative to PCr. (E) Difference between (C) and (D).

the exchange process. In terms of the two-compartment model, saturation of the *B* spins leads to $M_{zB}(0) = 0$ and $M_{zA}(0) = M_A^0$ such that Eqs. (3.12)–(3.18) simplify to

$$\frac{M_{zA}(t)}{M_{zA}^0} = \frac{1}{R_A T_{1A}} + \left[1 - \left(\frac{1}{R_A T_{1A}}\right)\right]e^{-R_A t} \tag{3.21}$$

In the majority of studies the multiple measurements needed to sample the time-dependent ST curve described by Eq. (3.21) are reduced to a single, steady-state measurement at very long saturation times. In that case, Eq. (3.21) reduces to

$$\frac{M_{zA}(\infty)}{M_{zA}^0} = \frac{1}{R_A T_{1A}} = \frac{1}{1 + k_{AB}T_{1A}} \tag{3.22}$$

Figure 3.19C–E shows a typical steady-state saturation recovery result obtained from rat skeletal muscle in which the γ-ATP resonance is selectively saturated. Saturation is never perfectly selective, i.e. when saturating the γ-ATP resonance there will also be some saturation of the PCr resonance due to "RF bleedover." If not compensated for, this unwanted saturation will lead to incorrect k_{AB} and T_{1A} values. Compensation can be achieved by performing a second experiment (Figure 3.19D) in which the saturation is performed at a resonance frequency "mirrored" with respect to the PCr resonance frequency (Figure 3.19C). In this experiment, no ST effects will occur, but there will be RF bleedover effects on the PCr signal. Subtracting the experiment will give a difference spectrum representing the MT without any effects of RF bleedover (Figure 3.19E). The magnetization transfer ratio (MTR) $M_{zA}(\infty)/M_{0zA}$ obtained from Figure 3.19C–E can be used to calculate the exchange rate according to Eq. (3.22), provided that T_{1A} is known. T_{1A} can be measured with a standard inversion recovery method during which the exchange partner B is continuously saturated.

3.3.3 Multiple Exchange Reactions

Selective perturbation of the γ-ATP resonance in a ^{31}P NMR spectrum allows the determination of the unidirectional rate constant from PCr to ATP, i.e. $k_{PCr-ATP}$. Determining the reverse rate constant, i.e. $k_{ATP-PCr}$, will result in the observation that $k_{PCr-ATP}[PCr] \neq k_{ATP-PCr}[ATP]$. This is because it was assumed that the PCr/ATP system can be described by a two-site chemical exchange reaction. However, for a complete description of the *in vivo* situation the PCr/ADP to Cr/ATP exchange reaction needs to be extended with an ADP/P_i to ATP reaction as catalyzed by ATPases and several other enzymes. Saturation of the PCr will lead to an underestimation of $k_{ATP-PCr}$, because the ATP is not only supplied with perturbed magnetization from PCr, but also with unperturbed, fully-relaxed magnetization from inorganic phosphate, P_i. This problem can be alleviated [70] by incorporating an additional saturation field. By continuously saturating the inorganic phosphate resonance, the ATP pool is no longer supplied by magnetization from the P_i pool. The three-site exchange has effectively been reduced to a two-site chemical exchange process. Rather than being a complication, the exchange reactions involving ATP and P_i holds unique information on the ATP synthesis rate. Following selective saturation of the γ-ATP resonance, any decrease in the inorganic phosphate P_i resonance is proportional to the ATP synthesis rate. Utilizing the improved sensitivity at high magnetic fields, Lei et al. [71] were able to detect significant changes in the small P_i resonance and showed that the calculated ATP synthesis rate is in excellent agreement with previous measurements of glucose consumption rates.

MT experiments have been performed during resting and stimulated conditions in several systems, like in animal skeletal muscle during muscle contraction [72, 73] and in the human cortex during visual stimulation [74]. The MT experiment has also been used to establish a clinically relevant relation between the cardiac post-ischemic creatine kinase reaction velocity and the severity of ischemia in perfused hearts [75]. By combining MT with magnetic resonance imaging techniques, Hsieh and Balaban [76] were able to spatially map the reaction velocity of creatine kinase. More recently the spatial distribution of creatine kinase activity was mapped based on MR fingerprinting [39]. The extended experimental duration of MT can be substantially reduced with more efficient acquisition methods [77]. MT experiments have also been combined with ^{13}C NMR to study the 2-oxoglutarate/glutamate exchange in rat brain *in vivo* [78, 79]. Extended reviews on MT methods with an emphasis on ^{31}P NMR can be found in the literature [80].

3.3.4 MT Contrast

In the preceding sections, MT was described between compounds in acid–base exchange reactions (used for pH determination as detailed in Chapter 2) and enzyme-catalyzed exchange reactions. In 1989, Wolff and Balaban reported a new MT mechanism between mobile, NMR observable water and bound, NMR unobservable water [81]. As shown in Figure 3.20, tissue

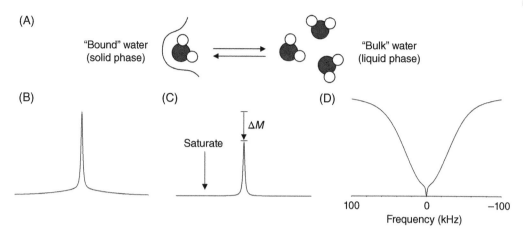

Figure 3.20 **Principle of magnetization transfer contrast.** (A) Water and metabolites are in rapid exchange between free, mobile, and bound, immobile states leading to (B) narrow and broad resonances, respectively. (C) Off-resonance saturation destroys the magnetization from immobile spins without directly affecting the signal from mobile spins. However, the magnetization from mobile spins is reduced indirectly due to the exchange mechanism between the two pools. (D) Repeating the measurement of (B and C) for different off-resonance frequencies creates a so-called z spectrum.

water exists in a number of environments with different rotational mobility. Water in the cytosol, extracellular fluid, and CSF is highly mobile with a very short rotation correlation time. Dipolar relaxation theory (Figure 3.4) predicts that this "bulk" water has a relatively long T_2 relaxation time constant, hence giving rise to a narrow NMR resonance. Water in protein hydration layers is relatively immobile with longer rotation correlation times. The T_2 relaxation of this "bound" water can be so short that the corresponding NMR resonance spans 10–100 kHz. All tissue water combined thus gives a NMR spectrum consisting of a narrow resonance from the mobile water on top of a very broad (NMR unobservable) resonance from the immobile water (Figure 3.20B). However, this immobile water pool can be studied through changes in the signal from the mobile water pool using the technique of off-resonance MT [81], in which a saturation pulse is applied at a frequency far enough off-resonance from the mobile spins such that the narrow resonance is not affected directly. In general, water in different compartments is in rapid exchange, either through proton exchange or cross-relaxation, such that saturation of the immobile spins leads to a signal decrease of the mobile spins (Figure 3.20C). When the experiment is repeated for different saturation frequency offsets, a so-called z-spectrum is obtained as shown in Figure 3.20D. When the saturation frequency is close to the mobile spins, the narrow resonance is directly affected according to

$$\frac{M_s}{M_0} = \frac{1 + \Delta\omega^2 T_2^2}{1 + \Delta\omega^2 T_2^2 + \gamma^2 B_1^2 T_1 T_2} \tag{3.23}$$

However, the direct saturation effect becomes negligible beyond a few kHz, such that any signal decrease can be attributed to MT from the immobile to the mobile water pools. Using equations similar to Eqs. (3.12)–(3.17), the z-spectrum can be quantitatively evaluated to obtain relaxation parameters of the immobile water protons, as well as the exchange rate between and the relative fractions of the pools.

Wolff and Balaban combined the off-resonance MT technique with MRI to obtain MT image contrast, different from conventional T_1, T_2, or proton density contrast. Figure 3.21A shows a

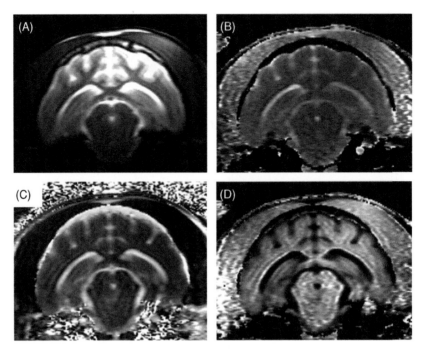

Figure 3.21 **Magnetization transfer contrast (MTC) imaging.** (A) Adiabatic spin-echo image (TR/TE = 5000/50 ms, surface coil transceiver) obtained on cat brain with kaolin-induced hydrocephalus. (B) T_1 relaxation time map, obtained by fitting five inversion recovery images (in the presence of off-resonance saturation) with varying recovery times. (C) Ratio image of the signal with and without 10 kHz off-resonance saturation. (D) Rate constant map as calculated from (B) and (C) according to Eq. (3.22). Rate constants vary between 0.2 and 0.8 s^{-1}. Cerebrospinal fluid exhibits a rate constant of almost zero. *Source:* Data from Kees P. J. Braun.

single-slice spin-echo image of cat brain with experimental hydrocephalus. Next, the same experiment was performed, but now with a 3 s saturation pulse ($B_1 = 8$ µT) irradiating 10 kHz off-resonance leading to the destruction of the broad resonance of immobile water spins and through chemical exchange to a reduction in the signal from mobile water spins. Figure 3.21C shows a calculated image of the ratio between saturation (M_s) and no saturation (M_0), also known as a MTR image. The CSF, having a low protein content, shows a ratio close to unity, whereas ratios in brain tissue vary between 0.4 and 0.9. A more quantitative representation is given by the rate constant of cross-relaxation, k, which can be calculated according to Eq. (3.22). Figure 3.21B shows a calculated T_{1A} map obtained with an inversion recovery sequence in the presence of off-resonance saturation. Figure 3.21D shows the calculated k map. The CSF now exhibits a cross-relaxation rate constant close to zero, while k ranges from 0.2 to 0.8 s^{-1} in brain tissue. The unique image contrast of MT imaging has found widespread applications in the clinical evaluation of neurological disorders, especially for multiple sclerosis and other white matter lesions and tumors.

In 1994, Dreher et al. [82] demonstrated that the off-resonance MT effect can also be observed for total creatine. This observation has been confirmed and extended to other metabolites including lactate and glutamate [83–87]. Figure 3.22A shows an off-resonance MT dataset for lactate in a C6 glioma, implanted in rat brain. The MT experiment was combined with spatial localization and spectral editing to eliminate signal from extracranial fat and mobile lipids in the tumor, respectively. Figure 3.22B shows the integrated lactate signal intensities

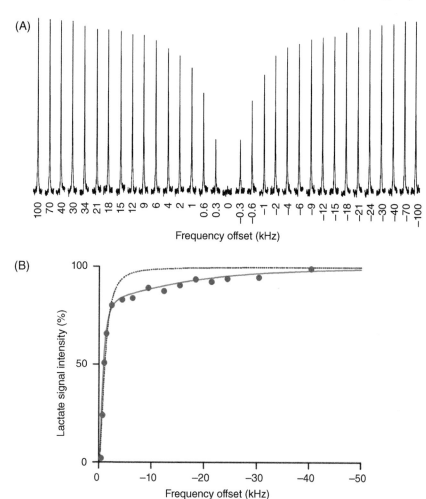

Figure 3.22 Magnetization transfer MR spectroscopy. (A) Experimental ^{1}H MRS of lactate obtained from a C6 glioma in rat brain *in vivo*. Each spectrum (32 averages, TR = 6000 ms) is acquired in the presence of a 3 s off-resonance saturation RF field ($\gamma B_1/2\pi$ = 450 Hz) applied at different frequencies between −100 and +100 kHz. Unambiguous lactate detection was achieved by a combination of spatial localization and spectral editing. (B) Integrated intensities (blue dots) of the lactate resonance shown in (A) as a function of the frequency offset of the saturation RF field. The red line depicts the theoretical curve for direct saturation. The blue curve is the best fit of the experimental data. *Source:* Data from Yanping Luo and Michael Garwood.

together with the theoretical curve for direct saturation. This curve was calculated according to Eq. (3.23) with $B_1 = 10\ \mu T$, $T_1 = 1730$ ms, and $T_2 = 202$ ms. A significant discrepancy between the theoretical curve for direct saturation and the experimentally measured data points indicates the presence of cross-relaxation between mobile and immobile (NMR invisible) metabolite pools. A fit of the experimental data (solid line) suggests a bound fraction of only 0.1% with an exchange rate 3.5 s^{-1}. Therefore, a small bound fraction may still cause a significant MT effect. The bound fraction of 0.1% will not affect the quantification of the lactate resonance. However, the MT effect will have consequences on metabolite quantification when water suppression is performed with long, low-power presaturation. In the presence of cross-relaxation as shown in Figure 3.22, the off-resonance saturation would lead to a significant reduction in the lactate resonance and, consequently to an underestimation of the lactate concentration, even though direct saturation effects are negligible.

Soon after the initial reports on metabolite MT effects, it became apparent that water is an important mediator in the observed signal modulations [84, 88–90]. Figure 3.23B and C shows ^{1}H NMR spectra from rat brain *in vivo* in the absence (Figure 3.23B) and presence (Figure 3.23C) of water saturation. Marked decreases in the methyl signals from total creatine and NAA can be observed. Repeating the experiment with difference saturation frequencies provides a z-spectrum for each metabolite (Figure 3.23D). In addition to the broad MTC effect between mobile and immobile spins observed for every metabolite, the z-spectrum for total creatine and NAA are characterized by a marked decrease upon saturation of the water spins at circa 4.65 ppm. Since the involved methyl protons are not in direct exchange with the water protons, the effects shown in Figure 3.23 must involve a combination of chemical exchange of water, protein, and exchangeable metabolite protons followed by inter- and intramolecular cross-relaxation with the non-exchangeable methyl protons (Figure 3.23A). The cross-relaxation mechanism is identical to the nOe discussed in Figures 3.5 and 3.6. The interaction between water and non-exchangeable metabolite protons is robust and can even be detected with a modest water perturbation, such as frequency-selective inversion (Figure 3.23E). Therefore, even pulsed water suppression techniques (see Chapter 6) can affect metabolite levels *in vivo*, although the commonly used CHESS method [91] was found to have minimal effects [84]. The complications of water–metabolite exchange phenomena in MR spectroscopy applications can be turned into an advantage when the same effect is used to indirectly detect metabolites via the high-intensity water signal.

3.3.5 Chemical Exchange Saturation Transfer (CEST)

Balaban and coworkers [92–95] demonstrated the detection of low-concentration metabolites with greatly enhanced sensitivity through a MT effect on the water which is now known as the CEST effect. The CEST effect is generally greatest and the specificity highest when the chemical exchange is in the slow to intermediate range, which is observed for many amide ($R(C{=}O)$—NH), amine (RR'—NH), and hydroxyl (R—OH) groups, whereby R represents other chemical groups with the molecule of interest. The chemical exchange between a compound and water is referred to as slow when $\Delta\nu\tau \gg 1$, where $\Delta\nu$ is the chemical shift difference between the two exchanging compounds and τ is the "lifetime" at either exchange site. Slowly exchanging compounds have two distinct chemical shifts, as opposed to a single averaged chemical shift for fast exchanging compounds as observed for inorganic phosphate (see Chapter 2). The "lifetime" τ is inversely proportional to the exchange rate and depends among other parameters upon the temperature and pH. Figure 3.24A shows a pulse-acquire ^{1}H NMR spectrum (no water suppression) of creatine dissolved in 50 mM phosphate buffer at 277 K. Besides the non-exchangeable methylene and methyl protons of creatine, the NMR spectrum contains a signal from the exchangeable creatine amine proton. The amine proton is clearly visible since the low temperature has pushed the system in the slow exchange regime. As a result, the z-spectrum (Figure 3.24B) shows only a modest CEST effect on the water upon saturation of the creatine amine proton. Note that in a typical z-spectrum, the water signal is referenced as 0 ppm (as opposed to the value of circa 4.65 ppm used in MRS). The CEST effect is often characterized by the MTR asymmetry factor, which compares the signals downfield and upfield from water in order to remove the symmetrical contributions from direct saturation and MTC. Increasing the temperature to 293 K (Figure 3.24C and D) and 310 K (Figure 3.24E and F) moves the system into the intermediate exchange regime, whereby the creatine amine signal becomes too broad for direct observation (Figure 3.24E). However, the increased chemical exchange rate leads to a greatly increased CEST

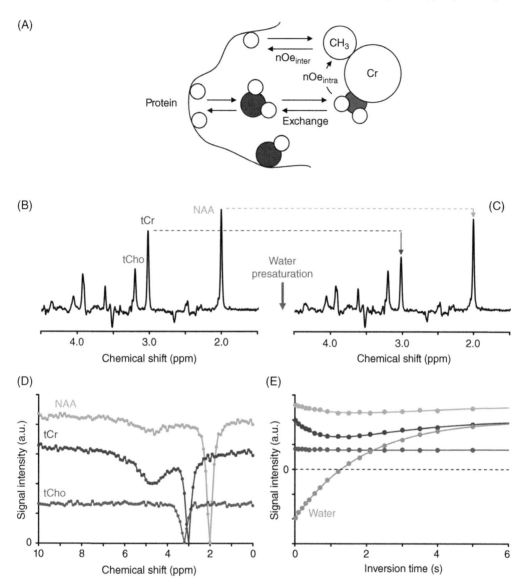

Figure 3.23 **Nuclear Overhauser effects in magnetization transfer MR spectroscopy.** (A) Model for magnetization transfer between water, exchangeable protons on proteins and non-exchangeable metabolite protons (methyl protons in creatine). Water magnetization can be transferred to the creatine methyl protons via a combination of chemical exchange and intra- and intermolecular nuclear Overhauser effects. (B, C) ^{1}H MR spectra of rat brain (9.4 T, TR/TE = 5000/68 ms) in (B) the absence and (C) the presence of low-power CW saturation (B_1 ~0.8 µT) of the water resonance. In both cases, the water resonance was minimized with MEGA water suppression. The creatine methyl and methylene protons and to a lesser degree the NAA methyl protons show a marked decrease in signal intensity due to nOe effects. (D) Acquisition of MR spectra with CW saturation at different chemical shifts shows a marked decrease in creatine (and NAA) intensity when the water resonance is saturated. In addition to the nOe effect, all metabolites show a significant MTC effect between bound and free metabolite pools spanning tens of kHz (see Figures 3.20 and 3.22). (E) Selective inversion of the water leads to a marked decrease in creatine (and NAA) signal intensity, roughly when the water signal crosses zero, further confirming a nOe effect between water and methyl protons.

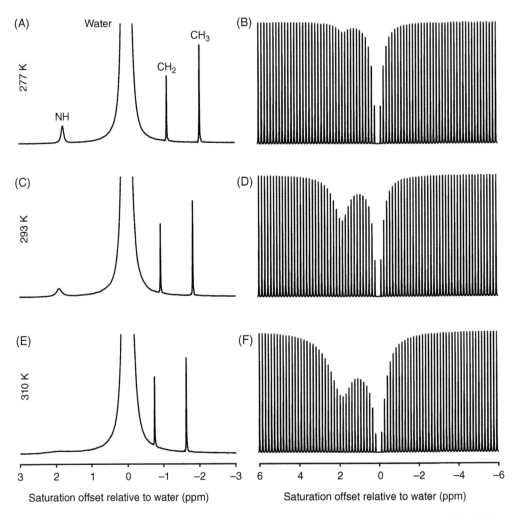

Figure 3.24 Chemical exchange saturation transfer (CEST) between water and creatine. (A) At lower temperature (277 K) the amine resonance of creatine at ~2 ppm downfield from water is readily observable with ¹H MRS. In other words, the chemical exchange between water and creatine is sufficiently slow to give two separate resonances. (B) As a consequence of the slow exchange, the water z spectrum only shows a modest decrease upon saturation of the creatine amine resonance. At higher temperatures (C, D) 293 K and (E, F) 310 K the spectroscopic detection of the creatine amine signal (C, E) becomes more difficult due to broader resonances as the result of increasing chemical exchange rates. (D, F) At the same time the CEST effect becomes more pronounced, leading to a substantial decrease in the water z spectrum upon saturation of the creatine amine resonance. Note that besides the temperature, the pH level also plays an important role in the magnitude of a CEST effect. All results were obtained with 20 mM creatine in phosphate buffer (pH 7.17).

effect on the water signal (Figure 3.24F). Detecting creatine through the CEST-mediated decrease in water signal intensity provides an increase in detection sensitivity of several orders of magnitude, thereby explaining the popularity of CEST for indirect metabolite detection. However, in human and animal tissues *in vivo*, the CEST effect is much more complicated and essentially combines all of the effects shown in Figures 3.20–3.24. Figure 3.25 shows a *z*-spectrum that can be expected *in vivo*. Over the frequency span covering ±50 kHz the *z*-spectrum will show two main effects (Figure 3.25A), namely the direct saturation of the water signal (green) and the MTC effect between mobile and immobile pools (blue). Zooming in to the

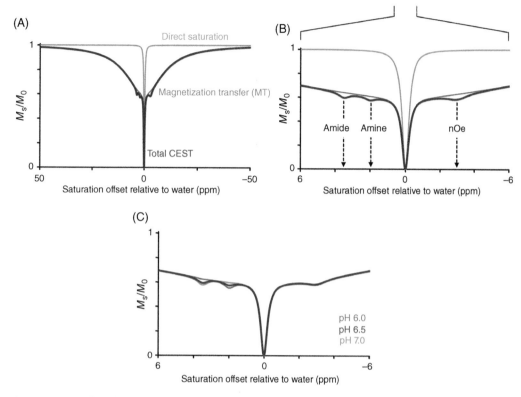

Figure 3.25 **Combined magnetization transfer effects observed *in vivo*.** (A) Many compounds, including water and most metabolites, display a "classical" magnetization transfer effect between immobile, bound spins and mobile, free spins giving rise to a broad MT effect (blue line). Near on-resonance application of the CW saturation pulse leads to a direct saturation curve given by Eq. (3.23) (green line). (B) Closer to water several additional features can be recognized that can be assigned to a CEST effect involving amide and amine protons at ~3.5 and ~2 ppm downfield from water, respectively. In certain cases an additional signal can be observed around 1 ppm downfield from water caused by a CEST effect involving hydroxyl protons from glucose or glycogen. A typical CEST spectrum from brain *in vivo* also displays upfield resonances that are typically assigned to nOe effects. (C) Most CEST signals display a strong dependence upon pH, as shown for the amide and amine signals. In addition to pH, temperature, saturation type, and amplitude all play a role in the relative amplitudes of the various CEST signals.

±3 kHz or ±6 ppm region of the *z*-spectrum around the water resonance at 0 ppm reveals several other effects related to CEST. Downfield from water at circa 3.5 and 2.0 ppm will be two signals corresponding to amide and amine protons, similar to that shown for creatine in water (Figure 3.24). In some cases, an additional signal may be observed circa 1 ppm downfield originating from the hydroxyl protons associated with glucose or glycogen. Upfield from water one or more signals can be observed that are generally designated as nOe peaks. These signals are related to MT between water and non-exchangeable protons, similar to the effects discussed in Figure 3.23. In the presence of nOe peaks, the use of MTR asymmetry measures is complicated such that analysis of the entire *z* spectrum is recommended. It should be realized that the CEST signals are dependent on many parameters, including pH (Figure 3.25C), temperature (Figure 3.24), saturation type, and saturation strength.

The CEST mechanism has been used to detect several metabolites such as creatine [96, 97], glutamate [98], glucose [99, 100], and glycogen [101] with greatly increased sensitivity. Whereas the prospects of obtaining high-resolution MR images of low-concentration metabolites are exciting, the opportunity should be met with a healthy dose of skepticism and caution. From Figures 3.20–3.25 it should be clear that MT and CEST effects *in vivo* contain many competing contributions that manifest themselves into a limited number of broad, overlapping signals originating from multiple different sources. In addition, the magnitude of MT and CEST effects depend on a wide range of experimental and biological factors. Caution should therefore be exercised when interpreting, for example, a glutamate CEST (gluCEST) image. Under the very best of conditions, the image should be interpreted as a glutamate-*weighted* CEST image as it would be impossible to exclude other competing contributions.

CEST and MTC MRI are undoubtedly exciting methods that will continue to provide novel and unique MR image contrast, whereby the prospects of adding a metabolic component to the arsenal of MR imaging modalities is a worthwhile goal. For extensive reviews on CEST and related phenomena the reader is referred to the literature [102–104].

3.4 Diffusion

3.4.1 Principles of Diffusion

Diffusion is the random translational (or Brownian) motion of molecules or ions that is driven by internal thermal energy. In 1855, Fick devised two differential equations that quantitatively described diffusion of molecules through ultrathin membranes [105, 106]. Fick's first law of diffusion states that the flux of particles (i.e. the number of particles that moves through a 2D plane per unit of time) at a given spatial position is directly proportional to the concentration gradient and the diffusion coefficient at that position. When combined with the conservation of mass, one arrives at Fick's second law of diffusion which predicts how the concentration of particles changes over time for a given concentration gradient and diffusion coefficient. The two diffusion laws were derived under the assumption of a net flux by diffusion over a concentration gradient. To describe diffusion in the absence of internal concentration gradients, like diffusion in pure water, it is convenient to introduce a probability function which gives the probability of a particle having moved over a certain distance and time. For isotropic diffusion, the probability function obeys Fick's second law of diffusion and predicts a Gaussian dependence on the displacement (Figure 3.26). As a random Brownian motion, the average displacement is zero. However, the average square displacement λ^2 associated with three-dimensional diffusion can be calculated according to the Einstein–Smoluchowski equation:

$$\lambda^2 = 6Dt \tag{3.24}$$

where D is the diffusion coefficient (in $\mu m^2 \ (ms)^{-1}$) and t is the diffusion time (in ms). Equation (3.24) highlights that displacement resulting from diffusion is simply related to the diffusion coefficient and that, for the case of unrestricted diffusion, the average square displacement increases linearly with the diffusion time t. For freely diffusing water at room temperature $(D \sim 2.2 \ \mu m^2 \ (ms)^{-1})$, the water molecules travel an average or root mean square distance of circa 25 μm in 50 ms. The average square displacements for diffusion in 1D and 2D spaces are given by 2Dt and 4Dt, respectively. Deviations from the Gaussian distribution function will arise when the translational, Brownian displacements are restricted by geometrical constraints.

3.4.2 Diffusion and NMR

Early on it was recognized that NMR is able to accurately measure the self-diffusion of molecules in isotropic liquids. In the classical paper on spin-echoes, Hahn [107] outlined the measurement of diffusion in the presence of a constant background gradient, which was

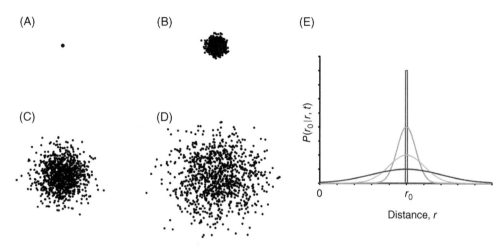

Figure 3.26 **Diffusion and spatial displacement.** Starting from a single point in space (A) the mean square displacement of diffusing particles increases linearly with time (B–D). The qualitative depiction of diffusion in (A–D) can be described quantitatively by a probability function *P* as is graphically depicted in (E). The Gaussian-distributed probability function describes the chance of a particle traveling from the initial position r_0 to position *r* during a diffusion time *t*. The situations sketched in (A–D) are represented by the blue, green, orange, and red probability functions.

subsequently modified and further developed by Carr and Purcell [108] and Woessner [109]. However, it was not until the development of a pulsed-gradient spin-echo method by Stejskal and Tanner [110] that NMR diffusion measurements became a routine technique to measure molecular self-diffusion coefficients.

Diffusion measurements by NMR rely on the principle of signal loss through diffusion-dependent phase dispersal as is qualitatively illustrated in Figure 3.27. Figure 3.27B shows transverse magnetization at time t_A prior to the application of a magnetic field gradient pulse (Figure 3.27A). In the absence of magnetic field inhomogeneity, all spins have the same phase. The magnetic field gradient gives each spin a position-dependent phase, given by $\phi(r) = 2\pi r G\delta$, where r (in cm) is the position along the direction of the gradient and G and δ are the magnetic field gradient strength (in $Hz\,cm^{-1}$) and duration (in s), respectively. In the absence of macro- or microscopic motion the spins reside at the same spatial location at time t_C (Figure 3.27D) such that application of an identical gradient with negative amplitude (Figure 3.27A) leads to a complete reversal of all gradient-induced, position-dependent phases (Figure 3.27G). Since the signal from the entire sample at time t_D originates from phase-coherent spins, it represents the maximum possible signal (Figure 3.27J). In the presence of macroscopic motion (e.g. linear flow) along the gradient direction (Figure 3.27E) all spins are spatially displaced by the same amount Δr. The total phase acquired by each spin represents the combined effects of the two magnetic field gradients and the displacement and is given by $\phi(r + \Delta r) = 2\pi r G\delta - 2\pi(r + \Delta r)G\delta = -2\pi\Delta r G\delta$. In other words, the position-dependent phase for flowing spins has been refocused, similar to the situation for stationary spins, whereby the macroscopic displacement has led to a global phase (Figure 3.27H and K). In the presence of diffusion (without macroscopic motion) the spins move to different spatial positions in a random manner (Figure 3.27F). Since the translational motion has perturbed the linear dependence of phase on spatial position, application of the second magnetic field gradient can no longer lead to a complete reversal of the phase acquired by the first magnetic field gradient. Therefore, the presence of diffusion leads to phase dispersal across the sample (Figure 3.27I), which in turn will lead to phase cancelation and hence signal loss when signal is acquired from the entire sample (Figure 3.27L). While the exact functional form for the

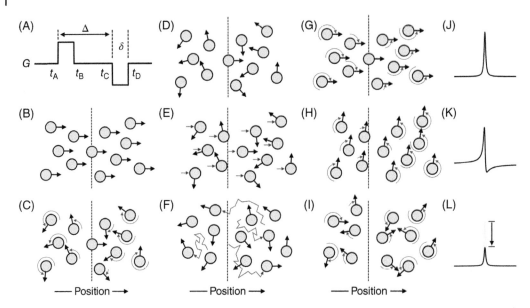

(A) (B) (C) (D) (E) (F) (G) (H) (I) (J) (K) (L)

—— Position ——→   —— Position ——→   —— Position ——→

Figure 3.27 **Diffusion, motion, and flow in the presence of magnetic field gradients.** (A) Bipolar gradients of duration δ and separation Δ. (B) At time t_A following excitation, but before the magnetic field gradients, the phase of spins at different spatial locations is (assumed) identical. (C) At time t_B the spins have acquired a position-dependent phase due to the first positive magnetic field gradient. (D) For static spins the situation at time t_C is identical to that at time t_B (ignoring relaxation effects). (E) In the presence of flow or macroscopic motion all spins move by the same amount during the period between t_B and t_C. (F) In the presence of diffusion, spins move randomly in different directions and over different distances. (G) The phase of static spins is perfectly refocused by the second magnetic field gradient of opposite sign, leading to (J) an in-phase MR spectrum of maximum intensity. (H) The gradient-induced phase dispersion of flowing or macroscopically displaced spins is also perfectly refocused, except for the phase caused by the net displacement. (K) The resulting MR spectrum has maximum intensity (due to the absence of phase dispersal), but comes with a nonzero global phase. (I) For diffusing spins the phase reversal achieved by the second magnetic field gradient is incomplete, leading to a residual phase dispersal across the sample and hence (L) a reduced, in-phase MR signal.

diffusion-dependent signal loss will be derived next, it is straightforward to see that the function will be dependent on the diffusion coefficient, the area of the magnetic field gradient (i.e. amplitude and duration) as well as on the separation between the magnetic field gradients. For example, increasing the delay Δ increases the diffusion time, which according to Eq. (3.24) increases the mean square displacement. As the spins move further away from the starting position, the phase reversal becomes worse, leading to more phase cancelation and hence more signal loss.

It should be noted that macroscopic motion does not lead to signal loss *per se*, but does lead to a global phase offset (Figure 3.27K). When the motion changes from scan to scan, the net phase $\Delta\phi$ will also change. During signal averaging in MRS this scan-to-scan phase variation can lead to signal loss, whereas for MRI applications any scan-to-scan phase instability will lead to image artifacts (ghosting) in the phase-encoding direction. This effect can be eliminated by storing and phase correcting each transient separately before summation for MRS. For MRI, the global phase offset can be minimized through the use of navigator echoes.

In order to quantitatively describe diffusion in the presence of time-varying magnetic field gradients, it is most convenient to follow the formalism outlined by Torrey [111], which was subsequently used by Stejskal and Tanner [110]. The solution to the Torrey–Bloch equations extended to include the effects of diffusion is given by

$$M_{xy}(b) = M_{xy}(0)e^{-bD}$$ (3.25)

with

$$b = \gamma^2 \int_0^t \left[\int_0^{t'} G(t'') dt'' \right]^2 dt'$$ (3.26)

Equation (3.25) predicts a simple, exponential decay of the detected signal due to diffusion-related signal loss. The diffusion coefficient D can be quantitatively measured from the detected signal $M_{xy}(b)$ as long as the "b-value" is quantitatively known. The b-value, as defined in Eq. (3.26), represents the amount of gradient-induced dephasing and can be quantitatively calculated for any MR sequence with any magnetic field gradient combination.

In his seminal paper of spin echoes [107], Hahn demonstrated that the diffusion-related signal loss in the presence of a constant amplitude background gradient field G_0 (Figure 3.28A) during the entire echo-time TE can be described by a b-value given by

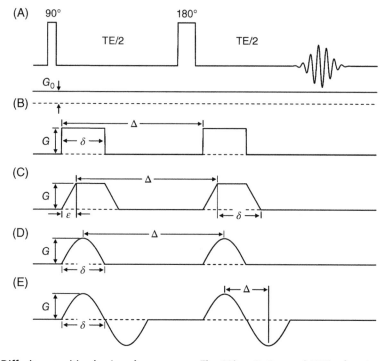

Figure 3.28 **Diffusion-sensitized spin-echo sequence.** The 90° excitation and 180° refocusing pulses form a spin-echo at TE. (A) In addition to a time-independent inhomogeneity G_0 in the main magnetic field, the pulse sequence can also be sensitized for diffusion by (B–D) the application of a pair of pulsed gradients surrounding the 180° pulse. Most commonly used gradient wave forms are (B) constant, (C) halfsine, and (D) trapezoidal. The exact definition of the pulse duration δ and the time between gradient pulses Δ are indicated. ε is the rise time of the trapezoidal ramp. Quantitative expressions for the corresponding b-values are given by Eqs. (3.27)–(3.31). (E) Diffusion-weighting can also be introduced by bipolar gradients which can be beneficial for diffusion cross-term elimination (see Exercises).

$$b = \frac{1}{12}\gamma^2 G_0^2 \text{TE}^3 \tag{3.27}$$

The background gradient field G_0 can arise from a macroscopically inhomogeneous magnetic field, or can be due to microscopic magnetic field gradients arising from local magnetic susceptibility differences. The TE^3 dependence of the spin-echo signal amplitude in the presence of magnetic field inhomogeneity and diffusion can lead to a significant underestimation of measured T_2 relaxation times when a Hahn spin-echo is used for refocusing. When the calculation as outlined for a regular spin-echo is repeated for a CPMG spin-echo pulse train with n 180° refocusing pulses, such that $\text{TE} = n\text{TE}_{\text{CPMG}}$, the b-value at the top of the spin-echo equals

$$b = \frac{1}{12n^2}\gamma^2 G_0^2 \text{TE}^3 \tag{3.28}$$

Therefore, the sensitivity of a CPMG spin-echo towards diffusion is greatly reduced and approaches zero when n → ∞ (or TE_{CPMG} → 0). This feature has been used by Michaeli et al. [61] to demonstrate that the decrease in observed T_2 relaxation times with increasing magnetic fields is largely due to diffusion through increased microscopic magnetic field gradients.

The use of a constant magnetic field gradient for diffusion weighting has several disadvantages related to decreased spectral resolution and undesirable slice effects. These problems were recognized and solved by Stejskal and Tanner [110] with the introduction of pulsed-gradient methods for the measurement of diffusion. Figure 3.28B–D shows three commonly used pulsed-gradient methods with constant, halfsine, and trapezoidal-shaped gradients for the measurement of diffusion as incorporated in a Hahn spin-echo sequence. For the constant amplitude gradient pair of duration δ and amplitude G, separated by a delay Δ (Figure 3.28B), the b-value is given by (see Exercises)

$$b = \gamma^2 G^2 \delta^2 \left(\Delta - \frac{\delta}{3} \right) \tag{3.29}$$

where b is expressed in $\text{s}\,\text{cm}^{-2}$, γG in $\text{Hz}\,\text{cm}^{-1}$, and Δ and δ in s. Even though this b-value is often given in the literature, it can never represent the real value, since it implies an infinitely short gradient rise time. More realistic (i.e. practically feasible) gradient waveforms are halfsine or trapezoidal-shaped gradients (Figure 3.28C and D). The b-values of the halfsine and trapezoidal gradient pairs are given by

$$b = \frac{4}{\pi^2}\gamma^2 G^2 \delta^2 \left(\Delta - \frac{\delta}{4} \right) \tag{3.30}$$

and

$$b = \gamma^2 G^2 \left[\delta^2 \left(\Delta - \frac{\delta}{3} \right) + \frac{\varepsilon^3}{30} - \frac{\delta\varepsilon^2}{6} \right] \tag{3.31}$$

respectively. ε is the gradient rise-time for the trapezoidal gradient waveform. Knowing the exact b-value for a given gradient pair combination allows the determination of the diffusion coefficient. Usually, the diffusion experiment is repeated with different b-values (either by changing G, δ, or Δ) to achieve a reasonable span of signal intensities. Figure 3.29A shows a series of diffusion-weighted ^{1}H NMR spectra from normal rat brain. Diffusion-sensitizing gradient pairs were positioned around each of the two 180° refocusing pulses in a PRESS

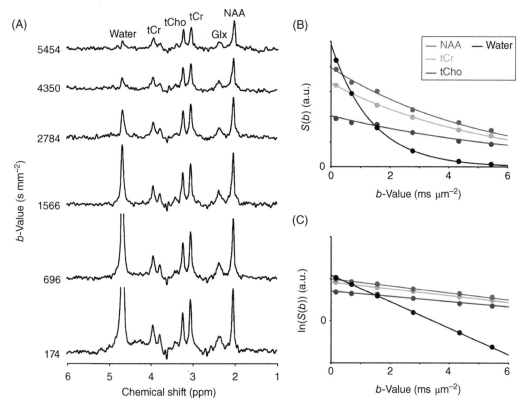

Figure 3.29 **Diffusion-weighted ^{1}H MR spectroscopy.** (A) ^{1}H MR spectra of rat brain *in vivo* acquired at 4.7 T (PRESS, TR/TE = 3000/144 ms, CHESS water suppression). The *b*-factors are 174, 696, 1566, 2784, 4350, 5454 s mm^{-2} (δ =6 ms and Δ =25 ms) with diffusion-weighting increasing from bottom to top. Observed resonances have been assigned to water, total creatine (tCr), choline-containing compounds (tCho), glutamate/ glutamine (Glx), and *N*-acetyl aspartate (NAA). Note that the water signal decreases substantially faster than the metabolites indicating a higher diffusion coefficient. All spectra were subject to the same phase correction. (B) Signal intensity of the major resonances shown in (A) and the exponential fit according to Eq. (3.25). (C) Natural logarithm of the signal intensities, whereby the diffusion coefficient is directly obtained from the slope.

localization method (see Chapter 6). Increasing the diffusion-weighting (by increasing the gradient amplitude) leads to increased signal attenuation. The diffusion-weighting increases from bottom to top. From the unsuppressed, residual water signal it can be seen that the diffusion of water is considerably faster than that of the metabolites, as expected. Fitting the experimental signals to Eq. (3.25) gives apparent diffusion constants for the major metabolites in the range of 0.15–0.17 μm^2 (ms)$^{-1}$, while the water apparent diffusion coefficient (ADC) amounts to 0.71 μm^2 (ms)$^{-1}$ (Figure 3.29B and C). For the majority of diffusion experiments, the *b*-value is incremented by changing the gradient amplitude, while keeping the duration and separation identical. The primary reasons for this choice are that for constant δ and Δ, (i) the echo-time can remain constant and (ii) the diffusion time remains constant, which is especially important when observing restricted diffusion. However, when the details of the restricting compartment are to be investigated a variation in the diffusion time is beneficial. In order to achieve the shortest possible gradient separation Δ, the magnetic field gradients can be executed as shown in Figure 3.28E.

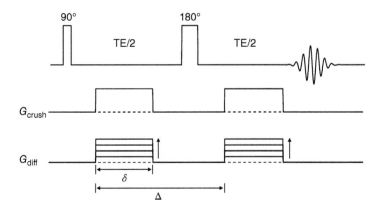

Figure 3.30 Diffusion cross-terms. A spin-echo with magnetic field gradients of constant amplitude G_{crush} used for removing unwanted coherences not properly refocused by the 180° pulse. When the sequence is extended with diffusion-sensitization gradients of variable amplitude G_{diff}, the resulting b-value will be characterized by contributions from the crusher and diffusion gradients, as well as from cross-terms between the two gradients. Cross-terms can be quantitatively accounted for using Eq. (3.26) or can be eliminated by modifying the pulse sequence such that spins are not being dephased by gradients other than the diffusion gradients during diffusion time Δ.

For accurate diffusion measurements, precise knowledge of the attenuation factor (i.e. b-value) is required. For simple, nonlocalized diffusion measurements the attenuation factor is given by Eqs. (3.29)–(3.31), or related expressions. However, for localized measurements (either imaging or localized spectroscopy) an accurate determination of the attenuation factor is more difficult, since all magnetic field gradients should be taken into account. Figure 3.30 shows a spin-echo pulse sequence incorporated with a pair of diffusion gradients (of duration δ, amplitude G_{diff}, and separated by a delay Δ). Furthermore, the sequence has two additional crusher magnetic field gradients to ensure proper refocusing by the 180° pulse. The b-value for the pulse sequence in Figure 3.30 is given by

$$b = \gamma^2 G_{diff}^2 \delta^2 \left(\Delta - \frac{\delta}{3} \right) + \gamma^2 G_{crush}^2 \delta^2 \left(\Delta - \frac{\delta}{3} \right) + 2\gamma^2 G_{diff} G_{crush} \delta^2 \left(\Delta - \frac{\delta}{3} \right) \tag{3.32}$$

The first term in Eq. (3.32) represents the attenuation term due to the diffusion gradients, while the second term is due to the crusher gradients. The last term is a so-called "cross-term" between the diffusion and crusher gradients. If not accounted for, the cross-terms can induce a significant error in the calculation of the diffusion coefficients. In many cases it is possible to redesign a pulse sequence such that cross-terms are minimized or even completely cancelled. Qualitatively, the presence of cross-terms can be determined when in between the time of dephasing and rephasing of one gradient pair combination, another gradient along the same or an orthogonal axis is applied. In some cases a different placement of magnetic field gradients within the sequence can eliminate cross-terms. In the case of Figure 3.30, simply shifting the magnetic field gradients to different times within the sequence does not eliminate cross-terms. A possible solution is provided by acquiring a second dataset with negative crusher gradients. While the first and second terms in Eq. (3.32) will be identical, the cross-term will be inverted due to the negative G_{crush} term. Adding the signals from the two experiments together eliminates the effect of the cross-terms.

Up to this point, the factors affecting the b-value have been described, since accurate knowledge of this value is essential for a proper interpretation of the diffusion experiment. It should be realized that the parameter of interest, i.e. the diffusion constant D, is also affected by several factors, especially when measurements in intact tissue are performed. The dependence of the diffusion constant of an isotropic medium (or a solute dissolved in an isotropic medium) on parameters like viscosity and temperature is described by

$$D = \frac{kT}{6\pi\eta R_H} \tag{3.33}$$

which is known as the Stokes–Einstein relation for translational diffusion. In Eq. (3.33), k is Boltzmann's constant ($1.380\,65 \cdot 10^{-23}\,\text{J K}^{-1}$), T is the absolute temperature, η is the viscosity, and R_H is the hydrodynamic radius. The hydrodynamic radius is a measure of the radius of a solute, including the surrounding hydration layers. Therefore, R_H can be much larger than would be expected on basis of the physicochemical structure of the solute. Note, that the viscosity in Eq. (3.33) itself also has a strong temperature dependence whereby the diffusion constant is not linear (as suggested by Eq. (3.33)) but exponentially dependent on the temperature. Figure 3.31 shows the temperature dependence of the diffusion constant of water [112].

Due to the noninvasive character of NMR, diffusion measurements can also be performed under *in vivo* conditions. The molecular displacement can be affected by a wide range of factors like the presence of cellular membranes, high concentrations of proteins, and the binding to macromolecules. Consequently, diffusion in the intracellular compartment may not be adequately described by the Stokes–Einstein relation (Eq. (3.33)). Furthermore, in the case of restricted diffusion (e.g. due to cellular compartmentalization), the measured diffusion constant may depend on the orientation of the diffusion-sensitizing gradients and on the specific timing parameters (i.e. δ and Δ) used. For these reasons, the molecular displacements measured *in vivo* do not represent pure molecular self-diffusion. To discriminate between molecular self-diffusion in isotropic media and the diffusive processes measured under *in vivo* conditions, the latter are described by an apparent diffusion coefficient (ADC). The measurement of water ADC values by imaging techniques is now frequently employed to discriminate ischemic brain lesions from normal tissue [113] and to access the anisotropy of tissue architecture [114].

Metabolite diffusion measurements are more demanding than water diffusion measurements. Besides being present at lower concentrations, metabolites also have substantially lower diffusion constants. This requires higher gradients and/or longer diffusion times for sufficient signal attenuation due to diffusion (i.e. *b*-values exceeding $10\,000\,\text{s mm}^{-2}$ or $10\,\text{ms μm}^{-2}$ are required). This is especially true for measurements involving nuclei with low gyromagnetic

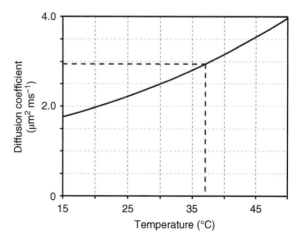

Figure 3.31 Temperature dependence of the water diffusion coefficient. In the physiological range, the diffusion coefficient of freely diffusion water is linearly dependent on the temperature. Note that the water apparent diffusion coefficient *in vivo* is several-fold lower due the hindered diffusion.

ratios. For example, for ^{31}P the gradients should be $(\gamma_H/\gamma_P)^2 \sim 6$ times stronger than for ^{1}H to obtain a similar signal attenuation [115–118].

Several experimental imperfections can influence the obtained ADC values. In general, gradient performance should be excellent in that the diffusion gradient pairs are well balanced and induce minimal eddy currents (i.e. time-varying magnetic fields following a gradient pulse). Both effects are easily verified by diffusion measurements on an *in vitro* sample. In the case of unbalanced gradient pairs or eddy currents, the phase and the shape of the resonance lines will change with increasing b-values, respectively. In Chapter 10, methods to minimize time-varying magnetic fields, like active screening and pre-emphasis, are described. The constant line widths and phases of the spectra shown in Figure 3.29A are a good indication that residual (time-varying) gradient effects are minimal.

Macroscopic motion and flow can also affect *in vivo* ADC measurements. A net displacement of spins (i.e. macroscopic motion) in the presence of a linear magnetic field gradient will cause a change in the phase of the entire NMR resonance. This is principally different from diffusion where the net displacement is zero, such that the ensemble average phase shift is also zero. However, macroscopic motion may also lead to signal attenuation on top of diffusion-related signal loss. For linear, macroscopic motion, the phase shift ϕ is identical for every position along the gradient direction and is given by

$$\phi = \gamma \delta \, \Delta \, Gv\cos \theta \tag{3.34}$$

where v is the linear velocity and θ is the angle between the gradient direction and the direction of motion. Equation (3.34) forms the basis for phase contrast MR angiography to obtain quantitative information about flow. From Eq. (3.34) it follows that coherent translational motion during a diffusion experiment does not have to lead to signal loss *per se*. However, during signal averaging (which will always be employed for metabolite ADC measurements), the phase shift may appear incoherent between acquisitions, thereby leading to signal loss. More complex motional patterns like turbulence lead to phase errors which are not refocused in individual acquisitions, such that signal loss can occur without signal averaging. In general, macroscopic motion-related signal losses *in vivo* are superimposed on diffusion-related signal losses, leading to an overestimation of the diffusion coefficients. Several methods are available to either minimize or compensate for macroscopic motion and flow [119–122]. Appropriate immobilization of the object under investigation (and particularly anesthetized laboratory animals) can minimize macroscopic motion-related signal loss to a large extent. Furthermore, physiological motion like, for instance, cardiac cycle-related CSF pulsations or respiration may be minimized by (a combination of) cardiac and respiratory triggering. Besides minimizing the motion-induced signal losses via a physiological approach, like immobilization or triggering, effects of motion can also be minimized by proper sequence programming. In principle, it is possible to utilize gradient moment nulling techniques to compensate for linear or higher order (i.e. accelerated) types of motion [119]. However, this tends to reduce the maximum attainable b-value. Any residual effects of macroscopic motion can be minimized by methods which compensate the acquired phase shift post-acquisition. Although not often used for spectroscopy applications, the use of navigator echoes is one of the most popular methods in MRI to compensate motion-related artifacts [122]. In this technique, an additional non-phase-encoded echo is acquired at each phase-encoding increment in a MRI sequence (for details of phase-encoding and image reconstruction see Chapter 4). Because the phase of these so-called navigator echoes should be identical in the absence of motion, deviations from a constant phase can be used to correct the image echoes for linear movements of the object. For spectroscopy, Posse et al. [120] have proposed a similar but slightly different method. By storing each acquired FID or echo separately, the FIDs can have an individual phase correction, after which they can be coherently summed as part of signal averaging without any motion-related signal loss.

3.4.3 Anisotropic Diffusion

In the preceding discussion the diffusion coefficient has been described as a scalar, i.e. identical diffusion along any spatial direction. This is a valid approximation if diffusion in an isotropic solution is considered, so that the diffusion has no directional preference. However, *in vivo* the presence of physical barriers (e.g. cell membranes, muscle fibers) may lead to anisotropic diffusion, i.e. the measured diffusion has a dependence on the direction of diffusion sensitization. This phenomenon has been observed in diffusion measurements of the spinal cord [123], brain white matter [114, 124], heart wall muscle [125–127], and skeletal muscle [115–118].

For a quantitative description of anisotropic diffusion, the classical Stejskal–Tanner relationship does no longer hold. The scalar diffusion coefficient D should be replaced by a rank two (i.e. three by three) diffusion tensor (or matrix) D, given by

$$D = \begin{pmatrix} D_{xx} & D_{xy} & D_{xz} \\ D_{yx} & D_{yy} & D_{yz} \\ D_{zx} & D_{zy} & D_{zz} \end{pmatrix} \tag{3.35}$$

The diagonal elements represent the diffusion along the x, y, and z axes in the laboratory frame, i.e. the frame in which the magnetic field gradients are applied. The off-diagonal elements represent the correlation between the diffusion in perpendicular directions. For isotropic diffusion, there is no correlation between diffusion in orthogonal directions and the off-diagonal elements are zero (see also Figure 3.32A for 2D diffusion). Furthermore, $D_{xx} = D_{yy} = D_{zz} = D$. For anisotropic diffusion one has to consider how the laboratory frame relates to the principal axes, i.e. the axes which coincide with the three main orthogonal directions of diffusion, of the cell (or tissue) frame of reference. In the principal axis system of the cell frame of reference, the diffusion tensor D' is given by

$$D' = \begin{pmatrix} D'_{xx} & 0 & 0 \\ 0 & D'_{yy} & 0 \\ 0 & 0 & D'_{zz} \end{pmatrix} \tag{3.36}$$

as is indicated for 2D diffusion in Figure 3.32B. However, the cell frame of reference almost never coincides with the laboratory frame. The diffusion tensor D in the laboratory frame is linked to D' by the relation

$$D = R^{-1} D' R \tag{3.37}$$

in which R is a rotation matrix (see Appendix A.1 and Chapter 5). When the cell frame of reference deviates from the laboratory frame, the off-diagonal elements in D are no longer zero and have to be taken into account for a complete description of anisotropic diffusion (Figure 3.32C). For uncharged particles, as water, the diffusion tensor D' (or D) is symmetric,

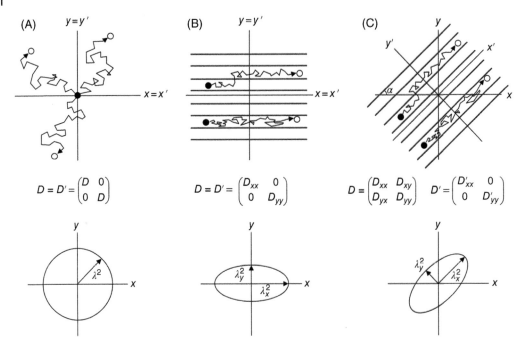

Figure 3.32 **Diffusion in the gradient and cell frames of reference.** (A) For isotropic, unrestricted diffusion in two spatial dimensions, the diffusion coefficient D and mean square displacement λ^2 are the same in all directions leading to a 2D diffusion tensor with all off-diagonal elements equal to zero. The 2D diffusion ellipsoid (bottom) will be round since $\lambda_x^2 = \lambda_y^2 = \lambda^2$. (B) For anisotropic, restricted diffusion the diffusion coefficients in the x and y directions are not equal to each other, leading to an elliptical displacement profile (bottom). However, since the gradient and cell frames of reference coincide, the off-diagonal elements of the 2D diffusion tensor are still all zero. (C) In the more general case where the gradient and cell frames of reference do not coincide, the off-diagonal elements become nonzero. However, the eigenvalues of the diffusion tensor provide the principle, orthogonal diffusivities, leading to a rotated, elliptical displacement profile (bottom).

i.e. $D_{xy} = D_{yx}$. Using the formalism of the diffusion tensor, the Stejskal–Tanner relationship of Eq. (3.25) can be rewritten as

$$\ln\left(\frac{S(b)}{S(0)}\right) = -\sum_{i=1}^{3}\sum_{j=1}^{3} b_{ij} D_{ij} \tag{3.38}$$

where b_{ij} is a component of the $\boldsymbol{b}$ matrix and D_{ij} is a component of the diffusion tensor $\boldsymbol{D}$. The b_{ij} terms can be analytically calculated using Eqs. (3.29)–(3.31). Note that cross-terms between perpendicular gradients are now appearing in b_{ij} components for which $i \neq j$. In order to quantitatively describe anisotropic diffusion at least seven experiments need to be performed (to obtain $D_{xx}, D_{xy}, D_{xz}, D_{yy}, D_{yz}, D_{zz}$, and $S(0)$), in which diffusion gradients are applied in various oblique directions. Note that for isotropic diffusion, i.e. $D_{ij} = 0$ for $i \neq j$ and $D_{ij} = D$ for $i = j$, Eq. (3.38) reduces to the classical Stejskal–Tanner equation (3.25) with $b = (b_{xx} + b_{yy} + b_{zz})/3$, showing that isotropic diffusion is simply a special case of a more general formulation of diffusion given by Eq. (3.38). Figure 3.33A and B shows the calculations of an ADC map obtained with diffusion-weighting along the x direction. The full diffusion tensor (Figure 3.33C) can be obtained by calculating ADC maps with diffusion weighting applied along six directions. The seven measurements represent the minimum number of experiments needed to construct the diffusion tensor. In reality, more experiments with diffusion weighting along different directions can

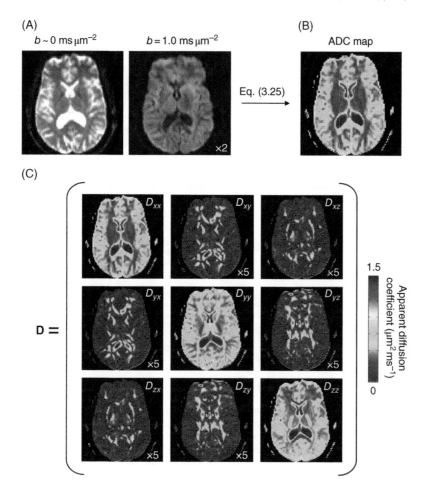

Figure 3.33 **Diffusion tensor imaging (DTI) of the human brain.** (A) A minimum of two MR images using low (left) and high (right) diffusion weighting allows (B) the calculation of a quantitative apparent diffusion coefficient (ADC) map. (C) Application of magnetic field gradients in at least six unique spatial directions allows the calculation of a diffusion tensor matrix in which the diagonal elements represent diffusion along the three principal axes and off-diagonal elements represent correlation between diffusion in orthogonal directions.

greatly improve the accuracy and sensitivity of the diffusion tensor. Some brain regions, such as cerebral gray matter and CSF have similar D_{xx}, D_{yy}, and D_{zz} values and small D_{xy}, D_{xz}, and D_{yz} values, indicating isotropic diffusion. However, in anisotropic tissues, like cerebral white matter, D_{xx}, D_{yy}, and D_{zz} are different and dependent on the orientation of the object relative to the magnetic field gradient direction. This observation of diffusion anisotropy in cerebral white matter is further confirmed by significant correlation between diffusion in orthogonal directions as described by D_{xy}, D_{xz}, and D_{yz}.

Although knowledge of the complete diffusion tensor provides insight into tissue orientation, for many applications the orientational dependence of the diffusion upon the gradient direction is of no interest and knowledge of the scalar diffusion coefficient suffices. For these applications one can use the average diffusion constant of three orthogonal directions, i.e. the trace of the diffusion tensor, since the trace of a matrix is invariant under rotations.

$$\mathrm{Tr}(\boldsymbol{D'}) = D'_{xx} + D'_{yy} + D'_{zz} = D_{xx} + D_{yy} + D_{zz} = \mathrm{Tr}(\boldsymbol{D}) = 3D_{av} \tag{3.39}$$

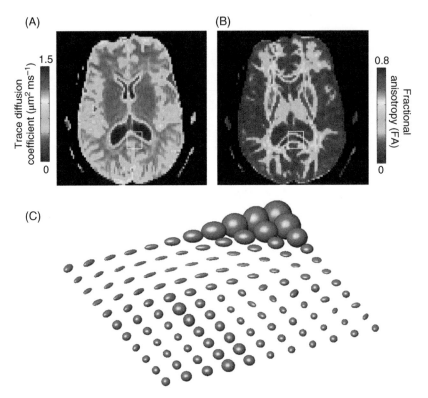

Figure 3.34 **Orientation dependent and independent parameters characterizing water diffusion *in vivo*.** (A) A diffusion trace image as calculated with Eq. (3.39) is orientation independent and primarily shows image contrast between hindered and free diffusion in brain and CSF, respectively. (B) A fractional anisotropy (FA) map emphasizes the orientation dependent diffusion, primarily providing image contrast based on the directional diffusion in white matter tracts. (C) Diffusion ellipsoid image extracted from the white box shows the features from (A) and (B) in more detail. Isotropic, but fast and slow diffusion in CSF and gray matter is represented by large and small diffusion spheres. Anisotropic diffusion in white matter is represented by ellipsoids that are elongated and angled in the direction of the white matter tracts, thereby allowing MR-based tractography.

Figure 3.34A shows a diffusion trace (D_{trace} or D_{av}) image calculated from the diffusion tensor in Figure 3.33C. All orientation-dependent contrast visible in the diffusion tensor images has disappeared and the remaining image contrast is largely due to diffusion coefficient differences between fast diffusing water in CSF and slower diffusing water in cerebral tissues. Note that there are several composite parameters that are rotationally invariant, of which the trace of the diffusion tensor is the most relevant and most commonly used.

For many other applications, like the study of cerebral white matter, diffusion anisotropy should be emphasized, rather than suppressed. Among several useful parameters [128] the fractional anisotropy (FA) is commonly used and is defined as

$$\text{FA} = \frac{\sqrt{3\left[\left(D'_{xx} - D_{\text{av}}\right)^2 + \left(D'_{yy} - D_{\text{av}}\right)^2 + \left(D'_{zz} - D_{\text{av}}\right)^2\right]}}{\sqrt{2\left(D'^2_{xx} + D'^2_{yy} + D'^2_{zz}\right)}} \tag{3.40}$$

where D'_{xx}, D'_{yy}, and D'_{zz} are the diffusion coefficients along the principal axis in the cell frame of reference and can be calculated as the eigenvalues of the diffusion tensor in the laboratory frame. Figure 3.34B shows the FA map as calculated from the diffusion tensor images shown in Figure 3.33C. Rather than eliminating the effects of diffusion anisotropy as was achieved for the diffusion trace (Figure 3.34A), the FA map clearly enhances those tissues that exhibit strong anisotropic diffusion behavior, i.e. white matter tracts.

As it is difficult to display diffusion tensor data with regular 2D images (at least six images are required as shown in Figure 3.33C), the concept of diffusion ellipsoid images has been proposed [114, 129]. A diffusion ellipsoid is a three-dimensional representation of the diffusion distance covered in space by molecules in a given diffusion time, t. Diffusion ellipsoids are easily constructed from the eigenvalues of the diffusion tensor (which correspond the D'_{xx}, D'_{yy}, and D'_{zz}) according to

$$\frac{x'^2}{2D'_{xx}t}+\frac{y'^2}{2D'_{yy}t}+\frac{z'^2}{2D'_{zz}t}=\left(\frac{x'}{\lambda_{x'}}\right)^2+\left(\frac{y'}{\lambda_{y'}}\right)^2+\left(\frac{z'}{\lambda_{z'}}\right)^2=1 \tag{3.41}$$

where λ corresponds to the 1D root mean square displacement along the principal diffusion axis (Eq. (3.24)). Figure 3.34C shows a diffusion ellipsoid image of a region in the human brain as indicated in Figure 3.34A and B. The degree of anisotropy is now embodied in the amount of eccentricity of the diffusion ellipsoid, whereas the bulk mobility of the diffusing water is related to the size of the ellipsoid. The preferred direction of diffusion is indicated by the orientation of the main axis of the diffusion ellipsoid. Therefore, in isotropic tissues, like gray matter and CSF, diffusion is similar in all directions and the diffusion ellipsoids are largely spherical. Since diffusion in CSF is faster than in gray matter, the diffusion spheroids in CSF are larger than in gray matter. In white matter, the diffusion ellipsoids are nonspherical with the preferred diffusion direction indicated by the most elongated ellipsoid dimension. Linking and tracing the preferred diffusion direction in adjacent ellipsoids (i.e. pixels) allows so-called white matter fiber tracking [130–135].

3.4.4 Restricted Diffusion

Since most membranes are semipermeable to water, diffusion anisotropy in diffusion tensor imaging (DTI) is largely caused by diffusion hindrance. Diffusion anisotropy has also been observed for metabolites in animal and human brain [121, 136–139] and skeletal muscle [115–118, 136]. However, unlike water, most metabolites are confined to the intracellular space on the time scale relevant for diffusion measurements. Therefore, in addition to diffusion hindrance due to intracellular organelles, metabolite diffusion is also affected by restricted diffusion caused by the boundaries of the cellular compartment.

Figure 3.35 shows the effect of restricted diffusion on the measured diffusion coefficient, D_{av}. In analogy to DTI, the orientational dependence of the measured diffusion coefficient can be eliminated by calculating the trace of the diffusion tensor, i.e. $D_{av} = (D_{xx} + D_{yy} + D_{zz})/3$. For unrestricted diffusion (Fig. 3.35A), the mean square displacement increases linearly with the diffusion time and as a result the measured diffusion coefficient is independent of the diffusion time. For restricted diffusion (Fig. 3.35B), however, the root mean square displacement increases linearly with diffusion time only up to the point that the mean square displacement becomes comparable with the size of the restricting compartment. For longer diffusion times, the mean square displacement levels off to a constant value (for a spherical restricting compartment) as the molecules bounce off the compartment walls (Fig. 3.35C). As a result, the measured ADC (Eq. (3.24)) decreases with increasing diffusion time (Fig. 3.35D). When performing metabolite diffusion measurements *in vivo* it is therefore crucial to state the diffusion time in order to allow comparison with other measurements. However, besides being an additional complication to *in vivo* diffusion measurements, the dependence of the ADC on the diffusion time can also be used to obtain information about the size of the restricting compartment. Following experiments on excised tissues, Moonen et al. [115] were the first to demonstrate restricted diffusion

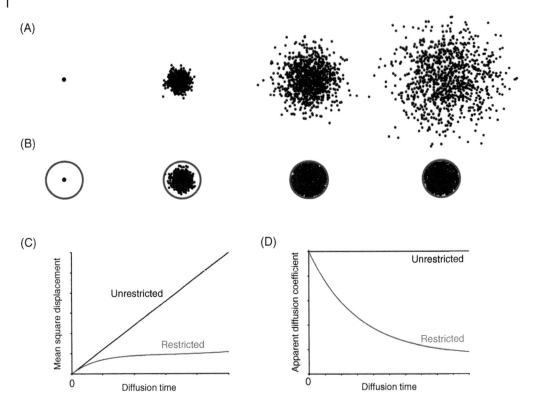

Figure 3.35 **Restricted diffusion.** (A) For free, unrestricted diffusion (i.e. in the absence of physical boundaries) the mean square displacement increases linearly with time (C), which according to Eq. (3.24) leads to a constant diffusion coefficient independent of the diffusion time (D). (B) In the presence of physical boundaries, the diffusion may appear free and unrestricted for short diffusion times. However, for longer diffusion times an increasing number of particles encounter the physical barrier, thereby leading to a decreased mean square displacement, which for a spherical compartment approaches the dimensions of the compartment. The restricted mean square displacement (C) leads to an increasingly smaller diffusion coefficient with increasing diffusion times (D).

of phosphocreatine in skeletal muscle. Quantitative studies on restricted metabolite diffusion were subsequently performed by van Gelderen et al. [117] and de Graaf et al. [118]. Figure 3.36 shows *in vivo* diffusion measurements of phosphocreatine and ATP on rat skeletal muscle. Clear diffusion anisotropy is visible when the magnetic field gradient orientation is changed relative to the muscle fibers. Diffusion-weighting along the muscle fibers (Figure 3.36A) leads to rapid signal decline for PCr and ATP with increasing *b*-value, indicating fast diffusion. Applying diffusion gradients perpendicular to the muscle fibers (Figure 3.36B) leads to greatly reduced diffusion due to restriction effects. Repeating the experiment at increasing diffusion times leads to a significant decline in the average (trace) diffusion coefficients (Figure 3.36D). Under the assumption of a cylindrical compartment, the curve in Figure 3.36D can be fitted with an analytical expression as given by Neuman [140]. Besides the size of the restricting compartment, the fitting results also indicated that the *in vivo* diffusion of PCr and ATP for zero diffusion time is close to that for free PCr and ATP in solution, indicating that *in vivo* diffusion is not fundamentally different from *in vitro* diffusion. *In vivo* diffusion is merely more complicated with (related) effects from diffusion hindrance, diffusion anisotropy, and restriction. More recently, diffusion-weighted ^{1}H MRS has been applied to study the morphology of brain cells, including neurons and astrocytes [141]. Additional information on the size and distribution of the restricting compartment can be obtained with double diffusion encoding methods [142].

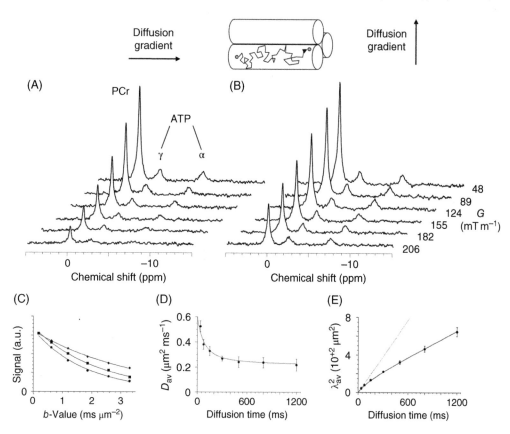

Figure 3.36 *In vivo* **diffusion-weighted** ^{31}P **NMR spectra of rat skeletal hind leg muscle.** The spectra were obtained with a 2.0 cm diameter surface coil and a diffusion-sensitized (selective) stimulated echo sequence (TR/TE/TM = 5000/20/60 ms) executed with adiabatic RF pulses (see Chapters 5 and 6). The *b*-factors are 93, 535, 1205, 1967, 2710, 3346, 4714, 5809, 6630 s mm^{-2} (δ = 9 ms and Δ = 73 ms) with increasing diffusion-weighting from top to bottom. Observed resonances have been assigned to phosphocreatine, and γ- and α-ATP. The β-ATP resonance is not visible due to the frequency-selective nature of the pulse sequence employed. All spectra were subject to the same phase correction. Diffusion-weighting along the direction of the main muscle fibers (A) leads to a significantly larger decrease in all signals as compared to when the diffusion-weighting is applied perpendicular to the main muscle fibers (B). (C) Integrated signal intensities for PCr as a function of the *b*-value (constant diffusion time = 75 ms). Diffusion anisotropy is evident from the curves for the different gradient directions. (D) Trace diffusion coefficient D_{av} for PCr as a function of the diffusion time. The solid line represents the best fit to the experimental data according to a theoretical model describing diffusion in infinitely long cylinders. (E) Trace mean square displacement λ_{av}^2 for PCr as a function of the diffusion time, as calculated with Eq. (3.24). The dotted lines indicate the mean square displacement for unrestricted diffusion. The difference between the dotted and solid lines is again indicative of diffusion restriction. *Source:* de Graaf et al. [118]. Reproduced with permission of Elsevier.

3.5 Dynamic NMR of Isotopically-Enriched Substrates

In Section 3.3, MT experiments were described in which endogenous compounds, like ATP and 2-oxoglutarate, were magnetically labeled by saturation or inversion. In the presence of a chemical exchange reaction or cross-relaxation, the magnetic label perturbs the reactions partner (e.g. PCr and glutamate) in a manner that is dependent on the exchange reaction rate, as well as the intrinsic T_1 relaxation parameters. While these experiments are completely noninvasive and provide valuable information on enzymes kinetics *in vivo*, they are limited to exchange reactions that are fast relative to the T_1 relaxation, i.e. $k_{AB} \gg 1/T_{1A,B}$. If this condition cannot be met, T_1 relaxation has already destroyed the magnetic label before it can reach the exchange partner. This therefore excludes the measurement of important, but relatively slow, metabolic pathways like glycolysis or the tricarboxylic acid (TCA) cycle.

The infusion of an exogenous compound that achieves a permanent magnetic perturbation offers an alternative to the previously described methods. Since the perturbation is permanent (with respect to T_1 relaxation), it allows the measurements of much slower reaction rates and metabolic fluxes. The primary example of these types of experiments is given by the intravenous infusion of ^{13}C-enriched substrates, like [1-^{13}C]-glucose or [2-^{13}C]-acetate, inhalation of ^{17}O-enriched oxygen gas or administration of ^{2}H or ^{15}N-enriched substrates. Since all dynamic NMR studies with isotopically enriched compounds follow similar rules and concepts, the most commonly used ^{13}C nucleus will be used as an example. Carbon-12 is the common isotope and is present at a natural abundance of 98.89%. However, since carbon-12 has no net nuclear spin it cannot be detected by NMR. Carbon-13 is the second-most abundant, stable, and nonradioactive isotope of carbon with a natural abundance of 1.11%. It has a nuclear spin ½ and is therefore NMR-visible (see Table 1.1). When 99$^+$% ^{13}C-enriched [1-^{13}C]-glucose is intravenously administered to the subject, human or animal, the [1-^{13}C]-glucose is taken up by the brain as if it was regular glucose. This is a fundamental advantage of NMR over similar experiments done with 2-deoxy-glucose and 2-fluoro-deoxy-gluxose in autoradiography and positron emission tomography (PET) [143, 144]. The addition of a single neutron to carbon-12 has a negligible effect on the biochemical properties (i.e. membrane transport, metabolic rates) of carbon-13. However, the modifications required for PET necessitates the use of correction terms to take differences between glucose and deoxy-glucose membrane transport into account. Another advantage of NMR is that while deoxy-glucose is metabolically inert, ^{13}C-glucose enters the normal metabolic pathways like glycolysis and the TCA cycle. The ^{13}C-label of [1-^{13}C]-glucose is then transferred to molecules further downstream, like [3-^{13}C]-pyruvate and lactate in glycolysis and [4-^{13}C]-glutamate in the TCA cycle. The next section will detail the flow of ^{13}C-label from glucose to metabolites, but for now it should be clear that following the flow of ^{13}C-label from a ^{13}C-labeled substrate to other "downstream" metabolic products allows the study of metabolic pathways *in vivo*.

In general, the detection of ^{13}C-label turnover involves a number of considerations and decisions that determine the exact information content of the experiment. These considerations/decisions can be divided into three groups.

1) **Choice of ^{13}C-labeled substrate.** While most studies are performed with [1-^{13}C]-glucose, several other substrates are available that provide complimentary information. In particular, [2-^{13}C]-acetate, [2-^{13}C]-glucose, and [1,6-^{13}C$_2$]-glucose/[U-^{13}C$_6$]-glucose are alternative substrates.

2) **Direct ^{13}C versus indirect ^{1}H-[^{13}C] detection.** ^{13}C-enriched compounds can be detected directly with ^{13}C NMR, or alternatively the protons attached to the ^{13}C nuclei can be detected with ^{1}H-[^{13}C]-NMR. While ^{1}H-[^{13}C]-NMR provides a higher sensitivity, direct ^{13}C NMR typically provides a higher information content, including the detection of isotopomer resonances.

3) **Choice of metabolic model.** The obtained ^{13}C turnover data must be fitted with a mathematical model in order to obtain information about the biochemical pathways. The complexity of the metabolic model (e.g. one, two, or three compartments) as well as the underlying assumptions (e.g. certain metabolic pools and fluxes are unknown and must be assumed) can have significant effects on the final metabolic flux values. The considerations involved with metabolic modeling are discussed in Section 3.5.2.

3.5.1   General Principles and Setup

Figure 3.37 shows the general setup of a ^{13}C NMR turnover study for animal and human subjects. Following preparations commonly employed in animal studies (general anesthesia and tracheotomy or intubation) two small lines are placed in the femoral artery and vein or tail vein for periodic sampling of arterial blood and intravenous infusion of substrates, respectively. In human studies, the two lines are typically placed in the antecubital vein in each arm. Signal is acquired with a ^{1}H-[^{13}C] RF coil for indirect detection (Figure 3.37A) or with a ^{13}C-[^{1}H] RF coil (Figure 3.37B) for direct ^{13}C detection. More information on ^{1}H-[^{13}C], ^{13}C-[^{1}H], and other RF coils can be found in Chapter 10.

The infusion protocol is typically designed to quickly reach a high and stabile isotopic enrichment of the substrate. Figure 3.37C and D shows a practical glucose infusion protocol for rat and human studies, respectively. The infusion protocols are designed to increase the intravenous glucose concentration from euglycemic levels (circa 4–5 mM following an overnight fast) to hyperglycemic levels (10–12 mM), while reaching a 50–70% fractional enrichment (FE) of the glucose within 1 min. However, exact knowledge of the time course of substrate concentration and isotopic enrichment is critical for a quantitative evaluation of ^{13}C labeling time courses. Without exact knowledge of the input function, the rate of ^{13}C-label appearance in metabolites like glutamate and glutamine cannot be interpreted since the entry rate of the ^{13}C-labeled substrate is unknown. The substrate concentration and isotopic enrichment time courses can be obtained through periodic sampling of arterial blood followed by quantitative analysis with a combination of methods, like glucosometer analysis, gas chromatography mass spectrometry (GC-MS), and high-resolution NMR spectroscopy (Figure 3.37E). Figure 3.37F shows typical blood glucose concentration and ^{13}C FE time courses. As an alternative to intravenous infusions, animal studies can be performed with intraperitoneal infusions, whereas in humans oral administration of glucose has been used [145].

In concert with the onset of infusion, NMR data acquisition (either direct or indirect) is started. The NMR spectra are typically acquired in blocks of 2–10 minutes in order to achieve sufficient signal-to-noise ratio per spectrum, while still allowing the detection of dynamic changes in the metabolite ^{13}C fraction. Figure 3.38A shows a typical ^{13}C NMR turnover dataset in rat brain following the intravenous infusion of [1,6-^{13}C$_2$]-glucose. Prior to infusion the ^{13}C NMR spectrum is, besides the 1.11% natural abundance, devoid of resonances. Immediately following the onset of infusion, the glucose resonances rapidly increase and remain relatively constant throughout the time course. Over time, the ^{13}C label can be observed in [4-^{13}C]-glutamate, [4-^{13}C]-glutamine, and several other metabolites. When the ^{13}C label has completed one revolution of the TCA cycle it appears in the C2 and C3 positions of the mentioned metabolites. The principles of ^{13}C and ^{1}H-[^{13}C] NMR methods are discussed in Chapter 8.

Following data acquisition, the NMR resonances must be converted to absolute concentrations and FEs (FE = ^{13}C/(^{12}C+^{13}C)). For ^{1}H-[^{13}C] NMR the calculation of FEs is typically more straightforward than for ^{13}C NMR, as ^{1}H-[^{13}C] NMR methods inherently detect the ^{13}C fraction, as well as the total pool of a given metabolite. For ^{13}C NMR the concentrations can be made relative to the 1.11% natural abundance signals or an assumed metabolite pool. Details on metabolite quantification are discussed in Chapter 9. Following this step, absolute metabolite turnover curves are obtained, as shown for [4-^{13}C]-glutamate and [4-^{13}C]-glutamine in Figure 3.38B, which will serve as input for the metabolic model.

3.5.2   Metabolic Modeling

Metabolic modeling is a set of procedures which express a collection of metabolic pathways as mathematical differential equations in order to obtain absolute fluxes through these pathways.

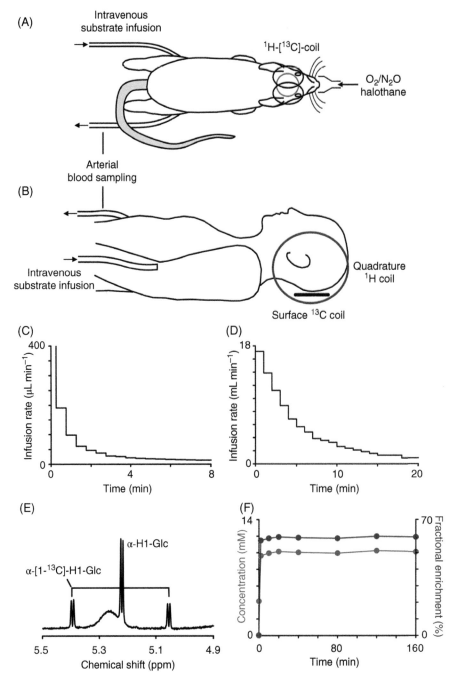

Figure 3.37 **Experimental setup for isotopic labeling studies.** (A) Indirect ^{1}H-[^{13}C]-NMR detection setup for rat brain and (B) direct ^{13}C-[^{1}H]-NMR detection setup for human brain. The indirect and direct detection schemes are readily changed by use of the proper RF coil setup. The ^{13}C-labeled substrate on rats is typically administered via intravenous infusion according to a protocol shown for ^{13}C-labeled glucose (0.75 M solution, 200 g animal) in (C) (initial infusion rate 1000 μl min^{-1} dropping to 14 μl min^{-1} after 8 min). Other ^{13}C-labeled substrates, such as acetate, require a modified infusion protocol. (D) Intravenous infusion protocol for ^{13}C-glucose on human subjects, aimed to rapidly raise the blood glucose level to ~10 mM and a >70% ^{13}C fractional enrichment. Oral administration of ^{13}C-glucose dissolved in water provides a completely noninvasive alternative to intravenous infusion. (E) ^{1}H NMR spectrum of blood plasma obtained at the end of a [1-^{13}C]-glucose infusion. The downfield α-H1-glucose resonance is readily detected at 5.22 ppm. In the absence of ^{13}C decoupling, α-[1-^{13}C]-H1-glucose appears as a doublet signal split by the ^{1}H—^{13}C heteronuclear scalar coupling. The ^{1}H NMR spectrum allows direct quantification of the glucose concentration and the ^{13}C fractional enrichment, as needed for quantitative metabolic modeling. (F) Measured blood glucose concentration (blue line) and ^{13}C fractional enrichment (red line) in an adult rat (200 g) during the infusion protocol shown in (C).

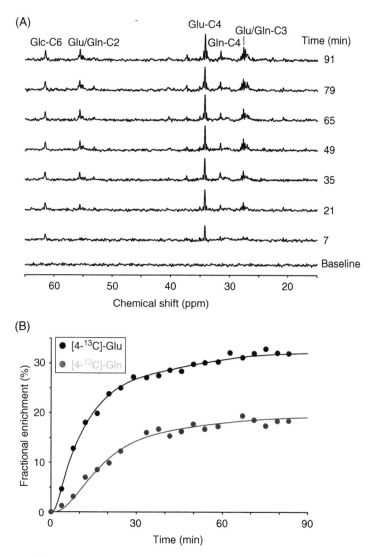

Figure 3.38 **Dynamic ^{13}C MRS *in vivo*.** (A) Time-resolved ^{1}H-decoupled, ^{13}C NMR spectra from rat brain *in vivo* following the intravenous infusion of [1,6-^{13}C$_2$]-glucose. Spectra are acquired with an adiabatic INEPT sequence (TR = 4000 ms, 200 μl, see also Chapter 8) at 7.05 T. (B) Turnover curves of [4-^{13}C]-glutamate and [4-^{13}C]-glutamine. Dots represent measured fractional enrichments as obtained through spectral fitting of the data shown in (A), whereas the solid line represents the best mathematical fit to a three-compartment (blood, glutamatergic neurons, and astroglia) metabolic model (see also Figure 3.40).

Some of the principles of metabolic modeling will first be demonstrated for a simple example, after which the complete metabolic model for cerebral energy metabolism and neurotransmitter cycling will be summarized.

Consider the linear three-step chemical reaction of a substrate S being converted to a product P, which is subsequently further metabolized to other products F.

$$S \xrightarrow{k_{S \to P}} P \xrightarrow{k_{P \to F}} F \qquad\qquad (3.42)$$

The reaction rate $k_{S \to P}$ (in s^{-1}) describes how fast an S molecule is converted to a P molecule and is determined by the enzyme characteristics, as well as thermodynamics. In this example it is assumed that the reaction is unidirectional, i.e. the reverse reaction ($P \to S$) does not occur. Reaction rates are very difficult to measure directly *in vivo* and do not give direct information about metabolism. Metabolic fluxes, V (in $mM \cdot s^{-1}$ or $mmol \cdot kg^{-1} \cdot s^{-1}$), are more relevant in the context of *in vivo* metabolic pathways and are defined as the amount of S that is converted to P per unit time, i.e.

$$V_{S \to P} = k_{S \to P} \cdot [S] \tag{3.43}$$

where $[S]$ is the concentration of substrate S (in mM or $mmol \cdot kg^{-1}$). The amount of product P changes over time according to the amount of S being converted to P minus the amount of P being converted to F. In other words, the size of the metabolic pool P changes according to the amount of S flowing to P (i.e. the "inflow") minus the flow of P to F (i.e. the "outflow"). Mathematically this can be expressed as a differential equation according to

$$\frac{d[P(t)]}{dt} = V_{S \to P} - V_{P \to F} = V_{in} - V_{out} \tag{3.44}$$

This type of equation is also referred to as a mass-balance equation since it balances ("tracks") all mass of compound P over time and over all possible reactions involving P. In many situations the metabolic pool (concentration) of compound P does not change with time, making $dP(t)/dt = 0$, which is referred to as a steady-state condition. It is now immediately clear that measuring the steady-state concentration of compound P does not give any insight into the formation and degradation (i.e. the "turnover") of compound P. The metabolic fluxes V_{in} and V_{out} can change dramatically, but as long as $V_{in} = V_{out}$ the concentration of compound P will remain constant.

The dynamic behavior of P can be studied in detail by introducing a marker in substrate S which can be followed over time when it flows to product P. Several techniques exist that introduce radioactive tracers to monitor metabolic pathways, including autoradiography, PET, and single photon emission computed tomography (SPECT). NMR spectroscopy is a unique technique to study metabolic pathways noninvasively *in vivo* since it utilizes nonradioactive substrates that are metabolized in the exact same manner as regular, non-labeled substrates. Carbon-13 is most frequently used as a NMR label for a number of reasons. Firstly, whereas carbon-12 does not have a nuclear magnetic moment, carbon-13 is a spin-1/2 nucleus and can therefore be detected by NMR spectroscopy. Secondly, most metabolically active compounds have a carbon backbone, making the label generally applicable. Thirdly, replacing the common ^{12}C isotope (98.89% natural abundance) with the ^{13}C isotope has negligible effects on metabolism.

Now reconsider the three-step chemical reaction of Eq. (3.42). Introduction of ^{13}C-labeled substrate S^* will lead to the formation of ^{13}C-labeled product P^*, which can be detected with ^{13}C or ^1H-[^{13}C]-NMR spectroscopy. Mathematically, the formation of ^{13}C-labeled product P^* can be described by

$$\frac{d[P^*(t)]}{dt} = \frac{[S^*(t)]}{[S]} V_{in} - \frac{[P^*(t)]}{[P]} V_{out} \tag{3.45}$$

This type of equation is referred to as an isotope-balance equation since it balances ("tracks") the isotope of compound P over time and over all possible reactions involving P. Together with the mass-balance equation (3.44) it gives a complete description of the total P and ^{13}C-labeled P^* metabolic pools. The ratio $[P^*]/[P]$ is typically referred to as the FE ($0 \leq \mathrm{FE} \leq 1$) of compound P.

When the FE of substrate S can be described by a (time-independent) step function (i.e. $[S^*(t)] = [S^*(0)]$), the time-dependence of $[P^*]$ under steady-state conditions (i.e. $V_{in} = V_{out}$) can be analytically determined as

$$\left[P^*(t) \right] = \left[P \right] \left(\frac{\left[S^* \right]}{\left[S \right]} \right) \left(1 - e^{-V_{in} t / [P]} \right) \tag{3.46}$$

It follows that $[P^*(t)]$ builds up as an exponential function to a final FE equal to ($[S^*]/[S]$) with a time-constant ($V_{in}/[P]$). Large metabolic pools will therefore turn over slower than small metabolic pools. When the substrate inflow cannot be described by a step function, Eq. (3.46) is more difficult to solve analytically and often a numerical approach is employed instead. This is mandatory for the description of metabolic pathways *in vivo*, where multiple pathways can flow into a single pool. Figure 3.39 shows the metabolic fate of ^{13}C isotopic label administered via ^{13}C-enriched glucose, acetate, or pyruvate. The glucose molecule is split into two pyruvate molecules in the glycolytic pathway, such that the glucose C1/C6, C2/C5, and C3/C4 positions end up in pyruvate C3, C2, and C1, respectively. Pyruvate is in rapid exchange with lactate and alanine. Acetate bypasses glycolysis and enters metabolism at the level of acetylCoA. Pyruvate enters the TCA cycle via acetylCoA, whereby the C1 position is cleaved off as carbon dioxide. The acetylCoA carbons merge with oxaloacetate in the TCA cycle to form citrate. The two ^{13}C labels present in acetylCoA end up in the C4 and C5 positions of glutamate and glutamine or continue down the TCA cycle whereby another carbon position is lost as carbon dioxide. The ^{13}C label that is left in oxaloacetate at the end of the first turn of the TCA cycle can merge with ^{13}C-labeled acetylCoA to form citrate that is ^{13}C labeled in multiple positions. These so-called isotopomers provide powerful, additional information on the ^{13}C label flow and can be observed and studied based on the ^{13}C–^{13}C homonuclear scalar coupling (see Table 2.6). Knowledge of the ^{13}C label flow as shown in Figure 3.39 is important for the metabolic modeling of *in vivo* metabolism. However, *in vivo* metabolism is typically more complicated due to the presence of distinct metabolic compartments. Figure 3.40 shows a four-compartment model (blood, astroglial, and two types of neurons) of cerebral metabolism that has been developed, updated, and refined over the last two decades [146, 147]. The metabolic model of Figure 3.40 together with the ^{13}C label flow of Figure 3.39 are sufficient to provide a quantitative, mathematical description of metabolism. For example, the mass-balance equation for the neuronal glutamate [Glu$_N$] and GABA [GABA$_N$] pools are given by

$$\frac{d[\mathrm{Glu_N}]}{dt} = V_{\mathrm{cycle,Glu/Gln}} + V_{x,\mathrm{2OG/Glu}} - \left(V_{\mathrm{cycle,Glu/Gln}} + V_{x,\mathrm{Glu/2OG}} \right) = 0 \tag{3.47}$$

$$\frac{d[\mathrm{GABA_N}]}{dt} = V_{\mathrm{GAD}} - \left(V_{\mathrm{shunt}} + V_{\mathrm{cycle,GABA/Gln}} \right) = 0 \tag{3.48}$$

Similar equations can be constructed for all other metabolic pools. For the isotope-balance equations one also has to consider the position of the carbon atom within a molecule, since different carbon positions will label at different rates depending on the substrate and metabolic pathways. For the C4 position of neuronal glutamate [Glu$_N$4] and the C2 position of neuronal GABA [GABA$_N$2], the isotope-balance equations are given by

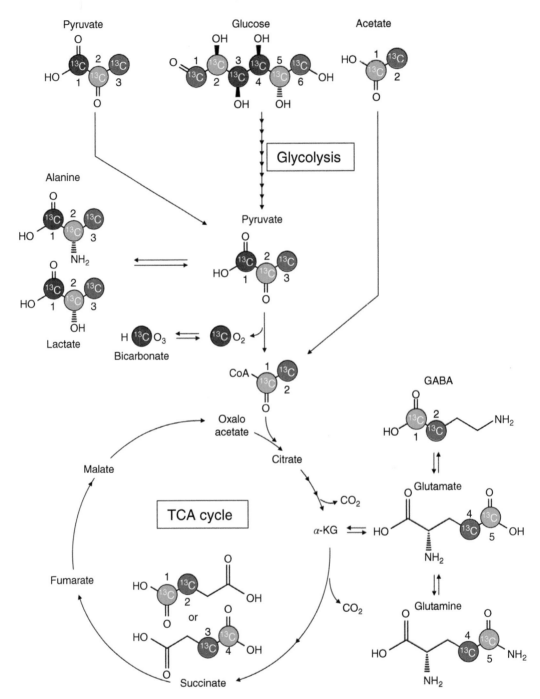

Figure 3.39 **^{13}C label flow through glycolysis and the tricarboxylic acid (TCA) cycle.** The intracellular pyruvate pool can receive ^{13}C label directly from ^{13}C-labeled pyruvate or indirectly via glycolysis from ^{13}C-labeled glucose. The pyruvate pool can be bypassed using ^{13}C-labeled acetate of which the ^{13}C label enters the indicated pathways at the level of acetyl CoA. Various carbon positions (red, green, or blue) have different metabolic fates, with the C1-pyruvate and C3/C4-glucose carbons ending up in bicarbonate during the pyruvate dehydrogenase (PDH) step. The C2-pyruvate and C2/C5-glucose carbons end up in the C2 position of lactate and alanine and after entering the TCA cycle can label glutamate and glutamine C5 and GABA C1. The C3-pyruvate and C1/C6-glucose carbons ultimately label glutamate and glutamine C4 and GABA C2 and are retained in the TCA cycle, where they can combine with ^{13}C-labeled acetyl-CoA to form compounds with ^{13}C labeling in multiple positions, i.e. isotopomers.

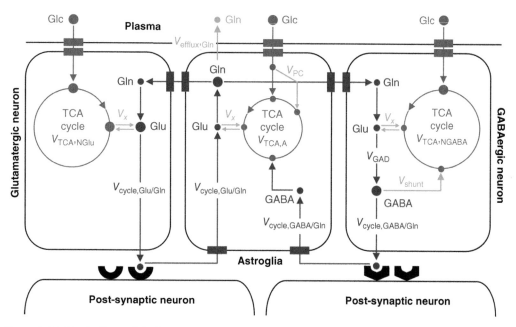

Figure 3.40 Metabolic model of cerebral metabolism. Four-compartment metabolic model comprising the blood, glutamatergic neurons, GABAergic neurons, and astroglia. Glucose enters the brain with the aid of glucose transporters in the blood–brain barrier. In the glycolytic pathway, glucose is broken down into two pyruvate molecules which enter the tricarboxylic acid (TCA) cycle. One of the TCA cycle intermediates, 2-oxoglutarate, is in rapid exchange with a large glutamate pool that can be observed with NMR. The TCA cycle flux can be obtained from the glutamate turnover and is denoted V_{TCA}. Due to compartmental localization of specific enzymes, the fate of glutamate differs in each of the three cellular compartments. In glutamatergic neurons, glutamate acts as an excitatory neurotransmitter and is released into the synaptic cleft in response to an action potential. Following interaction with post-synaptic receptors, the glutamate is taken up by the astroglia and converted to glutamine. Glutamine is ultimately transported back to the glutamatergic neuron, where it is converted to glutamate, thereby completing the so-called glutamate–glutamine neurotransmitter cycle. The flux through this cycle, $V_{cycle,Glu/Gln}$, can be obtained by following the glutamine turnover. In the GABAergic neuron, the glutamate is first converted to GABA, which is the primary inhibitory neurotransmitter. Similar to the glutamatergic neuron, a GABA–glutamine neurotransmitter exists between GABAergic neurons and astroglia. A metabolic pathway specific to astroglia is the carboxylation of pyruvate catalyzed by the astroglia-specific enzyme pyruvate carboxylase.

$$\frac{d\left[Glu_N 4^*\right]}{dt} = \frac{\left[Gln_N 4^*\right]}{\left[Gln_N\right]} V_{cycle,Glu/Gln} + \frac{\left[2OG_N 4^*\right]}{\left[2OG_N\right]} V_{x,2OG/Glu}$$

$$-\frac{\left[Glu_N 4^*\right]}{\left[Glu_N\right]}\left(V_{cycle,Glu/Gln} + V_{x,Glu/2OG}\right) \tag{3.49}$$

$$\frac{d\left[GABA_N 2^*\right]}{dt} = \frac{\left[Glu_N 4^*\right]}{\left[Glu_N\right]} V_{GAD} - \frac{\left[GABA_N 2^*\right]}{\left[GABA_N\right]}\left(V_{shunt} + V_{cycle,GABA/Gln}\right) \tag{3.50}$$

One noticeable exception to the relatively simple, linear metabolic fluxes is the transport of glucose across the blood–brain barrier. Glucose is transported from the blood into the brain by GLUT-1 transporters and can be described by a Michaelis–Menten kinetics [148–151]:

$$\frac{d\left[\text{Glc}_b\left(t\right)\right]}{dt} = \frac{V_{\text{max}}\left[\text{Glc}_p\left(t\right)\right]}{K_m + \left[\text{Glc}_b\left(t\right)\right]/V_d + \left[\text{Glc}_p\left(t\right)\right]} - \frac{V_{\text{max}}\left[\text{Glc}_b\left(t\right)\right]}{V_d(K_m + \left[\text{Glc}_p\left(t\right)\right]) + \left[\text{Glc}_b\left(t\right)\right]} - \text{CMR}_{\text{Glc}}$$

(3.51)

where $[\text{Glc}_p]$ and $[\text{Glc}_b]$ represent the time-dependent glucose concentration in plasma and brain, respectively. K_m is the Michaelis–Menten constant for GLUT-1 transporters, V_{max} the maximum attainable flux and V_d the brain glucose volume fraction. It should be noted that at higher magnetic fields, brain glucose levels and FEs can be measured directly in ${}^1\text{H}$ NMR spectra, such that any assumptions concerning glucose transport can be avoided (see Figures 3.38 and 8.24).

Mass and isotope balance equations similar to Eqs. (3.47)–(3.50) can be constructed for all metabolic pools, thereby making up the metabolic model. The model describing the metabolic pathways shown in Figure 3.40 is referred to as a four-compartment model, as it is composed of a blood compartment, and compartments for the glutamatergic neuron, GABAergic neuron, and astroglia. Using the blood or brain glucose levels and FEs, the metabolic pool sizes and the measured FE curves allows the calculation, through nonlinear least-squares optimization, of a number of important fluxes like the neuronal and astroglial TCA cycles and the glutamate–glutamine and GABA–glutamine neurotransmitter cycles. For more details on metabolic models and turnover data analysis the reader is referred to the literature [146, 147, 152].

Studies using different isotopes, such as ${}^2\text{H}$, ${}^{15}\text{N}$, or ${}^{17}\text{O}$, can be described by similar metabolic models as outlined for ${}^{13}\text{C}$ in Figures 3.39 and 3.40. However, each isotope in each position of a molecule has its own specific considerations. For example, ${}^2\text{H}$-enriched glucose will label the lactate and glutamate pools, similar to the labeling observed for ${}^{13}\text{C}$ MR studies. However, the ${}^2\text{H}$ label will be exchanged for ${}^1\text{H}$ nuclei in the later stages of the TCA cycle, such that ${}^2\text{H}$ labeling on the C3 position in glutamate and glutamine or the formation of ${}^2\text{H}–{}^2\text{H}$ isotopomers is not observed. ${}^{17}\text{O}$ MRS studies during and following the inhalation of ${}^{17}\text{O}$-enriched oxygen gas have to take into account that the ${}^{17}\text{O}$-labeled product, namely water, is not limited to the intracellular compartment. As a result, ${}^{17}\text{O}$-labeled water from other organs will enter the volume-of-interest (e.g. brain) and complicate the quantitative measurement of the cerebral metabolic rate of oxygen consumption (CMR_{O2}). As a result, dynamic ${}^{17}\text{O}$-based measurements of CMR_{O2} tend to use short periods (1–2 min) of ${}^{17}\text{O}_2$ inhalation in order to minimize the contribution of global, nonspecific water. However, the metabolic model still needs to be extended with terms for cerebral blood flow (CBF) to account for the non-negligible amount of ${}^{17}\text{O}$-labeled water coming from other organs.

3.5.3   Thermally Polarized Dynamic ${}^{13}\text{C}$ NMR Spectroscopy

3.5.3.1   [1-${}^{13}\text{C}$]-Glucose and [1,6-${}^{13}\text{C}_2$]-Glucose

The majority of ${}^{13}\text{C}$ NMR studies *in vivo* have used [1-${}^{13}\text{C}$]-glucose as the substrate. Besides the economic consideration that [1-${}^{13}\text{C}$]-glucose is the least expensive ${}^{13}\text{C}$-labeled form of glucose, the choice for [1-${}^{13}\text{C}$]-glucose is governed by the fact that the transfer of ${}^{13}\text{C}$ label to [4-${}^{13}\text{C}$]-glutamate is indicative of the most active energy-producing metabolic pathways, namely glycolysis and the neuronal TCA cycle (see Figure 3.39). While the labeling pattern of metabolites from [1,6-${}^{13}\text{C}_2$]-glucose is nearly identical to that from [1-${}^{13}\text{C}$]-glucose, the FE of pyruvate and hence of all subsequent metabolic pools will be twice as high, leading to an improved detection sensitivity. [1,6-${}^{13}\text{C}_2$]-glucose is more expensive than [1-${}^{13}\text{C}$]-glucose, but this is only a minor consideration when considering the small amounts of material used to study cerebral metabolism in mice or rats. For human experiments, [U-${}^{13}\text{C}_6$]-glucose is a less expensive alternative when performing ${}^1\text{H}$-[${}^{13}\text{C}$]-NMR. For direct ${}^{13}\text{C}$ NMR detection, [U-${}^{13}\text{C}_6$]-glucose is not recommended as the NMR spectrum will be complicated by the multiple homonuclear ${}^{13}\text{C}–{}^{13}\text{C}$

scalar couplings, giving rise to isotopomer resonances. The label transfer from $[1,6\text{-}^{13}C_2]$-glucose to $[4\text{-}^{13}C]$-glutamate is detectable within minutes following the start of glucose infusion. Provided that the exchange between the mitochondrial 2-oxoglutarate and cytosolic glutamate pools is rapid, the label transfer is indicative of the TCA cycle rate. Note that through simultaneous measurement of $[3\text{-}^{13}C]$-glutamate the TCA cycle rate can be determined even if the mitochondrial/cytosolic exchange rate is slow. Figure 3.38 further shows that, with a small time lag, label is transferred to $[4\text{-}^{13}C]$-glutamine. This has been interpreted as reflecting a glutamate–glutamine neurotransmitter cycle [153, 154]. Following 90 minutes of glucose infusion, a wide range of metabolites is labeled, including $[n\text{-}^{13}C]$-glutamate and glutamine (n = 2, 3, or 4), $[2\text{-}^{13}C]$-aspartate, $[3\text{-}^{13}C]$-aspartate, $[1,6\text{-}^{13}C_2]$-glucose, and $[n\text{-}^{13}C]$-GABA [n = 2, 3, or 4]. Besides the singlet ^{13}C NMR resonances, the NMR spectrum also contains several doublet and triplet resonances (Figure 3.41A). These so-called isotopomer resonances are due to homonuclear $^{13}C-^{13}C$ scalar coupling in molecules where two ^{13}C nuclei are immediately adjacent to each other, like in $[3,4\text{-}^{13}C_2]$-glutamate. This situation occurs when $[4\text{-}^{13}C]$-glutamate labeled in the first TCA cycle is labeled again at the C3 position in the second TCA cycle. The chance that this occurs increases with the duration of the infusion. The infusion for regular ^{13}C turnover studies is between 90 and 150 minutes, leading to several isotopomers, in particular $[2,3\text{-}^{13}C_2]$, $[3,4\text{-}^{13}C_2]$, and $[2,3,4\text{-}^{13}C_3]$-glutamate and glutamine (Figures 3.38 and 3.41A). However, for much longer infusion times, as required to detect ^{13}C-glycogen turnover [155–158], the ^{13}C NMR spectrum is dominated by isotopomer resonances [159]. Analysis of the exact isotopomer patterns can provide complementary information to the dynamic ^{13}C turnover curves as shown in Figure 3.29. Figure 3.42 summarizes the values of metabolic rates that have been measured in resting human [160–166] and anesthetized rat brain [146, 153, 167–173].

3.5.3.2 $[2\text{-}^{13}C]$-Glucose

In order to determine the rate of the glutamate–glutamine cycle from the glutamine synthesis rate, one must distinguish glutamine labeling via the glutamate–glutamine cycle from other metabolic pathways that may contribute to the flow through glutamine synthetase. The glutamate–glutamine cycle and anaplerosis are the only two pathways that provide carbon skeletons for glutamine synthesis. Glutamine efflux is the primary pathway of nitrogen removal from the brain. At steady state the concentration of glutamine remains constant. Therefore, any loss of glutamine by efflux to the blood must be compensated by synthesis *de novo* of glutamine through anaplerosis. For synthesis of glutamine *de novo* by anaplerosis, pyruvate derived from glucose is converted to oxaloacetate by carbon dioxide fixation (see Figure 3.40), a reaction catalyzed by the astroglial enzyme pyruvate carboxylase. Through the action of the TCA cycle, oxaloacetate is converted to 2-oxoglutarate, which is converted to glutamate. Astroglial glutamate is then converted to glutamine by glutamine synthetase. A limitation of using $[1\text{-}^{13}C]$-glucose as a labeled precursor to measure the glutamate–glutamine cycle flux is that ^{13}C label entering glutamine by this pathway cannot be easily distinguished from ^{13}C label that enters glutamine from the anaplerotic pathway or the astroglial TCA cycle. $[1\text{-}^{13}C]$-glucose labels $[4\text{-}^{13}C]$-glutamate through the action of pyruvate dehydrogenase and labels $[2\text{-}^{13}C]$-glutamate through pyruvate carboxylase. However, the carbon label of $[4\text{-}^{13}C]$-glutamate is quickly transferred to $[2\text{-}^{13}C]$-glutamate in subsequent turns of the overwhelming neuronal TCA cycle, thereby obscuring $[2\text{-}^{13}C]$-glutamate labeling by astroglial anaplerosis. Using $[2\text{-}^{13}C]$-glucose as a label precursor offers a highly sensitive alternative to measuring the fluxes though anaplerosis and the astroglial TCA cycle. $[2\text{-}^{13}C]$-glucose labels astroglial $[3\text{-}^{13}C]$-glutamate/glutamine through pyruvate carboxylase and labels $[5\text{-}^{13}C]$-glutamate/glutamine through the action of pyruvate dehydrogenase. However, the C5 position of TCA cycle

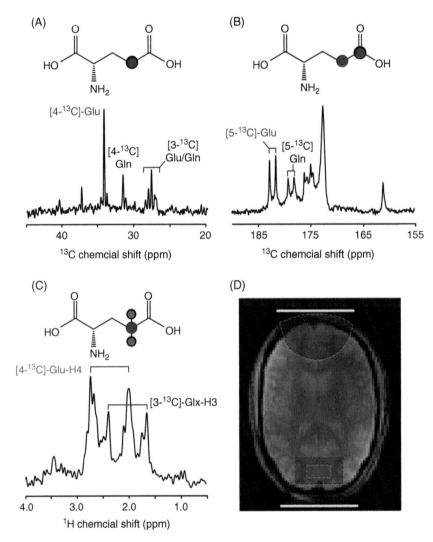

Figure 3.41 **Direct and indirect ^{13}C MRS on human brain *in vivo*.** (A) Polarization transfer-enhanced direct ^{13}C MRS of protonated carbon positions following administration of [1-^{13}C]-glucose. The detected [4-^{13}C]-glutamate resonance and carbon position are indicated in red. (B) Nuclear Overhauser enhanced direct ^{13}C MRS of non-protonated carbon positions following administration of [U-[13]C₆]-glucose. The detected C5 resonance and position in [4,5-[13]C₂]-glucose is indicated in red. (C) Indirect proton-observed, carbon-edited (POCE) MRS without broadband decoupling following the infusion of [U-[13]C₆]-glucose. The detected H4 proton resonance and position within [4-^{13}C]-glutamate are indicated in blue. (D) MRI showing the approximate detection volumes for (A–C). ^{13}C MRS in (A) was acquired from the lower red box in the occipital cortex, whereas ^{13}C MRS in (B) was acquired from the unlocalized, upper red area covering the frontal cortex. The indirect [1]H-[^{13}C]-MR spectra were acquired from the smaller blue box. The yellow lines indicate the approximate sizes and positions of the ^{1}H or ^{13}C detection coil.

intermediates does not lead to labeling in the C3 position of glutamine. Therefore, the accumulation of (astroglial) [3-^{13}C]-glutamine is a direct indication of anaplerotic activity, without contamination by neuronal TCA cycle metabolism. Under most conditions the anaplerotic pathway constitutes only a small fraction of the TCA cycle flux (Figure 3.42).

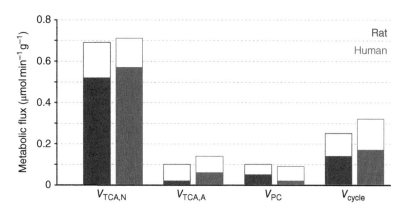

Figure 3.42 **Metabolic rates in human and rat brains**. Metabolic rates for human (blue) and rat (red) brains, representing a range of values found in the literature. Abbreviations: $V_{TCA,N}$, neuronal TCA cycle flux; $V_{TCA,A}$, astroglial TCA cycle flux; V_{PC}, pyruvate carboxylase flux; V_{cycle}, glutamate–glutamine neurotransmitter cycle.

3.5.3.3 [U-^{13}C$_6$]-Glucose

As mentioned earlier, using [U-^{13}C$_6$]-glucose for direct ^{13}C MR detection is not recommended due to the formation of extensive isotopomers that reduce the SNR and increase the spectral complexity. However, [U-^{13}C$_6$]-glucose (or [2-^{13}C]-glucose) can be used to detect the flow through metabolic pathways via the non-protonated carbon positions between 170 and 185 ppm [16, 174–176]. Since the non-protonated carbons only exhibit heteronuclear scalar coupling over two or three chemical bonds, the resonance splitting is very limited. As a result, low power decoupling can be used, which in turn leads to low RF power deposition. An additional advantage of non-protonated carbon detection is the absence of strong lipid resonances. Whereas [2-^{13}C]-glucose would provide singlet resonances for C1 and C5 glutamate and glutamine, [U-^{13}C$_6$]-glucose would lead to doublet formation due to the ^{13}C–^{13}C isotopomers. Figure 3.41B shows a typical ^{13}C MR spectrum of the non-protonated carbon positions in human frontal cortex, circa 60 min following the intravenous infusion of [U-^{13}C$_6$]-glucose. The main disadvantage of detecting non-protonated carbons are their long T_1 relaxation times of 10 s or more.

Indirect ^{1}H–[^{13}C] MR detection [162, 177–179] of [U-^{13}C$_6$]-glucose metabolism (Figure 3.41C) provides a higher-sensitivity alternative to direct ^{13}C MR detection without the complication of isotopomers. The higher sensitivity of proton detection is offset against a small spectral dispersion, leading to increased overlap of resonances. Chapter 8 discusses the relative merits of direct ^{13}C and indirect ^{1}H–[^{13}C] MR methods.

3.5.3.4 [2-^{13}C]-Acetate

Since glucose is transported and metabolized in both neurons and astroglia and extensive mixing of the label occurs through the glutamate–glutamine cycle, the TCA cycle fluxes in the neuronal and astroglial compartments are convolved. However, the formation of [4-^{13}C]-glutamate will be heavily weighted by the neuronal TCA cycle, making the glucose experiment relatively insensitive to the smaller astroglial TCA cycle flux. Neuronal and astroglial metabolism can be more directly distinguished by using acetate as a substrate. It has been shown that acetate is almost exclusively metabolized in astroglia due to a far greater transport capacity [180]. Therefore, [2-^{13}C]-acetate will be selectively transported into the astroglial compartment and converted to [2-^{13}C]-acetylCoA, which labels the astroglial TCA cycle intermediates. Next, the small (0.5–1.0 mM) astroglial glutamate pool is labeled after which the ^{13}C label arrives at the

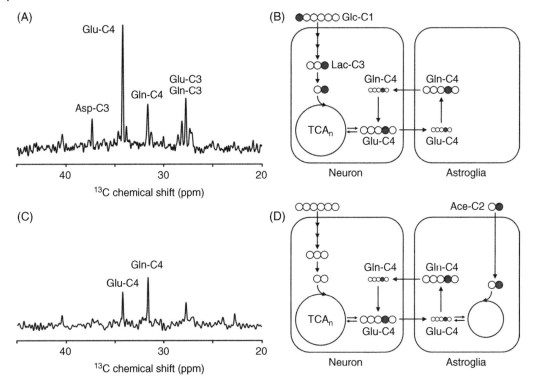

Figure 3.43 **Isotopic labeling strategies to probe cerebral metabolism.** (A, C) ^{13}C NMR spectra of human brain obtained 60 min following intravenous (A) [1-^{13}C]-glucose and (B) [2-^{13}C]-acetate administration. (B) The ^{13}C label from glucose first arrives in the neuronal glutamate pool, which after neurotransmission is converted to glutamine in the astroglial compartment. (D) [2-^{13}C]-acetate is first converted to glutamine in the astroglial before being transported to the neuronal compartment. While both substrates, glucose and acetate, probe identical metabolic pathways, they have different sensitivities towards the neuronal or astroglial TCA cycle flux or the glutamate–glutamine neurotransmitter cycle flux. The size of the glutamate and glutamine pools is indicated by the size of the circular carbon atoms. In (D) unlabeled glucose continues to be metabolized, thereby providing a dilution of the ^{13}C label in the neuronal glutamate pool.

large glutamine pool. The large neuronal glutamate pool is subsequently labeled from astroglial [4-^{13}C]-glutamine that is transported to the neuronal compartment as part of the glutamate–glutamine neurotransmitter cycle. [2-^{13}C]-Acetate experiments on human and animal brain have confirmed earlier observations made through the use of [1-^{13}C]-glucose and have determined that the astroglial TCA cycle is only a small fraction of the neuronal TCA cycle (Figure 3.42). Figure 3.43C shows a ^{13}C NMR spectrum acquired from human brain after 60 min following the onset of [2-^{13}C]-acetate infusion. Most noticeably is the different ratio between [4-^{13}C]-glutamate and [4-^{13}C]-glutamine when compared to using [1-^{13}C]-glucose as the substrate (Figure 3.43A). This can be understood from the fact that during [2-^{13}C]-acetate infusion the large astroglial glutamine pool is labeled first, while during [1-^{13}C]-glucose infusion the large neuronal glutamate pool is labeled first (Figure 3.43B and D).

While the exact formulation of metabolic models and the assumptions associated with them remain a point of discussion, the calculated metabolic fluxes obtained by different groups and with different substrates are remarkably similar (Figure 3.42). Rather than focusing on the relatively minor differences, the section on dynamic ^{13}C turnover is concluded with some of the novel findings and applications that have been reported.

One of the most interesting and significant findings obtained with dynamic ^{13}C NMR spectroscopy is that the glutamate–glutamine neurotransmitter cycle flux is linearly related to the TCA cycle flux [154, 181] over a wide range of cerebral activity. In essence, the ^{13}C NMR results support the notion that for every glutamate molecule that is released and recycled during neurotransmission a glucose molecule is oxidized in the TCA cycle. This mechanism provides a valuable link between function (neurotransmission) and energetics (glucose oxidation). Most of the commonly used high-resolution imaging modalities, like BOLD, CBF, or CBV MRI, or PET detect signal that is directly or indirectly related to energetics. With the metabolic link provided by ^{13}C NMR these high-resolution imaging results can potentially provide unique information on brain function. The relation between neurotransmitter cycling and glucose oxidation has since been extended to intense activation (seizures) [181–183], the separation of multiple cerebral tissue types [168] and the separation between excitatory and inhibitory neurotransmission [172, 173]. Furthermore, important links between neurotransmitter cycling, glucose oxidation, and the underlying neuronal firing have been obtained [182]. Besides ^{13}C NMR studies, dynamic metabolic pathways can also be studied with other nuclei, in particular deuterium [184, 221], nitrogen-15 [185–187], and oxygen-17 [188–190].

3.5.4 Hyperpolarized Dynamic ^{13}C NMR Spectroscopy

The low sensitivity of NMR originates from the low magnetic energy of nuclear spins compared to the thermal energy at room temperature, leading to an extraordinarily low spin polarization. The most straightforward, but arguably also the most expensive method to increase the spin polarization and hence the sensitivity is to increase the external magnetic field. However, even at a high magnetic field of 9.4 T, the spin-polarization is only 31 ppm. Improved RF coil design, like phased-array receivers and cooled coils, increases the detection sensitivity and decreases the noise, but do not fundamentally change the spin polarization. Therefore, the sensitivity potential of NMR, which will be reached when full polarization is achieved, remains largely dormant.

A number of methods have been proposed to enhance the nuclear polarization of spins to a significant fraction of unity. These hyperpolarization methods include a "brute force" approach [191–194], optical pumping of noble gases [195–197], para-hydrogen-induced polarization (PHIP) [198, 199], signal amplification by reversible exchange (SABRE) [200, 201], and dynamic nuclear polarization (DNP) [202, 203]. This section will briefly describe the principles of some of the hyperpolarization techniques. For more complete reviews the reader is referred to the literature [204–212].

3.5.4.1 Brute Force Hyperpolarization

The thermal equilibrium polarization of any sample increases with increasing magnetic field strength B_0 and with decreasing temperature T (see Eq. (1.4)). A relatively straightforward "brute force" approach to increase the polarization in a sample is therefore to place it in a strong magnetic field at a temperature close to 0 K. The spin polarization P obtained under very low temperatures is given by (see also Exercises)

$$P = \frac{n_\alpha - n_\beta}{n_\alpha + n_\beta} = \tanh\left(\frac{h\nu}{2kT}\right) \tag{3.52}$$

Figure 3.44 shows polarization curves for protons, carbon-13 nuclei, and electrons as a function of temperature in a magnetic field of 11.75 T (500 MHz for protons). Decreasing the temperature from body (310 K) to liquid helium (4 K) increases the polarization for protons circa 75 times to circa 0.3%. Even though this is a highly significant increase, the bulk of the spins do not contribute

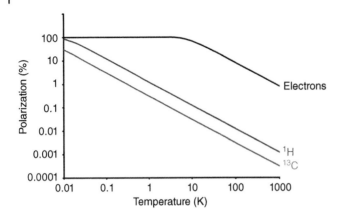

Figure 3.44 **Brute force hyperpolarization.** Spin polarization for ^1H, ^{13}C, and unpaired electrons as a function of temperature at a magnetic field of 11.75 T.

to the polarization until the temperature is decreased to the milliKelvin range close to absolute zero temperature. While working at liquid helium temperature is trivial, achieving temperatures close to absolute zero is challenging and in many cases impractical. Note that electrons have close to 100% polarization at 4 K due to their much larger gyromagnetic ratio ($\gamma_e = 658.21\gamma_H$).

While "brute force" approaches are relatively straightforward they are not used for *in vivo* NMR applications due to the extremely long T_1 relaxation time constants (hours to weeks) at very low temperatures.

3.5.4.2 Optical Pumping of Noble Gases

The term optical pumping refers to a process whereby (laser) light is used to excite or "pump" electrons from a low to a high energy state in atoms or molecules. In the case of hyperpolarized helium gas the excited electrons can be part of the substrate (metabstable ^3He) or part of an alkali metal vapor. In both cases the hyperpolarized electron spin state is subsequently transferred to the ^3He helium nucleus. The second approach, often referred to a spin exchange optical pumping (SEOP), is graphically shown in Figure 3.45. The optical cell contains the noble gas of interest (i.e. ^3He), buffer gas (^4He or N_2), and a small amount of vaporized alkali metal such as rubidium. Rubidium can be electronically hyperpolarized by irradiating the $5^{(2)}S_{1/2} \rightarrow 5^{(2)}P_{1/2}$ transition with resonant circularly polarized laser light of 795 nm. The electron spin polarization of the alkali metal vapor can be transferred to the nuclear spins of noble gas atoms like ^3He during collisions in which the Fermi contact mediates the spin exchange.

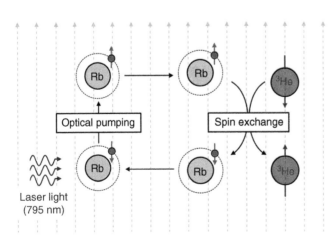

Figure 3.45 **Spin-exchange optical pumping (SEOP).** Optical pumping of rubidium (Rb) vapor with circularly polarized 795 nm laser light on the unresolved $5^2S_{1/2} \rightarrow 5^2P_{1/2}$ transition creates a large electron spin polarization in Rb atoms, which is in turn transferred to ^3He nuclei via spin exchange collisions in combination with hyperfine interactions.

3.5.4.3 Parahydrogen-induced Polarization (PHIP)

In 1986, Bowers and Weitekamp [198] theoretically predicted that the hydrogenation of small organic molecules with parahydrogen can lead to strong nonequilibrium polarizations of the protons in the hydrogenated compound. They confirmed their predictions experimentally in 1987 [199]. The novel polarization techniques were initially dubbed PASADENA and ALTADENA, but are now commonly known under their collective name PHIP.

Molecular hydrogen exists in two spin isomer states, orthohydrogen and parahydrogen (Figure 3.46A). Parahydrogen is the lowest energy state, a nuclear spin singlet, where the two proton spins are antiparallel to one another. Orthohydrogen, a nuclear spin triplet, exist in a higher energy state. As the isomers have different energies, the energy level populations are temperature dependent with a circa 25% mole fraction of parahydrogen at room temperature. Lowering the temperature to 20 K increases the molar fraction of parahydrogen to >99% (Figure 3.46B). However, the interconversion between ortho- and parahydrogen is normally very slow due to violation of symmetry selection rules. Fortunately, the addition of a paramagnetic catalyst can circumvent this selection rule, leading to the production of large quantities of parahydrogen in a matter of hours. Upon removal of the catalyst, and returning to ambient temperatures, the ortho/parahydrogen system only very slowly reestablishes thermodynamic equilibrium, thereby greatly facilitating storage and transport of parahydrogen.

When parahydrogen hydrogenates another molecule through a chemical reaction, the spin correlation between the two protons will initially be retained (Figure 3.46C and D). However, the symmetry of the hydrogen molecule will be broken due to scalar and dipolar couplings with other spins, as well as differences in chemical environment. Technically, the nuclei are not polarized, but are in a state of increased spin order. Nevertheless, the ordered spin state of each proton will give a highly enhanced NMR spectrum, consisting of characteristic anti-phase resonances.

While direct proton NMR observation of the PHIP effect is common in hydrogenation reactions in transition metal chemistry, for many *in vivo* applications it is desirable to transfer the hydrogen spin-state to an adjacent heteronucleus, like carbon-13. Firstly, heteronuclei often have a much longer T_1 relaxation time constant, thereby providing a greater window for efficient transfer of the compound and subsequent injection into the subject. Secondly, the natural abundance of carbon-13 is very low, such that injection of a hyperpolarized ^{13}C compound can be detected with minimal interference of "background" signals. As one of the first *in vivo* applications of hyperpolarization, ^{13}C MR angiography has provided images with excellent contrast-to-noise ratios [213]. Thirdly, the ordered spin-state of hydrogen does not provide a net polarization, resulting in anti-phase resonances. While this is not a problem for MR spectroscopy, it does lead to severe signal loss for MR imaging applications. Transfer to a heteronucleus can be accompanied by the generation of a net polarization on the carbon-13 nucleus (Figure 3.46E). Two commonly employed methods for transferring spin-order to heteronuclear spins are polarization transfer methods and so-called diabatic–adiabatic field cycling. The former methods are typically based on the INEPT sequence, which is commonly used for conventional ^{13}C NMR spectroscopy and will be discussed in Chapter 8. Diabatic–adiabatic field cycling rapidly (i.e. nonadiabatically) lowers the magnetic field to the range of μT, such that the heteronuclear $^1H–^{13}C$ spin-system is strongly coupled (i.e. has mixed spin-states) and dominated by the scalar coupling. The magnetic field is subsequently increased adiabatically, finally resulting in a net polarization of the carbon-13 nucleus. The main advantage of PHIP over DNP methods is the speed at which hyperpolarization is achieved. PHIP has been successful used on a number of compounds and has shown *in vivo* viability. However, the requirement of unsaturated precursor molecules with appropriate asymmetry presents a significant limitation, especially for biologically relevant compounds.

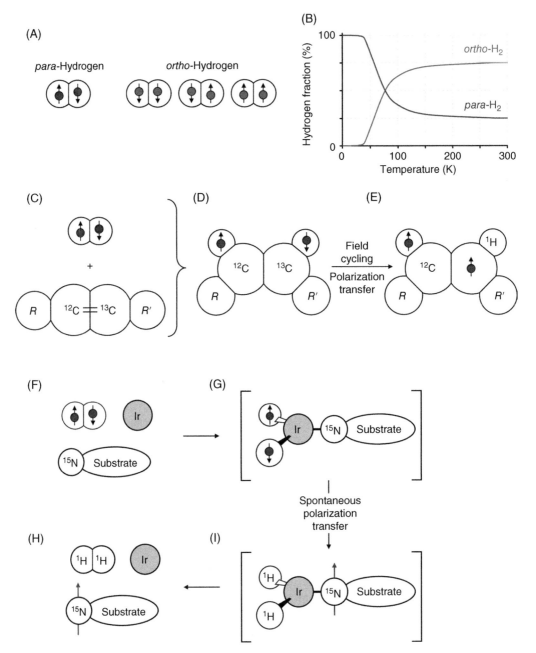

Figure 3.46 **Hyperpolarization based on parahydrogen.** (A) Four possible spin states of the hydrogen H_2 molecule, giving rise to 25% parahydrogen and 75% orthohydrogen at room temperature. (B) The fraction of parahydrogen can, with the aid of a catalyst, be increased to >99% when the temperature is lowered to <20 K. Following removal of the catalyst the system can be brought to room temperature where the para/orthohydrogen distribution returns only very slowly to the 25%/75% room temperature equilibrium. (C) When parahydrogen is combined with a compound containing double bonds, like alkenes, (D) a parahydrogen hydrogenated compound is formed in which the symmetry of the hydrogen nuclei is broken, thus leading to a greatly increased spin order, which is typically referred to as a parahydrogen induced polarization (PHIP). (E) In-phase signals with a longer T_1 relaxation time can be created by transferring the nonequilibrium proton spin order to a carbon-13 atom by field cycling or polarization transfer methods. (F–I) Signal amplification by reversible exchange (SABRE) also uses parahydrogen to achieve hyperpolarization, but does not have the requirement for a hydrogenation reaction as needed in PHIP. In SABRE an organometalic catalyst enables (G) the transient binding of parahydrogen and a substrate. (H) Following spontaneous polarization transfer from parahydrogen to the substrate, the substrate is released (I) making the catalyst available for another round of polarization exchange.

3.5.4.4 Signal Amplification by Reversible Exchange (SABRE)

The SABRE method [200, 201] is a relatively new hyperpolarization method that also uses parahydrogen. However, unlike PHIP, the parahydrogen during SABRE does not have to chemically react with the substrate molecule. Instead, the hyperpolarization is transferred via an intermediate complex between parahydrogen, substrate, and an organomettalic (iridium) catalyst (Figure 3.46F–I). The parahydrogen and substrate form a reversible bond with the iridium, whereby the polarization is spontaneously transferred to the substrate. Whereas the majority of SABRE applications have been performed in organic solvents, the method has shown viability in pure water [214]. SABRE will undoubtedly see continued development and may provide a promising method for affordable and efficient hyperpolarization of a wide range of compounds.

3.5.4.5 Dynamic Nuclear Polarization (DNP)

The method of DNP as described in this section is based on the "solid effect" described by Abragam and Proctor [215]. Inside a diamagnetic insulator with paramagnetic impurities, they observed a polarization much larger than the thermal equilibrium nuclear polarization when the insulator was exposed to microwave irradiation. A similar effect was predicted by Overhauser [10] and experimentally confirmed by Carver and Slichter [216] for metals five years earlier. A wonderful recollection of the early days of DNP is provided by Slichter [217].

To illustrate the principle of DNP, consider a group of nuclear spins resonating at Larmor frequency ν_N and a group of electron spins with Larmor frequency ν_e positioned in an external magnetic field B_0 (Figure 3.47A). At liquid helium temperatures (4 K) most of the electrons are in the low-energy α_e spin state, while the nuclear spins are distributed approximately equal over the α_N and β_N energy levels. Note that for an electron the magnetic moment in the low-energy α_e spin state is antiparallel to the applied magnetic field B_0 (Figure 3.47B). When the nuclei and electrons would be completely separated, nuclear transitions can be achieved in the radiofrequency range

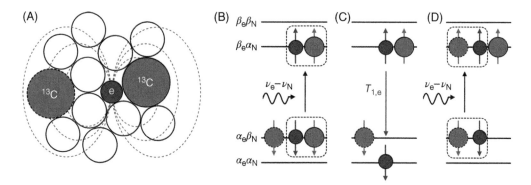

Figure 3.47 **Dynamic nuclear polarization (DNP).** (A) DNP is most efficient when the substrate is homogeneously mixed with a radical in an amorphous, noncrystalline, or "glassy" matrix such that the free electron has close interactions with nearby nuclear spins. (B) At low temperatures (<4 K) the free electrons have 100% polarization (i.e. all electron spins are in the low-energy α spin state) due to the large electron magnetic moment (see also Figure 3.44). Note that the electron spin orientation in the α spin state opposes the main magnetic field B_0. Selective irradiation of the microwave transition ($\nu_e - \nu_N$) "pumps" the double-quantum transition whereby both electron and nuclear spins flip, bringing the nuclear spin in the α spin state and hence increasing the nuclear spin polarization by one spin. (C) Rapid electron T_1 relaxation restores the electron thermal equilibrium situation, such that (D) the process in (B) can be repeated for another nuclear spin. The final nuclear spin polarization is determined by the increase due to pumping and a decrease due to nuclear spin T_1 relaxation.

at ν_N. Electronic transitions are, due to the much larger energy level difference, achieved in the microwave range at frequency ν_e. However, when the electron generates a finite magnetic field at the nuclei (Figure 3.47A), the two become dipolar-coupled. This so-called hyperfine interaction will lead to a mixing of the four pure states, which in turn will lead to a small, but nonvanishing probability of simultaneous electron and nuclear spin transitions or flips.

Now suppose that the system is irradiated at the microwave frequency $(\nu_e - \nu_N)$ and the electronic line width $\Delta \nu_e$ is small enough so that the transition $(\nu_e + \nu_N)$ does not occur (Figure 3.47B). The microwave irradiation results in a combined flip of the electron and nuclear spins, such that the nuclear spin polarization has been slightly increased. Rapid electron T_1 relaxation $(T_{1e} \sim 1\,ms)$ quickly flips the electron spin orientation (Figure 3.47C) back to the low-energy α_e spin state. The dipolar-coupled electron-nuclear system is then ready for another flip of the nuclear spin orientation (Figure 3.47D). Continuing this process will lead to a complete depletion of the antiparallel state, giving complete or 100% nuclear polarization.

It is easy to see that if a microwave frequency $(\nu_e + \nu_N)$ was used, all the nuclear spins would end up in an antiparallel state, giving a complete, but inverted nuclear polarization. In the example of Figure 3.47 it was assumed that the electron spin polarization P_e was complete, which is realistic at low temperatures (see Figure 3.44). However, when P_e is not complete, the final nuclear spin polarization P_N will be limited to $\pm P_e$. Typically the electron concentration is low (~0.001 electron spins per nuclear spin), so that one electron has to polarize not only the nuclear spins in the immediate vicinity, but also those that are further away. Resonant mutual flips or transitions between two neighboring nuclear spins, also referred to as spin diffusion, is the effective transport mechanism that supports this requirement. Furthermore, after each forced microwave-induced transition the electron spin must relax back into its thermal equilibrium before any of the nuclear spins in its sphere of influence relax through a nuclear relaxation mechanism. The final nuclear polarization thus becomes a balance between the rate at which the nuclear spins are being polarized and the rate at which the nuclear polarization undergoes T_1 relaxation. DNP has been shown in a wide range of molecules [205, 207, 209], whereby the enhanced polarization ranges from 5 to 40%.

The most commonly used and only FDA-approved substrate is $[1\text{-}^{13}C]$-pyruvate. Pyruvate has a long T_1 relaxation time constant, can be polarized up to 40%, and is biologically relevant as the final product of glycolysis and the substrate for subsequent degradation in the TCA cycle (Figure 3.48A). Figure 3.48B and C shows a typical application of hyperpolarized $[1\text{-}^{13}C]$-pyruvate to study tumor metabolism in a rat glioma model [218]. In the solid phase $[1\text{-}^{13}C]$-pyruvate (1 K temperature and several Tesla magnetic field) is hyperpolarized over the course of 30–120 min. Superheated water is then injected into the sample to dissolve the pyruvate and match the body temperature for subsequent injection. Following a rapid, high-concentration, intravenous bolus of $[1\text{-}^{13}C]$-pyruvate the metabolic fate of the substrate is followed by high temporal resolution, dynamic ^{13}C MRS (Figure 3.48B), or MRSI (Figure 3.48C). With global MRS (Figure 3.48B) it is clear that $[1\text{-}^{13}C]$-pyruvate is rapidly converted to $[1\text{-}^{13}C]$-lactate, $[1\text{-}^{13}C]$-alanine, and ^{13}C-bicarbonate. $[1\text{-}^{13}C]$-pyruvate hydrate $(HOO^{13}C{-}C(OH)_2{-}CH_3)$ forms spontaneously from $[1\text{-}^{13}C]$-pyruvate and has no biological relevance. The spatial distribution of $[1\text{-}^{13}C]$-pyruvate and $[1\text{-}^{13}C]$-lactate can be mapped with ultrafast MRSI methods (Figure 3.48C). As the hyperpolarized signal is a nonrenewable source of signal that *decreases* due to T_1 relaxation, special care has to be taken on how to most efficiently utilize the available signal. Most methods that detect hyperpolarized signals rely on low-nutation-angle excitation and ultrafast EPI or spiral-based MRSI. In addition, different nutation angles for substrate ($[1\text{-}^{13}C]$-pyruvate) and product ($[1\text{-}^{13}C]$-lactate) can prolong the presence of hyperpolarized signal. Other strategies to prolong signal lifetime is

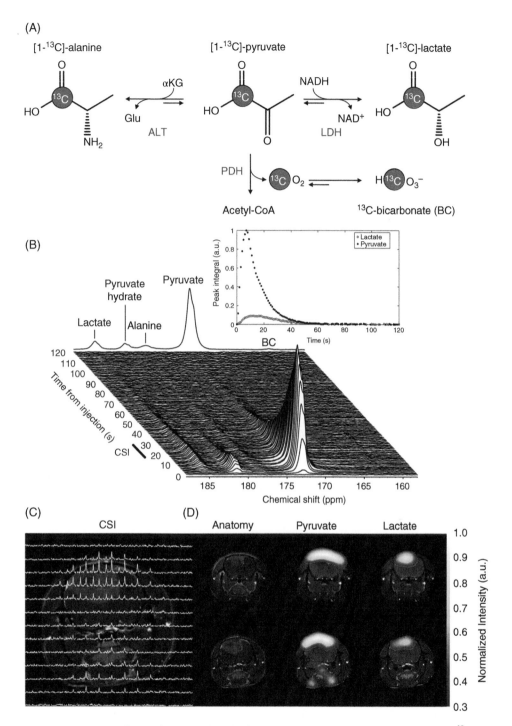

Figure 3.48 Dynamic nuclear polarization (DNP) of *in vivo* metabolism. (A) The hyperpolarized ^{13}C label of [1-^{13}C]-pyruvate can be converted to [1-^{13}C]-alanine or [1-^{13}C]-lactate by alanine transferase (ALT) or lactate dehydrogenase (LDH). Alternatively, [1-^{13}C]-pyruvate can be converted to acetyl-CoA by pyruvate dehydrogenase (PDH) whereby the ^{13}C label is lost as CO_2 and ultimately bicarbonate (BC). (B) Dynamic ^{13}C MRS on C6-glioma-bearing rat brain following the administration of 2 ml 75 mM hyperpolarized [1-^{13}C]-pyruvate. Visible signals are assigned to [1-^{13}C]-lactate (185 ppm), [1-^{13}C]-pyruvate hydrate (181 ppm), [1-^{13}C]-alanine (178 ppm), [1-^{13}C]-pyruvate (173 ppm), and ^{13}C-BC (162 ppm). ^{13}C MR spectra are acquired from the whole rat head with a surface coil and a 1 s time resolution. The inset shows the dynamics of the substrate [1-^{13}C]-pyruvate (solid circles) and the product [1-^{13}C]-lactate (open circles). (C) Representative ^{13}C MRSI or CSI superimposed on a T_1-weighted anatomical MRI. (D) Anatomical MRI and hyperpolarized [1-^{13}C]-pyruvate and [1-^{13}C]-lactate maps before (top) and 96 hours after radiotherapy (bottom). The radiotherapy led to reduced lactate levels. *Source:* Day et al. [218]. Reproduced with permission of John Wiley & Sons.

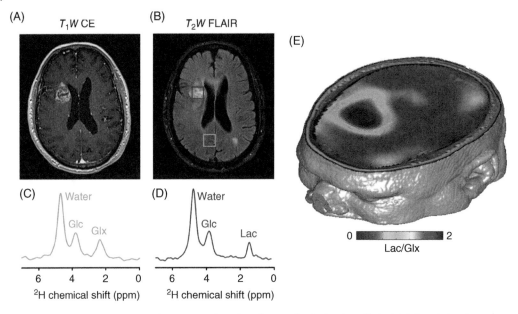

Figure 3.49 **Deuterium metabolic imaging (DMI) on human brain *in vivo*.** Clinical (A) T_1-weighted, contrast-enhanced (CE) and (B) T_2-weighted, fluid-attenuated inversion recovery (FLAIR) MR images of a patient with a glioblastoma multiforme (GBM) brain tumor. (C, D) Deuterium (^2H) MR spectra extracted from a 3D DMI dataset from the positions indicated in (B) circa 60 min following oral administration of [6,6'-$^2H_2$]-glucose. In addition to the natural abundance water signal, the ^2H MR spectrum from normal appearing brain tissue contains signals from the substrate, [6,6'-$^2H_2$]-glucose (Glc) and a downstream product, [4-^2H]-glutamate+glutamine (Glx) formed through the combined action of glycolysis and the TCA cycle. The ^2H MR spectrum from the tumor region is marked by increased lactate, characteristic of the Warburg effect. (E) A lactate/Glx ratio map provides a high contrast-to-noise measure of anaerobic over aerobic metabolism, providing unique metabolic information not available from clinical MR images. *Source:* Data from Henk De Feyter and Zachary Corbin.

to deuterate ^{13}C-labeled compounds, such as ^{13}C-glucose, to increase the ^{13}C T_1 relaxation time. Generally, the hyperpolarized signal has decayed to insignificant levels within one minute of injection.

Hyperpolarized MR of [1-^{13}C]-pyruvate in humans [219, 220] is still relatively rare, but initial results have shown metabolic effects in prostate cancer that were not observed with MRI methods. Hyperpolarization methods present unique research tools that allow the detection of metabolic processes that cannot be easily detected by other MR methods. However, the high cost, technical complexity, and supraphysiological conditions during substrate injection dampen the clinical utility and feasibility of hyperpolarized MR.

3.5.5 Deuterium Metabolic Imaging (DMI)

Carbon-13 NMR has traditionally been the standard tool to study metabolism *in vivo*. This is largely motivated by the fact that most biologically relevant molecules have a carbon backbone that can be tracked by ^{13}C NMR through multiple metabolic pathways. ^{13}C MRS has been successful in addressing several important aspects regarding cerebral metabolism, such as the intimate relationship between energy metabolism and neurotransmission [154, 181]. Unfortunately, ^{13}C MRS has never reached sufficient momentum to become a clinically viable technique. This must be attributed to the low sensitivity and long scan times associated with

^{13}C MRS, but also due to the substantial technical challenges related to RF coils, broadband decoupling, and spatial localization. Hyperpolarized ^{13}C MRS clearly tackled the sensitivity obstacle, but is limited in clinically viable substrates and further increases the technical challenges. With many pathologies, such as cancer, neurodegenerative diseases, and diabetes having a metabolic component, there is a clear need for metabolic imaging to supplemental structural MRI.

Deuterium metabolic imaging (DMI) is a novel modality with significant clinical potential [221]. Deuterium (^{2}H or D) is an isotope of hydrogen with a natural abundance of 0.015% and a nuclear spin of 1. Natural abundance ^{2}H NMR of *in vivo* tissues is characterized by sparse MR spectra containing a small HDO water signal. Intravenous or oral administration of a deuterated substrate, such as [6,6'-^{2}H$_2$]-glucose, leads to rapid ^{2}H labeling of lactate and glutamate/glutamine via glycolysis and the TCA cycle, similar to that observed using ^{13}C-labeled substrates (Figure 3.39). However, unlike ^{1}H or ^{13}C MR applications, ^{2}H MRS does not require water or lipid suppression, thereby making the method simple and robust. The sensitivity of ^{2}H MRS is relatively high due to a favorable T_1/T_2 ratio (T_1 between 100 and 400 ms for most metabolites, T_2 between 30 and 60 ms) that allows extensive signal averaging. Combining ^{2}H MRS with phase-encoding gradients allows DMI of dynamic ^{2}H metabolism in three dimensions. Figure 3.49 shows an example of DMI on a patient with a glioblastoma multiforme (GBM) tumor, whereby a lactate-over-glutamate (Lac/Glx) map provides high image contrast. While DMI is still in its infancy, results as shown in Figure 3.49 clearly show the clinical viability of the method. Similar to ^{13}C MR studies, DMI can utilize a wide range of ^{2}H-enriched substrates such as deuterated acetate and water.

Exercises

3.1   On a [2-^{13}C]-acetate sample at 7.05 T the proton frequency and carbon-13 frequencies were measured at 300.015 623 MHz and 75.229 674 MHz, respectively. In an *in vivo* ^{1}H NMR spectrum acquired from human brain at 4.0 T the proton frequency of glutamate-H4 was measured at 170.455 322 MHz. Calculate the *in vivo* carbon-13 frequency of glutamate-C4 at 4.0 T.

3.2   The 2-oxoglutarate/glutamate chemical exchange as catalyzed by glutamate dehydrogenase is characterized by a rate constant of 0.4 s^{-1}. Given the ^{13}C T_1 relaxation times of 1.5 and 1.7 s for 2-oxoglutarate and glutamate measured in the absence of saturation, calculate the signal decrease in glutamate upon continuous saturation of 2-oxogluterate.

3.3   **A** Given a longitudinal relaxation time constant T_1 of 4.0 s for free water ($\tau_c = 10^{-11}$ s) at 7.05 T, calculate the T_1 for bone ($\tau_c = 10^{-6}$ s) under the condition of pure dipolar relaxation.
      B Calculate the minimum T_1 relaxation time constant at 7.05 T as a result of pure dipolar relaxation.
      C Calculate the transverse relaxation time constants T_2 for water and bone at 7.05 T.
      D Calculate the T_1 and T_2s for water and bone at 17.6 T.

3.4   A NMR spectroscopist measures a sample of 100 mM H$_2$SO$_4$ and 100 mM glutathione in water with ^{1}H and ^{33}S NMR.
      A Describe the expected pulse-acquire ^{33}S NMR spectrum.
      B Describe the expected pulse-acquire ^{1}H NMR spectrum.

3.5 Inversion recovery is arguably the most robust method to measure the longitudinal relaxation time constant T_1. However, since inversion recovery is a relatively slow method, many alternative methods have been developed. Consider a pulse-acquire sequence executed with a repetition time TR. The T_1 relaxation time constant can be estimated after performing two experiments with nutation angles α_1 and $\alpha_2 = 2\alpha_1$ during which signals M_1 and M_2 are acquired.

A Derive an expression for T_1 as a function of α_1, α_2, M_1, M_2, and TR.

B Calculate T_1 when $\alpha_1 = 30°$, $\alpha_2 = 60°$, $M_1 = 100$, $M_2 = 150$, and TR = 500 ms.

C Discuss the pro's and con's of this method against an inversion recovery method and discuss how the performance of the two-point method can be optimized for a known T_1 range.

3.6 Consider a pulse-acquire experiment consisting of a RF pulse generating a nutation angle α followed by a recovery time TR.

A Starting with the Bloch equation for T_1 relaxation (i.e. Eq. (1.14)) derive an expression for the recovery of the longitudinal magnetization following a perturbation.

B Derive an expression for the steady-state longitudinal magnetization for a pulse-acquire sequence. Assume the absence of transverse magnetization prior to each excitation.

C Calculate after how many experiments the longitudinal magnetization is within 1% of the steady-state magnetization when $\alpha = 40°$ and TR = T_1.

D Suppose that 10 blocks of four averages are acquired sequentially in one experiment with $\alpha = 40°$ and TR = T_1 starting from an initial thermal equilibrium situation. Calculate the difference between the acquired signal in the first and the last block due to incomplete T_1 saturation during the first block.

E Derive the Ernst angle expression from Eq. (3.8).

F Calculate the Ernst angle for the excitation pulse of a spin-echo sequence with TR = $0.5T_1$. Assume negligible T_1 relaxation during the echo-time TE.

3.7 **A** Using Eq. (3.26) derive the analytical expression for the b-value of a pair of square diffusion gradients of amplitude G, duration δ, and separation Δ as incorporated in a Hahn spin-echo.

B For an echo-time TE of 20 ms and infinitely short RF pulses, calculate the maximum b-value achievable on a system with a maximum gradient strength of $40\,mT\,m^{-1}$ per direction. Assume infinitely fast gradient ramping.

C Recalculate the maximum b-value for the more realistic case in which the magnetic field gradients have a finite ramp time of 500 μs to maximum amplitude.

3.8 **A** Derive the b-value for a constant background gradient during a spin-echo sequence with echo time TE, i.e. Eq. (3.27).

B For a well-shimmed sample $G_0 = 1.0\,Hz\,mm^{-1}$. Calculate the echo-time by which the water signal ($D = 0.7\,\mu m^2\,ms^{-1}$) has decreased by 5% as a result of diffusion-related signal loss.

C Given a transverse relaxation time constant $T_2 = 70$ ms, calculate the overall signal loss.

D In order to reduce the overall signal loss, G_0 is increased such that the echo-time can be reduced. Discuss the pro's and con's of this approach over the more standard Stejskal–Tanner approach.

E Calculate the difference in signal intensity obtained for water at an echo-time of 100 ms when acquired with a standard Hahn spin-echo or a CPMG spin-echo pulse train with $TE_{CPMG} = 2$ ms. Assume $D = 0.7\,\mu m^2\,(ms)^{-1}$, $T_2 = 70$ ms, and a background magnetic field gradient $G_0 = 2500\,Hz\,mm^{-1}$.

3.9 Consider a stimulated echo sequence with two 15 ms 500 mT/m TE crusher gradients along the x direction, separated by 50 ms. Dephasing during the TM period (TM = 20 ms) is achieved with a 5 ms 400 mT m^{-1} TM crusher gradient along the y direction.

A Calculate the diffusion-weighting (i.e. "b-value") for the sequence using infinitely fast rise times.

B The creatine resonance at 3.03 ppm is detected with an intensity of 387 during the diffusion weighted experiment. In an earlier experiment with a lower diffusion weighting ($b = 120$ s mm^{-2}), the creatine intensity was 1045. Calculate the ADC of creatine.

C During a measurement with TM = 500 ms the creatine ADC is significantly larger than that calculated under B. Give the most likely explanation for this observation and a possible solution.

D Calculate the diffusion-weighting when the TE crusher gradients are applied simultaneously along all three spatial directions.

E Calculate the diffusion-weighting when the TM crusher gradient is changed to a 5 ms 500 mT m^{-1} gradient applied along the x and z directions.

3.10 **A** Derive Eq. (3.46) from Eq. (3.45).

B For a ^{13}C FE of compound S of 0.7, calculate the FE of compound P for $t \to \infty$.

C Given a FE for compound S of 0.7 and $[P] = 10$ mM, calculate the flux when $[P^*] = 3$ mM at $t = 12$ min.

D Given a FE for compound S of 0.7 and $[P] = 20$ mM, calculate the flux when $[P^*] = 6$ mM at $t = 12$ min.

3.11 Show, for 2D diffusion, that the trace of the diffusion tensor is rotationally invariant.

3.12 For an inversion recovery method, determine the inversion times that correspond to $M_z/M_0 = -0.9 + 0.1n$ ($0 \le n \le 18$) for $T_1 = 1000$ ms.

References

1 Pfeuffer, J., Tkac, I., Provencher, S.W., and Gruetter, R. (1999). Toward an *in vivo* neurochemical profile: quantification of 18 metabolites in short-echo-time ^{1}H NMR spectra of the rat brain. *J. Magn. Reson.* 141: 104–120.

2 Govindaraju, V., Young, K., and Maudsley, A.A. (2000). Proton NMR chemical shifts and coupling constants for brain metabolites. *NMR Biomed.* 13: 129–153.

3 Nicholson, J.K. and Wilson, I.D. (1989). High resolution proton magnetic resonance spectroscopy of biological fluids. *Prog. Nucl. Magn. Reson. Spectrosc.* 21: 449–501.

4 Preul, M.C., Caramanos, Z., Collins, D.L. et al. (1996). Accurate, noninvasive diagnosis of human brain tumors by using proton magnetic resonance spectroscopy. *Nat. Med.* 2: 323–325.

5 Hetherington, H.P., Pan, J.W., and Spencer, D.D. (2002). ^{1}H and ^{31}P spectroscopy and bioenergetics in the lateralization of seizures in temporal lobe epilepsy. *J. Magn. Reson. Imaging* 16: 477–483.

6 Petroff, O.A., Mattson, R.H., Behar, K.L. et al. (1998). Vigabatrin increases human brain homocarnosine and improves seizure control. *Ann. Neurol.* 44: 948–952.

7 Bloembergen, N., Purcell, E.M., and Pound, R.V. (1948). Relaxation effects in nuclear magnetic resonance absorption. *Phys. Rev.* 73: 679–712.

8 Solomon, I. (1955). Relaxation processes in a system of two spins. *Phys. Rev.* 99: 559–565.

9 Overhauser, A.W. (1953). Paramagnetic relaxation in metals. *Phys. Rev.* 89: 689–700.

10 Overhauser, A.W. (1953). Polarization of nuclei in metals. *Phys. Rev.* 92: 411–415.

11 Brown, T.R., Stoyanova, R., Greenberg, T. et al. (1995). NOE enhancements and T_1 relaxation times of phosphorylated metabolites in human calf muscle at 1.5 Tesla. *Magn. Reson. Med.* 33: 417–421.

12 Moonen, C.T.W., Dimand, R.J., and Cox, K.L. (1988). The noninvasive determination of linoleic acid content in human apidose tissue by natural abundance carbon-13 nuclear magnetic resonance. *Magn. Reson. Med.* 6: 140–157.

13 Gruetter, R., Novotny, E.J., Boulware, S.D. et al. (1992). Direct measurement of brain glucose concentrations in humans by ^{13}C NMR spectroscopy. *Proc. Natl. Acad. Sci. U. S. A.* 89: 1109–1112.

14 Gruetter, R., Novotny, E.J., Boulware, S.D. et al. (1994). Localized ^{13}C NMR spectroscopy in the human brain of amino acid labeling from D-[1-^{13}C]glucose. *J. Neurochem.* 63: 1377–1385.

15 Freeman, D.M. and Hurd, R. (1997). Decoupling: theory and practice. II. State of the art: *in vivo* applications of decoupling. *NMR Biomed.* 10: 381–393.

16 Li, S., Zhang, Y., Wang, S. et al. (2009). *In vivo* [13]C magnetic resonance spectroscopy of human brain on a clinical 3 T scanner using [2-^{13}C]-glucose infusion and low-power stochastic decoupling. *Magn. Reson. Med.* 62: 565–573.

17 Bachert-Baumann, P., Ermark, F., Zabel, H.J. et al. (1990). *In vivo* nuclear Overhauser effect in [31]P-(^{1}H) double-resonance experiments in a 1.5-T whole-body MR system. *Magn. Reson. Med.* 15: 165–172.

18 Bottomley, P.A. and Hardy, C.J. (1992). Proton Overhauser enhancements in human cardiac phosphorus NMR spectroscopy at 1.5 T. *Magn. Reson. Med.* 24: 384–390.

19 Evelhoch, J.L., Ewy, C.S., Siegfried, B.A. et al. (1985). [31]P spin-lattice relaxation times and resonance linewidths of rat tissue *in vivo*: dependence upon the static magnetic field strength. *Magn. Reson. Med.* 2: 410–417.

20 Mathur-De Vre, R., Maerschalk, C., and Delporte, C. (1990). Spin-lattice relaxation times and nuclear Overhauser enhancement effect for ^{31}P metabolites in model solutions at two frequencies: implications for *in vivo* spectroscopy. *Magn. Reson. Imaging* 8: 691–698.

21 Remy, C., Albrand, J.P., Benabid, A.L. et al. (1987). *In vivo* [31]P nuclear magnetic resonance studies of T_1 and T_2 relaxation times in rat brain and in rat brain tumors implanted to nude mice. *Magn. Reson. Med.* 4 (2): 144–152.

22 Blackledge, M.J., Oberhaensli, R.D., Styles, P., and Radda, G.K. (1987). Measurement of *in vivo* [31]P relaxation rates and spectral editing in human organs using rotating frame depth selection. *J. Magn. Reson.* 71: 331–336.

23 Thomsen, C., Jensen, K.E., and Henriksen, O. (1989). *In vivo* measurements of T_1 relaxation times of ^{31}P-metabolites in human skeletal muscle. *Magn. Reson. Imaging* 7: 231–234.

24 Buchthal, S.D., Thoma, W.J., Taylor, J.S. et al. (1989). *In vivo* T_1 values of phosphorus metabolites in human liver and muscle determined at 1.5 T by chemical shift imaging. *NMR Biomed.* 2: 298–304.

25 Gruetter, R., Boesch, C., Martin, E., and Wuthrich, K. (1990). A method for rapid evaluation of saturation factors in *in vivo* surface coil NMR spectroscopy using B_1-insensitive pulse cycles. *NMR Biomed.* 3: 265–271.

26 Hubesch, B., Sappey-Marinier, D., Roth, K. et al. (1990). P-31 MR spectroscopy of normal human brain and brain tumors. *Radiology* 174: 401–409.

27 Bottomley, P.A. and Ouwerkerk, R. (1994). The dual-angle method for fast, sensitive T_1 measurement *in vivo* with low-angle adiabatic pulses. *J. Magn. Reson. B* 104: 159–167.

28 Albrand, J.P., Foray, M.F., Decorps, M., and Remy, C. (1986). ^{31}P NMR measurements of T_2 relaxation times of ATP with surface coils: suppression of J modulation. *Magn. Reson. Med.* 3: 941–945.

29 Jung, W.I., Straubinger, K., Bunse, M. et al. (1992). ^{31}P transverse relaxation times of the ATP NMR signals of human skeletal muscle *in vivo*. *Magn. Reson. Med.* 28: 305–310.

30 Jung, W.I., Widmaier, S., Bunse, M. et al. (1993). ^{31}P transverse relaxation times of ATP in human brain *in vivo*. *Magn. Reson. Med.* 30: 741–743.

31 Straubinger, K., Jung, W.I., Bunse, M. et al. (1994). Spin-echo methods for the determination of ^{31}P transverse relaxation times of the ATP NMR signals *in vivo*. *Magn. Reson. Imaging* 12: 121–129.

32 Newcomer, B.R. and Boska, M.D. (1999). T_1 measurements of ^{31}P metabolites in resting and exercising human gastrocnemius/soleus muscle at 1.5 Tesla. *Magn. Reson. Med.* 41: 486–494.

33 Meyerspeer, M., Krssak, M., and Moser, E. (2003). Relaxation times of ^{31}P-metabolites in human calf muscle at 3 T. *Magn. Reson. Med.* 49: 620–625.

34 Lei, H., Zhu, X.H., Zhang, X.L. et al. (2003). *In vivo* ^{31}P magnetic resonance spectroscopy of human brain at 7 T: an initial experience. *Magn. Reson. Med.* 49: 199–205.

35 Lu, M., Chen, W., and Zhu, X.H. (2014). Field dependence study of *in vivo* brain ^{31}P MRS up to 16.4 T. *NMR Biomed.* 27: 1135–1141.

36 Christensen, K.A., Grant, D.M., Schulman, E.M., and Walling, C. (1974). Optimal determination of relaxation times of Fourier transform nuclear magnetic resonance. Determination of spin-lattice relaxation times in chemical polarized species. *J. Chem. Phys.* 78: 1971–1977.

37 Homer, J. and Beevers, M.S. (1985). Driven-equilibrium single-pulse observation of T_1 relaxation. A reevaluation of a rapid "new" method for determining NMR spin-lattice relaxation times. *J. Magn. Reson.* 63: 287–297.

38 Ma, D., Gulani, V., Seiberlich, N. et al. (2013). Magnetic resonance fingerprinting. *Nature* 495: 187–192.

39 Wang, C.Y., Liu, Y., Huang, S. et al. (2017). ^{31}P magnetic resonance fingerprinting for rapid quantification of creatine kinase reaction rate *in vivo*. *NMR Biomed.* 30 (12): doi: 10.1002/nbm.3786.

40 An, L., Li, S., and Shen, J. (2017). Simultaneous determination of metabolite concentrations, T_1 and T_2 relaxation times. *Magn. Reson. Med.* 78 (6): 2072–2081.

41 Bottomley, P.A., Foster, T.H., Argersinger, R.E., and Pfeifer, L.M. (1984). A review of normal tissue hydrogen NMR relaxation times and relaxation mechanisms from 1–100 MHz: dependence on tissue type, NMR frequency, temperature, species, excision, and age. *Med. Phys.* 11: 425–448.

42 Lynch, L.J. (1983). Water relaxation in heterogeneous and biological systems. In: *Magnetic Resonance in Biology*, vol. 2 (ed. J.S. Cohen), 248–304. New York: Wiley.

43 Fullerton, G.D. (1983). Physiological basis of magnetic relaxation. In: *Magnetic Resonance Imaging* (ed. D.D. Stark and W.G. Bradley), 88–108. St Louis: Mosby.

44 Bryant, R.G., Mendelson, D.A., and Lester, C.C. (1991). The magnetic field dependence of proton spin relaxation in tissues. *Magn. Reson. Med.* 21: 117–126.

45 Fischer, H.W., Rinck, P.A., van Haverbeke, Y., and Muller, R.N. (1990). Nuclear relaxation of human gray and white matter: analysis of field dependence and implications for MRI. *Magn. Reson. Med.* 16: 317–334.

46 Duewell, S.H., Ceckler, T.L., Ong, K. et al. (1995). Musculoskeletal MR imaging at 4 T and 1.5 T: Comparison of relaxation times and image contrast. *Radiology* 196: 551–555.

47 Diakova, G., Korb, J.P., and Bryant, R.G. (2012). The magnetic field dependence of water T_1 in tissues. *Magn. Reson. Med.* 68: 272–277.

48 Gideon, P. and Henriksen, O. (1992). *In vivo* relaxation of *N*-acetyl-aspartate, creatine plus phosphocreatine, and choline containing compounds during the course of brain infarction: a proton MRS study. *Magn. Reson. Imaging* 10: 983–988.

49 Toft, P.B., Christiansen, P., Pryds, O. et al. (1994). T_1, T_2, and concentrations of brain metabolites in neonates and adolescents estimated with H-1 MR spectroscopy. *J. Magn. Reson. Imaging* 4 (1): −5.

50 Kreis, R., Ernst, T., and Ross, B.D. (1993). Development of the human brain: *in vivo* quantification of metabolite and water content with proton magnetic resonance spectroscopy. *Magn. Reson. Med.* 30: 424–437.

51 Manton, D.J., Lowry, M., Blackband, S.J., and Horsman, A. (1995). Determination of proton metabolite concentrations and relaxation parameters in normal human brain and intracranial tumours. *NMR Biomed.* 8: 104–112.

52 Frahm, J., Bruhn, H., Gyngell, M.L. et al. (1989). Localized proton NMR spectroscopy in different regions of the human brain *in vivo*. Relaxation times and concentrations of cerebral metabolites. *Magn. Reson. Med.* 11: 47–63.

53 Brief, E.E., Whittall, K.P., Li, D.K., and MacKay, A. (2003). Proton T_1 relaxation times of cerebral metabolites differ within and between regions of normal human brain. *NMR Biomed.* 16: 503–509.

54 Ethofer, T., Mader, I., Seeger, U. et al. (2003). Comparison of longitudinal metabolite relaxation times in different regions of the human brain at 1.5 and 3 Tesla. *Magn. Reson. Med.* 50: 1296–1301.

55 Traber, F., Block, W., Lamerichs, R. et al. (2004). ^{1}H metabolite relaxation times at 3.0 Tesla: measurements of T_1 and T_2 values in normal brain and determination of regional differences in transverse relaxation. *J. Magn. Reson. Imaging* 19: 537–545.

56 Mlynarik, V., Gruber, S., and Moser, E. (2001). Proton T_1 and T_2 relaxation times of human brain metabolites at 3 Tesla. *NMR Biomed.* 14: 325–331.

57 Xin, L., Schaller, B., Mlynarik, V. et al. (2013). Proton T_1 relaxation times of metabolites in human occipital white and gray matter at 7 T. *Magn. Reson. Med.* 69: 931–936.

58 de Graaf, R.A., Brown, P.B., McIntyre, S. et al. (2006). High magnetic field water and metabolite proton T_1 and T_2 relaxation in rat brain *in vivo*. *Magn. Reson. Med.* 56: 386–394.

59 Cudalbu, C., Mlynarik, V., Xin, L., and Gruetter, R. (2009). Comparison of T_1 relaxation times of the neurochemical profile in rat brain at 9.4 Tesla and 14.1 Tesla. *Magn. Reson. Med.* 62: 862–867.

60 Lopez-Kolkovsky, A.L., Meriaux, S., and Boumezbeur, F. (2016). Metabolite and macromolecule T_1 and T_2 relaxation times in the rat brain *in vivo* at 17.2T. *Magn. Reson. Med.* 75: 503–514.

61 Michaeli, S., Garwood, M., Zhu, X.H. et al. (2002). Proton T_2 relaxation study of water, *N*-acetylaspartate, and creatine in human brain using Hahn and Carr-Purcell spin echoes at 4T and 7T. *Magn. Reson. Med.* 47: 629–633.

62 McConnell, H.M. (1958). Reaction rates by nuclear magnetic resonance. *J. Chem. Phys.* 28: 430–431.

63 Forsen, S. and Hoffman, R.A. (1963). A new method for the study of moderately rapid chemical exchange rates employing nuclear magnetic double resonance. *Acta Chem. Scand.* 17: 1787–1788.

64 Forsen, S. and Hoffman, R.A. (1963). Study of moderately rapid chemical exchange reactions by means of nuclear magnetic double resonance. *J. Chem. Phys.* 39: 2892–2901.

65 Brown, T.R. and Ogawa, S. (1977). ^{31}P nuclear magnetic resonance kinetic measurements on adenylatekinase. *Proc. Natl. Acad. Sci. U. S. A.* 74: 3627–3631.

66 Degani, H., Laughlin, M., Campbell, S., and Shulman, R.G. (1985). Kinetics of creatine kinase in heart: a ^{31}P NMR saturation- and inversion-transfer study. *Biochemistry* 24: 5510–5516.

67 Ren, J., Sherry, A.D., and Malloy, C.R. (2017). Efficient ^{31}P band inversion transfer approach for measuring creatine kinase activity, ATP synthesis, and molecular dynamics in the human brain at 7 T. *Magn. Reson. Med.* 78: 1657–1666.

68 Hsieh, P.S. and Balaban, R.S. (1988). Saturation and inversion transfer studies of creatine kinase kinetics in rabbit skeletal muscle *in vivo*. *Magn. Reson. Med.* 7: 56–64.

69 Buehler, T., Kreis, R., and Boesch, C. (2015). Comparison of ^{31}P saturation and inversion magnetization transfer in human liver and skeletal muscle using a clinical MR system and surface coils. *NMR Biomed.* 28: 188–199.

70 Ugurbil, K. (1985). Magnetization transfer measurements of individual rate constants in the presence of multiple reactions. *J. Magn. Reson.* 64: 207–219.

71 Lei, H., Ugurbil, K., and Chen, W. (2003). Measurement of unidirectional P_i to ATP flux in human visual cortex at 7 T by using *in vivo* ^{31}P magnetic resonance spectroscopy. *Proc. Natl. Acad. Sci. U. S. A.* 100: 14409–14414.

72 Gadian, D.G., Radda, G.K., Brown, T.R. et al. (1981). The activity of creatine kinase in frog skeletal muscle studied by saturation-transfer nuclear magnetic resonance. *Biochem. J.* 194: 215–228.

73 Brindle, K.M., Blackledge, M.J., Challiss, R.A., and Radda, G.K. (1989). ^{31}P NMR magnetization-transfer measurements of ATP turnover during steady-state isometric muscle contraction in the rat hind limb *in vivo. Biochemistry* 28: 4887–4893.

74 Chen, W., Zhu, X.H., Adriany, G., and Ugurbil, K. (1997). Increase of creatine kinase activity in the visual cortex of human brain during visual stimulation: a ^{31}P magnetization transfer study. *Magn. Reson. Med.* 38: 551–557.

75 Hamman, B.L., Bittl, J.A., Jacobus, W.E. et al. (1995). Inhibition of the creatine kinase reaction decreases the contractile reserve of isolated rat hearts. *Am. J. Physiol.* 269: H1030–H1036.

76 Hsieh, P.S. and Balaban, R.S. (1987). ^{31}P imaging of *in vivo* creatine kinase reaction rates. *J. Magn. Reson.* 74: 574–579.

77 Bottomley, P.A., Ouwerkerk, R., Lee, R.F., and Weiss, R.G. (2002). Four-angle saturation transfer (FAST) method for measuring creatine kinase reaction rates *in vivo. Magn. Reson. Med.* 47: 850–863.

78 Shen, J. (2005). *In vivo* carbon-13 magnetization transfer effect. Detection of aspartate aminotransferase reaction. *Magn. Reson. Med.* 54: 1321–1326.

79 Shen, J. and Xu, S. (2006). Theoretical analysis of carbon-13 magnetization transfer for *in vivo* exchange between alpha-ketoglutarate and glutamate. *NMR Biomed.* 19: 248–254.

80 Bottomley, P.A. (2016). Measuring biochemical reaction rates *in vivo* with magnetization transfer. *eMagRes* 5: 843–858.

81 Wolff, S.D. and Balaban, R.S. (1989). Magnetization transfer contrast (MTC) and tissue water proton relaxation *in vivo. Magn. Reson. Med.* 10: 135–144.

82 Dreher, W., Norris, D.G., and Leibfritz, D. (1994). Magnetization transfer affects the proton creatine/phosphocreatine signal intensity: *in vivo* demonstration in the rat brain. *Magn. Reson. Med.* 31: 81–84.

83 Roell, S.A., Dreher, W., Busch, E., and Leibfritz, D. (1998). Magnetization transfer attenuates metabolite signals in tumorous and contralateral animal brain: *in vivo* observations by proton NMR spectroscopy. *Magn. Reson. Med.* 39: 742–748.

84 de Graaf, R.A., van Kranenburg, A., and Nicolay, K. (1999). Off-resonance metabolite magnetization transfer measurements on rat brain *in situ. Magn. Reson. Med.* 41: 1136–1144.

85 Luo, Y., Rydzewski, J., de Graaf, R.A. et al. (1999). *In vivo* observation of lactate methyl proton magnetization transfer in rat C6 glioma. *Magn. Reson. Med.* 41: 676–685.

86 McLean, M.A., Simister, R.J., Barker, G.J., and Duncan, J.S. (2005). Magnetization transfer effect on human brain metabolites and macromolecules. *Magn. Reson. Med.* 54: 1281–1285.

87 McLean, M.A. and Barker, G.J. (2006). Concentrations and magnetization transfer ratios of metabolites in gray and white matter. *Magn. Reson. Med.* 56: 1365–1370.

88 Kruiskamp, M.J., de Graaf, R.A., van Vliet, G., and Nicolay, K. (1999). Magnetic coupling of creatine/phosphocreatine protons in rat skeletal muscle, as studied by ^{1}H-magnetization transfer MRS. *Magn. Reson. Med.* 42: 665–672.

89 Kruiskamp, M.J., de Graaf, R.A., van der Grond, J. et al. (2001). Magnetic coupling between water and creatine protons in human brain and skeletal muscle, as measured using inversion transfer ^{1}H-MRS. *NMR Biomed.* 14: 1–4.

90 Kruiskamp, M.J. and Nicolay, K. (2001). On the importance of exchangeable NH protons in creatine for the magnetic coupling of creatine methyl protons in skeletal muscle. *J. Magn. Reson.* 149: 8–12.

91 Haase, A., Frahm, J., Hanicke, W., and Matthaei, D. (1985). ^{1}H NMR chemical shift selective (CHESS) imaging. *Phys. Med. Biol.* 30: 341–344.

92 Wolff, S.D. and Balaban, R.S. (1990). NMR imaging of labile proton exchange. *J. Magn. Reson.* 86: 164–169.

93 Guivel-Scharen, V., Sinnwell, T., Wolff, S.D., and Balaban, R.S. (1998). Detection of proton chemical exchange between metabolites and water in biological tissues. *J. Magn. Reson.* 133: 36–45.

94 Ward, K.M., Aletras, A.H., and Balaban, R.S. (2000). A new class of contrast agents for MRI based on proton chemical exchange dependent saturation transfer (CEST). *J. Magn. Reson.* 143: 79–87.

95 Ward, K.M. and Balaban, R.S. (2000). Determination of pH using water protons and chemical exchange dependent saturation transfer (CEST). *Magn. Reson. Med.* 44: 799–802.

96 Kogan, F., Haris, M., Singh, A. et al. (2014). Method for high-resolution imaging of creatine *in vivo* using chemical exchange saturation transfer. *Magn. Reson. Med.* 71: 164–172.

97 Haris, M., Singh, A., Cai, K. et al. (2014). A technique for *in vivo* mapping of myocardial creatine kinase metabolism. *Nat. Med.* 20: 209–214.

98 Cai, K., Haris, M., Singh, A. et al. (2012). Magnetic resonance imaging of glutamate. *Nat. Med.* 18: 302–306.

99 Walker-Samuel, S., Ramasawmy, R., Torrealdea, F. et al. (2013). *In vivo* imaging of glucose uptake and metabolism in tumors. *Nat. Med.* 19: 1067–1072.

100 Xu, X., Chan, K.W., Knutsson, L. et al. (2015). Dynamic glucose enhanced (DGE) MRI for combined imaging of blood-brain barrier break down and increased blood volume in brain cancer. *Magn. Reson. Med.* 74: 1556–1563.

101 van Zijl, P.C., Jones, C.K., Ren, J. et al. (2007). MRI detection of glycogen *in vivo* by using chemical exchange saturation transfer imaging (glycoCEST). *Proc. Natl. Acad. Sci. U. S. A.* 104: 4359–4364.

102 Zhou, J. and van Zijl, P.C.M. (2006). Chemical exchange saturation transfer imaging and spectroscopy. *Prog. Nucl. Magn. Reson. Spectrosc.* 48: 109–136.

103 van Zijl, P.C. and Yadav, N.N. (2011). Chemical exchange saturation transfer (CEST): what is in a name and what isn't? *Magn. Reson. Med.* 65: 927–948.

104 van Zijl, P.C.M. and Shegal, A.A. (2016). Proton chemical exchange saturation transfer (CEST) MRS and MRI. *eMagRes* 5: 1307–1332.

105 Fick, A. (1855). On diffusion. *Poggendorf's Ann. Phys.* 94: 59–86.

106 Fick, A. On liquid diffusion. *Philos. Mag. J. Sci.* 1855 (10): 30–39.

107 Hahn, E.L. (1950). Spin echoes. *Phys. Rev.* 80: 580–594.

108 Carr, H.Y. and Purcell, E.M. (1954). Effects of diffusion on free precession in nuclear magnetic resonance experiments. *Phys. Rev.* 94: 630–638.

109 Woessner, D.E. (1961). Effects of diffusion in nuclear magnetic resonance spin-echo experiments. *J. Chem. Phys.* 34: 2057–2061.

110 Stejskal, E.O. and Tanner, J.E. (1965). Spin diffusion measurements: spin echoes in the presence of a time-dependent field gradient. *J. Chem. Phys.* 42: 288–292.

111 Torrey, H.C. (1956). Bloch equations with diffusion terms. *Phys. Rev.* 104: 563–565.

112 Le Bihan, D. (1995). *Diffusion and Perfusion Magnetic Resonance Imaging*. New York: Raven Press.

113 Moseley, M.E., Cohen, Y., Mintorovitch, J. et al. (1990). Early detection of regional cerebral ischemia in cats: comparison of diffusion- and T_2-weighted MRI and spectroscopy. *Magn. Reson. Med.* 14: 330–346.

114 Basser, P.J., Mattiello, J., and Le Bihan, D. (1994). MR diffusion tensor spectroscopy and imaging. *Biophys. J.* 66: 259–267.

115 Moonen, C.T., van Zijl, P.C., Le Bihan, D., and DesPres, D. (1990). *In vivo* NMR diffusion spectroscopy: ^{31}P application to phosphorus metabolites in muscle. *Magn. Reson. Med.* 13: 467–477.

116 Yoshizaki, K., Watari, H., and Radda, G.K. (1990). Role of phosphocreatine in energy transport in skeletal muscle of bullfrog studied by ^{31}P-NMR. *Biochim. Biophys. Acta* 1051: 144–150.

117 van Gelderen, P., DesPres, D., van Zijl, P.C., and Moonen, C.T. (1994). Evaluation of restricted diffusion in cylinders. Phosphocreatine in rabbit leg muscle. *J. Magn. Reson. B* 103: 255–260.

118 de Graaf, R.A., van Kranenburg, A., and Nicolay, K. (2000). *In vivo* ^{31}P-NMR diffusion spectroscopy of ATP and phosphocreatine in rat skeletal muscle. *Biophys. J.* 78: 1657–1664.

119 Pipe, J.G. and Chenevert, T.L. (1991). A progressive gradient moment nulling design technique. *Magn. Reson. Med.* 19: 175–179.

120 Posse, S., Cuenod, C.A., and Le Bihan, D. (1993). Human brain: proton diffusion MR spectroscopy. *Radiology* 188: 719–725.

121 Nicolay, K., Braun, K.P., de Graaf, R.A. et al. (2001). Diffusion NMR spectroscopy. *NMR Biomed.* 14: 94–111.

122 Ordidge, R.J., Helpern, J.A., Qing, Z.X. et al. (1994). Correction of motional artifacts in diffusion-weighted MR images using navigator echoes. *Magn. Reson. Imaging* 12: 455–460.

123 Moseley, M.E., Cohen, Y., Kucharczyk, J. et al. (1990). Diffusion-weighted MR imaging of anisotropic water diffusion in cat central nervous system. *Radiology* 176: 439–445.

124 Basser, P.J. and Jones, D.K. (2002). Diffusion-tensor MRI: theory, experimental design and data analysis – a technical review. *NMR Biomed.* 15: 456–467.

125 Scollan, D.F., Holmes, A., Winslow, R., and Forder, J. (1998). Histological validation of myocardial microstructure obtained from diffusion tensor magnetic resonance imaging. *Am. J. Physiol.* 275: H2308–H2318.

126 Hsu, E.W., Muzikant, A.L., Matulevicius, S.A. et al. (1998). Magnetic resonance myocardial fiber-orientation mapping with direct histological correlation. *Am. J. Physiol.* 274: H1627–H1634.

127 Scollan, D.F., Holmes, A., Zhang, J., and Winslow, R.L. (2000). Reconstruction of cardiac ventricular geometry and fiber orientation using magnetic resonance imaging. *Ann. Biomed. Eng.* 28: 934–944.

128 Le Bihan, D., Mangin, J.F., Poupon, C. et al. (2001). Diffusion tensor imaging: concepts and applications. *J. Magn. Reson. Imaging* 13: 534–546.

129 Basser, P.J., Mattiello, J., and Le Bihan, D. (1994). Estimation of the effective self-diffusion tensor from the NMR spin echo. *J. Magn. Reson.* 103: 247–254.

130 Douek, P., Turner, R., Pekar, J. et al. (1991). MR color mapping of myelin fiber orientation. *J. Comput. Assist. Tomogr.* 15: 923–929.

131 Nakada, T. and Matsuzawa, H. (1995). Three-dimensional anisotropy contrast magnetic resonance imaging of the rat nervous system: MR axonography. *Neurosci. Res.* 22: 389–398.

132 Conturo, T.E., Lori, N.F., Cull, T.S. et al. (1999). Tracking neuronal fiber pathways in the living human brain. *Proc. Natl. Acad. Sci. U. S. A.* 96: 10422–10427.

133 Mori, S., Crain, B.J., Chacko, V.P., and van Zijl, P.C. (1999). Three-dimensional tracking of axonal projections in the brain by magnetic resonance imaging. *Ann. Neurol.* 45: 265–269.

134 Jones, D.K., Simmons, A., Williams, S.C., and Horsfield, M.A. (1999). Non-invasive assessment of axonal fiber connectivity in the human brain via diffusion tensor MRI. *Magn. Reson. Med.* 42: 37–41.

135 Pajevic, S. and Pierpaoli, C. (1999). Color schemes to represent the orientation of anisotropic tissues from diffusion tensor data: application to white matter fiber tract mapping in the human brain. *Magn. Reson. Med.* 42: 526–540.

136 de Graaf, R.A., Braun, K.P., and Nicolay, K. (2001). Single-shot diffusion trace ^{1}H NMR spectroscopy. *Magn. Reson. Med.* 45: 741–748.

137 Ellegood, J., Hanstock, C.C., and Beaulieu, C. (2006). Diffusion tensor spectroscopy (DTS) of human brain. *Magn. Reson. Med.* 55: 1–8.

138 Upadhyay, J., Hallock, K., Erb, K. et al. (2007). Diffusion properties of NAA in human corpus callosum as studied with diffusion tensor spectroscopy. *Magn. Reson. Med.* 58: 1045–1053.

139 Palombo, M., Shemesh, N., Ronen, I., and Valette, J. (2017). Insights into brain microstructure from *in vivo* DW-MRS. *Neuroimage* doi: 10.1016/j.neuroimage.2017.11.028.

140 Neuman, C.H. (1974). Spin echo of spins diffusion in a bounded medium. *J. Chem. Phys.* 60: 4508–4511.

141 Palombo, M., Ligneul, C., and Valette, J. (2017). Modeling diffusion of intracellular metabolites in the mouse brain up to very high diffusion-weighting: diffusion in long fibers (almost) accounts for non-monoexponential attenuation. *Magn. Reson. Med.* 77: 343–350.

142 Shemesh, N., Ozarslan, E., Komlosh, M.E. et al. (2010). From single-pulsed field gradient to double-pulsed field gradient MR: gleaning new microstructural information and developing new forms of contrast in MRI. *NMR Biomed.* 23: 757–780.

143 Sokoloff, L., Reivich, M., Kennedy, C. et al. (1977). The [^{14}C]deoxyglucose method for the measurement of local cerebral glucose utilization: theory, procedure, and normal values in the conscious and anesthetized albino rat. *J. Neurochem.* 28: 897–916.

144 Reivich, M., Kuhl, D., Wolf, A. et al. (1979). The [18F]fluorodeoxyglucose method for the measurement of local cerebral glucose utilization in man. *Circ. Res.* 44: 127–137.

145 Mason, G.F., Falk Petersen, K., de Graaf, R.A. et al. (2003). A comparison of ^{13}C NMR measurements of the rates of glutamine synthesis and the tricarboxylic acid cycle during oral and intravenous administration of [1-^{13}C]glucose. *Brain Res. Brain Res. Protoc.* 10: 181–190.

146 Mason, G.F., Gruetter, R., Rothman, D.L. et al. (1995). Simultaneous determination of the rates of the TCA cycle, glucose utilization, alpha-ketoglutarate/glutamate exchange, and glutamine synthesis in human brain by NMR. *J. Cereb. Blood Flow Metab.* 15: 12–25.

147 Gruetter, R., Seaquist, E.R., and Ugurbil, K. (2001). A mathematical model of compartmentalized neurotransmitter metabolism in the human brain. *Am. J. Physiol. Endocrinol. Metab.* 281: E100–E112.

148 Mahler, H. and Cordes, E. (1971). *Biological Chemistry*. New York: Harper and Row.

149 Cunningham, V.J., Hargreaves, R.J., Pelling, D., and Moorhouse, S.R. (1986). Regional blood-brain glucose transfer in the rat: a novel double-membrane kinetic analysis. *J. Cereb. Blood Flow Metab.* 6: 305–314.

150 Gruetter, R., Ugurbil, K., and Seaquist, E.R. (1998). Steady-state cerebral glucose concentrations and transport in the human brain. *J. Neurochem.* 70: 397–408.

151 de Graaf, R.A., Pan, J.W., Telang, F. et al. (2001). Differentiation of glucose transport in human brain gray and white matter. *J. Cereb. Blood Flow Metab.* 21: 483–492.

152 Shulman, R.G. and Rothman, D.L. (2004). *Brain Energetics and Neuronal Activity*. Chichester: Wiley.

153 Sibson, N.R., Dhankhar, A., Mason, G.F. et al. (1997). *In vivo* ^{13}C NMR measurements of cerebral glutamine synthesis as evidence for glutamate-glutamine cycling. *Proc. Natl. Acad. Sci. U. S. A.* 94: 2699–2704.

154 Sibson, N.R., Dhankhar, A., Mason, G.F. et al. (1998). Stoichiometric coupling of brain glucose metabolism and glutamatergic neuronal activity. *Proc. Natl. Acad. Sci. U. S. A.* 95: 316–321.

155 Choi, I.Y., Tkac, I., Ugurbil, K., and Gruetter, R. (1999). Noninvasive measurements of [1-^{13}C]glycogen concentrations and metabolism in rat brain *in vivo*. *J. Neurochem.* 73: 1300–1308.

156 Choi, I.Y. and Gruetter, R. (2003). *In vivo* ^{13}C NMR assessment of brain glycogen concentration and turnover in the awake rat. *Neurochem. Int.* 43: 317–322.

157 Oz, G., Henry, P.G., Seaquist, E.R., and Gruetter, R. (2003). Direct, noninvasive measurement of brain glycogen metabolism in humans. *Neurochem. Int.* 43: 323–329.

158 Oz, G., Seaquist, E.R., Kumar, A. et al. (2007). Human brain glycogen content and metabolism: implications on its role in brain energy metabolism. *Am. J. Physiol. Endocrinol. Metab.* 292: E946–E951.

159 Sherry, A.D. and Malloy, C.R. (2016). Integration of ^{13}C isotopomer methods and hyperpolarization provides a comprehensive picture of metabolism. *eMagRes* 5: 885–900.

160 Shen, J., Petersen, K.F., Behar, K.L. et al. (1999). Determination of the rate of the glutamate/ glutamine cycle in the human brain by *in vivo* ^{13}C NMR. *Proc. Natl. Acad. Sci. U. S. A.* 96: 8235–8240.

161 Mason, G.F., Pan, J.W., Chu, W.J. et al. (1999). Measurement of the tricarboxylic acid cycle rate in human grey and white matter *in vivo* by ^{1}H-[^{13}C] magnetic resonance spectroscopy at 4.1T. *J. Cereb. Blood Flow Metab.* 19 (11): 1179–1188.

162 Pan, J.W., Stein, D.T., Telang, F. et al. (2000). Spectroscopic imaging of glutamate C4 turnover in human brain. *Magn. Reson. Med.* 44: 673–679.

163 Chen, W., Zhu, X.H., Gruetter, R. et al. (2001). Study of tricarboxylic acid cycle flux changes in human visual cortex during hemifield visual stimulation using ^{1}H-[^{13}C] MRS and fMRI. *Magn. Reson. Med.* 45: 349–355.

164 Chhina, N., Kuestermann, E., Halliday, J. et al. (2001). Measurement of human tricarboxylic acid cycle rates during visual activation by ^{13}C magnetic resonance spectroscopy. *J. Neurosci. Res.* 66: 737–746.

165 Lebon, V., Petersen, K.F., Cline, G.W. et al. (2002). Astroglial contribution to brain energy metabolism in humans revealed by ^{13}C nuclear magnetic resonance spectroscopy: elucidation of the dominant pathway for neurotransmitter glutamate repletion and measurement of astrocytic oxidative metabolism. *J. Neurosci.* 22: 1523–1531.

166 Mason, G.F., Petersen, K.F., de Graaf, R.A. et al. (2007). Measurements of the anaplerotic rate in the human cerebral cortex using ^{13}C magnetic resonance spectroscopy and [1-^{13}C] and [2-^{13}C] glucose. *J. Neurochem.* 100: 73–86.

167 Mason, G.F., Rothman, D.L., Behar, K.L., and Shulman, R.G. (1992). NMR determination of the TCA cycle rate and alpha-ketoglutarate/glutamate exchange rate in rat brain. *J. Cereb. Blood Flow Metab.* 12: 434–447.

168 de Graaf, R.A., Mason, G.F., Patel, A.B. et al. (2004). Regional glucose metabolism and glutamatergic neurotransmission in rat brain *in vivo*. *Proc. Natl. Acad. Sci. U. S. A.* 101: 12700–12705.

169 Sibson, N.R., Mason, G.F., Shen, J. et al. (2001). *In vivo* ^{13}C NMR measurement of neurotransmitter glutamate cycling, anaplerosis and TCA cycle flux in rat brain during [2-^{13}C]-glucose infusion. *J. Neurochem.* 76: 975–989.

170 Henry, P.G., Lebon, V., Vaufrey, F. et al. (2002). Decreased TCA cycle rate in the rat brain after acute 3-NP treatment measured by *in vivo* ^{1}H-[^{13}C] NMR spectroscopy. *J. Neurochem.* 82 (4): 857–866.

171 de Graaf, R.A., Brown, P.B., Mason, G.F. et al. (2003). Detection of [1,6-^{13}C$_2$]-glucose metabolism in rat brain by *in vivo* ^{1}H-[^{13}C]-NMR spectroscopy. *Magn. Reson. Med.* 49: 37–46.

172 Patel, A.B., de Graaf, R.A., Mason, G.F. et al. (2005). The contribution of GABA to glutamate/glutamine cycling and energy metabolism in the rat cortex *in vivo. Proc. Natl. Acad. Sci. U. S. A.* 102: 5588–5593.

173 van Eijsden, P., Behar, K.L., Mason, G.F. et al. (2010). *In vivo* neurochemical profiling of rat brain by ^{1}H-[^{13}C] NMR spectroscopy: cerebral energetics and glutamatergic/GABAergic neurotransmission. *J. Neurochem.* 112: 24–33.

174 Li, S., Yang, J., and Shen, J. (2007). Novel strategy for cerebral ^{13}C MRS using very low RF power for proton decoupling. *Magn. Reson. Med.* 57: 265–271.

175 Li, S., Zhang, Y., Wang, S. et al. (2010). ^{13}C MRS of occipital and frontal lobes at 3 T using a volume coil for stochastic proton decoupling. *NMR Biomed.* 23: 977–985.

176 Li, S., An, L., Yu, S. et al. (2016). ^{13}C MRS of human brain at 7 Tesla using [2-^{13}C]glucose infusion and low power broadband stochastic proton decoupling. *Magn. Reson. Med.* 75: 954–961.

177 Rothman, D.L., Behar, K.L., Hetherington, H.P. et al. (1985). ^{1}H-Observe/^{13}C-decouple spectroscopic measurements of lactate and glutamate in the rat brain *in vivo. Proc. Natl. Acad. Sci. U. S. A.* 82: 1633–1637.

178 Rothman, D.L., Hanstock, C.C., Petroff, O.A. et al. (1992). Localized ^{1}H NMR spectra of glutamate in the human brain. *Magn. Reson. Med.* 25 (1): 94–106.

179 De Feyter, H.M., Herzog, R.I., Steensma, B.R. et al. (2018). Selective proton-observed, carbon-edited (selPOCE) MRS method for measurement of glutamate and glutamine ^{13}C-labeling in the human frontal cortex. *Magn. Reson. Med.* 80: 11–20.

180 Waniewski, R.A. and Martin, D.L. (1998). Preferential utilization of acetate by astrocytes is attributable to transport. *J. Neurosci.* 18: 5225–5233.

181 Patel, A.B., de Graaf, R.A., Mason, G.F. et al. (2004). Glutamatergic neurotransmission and neuronal glucose oxidation are coupled during intense neuronal activation. *J. Cereb. Blood Flow Metab.* 24: 972–985.

182 Patel, A.B., de Graaf, R.A., Mason, G.F. et al. (2003). Coupling of glutamatergic neurotransmission and neuronal glucose oxidation over the entire range of cerebral cortex activity. *Ann. N. Y. Acad. Sci.* 1003: 452–453.

183 Patel, A.B., Chowdhury, G.M., de Graaf, R.A. et al. (2005). Cerebral pyruvate carboxylase flux is unaltered during bicuculline-seizures. *J. Neurosci. Res.* 79 (1–2): 128–138.

184 Lu, M., Zhu, X.H., Zhang, Y. et al. (2017). Quantitative assessment of brain glucose metabolic rates using *in vivo* deuterium magnetic resonance spectroscopy. *J. Cereb. Blood Flow Metab.* 37: 3518–3530.

185 Kanamori, K., Parivar, F., and Ross, B.D. (1993). A ^{15}N NMR study of *in vivo* cerebral glutamine synthesis in hyperammonemic rats. *NMR Biomed.* 6: 21–26.

186 Kanamori, K. and Ross, B.D. (1993). ^{15}N NMR measurement of the *in vivo* rate of glutamine synthesis and utilization at steady state in the brain of the hyperammonaemic rat. *Biochem. J.* 293: 461–468.

187 Shen, J., Sibson, N.R., Cline, G. et al. (1998). ^{15}N-NMR spectroscopy studies of ammonia transport and glutamine synthesis in the hyperammonemic rat brain. *Dev. Neurosci.* 20: 434–443.

188 Pekar, J., Ligeti, L., Ruttner, Z. et al. (1991). *In vivo* measurement of cerebral oxygen consumption and blood flow using ^{17}O magnetic resonance imaging. *Magn. Reson. Med.* 21: 313–319.

189 Fiat, D. and Kang, S. (1993). Determination of the rate of cerebral oxygen consumption and regional cerebral blood flow by non-invasive ^{17}O *in vivo* NMR spectroscopy and magnetic resonance imaging. Part 2. Determination of $CMRO_2$ for the rat by ^{17}O NMR, and $CMRO_2$, rCBF and the partition coefficient for the cat by ^{17}O MRI. *Neurol. Res.* 15: 7–22.

190 Zhu, X.H., Zhang, Y., Tian, R.X. et al. (2002). Development of ^{17}O NMR approach for fast imaging of cerebral metabolic rate of oxygen in rat brain at high field. *Proc. Natl. Acad. Sci. U. S. A.* 99: 13194–13199.

191 Johnson, R.T., Paulson, D.N., Giffard, R.P., and Wheatley, J.C. (1973). Bulk nuclear polarization of solid ^3He. *J. Low Temp. Phys.* 10: 35–58.

192 Frossati, G. (1998). Polarization of ^3He, D_2 (and possibly ^{129}Xe) using cryogenic techniques. *Nucl. Instrum. Meth. A* 402: 479–483.

193 Gadian, D.G., Panesar, K.S., Linde, A.J. et al. (2012). Preparation of highly polarized nuclear spin systems using brute-force and low-field thermal mixing. *Phys. Chem. Chem. Phys.: PCCP* 14: 5397–5402.

194 Hirsch, M.L., Kalechofsky, N., Belzer, A. et al. (2015). Brute-force hyperpolarization for NMR and MRI. *J. Am. Chem. Soc.* 137: 8428–8434.

195 Happer, W. (1972). Optical pumping. *Rev. Mod. Phys.* 44: 169–249.

196 Walker, T.G. and Happer, W. (1997). Spin-exchange optical pumping of noble-gas nuclei. *Rev. Mod. Phys.* 69: 629–642.

197 Kadlecek, S.J., Emami, K., Fischer, M.C. et al. (2005). Imaging physiological parameters with hyperpolarized gas MRI. *Prog. Nucl. Magn. Reson. Spectrosc.* 47: 187–212.

198 Bowers, C.R. and Weitekamp, D.P. (1986). Transformation of symmetrization order to nuclear-spin magnetization by chemical reaction and nuclear magnetic resonance. *Phys. Rev. Lett.* 57: 2645–2648.

199 Bowers, C.R. and Weitekamp, D.P. (1987). Parahydrogen and synthesis allow dramatically enhanced nuclear alignment. *J. Am. Chem. Soc.* 109: 5541.

200 Adams, R.W., Aguilar, J.A., Atkinson, K.D. et al. (2009). Reversible interactions with para-hydrogen enhance NMR sensitivity by polarization transfer. *Science* 323: 1708–1711.

201 Cowley, M.J., Adams, R.W., Atkinson, K.D. et al. (2011). Iridium *N*-heterocyclic carbene complexes as efficient catalysts for magnetization transfer from para-hydrogen. *J. Am. Chem. Soc.* 133: 6134–6137.

202 Abragam, A. and Goldman, M. (1978). Principles of dynamic nuclear polarization. *Rep. Prog. Phys.* 41: 395–467.

203 Ardenkjaer-Larsen, J.H., Fridlund, B., Gram, A. et al. (2003). Increase in signal-to-noise ratio of >10,000 times in liquid-state NMR. *Proc. Natl. Acad. Sci. U. S. A.* 100: 10158–10163.

204 Duckett, S.B. and Sleigh, C.J. (1999). Applications of the parahydrogen phenomenon: a chemical perspective. *Prog. Nucl. Magn. Reson. Spectrosc.* 34: 71–92.

205 Brindle, K.M., Bohndiek, S.E., Gallagher, F.A., and Kettunen, M.I. (2011). Tumor imaging using hyperpolarized ^{13}C magnetic resonance spectroscopy. *Magn. Reson. Med.* 66: 505–519.

206 Green, R.A., Adams, R.W., Duckett, S.B. et al. (2012). The theory and practice of hyperpolarization in magnetic resonance using parahydrogen. *Prog. Nucl. Magn. Reson. Spectrosc.* 67: 1–48.

207 Sriram, R., Kurhanewicz, J., and Vigneron, D. (2014). Hyperpolarized carbon-13 MRI and MRS studies. *eMagRes* 3: 1–14.

208 Kockenberger, W. (2014). Dissolution dynamic nuclear polarization. *eMagRes* 3: 161–170.

209 Brindle, K.M. (2015). Imaging metabolism with hyperpolarized ^{13}C-labeled cell substrates. *J. Am. Chem. Soc.* 137: 6418–6427.

210 Goodson, B.M., Whiting, N., Coffey, A.M. et al. (2015). Hyperpolarization methods for MRS. *eMagRes* 4: 797–810.

211 Comment, A. (2016). Dissolution DNP for *in vivo* preclinical studies. *J. Magn. Reson.* 264: 39–48.

212 Barskiy, D.A., Coffey, A.M., Nikolaou, P. et al. (2017). NMR hyperpolarization techniques of gases. *Chemistry* 23: 725–751.

213 Golman, K., Axelsson, O., Johannesson, H. et al. (2001). Parahydrogen-induced polarization in imaging: subsecond ^{13}C angiography. *Magn. Reson. Med.* 46: 1–5.

214 Zeng, H., Xu, J., McMahon, M.T. et al. (2014). Achieving 1% NMR polarization in water in less than 1 min using SABRE. *J. Magn. Reson.* 246: 119–121.

215 Abragam, A. and Proctor, W.G. (1958). Spin temperature. *Phys. Rev.* 109: 1441–1458.

216 Carver, T.R. and Slichter, C.P. (1953). Polarization of nuclear spins in metals. *Phys. Rev.* 92: 212–213.

217 Slichter, C.P. (2014). The discovery and renaissance of dynamic nuclear polarization. *Rep. Prog. Phy.* 77: 072501.

218 Day, S.E., Kettunen, M.I., Cherukuri, M.K. et al. (2011). Detecting response of rat C6 glioma tumors to radiotherapy using hyperpolarized [1-^{13}C]pyruvate and ^{13}C magnetic resonance spectroscopic imaging. *Magn. Reson. Med.* 65: 557–563.

219 Nelson, S.J., Kurhanewicz, J., Vigneron, D.B. et al. (2013). Metabolic imaging of patients with prostate cancer using hyperpolarized [1-^{13}C]pyruvate. *Sci. Transl. Med.* 5: 198ra108.

220 Cunningham, C.H., Lau, J.Y., Chen, A.P. et al. (2016). Hyperpolarized ^{13}C metabolic MRI of the human heart: initial experience. *Circ. Res.* 119: 1177–1182.

221 De Feyter, H.M., Behar, K.L., Corbin Z.A., Fulbright R.K., Brown P.B., McIntyre S., Nixon T.W., Rothman D.L., de Graaf R.A. (2018). Deuterium metabolic imaging (DMI) for MRI-based 3D mapping of metabolism *in vivo*. *Sci. Adv.* 4: eaat7314.

4

Magnetic Resonance Imaging

4.1 Introduction

Magnetic resonance imaging (MRI) [1, 2] is an essential part of any MR spectroscopy study. In its most basic form, MRI is used to provide the anatomical landmarks needed for MRS voxel placement. However, additional MRI scans are often beneficial for the interpretation of the spectroscopic results. For example, tissue segmentation based on quantitative T_1 mapping (Section 4.8.1) is useful to assess partial volume effects typically encountered with MRS voxels. While less critical for MRI, MRS data quality can be greatly improved by optimizing the B_0 and/or B_1 magnetic field homogeneity with MRI-based magnetic field maps (Sections 4.8.2 and 4.8.3). Furthermore, the metabolic information provided by MRS can be complemented with MR images depicting image contrast based on diffusion, blood flow, blood oxygenation, chemical exchange, and other processes.

Even when the primary focus is on MRS, it is advisable to have a thorough understanding of MRI acquisition and processing strategies in order to recognize potential artifacts and correctly interpret the complementary information provided by MRI. In addition, the principles of MRI also form the basis for MR spectroscopic imaging (MRSI) and will be the focus of Chapter 7.

4.2 Magnetic Field Gradients

The essential concept of MRI is that the resonance frequency ν is made position-dependent, such that after Fourier transformation the different frequencies correspond to spatial position rather than chemical shift. This can be accomplished by making the external magnetic field position-dependent with magnetic field gradients. The most commonly used magnetic field gradient is a magnetic field of which the amplitude varies linearly with position. Mathematically, a magnetic field gradient $G(r)$ whose amplitude varies linearly as a function of position r is given by

$$G(r) = \frac{dB(r)}{dr} = \text{constant} \tag{4.1}$$

where B is the magnetic field strength. The position r can represent any of the three orthogonal directions x, y, or z, or any vectorial combination thereof. Magnetic field gradients are generated by electrical currents in specially shaped coils within the bore of the magnet. Normally, there are three sets of gradients in a (MRI) magnet, the so-called X, Y, and Z gradients,

In Vivo *NMR Spectroscopy: Principles and Techniques*, Third Edition. Robin A. de Graaf.
© 2019 John Wiley & Sons Ltd. Published 2019 by John Wiley & Sons Ltd.

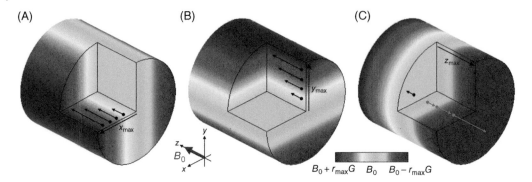

Figure 4.1 Magnetic fields of linear gradient coils. (A–C) Magnetic fields produced by (A) X, (B) Y, and (C) Z magnetic field gradients that add to the constant, static main magnetic field B_0. All magnetic gradient fields are zero in the gradient (and magnet) isocenter. Note that while the magnetic field strength increases linearly as a function of x, y, or z distance, the orientation of the magnetic field is always in the z direction (arrows), parallel to the main magnetic field.

corresponding to the direction along which the magnetic field strength changes. Chapter 10 will deal with the characteristics of gradient coils.

It is important to recognize the difference between the direction of a gradient and the direction of the magnetic field generated by that gradient. The magnetic field component that is relevant for NMR is by definition always directed along the z axis parallel to the main magnetic field B_0, *independent* of the gradient orientation. The direction of a gradient refers to the direction in which the strength of the magnetic field varies (Figure 4.1). The magnetic field gradient is positioned around the center of the magnet, i.e. a gradient adds to the main magnetic field on one side of the middle and subtracts from the static field on the other side. Therefore, the magnetic field strength of all gradients is zero in the magnet's isocenter.

The addition of a magnetic field gradient G to the static magnetic field B_0 generates a total magnetic field at position r given by

$$B(r) = B_0 + rG \tag{4.2}$$

When the position is expressed in meters (m), the gradient strength G is expressed in $T\,m^{-1}$ or $mT\,m^{-1}$. A commonly used non-SI unit for gradient strength is Gauss-per-centimeter or $G\,cm^{-1}$ in which 1 Tesla = 10 000 Gauss. The addition of a magnetic field gradient makes the Larmor resonance frequency linearly dependent on the spatial position according to

$$v(r) = \frac{\gamma}{2\pi} B_0 + \frac{\gamma}{2\pi} rG \tag{4.3}$$

Equation (4.3) reduces to $v(r) = v_0$ in the magnet isocenter ($r = 0$) or in the absence of a magnetic field gradient ($G = 0$). Equation (4.3) provides the basis for all MRI sequences, as well as most localized MRS methods. Using Eq. (4.3) all three dimensions of a three-dimensional object can be encoded. The specific encoding can be achieved in a number of ways, all of which use either the frequency or the phase of the signal to obtain spatial information.

4.3 Slice Selection

The spatial encoding of a three-dimensional object such as the human head is often reduced to a two-dimensional problem through the selection of a thin slice. A spatial slice can be selected by the combination of an RF pulse and a magnetic field gradient. A magnetic field gradient in

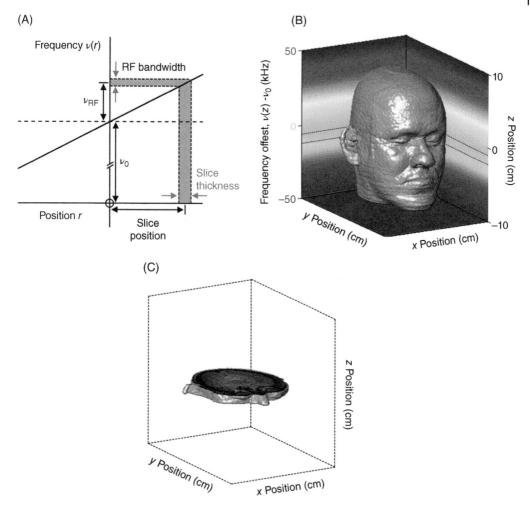

(A)

Frequency ν(r)

RF bandwidth

ν$_{RF}$

ν$_0$

Slice thickness

Position r

Slice position

(B)

Frequency offest, ν(z) -ν$_0$ (kHz)

y Position (cm)

x Position (cm)

z Position (cm)

(C)

y Position (cm)

x Position (cm)

z Position (cm)

Figure 4.2 **Principle of slice selection.** (A) A magnetic field gradient along any of the three Cartesian axes, or any vectorial combination thereof generates a linear relation between Larmor frequency and spatial position. (A–C) When spins are excited with a frequency-selective RF pulse in the presence of a magnetic field gradient, the frequency range will correspond to a spatial slice. The slice thickness and position are both dependent on the magnetic field gradient strength, whereby the slice thickness and position are also dependent on the RF pulse bandwidth and RF pulse offset, respectively. See text for more details.

the z direction creates a linear magnetic field distribution as function of the z position. As a consequence, each z position is characterized by a specific magnetic field and hence a specific resonance frequency, according to Eq. (4.3). When a RF pulse, specially designed to excite only a selective frequency range (see Chapter 5), is applied simultaneously with the magnetic field gradient, only a selective range of frequencies, and hence spatial positions, is excited (Figure 4.2).

The slice thickness is determined by two factors, the magnetic field gradient strength and the bandwidth of the RF pulse. A stronger magnetic field gradient creates a larger range of resonance frequencies across the sample (i.e. the slope of the diagonal line in Figure 4.2A increases), resulting in the selection of a thinner spatial slice when the RF pulse is kept constant (in terms of frequencies the slice thickness has not changed, since the bandwidth of the RF pulse remained constant). Vice versa, the same can be achieved by decreasing the RF pulse bandwidth (e.g. by increasing the pulse length), while keeping the magnetic gradient strength constant. On-resonance, both approaches lead to the same spatial slice. However, off-resonance, the former approach gives a smaller chemical shift displacement and is therefore the preferred method (see Chapter 6).

The spatial position of the slice is also determined by two factors, the magnetic field gradient strength and the transmitter frequency of the RF pulse, ν_{RF} (Figure 4.2A). The magnetic field gradient creates the spatial distribution of frequencies, but the transmitter frequency (and the bandwidth) of the RF pulse determines which frequencies are being excited. When the transmitter frequency equals the Larmor frequency, spins in the middle of the magnet ($\pm 0.5 \times$ bandwidth) are excited. To excite spins at a certain distance from the isocenter, the transmitter frequency should be adjusted according to Eq. (4.3).

Since the magnetic field gradient strength affects both the slice thickness and the slice position, it is a common practice to first determine the required gradient strength (for a given RF pulse length), after which the position is selected with the transmitter frequency of the RF pulse. For example, selection of a 2 mm slice, 5 mm out of the isocenter using a RF pulse with a bandwidth of 1000 Hz requires a gradient strength of

$$G_z = \frac{\text{RF bandwidth}}{\text{slice thickness}} = \frac{1.0}{0.2}\,\text{kHz cm}^{-1} = 5.0\,\text{kHz cm}^{-1} = 1.17\,\text{G cm}^{-1} = 11.7\,\text{mT m}^{-1} \quad (4.4)$$

The conversion from kHz cm^{-1} to G cm^{-1} is made for the proton-specific gyromagnetic ratio of $4257.6\,\text{Hz G}^{-1}$. With a fixed gradient strength and a constant RF pulse, the slice position can only be adjusted by changing the transmitter frequency according to

$$\nu_{RF} = \nu_0 + \text{slice position}\,(\text{cm}) \times G_z\,\left(\text{kHz cm}^{-1}\right) = \nu_0 + 2.5\,\text{kHz} \quad (4.5)$$

where ν_0 is the Larmor frequency in the absence of magnetic field gradients. Note that Eqs. (4.4) and (4.5) are modifications of Eq. (4.3), whereby units are in kHz and cm. While elimination of the gyromagnetic ratio makes the variables nucleus-specific, the use of uniform units (kHz and cm) for RF pulse frequency ν_{RF} (kHz), RF pulse amplitude $(\gamma/2\pi)B_{1max}$ (kHz), magnetic field strength $(\gamma/2\pi)B_0$ (kHz), and gradient strength $(\gamma/2\pi)G$ (kHz cm^{-1}) greatly simplify calculations and avoids common mistakes.

Slice selection can be performed for a single spatial slice, but can also be extended to select multiple spatial slices. The selection of multiple slices can be performed in a time-efficient manner in which signal from several slices is acquired during the recovery period of a single slice. This interleaved approach gives rise to multi-slice imaging [3]. Suppose the repetition time of a slice-selective imaging sequence is 2500 ms and the total acquisition time (from excitation to the end of acquisition) is 250 ms. The remaining 2250 ms can then be used to excite and acquire nine other spatial slices. Note that, in order to minimize cross-slice interference, all RF pulses in the MRI sequence need to be selective and need to select the same spatial slice. Even though all RF pulses are spatially selective, the acquisition of subsequent slices will lead to interference when the selected slices are close to each other (i.e. near-continuous coverage of the object under investigation). This is because the slice profile is not perfectly rectangular, leading to partial excitation and hence overlap of adjacent slices. For this reason, the slices are sometimes acquired in an interleaved manner (i.e. for 10 slices the acquisition order is 1, 3, 5, 7, 9, 2, 4, 6, 8, 10). In this manner, the effect of neighboring slices is significantly reduced (see also Exercises). An important point that should be noted is that the bandwidths of identical, frequency selective 90° and 180° RF pulses seldom have the same frequency profile, such that the selected slices are of different thickness (and/or shape).

For instance, the bandwidths of identical 90° and 180° sinc pulses differ by almost a factor of two (see Chapter 5). This has to be taken into account when quantitative, volumetric MRI measurements need to be performed.

4.4   Frequency Encoding

4.4.1   Principle

After selecting one or more spatial slices, the origin of the MR signal needs to be encoded in the two spatial dimensions within the slice. Again Eq. (4.3) provides the basis for spatial encoding. In the slice selection process, the distribution of spatially-dependent frequencies was maintained during the RF pulse in order to enable the RF pulse to select (e.g. excite) the desired frequencies. When, after excitation, the distribution of spatially-dependent frequencies is maintained during signal acquisition, the FID would hold information about the spatial origin of the signal.

Consider the object shown in Figure 4.3A consisting of three water-filled tubes. Signal obtained with a pulse-acquire sequence (Figure 4.3B and C) contains the Larmor resonance frequency of the water spins, but does not provide any information on the spatial distribution of the water. Spatial information can be encoded when a magnetic field gradient is applied imme-diately following excitation (Figure 4.3E). The magnetic field gradient causes the water spins in the three different tubes to attain different Larmor frequencies that are linearly dependent on spatial position (Figure 4.3D). Since the magnetic field gradient is applied during signal acquisi-tion, the different Larmor frequencies present in the sample are directly detected (Figure 4.3E).

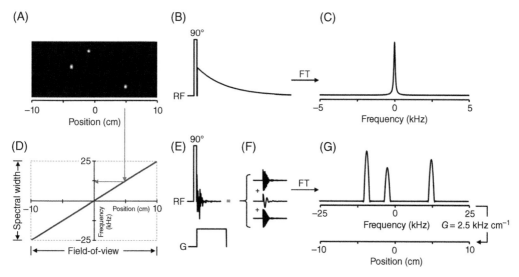

Figure 4.3 **Principle of frequency encoding.** (A) 2D object consisting of several small tubes. (B) A standard pulse-acquire experiment without magnetic field gradients provides (C) a spectrum with chemical information. (D, E) In order to obtain spatial information the pulse-acquire experiment can be extended with a magnetic field gradient such that each tube resonates with a different Larmor frequency. (F) Acquisition and (G) Fourier transformation of the multiple frequencies provides a spectrum in which the signal intensity corresponds to the proton density at every spatial location. With knowledge of the magnetic field gradient strength, the frequency axis can be converted to a spatial axis, thereby providing a 1D image of the object along the direction of the applied magnetic field gradient. The 1D image is commonly referred to as a readout profile.

Note that the total detected signal is simply the sum of the signals coming from the three different tubes (Figure 4.3F). Fourier transformation of the detected signal creates a MR "spectrum" that represents a 1D projection of the spin density distribution along the direction of the applied gradient (Figure 4.3G). The range of frequencies that are detected is set by the spectral width which, according to the Nyquist sampling theorem, is set by the reciprocal of the acquisition dwell-time (see Chapter 1). For MRI applications the frequency range (i.e. spectral width) is always converted to a position range (i.e. field-of-view or FOV) through division by the gradient strength (Figure 4.3G). The 1D image is typically referred to as a "readout" profile.

4.4.2 Echo Formation

While the simple pulse sequence shown in Figure 4.3E can explain the principles of frequency encoding, it is not an optimal choice for actual MRI studies. Hardware limitations on gradient switching make it impossible to instantaneously turn on the magnetic field gradient. As a result, the first few data points are sampled during a time-varying gradient which would lead to artifacts when a regular Fourier transformation is applied. As an alternative, the first few data points can be eliminated. Unfortunately, since the first few points contain the strongest MR signal, their elimination would greatly diminish the obtainable signal-to-noise ratio (SNR). The problem of finite gradient rise times is circumvented through echo formation with a gradient-echo sequence element (Figure 4.4).

A gradient-echo sequence consists, besides the excitation pulse, of two gradient pulses. One is applied prior to signal acquisition while the other, being of opposite sign and typically having twice the area, is applied during signal acquisition. Formation of the echo containing the spatially dependent frequencies can be explained in a number of ways. In Section 4.6, a concise *k*-space formalism applicable to any imaging sequence will be presented. For now, a more visual approach will be followed.

The function of the first (negative) magnetic field gradient in Figure 4.4B is to prepare the transverse magnetization for encoding of spatial information during signal acquisition. At position *r* the effect of the first gradient pulse is the generation of a phase shift given by

$$\phi_A(r,t) = \gamma r \int_0^t G_A(t')dt' \tag{4.6}$$

which for a constant-amplitude gradient (i.e. $G_A(t) = G_A$ when the gradient ramps in Figure 4.4 are of zero duration) simplifies to

$$\phi_A(r,t) = \gamma r G_A t \tag{4.7}$$

At the end of the first gradient pulse of duration T_A the transverse magnetization at each position *r* is encoded (i.e. prepared) with a specific phase shift $\phi_A(r, T_A)$. The second gradient is of opposite sign and the total area is twice that of the first gradient. Therefore, the total applied gradient in the middle of the second gradient is zero and consequently the acquired phase shift also equals zero. Since the transverse magnetization prior to signal acquisition had acquired a linear position-dependent phase shift during the first gradient, spins at different spatial positions have to rotate at different frequencies to fulfill the condition $\phi(r) = 0$ in the middle of the second gradient. The frequencies $\nu(r)$ are related to the initial acquired phase shifts according to

$$\nu(r) = \frac{d\phi(r,t)}{2\pi dt}, \quad \text{where } \phi(r,t) = \phi_A(r,T_A) + \gamma r \int_0^t G_B(t')dt' \tag{4.8}$$

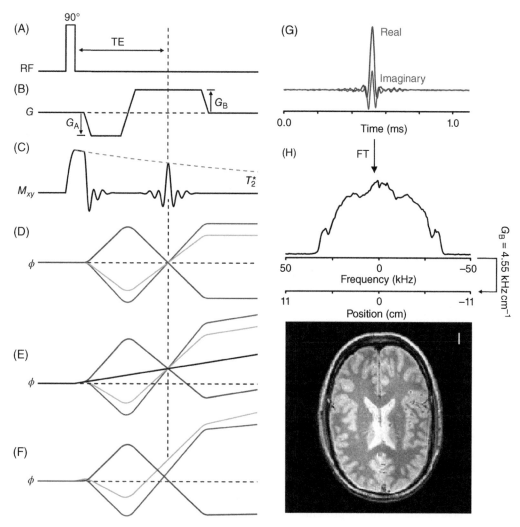

Figure 4.4 **Gradient-echo formation.** (A, B) Gradient echo pulse sequence consisting of an excitation pulse and two magnetic field gradients of opposite sign. MR signal acquisition occurs during the plateau of the second, positive magnetic field gradient. (C) Macroscopic magnetization as a function of RF and magnetic field gradients. While the initial FID signal is spatially encoded, it is not commonly used as the resulting readout profile would be distorted due to the time-varying spatial encoding during the gradient ramp. (D) Phase accumulation of spins at three spatial locations. Echo formation occurs when the phase accumulation is identical for all spins, which happens at echo-time TE when the positive-gradient area equals the negative-gradient area. (E) In the presence of a global frequency offset, echo formation still occurs at TE, albeit in the presence of a nonzero phase. (F) In the presence of magnetic field inhomogeneity (i.e. local frequency offsets) the residual phase at time of acquisition results in signal loss and (local) spatial displacement. (G) Gradient echo or "readout" signal acquired from human head. (H) Fourier transformation of the echo in (G) reveals the spin distribution along the frequency-encoding direction. The frequency axis is readily converted to a spatial axis by using Eq. (4.3). The spin distribution in the readout profile clearly shows the boundaries of the skull, as well as the separation between the two hemispheres as visible in the 2D reference image (I).

which for a constant-amplitude gradient (i.e. $G_B(t) = G_B$) simplifies to

$$\phi(r,t) = \phi_A(r, T_A) + \gamma r G_B t \qquad (4.9)$$

G_B is the amplitude of the second gradient, which is opposite to G_A such that $\phi(r, t)$ decreases over time. Maximum echo formation (i.e. $\phi(r) = 0$ for all positions r) occurs when $\phi_A(r, T_A) = -\phi_B(r, T_B)$ or, in other words, $G_A T_A = -G_B T_B$ (i.e. when the total, net gradient is zero). In the case of nonzero duration gradient ramps, the expression is more difficult but still relies on the principle that echo formation occurs when the phase induced by the first, negative gradient area is canceled by the second, positive gradient area for all spatial positions (see also Exercises). Since signal acquisition was performed during the second frequency encoding or "readout" gradient, the spatially dependent frequencies given by Eq. (4.8) are recorded and will give the spatial spin distribution upon Fourier transformation. Equations (4.6)–(4.9) and Figure 4.4D describe the phase evolution of spins at different spatial locations. In reality, signal is acquired from the entire sample simultaneously and phase evolution at individual positions can no longer be observed directly. Macroscopically, the phase dispersion of spins at different spatial locations leads to phase cancelation and hence signal loss. The phase dispersal and signal loss is generally larger with increasing total gradient area, but as long as signal has not been irreversibly lost due to T_2 relaxation, the phase dispersal and signal loss can be undone by reducing the gradient area. Maximum macroscopic signal is observed at the point of echo formation when the total gradient area is zero.

From the signal evolution $S(t)$ shown in Figure 4.4C it follows that while the FID signal is distorted by time-varying gradients, the echo is acquired during a period of constant gradient amplitude, thereby leading to an undistorted spatial profile following Fourier transformation. The outlined procedure is generally referred to as frequency encoding and is the most often employed method to obtain 1D spatial information.

In realistic samples, like the human head, externally applied magnetic field gradients are typically not the only perturbations to the homogeneous magnetic field B_0. Local magnetic field gradients and offsets are caused by differences in magnetic susceptibility (see Chapter 10), while magnet drift, respiration, and receiver offsets can cause a global frequency offset. Equations (4.6)–(4.9) and Figure 4.4A–D only considered the phase evolution due to externally applied magnetic field gradients. In the presence of a global magnetic field offset, the phase evolution changes as shown in Figure 4.4E. Besides the gradient-induced phases, the spins acquire an additional phase that increases linearly over time (Figure 4.4E, black line). As the magnetic field offset is never inverted, the phase accumulation continues over time, thereby making it impossible to achieve a zero phase for all spins. However, since it is still possible to achieve an equal (albeit nonzero) phase for all spins, perfect echo formation is still possible (Figure 4.4E). The magnetic field offset causes spins at different locations to attain different Larmor frequencies, which upon Fourier transformation, leads to a global, uniform image shift. In the presence of *local* magnetic field offsets (i.e. magnetic field inhomogeneity), it is no longer possible to achieve an equal phase for all spins at any point in time (Figure 4.4F). As a result, the phase dispersion leads to phase cancelation and hence signal loss. Furthermore, the additional frequency encoding caused by the local magnetic field offsets leads to shifted, incorrect spatial positions. In other words, magnetic field inhomogeneity leads to geometric image distortion in the frequency encoding direction. Consider a MRI sequence that acquires 200 pixels over a spectral width of 50 kHz, or 250 Hz per pixel. A local magnetic field offset of 250 Hz would then lead to a spatial misregistration of 1 pixel. While this effect can often be neglected, it becomes a significant consideration in the presence of metallic implants (with local offsets of several kHz) or when the (effective) spectral width becomes very small, as encountered for ultrafast MRI (see Section 4.7).

As shown in Figures 4.3 and 4.4, frequency encoding provides the spin density distribution along one spatial dimension. However, a one-dimensional projection does not provide information on the internal structure of two- or three-dimensional objects (Figure 4.4I). Therefore, a suitable method needs to be employed to identify the volume elements that span a multidimensional (e.g. two-dimensional) space. The described frequency encoding procedure provides a good method for imaging along one of the required spatial coordinates. The spatial distribution along the second dimension can be obtained in several ways. The first method employed [1] was the usage of a second frequency-encoding magnetic field gradient along an orthogonal direction executed in concert with the first frequency-encoding gradient. Since magnetic field gradients are vectors, the effect of two simultaneous gradients can be added in a vectorial manner. Therefore, two simultaneous magnetic field gradients of equal strength execute frequency encoding along an oblique orientation which makes an 45° angle with either gradient direction. The orientation of the oblique projection (given by angle θ relative to gradient G_1) can easily be varied by changing the relative amplitudes of the two magnetic field gradients, i.e. $\theta = \arctan(G_2/G_1)$. By making a large number of projections between $\theta = 0°$ and 180° the two-dimensional object can be reconstructed by the method of filtered back-projection, a technique originating from X-ray tomography. Although the technique of projection reconstruction started off the field of MRI [1], its popularity has decreased considerably and its applications are now limited to imaging of very short T_2 nuclei. For the majority of MRI methods, the second spatial dimension is characterized by phase encoding.

4.5 Phase Encoding

Figure 4.5A shows a gradient-echo MRI sequence employing frequency encoding with magnetic field gradient G_{freq} and phase encoding with magnetic field gradient G_{phase}. To obtain spatial information along the second dimension, a number of experiments need to be performed in which the amplitude of the phase-encoding gradient G_{phase} is incremented linearly. During each experiment the signal is acquired in the presence of a frequency-encoding gradient after which the resulting echo is stored in a 2D matrix, known as spatial frequency or k-space. Since the phase-encoding gradient is positioned before the acquisition window, it can only affect the phase of the signal. However, a phase that changes linearly with time (or with the amplitude of a phase-encoding gradient) makes up an indirect frequency $\nu = d\phi/(2\pi dt)$ that is independent of the frequencies acquired during frequency encoding.

Consider the three water-filled tubes from Figure 4.3A, whereby frequency encoding was used to resolve the signal distribution along the first spatial dimension (Figure 4.5D, middle trace with $G_{phase} = 0$). In the presence of a phase-encoding gradient (i.e. $G_{phase} \neq 0$) the signal from the tubes acquires a phase that is linearly dependent on their position along the phase-encoding gradient direction. For example, since the phase acquired by the left tube is much smaller than that acquired by the middle tube, it can be deduced that the left tube is much closer to the magnetic isocenter (in the phase-encoding direction) than the middle tube. Since the tubes have a non-negligible size, the acquired signal at stronger phase-encoding gradient amplitudes diminishes due to phase dispersal across a tube. Visually it is apparent from Figure 4.5C (and D) that the phase as a function of the phase-encoding gradient amplitude makes up a frequency that is different for the three tubes and that is directly proportional to their spatial position (Figure 4.5E). Fourier transformation of the phase-modulated signals with respect to the phase-encoding gradient amplitude generates a 2D image of the object (Figure 4.5E). Figures 4.3 and 4.5 provide an intuitive, albeit qualitative description of MR

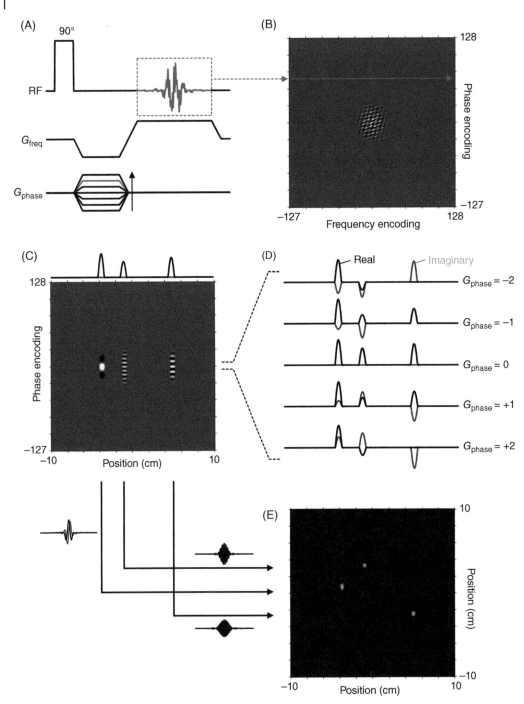

Figure 4.5 **Principle of phase encoding.** (A) Gradient echo pulse sequence extended with a phase-encoding gradient. (B) Data acquired during specific phase-encoding gradient amplitudes are stored as rows in a two-dimensional matrix which is typically referred to as *k*-space. (C, D) In image space, the phase of the 1D readout profile is modulated as a function of the phase-encoding gradient amplitude and position along the phase-encoding gradient direction. (E) The phase modulation makes up an apparent frequency, independent of the frequency encoding, which upon Fourier transformation provides the position along the phase-encoding direction.

imaging. In order to understand the origin of artifacts in MRI or the performance of more complicated imaging sequences, a more concise mathematical representation of MRI is needed. This is provided by the spatial frequency of k-space formalism.

4.6   Spatial Frequency Space

Slice selection, frequency encoding, and phase encoding are the standard procedures to independently encode all three spatial dimensions of an object. The spin warp MRI method as first described by Edelstein et al. [4] in 1980 combines slice selection, frequency encoding, and phase encoding into one pulse sequence (Figure 4.6A and B) and is still the most commonly used MR imaging method. The spin-warp sequence can be implemented as a spin-echo (Figure 4.6A) or gradient-echo (Figure 4.6B) variant, whereby the choice is primarily governed by the desired image contrast, RF power deposition, and image acquisition speed. Both variants are readily extended to allow multi-slice acquisition.

The qualitative description of frequency and phase encoding provided in Figures 4.3 and 4.5 will now be extended to a formal description of the data acquisition or k-space [5–7] that provides a simple Fourier transform relationship between the acquired signal and the MR image. In the presence of a time-dependent magnetic field gradient $G(t)$ the transverse magnetization M_{xy} acquired from a macroscopic sample is given by

$$M_{xy}\big(G(t)\big) = M_x\big(G(t)\big) + iM_y\big(G(t)\big) = \int_{-\infty}^{+\infty} M_0(r)e^{+i\gamma r \int_0^t G(t')dt'} dr \tag{4.10}$$

where $M_0(r)$ is the spin density at position r. Depending on the particular MRI sequence used, Eq. (4.10) should be multiplied with factors describing the effects of relaxation or diffusion. By defining a spatial frequency variable $k(t)$ as

$$k(t) = \gamma \int_0^t G(t')dt' \tag{4.11}$$

Eq. (4.10) reduces to

$$M_{xy}\big(k(t)\big) = \int_{-\infty}^{+\infty} M_0(r)e^{+ik(t)r} dr \tag{4.12}$$

Equation (4.12) can be recognized as an inverse Fourier transformation relation between the detected MR signal $M_{xy}(k(t))$ and the spatial spin density distribution $M_0(r)$. In other words, the spatial spin density distribution (i.e. the image) can be obtained through a Fourier transformation of the acquired signal (i.e. k-space). Note the similarity between the definition of the spatial frequency variable $k(t)$ in Eq. (4.11) and the phase acquired by a spin at position r during a gradient sequence in Eq. (4.6). The $k(t)$ parameter can thus be interpreted as the phase acquired by a spin placed at unit distance (for example, at 1 cm when the gradient is expressed in $G\,cm^{-1}$).

According to Eq. (4.12) the detected signal $M_{xy}(k(t))$ tends to zero when $k(t)$ goes to infinity. Under normal circumstances the decay of $M_{xy}(k(t))$ will be very rapid with increasing $k(t)$. However, as long as the magnetization has not been irreversibly dephased by T_2 relaxation, $M_{xy}(t)$ can always be recalled by letting the k-space trajectory return to low $k(t)$ values, i.e. a gradient echo can be generated when the total integrated gradient area (i.e. $k(t)$) tends

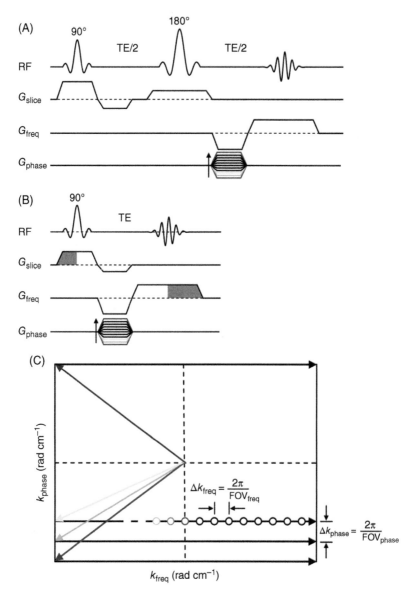

Figure 4.6 **Spin-echo and gradient-echo MRI.** (A) Spin-echo or spin-warp and (B) gradient-echo MRI sequences. In (A) the phase-encoding and negative frequency-encoding preparation gradients can be placed before the 180° refocusing pulse with inverted sign. As a rule of thumb, the slice selection gradients during the 90° and 180° pulses are not equal, as the bandwidths of the RF pulses are different. The negative gradient following the excitation pulse is required to refocus signal dephasing during excitation (see also Figure 5.9). The shaded gradients in (B) require balancing with gradients of opposite sign in order to allow steady-state free precession (SSFP). (C) Phase-encoded signals are stored as discrete points in *k*-space, whereby the spacing between adjacent points is inversely proportional to the FOV.

to zero. As an example of the structure of k-space, consider the spin-warp imaging sequence in Figure 4.6A or B. Figure 4.6C shows the corresponding k-space representation, in which the horizontal axis indicates k_{freq} (the k-parameter of the frequency-encoding direction) and the vertical axis indicates k_{phase} (the k-parameter of the phase-encoding direction). Immediately after excitation $k_{freq} = k_{phase} = 0$, since no gradient has been applied in either direction. In the first experiment, a maximum negative phase-encoding gradient will be applied (Figure 4.6C, red line) at the same time as the dephasing gradient in the frequency-encoding direction. According to the definition of the k-space variable (Eq. (4.11)) this gradient combination changes the k-space coordinate from the center to the lower left corner. However, at this point in the sequence no signal acquisition has been performed yet; the k-space coordinate is just being prepared to be at the correct k-space coordinate when data acquisition commences. During the acquisition period a positive frequency-encoding gradient is applied, thereby linearly increasing k_{freq} over time. At the top of the spin echo, $k_{freq} = 0$ and maximal signal will be observed (for that particular k_{phase}, which is nonzero in the first experiment). Since no phase-encoding gradient is applied during acquisition, the frequency-encoded data are stored along the horizontal line indicated in Figure 4.6C. In the following experiments k_{phase} is changed in discrete steps, such that k-space is "filled" with spatially encoded spin or gradient echoes according to parallel lines. In the middle experiment $k_{phase} = 0$ and maximum signal will be observed at the top of the echo ($k_{freq} = 0$). Because of the low k_{freq} and k_{phase} values, coordinates near the middle of k-space are called low spatial frequencies, while coordinates at the edges of k-space are referred to as high spatial frequencies. In general, it can be said that low spatial frequencies (which have the highest intensity) correspond to the overall shape of images, while high spatial frequencies represent more detailed features in the final image. Figure 4.7 shows an example of the relative importance of the different k-space coordinates on the final appearance of the image. An image reconstructed from low spatial frequencies only (Figure 4.7B) has approximately the same signal intensity as the image originating from the entire k-space data (Figure 4.7A). However, the image appears blurry (i.e. devoid of detailed spatial structures) because image detail is primarily contained in the missing spatial frequencies at higher k-space coordinates. When the image is reconstructed from high spatial frequencies only (Figure 4.7C), the resulting MR image has a lot of details and edges. Unfortunately, since the low spatial frequencies contain the bulk of the signal intensity, the SNR of the image is compromised. From Figure 4.7 it should be clear that a high-sensitivity, high-resolution MR image will need all of k-space.

The relation between gradient strength, gradient duration, spectral width, and FOV can be derived from the Nyquist sampling theory. A gradient-echo signal is, just as a regular FID, sampled at discrete time intervals Δt, also known as the dwell time. The Nyquist sampling theory states that a frequency is only accurately described when it is sampled at least twice per period, i.e. the phase difference between two sampling points should be maximally 2π over the entire FOV. This is equivalent in stating that signals are only accurately sampled when their frequency falls within the spectral bandwidth, which equals the reciprocal of the dwell time. The FOV in the frequency-encoding direction can then be expressed as

$$\text{FOV}(\text{cm}) = \frac{\text{spectral width}(\text{kHz})}{\text{gradient strength}(\text{kHz cm}^{-1})} = \frac{1}{\text{gradient strength}(\text{kHz cm}^{-1})\text{dwell time}(\text{ms})}$$

$$(4.13)$$

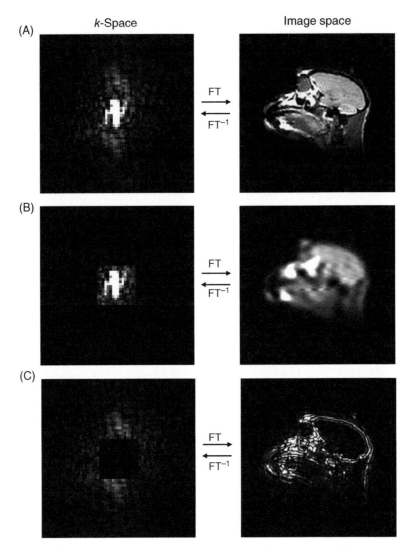

Figure 4.7 **Relationship between spatial frequency or *k*-space and image space.** (A) Fourier transformation of the entire *k*-space gives rise to a detailed image with high *S/N*. (B) Fourier transformation of only the low frequencies in *k*-space gives rise to a blurry image with almost the same *S/N* as in (A) but lacking significant details. (C) The higher spatial frequencies in *k*-space give rise to a very detailed image with a highly compromised *S/N*. Note that the vertical scale in (C) is twice that of (A) or (B).

Using the definition of *k*-space according to Eq. (4.11) the FOV can also be expressed in terms of the distance between *k*-space sampling points according to

$$\text{FOV (cm)} = \frac{2\pi}{\Delta k \left(\text{rad cm}^{-1}\right)} \qquad (4.14)$$

Equation (4.13) is only valid for the frequency-encoding direction in which the signal is sampled over a given receiver or spectral bandwidth. The phase-encoding direction has no associated spectral width, such that Eq. (4.14) must be used for the calculation of the FOV.

In that case the k-space separation Δk is given by $\Delta k = \gamma \Delta GT$, where T is the length of the (rectangular) phase-encoding gradient. In a typical MRI experiment of human brain, FOV = 24 cm, G = 5.0 mT m^{-1}, and $\gamma(^1\text{H})$ = 4257.6 Hz G^{-1}, resulting in Δt = 19.57 µs and a bandwidth of 51 091 Hz. Similar to 1D spectroscopic signals discussed in Chapter 1, MRI signals that lay outside the FOV and that therefore do not obey the Nyquist sampling criterion will be aliased. In the phase-encoding direction, aliased signals will fall on top of non-aliased signals, thereby compromising the image quality (see also Figure 4.10). Aliasing is typically not observed along the frequency encoding or readout direction, since the audio filters that prevent aliasing of noise also suppress any genuine signals outside the spectral width and therefore outside the FOV.

4.7 Fast MRI Sequences

The k-space formalism can be used to explain the principles of any gradient-based MRI method, including the ultrafast MRI pulse sequences. Fast imaging sequences are crucial for dynamic MRI studies, like functional MRI and for parametric MRI studies, like diffusion tensor imaging (DTI) or T_1 mapping. Rather than providing a complete overview of fast MRI methods, this section describes the features common to all ultrafast MRI methods, namely the fast and efficient traversal of k-space. For more extensive discussions of fast MRI methodology the reader is referred to the literature [8–11].

As illustrated in Figure 4.6, during regular spin-warp MR imaging 2D k-space is filled up one line per excitation (or repetition time TR). As a result, the experiment must be repeated N_p times, leading to a total image acquisition time of $N_p \times$ TR, where N_p is the number of phase-encoding lines. For a medium resolution 256×256 data matrix and a TR of 1000 ms, the MR image is acquired in circa 4 min. While this may be acceptable for certain anatomical MR scans, it is unsuitable for many other imaging studies, like fMRI. For a given image matrix size there are three straightforward ways by which the image acquisition time can be reduced, namely (i) reduction of the repetition time TR, (ii) reduction of the number of phase-encoding steps N_p, and (iii) increasing the number of k-space lines that can be acquired per TR.

4.7.1 Reduced TR Methods

Methods that achieve image acceleration by reducing the repetition time TR include fast low-angle shot (FLASH, [12]), steady-state free precession (SSFP, [13]), and their many variants [9–11]. Reducing the repetition time TR will proportionally decrease the image acquisition time and scan time reduction factors of 20–100× are common. All short TR methods establish an equilibrium condition between excitation and relaxation that is different from thermal equilibrium. The FLASH and SSFP-style sequences primarily differ in the magnetization components involved in establishing the new equilibrium condition. FLASH and its variants only use the longitudinal magnetization and purposely destroy transverse magnetization following each acquisition period, either by magnetic field gradient spoiling or RF spoiling [14]. SSFP methods attain a steady-state condition involving all three components of the magnetization vector. The transverse magnetization from one excitation can only constructively add to the signal of subsequent excitations when the gradient area along all three spatial dimensions over the repetition time TR is zero. The sequence shown in Figure 4.6B would not be suitable for SSFP-style signal acquisition due to the nonzero gradient areas along each spatial dimension (shaded areas). SSFP-style sequences therefore require additional magnetic field gradients to cancel out the non-balancing gradient parts. The advantage of SSFP-style sequences is that they can provide higher SNR and a wider range of image contrasts. FLASH-style sequences are often less sensitive to imperfections like magnetic field inhomogeneity and provide a more straightforward image contrast. Reviews on many aspects of FLASH and SSFP methods are available [9–11]. The general principles of steady-state conditions are described in more detail in Chapter 3.

4.7.2 Rapid k-Space Traversal

Increasing the fraction of k-space that can be sampled in a single TR forms the basis for the majority of ultrafast MRI methods, whereby so-called single-shot methods sample the entire k-space in a single TR. Prime examples of ultrafast MRI methods are echo-planar imaging (EPI, [15, 16]), spiral [17, 18], and rapid acquisition, relaxation enhanced (RARE, [19]) MRI. The basic MR pulse sequences for EPI and spiral MRI together with the k-space trajectories are shown in Figure 4.8.

The first part of a gradient-echo EPI sequence (Figure 4.8A) is identical to a standard gradient-echo MRI sequence (Figure 4.6B). Negative gradients along the frequency and phase-encoding direction (Figure 4.8A, red line) prepare the k-space coordinate to the lower left corner (Figure 4.8B, red line). Reversal of the frequency-encoding gradient reduces the k_x coordinate linearly, leading to k-space signal storage along the black horizontal line during

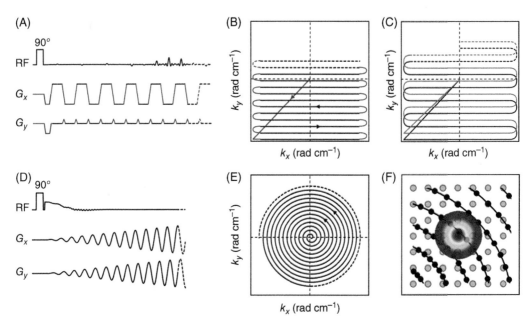

Figure 4.8 **Fast MRI based on efficient k-space traversal.** (A) Echo-planar imaging (EPI) sequence and (B) single and (C) two-shot k-space trajectories. The red lines in (A) and (B) correspond to preparation gradients, whereas the blue lines correspond to gradient switching. Without ramp sampling, signal is only acquired during the constant amplitude, black gradient periods. (C) In two-shot EPI every other k-space line in the phase-encoding direction is skipped, thus reducing the total EPI acquisition time twofold. The missing k-space lines are acquired during a second excitation. (D) Spiral MR imaging and (E) the corresponding k-space trajectory. (F) As many standard processing tools, such as Fourier transformation, require data on an equidistant Cartesian grid the spiral MRI data (black points) must be regridded to a rectangular grid (gray points). Regridding typically involves a Kaiser–Bessel kernel that calculates a point based on the intensity and phase of the surrounding acquired points.

signal acquisition. At the end of the acquisition window, the signal is macroscopically dephased due to the unbalanced second half of the frequency-encoding gradient. However, the signal has not been irreversibly lost due to the random microscopic processes underlying T_2 relaxation. Therefore, reducing the unbalanced gradient area and thus the k-space coordinate will lead to formation of a second echo. The second echo can be formed by inverting the frequency-encoding gradient sign (Figure 4.8A) and would hold the same spatial information as the first echo, albeit with a slightly increased T_2^* weighting due to the longer echo-time. Image acceleration can be achieved when this second echo is phase-encoded differently from the first echo. This is accomplished with a phase-encoding blip between the acquisition of the first and second echoes. The phase-encoding blip moves the k-space coordinate in the k_y direction, such that the second echo is stored along a separate horizontal line. It is now easy to recognize that multiple echoes can be formed by repeatedly inverting the sign of the frequency-encoding gradient. In conjunction with a phase-encoding gradient blip the EPI gradient pattern can traverse the entire k-space in a single TR. When the total gradient echo approaches zero (i.e. around the center of k-space) maximum intensity echo formation occurs, provided that the signal has not been irreversibly lost due to T_2 relaxation. Single-shot EPI traverses the entire k-space in a single experiment and is therefore the fastest option. However, single-shot EPI also has the highest sensitivity towards experimental imperfections, such as magnetic field inhomogeneity. Consider a practical example of acquiring a low-resolution 64×64 echo-planar image in a single shot. With an acquisition bandwidth of 200 kHz, the acquisition of a single echo takes 320 µs. With a 20 cm FOV the frequency-encoding gradient strength is 10 kHz cm^{-1}, which on state-of-the-art systems can be switched in circa 120 µs. The effective EPI echo-time TE occurs at the thirty-second echo, making TE = $32 \times (120 + 320 + 120)$ µs = 17.92 ms. For a medium-resolution 128×128 EPI dataset the echo-time becomes 56.32 ms. For longer TEs the SNR of the resulting EPI image can be severely reduced due to T_2 relaxation and display local signal drop-outs due to short T_2^* relaxation in areas of poor magnetic field inhomogeneity. Besides signal loss, single-shot EPI data is often affected by image distortion (Figure 4.9). In Figure 4.4E and F it was shown that a local magnetic field offsets lead to additional phase accrual over time, which in turn leads to a different local frequency and hence a different, incorrect spatial position. Using the example given above, in the frequency-encoding direction the signal is acquired in 640 µs for a 128×128 EPI dataset, which translates in a bandwidth of 200 kHz/128 = 1.563 kHz per pixel. A magnetic field offset of 150 Hz will thus result in a spatial displacement of less than 0.1 pixels in the frequency-encoding direction. Even though EPI does not have an actual bandwidth in the phase-encoding direction, one can calculate an effective bandwidth by the reciprocal of the inter-echo spacing. For a 128×128 dataset, the time between two echoes is $(120 + 640 + 120)$ µs = 880 µs, thereby making the effective bandwidth 1.136 kHz or only 8.88 Hz per pixel. A magnetic field offset of 150 Hz will thus result in a spatial displacement of more than 16 pixels in the phase-encoding direction (Figure 4.9C). Due to limitations on gradient strength, gradient slew rate, and peripheral nerve stimulation, the bandwidth for EPI in the phase-encoding direction will always be much lower than in the frequency-encoding direction. The geometric distortions can be greatly reduced by increasing the bandwidth per pixel through interleaved acquisition. In two-shot EPI (Figure 4.8C) half the number of echoes are acquired in the first experiment, whereas the remaining half is acquired during the second experiment. The price of a twofold reduced image acceleration is offset by a shorter echo-time TE and an increased bandwidth per pixel, leading to less geometric distortion. Figure 4.9E shows the greatly reduced geometric distortion obtained with a four-shot EPI sequence.

From Figure 4.8A–C it can be extrapolated that a wide range of gradient-encoding schemes can be used to achieve image acceleration, provided that the entire k-space is adequately sampled. One of the most efficient k-space sampling schemes is provided by spiral MRI in

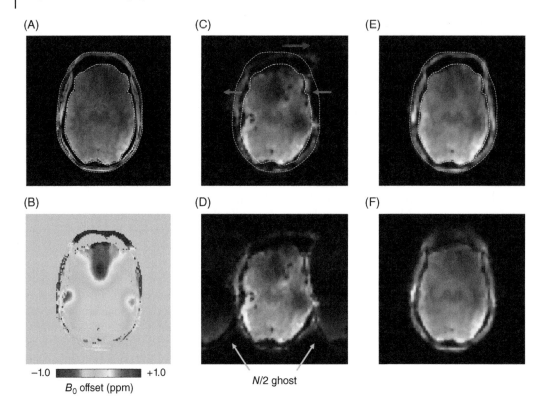

(A) (C) (E)

(B) (D) (F)

−1.0 ▮▮▮▮▮ +1.0
B_0 offset (ppm)

N/2 ghost

Figure 4.9 **Image artifacts in fast MRI.** (A) Reference MRI outlining the brain (yellow) and head (orange). (B) Magnetic field B_0 map indicating significant inhomogeneity of up to 1 ppm. (C) Single-shot echo-planar image showing image distortion that is directly proportional to the sign and magnitude of the magnetic field offset in (B). (D) Mismatch between the even and odd echoes in EPI lead to N/2 ghosting in the image domain. (E) Four-shot EPI shows greatly reduced image distortion. (F) Magnetic field inhomogeneity in spiral MRI leads to image blurring.

which two oscillating gradients (Figure 4.8D) sample k-space along a spiral trajectory. Since signal acquisition during the spiraling gradient is continuous without the need to wait for switching gradients, spiral MRI can be even faster than EPI. A consideration with spiral MRI is that the data is acquired and stored on a non-Cartesian grid (Figure 4.8F). Standard image processing using Fourier transformation requires data to be present on a Cartesian, equidistant grid. Therefore, spiral MRI data typically undergoes regridding in which the acquired data on the spiral grid is recalculated on a Cartesian grid (Figure 4.8F). Details on image regridding can be found in Ref. [20]. Similar to EPI, the total acquisition time of spiral MRI can be quite long, depending on the image resolution and spectral bandwidth. As a result, phase accrual due to magnetic field inhomogeneity leads to significant geometric distortion, which manifests itself as image blurring (Figure 4.9F) in areas of diverging magnetic field offsets. Similar to EPI, the image distortions in spiral MRI can be greatly reduced through interleaved, multi-shot acquisition.

Ultrafast MRI methods depend on high gradient fidelity such that the actual k-space trajectory is identical to the theoretical trajectory. However, a wide range of effects ranging from eddy currents to system delays to oblique slicing can lead to deviations of the actual k-space trajectory. When unaccounted for, these imperfections can lead to image artifacts such as the well-known N/2 ghost in EPI (Figure 4.9D), which is caused by a mismatch between the even and odd echoes. A wide variety of methods exist to characterize gradient infidelity [21] of which the use of magnetic field cameras to measure the exact k-space trajectory works exceptionally well ([22] and Chapter 10).

4.7.3 Parallel MRI

A completely different approach to scan time reduction is provided by parallel MRI methods which take advantage of the local spatial sensitivity information of an array of multiple receiver coils to partially replace spatial encoding by magnetic field gradients. Since signal can be acquired simultaneously (i.e. in parallel) by the multiple receiver coils, a reduction in k-space sampling directly results in a reduced image acquisition time while maintaining full spatial resolution. In order to explain the principles of parallel imaging, it is important to understand the effects of k-space undersampling and the resulting aliasing in the image domain. Consider the k-space and image space data in Figure 4.10A and B. The k-space is completely sampled according to the Nyquist sampling condition, meaning that the frequency at every spatial location is sampled at least twice (see also Chapter 1). As a result, the FOV in the resulting image is given by $2\pi/\Delta k$, whereby Δk is the distance between two adjacent k-space points in either of the two orthogonal directions. The image acquisition time can be reduced twofold by, for example, skipping every other k-space line along the phase-encoding direction (Figure 4.10C). In this case, Δk is twice as large, meaning that the FOV is twice as small (Figure 4.10D). Signals that are present outside the FOV are not properly sampled and will appear at an apparent, incorrect, or aliased frequency (Figure 4.10D, see also Chapter 1). Note that image aliasing only occurs along image dimensions that are obtained with phase encoding. Along the frequency-encoding direction, the audio filters that are associated with the acquisition bandwidth will suppress

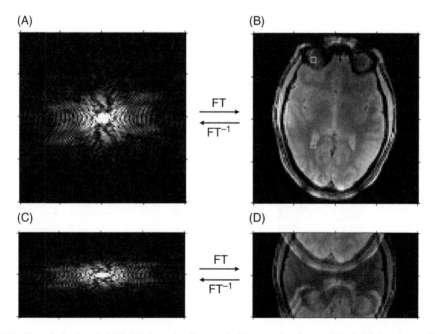

Figure 4.10 Signal aliasing in MRI. (A) Completely sampled k-space leading to (B) a full field-of-view (FOV) MR image. (C) k-space where every other phase-encoding step or k-space line has been omitted. The doubled k-space separation in (C) leads to a halved FOV in the image domain (D). As a consequence, uniquely encoded pixels separated by FOV/2 in (B, green pixels) will end up at the same spatial location following signal aliasing (D, red pixel).

signals outside the FOV (see also Chapter 10). The aliased MR image (Figure 4.10D) contains pixels representing the sum of aliased and non-aliased signals (i.e. the red box signal in Figure 4.10D is the sum of the two green boxes in Figure 4.10B). Without any additional information it is impossible to disentangle the aliased and non-aliased signals. The central concept of all scan time reduction methods based on parallel MRI is that additional information is provided by the signals acquired with multiple RF coils. When the additional information provided by the RF coils is sufficiently unique and complementary to the missing k-space lines, a full FOV image can be successfully reconstructed.

The most well-known and commonly used parallel MRI techniques are simultaneous acquisition of spatial harmonics (SMASH) [23], sensitivity encoding (SENSE) [24], and generalized autocalibrating partially parallel acquisitions (GRAPPA) [25]. All parallel MRI methods share the same basic idea of complementing incomplete k-space sampling with additional spatial information obtained from local RF coils. The methods mainly differ (i) in the required amount of prior knowledge on coil sensitivity profiles and (ii) in the manner by which the spatial information is combined. SMASH and SENSE both require accurate prior knowledge of the sensitivity profiles of each coil in a multi-coil receiver array. GRAPPA is a so-called autocalibration method, meaning that the coil sensitivity is part of and can be extracted from the acquired data.

4.7.3.1 SENSE

Consider the situation where the head of a human subject is placed in a four-element phased-array RF coil (Figure 4.11). Each RF coil receives signal from a localized region of the head and the four images could be processed and displayed individually (Figure 4.11B) or added

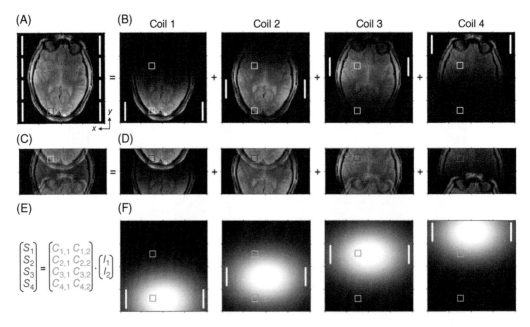

Figure 4.11 Principle of sensitivity encoding (SENSE). (A) Standard MR image reconstructed through summation of (B) signals acquired with four independent receiver coils. (C) Reduced FOV image and (D) the individual coil components. Each red pixel in (D) is the sum of the two yellow pixels in (B). Alternatively stated, each red pixel in (D) is the sum of the original (unknown) blue pixels in (A) multiplied by the sensitivity of the coil at those positions (green pixels in (F)). The previous statement in matrix form is shown in (E), which represent a set of four linear equations with two unknowns that are readily solvable by standard methods to provide the two image intensities I_1 and I_2.

together to provide a whole head image (Figure 4.11A). When every other k-space line is skipped, the image acquisition time is two times faster, but comes at the expense of signal aliasing (Figure 4.11C and D) similar to that observed with a single-channel RF coil (Figure 4.10). The key to parallel MRI is that the two aliased points are weighted by the sensitivity profile (Figure 4.11D) of each RF coil. Signal S_k detected on coil number k and spatial position (x, y) (red squares in Figure 4.11B) represents the sum of the object signal intensity I at spatial positions (x, y_m) (blue squares in Figure 4.11A) weighted or multiplied by the coil sensitivities $C_k(x,y_m)$ for each coil k at those positions (green squares in Figure 4.11F). In other words,

$$S_k\left(x,y\right)=\sum_{m=1}^{R}C_k\left(x,y_m\right)I\left(x,y_m\right) \tag{4.15}$$

where R represents the image acceleration factor. In the case of Figure 4.11, Eq. (4.15) expresses the fact that there are four measured values $S_k(x,y)$ and two unknown values $I(x,y_m)$. With knowledge of the sensitivity matrix C, Eq. (4.15) therefore describes an overdetermined set of linear equations that can be solved.

For the more general case of K coils and M spatial locations, K linear equations with M unknowns can be constructed and written in matrix form as

$$S=C\cdot I \tag{4.16}$$

where S is a $(K \times 1)$ vector and represents the measured complex signal intensity at a chosen pixel for each of the K coils. I is a $(M \times 1)$ vector and represents the signal in the full FOV image. C is a $(K \times M)$ sensitivity matrix describing the sensitivity of all K coils at all M spatial positions. Once the sensitivity matrix C is known, the unfolded image over the full FOV, I, can be calculated according to

$$I=\left(C^{H}C\right)^{-1}C^{H}\cdot S \tag{4.17}$$

where the superscript H indicates the transposed complex conjugate. Equation (4.17) assumes that there is no noise correlation between the coils. In reality, coil sensitivity profiles overlap and are not completely independent from one another, necessitating the extension of Eq. (4.17) with receiver noise correlation matrices [24].

A general requirement for a matrix inversion as shown in Eq. (4.17) is that the number of unique equations is equal or greater than the number of unknowns. In other words, the acceleration factor R can never be larger than the number of receiver coils K. However, the experimentally achievable acceleration factor is often much smaller than the number of receiver coils. If one would attempt to accelerate the data acquisition along the horizontal direction in Figure 4.11, the sensitivity profiles of the four RF coils would be too similar, resulting in four almost equal, or degenerate equations in Eq. (4.17). This would make the matrix inversion extremely sensitive to small variations in the data (i.e. noise), leading to noise amplification in the reconstructed image. The ability of a multielement receive array to uniquely encode the spatial positions is described by the geometry or g-factor [24]. The g-factor is calculated according to

$$g_m=\sqrt{\left(C^{H}C\right)^{-1}_{m,m}\left(C^{H}C\right)_{m,m}}\geq 1 \tag{4.18}$$

whereby m,m indicates the diagonal elements of the matrix. Similar to Eq. (4.17), the g-factor expressed by Eq. (4.18) assumes the absence of noise correlation between the coils. More

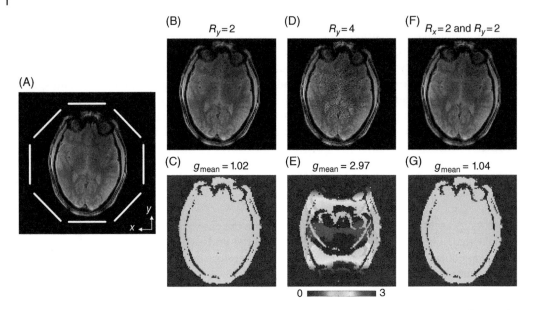

Figure 4.12 Coil geometry or *g*-map in accelerated imaging. (A) Geometric arrangement of an eight-element receiver coil relative to the human head. (B) MR image and (C) corresponding *g*-map obtained with a twofold acceleration in the *y* direction. (D, E) Fourfold acceleration in the *y* direction leads to (D) noise amplification as characterized by (E) high-intensity *g*-map. (F, G) The coil geometry in (A) does support a fourfold acceleration with minimal noise amplification when executed as a 2×2 acceleration in the *x* and *y* directions.

extensive equations are available when this assumption is not valid [24]. The *g*-factor is always greater or equal to 1.

The importance of the *g*-factor in relation to the acceleration factor R for a given coil array is shown in Figure 4.12. For a twofold acceleration in the *y* direction, the eight coils provide sufficiently unique encoding of all spatial positions to give a properly conditioned coil sensitivity matrix C. The matrix inversion of Eq. (4.17) leads to a constructed image with minimal noise amplification, characterized by a *g*-factor close to 1 (Figure 4.12B and C). For a fourfold acceleration in the *y*-direction, the eight coils do not provide adequate encoding, leading to degeneracy in the coil sensitivity matrix and hence noise amplification in the final image (Figure 4.12D and E). However, a fourfold acceleration is possible with the eight-coil setup of Figure 4.12, provided that the acceleration is split as twofold in *x* and twofold in *y* (Figure 4.12F and G).

The SNR of a parallel image reconstruction method, like SENSE, compared to the SNR of a non-accelerated image acquired with the same coil array is given by

$$\text{SNR}_{\text{accelerated}} = \frac{\text{SNR}_{\text{full}}}{g\sqrt{R}} \tag{4.19}$$

where the factor $\sqrt{R}$ originates from the fact that less time is spent on data acquisition. SENSE reconstruction relies on the availability of accurate coil sensitivity profiles as expressed in the matrix C. The sensitivity maps can be obtained in a number of ways. A popular choice is to divide the local coil images by an image acquired with a homogeneous volume coil from the same sample. Variations in proton density, RF transmit homogeneity, and relaxation are removed during the division, leaving only the variation caused by the coil receiver sensitivity. The coil sensitivity maps are typically smoothed to reduce noise contributions and extrapolated to ensure full coverage over the object.

4.7.3.2 GRAPPA

The main differences between the SENSE and GRAPPA parallel image reconstruction algorithms is that GRAPPA works in the k-space domain, whereas SENSE operates in the image domain. In addition, GRAPPA is a so-called autocalibration method which means that information on the coil sensitivity profiles is acquired as part of the data, thereby eliminating the need to acquire separate sensitivity maps. The GRAPPA reconstruction is based on using portions of acquired k-space to calculate portions that were not acquired. Figure 4.13 shows that in image space the presence of a local receiver coil leads to the multiplication (or weighting) of each spatial point by the sensitivity of the coil at that location (Figure 4.13E). According to Fourier theory (see Appendix A.3) a multiplication in one domain (image space) leads to a convolution in the complementary domain (k-space). A similar effect was observed during apodization in MR spectroscopy (see Chapter 1) where multiplication of the data with an exponentially decaying exponential in the time domain led to line broadening (i.e. convolution of the data with a Lorentzian line) in the frequency domain. For the example in Figure 4.13 the presence of a local receiver coil leads to a smearing of information across neighboring k-space locations (Figure 4.13F). Most importantly, since it originates from a convolution, the k-space

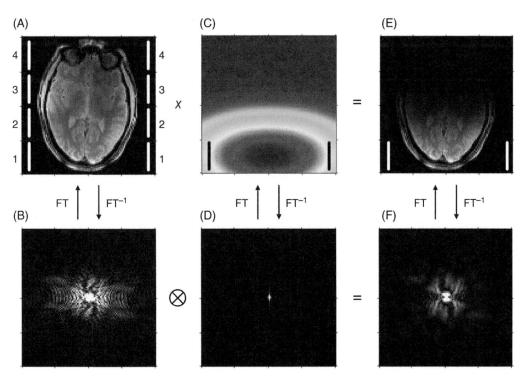

Figure 4.13 Coil sensitivity in image and k-space domains. (A) MR image and (B) corresponding k-space as acquired with a four-element receiver coil. (C) Sensitivity profile of receiver coil 1 in the image domain and (D) the corresponding k-space representation. (E) The MR image obtained by receiver 1 is the product of the total MR image (A) and the coil sensitivity profile (C). (F) The corresponding k-space is given by the *convolution* of the total k-space (B) and the k-space of a single receiver element (D). Each point in k-space relative to the surrounding points contains information about the coil sensitivity.

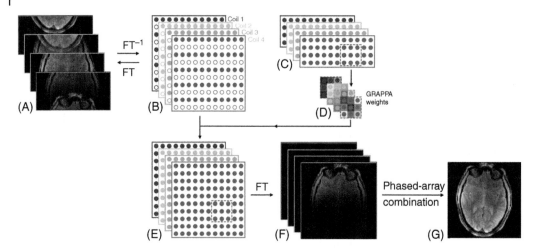

Figure 4.14 **Principle of generalized autocalibrating partially parallel acquisitions (GRAPPA).** (A) Accelerated, reduced field-of-view MR images and (B) the corresponding undersampled *k*-spaces. (C) A small part of *k*-space that is fully sampled can be used to determine or "calibrate" (D) the GRAPPA weights. (E) The GRAPPA weights can subsequently be used to calculate the missing *k*-space lines, after which (F) full field-of-view MR images can be reconstructed. (G) Standard phased-array coil combination strategies can be used to obtain the total MR image.

"smearing pattern" is identical for all *k*-space locations. In other words, some of the information about a given *k*-space point is also contained in the neighboring *k*-space points. This forms the basis for GRAPPA reconstruction in which the information contained in acquired *k*-space points can be used to recover the missing *k*-space data. Figure 4.14 shows the general workflow of a GRAPPA reconstruction. The multi-receiver MR images are acquired with a reduced FOV (Figure 4.14A), which is equivalent to stating that they were acquired with missing *k*-space lines (Figure 4.14B). As part of the GRAPPA acquisition scheme, a small part of *k*-space representing the autocalibration signal is fully sampled (Figure 4.14C). A small kernel (in Figure 4.14C a 3 × 3 grid) is used to establish the amplitude and phase relationship, also called the GRAPPA weights, between adjacent *k*-space points. The kernel is moved through the entire autocalibration part of *k*-space in order to get a reliable estimate of the GRAPPA weights. Once established, the kernel can be moved to the actual data where the missing *k*-space points can be calculated from the combination of acquired *k*-space points and GRAPPA weights. This will ultimately result in fully-sampled *k*-spaces for each coil (Figure 4.14E), which after Fourier transformation, can be combined into a full FOV MR image (Figure 4.14G).

While SENSE and GRAPPA approach the *k*-space undersampling problem from opposite directions, the resulting images are under most experimental conditions indistinguishable. The choice of one over the other is typically governed by personal preference or availability on a particular clinical MR platform.

4.8 Contrast in MRI

One of the most important tasks of MRI is to create appropriate contrast for the discrimination of areas of interest. In principle, any difference in the properties of spins can be used to create contrast. The most commonly used parameters are spin density, longitudinal relaxation time T_1, and transverse relaxation time T_2. A conventional spin-warp imaging sequence can exploit all these differences by proper adjustment of the experimental parameters TR (repetition time) and TE (echo time). For a spin-warp imaging sequence, the signal at the top of the echo is given by

$$M_{xy}\left(\text{TR,TE}\right) = M_0\left(1 - e^{-\text{TR}/T_1}\right)e^{-\text{TE}/T_2} \tag{4.20}$$

Herein, a 90° excitation pulse has been assumed and effects of the refocusing pulse (i.e. inversion of recovered longitudinal magnetization during TE/2) have been neglected. Figure 4.15A shows the signal dependence on TR (TE = 10 ms) for two pixels with white and gray matter pixels with $T_1 = 900$ and 1350 ms, respectively. The difference curve shows that maximum "T_1 contrast" is achieved when TR ~ 1200 ms. However, experimentally the image acquired with TR = 1200 ms (Figure 4.15B) is devoid of contrast, largely because the differences in proton density between gray and white matter cancel out the T_1 contrast. Figure 4.15C shows a MR image acquired with a long TR and short TE that reflects the proton density. The amount of T_1 contrast can be significantly enhanced by extending the sequence with an inversion pulse and a recovery delay TI, such that the signal at the top of the echo is given by

$$M_{xy}\left(\text{TI, TE}\right) = M_0\left(1 - 2e^{-\text{TI}/T_1}\right)e^{-\text{TE}/T_2} \tag{4.21}$$

where a long repetition time (TR > $5T_1$) and a perfect inversion pulse have been assumed. Figure 4.15D shows the signal dependence on TI for $T_1 = 900$ and 1350 ms, respectively. The difference curve again reveals that maximum T_1 contrast is generated at TI ~ 1200 ms. While the corresponding experimental image (Figure 4.15E) has improved gray/white matter con-

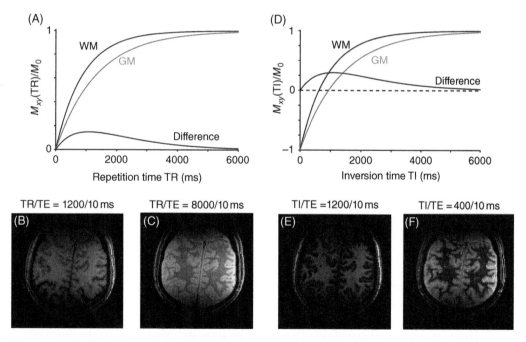

Figure 4.15 T_1, T_2, **and proton density MR image contrast.** (A) Saturation recovery signal curves for white matter (WM) and gray matter (GM) with T_1 values of 900 and 1350 ms, respectively. The difference between the curves indicates a maximum at TR ~ 1200 ms. Saturation recovery images from human brain acquired with (B) TR/TE = 1200/10 ms and (C) TR = 8000/10 ms. The image in (C) closely resembles a "proton-density" image. (D) Inversion recovery signal recovery curves for $T_1 = 900$ ms (WM) and $T_1 = 1350$ ms (GM). Inversion recovery images from human brain acquired with (E) TI/TE = 1200/10 ms and (F) TI/TE = 400/10 ms.

trast compared to Figure 4.15B, the cancelation of T_1 contrast by differences in proton density still largely prevents a clear discrimination between cerebral tissue types. However, for TI < 800 ms the gray/white matter T_1 contrast is inverted relative to the situation when TI > 1000 ms. Acquiring an image with TI = 400 ms therefore results in excellent image contrast (Figure 4.15F) since the T_1 and proton density differences between gray and white matter enhance each other. From Figure 4.15 it follows that the best image contrast is determined by an interplay between several parameters, like proton density, and T_1 and T_2 relaxation. As a result, a single "parameter-weighted" (e.g. "T_1-weighted") image cannot provide unambiguous information about that parameter and should be used with care (e.g. all "T_1-weighted" images in Figure 4.15 are also heavily weighted by proton density differences). A more quantitative approach in which one of the parameters (e.g. T_1) is uniquely determined independent of the other parameters offers a more objective strategy for tissue characterization and will be discussed next.

4.8.1 T_1 and T_2 Relaxation Mapping

Because MR images are always affected by a combination of T_1, T_2, spin density, and other effects, the generation of qualitative contrast on an unknown sample may result in ambiguous results. Furthermore, it is impossible to determine which effects are involved when signal changes during disease progression are observed. Therefore, it is often desirable to generate calculated images ("maps"), representing one of the parameters, from MRI data that are increasingly weighted for the parameter of choice. For instance, a T_1 map can be calculated from a series of images with increasing repetition (or inversion) time (Figure 4.16A). By analyzing each pixel in the series with a model describing the data, i.e. Eq. (4.21), the "T_1-weighted" images can be converted into quantitative T_1 and M_0 maps (Figure 4.16C). Besides the quantitative nature of the maps, they also generate new contrast, since the effects from other parameters (e.g. spin density) are entirely excluded.

Time constraints are critical for most clinical applications. For studies with an MRS focus, most experimental time should be used on the acquisition of spectroscopic data, whereas supporting MRI data should be acquired in the shortest amount of time. T_1 and T_2 measurements of complete relaxation curves can be prohibitively long and one typically has to resort to faster, more efficient methods. Dozens of methods and hundreds of papers have appeared on the fast measurement of T_1 relaxation time constants [26]. The T_1 measurement method of Look and Locker [27–29] is a popular choice to map longitudinal relaxation, without resorting to ultrafast MRI methods and their associated problems [30]. The measurement of T_2 or T_2^* relaxation is typically less time-consuming, as complete T_2 and T_2^* relaxation curves can be sampled through a multi-echo acquisition [31, 32]. More recently, a novel method for quantitative parameter mapping has been proposed. MR fingerprinting [33] aims at generating image contrast by rapidly varying the repetition time and nutation angle of the MRI sequence. In a short time span several hundred MR images are acquired, each with different weightings due to T_1 and T_2 relaxation and B_0 and B_1 magnetic fields. Since the data acquisition leads to image contrast that is a mix of many parameters, it is no longer possible to extract individual parameters through analytical expressions like those in Eqs. (4.20) and (4.21). However, MR signal evolution of water can be accurately described by the Bloch equations for any MR pulse sequence. The key to MR fingerprinting is the generation of a large dictionary that contains the temporal evolution of a spin for any possible combination of T_1, T_2, B_0, B_1, and any other parameter affecting the signal. The processing in MR fingerprinting then consists of finding, for a given MR image pixel, the best match between the experimentally measured signal evolution and one of the dictionary entries. MR fingerprinting is

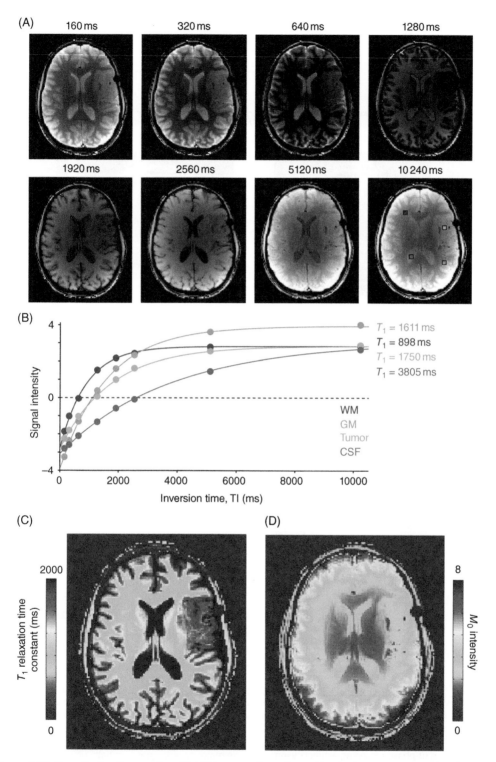

Figure 4.16 Quantitative T_1 mapping. (A) MR images of a patient with a brain tumor acquired with a Look-Locker implementation of inversion recovery. Inversion times are indicated above the images. (B) Signal recovery curves plotted for four pixels (GM, WM, CSF, and tumor) indicated in the final image of (A). The solid lines indicate the best mathematical fit, providing quantitative estimates for the relaxation time constant T_1 and the image intensity. (C, D) Repeating the mathematical fitting for each pixel gives (C) a quantitative T_1 map and (D) an intensity map. The intensity map has contributions from proton density M_0 and receiver coil sensitivity B_1^-.

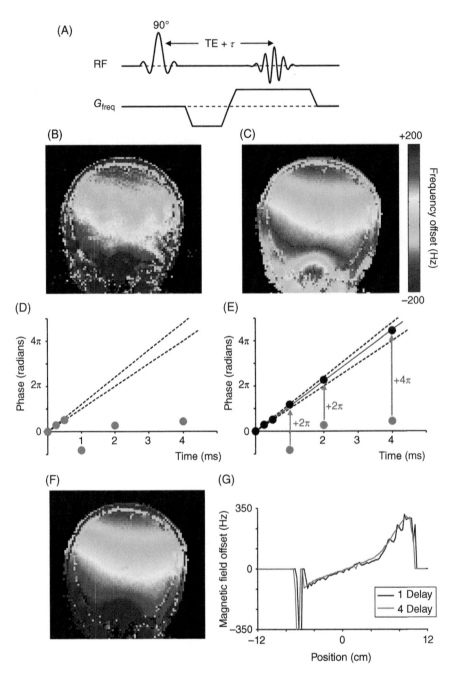

Figure 4.17 **Principle of magnetic field B_0 mapping.** (A) Basic gradient echo MR pulse sequence in which τ is a variable delay in addition to the constant echo-time TE. (B) For a short delay $\tau = 0.33$ ms, no phase wraps occur at the expense of a relatively low S/N. (C) For a long delay $\tau = 3.0$ ms, the S/N has greatly improved at the expense of phase-wrapping artifacts. The advantages of short and long delays can be combined as shown in (D, E) through the procedure of 1D temporal unwrapping. Acquiring several images with increasing delays τ allows the eliminating of phase-wrapping artifacts even when the majority of images is initially phase wrapped (MR pixels at $\tau = 1$, 2, and 4 ms in (D) are wrapped). Under the assumption that the first data point ($\tau > 0$) is not wrapped, a linear trend can be established (D). All other data points can then simply be unwrapped by multiples of 2π until they fall close to the line established by the previous points (E). Unwrapping followed by linear regression of a three delay dataset ($\tau = 0.33$, 1.0, and 3.0 ms) gives a B_0 map with excellent S/N and no phase wraps (F). (G) Comparison of a vertical trace through the field maps in (B) and (F) demonstrating the S/N advantage of additional longer delays.

capable of rapidly mapping several parameters in parallel and offers a real opportunity to bring quantitative parameter mapping to clinical MRI. More details of MR fingerprinting can be found in Chapter 3.

4.8.2 Magnetic Field B_0 Mapping

Knowledge of the magnetic field distribution within the object under investigation is important for a wide range of applications, including post-acquisition frequency correction in MRSI (see Chapter 7), and EPI image distortion correction and optimization of the magnetic field homogeneity through shimming (see Chapter 10). The magnetic field distribution is typically obtained from the additional phase evolution during a period of free precession. While the method can essentially be based on any MRI sequence, the principle will be demonstrated for a fast gradient-echo sequence as shown in Figure 4.17A. In the first (reference) experiment ($\tau = 0$), the signal at the top of the gradient echo at position r contains spatially dependent phases related to eddy currents, RF or B_1 magnetic fields, and B_0 magnetic field inhomogeneity. The multiple contributions to the observed phase makes it impossible to separate the phase associated with B_0 magnetic field inhomogeneity from the other phases. This problem is typically overcome by performing a second experiment whereby a small delay τ is inserted prior to acquisition to allow free precession under the influence of the local magnetic field B_0. Under the assumption that all other phases are constant between the two acquisitions, the phase difference is proportional to the B_0 magnetic field inhomogeneity according to

$$\Delta\phi = \arctan\left(\frac{R_1 I_2 - I_1 R_2}{R_1 R_2 + I_1 I_2}\right) = \gamma \Delta B_0 \tau \tag{4.22}$$

where R and I refer to the real and imaginary components of the observed signals, respectively, and 1 and 2 refer to the signals acquired with $\tau = 0$ and $\tau > 0$, respectively. The additional delay τ (typically between 0 and 5 ms) slightly decreases the signal intensity due to T_2 relaxation. However, since this only affects the signal intensity it has no consequence for the phase calculation according to Eq. (4.22). Dividing the phase $\Delta\phi$ by $2\pi\tau$ gives a quantitative B_0 map (Figure 4.18B). The most commonly encountered problem in B_0 mapping is so-called phase wrapping. The arctan function in Eq. (4.22) only calculates the phase between $-\pi$ and $+\pi$. When the real-phase evolution is very large (i.e. for $\Delta\omega/2\pi = 550$ Hz and $\tau = 4.0$ ms, $\Delta\varphi = 4.4\pi$), the calculated phase will wrap, through an integer number of 2π cycles, back into this range

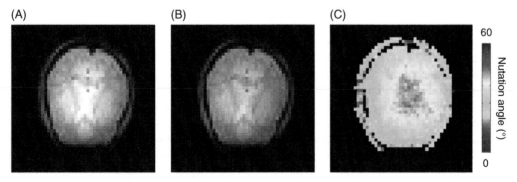

(A)   **(B)**   **(C)**

60

Nutation angle (°)

0

Figure 4.18 **Magnetic field B_1 mapping based on two nutation angles.** From the two acquired images (left, middle) a nutation angle map (right) can be generated which is directly proportional to the transmit B_1^+ magnetic field (see Exercises for more details).

(e.g. $4.4\pi - 2 \times 2\pi = 0.4\pi$). In general, phase wrapping leads to an incorrect estimation of the local magnetic field (e.g. 0.4π would be equivalent to a 50 Hz offset). However, in the presence of a wide range of magnetic field inhomogeneity, phase wrapping will also lead to discrete phase jumps (e.g. two adjacent points with real phases of 0.95π and 1.05π will be wrapped to 0.95π and -0.95π), making it impossible to accurately estimate global magnetic field homogeneity (Figure 4.17C). Phase wrapping can be avoided by chosen a very small evolution delay τ, such that the phase evolution remains with the $[-\pi \ldots +\pi]$ range. However, a small range of phases typically translates into a noisy B_0 map (Figure 4.17B). The problem of discrete phase jumps can also be solved through multidimensional spatial phase unwrapping algorithms (e.g. Ref. [34] and references therein). These algorithms are used in many different scientific disciplines and can be executed in a robust "black-box" implementation. However, these methods typically need to be tailored to the problem at hand and are computationally intensive. A simpler solution is to choose the evolution time τ small enough so that phase wrapping does not occur (e.g. the real phases are between $-\pi$ and $+\pi$, Figure 4.17B). Unfortunately, in the presence of noise the accuracy of the estimated offsets is not optimal when using a small τ delay. Accurate estimates of magnetic field offsets are obtained when the number of evolution delays as well as the delay durations are extended (see Figure 4.17F and G). While the acquired phase during the longer delays will likely be wrapped, the initial delay is chosen such that no phase wrap occurs (Figure 4.17D). Then, based on the linear phase-time trend established by the first delay, the subsequent delays can be unwrapped by adding or subtracting an integer number of 2π to the calculated phase (Figure 4.17E). Once all phases have been temporally unwrapped, a linear least-squares fit of the phase–time curve will provide a best estimate of the magnetic field offset (Figure 4.17F and G). Typical computation times for phase calculation, phase unwrapping, and linear least-squares fitting for a $128 \times 128 \times 64$ dataset is several seconds.

While the calculated B_0 map is, in principle, independent of the employed pulse sequence type (gradient-echo, spin-echo, EPI), the pulse sequence can nevertheless have a significant effect on the appearance of the B_0 map. Gradient-echo-based B_0 mapping can be fast and reliable, but the images may show areas of signal loss when strong magnetic field inhomogeneity is encountered due to phase cancelation during the initial echo-time TE. In these cases it is better to utilize spin-echo-based B_0 mapping, since phase evolution due to magnetic field inhomogeneity is refocused at the top of the echo. B_0 maps based on EPI can be acquired very rapidly. However, EPI image quality (signal loss and image distortion) is heavily influenced by magnetic field inhomogeneity and as such is less ideal to provide reliable B_0 maps.

The outlined steps represent standard B_0 mapping under the central assumption that the detected phase originates from a single chemical species, namely water. Outside the brain this assumption is violated due to the presence of lipids. The ^{1}H MR spectrum of lipids (Figure 2.35) displays many signals resonating at chemical shifts that are different from water. The main methylene CH_2 signal from lipids resonates around 1.3 ppm, or circa 435 Hz upfield from water at 3 T. For two adjacent MRI pixels in a homogeneous magnetic field that are dominated by water and lipids, respectively, a calculated B_0 map would display a steep magnetic field gradient of 435 Hz. When the incorrectly estimated B_0 map is used to establish shim correction currents, the actual magnetic field homogeneity would greatly deteriorate. The problem of lipids in B_0 mapping is present in all tissues containing fat, but is especially prevalent in human breast. Solutions to the lipid problem are found in eliminating the lipid signal or minimizing the phase evolution due to lipid–water chemical shift difference. Lipid signals can be removed prior to excitation by methods that are used for water suppression in ^{1}H MRS (see Chapter 6). As an alternative, the evolution time τ can be chosen such that the lipid methylene CH_2 chemical shift evolution is a multiple of 2π, e.g. $\tau = (1/435)$ Hz = 2.3 ms at 3.0 T. In a more advanced implementation, the evolution time can be finetuned by also taking the smaller lipid resonances (Figure 2.35) into consideration.

4.8.3 Magnetic Field B_1 Mapping

Spatial homogeneity of the magnetic field generated by RF coils is important for the unambiguous interpretation of images, optimizing sensitivity, and minimizing RF-related artifacts. However, inhomogeneous RF fields are frequently encountered in *in vivo* NMR applications at high magnetic fields (see Chapter 10) or with RF coils that are inherently inhomogeneous (e.g. surface coils). It therefore becomes important to measure the RF field quantitatively and a number of MRI techniques are available to achieve this goal. Most techniques rely on the relation between pulse length T, RF field amplitude B_1, nutation angle θ, and observed signal M_{xy}, which for a simple (fully relaxed) pulse-acquire experiment is given by

$$M_{xy} = M_0 \sin\theta, \quad \text{with } \theta = \gamma B_1 T \tag{4.23}$$

where M_0 is the thermal equilibrium magnetization. Acquiring two fully relaxed images with two different nutation angles θ_1 and $\theta_2 = 2\theta_1$ allows the calculation of the nutation angle according to

$$\theta_1 = \arccos\left(\frac{M_{xy}(\theta_2)}{2M_{xy}(\theta_1)}\right) \tag{4.24}$$

where $M_{xy}(\theta_1)$ and $M_{xy}(\theta_2)$ are the detected signals following nutation angles θ_1 and θ_2, respectively. The B_1 field can subsequently be calculated through Eq. (4.23). Since the method is based on only two measurements, the accuracy of the method is highest in the 80–100° range of nutation angles. Figure 4.19 shows an example of B_1 magnetic field mapping using a fast modification of the principle outlined by Eqs. (4.23) and (4.24) (see also Exercises).

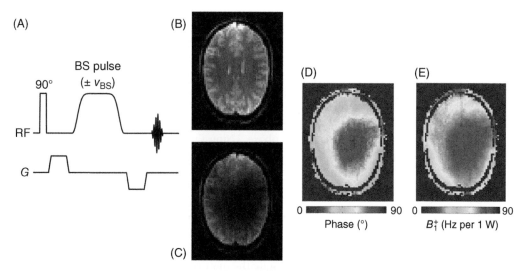

Figure 4.19 Magnetic field B_1 mapping based on Bloch–Siegert shifts. (A) Gradient echo MR sequence extended with an off-resonance Bloch–Siegert pulse. The actual frequency-encoding or readout gradients are not shown. Signal excited by the BS pulse is destroyed by the crusher gradients. (B, C) Real component of a MR image acquired in the presence of a Bloch–Siegert pulse applied at frequencies (B) $+\nu_{BS}$ or (C) $-\nu_{BS}$. (D) Phase difference between the MR images in (B) and (C). (E) Transmit B_1^+ map as calculated from the phase difference map in (D) according to Eq. (4.25).

An intrinsic assumption of pulse-angle-based B_1 mapping according to Eqs. (4.23) and (4.24) is the presence of fully relaxed longitudinal magnetization prior to each excitation. Under *in vivo* conditions the repetition time is often too short to fully satisfy this requirement, such that the detected signal is also affected by variations in T_1 relaxation, which would ultimately lead to an incorrect B_1 estimation. The dual refocusing echo acquisition mode (DREAM) method circumvents this problem by acquiring two signals with different B_1 weighting at the same time [35]. As will be discussed in Chapter 6, a sequence of three RF pulses can generate eight signals (three FIDs, four spin-echoes, and one stimulated echo). As each signal is generated by different RF pulse combinations, the various signals have a different dependence on the B_1 magnetic field amplitude. When two of the signals, e.g. an FID and a stimulated echo, can be detected separately, but simultaneously within the same acquisition window, a quantitative B_1 map can be calculated using their different B_1 dependencies (see Exercises).

A completely different approach to B_1 mapping is provided by the Bloch–Siegert (BS) method [36], in which a separate RF pulse, different from the MR excitation pulse, is used to generate phase-sensitive B_1-dependent image contrast (Figure 4.19). The basic BS B_1 mapping method is shown in Figure 4.19A. Following excitation, the magnetization evolves in the transverse plane and is detected as a MR gradient-echo signal. The BS pulse is applied far enough off-resonance such that it does not change the magnitude of the transverse magnetization. However, the BS pulse does affect the phase of the transverse magnetization according to

$$\phi_{BS} = B_{1,max}^2 \int_0^T \frac{\gamma^2 B_{1,normalized}^2(t)}{4\pi v_{RF}(t)} dt = B_{1,max}^2 K_{BS} \tag{4.25}$$

where the RF pulse frequency, v_{RF} (in Hz) is time-dependent for adiabatic RF pulses and time-independent for conventional, amplitude-modulated RF pulses. For a given RF pulse modulation $B_{1,normalized}(t)$ (unitless, scaled between -1 and $+1$) the integral in Eq. (4.25) reduces to a constant and the BS phase shift becomes only dependent on the maximum RF amplitude, $B_{1,max}$ (in T) and a RF pulse-specific constant, K_{BS} (in rad T^{-2}). In general, it is desirable to use RF pulses that maximize K_{BS}, while minimizing direct saturation of the water resonance [37].

In addition to the desired BS phase shift, the detected signal is also affected by phases from other sources, such as magnetic field inhomogeneity or eddy currents. It is therefore impossible to unambiguously determine the BS-induced phase from a single image. As a result, BS-based B_1 mapping always acquires two images acquired at offsets $+v_{RF}$ and $-v_{RF}$. Whereas the miscellaneous, non-RF-related phases are constant between the two scans, the BS-induced phase at the negative offset inverts relative to the positive offset. The difference between the two scans (Figure 4.19D) provides the RF-induced BS phase. Using Eq. (4.25) the BS phase can be converted to a B_1 map expressed in μT or Hz per unit input power (Figure 4.19E).

4.8.4 Alternative Image Contrast Mechanisms

Throughout the book a number of different image contrast mechanisms are discussed. Diffusion-weighted and diffusion-tensor imaging are introduced in Section 3.4, while magnetization transfer contrast (MTC) imaging and chemical exchange saturation transfer (CEST) MRI are discussed in Section 3.3. Despite the maturity of the MRI field, the techniques continued to see new innovations, including the discovery and implementation of novel contrast mechanisms. Among others, these include MR angiography [38–40], susceptibility-weighted MRI [41], quantitative susceptibility mapping (QSM, [42, 43]), diffusion kurtosis imaging [44], manganese-enhanced MRI (MEMRI) [45, 46], cardiac tagging [47], bolus tracking [48], and functional MRI based on blood-oxygenation-level dependent (BOLD, [49–52]) contrast, cerebral blood flow (CBF, [53]), or cerebral blood volume (CBV, [54–56]). Although many of these techniques can be used in combination with MRS studies and may help in the understanding of certain principles, they are outside the scope of this book. The interested reader is referred to some excellent books on MRI techniques [9–11]. Since fMRI based on the BOLD mechanism is such an important imaging modality with direct applications to fMRS [57–59], it will be briefly discussed here.

4.8.5 Functional MRI

Functional imaging techniques, NMR-based or otherwise, are aimed at the detection of the spatial distribution of neuronal activity in response to a stimulus (Figure 4.20). Although neuronal activity can be defined in several ways, it typically entails a number of events that include

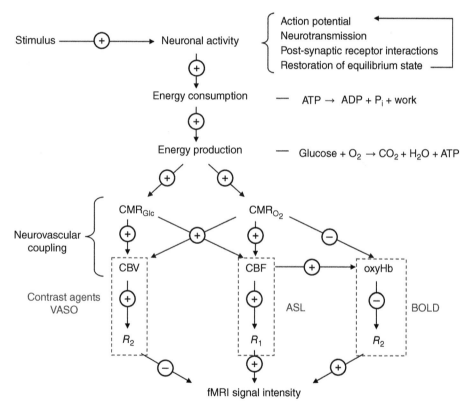

Figure 4.20 **Mechanisms underlying fMRI signal intensity changes.** The neuronal activity following a stimulus is associated with an increase in energy consumption, followed by an increase in energy production. Glucose oxidation is the primary mechanism for cerebral energy production, such that CMR_{Glc} and CMR_{O_2} must also increase in response to a stimulus. The glucose and oxygen must ultimately be supplied from the blood stream, leading to an increase in CBF and CBV. Since the CBF increases more than CMR_{O_2}, there will be a net increase in the amount of diamagnetic oxyhemoglobin. This will in turn lead to a decrease in the apparent transverse relaxation rate R_2 (or R_2^*), which will lead to a positive BOLD signal. NMR can also be sensitized for CBF and CBV after which stimulus-induced signal changes are observed through changes in the apparent longitudinal and transverse relaxation rates R_1 and R_2, respectively. It should be clear that fMRI techniques measure hemodynamic responses that, through a largely unknown neurovascular coupling, are related to the underlying electrical and metabolic processes. ^{13}C NMR can directly measure glucose oxidation and glutamatergic neurotransmission, thereby providing more direct information about the metabolic processes (see Chapter 3).

an action potential which leads to neurotransmission and post-synaptic receptor interactions. Following these events the equilibrium state must be reestablished, which includes the restoration of the intra/extracellular ion balance and recycling of neurotransmitters, such that the system is ready for another action potential. For most functional imaging methods the essential feature is that the restoration of an equilibrium state leads to an increase in energy consumption, which must ultimately also lead to an increase in energy production. Since the oxidation of glucose is the primary pathway for energy production in the brain, an increase in energy production must lead to an increase in the cerebral metabolic rates of glucose (CMR_{Glc}) as well as oxygen (CMR_{O_2}). The glucose and especially oxygen reserves inside the brain are very limited, such that increased demands for glucose and oxygen must be supplied by the blood, thus leading to an increase in CBF and CBV. The coupling between the metabolic parameters CMR_{Glc} and CMR_{O_2} and the hemodynamic parameters CBF and CBV is often referred to as the neurovascular coupling. The neurovascular coupling is still poorly understood. As a result, a significant fraction of the fMRI literature is dedicated to investigating the exact functional dependence of the measured parameters (CBF and CBV) on the parameters of interest (CMR_{Glc} and CMR_{O_2}). CBF and CBV can both be measured with NMR. CBF is most commonly measured through the use of arterial spin labeling (ASL) methods [53] in which the water spins in the carotid arteries are perturbed (saturated or inverted). During a delay period the perturbed water spins flow to the cerebral region of interest where they reduce the longitudinal magnetization. A difference image with and without ASL will show signal that is directly proportional to CBF, provided that the longitudinal T_1 relaxation is known. CBV can be measured through the use of intravascular contrast agents [54]. The intravascular (or "blood pool") contrast agent effectively eliminates the signal from the blood compartment. When the CBV increases, the amount of blood compartment within a MRI pixel increases, which leads to a decrease in signal intensity as the blood does not contribute any signal intensity. For human studies a completely noninvasive alternative for CBV measurements has been described that does not require a blood-pool agent [55, 56].

Even though NMR is able to measure CBF and CBV, these methods only represent a small fraction of the fMRI studies. An effect that is specific for NMR-based functional imaging can be found in the magnetic characteristics of hemoglobin. It has long been known that deoxyhemoglobin (i.e. hemoglobin without oxygen bound to it) is paramagnetic, whereas oxyhemoglobin is diamagnetic. As will be discussed in more detail in Chapter 10, the magnetic susceptibility difference between diamagnetic water and paramagnetic hemoglobin will lead to a disturbance of the magnetic field surrounding the red blood cells and blood vessels carrying the deoxyhemoglobin (Figure 4.21A). When CBF and CBV increase in response to a stimulus, the amount of deoxyhemoglobin inside red blood cells decreases, thereby leading to a reduction of the local magnetic field disturbance (Figure 4.21B). NMR sequences can be sensitized for local magnetic field distributions, for example, through the use of a gradient-echo MRI sequence (Figure 4.17A). During a resting condition in which no experimental stimuli are applied, the deoxyhemoglobin leads to a certain field disturbance which can be separated into intravascular, extravascular, static, and dynamic effects. Figure 4.21A shows a dynamic, extravascular effect in which a water molecule diffuses through the magnetic field disturbance caused by the deoxyhemoglobin. As the water diffuses it experiences a range of magnetic fields and hence resonance frequencies, which will lead to a certain amount of phase cancelation and thus signal loss in a gradient-echo MRI sequence. When during stimulation (Figure 4.21B) CBF and CBV increase and thus the amount of deoxyhemoglobin decreases, the phase dispersal of the diffusing water in the extravascular compartment will be less, such that more signal will be obtained. Note that the amount of deoxyhemoglobin in the blood only decreases because the increase in CBF exceeds the increase in CMR_{O_2}. Therefore, the so-called BOLD contrast [49–52] is typically positive when comparing an "activated" state with a "baseline" state

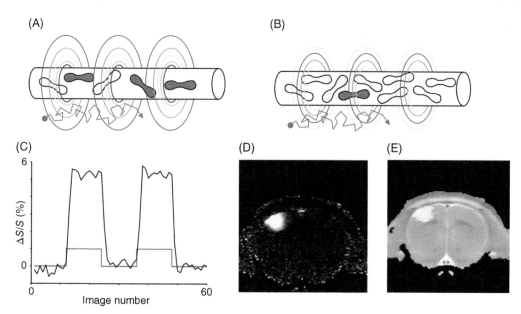

Figure 4.21 **Principle of blood-oxygen-level dependent (BOLD) fMRI.** (A) In a "resting" state the paramagnetic deoxyhemoglobin in red blood cells generate a significant magnetic field disturbance surrounding the cells and blood vessels. Through a number of intra- and extravascular effects this magnetic field inhomogeneity leads to signal loss in T_2 and T_2^*-weighted MRI sequences. (B) As a result of the interplay between CBF, CBV, and CMR_{O_2} changes following a stimulus, there is a net decrease in the amount of deoxyhemoglobin, and thus less signal loss during a T_2^*-weighted sequence, giving rise to a positive BOLD effect. (C) BOLD response during single forepaw stimulation in the rat at 9.4 T. Each image was acquired in circa 5 s with TE = 15 ms. A significant signal increase of 5–6% can be observed during the two stimulation blocks (gray line) relative to the baseline state. (D) Difference image, calculated as $(S_{\text{stimulated}} - S_{\text{baseline}})/S_{\text{baseline}}$, showing good localization of the forepaw region in the somatosensory cortex. (E) Cross-correlation map overlaid on an anatomical T_1 map. Only pixels with a cross-correlation coefficient of >0.8 are shown.

(Figure 4.21C–E). In reality, the signal is a complicated function of the various intra- and extravascular contributions, the magnetic field strength, echo-time, pulse sequence, blood vessel size and orientation, and other parameters. The interested reader is referred to the rich literature on fMRI [60–62]. For intracellular metabolites the situation is considerably simplified, as any effect must be attributed to a dynamic, extravascular contribution [57–59].

Exercises

4.1 Consider a spin-warp MRI sequence (TR – 90° – TE/2 – 180° – TE/2 – acquisition) and two tissue types with the following characteristics:
Tissue 1: $M_0 = 0.8$, $T_1 = 2.0$ s, $T_2 = 0.06$ s
Tissue 2: $M_0 = 0.7$, $T_1 = 1.0$ s, $T_2 = 0.03$ s
A At TR = 5.0 s calculate the echo-time TE which gives the largest absolute signal difference between the two tissues.
B At TE = 10 ms calculate the repetition time TR which gives the largest absolute signal difference between the two tissues.
C Calculate the echo-time TE at which the T_2-weighting cancels the combined proton-density/T_1-weighting at a repetition time TR of 2.0 s.

4.2 MRI signal is acquired with a gradient-echo sequence as 256 complex points over 1.28 ms during a readout gradient of 2.4466 G cm^{-1}. Signal excitation is achieved with a 1.5 ms sinc pulse (bandwidth = 3750 Hz). All gradient are ramped up and down between zero and their final amplitude in 500 μs.

A Calculate the FOV in the frequency-encoding ("readout") direction.

B For the same resolution and FOV as calculated under *A*, calculate the phase-encoding gradient increments (in G cm^{-1}) for a total phase-encoding gradient duration of 1.2 ms (including ramps).

C Calculate the slice selection gradient amplitude for 4.0 mm slices.

D For a maximum system gradient amplitude of 2.5 G cm^{-1} calculate the minimum achievable echo-time TE (from the middle of the excitation pulse to the top of the echo) for the parameters calculated under A–C. Hint: Magnetic field gradients in orthogonal directions can temporally overlap during certain periods within a sequence.

4.3 Consider the following sequences:
1) A multi-slice spin-echo sequence with TR/TE = 3000/12.5 ms.
2) A fast multi-slice gradient-echo method (FLASH) with TR = 10 ms, TE = 4 ms, and nutation angle $\alpha = 15°$.
3) A four-shot, multi-slice spin-echo EPI sequence with TR/TE = 3000/30 ms.

A For an image resolution of $32 \times 32 \times 32$ calculate the minimum acquisition times and determine the fastest sequence.

B Repeat the calculation for an image resolution of $128 \times 128 \times 128$.

C Calculate the relative SNR per unit time of the three methods for the middle k-space line for $T_1/T_2 = 1500/75$ ms. Assume negligible B_0 magnetic field inhomogeneity, identical acquisition bandwidths, and complete crushing of transverse magnetization between excitations. The effect of the 180° refocusing pulse on T_1 relaxation can be ignored.

4.4 A fast gradient-echo FLASH image is acquired as a 96×96 data matrix over a 19.2×19.2 cm FOV with 2.0 mm slices, TR/TE = 25/5 ms, 1 average, and nutation angle $\alpha = 20°$. Complete dephasing of transverse magnetization prior to each excitation can be assumed.

A For the corpus callosum with $T_1/T_2 = 1500/75$ ms, an absolute SNR of 40 was obtained. Which of the following modifications provides the largest and smallest improvement in absolute SNR?
- Increasing the FOV to 21.6×21.6 cm.
- Increasing the number of averages to 2.
- Decreasing the nutation angle to 15°.
- Increasing the repetition time to 100 ms.

B Which modification provides the largest improvement in SNR per unit of time?

C Calculate the Ernst angle and determine the increase in SNR relative to $\alpha = 20°$.

4.5 **A** Derive an expression for the longitudinal steady-state magnetization for a multi-slice gradient-echo sequence with *sequential* slice selection. Assume an in-slice nutation angle $\alpha 1$, an effective nutation angle $\alpha 2$ on either side of the slice due to an imperfect slice profile, and further assume that only adjacent slices affect each other.

B Derive an expression for the longitudinal steady-state magnetization for a multi-slice gradient-echo sequence with *interleaved* slice selection. Assume the same in-slice and out-of-slice nutation angles as in A.

C Calculate the signal gain of interleaved over sequential slice selection for a multi-slice gradient-echo sequence selecting 20 slices in 2000 ms for $T_1 = 1000$ ms and near-perfect slice profiles ($\alpha1 = 85°$, $\alpha2 = 10°$).

D Repeat the calculation in C for $\alpha1 = 75°$ and $\alpha2 = 35°$.

4.6 Two multi-slice, gradient-echo MR image datasets are acquired with TR/TE = 4000/4 ms and a 1 ms Gaussian excitation pulse with RF power settings of +6 dB and +12 dB. The absolute signal intensity of a specific pixel is 1025 and 768 in the two images, respectively.

A Calculate the nutation angle for the mentioned pixel position.

B Given the poorly defined slice profile of a Gaussian RF pulse (see also Chapter 5), there exists a range of nutation angles across the slice which will lead to an incorrect estimation. Based on simulations, a Gaussian pulse calibrated to give a 110° rotation on-resonance, gives the maximum amount of the signal from the entire slice. Recalculate the on-resonance nutation angle for the mentioned pixel position.

4.7 The magnetic field homogeneity is measured with a multi-slice gradient-echo sequence in which the echo-time is incremented from 4.0 ms to 5.0, 6.0, and 8.0 ms. For a given pixel, the acquired phases are calculated as −86°, +76°, −128°, and −171°, respectively.

A Under the assumption that the maximum magnetic field inhomogeneity is less than 500 Hz, calculate the frequency offset for the mentioned pixel position using (A) the first two data points and (B) all available data (Hint: Use a calculator).

B When the signal is acquired as a 64 × 64 matrix over a 50 kHz bandwidth, calculate the spatial displacement in the frequency- and phase-encoding directions for a magnetic field offset of 400 Hz.

4.8 **A** The Lissajou k-space trajectory is generated by the following gradient waveform combinations:

$$G_x(t) = \frac{G_{max}T}{v_x}\cos\left(\frac{2\pi v_x t}{T}\right) \quad \text{and} \quad G_y(t) = \frac{G_{max}T}{v_y}\cos\left(\frac{2\pi v_y t}{T}\right)$$

Draw the Lissajou k-space trajectory for $v_x = 9$ Hz, $v_y = 10$ Hz, and $0 \le t \le T$.

B The Rosette k-space trajectory is generated by the following gradient waveform combinations:

$$G_x(t) = \frac{G_{max}T}{v_1 + v_2}\cos\left(\frac{2\pi(v_1 + v_2)t}{T}\right) + \frac{G_{max}T}{v_1 - v_2}\cos\left(\frac{2\pi(v_1 - v_2)t}{T}\right)$$

$$G_y(t) = \frac{G_{max}T}{v_1 + v_2}\sin\left(\frac{2\pi(v_1 + v_2)t}{T}\right) - \frac{G_{max}T}{v_1 - v_2}\sin\left(\frac{2\pi(v_1 - v_2)t}{T}\right)$$

Draw the Rosette k-space trajectory for $v_1 = 10$ Hz, $v_2 = 1$ Hz, and $0 \le t \le T$.

C Discuss differences, advantages, and disadvantages of the alternative trajectories over standard rectangular k-space sampling.

4.9 Spatial mapping of the transmit B_1 field can be achieved with two fully relaxed measurements of two different excitation angles θ_1 and $\theta_2 = 2\theta_1$. A time-efficient alternative is given by the sequence: TR $- \theta -$ acquisition 1 $- \theta -$ acquisition 2.

   A Under the assumption that (i) TR $\gg T_1$ and (ii) acquisition time $1 \ll T_1$, derive the expression for the nutation angle as a function of the signals detected during acquisition times 1 and 2.

   B Determine the optimum nutation angle (i.e. the nutation angle that can be determined with the highest sensitivity). Note: Use standard error propagation theory and assume uncorrelated noise between the two measurements. Further note:

$$\frac{d(\arccos x)}{dx} = \frac{-1}{\sqrt{1-x^2}}.$$

   C Determine the optimum nutation angle for the standard, double-angle method in A.

4.10 During a gradient-echo sequence, the signal is excited with a homogeneous volume coil after which the signal is acquired with a two-element phased array receiver. The image acquisition is accelerated twofold by skipping every other line along the phase-encoding direction in k-space after which image reconstruction is executed according to the SENSE algorithm.

   A Derive an analytical expression for the signal intensities I_1 and I_2 in the non-aliased, full FOV image given the detected aliased signal S_1 and S_2 in the two receivers. The receiver sensitivity is described by a 2 × 2 matrix **C** for the two receivers and two positions (see also Figure 4.11E).

   B The receiver sensitivities at pixel location 1 are 10 and 2 and at pixel location 2 are 3 and 9. The signal intensities at the aliased pixel location are $S_1 = 280$ and $S_2 = 420$ for receivers 1 and 2, respectively. Calculate the signal intensities I_1 and I_2 at pixel locations 1 and 2 in the non-aliased, full FOV image.

   C Repeat the calculation in B when the receiver sensitivities at pixel location 1 are 10 and 2 and at pixel location 2 are 14 and 3. The signal intensities at the aliased pixel location are $S_1 = 280$ and $S_2 = 400$.

   D Repeat the calculation in B when the detected signal S_1 is 1% higher due to noise.

   E Repeat the calculation in C when the detected signal S_1 is 1% higher due to noise.

   F Draw a conclusion from the calculations performed in B–E regarding the suitability of a given receiver setup to achieve image acceleration.

4.11 **A** For two relaxation times T_{1A} and T_{1B} calculate the recovery delay TR that leads to the largest contrast in a standard gradient-echo sequence.

   B Show that a similar expression is valid for maximum T_2 contrast.

4.12 In general, T_1 relaxation time constants are measured with multipoint methods, such as inversion or saturation recovery. An alternative is provided by T_1 measurements based on two experiments with different repetition times. The signal intensity S_1 and S_2 of two spin-echo MRI experiments with repetition times TR1 and TR2 can be described by

$$S_1 = S_0 \left(1 - e^{-TR1/T_1}\right) e^{-TE/T_2}$$

$$S_2 = S_0 \left(1 - e^{-TR2/T_1}\right) e^{-TE/T_2}$$

A Derive an expression for the relaxation time T_1 when TR2 = 2TR1 and when TR2 = 3TR1.

B For the experiment in which TR2 = 2TR1 determine the optimal repetition time TR1 to determine a relaxation time constant T_1 of 1000 ms.

4.13 Consider a single-shot EPI sequence with the following parameters: 64 × 64 data matrix over a 20 cm FOV, imaging bandwidth = 100 kHz, and 100 μs gradient ramps.

A Calculate the effective spectral width across the FOV in the phase-encoding direction.

B Derive an expression for the spatial displacements along the frequency and phase-encoding directions during EPI assuming infinitely short gradient ramps.

C For a pixel with a frequency offset of +150 Hz, for example, due to residual magnetic field inhomogeneity, calculate the pixel shifts in the frequency- and phase-encoding directions.

4.14 Consider a 2D image with a 128 × 128 resolution over a 24 cm FOV.

A Calculate the linear phase that must be applied in k-space to shift the image by 1/3 of the FOV.

B Calculate the required first-order phase correction required to display the image as an in-phase, real image.

C Calculate the relative SNRs per unit time for a 64 × 64 image obtained by (i) downsampling the 128 × 128 image (i.e. adding 4 pixels) or (ii) acquiring a new image at a 64 × 64 resolution.

4.15 A 2 cm diameter cylinder filled with water (4 cm length) is placed in a 3 T magnet with its longest axis parallel to the main magnetic field.

A Draw the relative 1D spatial profiles when signal is acquired during a frequency-encoding gradient applied in the x, y, and z directions, respectively.

B Draw the spatial profile when signal is acquired in the presence of simultaneous x and y gradients.

Consider two 3 cm diameter spheres filled with water (H_2O, 4.7 ppm) and methanol (CH_3OH, 3.3 ppm), respectively, placed 5 cm apart (center-to-center) in the x direction of a 3 T magnet and being aligned in the y and z directions. FOV = 10 cm, dwell-time of 1 μs, 512 acquisition points.

C Draw the spatial profile when signal is acquired with a spin-echo sequence (TE = 8.4 ms) in the presence of frequency-encoding gradients in the x and y directions, respectively.

D Draw the spatial profile when signal is acquired with a gradient-echo sequence (TE = 8.4 ms) in the presence of frequency-encoding gradients in the x and y directions, respectively.

E Draw the spatial profile when the dwell-time is increased to 50 μs (spin-echo sequence) and the data are acquired at 11.7 T.

References

1 Lauterbur, P.C. (1973). Image formation by induced local interactions: examples employing nuclear magnetic resonance. *Nature* 242: 190–191.

2 Mansfield, P. and Grannell, P.K. (1973). NMR "diffraction" in solids? *J. Phys. C: Solid State Phys.* 6: L422–L427.

3 Feinberg, D.A., Crooks, L.E., Hoenninger, J.C. et al. (1986). Contiguous thin multisection MR imaging by two-dimensional Fourier transform techniques. *Radiology* 158: 811–817.

4 Edelstein, W.A., Hutchison, J.M.S., Johnson, G.A., and Redpath, T. (1980). Spin warp NMR imaging and applications to human whole body imaging. *Phys. Med. Biol.* 25: 751–756.

5 Ljunggren, S. (1983). A simple graphical representation of Fourier-based imaging methods. *J. Magn. Reson.* 54: 338–343.

6 Twieg, D.B. (1983). The k-trajectory formulation of the NMR imaging process with applications in analysis and synthesis of imaging methods. *Med. Phys.* 10: 610–621.

7 Mezrich, R. (1995). A perspective on k-space. *Radiology* 195: 297–315.

8 Wehrli, F.W. (1991). *Fast-scan Magnetic Resonance. Principles and Applications*. New York: Raven Press.

9 Haacke, E.M., Brown, R.W., Thompson, M.R., and Venkatesan, R. (1999). *Magnetic Resonance Imaging. Physical Principles and Sequence Design*. New York: Wiley-Liss.

10 Stark, D.D. and Bradley, W.G. (1999). *Magnetic Resonance Imaging*. New York: C. V. Mosby.

11 Bernstein, M.A., King, K.F., and Zhou, X.J. (2004). *Handbook of MRI Pulse Sequences*. Academic Press.

12 Haase, A., Frahm, J., and Matthaei, K.D. (1986). FLASH imaging: rapid NMR imaging using low flip angle pulses. *J. Magn. Reson.* 67: 258–266.

13 Gyngell, M.L. (1988). The application of steady-state free precession in rapid 2DFT NMR imaging: FAST and CE-FAST sequences. *Magn. Reson. Imaging* 6: 415–419.

14 Zur, Y., Wood, M.L., and Neuringer, L.J. (1991). Spoiling of transverse magnetization in steady-state sequences. *Magn. Reson. Med.* 21: 251–263.

15 Mansfield, P. (1977). Multiplanar image formation using NMR spin echoes. *J. Phys. C: Solid State Phys.* 10: L55–L58.

16 Stehling, M.K., Turner, R., and Mansfield, P. (1991). Echo-planar imaging: magnetic resonance imaging in a fraction of a second. *Science* 254: 43–50.

17 Ahn, C.B., Kim, J.H., and Cho, Z.H. (1986). High-speed spiral-scan echo planar NMR imaging. *IEEE Trans. Med. Imaging* MI-5: 2–7.

18 Meyer, C.H., Hu, B.S., Nishimura, D.G., and Macovski, A. (1992). Fast spiral coronary artery imaging. *Magn. Reson. Med.* 28: 202–213.

19 Hennig, J., Nauerth, A., and Friedburg, H. (1986). RARE imaging: a fast imaging method for clinical MR. *Magn. Reson. Med.* 3 (6): 823–833.

20 Jackson, J.I., Meyer, C.H., Nishimura, D.G., and Macovski, A. (1991). Selection of a convolution function for Fourier inversion using gridding. *IEEE Trans. Med. Imaging* 10: 473–478.

21 Duyn, J.H., Yang, Y., Frank, J.A., and van der Veen, J.W. (1998). Simple correction method for k-space trajectory deviations in MRI. *J. Magn. Reson.* 132: 150–153.

22 Wilm, B.J., Barmet, C., Pavan, M., and Pruessmann, K.P. (2011). Higher order reconstruction for MRI in the presence of spatiotemporal field perturbations. *Magn. Reson. Med.* 65: 1690–1701.

23 Sodickson, D.K. and Manning, W.J. (1997). Simultaneous acquisition of spatial harmonics (SMASH): fast imaging with radiofrequency coil arrays. *Magn. Reson. Med.* 38: 591–603.

24 Pruessmann, K.P., Weiger, M., Scheidegger, M.B., and Boesiger, P. (1999). SENSE: sensitivity encoding for fast MRI. *Magn. Reson. Med.* 42: 952–962.

25 Griswold, M.A., Jakob, P.M., Heidemann, R.M. et al. (2002). Generalized autocalibrating partially parallel acquisitions (GRAPPA). *Magn. Reson. Med.* 47: 1202–1210.

26 Kingsley, P.B. (1999). Methods of measuring spin-lattice (T_1) relaxation times: an annotated bibliography. *Concepts Magn. Reson.* 11: 243–276.

27 Look, D.C. and Locker, D.R. (1968). Nuclear spin-lattice relaxation measurements by tone-burst modulation. *Phys. Rev. Lett.* 20: 987–989.

28 Look, D.C. and Locker, D.R. (1970). Time saving in measurement of NMR and EPR relaxation times. *Rev. Sci. Instrum.* 41: 250–251.

29 Kaptein, R., Dijkstra, K., and Tarr, C.E. (1976). A single-scan Fourier transform method for measuring spin-lattice relaxation times. *J. Magn. Reson.* 24: 295–300.

30 Crawley, A.P. and Henkelman, R.M. (1988). A comparison of one-shot and recovery methods in T_1 imaging. *Magn. Reson. Med.* 7: 23–34.

31 Crawley, A.P. and Henkelman, R.M. (1987). Errors in T2 estimation using multislice multiple-echo imaging. *Magn. Reson. Med.* 4: 34–47.

32 Poon, C.S. and Henkelman, R.M. (1992). Practical T_2 quantitation for clinical applications. *J. Magn. Reson. Imaging* 2: 541–553.

33 Ma, D., Gulani, V., Seiberlich, N. et al. (2013). Magnetic resonance fingerprinting. *Nature* 495: 187–192.

34 Jenkinson, M. (2003). Fast, automated, *N*-dimensional phase-unwrapping algorithm. *Magn. Reson. Med.* 49: 193–197.

35 Nehrke, K. and Bornert, P. (2012). DREAM – a novel approach for robust, ultrafast, multislice B_1 mapping. *Magn. Reson. Med.* 68: 1517–1526.

36 Sacolick, L.I., Wiesinger, F., Hancu, I., and Vogel, M.W. (2010). B_1 mapping by Bloch-Siegert shift. *Magn. Reson. Med.* 63: 1315–1322.

37 Khalighi, M.M., Rutt, B.K., and Kerr, A.B. (2012). RF pulse optimization for Bloch-Siegert B_1^+ mapping. *Magn. Reson. Med.* 68: 857–862.

38 Wedeen, V.J., Meuli, R.A., Edelman, R.R. et al. (1985). Projective imaging of pulsatile flow with magnetic resonance. *Science* 230: 946–948.

39 Dumoulin, C.L. and Hart, H.R. Jr. (1986). Magnetic resonance angiography. *Radiology* 161: 717–720.

40 Nishimura, D.G. (1990). Time-of-flight MR angiography. *Magn. Reson. Med.* 14: 194–201.

41 Haacke, E.M., Xu, Y., Cheng, Y.C., and Reichenbach, J.R. (2004). Susceptibility weighted imaging (SWI). *Magn. Reson. Med.* 52: 612–618.

42 Wang, Y. and Liu, T. (2015). Quantitative susceptibility mapping (QSM): Decoding MRI data for a tissue magnetic biomarker. *Magn. Reson. Med.* 73: 82–101.

43 Liu, C., Li, W., Tong, K.A. et al. (2015). Susceptibility-weighted imaging and quantitative susceptibility mapping in the brain. *J. Magn. Reson. Imaging* 42: 23–41.

44 Jensen, J.H., Helpern, J.A., Ramani, A. et al. (2005). Diffusional kurtosis imaging: the quantification of non-gaussian water diffusion by means of magnetic resonance imaging. *Magn. Reson. Med.* 53: 1432–1440.

45 Lin, Y.J. and Koretsky, A.P. (1997). Manganese ion enhances T_1-weighted MRI during brain activation: an approach to direct imaging of brain function. *Magn. Reson. Med.* 38: 378–388.

46 Silva, A.C., Lee, J.H., Aoki, I., and Koretsky, A.P. (2004). Manganese-enhanced magnetic resonance imaging (MEMRI): methodological and practical considerations. *NMR Biomed.* 17: 532–543.

47 Axel, L. and Dougherty, L. (1989). MR imaging of motion with spatial modulation of magnetization. *Radiology* 171: 841–845.

48 Villringer, A., Rosen, B.R., Belliveau, J.W. et al. (1988). Dynamic imaging with lanthanide chelates in normal brain: contrast due to magnetic susceptibility effects. *Magn. Reson. Med.* 6: 164–174.

49 Ogawa, S., Lee, T.M., Kay, A.R., and Tank, D.W. (1990). Brain magnetic resonance imaging with contrast dependent on blood oxygenation. *Proc. Natl. Acad. Sci. U. S. A.* 87: 9868–9872.

50 Ogawa, S., Tank, D.W., Menon, R. et al. (1992). Intrinsic signal changes accompanying sensory stimulation: functional brain mapping with magnetic resonance imaging. *Proc. Natl. Acad. Sci. U. S. A.* 89 (13): 5951–5955.

51 Kwong, K.K., Belliveau, J.W., Chesler, D.A. et al. (1992). Dynamic magnetic resonance imaging of human brain activity during primary sensory stimulation. *Proc. Natl. Acad. Sci. U. S. A.* 89: 5675–5679.

52 Bandettini, P.A., Wong, E.C., Hinks, R.S. et al. (1992). Time course EPI of human brain function during task activation. *Magn. Reson. Med.* 25: 390–397.

53 Williams, D.S., Detre, J.A., Leigh, J.S., and Koretsky, A.P. (1992). Magnetic resonance imaging of perfusion using spin inversion of arterial water. *Proc. Natl. Acad. Sci. U. S. A.* 89: 212–216.

54 Mandeville, J.B., Marota, J.J., Kosofsky, B.E. et al. (1998). Dynamic functional imaging of relative cerebral blood volume during rat forepaw stimulation. *Magn. Reson. Med.* 39: 615–624.

55 Lu, H., Golay, X., Pekar, J.J., and Van Zijl, P.C. (2003). Functional magnetic resonance imaging based on changes in vascular space occupancy. *Magn. Reson. Med.* 50: 263–274.

56 Donahue, M.J., Lu, H., Jones, C.K. et al. (2006). Theoretical and experimental investigation of the VASO contrast mechanism. *Magn. Reson. Med.* 56: 1261–1273.

57 Zhu, X.H. and Chen, W. (2001). Observed BOLD effects on cerebral metabolite resonances in human visual cortex during visual stimulation: a functional ^{1}H MRS study at 4 T. *Magn. Reson. Med.* 46: 841–847.

58 Mangia, S., Garreffa, G., Bianciardi, M. et al. (2003). The aerobic brain: lactate decrease at the onset of neural activity. *Neuroscience* 118: 7–10.

59 Bednarik, P., Tkac, I., Giove, F. et al. (2015). Neurochemical and BOLD responses during neuronal activation measured in the human visual cortex at 7 Tesla. *J. Cereb. Blood Flow Metab.* 35: 601–610.

60 Moonen, C.T.W. and Bandettini, P.A. (2000). *Functional MRI* (ed. A.L. Baert, K. Sartor and J.E. Youker). New York: Springer.

61 Buxton, R.B. (2001). *An Introduction to Functional Magnetic Resonance Imaging: Principles and Techniques.* New York: Cambridge University Press.

62 Jezzard, P., Matthews, P.M., and Smith, S.M. (2003). *Functional MRI: An Introduction to Methods.* Oxford University Press.

5

Radiofrequency Pulses

5.1 Introduction

Radiofrequency (RF) pulses are required in all MR imaging and spectroscopy experiments. They are the primary tools used by the experimentalist to perform the desired spin manipulations like excitation, inversion, and refocusing. Short, constant-amplitude, or "square" RF pulses present the simplest option to achieve nonselective excitation of the spins. However, many *in vivo* NMR experiments require a certain degree of selectivity in which only spins in a limited range of frequencies are perturbed. Examples of this can be found in spatial slice selection, water suppression, and spectral editing. Frequency-selective RF pulses can be designed using Fourier transform theory, the Bloch equations, or elaborate optimization procedures. In other circumstances, the B_1-dependence of the nutation angle needs to be minimized in order to reduce artifacts and/or increase sensitivity. This can be achieved with composite RF pulses or with RF amplitude- and frequency-modulated (FM) adiabatic RF pulses.

In vivo NMR spectroscopy makes use of all features encountered in RF pulse design. In fact, successful *in vivo* NMR studies depend heavily on the thorough understanding and correct implementation of RF pulses. This chapter will deal with the description of RF pulses, emphasizing theoretical principles, practical utility, and experimental pitfalls. All RF pulses discussed in this chapter are available in PulseWizard, an open-source, Matlab-based RF generation and simulation tool that can be downloaded from the Wiley website (http://booksupport.wiley.com).

5.2 Square RF Pulses

An RF pulse is composed of a magnetic field rotating at or near the spin Larmor frequency. For the magnetic field strengths commonly used for NMR, the Larmor frequency falls in the radiofrequency or RF part of the electromagnetic spectrum. It should be realized that an RF pulse used for NMR is not the same as RF electromagnetic waves or radiation used by, for instance, commercial radio stations. The RF coils used in NMR are specifically designed to minimize the electric component, leaving only a rotating *magnetic* field. RF coils are said to operate in the "near field" region of the electromagnetic field. RF electromagnetic waves operate in the "far field" region in which electric and magnetic fields are closely associated with each other in a fixed amplitude and phase relationship. More details on RF coils and their magnetic and electric fields can be found in Chapter 10.

In Vivo *NMR Spectroscopy: Principles and Techniques*, Third Edition. Robin A. de Graaf.
© 2019 John Wiley & Sons Ltd. Published 2019 by John Wiley & Sons Ltd.

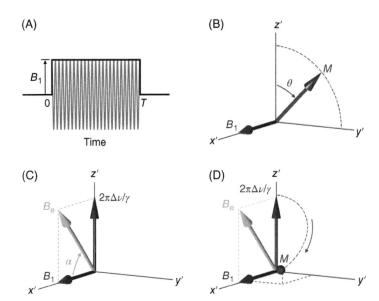

Figure 5.1 Magnetic fields and rotations during a square RF pulse. (A) A square RF pulse is a magnetic field oscillating at the Larmor frequency that is turned on with amplitude B_1 over a duration T. (B) In the rotating frame of reference $x'y'z'$ the static magnetic field B_0 vector is reduced to zero, making the RF magnetic field B_1 appear stationary along the x' axis. On-resonance spins rotate around B_1 over an angle θ according to Eq. (5.1). (C) For spins that are off-resonance by an amount $\Delta\nu$ the reduction of the static magnetic field vector is incomplete, thereby leading to a residual magnetic field vector equal to $2\pi\Delta\nu/\gamma$. The RF and off-resonance magnetic fields combine into an effective magnetic field B_e according to Eq. (5.2). (D) Off-resonance spins rotate around the effective field, leading to incomplete excitation and phase accumulation.

A square or "hard" RF pulse is generated by instantly turning on a time-varying current in an RF coil, leaving it on at constant amplitude for a duration T after which it is turned off. Even though in most publications only the RF pulse envelope is shown (Figure 5.1A), it should always be realized that the RF pulse is operating as a magnetic field rotating with a frequency close to the Larmor frequency and thus undergoing thousands to millions of revolutions during the duration of the pulse. The RF pulse is said to be "on resonance" when the RF pulse frequency equals the Larmor frequency of the spins. In that condition a square RF pulse with amplitude B_1 (in T) and duration T (in s) generates a nutation or flip angle θ (in radians) according to (Figure 5.1B)

$$\theta = \gamma B_1 T \tag{5.1}$$

where γ is the gyromagnetic ratio (in $\mathrm{rad\,T^{-1}\,s^{-1}}$, see Table 1.1). When the RF pulse rotates the longitudinal magnetization M_z into the transverse plane (M_{xy}), the 90° rotation is referred to as excitation. Inversion is the 180° rotation of M_z to $-M_z$ and requires double the RF pulse duration or double the RF pulse amplitude compared to a 90° excitation. An RF pulse that rotates the magnetization in the transverse plane by 180° is referred to as a refocusing pulse. A refocusing pulse is always an inversion pulse, but the opposite is not necessarily true (see Section 5.5.4).

Under many experimental conditions it is not possible to achieve the "on-resonance" condition for all spins in the sample, for example, in the presence of magnetic field inhomogeneity across the sample or when having spins with different chemical shifts. When the RF pulse is on-resonance for the N-acetyl aspartate (NAA) methyl spins at 2.01 ppm, it is automatically "off-resonance" for the creatine methyl spins at 3.03 ppm by an amount of 1.02 ppm or 304 Hz at 7.0 T. In a rotating frame of reference (rotating at the RF frequency ν) the magnetic field difference for off-resonance spins with Larmor frequency ν_0 appears as a vector of $2\pi(\nu_0 - \nu)/\gamma = 2\pi\Delta\nu/\gamma$ along the z' axis (Figure 5.1C). MRI coils and RF amplifiers used for human head applications are often limited to a maximum B_1 amplitude of 15–25 μT (or $(\gamma/2\pi)B_1 = 600$–1000 Hz for protons), making the B_1 magnetic field of comparable amplitude to the off-resonance magnetic field difference (Figure 5.1C). The two magnetic fields add as vectors to yield an effective field $\boldsymbol{B}_e$ whose amplitude and direction are given by

$$B_e = |\boldsymbol{B}_e| = \sqrt{B_1^2 + \left(\frac{2\pi\Delta\nu}{\gamma}\right)^2} \tag{5.2}$$

$$\alpha = \arctan\left(\frac{2\pi\Delta\nu}{\gamma B_1}\right) \tag{5.3}$$

whereby α gives the angle of the effective field relative to the transverse plane (Figure 5.1C). In the presence of off-resonance effects, the magnetization follows a complex path as it rotates around the effective field (Figure 5.1D). Rotations around the effective field are most easily described by rotations about the principal axis x', y', and z' in matrix notation. For a clockwise rotation around the x' axis over an angle θ, the rotation matrix $\boldsymbol{R}_x(\theta)$ is given by

$$\boldsymbol{R}_x(\theta) = \begin{pmatrix} 1 & 0 & 0 \\ 0 & \cos\theta & \sin\theta \\ 0 & -\sin\theta & \cos\theta \end{pmatrix} \tag{5.4}$$

The magnetization after the RF pulse can be calculated according to

$$\begin{pmatrix} M_x \\ M_y \\ M_z \end{pmatrix}_{\text{after}} = \boldsymbol{R}_x(\theta) \begin{pmatrix} M_x \\ M_y \\ M_z \end{pmatrix}_{\text{before}} \tag{5.5}$$

For an initial condition of thermal equilibrium (i.e. $(M_x, M_y, M_z) = (0, 0, M_0)$), Eq. (5.5) predicts excitation of the longitudinal magnetization along the y' axis according to $M_{y,\text{after}} = M_0\sin\theta$ and $M_{z,\text{after}} = M_0\cos\theta$. The matrices for rotations about the y' and z' axes are given by

$$\boldsymbol{R}_y(\theta) = \begin{pmatrix} \cos\theta & 0 & -\sin\theta \\ 0 & 1 & 0 \\ \sin\theta & 0 & \cos\theta \end{pmatrix} \tag{5.6}$$

$$\boldsymbol{R}_z(\theta) = \begin{pmatrix} \cos\theta & \sin\theta & 0 \\ -\sin\theta & \cos\theta & 0 \\ 0 & 0 & 1 \end{pmatrix} \tag{5.7}$$

These rotation matrices for clockwise rotations about the relevant axis (i.e. when viewed along this axis towards the origin) can be derived from the Bloch equations (Eqs. (1.9)–(1.11)) in the absence of relaxation (see also Exercises). The rotation matrices R_x, R_y, and R_z allow the description of the rotations of the magnetization around the effective field in consecutive steps by repeated application of the appropriate rotation matrix for non-tilted axes. For example, the rotations induced by an RF pulse along the x' axis over an angle θ with an effective field B_e tilted over α to the z' axis is equivalent to a rotation through $-\alpha$ about y' followed by a rotation through θ about x' and then followed by a rotation of $+\alpha$ about y' (Figure 5.1D). The corresponding rotation matrix for the rotation is the product of the three corresponding matrices given by Eqs. (5.5)–(5.7) and is given by

$$R_x(\theta,\alpha)=R_y(\alpha)R_x(\theta)R_y(-\alpha)$$

$$=\begin{pmatrix} \cos^2\alpha+\cos\theta\sin^2\alpha & \sin\alpha\sin\theta & \sin\alpha\cos\alpha(1-\cos\theta) \\ -\sin\alpha\sin\theta & \cos\theta & \cos\alpha\sin\theta \\ \sin\alpha\cos\alpha(1-\cos\theta) & -\cos\alpha\sin\theta & \sin^2\alpha+\cos\theta\cos^2\alpha \end{pmatrix} \quad (5.8)$$

The rotation matrix for an RF pulse with phase ϕ relative to the x' axis is obtained with an additional rotation about z', i.e.

$$R_x(\phi,\theta,\alpha)=R_z(\phi)R_x(\theta,\alpha)R_z(-\phi) \quad (5.9)$$

It is straightforward to confirm the validity of Eq. (5.8) by considering the case of a strong RF pulse ($|\gamma B_1|\gg 2\pi\Delta\nu$) such that off-resonance effects can be neglected. In that case, $\alpha=0°$ making $\cos\alpha=1$, $\sin\alpha=0$, $B_e=B_1$, and Eq. (5.8) then reduces to the rotation matrix given by Eq. (5.4). For the situation in which no B_1 field is present (i.e. $\cos\alpha=0$, $\sin\alpha=1$, and $B_e=2\pi\Delta\nu/\gamma$), Eq. (5.8) reduces to the description of the precession of transverse magnetization about z' given by Eq. (5.7).

The frequency dependence of the magnetization following an RF pulse of duration T applied along the x' axis can now be calculated by using the rotation matrices and Eq. (5.9). Starting with the thermal equilibrium magnetization M_0, the magnetization following the RF pulse is given by

$$M_x = M_0 \frac{2\pi\gamma B_1\Delta\nu}{\gamma^2 B_1^2+4\pi^2\Delta\nu^2}\left(1-\cos\sqrt{\gamma^2 B_1^2+4\pi^2\Delta\nu^2}T\right) \quad (5.10)$$

$$M_y = M_0 \frac{\gamma B_1}{\sqrt{\gamma^2 B_1^2+4\pi^2\Delta\nu^2}}\sin\sqrt{\gamma^2 B_1^2+4\pi^2\Delta\nu^2}T \quad (5.11)$$

$$M_z = \frac{M_0}{\gamma^2 B_1^2+4\pi^2\Delta\nu^2}\left(4\pi^2\Delta\nu^2+\gamma^2 B_1^2\cos\sqrt{\gamma^2 B_1^2+4\pi^2\Delta\nu^2}T\right) \quad (5.12)$$

Figure 5.2 shows the profiles of M_x, M_y, and M_z for a 90° nutation angle (applied along the x' axis) as calculated using Eqs. (5.10)–(5.12), respectively. On-resonance ($\Delta\nu=0$) Eqs. (5.10)–(5.12) reduce to $(M_x, M_y, M_z) = (0, M_0\sin\theta, M_0\cos\theta)$ confirming the rotation of magnetization in the yz plane (Figure 5.1B). Off-resonance two effects can be observed. Firstly, the amount of transverse magnetization $M_{xy}=(M_x^2+M_y^2)^{1/2}$ following excitation ($\theta=90°$) is always smaller for off-resonance spins compared to spins for which the on-resonance condition is satisfied. This can be problematic for applications that require uniform excitation of a large spectral width, such as encountered in ^{13}C MRS. For square RF pulses the only way to increase the

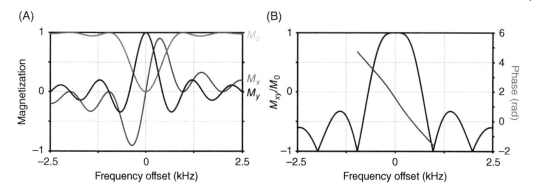

Figure 5.2 **Excitation profile of a square RF pulse.** (A) Magnetization components M_x, M_y, and M_z forming the excitation profile of a 250 μs square RF pulse as calculated with Eqs. (5.10)–(5.12). (B) Magnitude and phase profile following a 250 μs square 90° RF pulse.

excitation bandwidth is to decrease the pulse length T. In order to maintain a constant nutation angle θ the B_1 magnetic field amplitude must be increased proportionally. The B_1 amplitude that can be generated at the maximum output setting of the RF amplifier typically limits the achievable excitation bandwidth.

Secondly, on-resonance spins are excited along the $+y'$ axis generating only M_y without M_x. Off-resonance spins are excited into the transverse plane with both M_x and M_y components. In other words, off-resonance magnetization is excited with a nonzero phase that is approximately linear with the frequency offset (Figure 5.2B). This is explained by the fact that as soon as spins are rotated from their initial orientation parallel to the B_0 magnetic field, they will start to rotate around B_0 in addition to the rotation around B_1. The linear phase roll can be compensated with a first-order phase correction or through experimental design. For small frequency offsets the approximation $B_e \sim B_1$ holds and Eq. (5.10) yields $M_x/M_0 \approx 2\pi\Delta\nu/\gamma B_1$. Following an ideal, infinitely short 90° pulse and delay t, the transverse magnetization will precess in the transverse plane such that $M_x/M_0 = \sin(2\pi\Delta\nu t)$ which reduces to $2\pi\Delta\nu t$ for small frequency offsets. Equating these two expressions for M_x yields

$$t \sim \frac{1}{\gamma B_1} = \frac{2T}{\pi} \tag{5.13}$$

using Eq. (5.1) and $\theta = 90°$. From Eq. (5.13) it can be concluded that a square RF pulse of finite duration T can be approximated by an ideal, infinitely short RF pulse followed by a free precession delay equal to $2T/\pi$. The linear phase roll predicted by Eqs. (5.10)–(5.12) and shown in Figure 5.2 can be eliminated by subtracting the delay $2T/\pi$ from the first echo-time TE/2 period in a spin-echo sequence. Practical examples of this strategy to eliminate RF-induced phase rolls will be given in Chapter 6. For non-echo signals as acquired with, for example, a pulse-acquire sequence the linear phase roll can be corrected with first-order phase correction (see Chapter 1) or linear prediction methods (see Chapter 8).

Figure 5.3 shows the frequency profiles as calculated using Eqs. (5.10)–(5.12) for on-resonance nutation angles of 30°, 90°, and 180° at a constant maximum RF amplitude. The short duration of a 30° square RF pulse leads to a relatively uniform excitation across a large bandwidth. The six times longer pulse length required for a 180° inversion RF pulse leads to very poor off-resonance performance in which inversion is only achieved on-resonance. This in turn has sparked the development of adiabatic inversion pulses that are characterized by wide and

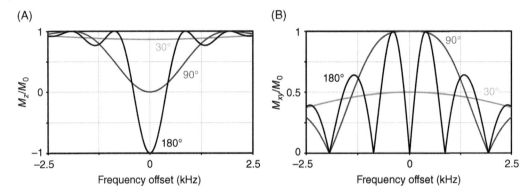

Figure 5.3 Frequency profiles of a square RF pulse. (A) Longitudinal magnetization M_z and (B) transverse magnetization M_{xy} following a 30°, 90°, and 180° square RF pulse. The RF pulses are executed with 167, 500, and 1000 μs pulse length and 500 Hz amplitude, respectively. Note that the desired rotations are only achieved close to on-resonance. As a result a 180° inversion pulse creates large amounts of transverse magnetization at almost every frequency offset.

uniform inversion bandwidths (see Section 5.5). Note that even though a 180° square RF pulse provides inversion on-resonance, it leads to significant excitation of magnetization at off-resonance frequencies. Magnetization excited by *any* inversion (or refocusing) pulse can lead to artifacts when insufficiently removed by magnetic field gradient crushers or phase cycling. This scenario is always encountered during single-volume localization as described in Chapter 6.

Equation (5.1) shows that the nutation angle θ of a square RF pulse can be adjusted by varying the RF pulse amplitude B_1, the RF pulse length T, or both. Since square RF pulses are typically used to achieve nonselective rotations over a wide frequency range, it is desirable to achieve the nutation angle at the shortest possible pulse length. This is normally achieved by setting the RF power amplifier to a high level, close to the maximum amplifier output which then generates the maximum possible RF pulse amplitude B_{1max}.

When NMR signal is acquired with increasing RF pulse lengths and a fixed RF pulse amplitude B_{1max}, a calibration curve as shown in Figure 5.4A will be obtained, in which the signal intensity is proportional to $\sin(\gamma B_{1max}T)$. The highest signal corresponds to a 90° nutation angle, after which the signal reduces to zero and becomes maximum negative for 180° and 270° nutation angles, respectively. The signal shown in Figure 5.4A is obtained with a homogeneous RF coil and a long repetition time TR with respect to the T_1 relaxation time constant of the sample. The experiment needs to be performed on-resonance, to avoid undesirable off-resonance phase shifts (Figure 5.2). Furthermore, when the calibration experiment is performed off-resonance, the highest signal will be an overestimation of the actual on-resonance 90° pulse length. Figure 5.4B shows the situation when the repetition time of the calibration experiment is comparable to T_1 (TR ~ T_1). The curve can no longer be described by a simple sinusoidal function, but needs to be described by Eq. (3.8). When the highest signal is taken as the 90° nutation angle, a significant underestimation of the actual 90° pulse length results. In principle, the obtained calibration curve could still be used to retrieve the actual 90° nutation angle by fitting the curve with Eq. (3.8). However, this requires an accurate knowledge of the T_1 relaxation time, which is seldom true prior to a calibration experiment. Note that the pulse length corresponding to a 180° pulse is relatively insensitive to the repetition time making it, in principle, suitable for pulse calibration. However, in situations of significant B_1 magnetic field inhomogeneity (as, for example, encountered with surface coils), the zero-crossing in the calibration curve corresponding to a 180° nutation angle may no longer be observed. Figure 5.4C

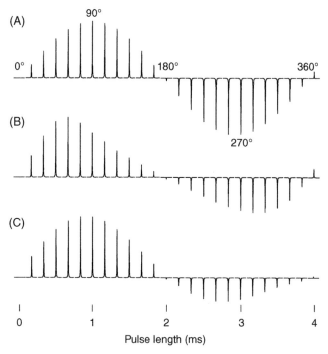

Figure 5.4 Nutation angle calibration of a square RF pulse. (A) On-resonance ($\Delta\nu = 0$), under conditions of complete longitudinal relaxation, the transverse magnetization follows a sinusoidal modulation as a function of pulse length according to Eq. (5.11). (B) Under conditions of rapid pulse repetition (TR $\sim T_1$), the observed signal intensity becomes dependent upon T_1, TR, and the nutation angle according to Eq. (3.8). (C) In the presence of RF transmit B_1^+ magnetic field inhomogeneity, the observed signal intensity results from a variety of different nutation angles making accurate pulse length calibration difficult.

shows the calibration curve in case the B_1 magnetic field inhomogeneity is ±50% of the nominal B_1 field strength. It can be seen that especially for longer pulse lengths (corresponding to ≥180° nutation angles in a homogeneous B_1 field), B_1 magnetic field inhomogeneity results in severe signal loss. Even though the 180° nutation angle can still be determined from Figure 5.4C, it will be impossible for larger B_1 inhomogeneity. While an RF pulse calibration as shown in Figure 5.4A is conceptually simple and standard for homogeneous RF probes encountered in high-resolution NMR, most RF pulse calibrations performed *in vivo* use MRI-based B_1 maps in order to deal with spatial heterogeneity and provide localized RF calibration. MRI-based B_1 mapping is discussed in Chapter 4.

5.3   Selective RF Pulses

A square RF pulse is the workhorse for high-resolution, liquid-state NMR where most applications require nonselective excitation over a wide spectral range. However, square RF pulses are not commonly used for *in vivo* NMR since most applications require frequency-selective excitation. Frequency-selectivity can, to a certain degree, be achieved by prolonging the duration of a square RF pulse. However, even though most of the excited spins are within a selective frequency band, there is also significant excitation of spins off-resonance, due to the sinc-like excitation profile (see Figures 5.2 and 5.3). With the development of MRI and localized MRS methods came the demand for RF pulses with well-defined, selective frequency profiles. Here, the principles of frequency-selective RF pulses are described, as well as their design and optimization.

5.3.1 Fourier-transform-based RF Pulses

Selective RF pulses can be designed based on the Fourier transformation. Figure 5.5 shows that Fourier transformation of a square time domain function yields a frequency domain function that is described by a "sine cardinal" or "sinc" function. The sinc function is a reasonable approximation of the exact frequency response of a square RF pulse given by Eqs. (5.10)–(5.12) (see Exercises). Since the Fourier transformation is a reversible operation, the inverse Fourier transformation of the sinc function in the frequency domain produces a square function in the time domain (Figure 5.5A). If the sinc function is applied as the RF modulation function in the time domain it can be expected to produce a selective frequency profile resembling a square function. Unfortunately, as the sinc function stretches over the span $[-\infty \ldots +\infty]$ it is an impractical RF modulation function. A sinc RF pulse is therefore always executed as a sinc function truncated at the nth zero crossing according to

$$B_1(t) = B_{1\max} f_B(t) = B_{1\max} \frac{\sin(2\pi nt/T)}{(2\pi nt/T)} \quad \text{for } -\frac{T}{2} \leq t \leq +\frac{T}{2} \tag{5.14}$$

with $B_1(0) = B_{1\max}$. Figure 5.5C shows a "five-lobe" sinc RF pulse truncated at the third zero crossing ($n = 3$) and the frequency profile obtained by Fourier transformation (Figure 5.5D). The frequency response does resemble selective excitation according to a square function. Unfortunately, the frequency profile is compromised by wiggles that originate from the required truncation of the infinite sinc function. The frequency profile in Figure 5.5D is an accurate reflection of the profile obtained experimentally when the nutation angle is low ($\theta < 30°$).

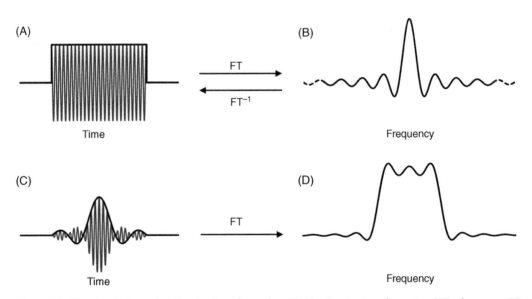

Figure 5.5 RF pulse design using Fourier transformation. (A) The Fourier transformation (FT) of a square RF pulse envelope predicts (B) a "sinc" excitation profile for small nutation angles. (C) An RF pulse executed with a truncated sinc envelope achieves (D) a frequency-selective excitation profile. Note that a sinc pulse refers to the RF *envelope* function; the magnetic field always oscillates close to the spin Larmor frequency.

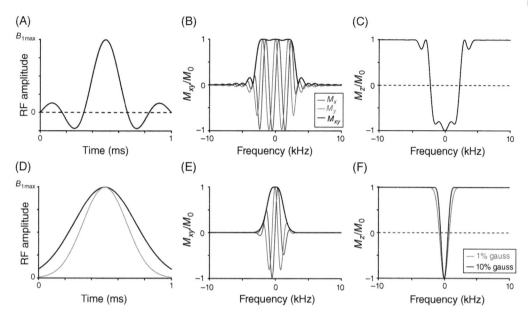

Figure 5.6 **Frequency profiles of sinc and Gaussian RF pulses.** (A) Five-lobe sinc ($n = 3$) RF pulse and the (B) excitation and (C) inversion profiles. Spin evolution during the RF pulse leads to a linear phase roll across the excitation profile. (D) Gaussian RF pulses truncated at an initial amplitude of 1 or 10%. (E) Excitation profile for a Gaussian RF pulse truncated at 1% and (F) inversion profiles for both Gaussian RF pulses. For all simulations the maximum RF amplitude B_{1max} was scaled to achieve the desired on-resonance nutation for a 1 ms pulse length.

For larger nutation angles, the exact frequency profile needs to be calculated with the Bloch equations (or rotation matrices in the absence of relaxation). Figure 5.6 shows the excitation (Figure 5.6B) and inversion (Figure 5.6C) profiles for a "five-lobe" sinc RF pulse. Significant deviations from the Fourier-transform-based frequency profile can be observed, caused by the fact that the Fourier transformation is a linear operation ($\mathrm{FT}(A) + \mathrm{FT}(B) = \mathrm{FT}(A + B)$), whereas signal excitation in NMR is nonlinear (e.g. $M_{xy}(\theta = 40°) + M_{xy}(\theta = 60°) \neq M_{xy}(\theta = 100°)$). This means that Fourier transformation does not accurately reflect the NMR response to an RF pulse, except for very small nutation angles where the NMR response is roughly linear with the nutation angle (i.e. $\sin\theta \sim \theta$ for $\theta < 30°$). Despite this limitation, the sinc RF pulse derived from Fourier transformation is still used today, especially for nutation angles below 90°.

Another popular RF pulse that is based on Fourier transform theory is the Gaussian RF pulse [1]. Since a Gaussian function is its own Fourier transformation, the frequency domain profile of a Gaussian RF pulse is well behaved without the wiggles common to sinc-based excitation. A Gaussian RF pulse is described by

$$B_1(t) = B_{1max} e^{-\beta(2t/T)^2} \qquad \text{for} -\frac{T}{2} \leq t \leq +\frac{T}{2} \tag{5.15}$$

where β determines the truncation of the asymptotic Gaussian function, and is given by $\beta = -\ln(B_1(T/2)/B_{1max})$. Figure 5.6E and F shows the excitation and inversion profiles of a Gaussian RF pulse. Gaussian RF pulses are popular in MRS applications since they achieve excitation and inversion over a small, frequency-selective bandwidth for relatively short pulse lengths. This is advantageous in water and lipid suppression to reduce T_1 relaxation during the RF pulse (Chapter 6) and to minimize the echo-time in spectral editing (Chapter 8).

5.3.2 RF Pulse Characteristics

The family of RF pulses is, in principle, infinitely large and each RF pulse produces a unique frequency profile. In order to provide an objective evaluation of the characteristics and performance of an RF pulse a number of practical parameters are introduced. Since the bandwidth and pulse length of an RF pulse are inversely related, RF pulses are often characterized by the product of bandwidth BW and pulse length T, which is also known as an R-value. The bandwidth is being measured as the frequency width at half maximum (Figure 5.7B) of either M_{xy} or M_z, depending on whether an excitation or inversion pulse is considered. This product is constant for a given RF pulse and nutation angle and is summarized in Table 5.1 for some RF pulses useful for *in vivo* NMR. The R value is a convenient parameter for calculations in many experiments. For example, suppose that the required gradient strength to select a 0.5 cm slice with a 1.0 ms 90° Gaussian pulse needs to be calculated. The combination of Eq. (4.4) and the R value of Table 5.1 yields a direct answer, since the bandwidth of the pulse is

$$\mathrm{BW} = \frac{R}{T} = 2.70\,\mathrm{kHz} \tag{5.16}$$

after which the gradient strength can be calculated according to

$$G = \frac{\mathrm{BW}}{\text{slice thickness}} = 5.40\,\mathrm{kHz\,cm^{-1}} = 12.7\,\mathrm{mT\,m^{-1}} \tag{5.17}$$

where the conversion from kHz to mT was made with the proton gyromagnetic ratio of $42.567\,\mathrm{kHz\,mT^{-1}}$. From this example it should be clear that a low R value is desirable when gradient strength is limited. However, when sufficient gradient strength is available RF pulses

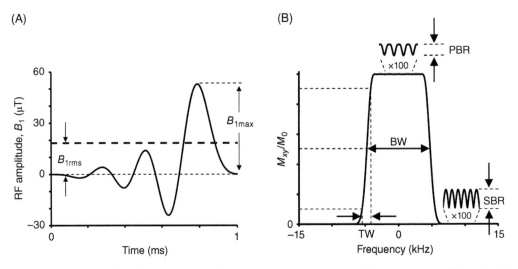

Figure 5.7 **Definition of useful RF pulse and frequency profile parameters.** (A) The maximum RF amplitude B_{1max} and the root-mean-squared RF amplitude B_{1rms} are important parameters characterizing the power requirements of an RF pulse. Both amplitudes scale linearly with the inverse of the pulse length. (B) The bandwidth BW is often the most critical parameter describing a frequency profile and is typically measured at $M_{xy}/M_0 = 0.5$ for excitation and $M_z/M_0 = 0$ for inversion profiles. Additional parameters are transition width TW, in-slice or pass band ripple (PBR), and out-of-slice or stop band ripple (SBR). For a given RF pulse, BW and TW scale linearly with the inverse of the pulse length.

Table 5.1 Characteristics of amplitude-modulated RF pulses.[a]

Pulse name	Nutation angle θ (°)	R_{BW}[b]	R_{TW}[c]	Pulse integral	RMS pulse integral	f[d] (%)
Square	90	1.37	0.61	1.0000	1.0000	50
Square	180	0.80	0.60	1.0000	1.0000	—
Sinc	90	5.95	0.95	0.1775	0.4011	50
Sinc	180	4.45	2.27	0.1775	0.4011	—
Gaussian, 1%	90	2.70	1.86	0.4116	0.5401	50
Gaussian, 1%	180	1.53	1.34	0.4116	0.5401	—
Gaussian, 10%	90	2.00	1.14	0.5650	0.6416	50
Gaussian, 10%	180	1.15	0.95	0.5650	0.6416	—
Hermitian	90	5.43	2.56	0.1967	0.3956	50
Hermitian	180	3.49	1.81	0.1541	0.3804	—
SLR linear phase	90	9.62	1.92	0.0984	0.3034	50
SLR minimum phase	90	9.48	1.67	0.1113	0.3462	23.6
SLR linear phase	180	10.30	1.97	0.0304	0.1816	—
SLR minimum phase	180	10.02	1.44	0.0418	0.2452	—

[a] All RF pulses can be found in PulseWizard (http://booksupport.wiley.com)
[b] R_{BW} = bandwidth (kHz) × pulse length (ms) measured at $M_{xy}/M_0 = 0.50$ for 90° and $M_z/M_0 = 0.0$ for 180° RF pulses.
[c] R_{TW} = transition width (kHz) × pulse length (ms) measured at $0.05 \le M_{xy}/M_0 \le 0.95$ for 90° and $-0.90 \le M_z/M_0 \le 0.90$ for 180° RF pulses.
[d] f = pulse length fraction (%) requiring refocusing of chemical shifts and magnetic field gradients.

with higher R values are recommended, since they can minimize chemical shift displacement artifacts in spatial localization sequences (see Chapter 6). A Gaussian pulse is very popular for water suppression and spectral editing due to its relatively low R value in combination with reasonable frequency-selectivity. Selective refocusing of a 200 Hz bandwidth can be achieved with a 7.7 ms Gaussian pulse, whereas a "five-lobe" sinc pulse ($n = 3$) would require a 22.3 ms pulse length.

The frequency profile of a selective RF pulse is, in addition to the bandwidth BW, characterized by a transition width TW between frequencies experiencing the desired rotation and frequencies that are not affected by the RF pulse (Figure 5.7B). The product of transition width and pulse length, R_{TW}, is a convenient parameter characterizing the selectivity of an RF pulse. Low R_{TW} values and high R_{BW}/R_{TW} ratios are desirable as they most closely resemble an ideal, square frequency profile. However, in reality the transition width is inversely linked to the amount of out-of-slice or stop-band ripples (SBR, Figure 5.7B) and in-slice or pass-band ripples (PBR, Figure 5.7B), whereas the maximum R_{BW} is limited by the achievable RF amplitude B_{1max}.

Besides the R values, Table 5.1 also shows pulse integral and pulse power parameters. The pulse integral is defined as the integral over the phase-sensitive RF pulse shape normalized to the range [0 ... 1] such that the pulse integral of a square RF pulse equals 1. The pulse integral is a convenient parameter to calculate the required RF pulse amplitude B_{1max} for a given pulse length T and nutation angle θ (in radians) according to

$$\theta = \gamma \int_0^T B_1(t)\,dt = \gamma B_{1max} \int_0^T f_B(t)\,dt = \gamma B_{1max}[\text{pulse integral}]T \tag{5.18}$$

For example, the pulse integral of a normalized, five-lobe sinc function is only 17.8% of a square function. In order to maintain the same nutation angle, a sinc RF pulse therefore needs $100/17.8 = 5.62$ times more amplitude than a square RF pulse of equal length. On most MR scanners, the power of a shaped RF pulse is calculated from the calibrated power of a square RF pulse. Suppose that a 1.0 ms square RF pulse is calibrated to provide an on-resonance 90° excitation at an RF amplifier output setting of 200 W. The RF power requirement of a 2.0 ms 90° sinc pulse can then be calculated using Eq. (5.18) and Table 5.1 according to

$$\frac{\gamma B_1}{2\pi}(\text{square}) = \frac{90^\circ}{360^\circ \times 1\,\text{ms}} = 0.25\,\text{kHz} \tag{5.19}$$

$$\frac{\gamma B_1}{2\pi}(\text{sinc}) = \frac{90^\circ}{360^\circ \times 0.178 \times 2\,\text{ms}} = 0.70\,\text{kHz} \tag{5.20}$$

The 2.81-fold increase in RF amplitude can be achieved by increasing the RF current I (in A) proportionally. However, since power P (in W) scales according to I^2R with R being the resistance (in Ω), the power level needed from the RF amplifier is given by

$$P(\text{sinc}) = (2.81)^2 \times 200\,\text{W} = 1579\,\text{W} \tag{5.21}$$

On most MR scanners the calculations in Eqs. (5.19)–(5.21) are performed automatically without user interaction. However, it is advised to become familiar with RF pulse power calculations, as the peak power requirements of shaped RF pulses can quickly approach and exceed maximum RF amplifier output and SAR power deposition guidelines (see Exercises). In cases where the RF power cannot be automatically calibrated by the system, the power of the shaped RF pulse can be manually calibrated (Figure 5.8). This situation can be encountered with highly inhomogeneous RF coils, adiabatic RF pulses, or with non-proton nuclei. Under most conditions, the bandwidth of the shaped RF pulse should remain constant such that the pulse length should stay constant. In that case the RF pulse is calibrated by varying the RF amplitude (Figure 5.8). Since most RF amplifiers operate on a logarithmic decibel

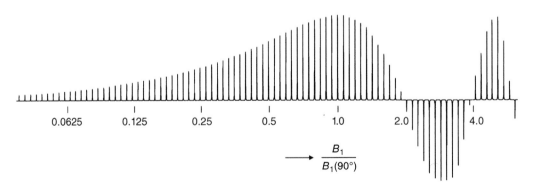

Figure 5.8 **Nutation angle calibration of an amplitude-modulated RF pulse.** The nutation angle of a fixed duration RF pulse can be calibrated by increasing the RF amplifier output in regular increments of 0.5 dB. The observed signal intensity deviates from a sinusoidal dependence due to the nonlinear decibel scale. Note that a doubling of the nutation angle is achieved every 6 dB.

scale, the calibration curve deviates from a simple sinusoidal curve (compare Figure 5.4) whereby the RF amplitude B_1 scales according to

$$B_1 = B_1\left(90°\right)10^{\Delta dB/20} \tag{5.22}$$

where ΔdB is the change in transmitter output in dB units. For a 6 dB change, the B_1 magnetic field amplitude and thus the nutation angle will be doubled. Similar to a square RF pulse (Figure 5.4), the first signal maximum encountered with increasing RF amplitude corresponds to a 90° nutation, whereas the first zero crossing indicates a 180° nutation.

The electric field associated with the magnetic field necessary to rotate nuclear spins is the dominant source of RF heating in conductive samples. The calculation of the magnitude and spatial distribution of electric fields is complex and requires electromagnetic simulations based on Maxwell's equations, which are discussed in Chapter 10. However, since tissue heating is *proportional* to the amount of RF power transmitted into the sample, a useful measure for the relative RF power demands of an RF pulse is provided by the root mean square (rms) RF amplitude B_{1rms}, which is proportional to the rms pulse integral according to

$$B_{1rms} = B_{1max}\sqrt{\int_0^1 f_B^2(t)dt} = B_{1max}\left[\text{rms pulse integral}\right] \tag{5.23}$$

The square of B_{1rms} is then proportional to the average power according to $P \propto (B_{1rms})^2 T$. From Table 5.1 it follows that a square pulse is optimal in terms of required peak amplitude B_{1max} and average power $(B_{1rms})^2 T$. In other words, the price for improved selectivity is increased RF amplitude and power requirements. Note that doubling the pulse length of a given RF pulse decreases the average power twofold, thereby presenting a straightforward method to reduce tissue heating. The reduced power requirements come at the price of a decreased bandwidth. It should further be noted that for conventional, amplitude-modulated RF pulses B_{1max} increases linearly with increasing R value (and hence the bandwidth).

The frequency profile of a square excitation RF pulse exhibits a linear phase variation due to the fact that magnetization evolves in the transverse plane during the RF pulse (Figure 5.2). The time that the magnetization evolves in the transverse plane during a shaped, frequency-selective RF pulse depends on the exact modulation function. For time-symmetric RF pulses, like sinc and Gaussian, the magnetization spends effectively 50% of the pulse length T in the transverse plane, thereby acquiring a linear, frequency-dependent phase according $2\pi\Delta\nu T/2$, whereby $\Delta\nu$ is the off-resonance frequency difference. When the excitation pulse is part of a spin-echo sequence, as is the case for PRESS localization (see Chapter 6), the linear phase roll can be removed by incorporating 50% of the excitation pulse length into the first echo-time. In other words, the echo-time should be counted from the center of the RF pulse instead of the end. For time-asymmetric RF pulses the pulse length fraction f that represents the effective time of phase evolution is obtained through simulations (Table 5.1). A spatial phase roll is observed in addition to a spectral phase roll, when the excitation pulse is executed in the presence of a magnetic field gradient in order to select a spatial slice (Figure 5.9). Before the RF pulse, the magnetization is along the z' axis parallel to the main magnetic field (Figure 5.9A). When the RF pulse is executed, the magnetization becomes partially perpendicular to the main magnetic field and starts to precess around it. In the presence of a magnetic field gradient, spins at different positions within the slice acquire a different amount of phase (Figure 5.9B). At the end of the RF pulse the magnetization is excited, but is also dephased, giving little to no macroscopic signal (Figure 5.9C). The gradient-induced dephasing can be reversed by applying a gradient-refocusing pulse

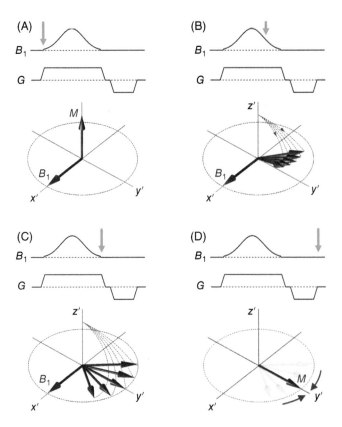

Figure 5.9 **Phase evolution during RF pulses.** (A) As the longitudinal magnetization is rotated towards the transverse plane, (B) spins acquire phase under the effects of chemical shifts and magnetic field gradients. (C) At the end of the excitation pulse, the magnetization is dephased along the slice selection direction giving no detectable, macroscopic signal. (D) The gradient-induced dephasing can be rephased by a negative rewinding gradient with an area corresponding to 50% of the positive, slice selection gradient area for time-symmetric RF pulses. Phase accrual due to chemical shifts can be removed with a spin-echo sequence when the effective time of excitation is set at 50% of the pulse length. Time-asymmetric RF pulses require a different amount of compensation (see Table 5.1).

following the RF pulse of opposite sign and with an area that equals the pulse length fraction f (which equals 50% for the Gaussian pulse in Figure 5.9). Time-symmetric 180° refocusing pulses do not require gradient refocusing as the dephasing during the first half of the RF pulse is automatically refocused during the second half.

5.3.3   Optimized RF Pulses

RF pulses shaped according to square, sinc, or Gaussian modulation functions are very useful and continue to be used in modern MR experiments. However, many aspects of these basic RF pulses can be improved in terms of frequency-selectivity, bandwidth per unit input power, or immunity towards imperfections such as RF field inhomogeneity. While the Fourier transformation has been essential for the design of these basic RF pulses, it does not provide the flexibility to incorporate other design criteria. The Bloch equations (Eqs. (1.9)–(1.11)) provide a complete description of NMR, including the effects of RF pulses. For a given RF pulse the Bloch equations can be solved numerically to reveal the excitation or inversion profile.

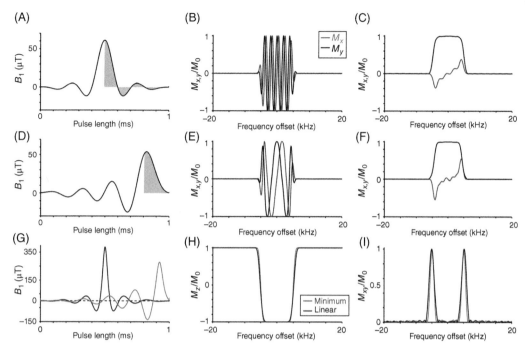

Figure 5.10 **RF pulses optimized with the Shinnar–Le Roux (SLR) algorithm.** (A) Linear phase and (D) minimum phase excitation pulses and (G) linear and minimum phase inversion pulses with a product of bandwidth and pulse length (*R*-value) of circa 10. Excitation profiles (B, E) before and (C, F) after compensation of the linear phase roll. Note that the minimum phase pulse required a much smaller compensation of ~60° per kHz compared to the linear phase pulse (~180° per kHz). (H) Inversion profile and (I) associated excitation profile of the SLR-optimized inversion pulses (G). Note the narrower transition width for the asymmetric, minimum phase inversion pulse.

In the absence of relaxation the Bloch equations reduce to 3×3 rotation matrices, providing a straightforward method for the numerical calculation of the "forward" solution. The "inverse" problem, however, in which the RF pulse is calculated from a given frequency profile is much more difficult. For small nutation angles, the inverse problem can be solved by taking the inverse Fourier transformation of the frequency profile, resulting in sinc and Gaussian RF pulses. Unfortunately, this approach does not work well for larger nutation angles, due to nonlinearities in the Bloch equations that are not accounted for during a linear Fourier transform operation. In the large-nutation-angle regime, improved RF pulses can be obtained through iterative numerical optimization methods, like gradient descent algorithms, optimal control theory [2–5], simulated annealing [6], and neural networks [7]. These optimization have resulted in a number of greatly improved RF pulses, like the optimized 180° pulse by Mao et al. [2] and self-refocused RF pulses by Geen and Freeman [6, 8]. A robust and elegant method for RF pulse optimization is the Shinnar–Le Roux (SLR) algorithm [9–14]. The SLR algorithm reformulates RF pulse optimization in terms of digital filter design. Considerations that are typical for digital filters, such as PBR and SBR, bandwidth and phase behavior, now become the inputs for RF pulse design. Figure 5.10 shows three examples of SLR-optimized RF pulses. Figure 5.10A shows a time-symmetric SLR pulse that produces a frequency-selective excitation profile with a linear phase roll (Figure 5.10B). After removing the linear phase roll, the profile shows a near-constant phase across the excitation bandwidth (Figure 5.10C). Figure 5.10D shows a minimum-phase, time-asymmetric SLR pulse. The linear phase roll across the excitation profile is much less because the spins are excited onto the transverse plane towards the end of the pulse (Figure 5.10E). The advantages of minimum phase excitation pulses are (i) shorter minimum echo-times and (ii) less gradient rephasing requirements (see Figure 5.9). The need for minimum phase refocusing or inversion pulses appears less obvious, as either the entire RF pulse or none of the RF pulse is part of the echo time. However, an additional feature of time-asymmetric RF pulses is that they can have a narrower transition width than an equivalent time-symmetric RF pulse. This feature can become important in, for example, spectral editing, where a narrower transition zone can reduce the amount of unwanted coediting (see Chapter 8). Figure 5.10G shows time-symmetric and minimum-phase inversion pulses optimized with the SLR algorithm. Both RF pulses provide inversion profiles (Figure 5.10H) that are greatly improved compared to sinc and Gaussian profiles (compare Figure 5.6), whereby the minimum phase inversion pulse provides a narrower transition width or R_{TW} value. Note that despite the excellent frequency profile, a frequency-selective inversion pulse *always* generates significant amounts of unwanted transverse magnetization (Figure 5.10I) that can lead to artifacts if not properly removed. RF pulses based on the SLR algorithm can be generated in PulseWizard, an open-source, Matlab-based RF pulse generation and simulation tool that can be downloaded from the Wiley website (http://booksupport.wiley.com).

In general, improved selectively comes at the price of increased RF power requirements. While some RF pulse optimization algorithms can accommodate minimization of the RF peak and average powers, many others do not specifically consider RF power. When a given RF pulse exceeds the RF power capabilities of the system, the user is often limited to (i) increase the pulse length or (ii) use a different, lower-power RF pulse, typically with a lower bandwidth. An alternative method to reduce the RF power requirements of a given RF pulse is provided by the variable rate selective excitation (VERSE) algorithm [15–17]. The VERSE algorithm rests on the fact that for on-resonance spins the rotation around a magnetic field is proportional to the magnetic field strength and duration. Therefore, the same on-resonance rotation can be achieved by many different combinations of magnetic field strength and duration, provided that their product is constant. Figure 5.11 shows the case of a frequency-selective RF pulse in the presence of a constant amplitude magnetic field gradient (blue lines). The maximum RF amplitude at the center of the pulse can be reduced by distributing the RF more uniformly over the duration of the pulse. This can be achieved by replacing the constant

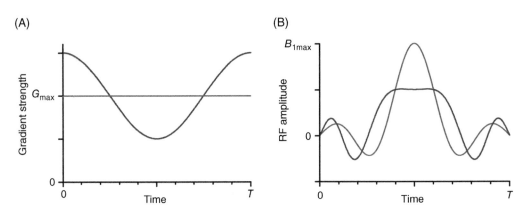

Figure 5.11 Variable rate selective excitation (VERSE) pulse optimization. (A) The original sinc pulse excites a spatial slice in the presence of a constant magnetic field gradient (blue line). The VERSE gradient waveform was chosen as $G(t) = G_{max}(0.5\cos(2\pi t/T) + 1)$ with $0 \le t \le T$ (red line). (B) Recalculating the RF waveform for the time-dependent gradient leads to a more uniform RF pulse that can be executed at a lower RF peak amplitude. The average and peak power of the VERSE pulse (red line) are only 26 and 64% of the original pulse (blue line), respectively.

amplitude gradient with, for example, a sinusoidally varying gradient (Figure 5.11, red line). The new RF pulse as calculated with the VERSE algorithm based on the modified gradient waveform is indeed more uniformly distributed, thereby leading to lower maximum and rms RF amplitudes (Figure 5.11B, see also Exercises). Note that the VERSE algorithm is only valid for on-resonance spins. Off-resonance, the slice profile is slightly degraded, although it typically remains acceptable for the chemical shift range encountered in *in vivo* ^{1}H MRS. An additional advantage of gradient modulation is the reduced chemical shift displacement. This has led to FOCI and GOIA RF pulses that use gradient modulation to increase the effective pulse bandwidth without increasing the RF peak amplitude. These gradient modulation methods are discussed in Chapter 6.

5.3.4 Multifrequency RF Pulses

For many applications, like spectral editing, multivolume localization, and signal suppression, it may be desirable to utilize RF pulses that achieve a specific rotation at multiple, selective frequency bands. One way to generate a multifrequency RF pulse relies on the complex-valued addition of RF pulses with shifted frequency bands (Figure 5.12). A regular single-frequency 180° RF pulse inverts magnetization over a selective frequency range centered on the Larmor frequency (Figure 5.12A and B). The selective inversion band can be shifted $\Delta\nu$ Hz off-resonance by changing the RF pulse center frequency, which is equivalent to the addition of a phase ramp $\phi(t)$ during the RF pulse given by

$$\phi(t) = 2\pi \int_0^t \Delta\nu \mathrm{d}t = 2\pi\Delta\nu t \tag{5.24}$$

An RF pulse that inverts simultaneously at both frequencies can be generated by adding the complex-valued RF pulse shapes of Figure 5.12A and C, given the multifrequency pulse shown in Figure 5.12E. Note that the peak amplitude of the combined RF pulse is twice as high as the separate single-banded RF pulses. This can pose problems when more than two bands need to be selected as, for instance, in Hadamard encoding (see Chapter 7). While several solutions to this problem have been proposed [18], restrictions on available RF peak power have limited the applications of multifrequency RF pulses to low-order Hadamard matrices ($n \leq 8$). In the specific case of two frequency bands generated with time-asymmetric RF pulses, the peak amplitude of the multifrequency pulse can be significantly reduced by time inverting one of the two pulses before complex addition (Figure 5.12G and H).

An assumption underlying the simple addition of two RF pulses as outlined in Figure 5.12 is that the rotations required to invert one frequency band do not affect the rotations that lead to the inversion of the other frequency band. This assumption is valid when the frequency band separation (BS) is much larger than the frequency bandwidth (BW) as was the case in Figure 5.12. However, when the frequency separation between the inversion bands gets smaller, the frequency profile quickly deteriorates. Figure 5.13A shows a double-frequency inversion pulse based on an optimized SLR inversion pulse. Upon decreasing the frequency BS the inversion profiles quickly deteriorate with incomplete inversion and additional unwanted excitation. Note that even though the inversion profile for the longitudinal magnetization appears reasonable for BS/BW > 6, the pulse generates significantly more transverse magnetization than an ideal multifrequency pulse. This puts higher demands on signal dephasing by magnetic field gradients.

Steffen et al. [19] have presented an elegant method to reduce the deterioration of multifrequency RF pulses applied at nearby frequencies. Essentially by tracking the phase evolution generated by pulse 1 at the frequency of pulse 2 and applying this as a correction, they were able to greatly improve the frequency profile of multifrequency RF pulses (Figure 5.13B).

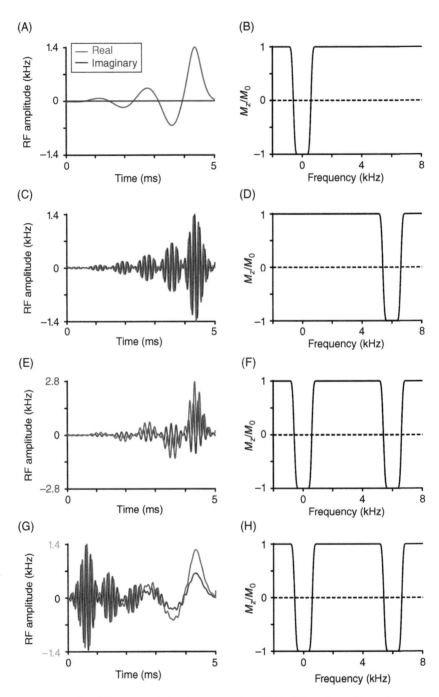

Figure 5.12 **Principle of multifrequency RF pulse generation.** (A) An SLR-optimized inversion pulse produces (B) a frequency-selective inversion profile centered around 0 Hz. (C) In the presence of a linear phase ramp, the SLR inversion pulse becomes complex (i.e. having real B_{1x} [blue line] and imaginary B_{1y} [red line] modulations) leading to (D) a frequency-shifted inversion profile. Complex addition of the pulses in (A) and (C) produces (E) a multifrequency RF pulse that produces (F) simultaneous selective inversion bands at two different frequencies. Note that the maximum RF amplitude of the multifrequency RF pulse is twice as high as that of each single-frequency pulse. (G) When the RF pulse in (C) is time-reversed before complex addition, the resulting multifrequency RF pulse produces (H) the same frequency profile at the maximum RF amplitude of a single-frequency pulse.

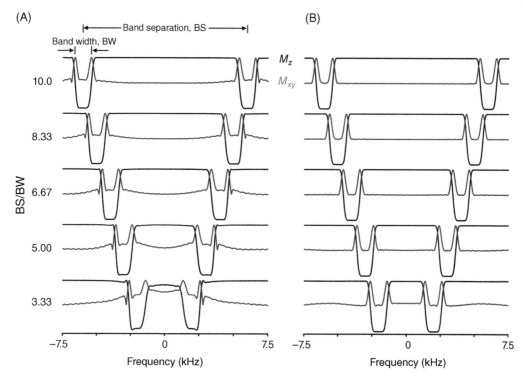

Figure 5.13 Frequency band separation for multifrequency RF pulses. (A) As the band separation (BS) between two selective inversion bands is reduced relative to the band width (BW), the rotations necessary for one band increasingly affect the rotations for the other band. This leads to increasing distortions (incomplete inversion, increased excitation) to the frequency profiles of either band with decreasing BS/BW. (B) By determining and correcting the mutual interactions between the two inversion bands, the ideal inversion profiles can be maintained for much smaller BS/BW.

Similar improvements can be achieved by incorporating the positions of the frequency bands in a pulse optimization algorithm.

An older and different class of multifrequency RF pulses is provided by "delays alternating with nutation for tailored excitation" or DANTE [20–22]. In the DANTE pulse family a given RF pulse with a desired frequency profile is broken up into smaller RF pulse segments of length $t_{segment}$. When the pulse segments are interleaved with delays t_{delay}, the resulting DANTE pulse produces a multifrequency or multiband excitation profile. The shape of the individual frequency bands is simply given by the profile of the non-segmented parent pulse. The separation between adjacent bands is given by the reciprocal of the interpulse segment delay, i.e. $(t_{segment} + t_{delay})^{-1}$. The overall multifrequency profile is convolved with the frequency profile of an individual RF pulse segment, such that higher frequency bands have generally less intensity. More recently, the DANTE principle has been repurposed under the acronym PINS (power independent of number of slices) for use in simultaneous multiband, accelerated imaging [23].

5.4 Composite RF Pulses

For all RF pulses discussed in the previous paragraphs, the on-resonance nutation angle is given by Eq. (5.1) and is linearly dependent on the RF amplitude B_1. When the RF field generated by the coil is spatially inhomogeneous, this immediately translates to the fact that a range

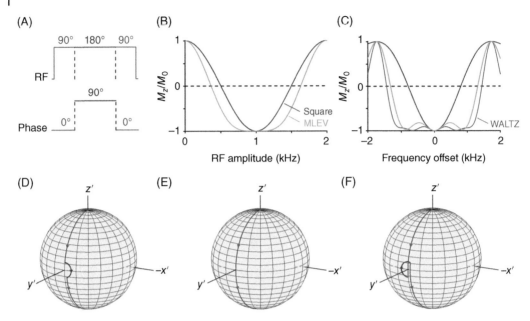

Figure 5.14 Performance of composite RF pulses. (A) A MLEV composite inversion pulse is composed of a 180° square pulse flanked by two phase-shifted 90° square pulses. (B) On-resonance inversion efficiency as function of the RF amplitude for square (red line) and composite MLEV (green line) RF pulses. (C) Frequency profiles for square (red line) and composite MLEV (green) and composite WALTZ 90°(+x)180°(−x)270°(+x) (blue) RF pulses. For all pulses the RF amplitude was adjusted to give complete inversion on-resonance. (D, E) Rotations during the composite MLEV pulse when the RF amplitude is (D) 10% lower than, (E) at, and (F) 10% higher than the nominal RF amplitude. Blue and red lines indicate the rotations during the two 90° and single 180° RF pulses, respectively.

of different nutation angles will be produced. Besides signal loss, the variation in nutation angles can also lead to artifacts when accurate nutation angles are required. For this purpose, so-called composite RF pulses have been developed which have a certain degree of insensitivity to the RF amplitude [24–31]. Composite pulses are, as the name suggests, composed of several closely spaced RF pulses with variable nutation angles and/or phases. The overall effect of a composite RF pulse is identical to a single hard pulse, except that it is less sensitive to imperfections. The first composite pulses were developed intuitively (assisted by computer simulations) by considering the trajectory of magnetization at different RF amplitudes. The degree of compensation for imperfections was therefore determined by the limited human ability to visualize three-dimensional rotations. As a consequence, many of the earlier developed composite pulses compensated for one imperfection (e.g. variation in RF amplitude), while the performance towards another parameter (e.g. resonance offset) degraded. Nevertheless, some of these early efforts provided composite pulses with dual compensation. For example, the 180° composite pulse $90°_{+x}180°_{+y}90°_{+x}$ (Figure 5.14A) is widely used for wideband inversion with insensitivity to variations in RF amplitude and resonance offset. Figure 5.14B shows a simulation of the longitudinal magnetization after this composite pulse as a function of the RF amplitude, while Figure 5.14C gives the longitudinal magnetization as a function of frequency offset. In both figures, the performance of a conventional "square" 180° pulse is also shown. Even though the composite pulse is twice as long as the conventional 180° pulse, it inverts a much wider bandwidth with more insensitivity to RF amplitude variations. Figure 5.14D–F shows trajectories of the magnetization (initially along z) as a function of the RF amplitude. When the RF amplitude is 10% below the nominal value for 90° and 180° pulses, the composite pulse is given by $81°_{+x}162°_{+y}81°_{+x}$. Although the first pulse does not properly excite the magnetization onto the transverse plane (Figure 5.14D), the following $162°_{+y}$ pulse rotates it through the transverse plane (Figure 5.14E), after which the last $81°_{+x}$ pulse will rotate the magnetization close to the $-z$ axis. When the RF amplitude is too high, the opposite will happen, the first pulse will rotate the magnetization too far, after which the second pulse will reverse the imperfection so that the final nutation angle will be close to the $-z$ axis. The overall nutation angle of the composite pulse is therefore largely determined by the outer two pulses, whereas the inner pulse provides the compensation for imperfections. Similar arguments can also be used for explaining the compensation for magnetization vectors at different frequency offsets. Figure 5.14C shows the off-resonance performance of another important composite pulse, $90°_{x}180°_{-x}270°_{x}$. The composite pulses $90°_{x}180°_{y}90°_{x}$ (MLEV) and $90°_{x}180°_{-x}270°_{x}$ (WALTZ) are very popular for heteronuclear broadband decoupling applications and are discussed in greater detail in Chapter 8. Note that while the off-resonance performance of WALTZ is superior to that of MLEV, the sensitivity of WALTZ towards B_1 magnetic field inhomogeneity is identical to that of a conventional square pulse.

A more systematic procedure in the development of composite RF pulses was proposed by Levitt and Ernst [28]. Their recursive expansion procedure allows the generation of composite RF pulses with arbitrary nutation angle which compensate to any desired degree for the effects of RF inhomogeneity. Unfortunately, the compensation towards RF inhomogeneity is achieved at the expense of severe bandwidth reduction. Garwood and Ke [30] have shown that symmetrization of the composite pulses obtained with the recursive expansion procedure restores the off-resonance characteristics, while the compensation with respect to RF inhomogeneity is retained. The composite pulse $90°_{x}180°_{y}90°_{x}$, developed based on intuitive arguments, is part of the family of pulses obtained through the symmetrical recursive expansion procedure.

RF pulses optimized with algorithms based on optimal control theory [4] can be viewed as an extreme form of composite RF pulses in which hundreds of small RF pulse segments with different amplitude and phase achieve an overall nutation angle with extreme tolerance to frequency offsets and incorrect RF amplitude calibration. An universally applicable set of excitation, refocusing, and inversion RF pulses have been designed [32], which lend themselves perfectly for complete automation in high-resolution NMR. These universal RF pulses have not been applied to *in vivo* NMR, presumably due to the lower available B_1 amplitude and hence longer minimum pulse lengths. An alternative method for the compensation of imperfections is the use of adiabatic RF pulses. Adiabatic RF pulses are amplitude- and frequency-modulated pulses with an inherent insensitivity to imperfections.

5.5 Adiabatic RF Pulses

The principle of adiabatic rotations is as old as NMR itself and was used by Bloch [33] and Purcell [34] in the form of continuous wave (CW) excitation. Adiabatic RF pulses differ from conventional RF pulses in that they employ frequency modulation in addition to amplitude modulation. By purposely changing the transmitter offset frequency from far off-resonance to on-resonance, adiabatic RF pulses achieve an extreme tolerance towards variations in B_1 magnetic field strength and frequency offsets. For *in vivo* NMR applications adiabatic RF pulses were first used in combination with surface coil transceivers [35, 36]. Surface coils can be placed close to the object under investigation, thereby providing a sensitive receive element. However, when the same surface coil is also used for transmission of conventional RF pulses, the inhomogeneous B_1 magnetic field leads to a range of nutation angles and hence to signal loss (Figure 5.15A). When adiabatic RF pulses are transmitted, the nutation angle is largely independent of the B_1 magnetic field, thereby leading to enhanced sensitivity (Figure 5.15B).

(A)   (B)

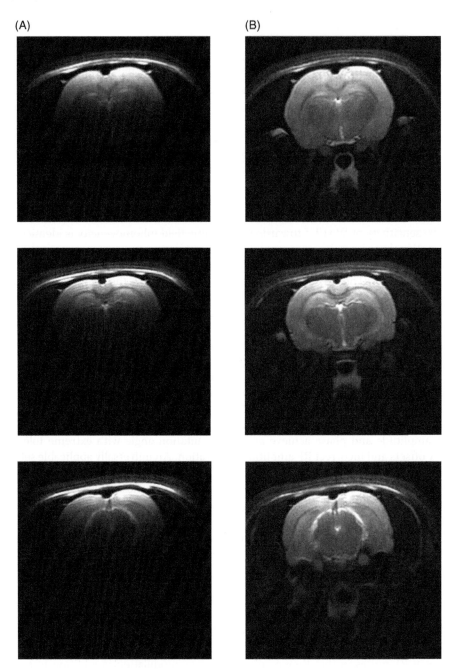

Figure 5.15 **Adiabatic RF pulses in MRI.** Surface coil MR images of rat brain (1.5 mm slice thickness) as acquired with (A) conventional sinc and (B) adiabatic RF pulses. The insensitivity of adiabatic RF pulses towards the RF amplitude eliminates all signal losses during RF transmission.

While the immunity towards RF field inhomogeneity sparked the initial development of adiabatic RF pulses for *in vivo* NMR, the continued use of these RF pulses is often based on some of the other favorable features, such as tolerance towards frequency offsets and potential for robust pulse sequence automation.

5.5.1 Rotating Frame of Reference

In a standard rotating frame of reference, the frame rotates at the constant frequency of the RF pulse, such that the B_1 field vector appears static along the x' axis (Figure 1.5). Since adiabatic RF pulses modulate the frequency during the RF pulse, the standard rotating frame of reference is typically changed to a frequency-modulated (FM) reference frame. The FM frame changes its speed of rotation to match the RF pulse frequency. As a consequence, the B_1 field vector appears static along the x' axis of the FM frame. Spins are off-resonance during the majority of an adiabatic RF pulse and experience a magnetic field vector along the z' axis equal to $\nu(t) - \nu_0$, where $\nu(t)$ represents the frequency modulation function. In general, the amplitude $B_1(t)$ and frequency $\Delta\nu(t)$ modulations of adiabatic RF pulses can be described by modulation functions $f_B(t)$ and $f_\nu(t)$ as

$$B_1(t) = B_{1\max} f_B(t) \tag{5.25}$$

$$\nu(t) - \nu_c = \Delta\nu(t) = \nu_{\max} f_\nu(t) \tag{5.26}$$

where $B_{1\max}$ and $\nu_{\max}$ are the maximum RF amplitude and maximum frequency offset (in Hz), respectively, and ν_c is the RF carrier frequency (in Hz). One of the most common adiabatic RF pulses is the hyperbolic secant (HS1) pulse [37, 38] for which the modulations are given by

$$B_1(t) = B_{1\max} \operatorname{sech}\left(\beta\left(1 - \frac{2t}{T}\right)\right) \tag{5.27}$$

$$\Delta\nu(t) = \nu_{\max} \tanh\left(\beta\left(1 - \frac{2t}{T}\right)\right) \tag{5.28}$$

where T is the pulse length and $0 \le t \le T$. β is a dimensionless truncation factor which is typically defined as $\operatorname{sech}(\beta) = 0.01$. Figure 5.16A and B shows the RF modulations as defined by Eqs. (5.27) and (5.28). An HS1 pulse achieves inversion of the magnetization when the pulse is executed over the full duration. In that case it belongs to the family of adiabatic full passage (AFP) pulses. When a HS1 pulse is only executed from 0 to $T/2$, it achieves excitation and can then be classified as an adiabatic half passage (AHP) pulse. On most modern spectrometers adiabatic RF pulses are implemented as amplitude- and *phase*-modulated RF pulses. The frequency and phase modulations are closely related as phase is the time integral of frequency according to

$$\phi(t) = 2\pi \int_0^t \Delta\nu(t)\,\mathrm{d}t \tag{5.29}$$

which, for a HS1 pulse evaluates to

$$\phi(t) = \frac{\pi\nu_{\max}T}{\beta}\ln\left[\frac{\operatorname{sech}(\beta(1 - 2t/T))}{\operatorname{sech}(\beta)}\right] \tag{5.30}$$

Figure 5.16C shows the HS1 phase modulation given by Eq. (5.30). The derivation of a closed-form expression for the phase modulation based on other frequency modulation functions $f_\nu(t)$ may not always be possible. In general, a closed-form expression is not required as the phase modulation can be obtained through numerical integration according to Eq. (5.29). In a standard, constant frequency, phase-modulated (PM) reference frame the trajectory of the RF field can be described by two magnetic fields $B_{1x}(t)$ and $B_{1y}(t)$, along the x and y axes,

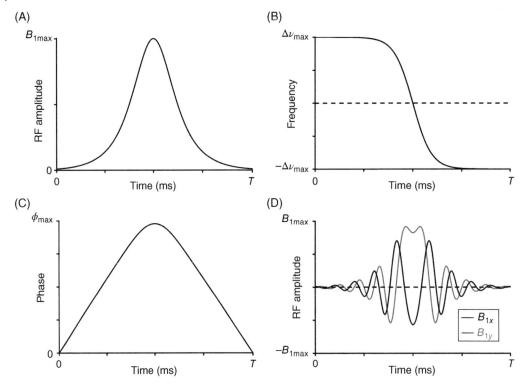

Figure 5.16 **Hyperbolic secant adiabatic RF pulse.** (A) Amplitude and (B) frequency modulation functions for a hyperbolic secant (HS1) adiabatic RF pulse. (C) Phase modulation function as calculated from (B) using Eq. (5.29). (D) Real (black, B_{1x}) and imaginary (gray, B_{1y}) components of the RF modulation.

respectively, given by $B_{1x}(t) = B_1(t)\cos\phi(t)$ and $B_{1y}(t) = B_1(t)\sin\phi(t)$ and are shown in Figure 5.16D. In the PM frame the phase and thereby the orientation of the B_1 field is not constant, leading to a complicated trajectory of the effective field $B_e(t)$ vector. Whereas adiabatic RF pulses are experimentally executed with amplitude and phase modulation, the principles of adiabatic RF pulses and their effect on the magnetization are best understood in the FM frame with amplitude and frequency modulation. Figure 5.17A shows the $B_1(t)$ and $\Delta\nu(t)$ magnetic field vectors in the FM frame, together with the effective field $B_e(t)$ vector. The FM frame in Figure 5.17A is similar to the standard rotating frame in Figure 5.1, with the exception that $\Delta\nu(t)$ is time-dependent for adiabatic RF pulses. The rate of change of $B_e(t)$ is equal to $d\alpha(t)/dt$, where $\alpha(t)$ is the angle between $B_e(t)$ and the x' axis (Figure 5.17A). In a rotating frame of reference $x''y''z''$ in which the orientation of $B_e(t)$ is always along z'', the rate of change in $B_e(t)$ can be represented as a vector along the y'' axis (Figure 5.17B). The $B_e(t)$ and $d\alpha(t)/dt$ vectors together make up a total effective field $B_e'(t)$ around which the magnetization is precessing. The condition for a successful adiabatic rotation is to ensure a slow rate of change $d\alpha(t)/dt$ relative to the magnitude of $B_e(t)$, such that $B_e'(t) \sim B_e(t)$ and the magnetization follows the effective field throughout the RF pulse. This is typically referred to as the adiabatic condition.

5.5.2   Adiabatic Condition

The rotations of the initial longitudinal magnetization and magnetic fields during a hyperbolic secant AFP pulse are shown in Figure 5.18 for different peak RF amplitudes, B_{1max}. At the onset of the pulse, the effective field $B_e(0)$ is equal to the frequency offset ν_{max} along the $+z'$ axis.

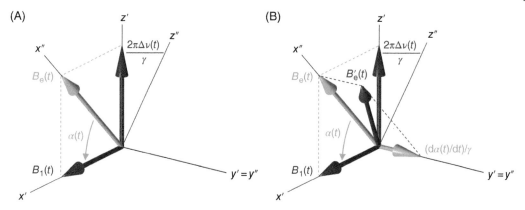

Figure 5.17 **Magnetic field vectors in rotating frames used to describe adiabatic RF pulses.** (A) The frequency frame $x'y'z'$ precesses at the instantaneous frequency of the RF pulse making the $B_1(t)$ orientation (arbitrarily chosen along x') stationary. (B) The second rotating frame $x''y''z''$ rotates about y' with angular velocity $(d\alpha(t)/dt)$ of the $B_e(t)$ rotation in the frequency frame. In the second rotating frame, the $B_e(t)$ orientation is static, leading to an additional vector $([d\alpha(t)/dt]/\gamma)$ along y'' as $B_e(t)$ is rotating. The adiabatic condition is satisfied when the additional vector $([d\alpha(t)/dt]/\gamma)$ is much smaller than the magnitude of $B_e(t)$, such that $B_e'(t) \sim B_e(t)$. This is generally achieved by going slow, using sufficient RF amplitude or both.

In order to work with consistent units, either B_1 needs to be multiplied with or $\Delta\nu$ needs to be divided by $(\gamma/2\pi)$. The longitudinal magnetization is parallel to $B_e(0)$. As the pulse is executed according to the modulations shown in Figure 5.16A and B, the frequency offset decreases and the RF field along the x' axis increases, rotating the effective field $B_e(t)$ away from the $+z'$ axis. For the lowest RF amplitude (Figure 5.18A and B) it can be seen that the effective field $B_e(t)$ rotates too fast midway the pulse, thereby generating a significant $(d\alpha/dt)$ vector. As a result, the effective field $B_e'(t)$ deviates substantially from $B_e(t)$, causing the magnetization to follow an erratic trajectory that does not lead to inversion (Figure 5.18B). An adiabatic inversion only occurs when the magnetization can follow the effective field $B_e(t)$, which happens when $B_e(t) \sim B_e'(t)$ throughout the pulse. This can be expressed by the inequality

$$\left| B_e\left(t\right) \right| = \sqrt{ B_1^2\left(t\right) + \left(\frac{\Delta\omega\left(t\right)}{\gamma} \right)^2 } \gg \left| \frac{d\alpha\left(t\right)}{\gamma dt} \right| \tag{5.31}$$

$$\text{with} \quad \alpha\left(t\right) = \text{atan}\left(\frac{\Delta\omega\left(t\right)}{\gamma B_1\left(t\right)} \right) \tag{5.32}$$

Equation (5.31) is commonly referred to as the "adiabatic condition" and has been extensively used in the design and optimization of modulation functions for AFP pulses [39, 40]. From Figure 5.18A and B it is clear that inequality (Eq. (5.31)) is not satisfied during a significant part of the RF pulse when $|d\alpha/\gamma dt| > |B_e|$. Increasing the maximum RF amplitude of the HS AFP pulse twofold (Figure 5.18C) improves the situation by slowing the rate of change of $B_e(t)$ to the point that $|d\alpha/\gamma dt| < |B_e|$ throughout the pulse. The magnetization can follow the effective field $B_e(t)$ for most of the pulse duration (Figure 5.18D), leading to a near-complete inversion (Figure 5.18D). From the rotations shown in Figure 5.18A–D it can be extrapolated that the inversion performance of an AFP pulse is independent of the RF amplitude as long as the adiabatic condition, Eq. (5.31), is satisfied and the magnetization follows the effective field. This is confirmed in Figure 5.18E and F where a further doubling of the maximum RF amplitude leads to an even better adherence of the adiabatic condition, making $B_e \sim B_e'$ throughout the

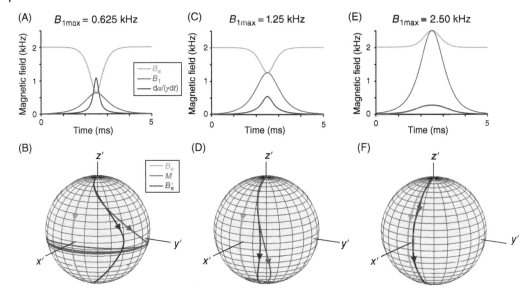

Figure 5.18 Fulfillment of the adiabatic condition for different RF amplitudes. Temporal evolution of $B_e(t)$, $B_1(t)$, and $(d\alpha(t)/dt)/\gamma$ during a 5 ms HS1 pulse ($R = 20$) at maximum RF amplitudes B_{1max} of (A, B) 0.625 kHz, (C, D) 1.25 kHz, and (E, F) 2.50 kHz. At an RF amplitude of 0.625 kHz the adiabatic condition of Eq. (5.31) is (A) not sufficiently satisfied leading to (B) a deviation between the $B_e(t)$ and magnetization M trajectories that ultimately results in an incomplete inversion of M at the end of the RF pulse. At RF amplitudes of (C, D) 1.25 kHz and (E, F) 2.50 kHz the adiabatic condition is satisfied throughout the pulse, making $B_e(t) \sim B'_e(t)$. As a result, the magnetization closely follows $B_e(t)$ leading to a near-complete inversion at the end of the RF pulse. Note that in (B, D, F) the effective fields are shown as $B_e/|B_e|$ and $B'_e/|B'_e|$ at all times for display purposes, such that they follow a path on the spherical surface. In reality, B_e and B'_e do not follow a spherical path and run far outside the unit sphere.

pulse and leading to inverted magnetization at the end of the pulse (Figure 5.18F). Any adiabatic pulse has a minimum RF amplitude threshold that needs to be exceeded before the adiabatic condition is properly satisfied. However, once above the minimum threshold most adiabatic RF pulses do not show an upper limit over the RF amplitude ranges encountered *in vivo*.

Using arguments similar to those shown in Figure 5.18, it is straightforward to show that HS1 AFP pulses only achieve inversion for frequency offsets ν that satisfy $-\nu_{max} \leq \nu \leq \nu_{max}$, i.e. the inversion profile is frequency selective. For frequency offsets $\nu > \nu_{max}$ the effective field $B_e(t)$ rotates away from the $+z'$ axis as $\Delta\nu(t)$ decreases and $B_1(t)$ increases. However, as $\nu(t) - \nu > 0$ throughout the pulse, the effective field never crosses the transverse plane. Instead, $B_e(t)$ rotates back towards the $+z'$ axis. When the adiabatic condition is adequately satisfied, the magnetization follows $B_e(t)$ as it rotates away from and then back to the $+z'$ axis, leading to an effective rotation of $0°$ at the end of the pulse.

5.5.3 Modulation Functions

The amplitude- and frequency-modulation functions for adiabatic RF pulses given by Eqs. (5.25) and (5.26) span, in principle, an infinitely large space. Simple modulation functions as used for chirp pulses were chosen based on intuition, whereas the hyperbolic secant modulation functions were derived as an analytical solution to the Bloch equations [37, 41]. A large number of modulation functions have been optimized numerically based on the adiabatic condition of Eq. (5.31). It is important to realize that the modulation functions are the primary factor in determining the overall performance of an adiabatic RF pulse in terms of minimum threshold RF amplitude, frequency selectivity, and attainable bandwidth. Adiabatic RF pulses based on tanh/tan modulation generally give the best performance for nonselective excitation and inversion of wide frequency ranges and are given by

$$B_1(t) = B_{1,\max} \tanh\left(\xi\left(1 - \left|1 - \frac{2t}{T}\right|\right)\right) \tag{5.33}$$

and

$$\Delta\nu(t) = \nu_{\max} \frac{\tan\left(\kappa(1 - 2t/T)\right)}{\tan(\kappa)} \qquad 0 \leq t \leq T \tag{5.34}$$

which for $\xi = 10$ and $\kappa = \text{atan } [20]$ closely resembles an AFP pulse with numerically optimized modulation (NOM) functions optimized to give the largest bandwidth at the lowest RF amplitude [30, 39].

Tannus and Garwood used the adiabatic condition of Eq. (5.31) to derive a general class of adiabatic modulation functions that satisfy the adiabatic condition equally for all frequencies $|\nu| \leq |\nu_{\max}|$. The details of these so-called offset-independent adiabaticity (OIA) modulation functions can be found in Refs. [40, 42]. One of the most useful groups of OIA pulses is the HSn family, which can be described by

$$B_1(t) = B_{1\max} \operatorname{sech}\left(\beta\left(1 - \frac{2t}{T}\right)^n\right) \tag{5.35}$$

$$\Delta\nu(t) = \Delta\nu_{\max}\left[1 - 2\left(\frac{\int_0^t f_B^2(t)\mathrm{d}t}{\int_0^T f_B^2(t)\mathrm{d}t}\right)\right] \tag{5.36}$$

where n is an integer. Note that for $n = 1$ the pulse reduces to the HS1 pulse described by Eqs. (5.27) and (5.28). Once the adiabatic condition is satisfied, OIA pulses have some common features that include a constant inversion profile for $|\nu| \leq |\nu_{\max}|$ and an identical average B_1 field, $B_{1\mathrm{rms}}$ for the minimum $B_{1\max}$, necessary to achieve inversion. The point at which the adiabatic condition is satisfied varies widely among OIA and other AFP pulses. Figure 5.19 shows the performance of three AFP pulses over a range of B_1 amplitudes and frequency offsets. Figure 5.19C–E and H–J shows the performance of the OIA pulses HS1 (Figure 5.19A and B) and HS8 (Figure 5.19F and G) with different R-values. Figure 5.19M–O shows the performance of a tanh/tan AFP pulse (Figure 5.19K and L) given by Eq. (5.33) and (5.34). From Figure 5.19 it follows that HS1 and HS8 pulses both achieve OIA at $R \geq 40$. The HS8 pulse achieves good performance at $B_{1\max} \sim 1.3\,\mathrm{kHz}$, whereas the HS1 pulse requires $B_{1\max} \sim 2.5\,\mathrm{kHz}$ to achieve the same inversion. When the different amplitude modulation functions are taken into account, the average RF amplitude $B_{1\mathrm{rms}}$ is roughly similar for both pulses. The HS8 pulse is therefore a good choice when the maximum RF amplitude is limited, whereas the HS1 provides a sharper inversion profile with narrower transition bands. At low R-values, the HS8 pulse does not adequately satisfy the adiabatic condition, making the HS1 pulse the only valid choice. The tanh/tan AFP pulse behaves very differently from the HS1 and HS8 pulses. Even though the minimum R-value to achieve adequate performance is much higher, the pulse performs well at relatively low $B_{1\max}$. In addition, whereas HS1 and HS8 achieve frequency-selective inversion independent of the

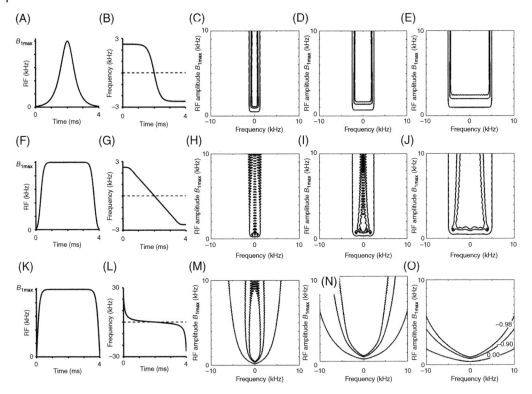

Figure 5.19 **Effect of AFP modulation functions on inversion performance.** (A, F, K) RF amplitude and (B, G, L) frequency modulation functions for (A, B) HS1, (F, G) HS8, and (K, L) NOM AFP pulses. Inversion profiles as a function of RF amplitude B_{1max} and frequency offset for (C–E) HS1, (H–J) HS8, and (M–O) NOM AFP pulses. HS1 and HS8 AFP pulses were executed with (C, H) $R = 10$, (D, I) $R = 20$, and (E, J) $R = 40$ whereas the NOM pulse was executed with (M) $R = 100$, (N) $R = 200$, and (O) $R = 400$. Contour lines are shown for $M_z/M_0 = -0.98$ (inner line), -0.90 (middle line), and 0.00 (outer line).

applied RF amplitude, the inversion bandwidth of the tanh/tan AFP pulse increases with increasing B_{1max}. This is because the tanh/tan AFP pulse does not exhibit OIA. As a result, the tanh/tan AFP pulse is an excellent choice for inversion over wide bandwidths, whereas the HS1 and HS8 are better choices for frequency-selective inversion as required in spatial localization. The RF amplitude and frequency offset performance of a given AFP pulse is not easily deduced from visual inspection of the $B_1(t)$ and $\Delta\nu(t)$ modulation functions. Instead, the pulse performance should be calculated using the Bloch equations (see Figure 5.19 and PulseWizard) or experimentally calibrated.

5.5.4 AFP Refocusing

AFP pulses demonstrate a superb performance for inverting longitudinal magnetization ($M_z \rightarrow -M_z$). However, other rotations like excitation ($M_z \rightarrow M_x$) and refocusing ($M_x + iM_y \rightarrow M_x - iM_y$) are equally important for a wide range of experiments. The primary difference between M_z and M_{xy} during an AFP pulse is that M_z is initially parallel to $B_e(0)$, whereas M_{xy} is initially perpendicular to $B_e(0)$. When the effective field $B_e(t)$ rotates slowly relative to its amplitude, i.e. when $|d\alpha(t)/\gamma dt| \ll |B_e(t)|$, M_z remains parallel to $B_e(t)$ which ultimately leads to spin inversion. Similarly, provided that the adiabatic condition is satisfied, M_{xy} will remain perpendicular to $B_e(t)$ throughout the pulse, leading to an inversion of the xy plane, i.e. refocusing. However, the refocusing

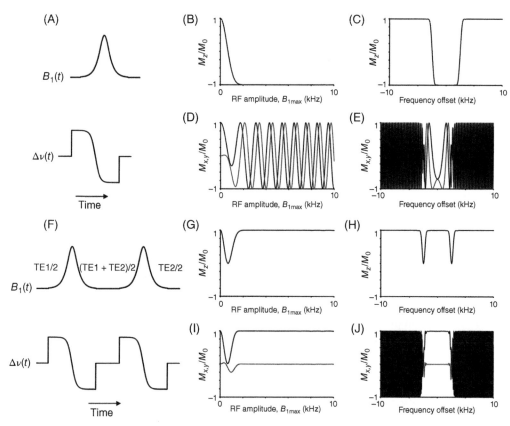

Figure 5.20 **Inversion and refocusing properties of AFP pulses.** (A) A single AFP pulse achieves (B) B_1-independent inversion above a minimum threshold B_1 and (C) frequency-selective inversion. (D, E) Using a single AFP pulse for refocusing leads to a (D) B_1 and (E) frequency-dependent phase. (F) A double AFP pulse sequence achieves an identity or 360° rotation that is (G) B_1-independent and (H) frequency-selective. Using the AFP pulse pair for refocusing eliminates the (I) B_1 and (J) offset-dependent phase roll. Note that the two AFP pulses need to be executed as a double spin-echo in order to obtain proper refocusing of magnetic field inhomogeneity and chemical shifts. Each AFP pulse should be surrounded by unique magnetic gradient crushers to remove unwanted coherences (see Figure 6.18).

properties are compromised as the transverse magnetization has acquired a phase angle $\beta(t)$ due to precession around $B_e(t)$ given by

$$\beta(t) = 2\pi \int_0^t B_e(t) dt \qquad (5.37)$$

where $B_e(t)$ is the effective magnetic field given by Eq. (5.31). Figure 5.20 shows the effects of an AFP pulse on the three magnetization components M_x, M_y, and M_z. The AFP pulse achieves B_1 insensitive (Figure 5.20B) and frequency-selective (Figure 5.20C) inversion when starting with M_z prior to the pulse. Using M_x as a starting point prior to the AFP pulse will lead to B_1- and offset-dependent signal dephasing (Figure 5.20D and E) according to Eq. (5.37). The presence of RF inhomogeneity and different frequencies will lead to phase cancelation and signal loss. In other words, a single AFP pulse is a good inversion pulse, but it is a poor choice for signal refocusing as required in a spin-echo sequence.

Conolly et al. [16, 43] were the first to recognize that the application of additional AFP pulses could rewind the phase acquired during the first AFP pulse. After the initial description of a 3π pulse, consisting of three AFP pulses of various lengths, Conolly et al. [16] subsequently proposed the use of two identical AFP pulses to rewind undesirable phase accumulation. Figure 5.20F shows the pulse sequence element in which two identical AFP pulses are surrounded by delays to form a double spin-echo. The phase accumulated during the first pulse according to Eq. (5.37) is canceled by the second RF pulse, such that the transverse magnetization phase is constant as a function of RF amplitude (Figure 5.20I) and frequency offset (Figure 5.20J). For the longitudinal magnetization the AFP pulse pair acts as a 2π pulse, bringing the magnetization back to the $+z'$ axis when the RF pulse satisfies the adiabatic condition. The AFP-based double spin-echo element provides very robust slice selection when the pulses are executed during magnetic field gradients as done during LASER localization (see Chapter 6).

It should be noted that the evaluation of *any* refocusing pulse, adiabatic or conventional, requires two simulations starting with M_x or M_y prior to the RF pulse. The simulation with M_x as a starting point may show the rotation $(M_x \rightarrow -M_x)$ within the frequency band selected by the RF pulse, leading to the conclusion that proper refocusing has been achieved. However, if the M_y simulation shows the rotation $(M_y \rightarrow -M_y)$ the pulse may not necessarily have achieved refocusing. The two independent simulations with M_x or M_y as starting point can be combined into a refocused component and associated phase [44, 45], that provide an objective evaluation of refocusing pulses (see Exercises).

5.5.5 Adiabatic Plane Rotation of Arbitrary Nutation Angle

As noted, AFP pulses are an excellent choice for spin inversion. While a single AFP pulse is not suitable for spin refocusing, the combination of two AFP pulses allows the formation of a perfectly refocused, frequency-selective spin-echo. When an AFP pulse is only executed until its midpoint in time, the pulse achieves B_1-insensitive excitation albeit forfeiting frequency-selectivity. Unfortunately, the AFP pulse is not suitable for any other rotations that, for example, require plane rotations or arbitrary nutation angles. The B_1-*Insensitive Rotation* using *4* segments (BIR-4) pulse is a plane rotation pulse that can achieve arbitrary nutation angles with high immunity to both RF inhomogeneity and resonance offsets. BIR-4 was derived from composite pulses, which are composed of multiple hard RF pulses that compensate each other's imperfections. Garwood and Ke [30] generated a time-symmetric, composite pulse $90°_{0°}180°_{180°+\theta/2}90°_{0°}$ which can achieve an arbitrary nutation angle θ simply by changing the phase of the center 180° pulse. The immunity towards RF inhomogeneity and frequency offset could be greatly enhanced by replacing the hard RF pulses with AFP pulse segments as shown in Figure 5.21. Without the phase offset for the middle two 90° segments the pulse would simply generate an identity (or 360°) rotation. However, with a phase offset of 180° + $\theta/2$ for the middle two segments an arbitrary nutation angle θ can be achieved. In addition, the time symmetry of BIR-4 ensures that the net rotation angle of the magnetization around the effective field according to Eq. (5.37) is zero, thus making BIR-4 a general purpose, plane rotation pulse. BIR-4 has found applications in MRI [46, 47], polarization transfer [48, 49], spectral editing [50, 51], T_1 relaxation measurements [52, 53], and saturation transfer [54]. An example of the plane rotation capabilities and the immunity towards RF inhomogeneity of BIR-4 is shown in Figure 5.22, where 180° BIR-4 pulses are used in surface coil multi-echo imaging to get artifact-free T_2-sensitized images.

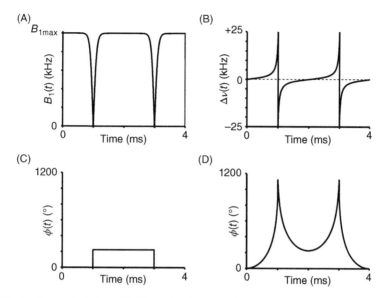

Figure 5.21 **BIR-4 pulse modulations.** (A) RF amplitude and (B) frequency modulations of a BIR-4 pulse, based on the modulation functions given by Eqs. (5.33) and (5.34). (C) A phase offset of the middle two segments by $180° + \theta/2$ lead to a final nutation angle of θ. (D) Phase modulation obtained by numerical integration of the frequency modulation in (B) and the addition of the discrete phase jumps $180° \pm \theta/2$ as indicated in (C).

Echo number

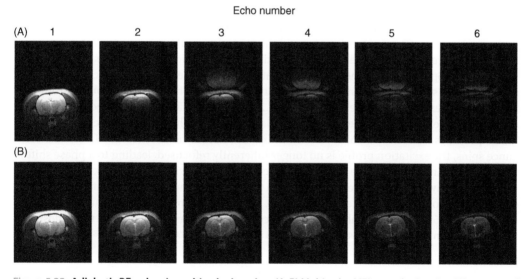

Figure 5.22 **Adiabatic RF pulses in multi-echo imaging.** (A, B) Multi-echo MRI on rat brain using (A) square and (B) adiabatic 180° BIR-4 RF pulses for refocusing. RF pulse irradiation and signal acquisition were performed with a surface coil (Ø 20 mm). In both cases, the first image (echo number $n = 1$) was generated with the adiabatic slice-selective pulse sequence of Figure 5.20F (TR/TE = 3000/35 ms, 2 mm slice thickness). The subsequent echoes are sampled at TE = $(35 + (n - 1)15)$ ms, with $n = 2$–6. (A) The sensitivity of square-refocusing pulses towards B_1^+ transmit inhomogeneity leads to signal loss (for $n \geq 2$) and image artifacts (mirror images for $n \geq 3$). (B) The insensitivity towards B_1^+ inhomogeneity combined with the refocusing capabilities of 180° BIR-4 pulses give artifact-free images without the need for phase cycling or magnetic field gradients, in which the signal decrease in subsequent images is purely caused by T_2 relaxation (ignoring diffusion).

5.6 Multidimensional RF Pulses

An RF pulse executed in the presence of a constant-amplitude gradient selects a slice along one spatial dimension. Spatial localization along multiple spatial dimensions is achieved by executing multiple RF pulses in concert with constant-amplitude gradients along different spatial directions. Well-known examples of 3D voxel selection are the PRESS, STEAM, and LASER spatial localization methods, as will be discussed in Chapter 6. Multidimensional spatially selective RF pulses aim at selecting more than one spatial dimension with a single RF pulse. The concept of multidimensional RF pulses was proposed by Bottomley and Hardy [55, 56] in 1987. Two years later, Pauly et al. [57] provided a convenient notation based on the k-space formalism used in MRI (see Chapter 4) for the description and further development of multidimensional RF pulses. Within the small-nutation-angle approximation of k-space the magnetization excited by a gradient-modulated RF pulse is given by

$$M_{xy}(r) = \gamma M_0 \int_k W(k) S(k) e^{-ikr} dk \qquad (5.38)$$

where $M_{xy}(r)$ represents the excited magnetization at position r (i.e. the multidimensional excitation profile), $W(k)$ and $S(k)$ are k-space weighting and sampling functions, and k is the standard k-space parameter. Note the similarity of Eq. (5.38) and the k-space description of MRI given by Eq. (4.12). Equation (5.38) states that the excitation profile $M_{xy}(r)$ is the Fourier transformation of k-space weighted by the RF pulse. Once a k-space sampling function or trajectory has been selected, the RF pulse weighting function $W(k)$ can be calculated from the inverse Fourier transformation of the excitation profile. The requirements for the k-space trajectory are identical to those for MRI, i.e. the trajectory should uniformly cover the part of k-space where the spatial frequency weighting function has significant energy and it should cover that region with sufficient density to avoid aliasing. A spiral k-space sampling function as shown in Figure 5.23A as achieved by two oscillating magnetic field gradients (Figure 5.23B) represents a popular choice. The k-space weighting function, i.e. the RF pulse, can be obtained analytically for simple excitation profiles [58] or numerically for complicated, but anatomically more relevant excitation profiles [59]. Figure 5.23C shows the RF waveform required to select a 2D Gaussian excitation profile (Figure 5.23D). Note that for the on-resonance spins the excitation profile is inherently refocused with maximum M_y in the absence of M_x. For off-resonance spins, the excitation profile is no longer inherently refocused, leading to a phase shift in the spectral dimension. However, for more complicated excitation profiles, off-resonance modulation may have a more delirious effect in that the excitation profile is shifted and broadened, leading to significant localization errors. While multidimensional RF pulses have been used for MRS [59], the undesirable off-resonance effects have largely prevented the widespread implementation of multidimensional RF pulses in *in vivo* MRS applications.

5.7 Spectral–Spatial RF Pulses

Using the same k-space interpretation of small-nutation excitation presented for multidimensional RF pulses, it is possible to design RF pulses that are simultaneously spatially and spectrally selective. These so-called spectral–spatial pulses eliminate the chemical shift artifact (see Chapter 6) and can substitute combinations of conventional RF pulses which produce spatial and spectral selectivity (e.g. water suppressed, localized ^{1}H MRS or fat/water imaging). A concise theoretical description as well as experimental applications are given by Meyer et al. [60] and

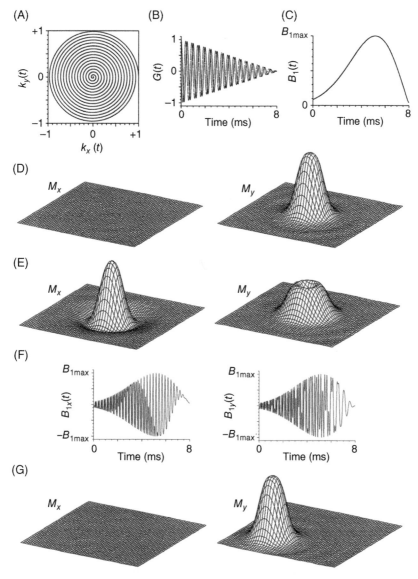

Figure 5.23 **Two-dimensional excitation with gradient-modulated RF pulses.** (A) Sixteen-turn spiral k-space sampling scheme and (B) the corresponding constant angular rate magnetic field gradients (G_x = solid line, G_y = dashed line). (D) Excitation of a symmetrical, Gaussian profile in the gradient isocenter requires (C) a real-valued RF modulation. Note that the excitation profile is inherently refocused. (E) Excitation profile 100 Hz off-resonance. The profile is no longer refocused and is slightly broadened. (G) Excitation out of the gradient isocenter (or excitation of asymmetrical profiles) is accompanied by (F) a complex RF modulation in analogy with conventional, frequency-shifted RF pulses (Figure 5.12C).

Spielman et al. [61]. Figure 5.24 shows a typical example in which an RF pulse executed during an oscillating gradient (Figure 5.24A) leads to an excitation profile that is selective in both the spatial and spectral domains. The slice profile in the spatial domain is given by the individual sub-pulses, whereas the spectral profile is given by the overall pulse modulation. To achieve more rectangular slice profiles, the RF pulse can be modulated according to a sinc function or other optimized functions. In the spectral dimension, several side lobes can be observed due to discrete sampling

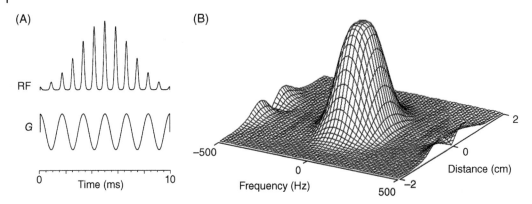

Figure 5.24 Spectral–spatial RF pulses. (A) RF and gradient modulations for an RF pulse with selectivity along the spatial and spectral domains. (B) Simulated excitation profile of the spectral–spatial pulse shown in (A). The overall pulse modulation leads to a Gaussian excitation profile in the spectral domain, whereas the individual pulse modulations (during constant gradients) lead to a Gaussian profile in the spatial domain. In the spectral domain aliasing side lobes are observed, which are absent in the spatial domain.

of the spectral k-space domain (i.e. one pulse for each half cycle of the oscillating magnetic field gradient). Because the spatial k-space domain is sampled continuously, no spatial side lobes are observed in the spatial dimension. The spectral selectivity can be used for selective excitation of one resonance (e.g. NAA) while simultaneously suppressing another (e.g. water). Because of the narrow bandwidth in the spectral dimension, which is caused by the long pulse length dictated by the maximum achievable slew rate, only resonances which resonate close to each other (e.g. NAA and lactate) can be simultaneously excited. Note that spectral–spatial pulses are not inherently refocused. In the limit of the small-nutation-approximation the acquired phase in both the spectral and spatial dimension is linear. Thus, the spatial phase dispersion can be corrected by gradient reversal and the spectral response can be refocused by a 180° RF pulse. Recently, spectral–spatial RF pulses have seen increased usage with hyperpolarized ^{13}C MR applications [62, 63]. RF pulses are designed with multiple frequency bands in order to allow the excitation of multiple ^{13}C hyperpolarized resonances, each with their own optimal nutation angle.

Exercises

5.1 Longitudinal magnetization can be "excited" into the transverse plane by a 90° (or $\pi/2$) pulse.
 A If the pulse length of the square 90° pulse is 1.0 ms, what is the required B_1 magnitude in μT and in Hz to achieve excitation?
 B How many Larmor precession cycles will occur in the laboratory frame at $B_0 = 3.0$ T during the 90° excitation pulse?

5.2 A birdcage RF coil generates a B_1 field of 37.6 μT at the maximum amplifier output. Calculate the shortest possible pulse length to achieve signal inversion with a "square" RF pulse.

5.3 Consider a multi-slice spin-warp sequence with 1 ms $SLR_{linear-phase}$ ($R \sim 10$) and 2 ms $SLR_{linear-phase}$ ($R \sim 10$) pulses for excitation and refocusing, respectively (see Table 5.1).
 A Calculate the slice selection gradient strengths (in $Hz\,cm^{-1}$) during each pulse in order to select 2 mm thick slices.

B Assuming constant-slope (150 μs from 0 to 100%) gradients, calculate the 90° slice-refocusing gradient strength for a plateau duration of 0.5 ms.

C Repeat the calculation under B when excitation is performed with a minimum-phase SLR ($R \sim 10$, Table 5.1) excitation pulse.

5.4 Derive an expression for the on-resonance B_1-dependence of the longitudinal magnetization following a composite MLEV pulse ($90°_{+x}180°_{+y}90°_{+x}$).

A Calculate the range of nutation angle (between 0° and 180°) for which the longitudinal magnetization is at least 99% inverted (i.e. $M_z/M_0 < -0.98$) following a MLEV pulse.

B Explain the mechanism of compensation in terms of M_x, M_y, and M_z components following the MLEV pulse.

C Calculate the B_1 insensitivity of MLEV relative to a square pulse for the 99% inversion interval.

D Show that the on-resonance nutation angle generated by a composite WALTZ pulse ($90°_{+x}180°_{-x}270°_{+x}$) has the same sensitivity towards RF inhomogeneity as a regular 180° pulse.

5.5 Consider a PRESS sequence executed with 2 ms sinc-shaped RF pulses for excitation and refocusing.

A Calculate the relative change in RF power deposition when the pulse lengths are increased to 4 ms.

B Calculate the relative change in RF power deposition when the pulses are changed to linear-phase SLR excitation and refocusing pulses with $R = 10$ (Table 5.1).

C Given a maximum RF amplitude and pulse length of 1500 Hz and 5 ms, respectively, calculate the RF pulse combination (using the pulses tabulated in Table 5.1) that gives the smallest chemical shift displacement.

D Calculate the relative change in RF power deposition when the echo- and repetition times are reduced and increased by a factor of 2, respectively.

5.6

A A 2 ms square pulse was calibrated to produce a 90° nutation angle at a RF power of +23 dB. Calculate the RF power setting for a 1 ms sinc pulse producing a 180° nutation angle.

B A given RF coil was calibrated to produce a 1310 Hz B_1 magnetic field at a RF power setting of 10 dB. Calculate the RF power setting for a 1 ms five-lobe sinc inversion pulse.

5.7 Consider a 10 ms AFP pulse with sech/tanh modulation and $R = 60$. Over the B_1 amplitude range in which the adiabatic condition is satisfied, qualitatively describe what happens to the longitudinal and transverse magnetization during the pulse (starting from a thermal equilibrium state) for

A a frequency offset of +30 kHz

B a frequency offset of +3 kHz

C a frequency offset of +0.3 kHz and

D a frequency offset of −3 kHz.

E Describe how the rotations at each frequency can be used for specific applications and discuss if the same could be achieved with AFP pulses based on tanh/tan modulation.

5.8 A PRESS sequence is executed with 3 ms optimized SLR90 ($R = 12$) and SLR180 ($R = 12$) excitation and refocusing pulses, respectively. Signal is acquired from a $2 \times 2 \times 2 \, \text{cm} = 8 \, \text{ml}$ volume in the human occipital cortex with TR = 1500 ms and TE = 20 ms.

    A Given a maximum gradient amplitude of $40 \, \text{mT} \, \text{m}^{-1}$ and constant-time gradient ramping of 500 μs, calculate the amplitude of a trapezoidal slice selection refocusing gradient with a 1 ms plateau period.

    B Recalculate the slice selection-refocusing gradient amplitude when the sequence is executed with constant-slope gradient ramping of $80 \, \text{mT} \, \text{m}^{-1} \, \text{ms}^{-1}$.

    C Repeat the calculations under A and B for a minimum phase SLR90 ($R = 12$) pulse for which the effective dephasing pulse length equals $T/10$.

5.9 The rotations induced by a RF pulse along the x' axis over an angle θ with an effective field B_e tilted over α to the z' axis is equivalent to a rotation through $-\alpha$ about y' followed by a rotation through θ about x' and then followed by a rotation of $+\alpha$ about y', resulting in a total rotation matrix given by Eq. (5.8). Demonstrate that the same total rotation can also be achieved by rotating through $90° - \alpha$ about y' followed by a rotation through θ about z' and finished by a rotation of $\alpha - 90°$ about y'.

5.10

    A In PulseWizard (http://booksupport.wiley.com) generate an AFP inversion pulse based on sech/tanh modulations and $R = 10$.

    B Using the AFP pulse generated under A, determine the minimum RF magnetic field amplitude (in Hz) for which the pulse achieves 95% inversion (i.e. M_0 is inverted to $-0.9 M_0$). Assume a 2 ms pulse length.

    C Simulate the frequency profile of the AFP pulse at the RF amplitude determined under B and determine the bandwidth of the inversion profile at $M_z = 0$.

    D Repeat the simulation under C for RF amplitude that are 2, 4, and 8 times larger than the minimum threshold. Comment on the result in light of *in vivo* MRS application using surface coil transception.

References

1 Bauer, C., Freeman, R., Frenkiel, T. et al. (1984). Gaussian pulses. *J. Magn. Reson.* 58: 442–457.

2 Mao, J., Mareci, T.H., Scott, K.N., and Andrew, E.R. (1986). Selective inversion radiofrequency pulses by optimal control. *J. Magn. Reson.* 70: 310–318.

3 Conolly, S., Nishimura, D., and Macovski, A. (1986). Optimal control solutions to the magnetic resonance selective excitation problem. *IEEE Trans. Med. Imaging* MI5: 106–115.

4 Skinner, T.E., Reiss, T.O., Luy, B. et al. (2003). Application of optimal control theory to the design of broadband excitation pulses for high-resolution NMR. *J. Magn. Reson.* 163: 8–15.

5 Janich, M.A., Schulte, R.F., Schwaiger, M., and Glaser, S.J. (2011). Robust slice-selective broadband refocusing pulses. *J. Magn. Reson.* 213: 126–135.

6 Geen, H., Wimperis, S., and Freeman, R. (1989). Band-selective pulses without phase distortion. A simulated annealing approach. *J. Magn. Reson.* 85: 620–627.

7 Gezelter, J.D. and Freeman, R. (1990). Use of neural networks to design shaped radiofrequency pulses. *J. Magn. Reson.* 90: 397–404.

8 Geen, H. and Freeman, R. (1991). Band-selective radiofrequency pulses. *J. Magn. Reson.* 93: 93–141.

9 Shinnar, M., Eleff, S., Subramanian, H., and Leigh, J.S. (1989). The synthesis of pulse sequences yielding arbitrary magnetization vectors. *Magn. Reson. Med.* 12: 74–80.

10 Shinnar, M., Bolinger, L., and Leigh, J.S. (1989). The use of finite impulse response filters in pulse design. *Magn. Reson. Med.* 12: 81–87.

11 Shinnar, M., Bolinger, L., and Leigh, J.S. (1989). The synthesis of soft pulses with a specified frequency response. *Magn. Reson. Med.* 12: 88–92.

12 Shinnar, M. and Leigh, J.S. (1989). The application of spinors to pulse synthesis and analysis. *Magn. Reson. Med.* 12: 93–98.

13 Pauly, J., Le Roux, P., Nishimura, D., and Macovski, A. (1991). Parameter relations for the Shinnar-Le Roux selective excitation pulse design algorithm. *IEEE Trans. Med. Imaging* 10: 53–65.

14 Matson, G.B. (1994). An integrated program for amplitude-modulated RF pulse generation and re-mapping with shaped gradients. *Magn. Reson. Imaging* 12: 1205–1225.

15 Conolly, S., Nishimura, D., Macovski, A., and Glover, G. (1988). Variable-rate selective excitation. *J. Magn. Reson.* 78: 440–458.

16 Conolly, S., Glover, G., Nishimura, D., and Macovski, A. (1991). A reduced power selective adiabatic spin-echo pulse sequence. *Magn. Reson. Med.* 18: 28–38.

17 Slotboom, J., Vogels, B.A.P.M., de Haan, J.G. et al. (1994). Proton resonance spectroscopy study of the effects of L-ornithine-L-aspartate on the development of encephalopathy using localization pulses with reduced specific absorption rate. *J. Magn. Reson. B* 105: 147–156.

18 Goelman, G. (1997). Two methods for peak RF power minimization of multiple inversion-band pulses. *Magn. Reson. Med.* 37: 658–665.

19 Steffen, M., Vandersypen, L.M.K., and Chuang, I.L. (2000). Simultaneous soft pulses applied at nearby frequencies. *J. Magn. Reson.* 146: 369–374.

20 Bodenhausen, G., Freeman, R., and Morris, G.A. (1976). Simple pulse sequence for selective excitation in Fourier transform NMR. *J. Magn. Reson.* 23: 171–175.

21 Morris, G.A. and Freeman, R. (1978). Selective excitation in Fourier transform nuclear magnetic resonance. *J. Magn. Reson.* 29: 433–462.

22 Wu, X.L., Xu, P., and Freeman, R. (1989). Selective excitation with the DANTE sequence. The baseline syndrome. *J. Magn. Reson.* 81: 206–211.

23 Norris, D.G., Koopmans, P.J., Boyacioglu, R., and Barth, M. (2011). Power independent of number of slices (PINS) radiofrequency pulses for low-power simultaneous multislice excitation. *Magn. Reson. Med.* 66: 1234–1240.

24 Levitt, M.H. and Freeman, R. (1979). NMR population inversion using a composite pulse. *J. Magn. Reson.* 1979 (33): 473–476.

25 Freeman, R., Kempsell, S.P., and Levitt, M.H. (1980). Radiofrequency pulse sequences which compensate their own imperfections. *J. Magn. Reson.* 38: 453–479.

26 Levitt, M.H. (1982). Symmetrical composite pulse sequences for NMR population inversion. I. Compensation of radiofrequency field inhomogeneity. *J. Magn. Reson.* 48: 234–264.

27 Levitt, M.H. (1982). Symmetrical composite pulse sequences for NMR population inversion. II. Compensation for resonance offset. *J. Magn. Reson.* 50: 95–110.

28 Levitt, M.H. and Ernst, R.R. (1983). Composite pulses constructed by a recursive expansion procedure. *J. Magn. Reson.* 55: 247–254.

29 Levitt, M.H. (1986). Composite pulses. *Prog. Nucl. Magn. Reson. Spectrosc.* 18: 61–122.

30 Garwood, M. and Ke, Y. (1991). Symmetric pulses to induce arbitrary flip angles with compensation for RF inhomogeneity and resonance offsets. *J. Magn. Reson.* 94: 511–525.

31 Wimperis, S. (1994). Broadband, narrowband and passband composite pulses for use in advanced NMR experiments. *J. Magn. Reson. A* 109: 221–231.

32 Nimbalkar, M., Luy, B., Skinner, T.E. et al. (2013). The fantastic four: a plug "n" play set of optimal control pulses for enhancing NMR spectroscopy. *J. Magn. Reson.* 228: 16–31.

33 Bloch, F., Hansen, W.W., and Packard, M.E. (1946). The nuclear induction experiment. *Phys. Rev.* 70: 474–485.

34 Purcell, E.M., Torrey, H.C., and Pound, R.V. (1946). Resonance absorption by nuclear magnetic moments in a solid. *Phys. Rev.* 69: 37–38.

35 Bendall, M.R. and Pegg, D.T. (1986). Uniform sample excitation with surface coils for *in vivo* spectroscopy by adiabatic rapid half passage. *J. Magn. Reson.* 67: 376–381.

36 Garwood, M., Ugurbil, K., Rath, A.R. et al. (1989). Magnetic resonance imaging with adiabatic pulses using a single surface coil for RF transmission and signal detection. *Magn. Reson. Med.* 9: 25–34.

37 Silver, M.S., Joseph, R.I., and Hoult, D.I. (1984). Highly selective $\pi/2$ and π pulse generation. *J. Magn. Reson.* 59: 347–351.

38 Hioe, F.T. (1984). Solution of Bloch equations involving amplitude and frequency modulations. *Phys. Rev. A* 30: 2100–2103.

39 Ugurbil, K., Garwood, M., and Rath, A.R. (1988). Optimization of modulation functions to improve insensitivity of adiabatic pulses to variations in B_1 magnitude. *J. Magn. Reson.* 80: 448–469.

40 Tannus, A. and Garwood, M. (1996). Improved performance of frequency-swept pulses using offset-independent adiabaticity. *J. Magn. Reson. A* 120: 133–137.

41 Silver, M.S., Joseph, R.I., and Hoult, D.I. (1985). Selective spin inversion in nuclear magnetic resonance and coherent optics through an exact solution of the Bloch-Riccati equation. *Phys. Rev. A* 31: 2753–2755.

42 Tannus, A. and Garwood, M. (1997). Adiabatic pulses. *NMR Biomed.* 10 (8): 423–434.

43 Conolly, S., Nishimura, D., and Macovski, A. (1989). A selective adiabatic spin-echo pulse. *J. Magn. Reson.* 83: 324–334.

44 Bendall, M.R. and Pegg, D.T. (1985). Comparison of depth pulse sequences with composite pulses for spatial selection in *in vivo* NMR. *J. Magn. Reson.* 63: 494–503.

45 Bendall, M.R., Garwood, M., Ugurbil, K., and Pegg, D.T. (1987). Adiabatic refocusing pulse which compensates for variable RF power and off-resonance effects. *Magn. Reson. Med.* 4: 439–499.

46 Staewen, R.S., Johnson, A.J., Ross, B.D. et al. (1990). 3-D FLASH imaging using a single surface coil and a new adiabatic pulse, BIR-4. *Invest. Radiol.* 25: 559–567.

47 de Graaf, R.A., Rothman, D.L., and Behar, K.L. (2003). Adiabatic RARE imaging. *NMR Biomed.* 16 (1): 29–35.

48 Merkle, H., Wei, H., Garwood, M., and Ugurbil, K. (1992). B_1-insensitive heteronuclear adiabatic polarization transfer for signal enhancement. *J. Magn. Reson.* 99: 480–494.

49 Shen, J., Petersen, K.F., Behar, K.L. et al. (1999). Determination of the rate of the glutamate/glutamine cycle in the human brain by *in vivo* ^{13}C NMR. *Proc. Natl. Acad. Sci. U. S. A.* 96: 8235–8240.

50 Garwood, M. and Merkle, H. (1991). Heteronuclear spectral editing with adiabatic pulses. *J. Magn. Reson.* 94: 180–185.

51 de Graaf, R.A., Luo, Y., Terpstra, M., and Garwood, M. (1995). Spectral editing with adiabatic pulses. *J. Magn. Reson. B* 109 (2): 184–193.

52 Sakuma, H., Nelson, S.J., Vigneron, D.B. et al. (2005). Measurement of T_1 relaxation times in cardiac phosphate metabolites using BIR-4 adiabatic RF pulses and a variable nutation method. *Magn. Reson. Med.* 29: 688–691.

53 Bottomley, P.A. and Ouwerkerk, R. (1994). The dual-angle method for fast, sensitive T_1 measurement *in vivo* with low-angle adiabatic pulses. *J. Magn. Reson. B* 104: 159–167.

54 Bottomley, P.A., Ouwerkerk, R., Lee, R.F., and Weiss, R.G. (2002). Four-angle saturation transfer (FAST) method for measuring creatine kinase reaction rates *in vivo*. *Magn. Reson. Med.* 47: 850–863.

55 Bottomley, P.A. and Hardy, C.J. (1987). Two-dimensional spatially selective spin inversion and spin-echo refocusing with a single nuclear magnetic resonance pulse. *J. Appl. Phys.* 62: 4284–4290.

56 Bottomley, P.A. and Hardy, C.J. (1987). PROGRESS in efficienct three-dimensional spatially localized *in vivo* ^{31}P NMR spectroscopy using multidimensional spatially selective (ρ) pulses. *J. Magn. Reson.* 74: 550–556.

57 Pauly, J., Nishimura, D., and Macovski, A. (1989). A linear class of large-tip-angle selective excitation pulses. *J. Magn. Reson.* 82: 571–587.

58 Pauly, J., Nishimura, D., and Macovski, A. (1989). A k-space analysis of small-tip-angle excitation. *J. Magn. Reson.* 81: 43–56.

59 Qin, Q., Gore, J.C., Does, M.D. et al. (2007). 2D arbitrary shape-selective excitation summed spectroscopy (ASSESS). *Magn. Reson. Med.* 58: 19–26.

60 Meyer, C.H., Pauly, J.M., Macovski, A., and Nishimura, D.G. (1990). Simultaneous spatial and spectral selective excitation. *Magn. Reson. Med.* 15: 287–304.

61 Spielman, D., Meyer, C., Macovski, A., and Enzmann, D. (1991). ^{1}H spectroscopic imaging using a spectral-spatial excitation pulse. *Magn. Reson. Med.* 18: 269–279.

62 Lau, A.Z., Chen, A.P., Hurd, R.E., and Cunningham, C.H. (2011). Spectral-spatial excitation for rapid imaging of DNP compounds. *NMR Biomed.* 24: 988–996.

63 Marco-Rius, I., Cao, P., von Morze, C. et al. (2017). Multiband spectral-spatial RF excitation for hyperpolarized [2-^{13}C]dihydroxyacetone ^{13}C-MR metabolism studies. *Magn. Reson. Med.* 77: 1419–1428.

6

Single Volume Localization and Water Suppression

6.1 Introduction

Restricting signal detection to a well-defined region-of-interest (ROI) is crucial for meaningful *in vivo* NMR spectroscopy. First and foremost, spatial localization is used to remove unwanted signals from outside the ROI, like extracranial lipids in MRS applications of the brain. However, spatial localization has many additional benefits, in particular in managing tissue and magnetic field heterogeneity. For example, at a macroscopic level the brain can be segmented into gray matter, white matter, and cerebrospinal fluid (CSF), each of which has unique metabolic profiles. Careful positioning of the localized volume can minimize "partial volume effects" (i.e. contamination of signal from one compartment by signal from another compartment), thereby providing a more genuine tissue characterization. Additional benefits of spatial localization originate from the fact that variations in B_0 and B_1 magnetic fields are greatly reduced over the small localized volume, thereby providing narrower spectral lines and more uniform signal excitation and reception, respectively.

Figure 6.1A and C shows unlocalized ^{1}H and ^{31}P NMR spectra from newborn piglet brain, respectively. The ^{1}H NMR spectrum is acquired with a combination of water suppression techniques (as will be discussed in Section 6.3). Figure 6.1D and F shows the corresponding localized ^{1}H and ^{31}P NMR spectra. Clearly, the unlocalized ^{1}H NMR spectrum is contaminated by large signals from extracranial lipids, such that cerebral metabolite resonances cannot be observed. Furthermore, because the B_0 magnetic field homogeneity over a large unlocalized area is difficult to optimize, the resonances are broad, thereby further reducing the spectral resolution and degrading the water suppression. Upon spatial localization, the overwhelming lipid resonances are removed and the B_0 magnetic field homogeneity has improved, giving a recognizable ^{1}H NMR spectrum of brain. Besides the broad ^{31}P NMR signal from bone, the unlocalized and localized ^{31}P NMR spectra do not differ as much as the corresponding ^{1}H NMR spectra. However, the ratio between phosphocreatine and ATP has been reduced upon localization, indicating that ^{31}P signals from extracranial muscle tissue (with a high PCr/ATP ratio) have been excluded. Furthermore, as in the case of ^{1}H MRS, all the ^{31}P NMR resonances are narrower in the localized spectrum. It can be concluded from Figure 6.1 that spatial localization does not only allows the unambiguous detection of metabolites from a well-defined spatial volume, but also increases the spectral resolution by (i) narrower resonances, (ii) exclusion of broad unwanted resonances, and (iii) improved water suppression (for ^{1}H MRS).

Over the last three decades, a wide variety of spatial localization techniques have been developed. To avoid getting lost in the maze of acronyms, they are divided into single- and multi-voxel

In Vivo NMR Spectroscopy: Principles and Techniques, Third Edition. Robin A. de Graaf.
© 2019 John Wiley & Sons Ltd. Published 2019 by John Wiley & Sons Ltd.

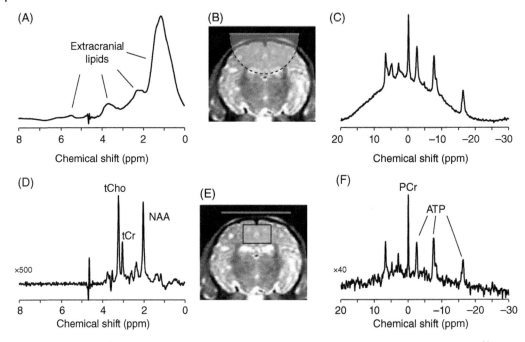

Figure 6.1 **^{1}H and ^{31}P NMR spectra from newborn piglet brain** *in vivo*. (A) Nonlocalized ^{1}H and (C) ^{31}P NMR spectra obtained from (B) a 3D volume spanning the sensitive area of the surface coil (shaded yellow). The spectra are characterized by the presence of unwanted signal from skull, bone, and lipids and broad resonances due to the low magnetic field homogeneity over a large volume. (D) PRESS-localized (Section 6.2.5, TE = 144 ms) ^{1}H and (F) ISIS-localized (Section 6.2.1) ^{31}P NMR spectra obtained from (E) a 1.0 ml volume placed in the cortex. The improved magnetic field homogeneity and absence of complicating signals leads to high-quality NMR spectra from a well-defined spatial location.

techniques. Single-voxel techniques, in which signal is acquired from a single spatial volume will be described in this chapter, while multi-voxel techniques are discussed in Chapter 7.

Spatially-dependent magnetic field gradients are at the heart of spatial localization methods. Several of the earlier methods utilize magnetic field gradients in the excitation RF field B_1 [1–9], based on the classical rotating frame zeugmatography experiment by Hoult [1]. However, the localized volume obtained with B_1-gradient-based localization techniques is typically inferior to that obtained with B_0-gradient-based methods. As a result, almost all spatial localization requirements are currently addressed with B_0-gradient-based methods, as will be described in the first part of this chapter. For more information on B_1-based localization, the reader is referred to the literature [1–10].

6.2 Single-volume Localization

All localization methods based on B_0 magnetic field gradients rely on the selection of a spatially selective slice by the application of a frequency-selective RF pulse in the presence of a magnetic field gradient, identical to slice selection in MRI (Figure 6.2A). Following 1D slice selection a second RF pulse executed during a magnetic field gradient along an orthogonal spatial direction is used to select a 2D column (Figure 6.2B). Finally, a third RF pulse is used along the remaining orthogonal gradient direction to reduce the 2D column to a 3D voxel (Figure 6.2C).

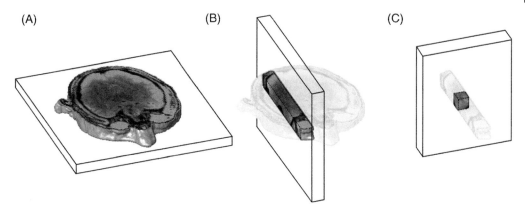

Figure 6.2 **Principle of 3D voxel selection.** Single voxel selection based on magnetic field gradients depends on the consecutive selection of (A) a 1D plane, (B) a 2D column, and finally (C) a 3D voxel along three orthogonal directions in space. Various localization methods differ in the type of RF pulses (excitation, inversion, refocusing) used to select the three spatial planes. A critical aspect of all localization methods is the elimination of unwanted signals (light shaded areas in B and C), while simultaneously retaining signal in the 3D voxel. The voxel can be rotated and translated by changing the relative gradient amplitudes during the RF pulses and the frequency of the RF pulses, respectively (see also Figure 4.2).

The 3D voxel can be rotated and translated by changing the magnetic field gradient orientation and RF frequency offset, respectively. Many RF pulse and magnetic field gradient combinations exist to select a three-dimensional volume. They can be divided into techniques that leave the magnetization in the volume-of-interest unperturbed during the localization procedure (here after referred to as outer volume suppression, OVS) and those that selectively perturb the magnetization in order to remove unwanted magnetization outside the volume-of-interest (hereafter referred to as single-voxel or single-volume localization). Since OVS is most commonly used for the removal of extracranial lipid signals in MR spectroscopic imaging, the principles of OVS will be discussed in Chapter 7.

6.2.1   Image Selected *In Vivo* Spectroscopy (ISIS)

The Image Selected *In Vivo* Spectroscopy (ISIS) localization method, as first described by Ordidge et al. [11] achieves full 3D localization in eight scans. The principle of ISIS localization is depicted in Figure 6.3. ISIS employs frequency-selective inversion pulses in the presence of a magnetic field gradient followed by nonselective excitation of the entire sample. For a single slice, the inversion pulse is turned on or off on alternate acquisitions such that following excitation the magnetization within the slice of interest starts along the $-y$ or $+y$ axis, respectively. The unwanted magnetization outside the slice is unperturbed by the inversion pulse such that it is excited along the $+y$ axis in both experiments. The 1D slice is calculated as the difference between the two experiments, retaining the magnetization of interest while simultaneously subtracting out the unwanted signals (Figure 6.3C). The localization can be extended to three dimensions by extending the number of inversion pulse on/off combinations along orthogonal magnetic field gradient directions. When zero or an even number of 180° pulses are executed, the desired magnetization in the cross-section of the three selected slices ends up along the positive longitudinal axis and following a $90°_x$ excitation pulse will end up along the $+y$ axis. During a scan with an odd number of 180° pulses, the desired magnetization ends up along the negative longitudinal axis and is excited to the $-y$ axis. Adding and subtracting the individually stored scans with even and odd number of 180° pulses, respectively, will constructively

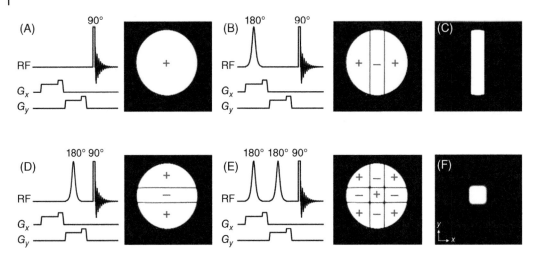

Figure 6.3 **Principle of ISIS localization.** (A, B) ISIS localization of a 1D plane requires two scans, (A) without and (B) with inversion of the slice. (C) The difference between (A) and (B) selects the slice by subtracting out the unwanted signals. (D, E) Extension to a 2D column requires two additional scans, whereby (F) the volume is selected as (A) − (B) − (D) + (E). Extension to a 3D voxel requires an additional four scans for a total of eight scans (not shown). The subtraction and addition of the various scans is typically part of the sequence phase cycle. The red + and blue − signs indicate the related sign of the magnetization at that spatial location.

accumulate signal from the desired location while destructively canceling signal from all other locations. Note that in practice the add–subtract scheme is achieved by cycling the receiver phase, making the procedure invisible to the user. Figure 6.3D–F shows the extension to a 2D column, requiring 2×2 scans with the inversion pulse combinations shown. Further extension to a 3D voxel will require a total of eight ($=2 \times 2 \times 2$) scans each with a unique combination of zero to three inversion pulses. Note that slice selection gradients are present even in the absence of an inversion pulse in order to keep eddy currents identical between all eight scans.

For ^{1}H MRS, ISIS is rarely executed as the basic scheme depicted in Figure 6.3. Besides the requirements of water suppression, a potential problem inherent to ISIS is that in each acquisition signal from the entire sensitive coil volume is obtained. When the VOI is small compared to the entire excited volume, even small subtraction errors could lead to large contamination artifacts. Furthermore, in areas where the B_0 magnetic field homogeneity, and hence the water suppression is poor, the large unsuppressed water resonance could lead to receiver dynamic range problems during the individual eight scans (even when the water is perfectly suppressed in the final spectrum). In addition, manual or iterative shimming on the localized volume is tedious, as complete 3D localization is only achieved after eight scans. All these potential problems can be significantly reduced when ISIS is combined with a (separate) method for OVS. Connelly et al. [12] originally proposed to combine ISIS with so-called noise pulses [13], giving rise to the Outer volume Suppressed Image Related *In vivo* Spectroscopy (OSIRIS) technique. The frequency domain excitation profile of a noise pulse consists of a selective null-band in which magnetization remains unperturbed, together with noise (i.e. random dephasing of magnetization) outside this null-band. This combination leads to a significant suppression of the magnetization outside the VOI *in a single scan*, such that receiver gain and motion problems are minimized and manual shimming becomes feasible. While noise pulses were initially proposed, essentially any kind of OVS can be used, including techniques that combine the ISIS and OVS principles into one pulse [14, 15]. In order to achieve a short echo-time, the SPECIAL

[16] localization method uses 1D ISIS to select a slice in one of the three spatial direction. Sensitivity to motion and subtraction artifacts is reduced by selecting the other two voxel dimensions with (single-scan) excitation and refocusing pulses.

Even though ISIS can theoretically give perfect localization in eight scans, the localization accuracy can be compromised by a number of factors. When the *in vivo* MRS experiment is executed with surface coils (for increased sensitivity), B_1 magnetic field inhomogeneity can degrade the performance (i.e. slice profile) of the inversion and excitation pulses. While this problem can lead to greatly reduced sensitivity and increased contamination from unwanted signals, it is easily alleviated by using adiabatic RF pulses. As a result, ISIS is typically executed with AFP pulses for selective inversion and AHP or BIR-4 pulses for excitation (see Chapter 5).

Another potential source of signal loss and contamination is caused by so-called "T_1 smearing" [17–19]. When the longitudinal T_1 relaxation is incomplete between subsequent acquisitions within the eight ISIS add–subtract cycles, the longitudinal magnetization prior to a given inversion (and acquisition) depends on the particular order in which the add–subtract scheme is executed. Because the longitudinal magnetization is not the same for all eight acquisitions, the cancelation of unwanted signals is incomplete. The degree of contamination due to "T_1 smearing" depends on the repetition time relative to the T_1 relaxation time, the degree of B_1 inhomogeneity and the order in which the eight ISIS acquisitions are performed [17–19]. "T_1 smearing" can be effectively minimized by the application of an adiabatic, post-acquisition saturation pulse that ensures an identical longitudinal magnetization for all acquisitions.

ISIS in combination with a pulse-acquire detection module is a popular method for localized ^{31}P MRS of short T_2 species such as ATP. However, T_2 relaxation during the inversion pulses can degrade the localization performance of ISIS with decreased signal within the VOI and increased contamination from outside the VOI. While T_2 relaxation is unavoidable, the effect is exacerbated at higher RF amplitudes as the magnetization spends a greater amount of time in the transverse plane.

While ISIS is still sporadically used, its popularity has greatly diminished due to the availability of other localization methods that achieved high-quality spatial localization in a single scan.

6.2.2 Chemical Shift Displacement

The application of a magnetic field gradient G (in $\mathrm{T\,m^{-1}}$) makes the spin Larmor frequency, ν (in MHz), linearly dependent on the spatial position r (in m) according to

$$\nu(r) = \nu_0 + \left(\frac{\gamma}{2\pi}\right) r G \tag{6.1}$$

where ν_0 represents the Larmor frequency of the spins in an external magnetic field B_0 in the absence of a magnetic field gradient. The spatial position r can represent any of the Cartesian directions or any vectorial combination thereof. An RF pulse with center frequency ν_{RF} will select a slice centered at position r according to

$$r = \frac{2\pi(\nu_{RF} - \nu_0)}{\gamma G} \tag{6.2}$$

When the RF frequency ν_{RF} equals the Larmor frequency ν_0 of the spins in the absence of a magnetic field gradient, a slice is selected in the magnet isocenter ($r = 0$). When $\nu_{RF} > \nu_0$ a slice will be selected on the positive side of the corresponding magnet direction ($r > 0$), provided G

is a positive gradient. When the sample also contains spins with a Larmor frequency $\nu_0 + \Delta\nu$, the same RF pulse selects a slice that is offset by an amount Δr according to

$$\Delta r = -\frac{2\pi\Delta\nu}{\gamma G} \tag{6.3}$$

The selection of different spatial positions for spins with different Larmor frequencies is referred to as the chemical shift displacement artifact (CSDA) or error (CSDE). Figure 6.4 demonstrates the effect of the CSDA in ^{1}H MRS. According to Eq. (6.3) the CSDA can only be

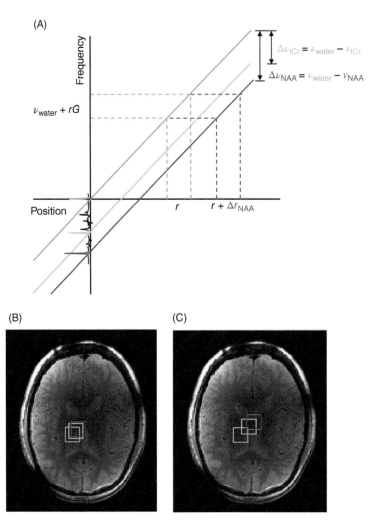

Figure 6.4 Chemical shift displacement in localized MRS. (A) A frequency-selective RF pulse with frequency $\nu_{water} + rG$ (in Hz) during a magnetic field gradient G (in Hz cm^{-1}) selects, for the on-resonance water signal (green line), a spatial slice centered at position r (in cm). The same RF pulse selects, for the off-resonance NAA signal (red line), a slice that is centered at position $r + \Delta r$ whereby Δr is the chemical shift displacement given by Eq. (6.4). (B) An example of chemical shift displacement for water (green), total creatine (orange), and NAA (red) at 3 T. Displacement in the third spatial dimension is not shown. (C) For a localization method with identical RF bandwidths as used in (B), the chemical shift displacement at 7 T is increased $(7/3) = 2.33$ times along *each* spatial dimension.

reduced by increasing the gradient strength G which, in order to keep a constant slice thickness, demands a proportionally increased RF pulse bandwidth. Equation (6.3) is often rewritten in a more useful form relating the CSDA Δr_{CSDA} to the RF pulse bandwidth, BW (in Hz), and slice thickness, Δr_{slice}, according to

$$\Delta r_{CSDA} = -\frac{\Delta \nu}{BW}\Delta r_{slice} \tag{6.4}$$

Equation (6.4) demonstrates that the CSDA can indeed only be minimized by increasing the RF pulse bandwidth BW. However, for amplitude-modulated RF pulses the bandwidth is linearly proportional to the maximum RF amplitude B_{1max}. Especially on human MR scanners the maximum achievable RF amplitude is limited by the available hardware and power deposition considerations. For example, on a human 3 T MR scanner the maximum RF amplitude may be limited to 35 µT, corresponding to $(\gamma/2\pi)B_{1max}$ of 1500 Hz for protons. From Table 5.1 it follows that a sinc 180° pulse executed at the maximum RF amplitude requires a 1.9 ms duration to refocus a bandwidth of circa 2.4 kHz. The chemical shift difference between water and N-acetyl aspartate (NAA) is circa 2.7 ppm corresponding to 346 Hz at 3 T. The combination of the limited RF pulse bandwidth and the chemical shift difference leads to a 14.4% displacement of the NAA slice compared to the water slice, according to Eq. (6.4). When the same RF pulse is used to select the spatial slice in all three directions, the 3D voxels for NAA and water only overlap by 63% (Figure 6.4B).

The CSDA artifact increases at higher magnetic fields since chemical shift differences (in Hz) scale linearly with magnetic field strength. In addition, the maximum achievable RF amplitude B_{1max} is further limited by MR hardware, RF field inhomogeneity, and RF power deposition, such that B_{1max} on a human 7 T MR scanner typically does not exceed 25 µT. The lower RF amplitude forces the sinc 180° pulse to be executed with a 2.7 ms length leading to a refocusing bandwidth of 1.7 kHz. The 2.7 ppm chemical shift difference, corresponding to 805 Hz at 7 T, now leads to a 48.5% displacement between the NAA and water slices (Figure 6.4C). For a 3D voxel selected by the same RF pulse this translates to a 14% overlap between the NAA and water volumes.

The CSDA leads to a variety of problems of which misinterpretation of the voxel composition is probably most problematic as it can go undetected when only evaluating MR spectra post-acquisition. In the case of a significant CSDA (Figure 6.4C), effects of magnetic field B_0 or B_1 inhomogeneity may be observed. For example, the magnetic field B_0 homogeneity is often optimized on the water signal with a combination of B_0 mapping and iterative algorithms. A situation could arise in which a high magnetic field homogeneity is achieved for the water resonance, but whereby the resonance lines for the metabolites are broad as they originate from spatial positions with a suboptimal homogeneity.

When performing signal localization in areas close to the skull, the CSDA can lead to severe lipid contamination even though the localization method works perfect for water and metabolites close to water. Figure 6.5 shows a practical example of the water–lipid CSDA in proton MRS of brain. When the RF pulse offsets needed to select the localized volume are calculated relative to the water resonance frequency, a significant amount of extracranial lipids is observed (Figure 6.5A). When the volume placement is based on the NAA methyl resonance at 2.01 ppm, the chemical shift displacement for the lipids is much smaller and no significant extracranial lipid resonances are observed (Figure 6.5B). Alternatively, the sign of the slice selection gradient could have been inverted, leading to a CSDA in the opposite direction away from the skull. Note that these solutions to reduce the effect of the CSDA are bandages at best and that the actual CSDA has not changed.

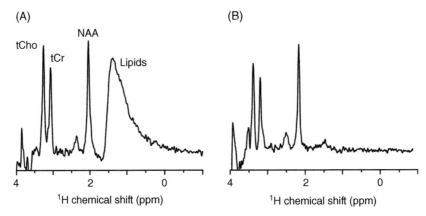

Figure 6.5 **Chemical shift displacement artifacts in localized ^{1}H MRS from newborn piglet brain *in vivo*.** (A) Localized ^{1}H NMR spectrum obtained from a 1.0 ml voxel positioned within the brain close to the skull (Figure 6.1E). RF frequencies were calculated and signal was acquired relative to the water resonance. Because of the circa 700 Hz frequency difference between water and lipids, the chemical shift displacement causes contamination from extracranial lipid signals. (B) ^{1}H NMR spectrum from the same localized volume as in (A), but now calculated and acquired relative to the NAA resonance. The reduced frequency difference between NAA and lipids (circa 100 Hz) resulted in minimal lipid contamination.

As mentioned earlier, the CSDA can only be reduced by increasing the magnetic field gradient in concert with a proportionally increased RF bandwidth. The bandwidth of conventional, amplitude-modulated RF pulses increases linearly with the RF amplitude, leading to a large CSDA as shown in Figure 6.4. The bandwidth of amplitude- and frequency-modulated adiabatic RF pulses increases as the square of the RF amplitude [20], explaining the surge in adiabatic localization methods such as localization by adiabatic selective refocusing (LASER) and semi-LASER which will be discussed in Section 6.2.7. For both conventional and adiabatic RF pulses the limitations related to RF amplitude and RF power deposition are typically encountered well before any gradient strength limitations, especially on human MR systems. This observation has led to the development of gradient-modulated RF pulses, which can achieve high magnetic field gradient amplitude at moderate RF peak amplitude.

Frequency-Offset Corrected Inversion (FOCI) pulses [21–24] were specifically designed to reduce the chemical shift displacement by increasing the effective RF pulse bandwidth through magnetic field gradient modulation. Starting with a regular AFP pulse executed with hyperbolic secant modulation, as described in Section 5.5, a FOCI pulse is created by multiplying the AFP RF, frequency, and gradient modulation by a function $A(t)$ which maximizes the RF amplitude modulation over the range $|t| > (1/\beta)\mathrm{asech}(1/f)$ according to

$$B_1(t) = A(t) B_{1\max} \mathrm{sech}(\beta t) \tag{6.5}$$

$$\Delta v(t) = A(t) \Delta v_{\max} \tanh(\beta t) \tag{6.6}$$

$$G(t) = A(t) G_{\max} \tag{6.7}$$

with

$$A(t) = \cosh(\beta t) \quad \text{for } |t| < \left(\frac{1}{\beta}\right)\mathrm{asech}\left(\frac{1}{f}\right)$$

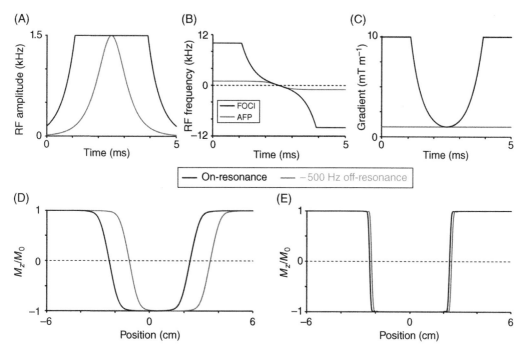

Figure 6.6 Frequency-offset corrected inversion (FOCI) pulses. (A) Amplitude, (B) frequency, and (C) magnetic field gradient modulation of AFP (HS1 modulation, gray lines) and FOCI (black lines) pulses. (D, E) Simulated spatial inversion profile of (D) an AFP pulse ($\Delta\nu_{max} = 1.0\,kHz$, $T = 5\,ms$, $G_{max} = 1.0\,mT\,m^{-1} = 0.426\,kHz\,cm^{-1}$) and (E) a FOCI pulse ($f = 10$) on-resonance (black lines) and $-500\,Hz$ off-resonance (gray lines). The increased frequency modulation range and associated magnetic field gradient amplitude of FOCI greatly reduces the chemical shift displacement.

$$A(t) = f \quad \text{for} \quad \left(\frac{1}{\beta}\right)\text{asech}\left(\frac{1}{f}\right) \le |t| \le 1$$

where $-1 \le t \le 1$. The FOCI factor f is typically set to 10, thereby increasing the RF bandwidth and decreasing the CSDA 10-fold (Figure 6.6D and E). Notice that the maximum RF amplitude of the FOCI pulse is equal to the original AFP pulse, such that the RF peak power requirements are unchanged. However, the area of the FOCI RF waveform is larger than the AFP RF waveform, such that the increased bandwidth comes at the cost of an increased average RF power and thus increased RF power deposition. FOCI pulses are seeing increasing usage, especially at higher magnetic fields. The gradient-offset independent adiabaticity (GOIA) method [25] is an alternative algorithm to design gradient-modulated adiabatic pulses based on the adiabatic condition.

6.2.3 Coherence Selection

Spatial localization relies on the ability to discriminate signals from inside and outside the volume of interest (VOI). Practically this is achieved by making the spin rotations in the VOI unique and different from all other spatial locations. For ISIS localization the VOI was differentiated from all other locations by an eight-step add–subtract scheme used in combination with slice-selective inversion pulses. For the single-shot localization methods (Sections 6.2.4–6.2.7) the separation of desired from undesired magnetization is based on coherence selection using phase cycling or magnetic field gradients. In Chapter 1, coherence was described as a collection of spins that have attained partial phase coherence leading to a macroscopic magnetization vector. Coherence includes detectable transverse magnetization, but also undetectable multiple-quantum-coherence (see Chapter 8). Due to the absence of phase coherence, longitudinal magnetization is generally not considered coherence. Coherence selection is aimed at selecting a particular coherence component, while simultaneously eliminating all others. Coherence selection relies on the fact that various coherences respond differently to RF pulses, RF phases, and magnetic field gradients. Here, the principles of coherence selection based on phase cycling and magnetic field gradients are described. A more in-depth discussion on coherences can be found in Chapter 8 and in the literature [26].

6.2.3.1 Phase Cycling

The aim of phase cycling is to retain signals that experience a certain manipulation (e.g. refocusing) while canceling out signals that do not experience it. A classic example of phase cycling is the removal of imperfections in a 180° refocusing pulse as shown in Figure 6.7. When the refocusing pulse is designed to select a spatial slice (Figure 6.7B), the spins in the sample can undergo a number of different rotations. Within the slice the spins experience 180° refocusing (red), provided that the RF pulse nutation angle is calibrated correctly. Outside the slice, spins do not experience any rotation (green), whereas spins in the transition zone experience partial refocusing and therefore also partial excitation (blue). When the spin-echo sequence of Figure 6.7A is executed without phase cycling, the detected signal can arise from all positions excited by the 90° pulse. In addition, the detected signal is affected by a range of contrasts as the refocused spin-echo signal (red) is T_2-weighted over an echo-time TE, the non-refocused signal (green) is T_2^*-weighted over TE, and the excited signal (blue) is T_2^*-weighted over TE/2. Phase cycling of the 180° refocusing pulse can be used to remove unwanted signals (blue, green) to achieve unambiguous slice selection with uniform T_2 contrast. During phase cycling, the 180° pulse is applied along a different Cartesian axis in subsequent experiments. In Figure 6.7C and D, the 180° pulse was applied along the $+x'$ and $-x'$ axes, respectively, resulting in cancelation of the excited signal (blue) when the two experiments are added together (Figure 6.7E). Note that this two-step phase cycle is not able to remove the non-refocused signal (green). The 180° refocusing pulse can also be applied along the $+y'$ (Figure 6.7F) and $-y'$ (Figure 6.7G) axes, again leading to the removal of the excited signal upon addition (Figure 6.7H). When the four steps are combined by subtracting the result of Figure 6.7E from that in Figure 6.7H, the non-refocused signal (green) is canceled out leaving only the desired, refocused signal (Figure 6.7I). This four-step phase cycle on 180° refocusing pulses is referred to as EXORCYCLE [27]. In practice, addition of signals (and cancelation of unwanted signals) happens naturally as part of signal averaging, whereby the receiver phase is changed by 180° when signal subtraction is required. Phase cycling has traditionally been the method of choice to separate signals that experience different RF pulse manipulations. However, the need for multiple transients (4^n transients for n 180° pulses) is incompatible with the single-scan methods (STEAM, PRESS, and LASER) that are commonly used for *in vivo* MRS. Fortunately, dephasing of unwanted signals by magnetic field gradients is a powerful, single-scan alternative to phase cycling that often provides superior suppression.

6.2.3.2 Magnetic Field Gradients

Figure 6.8 shows a typical implementation of magnetic field gradients to ensure that only signal that is refocused by the 180° pulse is detected. The 180° refocusing pulse is surrounded by two magnetic field gradient pulses of equal area and direction. Signal excited by the 90° pulse is dephased by the first magnetic field gradient (Figure 6.8C) as spins at different positions in the

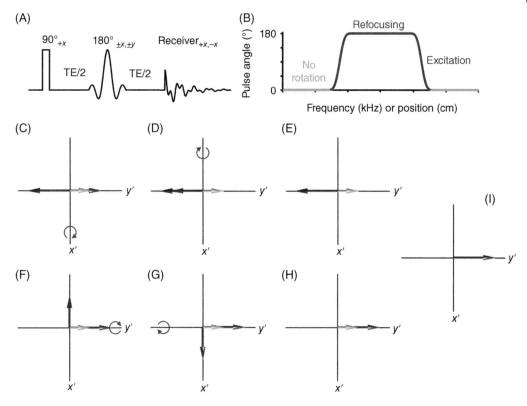

Figure 6.7 **Principle of phase cycling.** (A) Spin-echo pulse sequence. In order to remove signals that do not experience complete refocusing, the phase of the 180° pulse is cycled as to apply the RF pulse along the +*x'*, −*x'*, +*y'*, and −*y'* axes in subsequent scans. In concert with the phase of the 180° pulse, signal reception is performed along the +*x'*, +*x'*, −*x'*, and −*x'*, respectively. (B) Nutation angle profile for the frequency-selective 180° pulse indicates frequency ranges experiencing refocusing (red), excitation (blue), and no rotation (green). When the frequency-selective 180° pulse is executed during a magnetic field gradient the different regions correspond to various spatial positions. (C, D) Magnetization vectors representing the three frequency ranges in (B) immediately after the 180° pulse executed along (C) +*x'* and (D) −*x'*. (E) The sum of (C) and (D) removes signal excited by the 180° pulse (blue). (F, G) Magnetization vectors immediately after the 180° pulse executed along (F) +*y'* and (G) −*y'*. (H) The sum of (F) and (G) again removes signal excited by the 180° pulse (blue). (I) The difference between (E) and (H) (i.e. (F) + (G) − (C) − (D)), as experimentally achieved by phase cycling the receiver, also removes signal not refocused by the 180° pulse (green), while simultaneously retaining the refocused signal (red).

slice experience different magnetic fields. The 180° refocusing pulse rotates the transverse plane around the *y'* axis, thereby inverting the phase accumulated during the first TE/2 period and magnetic field gradient (Figure 6.8D). At the same time spins residing in the slice transition zone (blue, see also Figure 6.7B) are excited long the −*x'* axis. As the phase accumulation during the second magnetic field gradient is equal to that acquired during the first magnetic field gradient, the spins that have experienced a 180° refocusing will have a net phase accumulation of zero, leading to complete signal recovery (Figure 6.8E). The unwanted signal that is excited by the 180° pulse did not experience the first magnetic field gradient and is therefore dephased by the second magnetic field gradient in a single scan. Signal that was excited by the initial 90° pulse and not refocused by the 180° pulse (not shown) will dephase by both magnetic field gradients.

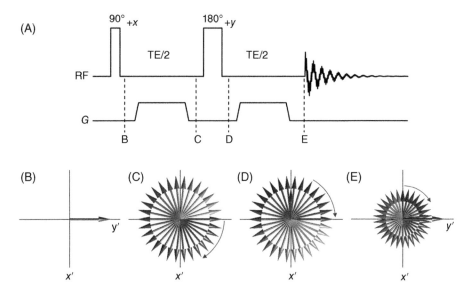

Figure 6.8 **Coherence selection with magnetic field gradients.** (A) Spin-echo pulse sequence in which the 180° refocusing pulse is surrounded by two equal-area magnetic field gradients. (B–E) State of the macroscopic magnetization vector at the time points indicated in (A). At time (C) the magnetic field gradient has dephased all magnetization residing in the transverse plane. Following refocusing (D) and rephasing of the spins by a second magnetic field gradient (E) the desired signal (red vector) can be detected. Signal excited by the 180° RF pulse (blue vector) did not experience the first magnetic field gradient and is therefore dephased by the second magnetic field gradient. Signal excited by the initial 90° RF pulse that does not experience the 180° RF pulse is dephased continuously by T_2^* relaxation as well as by *both* magnetic field gradients (not shown).

It is important to remember that magnetic field gradients *dephase* unwanted signals, rather than eliminate them. This means that as long as the signal has not irreversibly decayed due to T_2 relaxation the unwanted signal can, perhaps unwillingly, be recalled by magnetic field gradients that cancel the signal dephasing. Gradient-recalled rephasing of unwanted signals can become a significant problem in sequences with many RF pulses and gradients. Two simple rules can be used to minimize gradient-recalled rephasing: (i) Placement of magnetic field gradients along orthogonal spatial directions cannot lead to signal rephasing. (ii) When magnetic field gradients must be used along the same spatial direction, a doubling of the gradient area prevents signal rephasing. While these simple rules are a good starting point for gradient placement in multi-pulse MR sequences, other effects such as local, sample-induced magnetic field inhomogeneity gradients (see Section 6.2.5 and Figure 6.15) make the minimization of signal rephasing an empirical process that is best performed *in vivo*. Phase cycling removes unwanted signals through subtraction, thus preventing the possibility of signal refocusing. While phase cycling by itself is generally not sufficient for *in vivo* MRS, phase cycling in conjunction with magnetic field gradient dephasing can be a very effective method to achieve high-quality localization.

6.2.4 STimulated Echo Acquisition Mode (STEAM)

STimulated Echo Acquisition Mode (STEAM) is a localization method [28–34] capable of complete 3D localization in a single acquisition, making it a so-called single-shot or single-scan localization technique. Before describing the STEAM localization technique it is informative to consider the basic pulse sequence (without magnetic field gradients) consisting of three RF

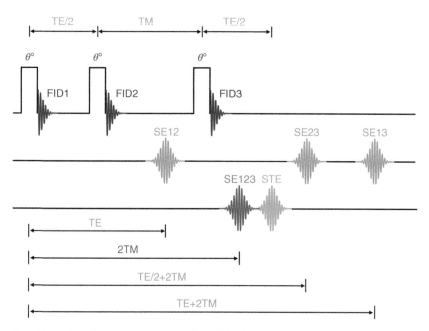

Figure 6.9 Echo formation during a sequence of three RF pulses. In principle, three FIDs (FID1, FID2, and FID3), four spin echoes (SE12, SE13, SE23, and SE123), and one stimulated echo (STE) can be generated. The echoes appear at distinct temporal positions, depending on the TE and TM periods. The relative amplitudes of the echoes are determined by the nutation angles θ of the RF pulses. Only two signals, STE and SE123, are generated through a combination of all three RF pulses and form the target for STEAM and PRESS localization, respectively. A significant part of robust signal localization is the successful removal of the unwanted FIDs and echoes by phase cycling and/or magnetic field gradients.

pulses, each with a nominal nutation angle of 90° (Figure 6.9). As early as 1950, Hahn described that this pulse sequence generates five echoes and three FIDs [35]. When the first two pulses are separated by a delay TE/2 and the last two pulses (i.e. pulses 2 and 3) are separated by a delay TM, then four spin-echoes can be formed. The spin-echo formed by pulses 1 and 2 (SE12) occurs at a position TE after the initial excitation pulse, SE13 at TE + 2TM, SE23 at position TE/2 + 2TM, and S123 (due to refocusing by pulses 2 and 3 of magnetization excited by pulse 1) occurs at position 2TM. Besides these four spin-echoes a so-called stimulated echo (STE) is formed at position TE + TM (or after a delay TE/2 following the last 90° pulse). Figure 6.10 shows an experimental verification of the generated echoes on a sample containing water with TE = 200 ms and TM = 20 ms. The echoes were formed in the presence of a low-amplitude magnetic field gradient present during the entire pulse sequence. The positions of the different echoes depend on the timing parameters TE and TM, while the relative amplitudes depend on the nutation angles of the RF pulses. In most imaging and spectroscopy experiments, the STE is the signal of interest. (The spin-echo SE123 is the signal of interest for PRESS, see next section.) Therefore, the FID and spin-echo signals need to be eliminated in order to obtain unambiguous information from the STE. This can be achieved by phase cycling the individual RF pulses (Figure 6.10B and C), at the expense of losing the single-scan nature of STEAM and increasing its sensitivity to motion and subtraction artifacts. As an alternative, the unwanted FID and spin-echo signals can be eliminated with the use of magnetic field gradients (Figure 6.10D and E). By placing a "crusher" gradient in the TM period, all spin-echoes are eliminated leaving only the STE and a FID component arising from the last excitation pulse (Figure 6.10D). The FID component can easily be eliminated by placing identical "crusher"

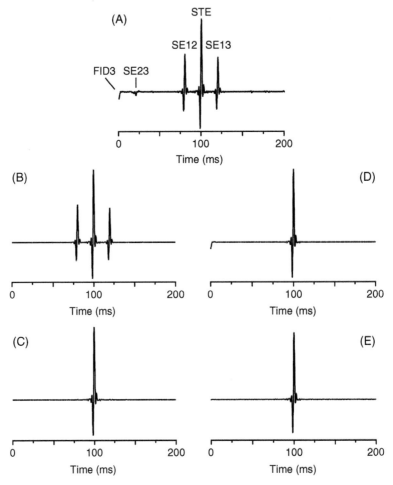

Figure 6.10 Experimental verification of echo formation and destruction during a sequence of three RF pulses on a water-filled sphere. For the particular sequence used, TE = 200 ms and TM = 20 ms with nominal nutation angles of 60°. The echoes were acquired immediately after the last RF pulse in the presence of a low-amplitude magnetic field gradient (which was present during the entire sequence). (A) Echoes without any phase cycling or gradient crushers. Besides the stimulated echo STE, the spin echoes SE12 and SE13 are especially intense. (B) A two-step phase cycle of second and third RF pulses (e.g. 60°$_{+x}$ and 60°$_{-x}$ with constant phase reception) removes the FID3 and SE23 signals. (C) Signal following a complete eight-step phase cycle of all three RF pulses. All unwanted echoes have been canceled, leaving only the STE. (D) Echo formation in the presence of a TM gradient crusher. Besides the desired STE, only an unwanted FID arising from the last RF pulse remains. (E) Complete removal of all unwanted coherences is achieved by a combination of TE and TM crushers in a single acquisition.

gradients in both TE/2 periods. With these additional TE and TM crusher gradients, Figure 6.10E shows that a clean selection of the STE can achieved in a single scan. In the presence of strong B_0 magnetic field inhomogeneity at higher magnetic field strengths, it is possible to observe so-called multiple spin-echoes at multiples of TM [36, 37]. Although these additional spin-echoes can be used for some applications, they are, in general, not desirable. Phase cycling may not be able to remove these unwanted echoes, but magnetic field gradients can still produce a clean selection of the STE.

Using rotation matrices (see Exercises) it can be shown that with these TE and TM crushers only signal arises which was along the longitudinal axis during the TM period (provided that the TE and TM crushers are of sufficient strength to achieve complete dephasing of wanted signals). The amplitude of the STE equals

$$M_{xy} = \frac{1}{2} M_0 \sin\theta_1 \sin\theta_2 \sin\theta_3 e^{-TM/T_1} e^{-TE/T_2} e^{-bD} \tag{6.8}$$

where θ_i (i = 1, 2, or 3) represents the nutation angle of RF pulse i. Equation (6.8) shows that the amplitude of a STE (and of STEAM) is theoretically only 50% of a spin-echo. Later, it will be shown that, in practice, this does not always hold due to the different sensitivities of the localization sequence to relaxation, scalar coupling, and diffusion. Figure 6.11 graphically demonstrates the formation of a STE under the influence of TE and TM magnetic field gradients. Equation (6.8) further shows that longitudinal T_1 relaxation only affects the final STE during the TM period (assuming a long repetition time), while transverse T_2 relaxation only plays a role during the TE period. The last term in Eq. (6.8) describes the effect of diffusion on the STE whereby the diffusion-weighting factor b is described in Chapter 3. The separation between the two diffusion gradients is at least equal to TM, making the diffusion b factor large even when gradient strength G or gradient duration δ are small. Since the magnetization during TM is not influenced by T_2 relaxation, STEs offer a good method of measuring diffusion constants of short T_2 species (for which the longitudinal T_1 relaxation time is relatively long). In Chapter 3 it was discussed how a STE sequence can be used to study diffusion restriction in skeletal muscle by varying the TM period. In this way, the diffusion weighting increases, while the signal loss due to T_2 relaxation remains constant. The response of scalar-coupled spins to the STEAM sequence can be complicated, because the multiple 90° pulses can create multiple quantum coherences and induce polarization transfer effects. These effects and their utilization in spectral editing are described in Chapter 8.

The preceding discussion has been limited to the basic three-pulse STE sequence. The STEAM localization sequence is a direct derivation of the basic pulse sequence in which the nonselective square RF pulses have been replaced by frequency-selective RF pulses in the presence of magnetic field gradients (Figure 6.12). Since transverse magnetization is dephased during part of the excitation pulse length, gradient refocusing lobes are required following slice selection. Improper balancing of the slice selection-refocusing gradients can lead to incorrect echo formation, distorted line shapes, and signal loss. To reduce the minimum attainable echo time, inherently refocused 90° RF pulses can be employed, as shown. However, standard, time-symmetrical 90° RF pulses can be used for longer echo-times. On state-of-the-art preclinical MR systems with fast and strong magnetic field gradients it has been shown that excellent proton NMR spectra can be obtained at TE = 1 ms [38], whereas on human MR systems echo-times of 5–10 ms can be achieved [39].

6.2.5 Point Resolved Spectroscopy (PRESS)

The Point Resolved Spectroscopy (PRESS) localization method [40–43] is a so-called double spin-echo method, in which slice-selective excitation is combined with two slice-selective refocusing pulses (Figure 6.13). When the first 180° pulse is executed after a time TE1/2 following the excitation pulse, a spin-echo is formed at time TE1. The second 180° pulse refocuses this spin-echo during a delay TE2, such that the final spin-echo is formed at the PRESS echo-time TE = TE1 + TE2. The first echo contains signal from a column which is the intersection between the two orthogonal slices selected by the 90° pulse and the first 180° pulse. The second

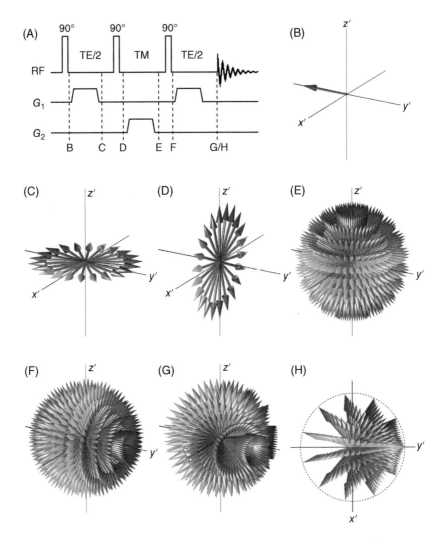

Figure 6.11 Stimulated echo formation by three 90° RF pulses in combination with TE and TM crusher gradients. (A) Stimulated echo pulse sequence. (B–H) State of the macroscopic magnetization vector at the time points indicated in (A). Magnetization following an (B) initial 90°$_x$ excitation pulse, (C) the first TE crusher, and (D) the second 90°$_{+x}$ RF pulse. (E) Application of a TM crusher gradient distributes the magnetization across a 3D sphere, after which (F) the third RF pulse rotates the entire sphere around the +x′ axis. At this point of time, no signal can be detected as the magnetization is completely dephased. (G) In the second TE/2 period, chemical shifts, magnetic field inhomogeneity, and dephasing induced by the first TE crusher are refocused. However, the unbalanced TM crusher prevents a complete refocusing, leading to a 50% signal loss (see also Exercises). (H) Top view of (G) revealing a net macroscopic magnetization along the +y′ axis.

spin-echo only contains signal from the intersection of the three planes selected by the three pulses resulting in the desired volume. Signal outside the VOI is either not excited or not refocused, leading to rapid dephasing of signal by the "TE crusher" magnetic field gradients (see Figure 6.13). PRESS is a popular localization method at lower magnetic fields due to the increased SNR compared to STEAM. However, due to limitations in RF peak power, especially on human MR systems, the bandwidth of conventional 180° refocusing pulses is typically limited to 1–2 kHz. At higher magnetic fields this would lead to unacceptable chemical shift

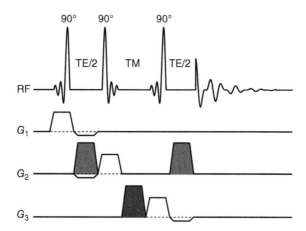

Figure 6.12 **STEAM pulse sequence for 3D spatial localization.** The square RF pulses of Figure 6.11A have been replaced by frequency-selective RF pulses in the presence of magnetic field gradients to select three orthogonal spatial slices. Refocusing gradients are required to rephase the phase evolution during the excitation pulses (see Chapter 5). These gradients may also be positioned with opposite sign on the other side of the TM period. The TE and TM crusher gradients should be of sufficient strength to guarantee the complete removal of unwanted coherences, which is synonymous with accurate localization.

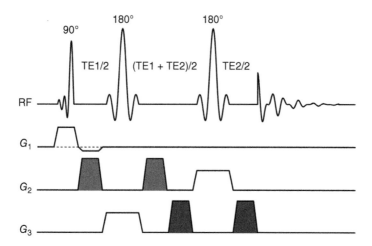

Figure 6.13 **PRESS pulse sequence for 3D spatial localization.** Two pairs of TE crusher gradients (different in direction and/or amplitude) surrounding the 180° refocusing pulses ensure the selection of the desired coherences, while simultaneously destroying all others. Echo formation for spins inside the 3D localized volume occurs at a total echo-time TE = TE1 + TE2.

displacement. Fortunately, the adiabatic LASER and semi-LASER methods provide high bandwidth alternatives to PRESS for single voxel localization at high magnetic fields and will be discussed in Section 6.2.7.

6.2.6 Signal Dephasing with Magnetic Field Gradients

The primary reason for the appearance of spectral artifacts in single voxel localization, such as spurious and variable signals, incomplete water suppression, and lipid contamination, is given by an incomplete dephasing of unwanted coherences. As was shown for a STE pulse sequence

(Figure 6.9), a three RF pulse sequence can generate three FID signals, four spin echoes, and a STE. During a STEAM experiment all coherences, except the STE, must be destroyed (e.g. by magnetic field gradient dephasing), while in a PRESS sequence the spin echo arising from the center of the three intersecting, orthogonal planes must be selected. Dephasing of transverse magnetization with magnetic field gradients is the method of choice for the discrimination between wanted and unwanted signals. However, a careful analysis of STEAM and PRESS shows that not all unwanted signals are dephased by the same amount. Figure 6.14A gives an overview of the amount of dephasing for the seven types of echoes and FIDs encountered in STEAM and PRESS. For example, the FID signal originating from the first 90° pulse is dephased by all magnetic field gradients. The limited crushing capability for the FID originating from the last RF pulse is common to both STEAM and PRESS, whereas the low magnetic field gradient dephasing for the spin echo between RF pulses 2 and 3 is unique to the PRESS sequence. Note that the TM crushing gradient G_3 in STEAM is extremely effective. All unwanted coherences, except the FID from the last pulse, are suppressed. Further note that some unwanted coherences are refocused by inappropriate G_2 and G_3 gradient combinations. For example, PRESS will detect an unwanted STE (STE123) on top of the spin-echo (SE123) when $G_2 = G_3$. This problem is readily alleviated when the crusher gradient pairs are applied along different (orthogonal) spatial axes.

A practical consideration when using magnetic field gradients for signal dephasing or "crushing" is the minimum gradient area that is required to achieve a certain level of signal suppression. The detectable transverse magnetization $M_x + iM_y$ following the application of a magnetic field gradient with amplitude G and duration t is given by

$$M_x(t) + iM_y(t) = \frac{1}{2r_{max}} \int_{-r_{max}}^{+r_{max}} M_0(r)e^{+i\gamma Grt} dr \tag{6.9}$$

where $M_0(r)$ represents the position-dependent spin density. For a uniform sample (i.e. $M_0(r) = 1$) spanning from $-r_{max}$ to $+r_{max}$ the real part of Eq. (6.9) evaluates to

$$M_x(t) = \frac{\sin(\gamma Gr_{max}t)}{\gamma Gr_{max}t} \tag{6.10}$$

with the envelope function given by $1/(\gamma Gr_{max}t)$. Equations (6.9) and (6.10) are shown in Figure 6.14B. A 10 cm sample ($r_{max} = 5$ cm) subjected to a 2 ms magnetic field gradient requires a gradient strength of 37.4 mT m^{-1} to achieve a suppression factor of 1000. The experimental suppression achieved *in vivo* is determined by the externally applied magnetic field gradients, as well as the internal magnetic field distribution. Figure 6.15B shows an axial magnetic field B_0 map through the human brain, displaying areas of poor magnetic field homogeneity with local magnetic field gradients of up to 100 Hz cm^{-1} (Figure 6.15C, red line). Figure 6.15D shows the nine spatial areas that are selected with PRESS localization composed of slice-selective 90° excitation and slice-selective 180° refocusing along orthogonal axes (the slice in the third dimension is not shown). The VOI (Figure 6.15D, red) experiences both RF pulses, whereas four spatial areas experience only one of the two RF pulses (Figure 6.15D, blue) and the remaining volumes do not experience any RF pulse (Figure 6.15D, green). For the slice order shown, the poorly shimmed brain regions are excited by the 90° pulse and need to be dephased by the crusher gradients surrounding the 180° refocusing pulse. When the local magnetic field inhomogeneity opposes the external magnetic field gradients, there may be a point in time where the dephasing of the external and internal fields cancel, leading to signal rephasing. With TE crushers executed as 1 ms 10 mT m^{-1} (=4257 Hz cm^{-1} for protons) gradients, the internal

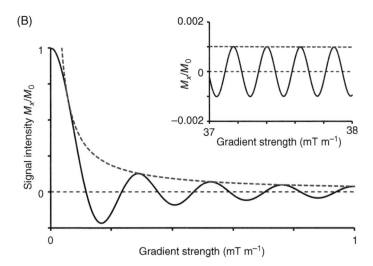

(A) Signal	STEAM	PRESS
FID1	$2G_2 + G_3$	$2G_2 + 2G_3$
FID2	$G_2 + G_3$	$G_2 + 2G_3$
FID3	G_2	G_3
SE12	G_3	$2G_3$
SE13	G_3	$2G_2$
SE23	$G_2 - G_3$	G_2
SE123	$2G_2 - G_3$	0
STE123	0	$G_2 - G_3$

Figure 6.14 **Magnetic field gradient crusher efficiency.** (A) Suppression efficiency of different signals encountered in STEAM and PRESS. The gradient terms refer to those displayed in Figures 6.12 and 6.13 for STEAM and PRESS, respectively. Note that there is strong potential for unwanted signal detection when gradients G_2 and G_3 are along the same spatial direction. For example, when $G_2 = G_3$ the spin-echo SE23 is refocused in addition to the stimulated echo STE in STEAM. Gradient combinations applied along orthogonal axes can never lead to unwanted signal refocusing. (B) Suppression efficiency for a 2.0 ms magnetic field gradient as applied over a 10 cm wide object of uniform density. A magnetic field gradient amplitude of $37.4\,\text{mT}\,\text{m}^{-1} = 3.74\,\text{G}\,\text{cm}^{-1}$ is required to suppress the signal a 1000-fold. Note that the envelope function (gray dotted line) should be used to evaluate the suppression factor, since the rapid oscillations of the signal (black solid line) are highly dependent on sample geometry, uniformity, and local magnetic field gradients. Note that a smaller sample needs a proportionally larger gradient crusher amplitude to achieve the same level of suppression.

magnetic field inhomogeneity gradient of $-100\,\text{Hz}\,\text{cm}^{-1}$ will achieve complete cancelation at 85 ms, possibly leading to the appearance of unwanted (water) signal in the acquisition window. Figure 6.15F and G shows a practical example of this phenomenon on human brain. The appearance of a significant amount of refocused water between 100 and 250 ms will result in the appearance of "ghosting" artifacts in the spectrum (Figure 6.15G). In this case the ghosting

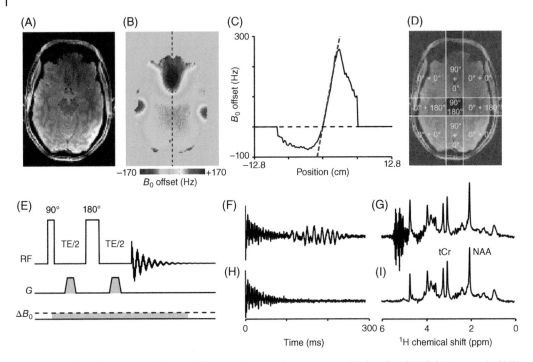

Figure 6.15 **Spatial origins of different FID and echo signals encountered in localized MRS.** (A) Anatomical MRI and (B) magnetic field B_0 map from the human head. (C) 1D B_0 map trace in the anterior–posterior direction (along black dotted line in (B)) reveals local magnetic field gradients in the frontal cortex of up to 100 Hz cm^{-1} (red dotted line). (D) Spatial distribution of nutation angles arising from the first two pulses in a PRESS sequence. For example, the SE123 signal corresponds to the voxel location at the intersection of the three planes (red area), where the FID1 signal originates from the blue areas labeled 90° + 0°. (E) Spin-echo sequence with TE crusher gradients in the presence of a negative local magnetic field inhomogeneity gradient ΔB_0. Frequency-selective RF shapes and slice-selection gradients have been omitted for simplicity. When the local gradient opposes the TE crusher gradients there will be a time when unwanted signals such as FID1 and FID2 are insufficiently dephased, leading to (F) delayed echoes in the time domain and (G) spurious signals in the frequency domain. Refocusing of only a small amount of water or lipids can quickly overwhelm the NMR spectrum. (H, I) Increasing the TE crusher amplitude, changing the TE crusher sign, or changing the slice order and orientation can all be used to prevent unwanted signal refocusing.

signals appear mainly downfield from water. For other volumes and subjects the ghosting signal can also appear upfield, thereby obscuring the metabolite signals [44, 45]. A straightforward method to remove water ghosting signals is to increase the crusher gradient strength, as to push the temporal point of water refocusing outside the acquisition window. However, especially on human MR systems the magnetic field gradient strength may not be sufficient to completely remove the ghosting artifact. In that case, the slicing order may be changed to achieve improved dephasing of a certain area. In the case of Figure 6.15D a reversal of the 90° and 180° slicing order would lead to improved performance, since the inhomogeneous frontal area is not excited. For *in vivo* ^{1}H MRS studies on the human brain, it is important to recognize magnetic field inhomogeneity and motion artifacts due to swallowing, primarily in sagittal (between both eyes, or slices from left to right) and coronal (slices from the back to the front) slice orientations. Water in the sinuses and mouth are another source for artifacts in localized proton MRS. In transverse or axial (through both eyes, or slices from top to bottom) slices through the brain, internal motion is in general minimal. Note that a transverse orientation in animal studies is in the direction of a human coronal orientation and vice versa, due to the different positioning of animals in the magnet. Unwanted signal rephasing can also be improved by changing the sign and orientation of the magnetic field crusher gradients, such that the external dephasing works in concert with the internal gradients.

Phase rotation is an elegant method to remove spurious signals in the absence of strong magnetic field gradients [46–48]. The technique relies on the fact that spins in the VOI respond differently to RF pulse phase changes than spins outside the VOI. By systematically incrementing the phase of the RF pulses and storing the data separately, a 2D FT can separate the VOI signal from all other signals as an apparent frequency along the second dimension. Phase rotation has been shown to provide excellent localization for STEAM and PRESS at short TE. While it is not a popular technique, it is well described and characterized and may prove important when magnetic field gradients provide an insufficient level of signal dephasing.

It should be emphasized that crusher gradients can typically not be optimized on phantoms due to the absence of internal magnetic field gradients of correct amplitude and orientation. In addition, the T_2 relaxation of undoped (pure) water is typically so long that unwanted echoes can form that would never be observed *in vivo*. It is recommended that crusher gradients are optimized *in vivo* and that caution should be used with any modification that could modify internal magnetic field gradients, such as head rotation and the presence of surgical clips or prosthetics.

A method that allows convenient visualization of the localized volume is shown in Figure 6.16. Any MRS pulse sequence can be extended with MR imaging gradients as shown for PRESS

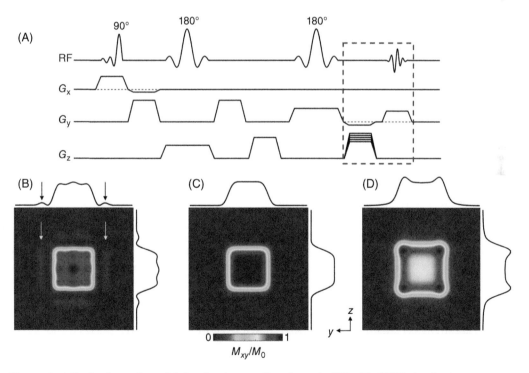

Figure 6.16 **Evaluation of spatial localization quality through MRI.** (A) PRESS localization sequence extended with frequency- and phase-encoding MRI gradients (red dotted box). (B) Image of the localized volume selected by the two 180° refocusing pulses in PRESS when executed with five-lobe sinc pulses (see Chapter 5). The imperfect frequency profile of sinc-refocusing pulses translates directly into a suboptimal voxel shape, characterized by out-of-volume signal refocusing (arrows). (C, D) Image of a PRESS localized volume when executed with Shinnar–Le Roux (SLR)-refocusing pulses (see Chapter 5) calibrated (C) to give 180° refocusing on-resonance and (D) at a 3 dB (~30% lower nutation angle) lower power setting.

localization (Figure 6.16A). The resulting MR image readily reveals imperfections in the localized volume due to, for example, the use of non-optimized RF pulses (Figure 6.16B). Replacing the refocusing pulses with SLR-optimized RF pulses (see Chapter 5) reveals greatly improved localization accuracy (Figure 6.16C). However, the spatial profile is highly sensitive towards incorrect RF power settings (Figure 6.16D). Extending the localization method with MR imaging gradients can also provide a visual means to localize the origin of insufficiently dephased signal, such as shown in Figure 6.15F. The acquisition of a full 2D MR image is often not required and a 1D "readout" profile can be used to get a first impression on the quality of spatial localization. Another strategy to track the origins of unwanted signal is to "turn off" certain RF pulses in a given pulse sequence. For example, the power of the excitation pulse in a PRESS sequence can be reduced to zero. When an artifactual signal persists, it is an indication that the signal has been excited by an imperfect 180° pulse.

From Figure 6.16C and D it is clear that the localization quality of spatial localization methods based on conventional, amplitude-modulated RF pulses (STEAM, PRESS, SPECIAL) has a dependence on the transmit RF field amplitude. The RF amplitude is typically calibrated using MRI-based B_1 mapping (see Chapter 4). For MRS applications, the time-consuming B_1 mapping can be replaced with a fast, two-point measurement as shown for Bloch–Siegert (BS)-based B_1 mapping using STEAM localization (Figure 6.17A). The measurement is limited to the acquisition of two water spectra (Figure 6.17B) that have been sensitized to provide a B_1-dependent BS shift (see Chapter 4). The spectroscopic BS measurement is robust towards the RF power setting of the BS pulses (Figure 6.17C), towards the nutation angle of the STEAM sequence (Figure 6.17D) and towards magnetic field offsets as encountered in regions of low magnetic field homogeneity (Figure 6.17E). Overall the spectroscopic BS method is sufficiently robust to be used before shimming and before global RF power adjustments, thereby providing reliable estimates of the local RF transmit field.

6.2.7 Localization by Adiabatic Selective Refocusing (LASER)

LASER is a single-scan 3D localization method executed with adiabatic excitation and refocusing RF pulses. It is a modification of the previously described SADLOVE technique [49] and is based on the well-known principle of frequency-selective refocusing with pairs of AFP pulses [50] as described in Section 5.5. The basic LASER sequence is depicted in Figure 6.18A. In short, a single AFP pulse in the presence of a magnetic field gradient can achieve slice-selective refocusing. However, the frequency-modulation of the pulse induces a nonlinear B_1 and position-dependent phase across the slice which will lead to severe signal cancelation. In Chapter 5 it was shown that a second, identical AFP pulse can refocus the nonlinear phase such that perfect refocusing can be achieved. Since frequency-selective adiabatic excitation pulses remain elusive, the entire sample is excited with a nonselective adiabatic excitation pulse after which three pairs of AFP pulses achieve 3D localization by selectively refocusing three orthogonal slices. The advantages of the LASER technique over STEAM and PRESS are twofold, namely (i) the method is completely adiabatic and (ii) by employing high-bandwidth AFP pulses the localization can be extremely well defined, both in terms of minimal chemical shift displacement, as well as sharpness of the localization edges (Figure 6.18C–E). The localization can be further improved by using FOCI pulses for refocusing [23]. The combination of strong magnetic field gradients and efficient RF coils on preclinical scanners allows LASER to be executed with short echo-times on the order of 10–15 ms. The lower available RF amplitudes on human clinical scanners together with strict limits on RF power deposition lengthen the minimum echo-time of LASER to more than 50 ms. In order to reduce the minimum echo-time and RF deposition, the semi-LASER method [24, 51, 52] has been proposed in which one pair of AFP pulses is

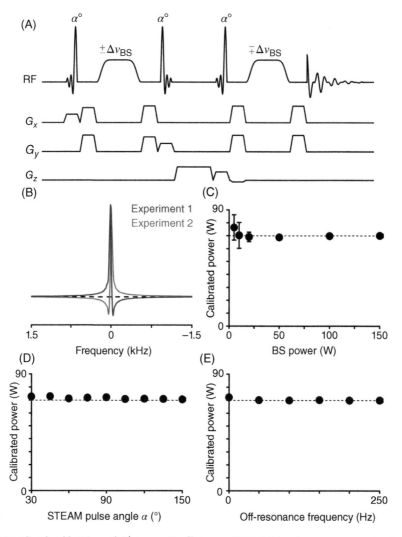

Figure 6.17 Localized calibration of B_1^+ transmit efficiency. (A) STEAM pulse sequence extended with two Fermi pulses to induce a B_1^+-dependent Bloch–Siegert (BS) shift. The BS pulses before and after the TM period are transmitted at opposite off-resonance frequencies (±2.0 kHz) in order to preserve the BS phase shift. (B) Localized water spectra acquired during two scans in which the frequency order of the two BS pulses in (A) were reversed. The phase difference between the two scans is directly proportional to the local B_1^+ transmit field according to Eq. (4.25). (C) Calibrated RF power (in W for a 0.5 ms 90° pulse) as a function of the BS pulse power. While for very low powers the BS phase is small and dominated by noise, the calibration power quickly converges to the "ground truth" value of 70 W. (D) Calibrated RF power as a function of the STEAM nutation angle α. The calibrated power determination remains robust despite an eightfold variation in S/N. (E) Calibrated RF power as a function of the water off-resonance frequency. While for the 250 Hz data point the BS pulses are applied at −1.75 kHz and +2.25 kHz, the BS phase difference and calculated B_1^+ transmit efficiency are still accurate. Note that in (D) and (E) the error on the calibrated power falls within the round extent of each data point.

replaced with a single slice-selective excitation pulse (Figure 6.18B). The sequence can achieve an echo time of 30 ms at modest maximum RF amplitude (23 µT) at 7 T [53]. Semi-LASER has demonstrated excellent test–retest reproducibility within and between different MR groups. As a result, the semi-LASER sequence is quickly gaining popularity as a robust, clinical localization method for quantitative ^{1}H MRS.

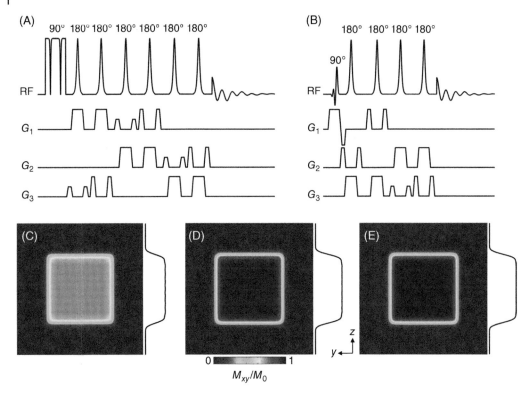

Figure 6.18 3D spatial localization based on adiabatic refocusing. (A) LASER and (B) semi-LASER pulse sequences. TE crushers are placed along orthogonal axes or in a 1 : 2 ratio to avoid refocusing of unwanted signals. (C–E) MR images of the localized volume selected by two pairs of 180° refocusing pulses at three different RF power settings. (D) AFP pulses are executed at the minimum RF power level needed to provide refocusing. (C, E) AFP pulses are executed at (C) a 3 dB lower power setting (~30% lower) and at (E) a 6 dB higher power setting (50% higher) than in (D). The voxel localization is essentially independent of the RF amplitude above the minimum RF threshold. Note that at lower RF power levels the voxel definition is still excellent, albeit at a reduced intensity.

At first sight LASER may not appear suitable for the detection of scalar-coupled spin-systems due to the requirement of six AFP-refocusing pulses. Scalar coupling evolution during a regular spin-echo can lead to significant signal loss for strongly-coupled spin-systems like glutamate and glutamine even for modest echo-times (10–40 ms). Fortunately, LASER is not a regular spin-echo method, but is closer related to the Carr–Purcell–Meiboom–Gill (CPMG) multiple-spin-echo pulse train [54, 55]. It is well known that evolution due to scalar coupling during a CPMG pulse train can be inhibited [56, 57] when

$$\sqrt{\Delta\nu_{AB}^2 + J_{AB}^2}\, TE_{CPMG} \ll 1 \tag{6.11}$$

where $\Delta\nu_{AB}$ and J_{AB} are the frequency separation and scalar coupling constant between spins A and B, respectively, and TE_{CPMG} is the echo-time surrounding one 180° pulse, which equals TE/6 for LASER as shown in Figure 6.18A. Equation (6.11) cannot be readily satisfied for weakly-coupled spin-systems like lactate, since the large frequency separation between lactate-H2 and H3 demands an unrealistically short echo-time. However, for strongly-coupled spin-systems like glutamate, even modest LASER echo-times up to 40 ms satisfy the condition set by Eq. (6.11) leading to greatly enhanced signal detection. Figure 6.19 compares glutamate

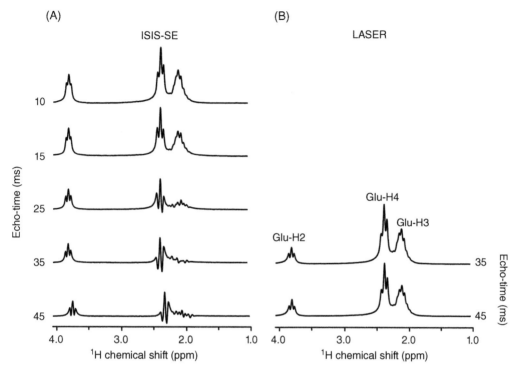

Figure 6.19 **Scalar coupling evolution of glutamate as a function of echo-time.** (A, B) ^{1}H NMR spectra of glutamate acquired with (A) a regular spin-echo (SE) sequence preceded by 3D ISIS localization and (B) a 3D LASER sequence. While the minimum echo-time for LASER is much longer than for ISIS-SE, the scalar coupling evolution for a strongly-coupled spin-system like the glutamate H3 and H4 protons is greatly reduced due to the CPMG character of the sequence.

detection with LASER and a conventional spin-echo method for increasing echo-times. While the minimum LASER echo-time on this particular system was limited to 35 ms (Figure 6.19B), the signal intensity for glutamate-H3 and H4 corresponds to that obtained at TE = 10 ms during a regular spin-echo acquisition (Figure 6.19A). Increasing the LASER echo-time to 45 ms has minimal effects on the spectral appearance, while the spin-echo glutamate intensity has essentially reduced to zero. Therefore, despite the increased echo-time, LASER is an excellent method to detect strongly-coupled spin-systems like glutamate and glutamine.

6.3 Water Suppression

Water is the most abundant compound in mammalian tissue and as a result the proton NMR spectrum of almost all tissues is dominated by a resonance at circa 4.65 ppm originating from the two protons of water (e.g. Figure 6.20A). In some tissues, like muscle and breast, the second-most abundant compounds are a wide variety of lipids which give rise to multiple resonances in the 1–2 ppm range (see Chapter 2). The concentration of metabolites is often >10 000 lower than that of water. Modern 16-bit analog-to-digital converters (ADCs, see Chapter 10) are able to adequately digitize the low metabolite resonances in the presence of a large water resonance without degrading the metabolite signal-to-noise ratio (SNR, Figure 6.20B). However, the presence of a large water resonance leads to baseline distortions and spurious signals due to

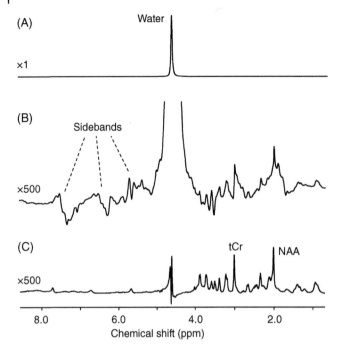

(A)

Water

×1

(B)

Sidebands

×500

(C)

×500

tCr

NAA

8.0   6.0   4.0   2.0

Chemical shift (ppm)

Figure 6.20 Demonstration of the necessity of water suppression *in vivo*. (A) A 3D localized ^{1}H NMR acquired from brain is dominated by the singlet resonance of water. (B) While the dynamic range of modern AD converters is sufficient to detect small metabolite signals in the presence of a large water signal, small (typically <1%) vibration-induced sidebands of the water obscure the metabolite resonances. (C) Removal of the water resonance (and consequently any associated sidebands) results in an artifact-free ^{1}H NMR spectrum that allows reliable detection and quantification of metabolites.

vibration-induced signal modulation (Figure 6.20B) which in turn make the detection of metabolites unreliable [58–60]. Suppression of the water signal eliminates baseline distortions and spurious signals, leading to a reliable and consistent detection of metabolite spectra (Figure 6.20C).

The suppression of a particular resonance in an NMR spectrum requires a difference in a property between the molecule of interest and the compound that is interfering with detection (i.e. water). This property need not be a magnetic one, but it must be reflected in a NMR observable parameter. Thus, besides differences in chemical shift, scalar coupling, and T_1 or T_2 relaxation, properties such as diffusion and exchange can also be exploited. Although there is not a universal technique there are some criteria by which the existing water suppression methods can be evaluated. These criteria include: (i) degree of suppression; (ii) insensitivity to RF imperfections (inhomogeneity); (iii) ease of phasing the spectra; (iv) insensitivity to relaxation effects; (v) perturbation of other resonances; and (vi) detection of resonances near or at the water resonance frequency.

The existing water suppression techniques can be divided into four groups, namely: (i) methods that employ frequency-selective excitation and/or refocusing; or (ii) utilize differences in relaxation parameters; (iii) spectral editing methods like polarization transfer; and (iv) other methods, including software-based water suppression. These four groups will be discussed in detail in the next paragraphs.

6.3.1 Binomial and Related Pulse Sequences

One of the most obvious choices for water suppression is to utilize the difference in chemical shift (resonance frequency) between water and the other resonances. These methods are most widely used and in general they satisfy most of the above mentioned criteria, often providing suppression factors of 1000-fold or more. However, criteria (vi) can never be fulfilled, i.e. since there is no chemical shift difference, resonances at or near the water resonance will also be suppressed.

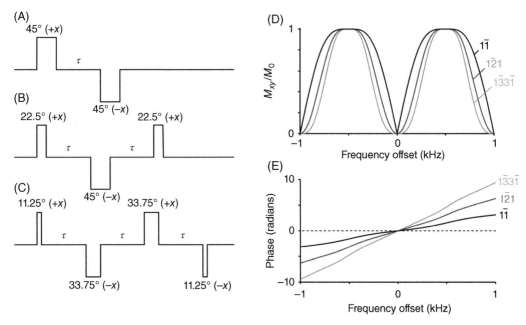

Figure 6.21 **Binomial water suppression sequences.** (A) First-, (B) second-, and (C) third-order binomial water suppression sequences where the nutation angles of the individual RF pulses are scaled according to a binomial distribution. In shorthand notation, the pulses are indicated by $1\bar{1}$, $1\bar{2}1$, and $1\bar{3}3\bar{1}$ where the sum of the numbers represents the total nutation angle $1/(2\tau)$ Hz off-resonance. The bar represents a phase inversion such that the on-resonance nutation angle is always zero. (D) Excitation profile and (E) phase profile of the first-, second-, and third-order binomial sequences ($\tau = 1$ ms).

Among the first water suppression methods to be used in FT NMR spectroscopy are the sequences of short high-amplitude RF pulses interleaved with delays to achieve frequency-selective excitation [61–67]. The sequences are designed in such a manner that metabolite resonances are rotated to the transverse plane for detection, while the water magnetization is returned to the longitudinal axis at the end of the sequence, making it unobservable. The pulse trains of which the nutation angles are given by a binomial expansion [62, 64, 66, 67] were among the first described and Figure 6.21A–C shows the first three binomial RF pulse trains, together with their excitation amplitude (Figure 6.21D) and phase behavior (Figure 6.21E). In all cases the sum of the nutation angles adds up to zero, whereas the sum of the absolute-valued nutation angles add up to 90° for an excitation binomial pulse. For an inversion or refocusing binomial pulse the nutation angles in Figure 6.21A–C need be doubled, thus adding up to 180° in absolute value. On-resonance magnetization does not precess in the rotating frame during the delay τ, such that the total nutation angle is 0°. Magnetization that is off-resonance by an amount $1/(2\tau)$ acquires a phase of 180° during each delay τ, such that the total nutation angle adds up to 90° for excitation or 180° for inversion/refocusing. In other words, on-resonance spins (from water) are not excited, whereas spins off-resonance by $1/(2\tau)$ are fully excited, inverted, or refocused. Spins with other frequencies are excited by a smaller amount according to the curves shown in Figure 6.21D. The curves are described by $M_{xy}(\nu) = M_0\sin^n(\pi\nu\tau)$, whereby n is a positive, integer number representing the binomial order. It follows that for increasing n the suppression frequency range becomes broader (better water suppression) but comes at the price of a narrower frequency range experiencing full excitation (less excitation of metabolites). An additional, noteworthy feature of binomial excitation pulses is the

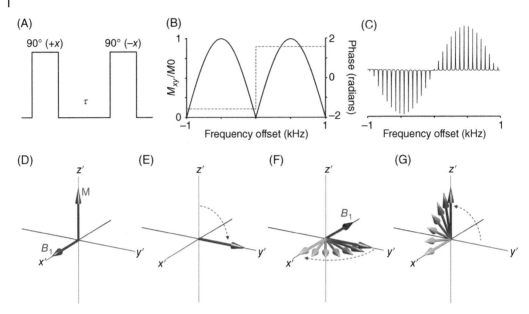

Figure 6.22 **Jump-return water suppression method.** (A) Jump-return sequence providing maximum excitation 1/(4τ) Hz off-resonance. (B) Simulated and (C) experimental excitation profile of a jump-return sequence with τ =0.5 ms. The phase is constant across the excitation profile, in stark contrast to the linear phase encountered with binomial sequences (Figure 6.21E). (D–G) Rotations of a macroscopic magnetization vector (D) immediately before and (E) after the first 90° RF pulse. (F) During the delay τ, spins acquire a phase proportional to their Larmor frequency, such that on-resonance spins (red) remain along the *y′* axis. (G) Distribution of spins following the second 90° RF pulse, showing full excitation of spins 1/(4τ) Hz off-resonance (green) and no excitation for on-resonance spins (red). Spins at intermediate frequencies are excited according to a sinusoidal function with a constant phase along the +*x′* axis.

(approximately) linear phase across the profile (Fig. 6.21E) which can be removed by incorporating half of the binomial pulse duration in the first TE/2 period of a spin-echo sequence.

The jump-return (JR) pulse sequence ([63, 65], Figure 6.22A) is closely related to binomial RF pulses and consists of 90°(+*x*) − τ − 90°(−*x*), in which τ = 1/(4Δ*ν*) for complete excitation Δ*ν* Hertz off-resonance. The simulated (Figure 6.22B) and experimental (Figure 6.22C) excitation profiles highlight the main advantage of the JR sequence over the binomial pulse trains in that the phase is virtually constant across the excitation profile (except from a phase inversion on-resonance). Figure 6.22D–G shows the rotations of magnetization at different frequencies during a JR sequence. Since the final 90° pulse rotates the entire *xy* plane into the *xz* plane (Figure 6.22G), the phase is constant for all frequency offsets. The JR sequence is readily converted to an adiabatic version, known as a Solvent Suppression Adiabatic Pulse (SSAP, [14, 68]).

All binomial and JR pulse sequences are shown under the assumption of instantaneous rotation by infinitely short RF pulses. In reality, the maximum RF amplitude is limited such that the RF pulse duration becomes finite. In order to attain fidelity of the excitation profile, the interpulse delay τ should be shortened according to Eq. (5.13) to take phase accumulation during the RF pulse into account (see Exercises). Binomial RF pulses provide very robust water suppression independent of T_1 relaxation and RF inhomogeneity and have been successfully used in a range of applications. However, their semi-selective excitation profile is often deemed undesirable such that their use is overshadowed by frequency-selective methods that do not perturb the metabolite signals.

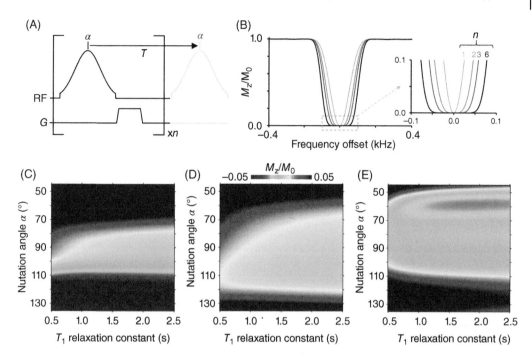

Figure 6.23 Water suppression through frequency-selective excitation. (A) General water suppression element consisting of a frequency-selective RF pulse of nutation angle α followed by a magnetic field crusher gradient. The element is often repeated n times with an interpulse delay T. (B) Suppression profiles of a 10 ms Gaussian pulse (10% truncation level, see Chapter 5) repeated 1, 2, 3, or 6 times in the absence of T_1 relaxation. (C–E) Water suppression efficiency for a range of nutation angles α and T_1 relaxation times for (C, D) CHESS executed with (C) three and (D) six RF pulses ($T = 12.5$ ms) and (E) WET executed with four RF pulses with relative angles α, 1.25α, 0.85α, and 1.98α ($T = 60$ ms).

6.3.2 Frequency-Selective Excitation

Frequency-selective, shaped RF pulses were discussed in Chapter 5 and their use in combination with magnetic field gradients has been demonstrated for slice selection in MRI and volume selection in MRS. The frequency-selectivity of shaped RF pulses is readily utilized in water suppression through the pulse sequence element shown in Figure 6.23A. In essence, the selective RF pulse excites the water onto the transverse plane after which all coherences are dephased by the following magnetic field gradient. This sequence was originally described by Haase et al. [69] and was named chemical shift selective (CHESS) water suppression. A similar sequence was described by Doddrell and coworkers [70] given the name suppression by mistimed echo and repetitive gradient episodes (SUBMERGE). CHESS can precede any pulse sequence, since it leaves the metabolite resonances unperturbed. Most often, Gaussian-shaped RF pulses are used for frequency-selective excitation because of their favorable bandwidth–pulse length product and the well-defined frequency profile. The suppression efficiency of CHESS clearly depends on the ability of the RF pulse to reduce the longitudinal magnetization to zero for all water spins in the ROI. The main factors that complicate this requirement are (i) signal recovery due to T_1 relaxation, (ii) incomplete signal excitation due to inhomogeneity in the transmit magnetic field, and (iii) incomplete signal excitation due to a spread of frequencies in the water resonance. These three complications are typically minimized by repeating the CHESS element multiple times. Figure 6.23B shows that repeated application of a CHESS element broadens the frequency range over which suppression is achieved. Traditionally, CHESS has been executed with nominal 90° excitation pulse and repeated three times to decrease the sensitivity towards T_1 relaxation and B_1 magnetic field inhomogeneity (Figure 6.23C). However, T_1 relaxation between the elements leads to significant signal recovery, especially when the excitation pulse deviates from 90°. Extending the three-element CHESS to a six-element CHESS sequence does improve the immunity towards imperfections (Figure 6.23D). However, since the residual magnetization is positive for all (T_1, B_1) combinations, the experimentally achievable water suppression may be worse than for the three-element CHESS method. A more systematic approach to improve CHESS water suppression is provided by the water suppression enhanced through T_1 effects (WET) method [71]. Figure 6.23E shows the performance of a four-element WET sequence, whereby the nutation angle of the four RF pulses are optimized as to compensate for T_1 relaxation in between the pulses. The performance of a four-element WET method greatly exceeds that of a standard three-element CHESS sequence (Figure 6.23C) and even challenges that of a six-element CHESS sequence (Figure 6.23D). In Section 6.3.4 it will be shown that rather than being a complication, T_1 relaxation can be utilized as an integral part of a water suppression method to achieve near-perfect removal of the water resonance.

Similar to spatial localization sequences, the presence of multiple RF pulses interspersed with magnetic field gradients generates the potential for unwanted echoes due to unintended signal rephasing. The most common way to minimize unwanted echoes is to place magnetic field gradients along orthogonal axes where possible. However, with more than three CHESS pulses multiple magnetic field gradients will appear along the same axes. In that case, unwanted echoes can be eliminated by increasing the gradient amplitude by a factor of two during subsequent CHESS elements (i.e. applying gradients in a 1 : 2 : 4 : 8 ratio, whereby 1 represents the gradient area for complete dephasing) [72].

When the selective RF pulse shown in Figure 6.23A is a long (in the order of T_1), low-power, constant amplitude irradiation field, the sequence element corresponds to presaturation [73]. During RF presaturation the field is continuously applied at the water frequency in order to reach a (hypothetical) situation of equal spin populations (i.e. completely dephase the water magnetization in a plane perpendicular to the excitation axis). In practice, complete saturation cannot be obtained, since saturation is competing with T_1 relaxation. The best results are obtained when irradiation is maintained for $2-3T_{1water}$. In practice, the saturation pulse length is chosen, after which the pulse power is optimized with respect to optimal water suppression (for a given repetition time TR). Presaturation is a popular water suppression technique in high-resolution liquid-state NMR and it has received some attention in early *in vivo* NMR studies. This was mainly governed by the ease of implementation and reasonable suppression factor (~1000) of presaturation. However, the disadvantages of presaturation have dramatically reduced its use for *in vivo* MRS. Excessive heat deposition is a consideration, but presaturation may also obscure actual metabolite concentrations by off-resonance magnetization transfer between NMR invisible ("bound" or immobile) metabolites and NMR visible, mobile metabolites (see Chapter 3). Furthermore, presaturation does, in general, not allow (or at least complicates) the observation of exchangeable protons. Finally, presaturation requires a high B_0 magnetic homogeneity, although this is a requirement for all chemical shift-based water suppression techniques.

The absence of a frequency-selective adiabatic excitation pulse has prevented a direct adiabatic analogue of CHESS. However, de Graaf and Nicolay [74] realized that a frequency-selective inversion, as achieved by a hyperbolic secant AFP pulse, is always accompanied by frequency-selective excitation at the two frequencies where the longitudinal magnetization becomes zero. Since the inversion bandwidth of AFP pulses is constant over a large B_1 range, the points of excitation are essentially adiabatic and can be used in a CHESS-type water suppression sequence, dubbed SWAMP for Selective Water Suppression with Adiabatic-Modulated

Pulses [74]. Using four or six AFP pulses interleaved with magnetic field crusher gradients typically reduces the residual water well below the metabolite resonances, without subject-dependent power adjustments. The insensitivity of SWAMP towards variations in T_1 relaxation can be further improved by adjustment of frequency offsets and inter-pulse delays [75].

6.3.3   Frequency-Selective Refocusing

As an alternative to suppressing the water resonance prior to excitation, suppression can also be achieved during a subsequent spin-echo period. The main advantage of water suppression following excitation is that the suppression efficiency is not degraded by T_1 relaxation. The main drawback of frequency-selective refocusing over excitation methods is that the minimum echo-time is prolonged due to the requirement for frequency-selective 180° pulses. While this prevents short-TE MRS, it does not impede experiments at longer echo-times, like spectral editing (see Chapter 8). Three spin-echo-based methods are WATERGATE [76], excitation sculpting [77], and MEGA [78] of which the basic MEGA sequence is shown in Figure 6.24. In essence, all three methods rely on the fact that the water is selectively dephased, while the metabolites of interest are rephased during the spin-echo period. For instance, consider the MEGA sequence of Figure 6.24A. All resonances are excited by a nonselective 90° pulse. In the transverse plane, the spins experience two orthogonal magnetic field gradients G_1 and G_2. The essential "trick" of all three water suppression techniques is that the water resonance experiences an even number of refocusing pulses during two equal magnetic field gradients, while the metabolites are refocused by only one 180° pulse. This will result in a dephasing of the water by the orthogonal magnetic field gradients, while the metabolites can be observed. A detailed theoretical treatment of these techniques using the product operator formalism is given elsewhere [77, 78]. From these calculations it follows that residual, unsuppressed signal of excitation sculpting and MEGA is proportional to $\cos^4(\theta/2)$, where θ is the nutation angle of the selective-refocusing pulses (Figure 6.24B). Both excitation sculpting and MEGA can be transformed to adiabatic sequences, thereby providing complete B_1-insensitive water suppression. This is achieved by replacing the selective-refocusing pulses with adiabatic full passage pulses and the nonselective pulses with BIR-4 pulses. Figure 6.24C shows an *in vivo* evaluation of the B_1 insensitivity of MEGA water suppression. Variation of the RF amplitude by a factor of >10 has negligible effect on the spectral appearance. The MEGA sequence is quickly becoming a popular method for spectral editing and will be discussed in more detail in Chapter 8.

6.3.4   Relaxation-Based Methods

Techniques exploiting chemical shift differences are currently most often used for *in vivo* water suppression. However, for some specific applications, longitudinal and/or transverse relaxation may be used to discriminate between water and other resonances. When the water and metabolite relaxation times are sufficiently different, it is even possible to suppress the water signal and observe metabolites in the close proximity to the water resonance. Most relaxation-based methods make use of differences in longitudinal T_1 relaxation. Water eliminated Fourier transform spectrum (WEFT) is among the oldest T_1-based water suppression methods [79–81]. The pulse sequence of WEFT is identical to an inversion recovery sequence used for T_1 relaxation time measurements. It consists of a nonselective 180° pulse followed by an adjustable delay t and a magnetic field gradient to remove transverse components during t. After the delay t, the magnetization is excited by a nonselective 90° pulse and acquired. The delay t is chosen so that the water magnetization has recovered to its null point (i.e. $t = T_{1,\text{water}} \cdot \ln(2)$), whereas the metabolites have partially recovered. In order to ensure near-complete recovery of

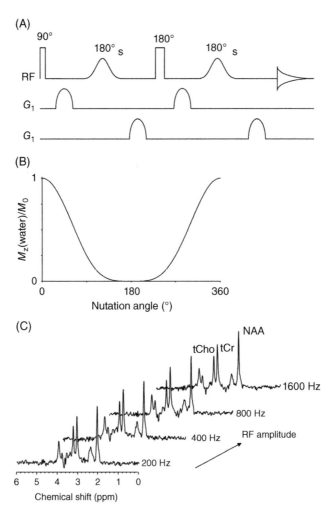

Figure 6.24 Water suppression through frequency-selective refocusing. (A) RF and gradient combinations as used for MEGA water suppression. (B) Residual water signal as a function of nutation angle for conventional frequency-selective RF pulses. (C) ^{1}H NMR spectra obtained from rat brain *in vivo* (TR/TE = 3000/144 ms, 125 µl) with MEGA water suppression executed with adiabatic full passage pulses. The water suppression remains excellent despite an eightfold variation in the RF amplitude.

the magnetization, the repetition time TR between successive excitations needs to be $4-5T_{1,water}$. Unfortunately, this would lead to excessive measurement time, since the *in vivo* T_1 relaxation time of water can be 1.5–2.0 s (see Chapter 3). It is possible to operate in a steady-state mode, such that $TR < 4-5T_1$, making the simple expression for the water null point more complicated and TR dependent. However, the exact dependency need not be known if the delay t is optimized empirically.

A comparable method is the so-called driven equilibrium Fourier transform (DEFT) technique [82–84] given by the sequence $90° - t - 180° - t - 90°$. The magnetization is excited by the first pulse. Longitudinal magnetization that has recovered during the t period is inverted by the 180° pulse. The water resonance can be nulled just prior to the last 90° pulse by proper choice of t. The metabolites are supposed to have relaxed back during $2t$ and will be excited by the last 90° pulse. The advantage of DEFT over WEFT is that the choice of t is less critical, since it is only necessary to remove water that relaxed back during the first t delay. Furthermore, phase cycling during DEFT may improve suppression.

The main disadvantages of WEFT and DEFT relate to the degree of water suppression and metabolite signal recovery. Since the difference in *in vivo* T_1 relaxation time between water and metabolites is small, good water suppression is always accompanied by a severe reduction in sensitivity of the metabolite resonances since the metabolites are incompletely recovered during the delay t. Further, *in vivo* T_1 relaxation times of water are very heterogeneous (see Chapter 3). This results in incomplete water suppression, because the delay t can only be optimized for a single T_1 relaxation time. Finally, T_1-based water suppression makes quantification of metabolite resonances cumbersome. T_1-based signal suppression is more commonly used to suppress lipid resonances in MR spectroscopic imaging, whereby the larger T_1 difference between metabolites and lipids reduces some of the aforementioned complications. Lipid suppression is discussed in detail in Chapter 7.

The performance of all water suppression techniques that rely on selective removal of the water resonance prior to signal excitation is degraded by variations in T_1 relaxation and B_1 transmit homogeneity (Figure 6.23). Several methods have been reported [38, 39, 71] in which the T_1 and B_1 sensitivity of CHESS-based water suppression is decreased by optimizing the nutation angles of subsequent CHESS elements and/or optimizing the inter-pulse delay. One of these methods, variable pulse powers and optimized relaxation delays (VAPOR, [38]), makes T_1 relaxation in between the RF pulses an integral part of the sequence to provide excellent water suppression with a large insensitivity towards T_1 and B_1-inhomogeneity. The VAPOR method is depicted in Figure 6.25A. It consists of seven frequency-selective RF pulses interspersed with optimized T_1 recovery delays. Figure 6.25B shows the trajectories of the longitudinal magnetization under the influence of B_1 magnetic field inhomogeneity and T_1 relaxation. The interplay between repeated excitation and T_1 relaxation gradually decreases the longitudinal magnetization to zero. Figure 6.25C shows the large immunity of VAPOR towards T_1

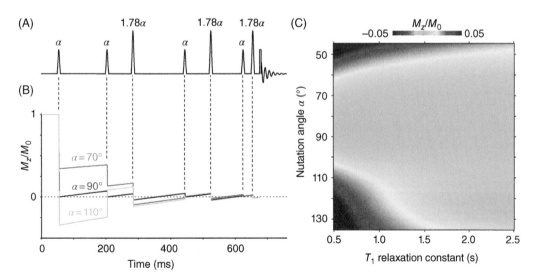

Figure 6.25 **Water suppression with variable pulse powers and optimized relaxation delays (VAPOR).** (A) RF pulse and delay sequence used in seven-element VAPOR. Inter-pulse delays are given by 150, 80, 160, 80, 100, 30, and 26 ms. (B) Trajectories of the longitudinal magnetization M_z for $\alpha = 70°$, 90°, and 110° ($T_1 = 1500$ ms) all end up close to zero at the end of the sequence. (C) VAPOR water suppression efficiency for a range of nutation angles α and T_1 relaxation times. The suppression efficiency can be directly compared to the CHESS and WET methods shown in Figure 6.23.

relaxation and magnetic field B_1 inhomogeneity. Especially, when switching between phantoms and *in vivo* tissues, the VAPOR water suppression can be fine-tuned by adjusting the delay between the final water suppression pulse and the excitation pulse. Recognizing that the residual water signal always increases from the negative z axis during the last delay, the zero crossing delay can be optimized by increasing (for longer T_1s in a phantom) or decreasing (for shorter *in vivo* T_1s) the final delay. An eight-element VAPOR method has been described for increased performance [39]. The long duration of a VAPOR sequence provides ample time for spins to undergo chemical exchange. As a result, resonances of water-exchangeable protons downfield from water are greatly suppressed during VAPOR water suppression. When these signals are the target of investigation, it is advisable to use non-water-suppressed MRS as discussed next.

6.3.5 Non-water-suppressed NMR Spectroscopy

The majority of ^{1}H MRS studies have focused on selectively suppressing the water resonance to unambiguously observe the metabolite resonances. However, detection of the water resonance can be very useful for a number of operations, including (i) internal concentration

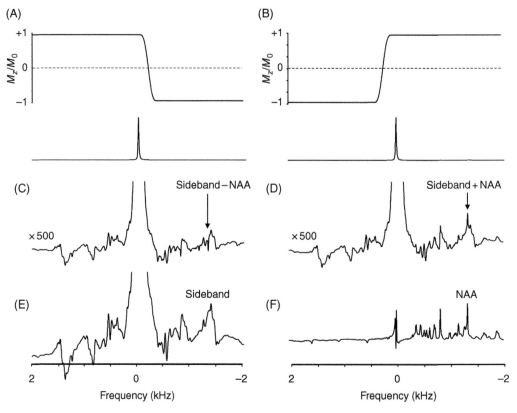

Figure 6.26 Water suppression through metabolite cycling. (A, B) Metabolite-cycled MRS is based on two scans during which the signals (A) upfield and (B) downfield from water are selectively inverted. (C, D) In both non-water-suppressed scans the vibration-related sidebands associated with the water overwhelm the smaller metabolite signals. However, since the water signal and related sidebands are *identical* in both scans they are suppressed in the difference spectrum (F) = (B) − (A), thereby revealing the metabolite resonances. (E) The sum of (A) and (B) provides the water signal and associated sidebands which can be used for B_0 eddy current correction. Since the water resonance is not suppressed during acquisition, signal from water-exchangeable protons in the downfield region of the spectrum are greatly enhanced (see also Figure 2.1).

referencing, (ii) eddy current and vibration correction, (iii) automated phasing, (iv) addition of multiple receivers, and (v) magnetic field drift monitoring and correction.

While arguments concerning limited ADC dynamic range are no longer valid with modern hardware, the primary reason against non-water-suppressed ^{1}H MRS is the interference of strong spectral sidebands (Figure 6.20) arising from the water due to mechanical vibrations [58–60]. Several reports have appeared addressing this problem using 2D approaches [85], sideband symmetry arguments [86], use of long echo-times [58], and hardware solutions to compensate the B_0 effect of the vibrations [60]. However, none of these reports is generally applicable under all experimental conditions. A modification of the method described by Dreher and Leibfritz [87] is depicted in Figure 6.26. It involves the acquisition of two separate scans in which either the upfield (−1.6 to 4.4 ppm) metabolite resonances (Figure 6.26A) or the downfield (5.0 to 11.0 ppm) metabolite resonances (Figure 6.26B) are inverted. The large (unsuppressed) water resonance, as well as the water-related sidebands are not inverted in either scan. The difference between the two scans therefore results in a water-subtracted (or suppressed) metabolite spectrum without any interfering water-related sidebands. The sum of the two acquisitions gives the water and water-related sidebands and can be used for eddy current and vibration correction, phase correction, or internal concentration referencing. The method outlined in Figure 6.26 is typically referred to as water-metabolite or metabolite-cycled MRS. Unlike the methods described in the previous sections, the metabolite-cycled MRS only minimally perturbs the water resonance. As a result, water-exchangeable protons are detected at greatly increased sensitivity, as can be seen in the example spectrum of Figure 2.1.

Exercises

6.1 Consider the PRESS localization method with the following parameters:
TR = 2500 ms, TE = 288 ms (TE1 = TE2 = 144 ms), 2 ms SLR90 ($R = 5.6, B_1(90) = 750$ Hz) excitation, 4 ms sinc180 ($R = 5.8, B_1(90) = 1350$ Hz) refocusing, $B_0 = 3$ T, voxel size = $3 \times 3 \times 3$ cm.

A Calculate the percentage spatial overlap of voxel locations for the NAA methyl (2.01 ppm), creatine methyl (3.03 ppm), and choline methyl (3.22 ppm) resonances. Assume perfectly rectangular slice profiles along all dimensions.

B Given the limitations of the RF and gradient power supplies ($B_{1max} = 2500$ Hz, $G_{max} = 20$ mT m^{-1}), calculate the percentage voxel overlap and pulse lengths that give the highest amount of spatial voxel overlap.

C In the case of lactate (assume two-spin system with $J = 6.9$ Hz) calculate the amount of observable lactate as percentage of the amount that would have been observed in the absence of a chemical shift artifact. Use the RF and gradient combination calculated under A and neglect the chemical shift displacement during excitation. See Chapter 8 for more details on scalar coupling evolution during slice selective RF pulses.

D Repeat the calculation for lactate detection at TE = 576 ms.

6.2 Consider a CHESS-2 sequence prior to a nonselective excitation executed with two 10 ms Gaussian pulses, each followed by a 20 ms crusher gradient. Assume the first pulse is calibrated to give a 90° excitation.

A In the absence of B_1 magnetic field inhomogeneity calculate the optimal nutation angle of the second pulse that gives the best suppression of brain gray matter water ($T_1 = 1250$ ms). Neglect T_1 relaxation during the RF pulses and assume a long repetition time (TR > $5T_1$).

B Calculate the relative water intensity (M_z/M_0) for the optimized sequence calculated under A when the B_1 field is off by –10% and +10% due to a miss calibration or B_1 inhomogeneity.

C For a voxel composed of 60% gray matter (50 M water, T_1 = 1250 ms) and 40% white matter (48 M water, T_1 = 1050 ms) calculate the water resonance intensity relative to the creatine methyl resonance (10 mM, T_1 = 1400 ms) for the optimized sequence calculated under A.

D Repeat the calculation under C for a miss calibration of the B_1 field by –10% and +10%.

6.3 **A** Under the assumption that sufficient inhibition of scalar coupling evolution occurs when $(\Delta\nu^2 + J^2)^{1/2}\text{TE}_{\text{CPMG}} = 0.2$ calculate the maximum TE_{CPMG} for citrate at 1.5 T.

B Repeat the calculation for a magnetic field strength of 7.0 T.

C Repeat the calculation for lactate at 7.0 T.

6.4 **A** Derive an expression for the residual longitudinal magnetization M_z following n CHESS elements during the TR period of a PRESS sequence in the presence of T_1 relaxation. Assume CHESS elements of equal length Δt and $\text{TR} \gg T_1$.

B Derive an expression for the residual longitudinal magnetization M_z following $n1$ and $n2$ CHESS elements during the TR and TM periods of a STEAM sequence in the presence of T_1 relaxation. Assume CHESS elements of lengths Δt, $\text{TR} \gg T_1$, and TM equals the length of $n2$ CHESS elements.

C Calculate the residual signals following the PRESS sequence for CHESS ($n = 3$) nutation angles of 70°, 90°, and 110° in the absence of T_1 relaxation. Repeat the calculation for a T_1 relaxation time of 1500 and 25 ms CHESS elements. Assume $\text{TR} \gg T_1$.

D Calculate the residual signals following the STEAM sequence for CHESS ($n = 3$) nutation angles of 70°, 90°, and 110° in the presence of T_1 relaxation (T_1 = 1500 ms, 25 ms CHESS elements, TM = 75 ms). Repeat the calculation for a TM period of 250 ms. Assume $\text{TR} \gg T_1$.

6.5 The excitation profile for a Gaussian-shaped RF pulse is given by:

$$M_{xy}(\delta) = M_0 e^{-2(\delta - 4.7)^2}$$

where δ is the chemical shift is ppm.

A In the absence of T_1 relaxation, calculate the signal intensity for the total creatine resonance at 3.93 ppm when a single pulse is used for CHESS water suppression followed by nonselective excitation.

B Repeat the calculation when the CHESS element is repeated six times prior to nonselective excitation in order to improve the water suppression.

C Summarize the conclusions of the calculation performed under A and B and explain how finite metabolite T_1 relaxation would affect the results.

6.6 **A** During a comparison between STEAM and LASER localization it is noticed that the localized STEAM spectra display a 10-fold lower SNR. Name at least three possible reasons for the large difference in SNR and suggest means to eliminate them.

B During the development of a novel localization method, the localized volume is visualized by incorporating frequency- and phase-encoding gradient into the sequence. The image shows strong ripples across and outside the localized volume, indicating an inadequate localization performance. However, a spectroscopic experiment on a two-compartment water–lipid phantom yields a ^{1}H NMR spectrum from the water compartment devoid of any lipid contamination. Explain these apparently contradicting results and suggest experimental modifications to test the answer.

C The T_2 relaxation time for glycine in an *in vitro* sample doped with $MnCl_2$ is determined as 150 ± 8 ms. The glycine line width as measured from ^{1}H NMR spectra acquired from $5 \times 5 \times 5$ mm = 125, 64, 27, 8 μl and $1 \times 1 \times 1$ mm = 1 μl volumes are given by 5.0, 3.8, 2.5, 2.6, and 3.6 Hz, respectively. Give an explanation for the variation in spectral line width for the different volume sizes.

6.7 A For a 2D ISIS sequence verify that all signal outside the VOI is canceled at the end of the four required scans.

B Discuss if the presence of short T_1 relaxation times would compromise the localization quality.

6.8 A For a regular spin-echo sequence show that a simple two-step phase cycle on the 180° pulse $(+x, -x)$ in conjunction with a constant receiver phase is sufficient to eliminate imperfections excited by the refocusing pulse (i.e. when the nutation angle deviates from 180°). Assume a perfect 90° nutation angle for the excitation pulse.

C Show that the two-step phase-cycle under (A) is insufficient when the excitation pulse has a nutation angle different from 90°. Suggest an extended phase cycle for the 180° pulse that will be valid for any nutation angle of the excitation pulse.

D Determine the number of phase cycle steps required for a PRESS sequence.

E Show that the placement of two equal magnetic field "crusher" gradients on either side of the 180° pulse is equivalent to the extended phase cycle derived under (B).

6.9 A For a stimulated-echo sequence (without magnetic field gradients) with TE = 20 ms and TM = 200 ms calculate the number of signals that can be expected following the final RF pulse.

B With the insertion of equal-area magnetic field crusher gradients in the two TE/2 and one TM period, calculate the number of signals that can be expected following the final TE crusher gradient.

C Recalculate the answer when the TM crusher gradient is made twice the area of a single TE crusher gradient.

D Discuss the spatial origin of the detected signal obtained with a STEAM sequence (see Figure 6.12) for the TE/TM crusher gradient combinations discussed under B and C.

6.10 A Under the assumption of infinitely short RF pulses, calculate the intra-pulse delay for a JR pulse to achieve excitation at ± 250 Hz off-resonance.

B Calculate the required pulse length when the maximum available B_1 magnetic field strength equals 2.0 kHz.

C Recalculate the intra-pulse delays obtained under (A) to account for chemical shift evolution during the RF pulses.

D Derive an expression for the evolution of a scalar-coupled two-spin-system AX resonating at frequencies ν_A and ν_X immediately following the refocusing pulse during a spin-echo sequence with echo-time TE and a JR-refocusing pulse with "null" and maximum refocusing frequencies of ν_{null} and ν_{max}, respectively. Assume ideal (infinitely short) RF pulses. See chapter 8 for details on scalar coupling evolution.

6.11 **A** Derive an expression for the evolution of the A spins in an A_3X spin-system during a STE sequence with echo time TE and mixing time TM. Assume ideal RF pulses (without slice selection) and complete dephasing by the TE and TM magnetic field "crusher" gradients.

B For lactate at 9.4 T ($^3J_{HH} = 6.9$ Hz) calculate the shortest TM delays which give the minimum and maximum amounts of detectable signal. Consider lactate an A_3X spin-system.

6.12 **A** For a magnetic field "crusher" gradient of 2.0 ms duration, calculate the required amplitude (in mT m^{-1}) to suppress the transverse magnetization by a factor of 100. Assume a uniform sample of length $L = 5$ cm.

B A post-surgery metal clip in the skull generates a local magnetic field disturbance of −45 Hz/mm. During PRESS localization, unwanted signal in that spatial area is dephased by a 2 ms magnetic field crusher gradient of 30 mT/m. Calculate the time following signal dephasing at which unwanted echo formation can occur.

References

1 Hoult, D.I. (1979). Rotating frame zeugmatography. *J. Magn. Reson.* 33: 183–197.

2 Garwood, M., Schleich, T., Ross, B.D. et al. (1985). A modified rotating frame experiment based on a Fourier series window function. Application to *in vivo* spatially localized NMR spectroscopy. *J. Magn. Reson.* 65: 239–251.

3 Garwood, M., Schleich, T., Bendall, M.R., and Pegg, D.T. (1985). Improved Fourier series windows for localization in *in vivo* NMR spectroscopy. *J. Magn. Reson.* 1985: 510–515.

4 Hoult, D.I., Chen, C.N., and Hedges, L.K. (1987). Spatial localization by rotating frame techniques. *Ann. N. Y. Acad. Sci.* 508: 366–375.

5 Blackledge, M.J., Rajagopalan, B., Oberhaensli, R.D. et al. (1987). Quantitative studies of human cardiac metabolism by ^{31}P rotating-frame NMR. *Proc. Natl. Acad. Sci. U. S. A.* 84: 4283–4287.

6 Styles, P., Blackledge, M.J., Moonen, C.T., and Radda, G.K. (1987). Spatially resolved ^{31}P NMR spectroscopy of organs in animal models and man. *Ann. N. Y. Acad. Sci.* 508: 349–359.

7 Garwood, M., Robitaille, P.M., and Ugurbil, K. (1987). Fourier series windows on and off resonance using multiple coils and longitudinal magnetization. *J. Magn. Reson.* 75: 244–260.

8 Styles, P. (1992). Rotating frame spectroscopy and spectroscopic imaging. In: *NMR Basic Principles and Progress* (ed. P. Diehl, E. Fluck, H. Gunther, et al.), 45–66. Berlin: Springer-Verlag.

9 Hendrich, K., Hu, X., Menon, R.S. et al. (1994). Spectroscopic imaging of circular voxels with a two-dimensional Fourier-series window technique. *J. Magn. Reson. B* 105: 225–232.

10 Hoult, D.I. (1980). Rotating frame zeugmatography. *Philos. Trans. R. Soc. Lond. B Biol. Sci.* 289 (1037): 543–547.

11 Ordidge, R.J., Connelly, A., and Lohman, J.A. (1986). Image-selected *in vivo* spectroscopy (ISIS). A new technique for spatially selective NMR spectroscopy. *J. Magn. Reson.* 66: 283–294.

12 Connelly, A., Counsell, C., Lohman, J.A., and Ordidge, R.J. (1988). Outer volume suppressed image related *in vivo* spectroscopy (OSIRIS), a high-sensitivity localization technique. *J. Magn. Reson.* 78: 519–525.

13 Ordidge, R.J. (1987). Random noise selective excitation pulses. *Magn. Reson. Med.* 5: 93–98.

14 de Graaf, R.A., Luo, Y., Terpstra, M. et al. (1995). A new localization method using an adiabatic pulse, BIR-4. *J. Magn. Reson. B* 106: 245–252.

15 de Graaf, R.A., Luo, Y., Garwood, M., and Nicolay, K. (1996). B1-insensitive, single-shot localization and water suppression. *J. Magn. Reson. B* 113 (1): 35–45.

16 Mlynarik, V., Gambarota, G., Frenkel, H., and Gruetter, R. (2006). Localized short-echo-time proton MR spectroscopy with full signal-intensity acquisition. *Magn. Reson. Med.* 56: 965–970.

17 Lawry, T.J., Karczmar, G.S., Weiner, M.W., and Matson, G.B. (1989). Computer simulation of MRS localization techniques: an analysis of ISIS. *Magn. Reson. Med.* 9 (3): 299–314.

18 Burger, C., Buchli, R., McKinnon, G. et al. (1992). The impact of the ISIS experiment order on spatial contamination. *Magn. Reson. Med.* 26 (2): 218–230.

19 Matson, G.B., Meyerhoff, D.J., Lawry, T.J. et al. (1993). Use of computer simulations for quantitation of 31P ISIS MRS results. *NMR Biomed.* 6 (3): 215–224.

20 Kupce, E. and Freeman, R. (1995). Stretched adiabatic pulses for broadband spin inversion. *J. Magn. Reson. A* 117: 246–256.

21 Ordidge, R.J., Wylezinska, M., Hugg, J.W. et al. (1996). Frequency offset corrected inversion (FOCI) pulses for use in localized spectroscopy. *Magn. Reson. Med.* 36: 562–566.

22 Payne, G.S. and Leach, M.O. (1997). Implementation and evaluation of frequency offset corrected inversion (FOCI) pulses on a clinical MR system. *Magn. Reson. Med.* 38: 828–833.

23 Kinchesh, P. and Ordidge, R.J. (2005). Spin-echo MRS in humans at high field: LASER localisation using FOCI pulses. *J. Magn. Reson.* 175 (1): 30–43.

24 Sacolick, L.I., Rothman, D.L., and de Graaf, R.A. (2007). Adiabatic refocusing pulses for volume selection in magnetic resonance spectroscopic imaging. *Magn. Reson. Med.* 57: 548–553.

25 Tannus, A. and Garwood, M. (1997). Adiabatic pulses. *NMR Biomed.* 10 (8): 423–434.

26 Keeler, J. (2005). *Understanding NMR Spectroscopy*. New York: Wiley.

27 Bodenhausen, G., Freeman, R., and Turner, D.L. (1977). Suppression of artifacts in two-dimensional *J* spectroscopy. *J. Magn. Reson.* 27: 511–514.

28 Frahm, J., Merboldt, K.D., and Hanicke, W. (1987). Localized proton spectroscopy using stimulated echoes. *J. Magn. Reson.* 72: 502–508.

29 Granot, J. (1986). Selected volume excitation using stimulated echoes (VEST). Application to spatially localized spectroscopy and imaging. *J. Magn. Reson.* 70: 488–492.

30 Kimmich, R. and Hoepfel, D. (1987). Volume-selective multipulse spin-echo spectroscopy. *J. Magn. Reson.* 72: 379–384.

31 Frahm, J., Bruhn, H., Gyngell, M.L. et al. (1989). Localized high-resolution proton NMR spectroscopy using stimulated echoes: initial applications to human brain *in vivo*. *Magn. Reson. Med.* 9: 79–93.

32 Moonen, C.T.W., Von Kienlin, M., van Zijl, P.C.M. et al. (1989). Comparison of single-shot localization methods (PRESS and STEAM) for *in vivo* proton NMR spectroscopy. *NMR Biomed.* 2: 201–208.

33 van Zijl, P.C., Moonen, C.T., Alger, J.R. et al. (1989). High field localized proton spectroscopy in small volumes: greatly improved localization and shimming using shielded strong gradients. *Magn. Reson. Med.* 10 (2): 256–265.

34 Bruhn, H., Frahm, J., Gyngell, M.L. et al. (1991). Localized proton NMR spectroscopy using stimulated echoes: applications to human skeletal muscle *in vivo*. *Magn. Reson. Med.* 17: 82–94.

35 Hahn, E.L. (1950). Spin echoes. *Phys. Rev.* 80: 580–594.

36 Deville, G., Bernier, M., and Delrieux, J.M. (1979). NMR multiple spin echoes in solid ^{3}He. *Phys. Rev. B* 19: 5666–5688.

37 Mori, S., Hurd, R.E., and van Zijl, P.C. (1997). Imaging of shifted stimulated echoes and multiple spin echoes. *Magn. Reson. Med.* 37: 336–340.

38 Tkac, I., Starcuk, Z., Choi, I.Y., and Gruetter, R. (1999). *In vivo* ^{1}H NMR spectroscopy of rat brain at 1 ms echo time. *Magn. Reson. Med.* 41: 649–656.

39 Tkac, I., Andersen, P., Adriany, G. et al. (2001). *In vivo* ^{1}H NMR spectroscopy of the human brain at 7 T. *Magn. Reson. Med.* 46: 451–456.

40 Bottomley, P.A. (1984). Selective volume method for performing localized NMR spectroscopy. US patent 4,480,228.

41 Bottomley, P.A. (1987). Spatial localization in NMR spectroscopy *in vivo*. *Ann. N. Y. Acad. Sci.* 508: 333–348.

42 Jung, W.I. (1996). Localized double spin echo proton spectroscopy part I: basic concepts. *Concepts Magn. Reson.* 8 (1): 17.

43 Jung, W.I. (1996). Localized double spin echo proton spectroscopy part II: weakly coupled homonuclear spin systems. *Concepts Magn. Reson.* 8: 77–103.

44 Ernst, T. and Chang, L. (1996). Elimination of artifacts in short echo time ^{1}H MR spectroscopy of the frontal lobe. *Magn. Reson. Med.* 36: 462–468.

45 Kreis, R. (2004). Issues of spectral quality in clinical [1]H-magnetic resonance spectroscopy and a gallery of artifacts. *NMR Biomed.* 17: 361–381.

46 Hennig, J. (1992). The application of phase rotation for localized *in vivo* proton spectroscopy with short echo times. *J. Magn. Reson.* 96: 40–49.

47 Knight-Scott, J., Shanbhag, D.D., and Dunham, S.A. (2005). A phase rotation scheme for achieving very short echo times with localized stimulated echo spectroscopy. *Magn. Reson. Imaging* 23: 871–876.

48 Ramadan, S. (2007). Phase rotation in *in vivo* localized spectroscopy. *Concepts Magn. Reson. A* 30: 147–153.

49 Slotboom, J., Mehlkopf, A.F., and Bovee, W.M. (1991). A single-shot localization pulse sequence suited for coils with inhomogneous RF fields using adiabatic slice-selective RF pulses. *J. Magn. Reson.* 95: 396–404.

50 Conolly, S., Glover, G., Nishimura, D., and Macovski, A. (1991). A reduced power selective adiabatic spin-echo pulse sequence. *Magn. Reson. Med.* 18: 28–38.

51 Scheenen, T.W., Heerschap, A., and Klomp, D.W. (2008). Towards ^{1}H-MRSI of the human brain at 7T with slice-selective adiabatic refocusing pulses. *MAGMA* 21: 95–101.

52 Scheenen, T.W., Klomp, D.W., Wijnen, J.P., and Heerschap, A. (2008). Short echo time ^{1}H-MRSI of the human brain at 3T with minimal chemical shift displacement errors using adiabatic refocusing pulses. *Magn. Reson. Med.* 59: 1–6.

53 van de Bank, B.L., Emir, U.E., Boer, V.O. et al. (2015). Multi-center reproducibility of neurochemical profiles in the human brain at 7 T. *NMR Biomed.* 28: 306–316.

54 Carr, H.Y. and Purcell, E.M. (1954). Effects of diffusion on free precession in nuclear magnetic resonance experiments. *Phys. Rev.* 94: 630–638.

55 Meiboom, S. and Gill, D. (1958). Modified spin-echo method for measuring nuclear relaxation times. *Rev. Sci. Instrum.* 29: 688–691.

56 Allerhand, A. (1966). Analysis of Carr-Purcell spin-echo NMR experiments. I. The effects of homonuclear coupling. *J. Chem. Phys.* 44: 1–9.

57 Hennig, J., Thiel, T., and Speck, O. (1997). Improved sensitivity to overlapping multiplet signals in *in vivo* proton spectroscopy using a multiecho volume selective (CPRESS) experiment. *Magn. Reson. Med.* 37: 816–820.

58 van der Veen, J.W., Weinberger, D.R., Tedeschi, G. et al. (2000). Proton MR spectroscopic imaging without water suppression. *Radiology* 217: 296–300.

59 Clayton, D.B., Elliott, M.A., Leigh, J.S., and Lenkinski, R.E. (2001). [1]H spectroscopy without solvent suppression: characterization of signal modulations at short echo times. *J. Magn. Reson.* 153: 203–209.

60 Nixon, T.W., McIntyre, S., Rothman, D.L., and de Graaf, R.A. (2008). Compensation of gradient-induced magnetic field perturbations. *J. Magn. Reson.* 192: 209–217.

61 Morris, G.A. and Freeman, R. (1978). Selective excitation in Fourier transform nuclear magnetic resonance. *J. Magn. Reson.* 29: 433–462.

62 Sklenar, V. and Starcuk, Z. (1982). 1-2-1 pulse train: a new effective method of selective excitation for proton NMR in water. *J. Magn. Reson.* 50: 495–501.

63 Plateau, P. and Gueron, M. (1982). Exchangeable proton NMR without baseline distortion using a new strong-pulse sequence. *J. Am. Chem. Soc.* 104: 7310–7311.

64 Hore, P.J. (1983). Solvent suppression in Fourier transform nuclear magnetic resonance. *J. Magn. Reson.* 55: 283–300.

65 Bleich, H. and Wilde, J. (1984). Solvent resonance suppression using a sequence of strong pulses. *J. Magn. Reson.* 56: 154–155.

66 Starcuk, Z. and Sklenar, V. (1985). New hard pulse sequences for solvent signal suppression in Fourier transform NMR. I. *J. Magn. Reson.* 61: 567–570.

67 Starcuk, Z. and Sklenar, V. (1986). New hard pulse sequences for solvent signal suppression in Fourier tranform NMR. II. *J. Magn. Reson.* 66: 391–397.

68 Ross, B.D., Merkle, H., Hendrich, K. et al. (1992). Spatially localized *in vivo* ^{1}H magnetic resonance spectroscopy of an intracerebral rat glioma. *Magn. Reson. Med.* 23 (1): 96–108.

69 Haase, A., Frahm, J., Hanicke, W., and Matthaei, D. (1985). ^{1}H NMR chemical shift selective (CHESS) imaging. *Phys. Med. Biol.* 30: 341–344.

70 Doddrell, D.M., Galloway, G.J., Brooks, W.M. et al. (1986). Water signal elimination *in vivo*, using "suppression by mistimed echo and repetitive gradient episodes". *J. Magn. Reson.* 70: 176–180.

71 Ogg, R.J., Kingsley, P.B., and Taylor, J.S.W.E.T. (1994). a T_1- and B_1-insensitive water-suppression method for *in vivo* localized ^{1}H NMR spectroscopy. *J. Magn. Reson. B* 104: 1–10.

72 Moonen, C.T.W. and van Zijl, P.C.M. (1990). Highly effective water suppression for *in vivo* proton NMR spectroscopy (DRYSTEAM). *J. Magn. Reson.* 88: 28–41.

73 Hoult, D.I. (1976). Solvent peak saturation with single phase and quadrature Fourier transformation. *J. Magn. Reson.* 21: 337–347.

74 de Graaf, R.A. and Nicolay, K. (1998). Adiabatic water suppression using frequency selective excitation. *Magn. Reson. Med.* 40: 690–696.

75 Starcuk, Z. Jr., Starcuk, Z., Mlynarik, V. et al. (2001). Low-power water suppression by hyperbolic secant pulses with controlled offsets and delays (WASHCODE). *J. Magn. Reson.* 152 (1): 168–178.

76 Piotto, M., Saudek, V., and Sklenar, V. (1992). Gradient-tailored excitation for single-quantum NMR spectroscopy of aqueous solutions. *J. Biomol. NMR* 2 (6): 661–665.

77 Hwang, T.L. and Shaka, A.J. (1995). Water suppression that works. Excitation sculpting using arbitrary waveforms and pulsed field gradients. *J. Magn. Reson. A* 112: 275–279.

78 Mescher, M., Tannus, A., O'Neil Johnson, M., and Garwood, M. (1996). Solvent suppression using selective echo dephasing. *J. Magn. Reson. A* 123: 226–229.

79 Patt, S.L. and Sykes, B.D. (1972). Water eliminated Fourier transform NMR spectroscopy. *J. Chem. Phys.* 56: 3182–3184.

80 Benz, F.W., Feeney, J., and Roberts, G.C.K. (1972). Fourier transform proton NMR spectroscopy in aqueous solution. *J. Magn. Reson.* 8: 114–121.

81 Gupta, R.K. (1976). Dynamic range problem in Fourier transform NMR. Modified WEFT pulse sequence. *J. Magn. Reson.* 24: 461–465.

82 Becker, E.D., Ferretti, J.A., and Farrar, T.C. (1969). Driven equilibrium Fourier transform spectroscopy. A new method for nuclear magnetic resonance signal enhancement. *J. Am. Chem. Soc.* 91: 7784–7785.

83 Shoup, R.R. and Becker, E.D. (1972). The driven equilibrium Fourier transform technique: an experimental study. *J. Magn. Reson.* 8: 298–310.

84 Hochmann, J. and Kellerhals, H. (1980). Proton NMR on deoxyhemoglobin: use of a modified DEFT technique. *J. Magn. Reson.* 38: 23–39.

85 Hurd, R.E., Gurr, D., and Sailasuta, N. (1998). Proton spectroscopy without water suppression: the oversampled J-resolved experiment. *Magn. Reson. Med.* 40: 343–347.

86 Serrai, H., Clayton, D.B., Senhadji, L. et al. (2002). Localized proton spectroscopy without water suppression: removal of gradient induced frequency modulations by modulus signal selection. *J. Magn. Reson.* 154: 53–59.

87 Dreher, W. and Leibfritz, D. (2005). New method for the simultaneous detection of metabolites and water in localized *in vivo* ^{1}H nuclear magnetic resonance spectroscopy. *Magn. Reson. Med.* 54: 190–195.

7

Spectroscopic Imaging and Multivolume Localization

7.1 Introduction

The localization techniques discussed in Chapter 6 allow the detection of signal from a single volume. The advantages of single-volume MRS are that (i) the volume is typically well-defined with minimal contamination (e.g. extracranial lipids in brain MRS), (ii) the magnetic field homogeneity across the volume can be readily optimized, leading to (iii) improved water suppression and spectral resolution. The main disadvantage of single-volume localization methods is that no signal is acquired from large parts of the object, thereby potentially missing important areas of interest.

Multivolume or multi-voxel localization or magnetic resonance spectroscopic imaging (MRSI) allows the detection of localized MR spectra from a multidimensional array of locations. While technically more challenging, due to (i) significant magnetic field inhomogeneity across the entire object, (ii) spectral degradation due to intervoxel contamination, (iii) long data acquisition times, and (iv) processing of large multidimensional datasets, MRSI can detect metabolic profiles from multiple spatial positions, thereby offering an unbiased characterization of the entire object under investigation. This chapter will review the principles of multivolume localization techniques and, in particular, of MRSI.

7.2 Principles of MRSI

MRSI can be based on B_0 or B_1 magnetic field gradients, similar to single-voxel methods. However, with the exception of a limited number of specialized applications, B_1-based rotating frame spectroscopic imaging has essentially been replaced with B_0-based spectroscopic imaging, which will be the topic for the rest of this chapter. The interested reader is referred to the literature [1–4] for more details on rotating frame spectroscopic imaging.

The principles of MRSI [5, 6] are very similar to phase-encoding in magnetic resonance imaging (MRI). Any MR spectroscopy pulse sequence can be converted to an MRSI sequence by inserting phase-encoding gradients during one of the free evolution periods. However, technical complications typically demand additional sequence modifications, as will be discussed. Following the first *in vitro* [5, 6] and *in vivo* [7, 8] demonstrations, MRSI has seen continued development and is now routinely used in clinical research.

Before proceeding with a formal, quantitative *k*-space description of MRSI, a qualitative and intuitive explanation of MRSI will be provided. Consider the basic spin-echo sequence of Figure 7.1B and the four-tube phantom containing various chemicals (Figure 7.1A). For simplicity, the tubes are considered infinitesimally narrow such that the gradient-induced phase dispersal *across* a tube can be ignored. A regular, spin-echo 1D MR spectrum (Figure 7.1C) reveals the chemical content of the phantom, but does not provide any spatial information. Application of a magnetic field gradient during the second TE/2 period (Figure 7.1D) adds spatial encoding to the acquired MR signal. During the magnetic field gradient *G*, the magnetic field and hence the spin Larmor frequency *v* becomes linearly dependent on position *r* according to

$$v(r) = \frac{\omega(r)}{2\pi} = \frac{\gamma r G}{2\pi} \tag{7.1}$$

Transverse magnetization will have acquired a phase at the end of a constant-amplitude gradient pulse of duration *T* given by

$$\phi(r) = \int_0^T \omega(r)dt = \gamma r G T \tag{7.2}$$

Since the magnetic field gradient is only executed during one of the TE/2 periods, the phase shift given by Eq. (7.2) will not be refocused and will therefore encode the acquired free induction decay (FID) or echo with spatial information. Figure 7.1D shows spin-echo acquisitions with different phase-encoding gradient amplitudes, together with the corresponding MR spectra following Fourier transformation (Figure 7.1E). It follows that different chemicals have varying phase accumulations according to the spatial position of the tube in which they reside. The total creatine (tCr) resonances do not acquire any gradient-induced phase in any of the experiments, which can only be explained when the tube containing tCr resides in the gradient isocenter ($r = 0$), which normally coincides with the magnet isocenter. The resonances of *N*-acetyl aspartate (NAA) and choline (Cho) display phase accumulation in opposite directions, which can be explained when their respective tubes reside on opposite sides of the magnetic isocenter. Finally, glutamate (Glu) resonances display an increased phase accumulation relative to NAA, such that the Glu tube must be positioned further away from the magnetic isocenter than the tube containing NAA. While the data in Figure 7.1 clearly show the position-dependent phase accumulation, a visual interpretation and extraction of information as described here is almost never possible since it requires well-separated, discrete phantoms containing a single metabolite in each tube. Fortunately, there is a tool that can be used to extract the spatial information automatically even on complex biological samples. When the signal intensities of NAA and Cho resonances are plotted as a function of the phase-encoding gradient amplitude (Figure 7.1G), it becomes apparent that the position-dependent phase accumulation makes up an apparent frequency $v(r)$ according to Eq. (7.1). Application of a Fourier transformation with respect to the gradient amplitude can be used to separate the spatial positions of the various signals (Figure 7.1F). The 2D plot of Figure 7.1F reveals that the tCr signal comes from a tube in the magnet isocenter. NAA and Glu reside on one side of the magnet, whereas the Cho-containing tube is on the opposite side. The data shown in Figure 7.1F are referred to as a 1D MR spectroscopic image since one spatial axis has been resolved.

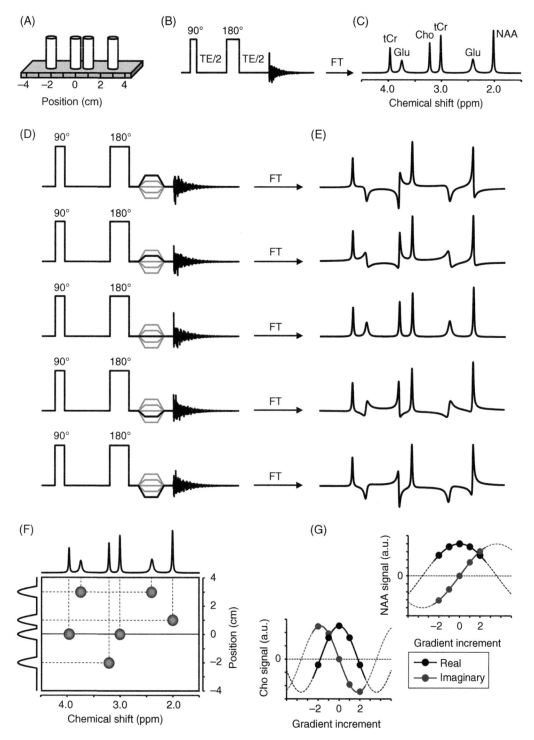

Figure 7.1 Principle of MR spectroscopic imaging. (A) Object consisting of four narrow tubes containing different chemicals in water. (B) A regular spin-echo sequence without phase-encoding gradients provides (C) a spectrum with resolved chemical shifts, but without spatial information. (D) Spin-echo sequence with a linearly incrementing phase-encoding gradient along the direction of the object provides (E) spectra in which the phase of various resonances is proportional to the position of the tube along the phase-encoding direction. (G) The phase-sensitive signal of the NAA and choline as a function of the phase-encoding gradient makes up an apparent frequency, which upon Fourier transformation (F) reveals the spatial position.

7.3 *k*-Space Description of MRSI

Figure 7.1 provides an intuitive, but qualitative description of the basic MRSI method. A quantitative description of MRSI is based on the *k*-space formalism as introduced in Chapter 4. The total acquired time-domain signal $S(t)$ represents the sum of signal from elementary volume elements $s(r, t)dr$ from each point r in the sample according to

$$S(t) = \int_{-\infty}^{+\infty} s(r,t)\,dr \tag{7.3}$$

The total spectrum of the sample $F(\omega)$ is obtained by Fourier transformation of $S(t)$ and equals the sum of spectra from elementary volume elements $f(r, \omega)dr$:

$$F(\omega) = \int_{-\infty}^{+\infty} S(t)e^{-i\omega t}\,dt = \int_{-\infty}^{+\infty} f(r,\omega)\,dr \tag{7.4}$$

Up to this point there is no difference between Eq. (7.4) and the Fourier transformation given by Eq. (1.20). However, as was already intuitively demonstrated in Figure 7.1, the spatial distribution of $F(\omega)$, i.e. $f(r, \omega)$ can be obtained by application of a phase-encoding gradient, which induces a phase shift on each elementary volume element according to

$$f'(r,\omega) = f(r,\omega)e^{i\gamma rGt} \tag{7.5}$$

such that the entire spectrum can be written as

$$F(G,\omega) = \int_{-\infty}^{+\infty} f(r,\omega)e^{i\gamma rGt}\,dr \tag{7.6}$$

The introduction of the *k*-space formalism, $k = \gamma Gt$, converts Eq. (7.6) to

$$F(k,\omega) = \int_{-\infty}^{+\infty} f(r,\omega)e^{ikr}\,dr \tag{7.7}$$

Clearly, the phase-modulated spectra of the entire sample $F(k, \omega)$ represents the inverse Fourier transformation of the spectra $f(r, \omega)$ from the individual volume elements. Therefore, $f(r, \omega)$ can be obtained by Fourier transformation of $F(k, \omega)$ according to

$$f(r,\omega) = \int_{-\infty}^{+\infty} F(k,\omega)e^{-ikr}\,dr \tag{7.8}$$

where r can represent any direction in 3D space. The general 1D case expressed by Eq. (7.8) can be extended to the three Cartesian directions by applying three orthogonal gradients independently in subsequent experiments. The volume elements are then calculated with a 4D Fourier transformation. The first spectral Fourier transformation of a complete 4D MRSI dataset $s(k_x, k_y, k_z, t)$ yields the spectra $F(k_x, k_y, k_z, \omega)$ after which a 3D spatial FT gives the spectra $f(x, y, z, \omega)$ from the spatial positions x, y, and z.

While Eq. (7.8) suggests that the signal is continuously sampled in *k*-space, in reality the signal is only acquired at discrete *k*-space positions. The spatially resolved spectra are obtained through a discrete Fourier transformation. Following the acquisition of N *k*-space samples, most discrete Fourier transform algorithms provide as output N spatial positions located between $-N/2$ and $N/2 - 1$. As a result, for even N the zero spatial frequency point is not centered in space, but appears at a position -0.5. This is referred to as a half-pixel shift and should be taken into account when accurate localization or co-registration between high-resolution MRI and low-resolution MRSI data is required.

The increments in gradient area, which are typically achieved by increasing the amplitude, determine the digitization rate in the spatial frequency or k-space domain. By analogy with the acquisition of a one-dimensional spectrum, this digitization is governed by the Nyquist sampling criterion, i.e. the maximal phase shift difference between two gradient increments over the entire field-of-view (FOV) equals 2π in order to avoid aliasing.

$$2\pi = \gamma \text{FOV} \Delta Gt, \quad \text{or FOV} = \frac{2\pi}{\gamma \Delta Gt} = \frac{2\pi}{\Delta k} \tag{7.9}$$

The nominal voxel size ΔV is directly related to the FOV and the number of phase-encoding increments N and is given by

$$\Delta V = \frac{\text{FOV}}{N} \tag{7.10}$$

In practice, the minimum voxel size is, besides the FOV, determined by the allowable measurement time and sensitivity. For conventional MRSI, the encoding of $N_1 \times N_2 \times N_3$ volume elements (voxels) requires $N_1 \times N_2 \times N_3$ separate acquisitions, each with a repetition time TR.

7.4 Spatial Resolution in MRSI

The nominal voxel size in an MRSI experiment is simply given by Eq. (7.10), i.e. the entire FOV divided by the number of phase-encoding gradient increments. However, the actual voxel size can substantially deviate from this nominal value due to the characteristics of the Fourier transformation [9–14].

A time domain signal $s(k)$ measured over an infinitely long period of time will produce a Dirac delta function or a single frequency upon Fourier transformation (Figure 7.2A and B). However, under realistic conditions $s(k)$ is sampled over a finite time, described by the sampling function $F_{\text{sample}}(k)$. The Fourier transformation must therefore be made over the product of $s(k)$ and $F_{\text{sample}}(k)$, leading to a convolution of $s(r)$ and $F_{\text{sample}}(r)$ in the spatial domain (see also Appendix A.3). For a constant grid of N samples (i.e. phase-encoding steps) $F_{\text{sample}}(k)$ is given by a sinc function (Figure 7.2D, see Exercises). This is the case for MRS, MRI, MRSI, and, in general, all techniques requiring Fourier transformation. The Fourier transformation of the sampling points, $F_{\text{sample}}(k)$ (or sampling grid when more dimensions are involved), is often referred to as the point spread function (PSF). Although the PSF always influences the Fourier transformed data, it has not been mentioned earlier, because the PSF is not a dominant factor for MRS. Provided that the signal has decayed to zero at the end of the acquisition period, the PSF of a time-domain FID is much narrower than the typical line widths that are dominated by T_2^* relaxation. The PSF only becomes apparent when the acquisition time is too short, leading to a truncation artifact in the NMR spectrum. However, with MRSI the effects of the PSF become more pronounced because of (i) the limited number of phase-encoding increments (and consequently the limited number of k-space samples) and (ii) the spatial time domain data are not influenced by parameters such as T_2^* (since T_2^* effects are refocused in a spin-echo

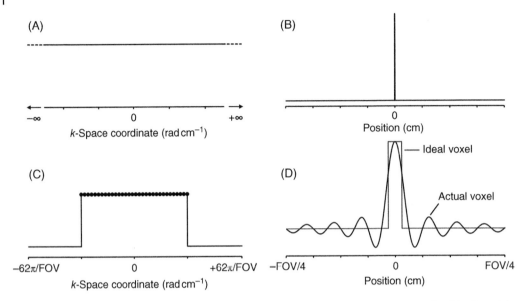

(A)

$-\infty$ 0 $+\infty$

k-Space coordinate (rad cm^{-1})

(B)

0

Position (cm)

— Ideal voxel

Actual voxel

(C)

-62π/FOV   0 $+62\pi$/FOV

k-Space coordinate (rad cm^{-1})

(D)

$-$FOV/4   0   FOV/4

Position (cm)

Figure 7.2 **Spatial point spread function.** (A) A *k*-space signal sampled from $-\infty$ to $+\infty$ provides (B) a Dirac delta function upon Fourier transformation. (C) With MRSI *k*-space is sampled over a limited range, which leads to (D) a "sinc" spatial response or spatial point spread function (PSF) upon Fourier transformation. Note that the PSF extends far beyond the ideal or nominal voxel dimension, which is simply given by the field-of-view divided by the number of *k*-space or phase-encoding samples (i.e. Eq. (7.10)).

sequence or are identical for all phase-encoding increments in a pulse-acquire sequence). Figure 7.2D shows the PSF of a *k*-space region sampled by 32 points (Figure 7.2C). Three noticeable features can be extracted from Figure 7.2. Firstly, while the nominal resolution in MRSI is given by the FOV divided by the number of phase-encoding increments *N*, i.e. Eq. (7.10), the actual resolution as defined as the full width at half maximum (FWHM) of the PSF is 1.21(FOV/*N*). In other words, the actual voxel size is 21% larger than the nominal voxel size. Secondly, due to residual phase dispersal across the voxel, the PSF is not uniform across the nominal voxel size leading to the observation that only 87.3% of the observed signal originates from the desired (or intended) spatial location [15]. Thirdly, the remaining 12.7% of the signal is spread to adjacent voxels, a phenomenon referred to as signal leakage or bleeding. Similarly, voxels at other positions contaminate the voxel displayed in Figure 7.2D. The exact contribution of the PSF to remote voxels depends on the object under investigation and the relative positions to the points of the spatial grid. In the best case, as encountered for a homogeneous sample with homogeneous B_0 and B_1 magnetic fields, the positive and negative "sinc" lobes cancel out perfectly. Unfortunately for most realistic samples, the spin density distribution, nor the magnetic field B_0 and B_1 distributions are homogeneous. This leads to an imperfect cancelation of the side lobes, which in turn leads to signal contamination, potentially from locations far removed from the voxel of interest. While the PSF plays a central role in all MRSI studies, its effects are particularly obvious in ^{1}H MRSI of the brain, where strong lipid signals from outside the brain can contaminate spectra from within the brain. Standard MRSI acquisition methods on the brain therefore include lipid suppression modules that will be discussed in Section 7.6. Note that for 2D and 3D MRSI grids 76.2 and 66.7% of the signal originates from the theoretical square or cubic voxel position, respectively.

The PSF can be artificially improved by applying apodization functions in analogy to those used for MRS [9, 16]. However, in the case of MRSI the apodization functions need to be applied in the spatial frequency or k-space domain. Since MRSI data are normally acquired with the spatial frequencies centered on the origin of k-space, the apodization function need to be symmetrical with respect to the origin. Figure 7.3 shows the effects of some commonly used apodization functions on the PSF. Figure 7.3B shows the PSF without any apodization of k-space (Figure 7.3A). Figure 7.3D shows the PSF after the application of a cosine apodization function according to

$$W(k) = \cos\left(\frac{\pi k}{2k_{\max}}\right) \quad \text{for} -k_{\max} \leq k \leq k_{\max} \tag{7.11}$$

where $k_{\max}$ is the maximum sampled position in k-space. The ripples are significantly reduced at the expense of a slight increase in the width (and integrated area) of the main lobe, i.e. the actual, localized volume increases. The ripples can be reduced further by using apodization which is described by a Gaussian function (Figure 7.3E):

$$W(k) = e^{-4(k/k_{\max})^2} \quad \text{for} -k_{\max} \leq k \leq k_{\max} \tag{7.12}$$

The reduction of ripples is again accompanied by an increase in the frequency width at half maximum (FWHM), leading to a decreased spatial resolution. Theoretically, the optimal filter in terms of maximal ripple reduction versus minimal FWHM is given by the Dolph–Chebyshev windowing function [16]. However, this function is rarely used as it has a dependence on the number of sampling points and is relatively cumbersome to calculate. An analytical function that closely approximates the performance of the Dolph–Chebyshev filter is given by the Hamming function (Figure 7.3G):

$$W(k) = 0.54 + 0.46\cos\left(\frac{\pi k}{k_{\max}}\right) \quad \text{for} -k_{\max} \leq k \leq k_{\max} \tag{7.13}$$

Note that the FWHM of the 1D PSFs without and with cosine, Gaussian and Hamming filtering are 21, 69, 119, and 85% broader than the ideal or nominal volume. The maximum amplitude of the side lobes are reduced to 21.8, 7.1, 0.3, and 0.8% of the main PSF signal, respectively.

The application of post-acquisition spatial frequency apodization functions is not optimal in terms of time-efficiency and consequently sensitivity [12]. This is because all k-space samples are acquired with the same number of acquisitions, after which post-acquisition apodization reduces the signal of the high-frequency k-space coordinates. In Section 7.5, methods are described that can achieve k-space apodization during acquisition.

7.5 Temporal Resolution in MRSI

Conventional MRSI sequences measure a single k-space point per repetition time. For a 3D phase-encoded MRSI experiment, the total measurement time $T_{\text{measurement}}$ is given by $\text{NA} \times N_x \times N_y \times N_z \times \text{TR}$, whereby N_n ($n = x, y, z$) are the number of phase-encoding increments, NA the number of averages, and TR the repetition time. For a medium-resolution 3D dataset ($N_x = N_y = N_z = 16$), NA = 1 and a repetition time that is such that T_1-weighting is not a major concern (e.g. 2000 ms at 1.5 T), the measurement time will be 136 min, well outside

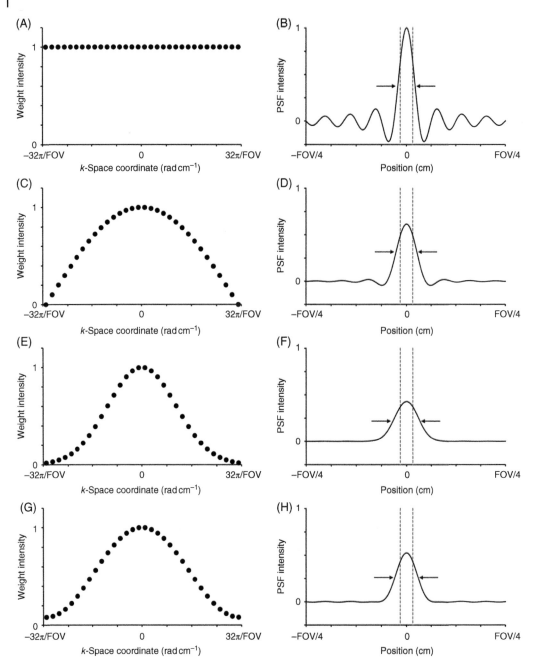

Figure 7.3 **Post-acquisition *k*-space apodization.** (A, C, E, G) *k*-space signals and (B, D, F, H) corresponding point spread functions when the *k*-space signals are (A, B) not apodized, (C, D) apodized by a cosine function (Eq. (7.11)), (E, F) apodized by a Gaussian function (Eq. (7.12)), and (G, H) apodized by a Hamming function (Eq. (7.13)). The dotted lines indicate the ideal or nominal voxel width, whereas the arrows indicate the actual voxel width at half maximum.

acceptable MR examination times for human subjects. Even a moderately high-resolution 2D dataset ($N_x = N_y = 32$) acquired in 34 min will challenge subject compliance limits, especially in patient populations. For most applications it is therefore crucial to increase the temporal resolution of MRSI. Techniques that achieve an increase in temporal resolution can be divided into three groups, namely (i) conventional methods, requiring minimal additional pre- or post-processing, (ii) methods based on fast MRI sequences that typically require reordering and sometimes re-gridding of k-space, and (iii) methods that use prior knowledge to minimize and optimize k-space sampling.

7.5.1 Conventional Methods

7.5.1.1 Circular and Spherical k-Space Sampling

As was demonstrated in Chapter 4, the low-spatial-frequency coordinates of k-space generally contribute to the bulk of the observed signal with minimal information about the object shape and boundaries. The high-spatial-frequency coordinates of k-space hold information about the detailed features of an object, like boundaries, but contribute minimally to the bulk signal. Because of sensitivity restrictions, MRSI is inherently a low to medium-spatial-resolution technique and as such the decision can be made to emphasize the low k-space coordinates over the higher ones in order to favor sensitivity over detail and resolution. Circular 2D or spherical 3D k-space sampling [17, 18] achieves this goal by simply not acquiring the high k-space coordinates (Figure 7.4A). This reduced k-space sampling leads to a circa 21.5 and 47.6% reduction in measurement time for 2D and 3D MRSI, respectively. The non-acquired k-space coordinates are typically replaced by zeros during post-acquisition processing, such that standard Fourier transformation can be used to process the MRSI data. However, ignoring the higher k-space coordinates will have an effect on the PSF of the experiment. Figs. 7.4B–D compare the 2D PSF of a regular, rectangular sampled k-space with that of a circularly sampled k-space, as shown in Figure 7.4A. As expected, the main PSF lobe of the circularly sampled k-space grid is wider than the conventional PSF lobe, leading to an effective volume that is circa 55% larger than the nominal volume (and circa 29% larger when compared to the volume obtained with standard sampling). In addition to the wider main lobe, the contamination to other voxels is significantly reduced by circular k-space sampling.

7.5.1.2 k-Space Apodization During Acquisition

In Section 7.4, it was demonstrated that post-acquisition k-space apodization can improve the PSF by suppressing voxel bleeding at the expense of a broader main lobe. However, performing this operation post-acquisition is not optimal in terms of sensitivity per unit time, since the high k-space coordinates are suppressed by the apodization [12]. A better approach would be to spend less time acquiring the high k-space coordinates and more time on acquiring the low k-space coordinates. This can be achieved by (i) performing fewer signal averages or (ii) introduce more T_1-weighting at high k-space coordinates. The first option is only applicable when signal averaging is required, which typically limits it to low-sensitivity nuclei, such as ^{31}P MRSI. The sensitivity of ^{1}H MRSI is typically sufficient in a single average. Figure 7.5 shows the principle of k-space weighting during signal acquisition for a 1D MRSI experiment. Instead of acquiring all 16 k-space coordinates with the same number of averages (NA = 3), the number of averages can be varied between NA = 1 (outer edges of k-space) and NA = 5 (middle of k-space) to approximate a Gaussian apodization function according to Eq. (7.12). Figure 7.5B shows the greatly improved PSF for the weighted k-space acquisition, with only minor differences due to the stepwise approximation of a continuous Gaussian function.

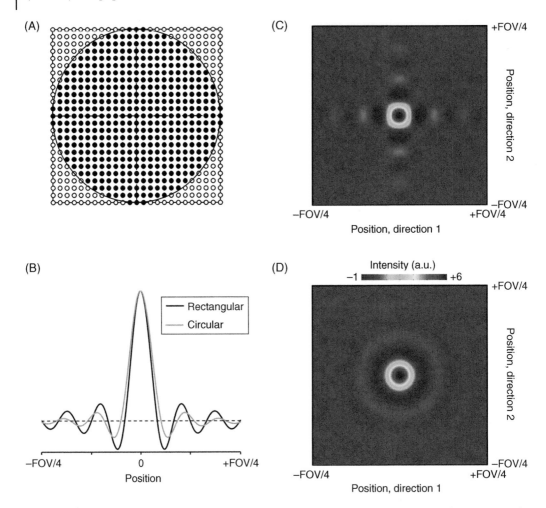

Figure 7.4 Circular k-space encoding. (A) Circular k-space sampling over a 25×25 grid. During regular k-space sampling all k-space positions are acquired, whereas circular k-space sampling only acquires the black k-space positions. The missing k-space positions are substituted with zeroes during post-acquisition processing. (B) 1D and (C, D) 2D point spread functions for (B, C) regular and (B, D) circular k-space sampling.

While the acquisition k-space apodization shown in Figure 7.5A and B improves the PSF and the sensitivity, it did not result in increased temporal resolution since the total number of averages remained constant.

An alternative to k-space weighting by variable signal averaging was proposed by Kuhn et al. [19]. Different k-space coordinates can be weighted differently by employing variable repetition time acquisitions. The amount of signal at position k, relative to $k = 0$ essentially determines the weighting (or apodization) function according to

$$W(k) = \frac{M_{xy}(\text{TR},k)}{M_{xy}(\text{TR},k=0)} = \frac{1-e^{-\text{TR}(k)/T_1}}{1-e^{-\text{TR}(k=0)/T_1}} \tag{7.14}$$

where the right-hand side of Eq. (7.14) is valid for 90° excitations. The desired apodization function $W(k)$ can be chosen according to Eqs. (7.11)–(7.13) after which Eq. (7.14) allows the

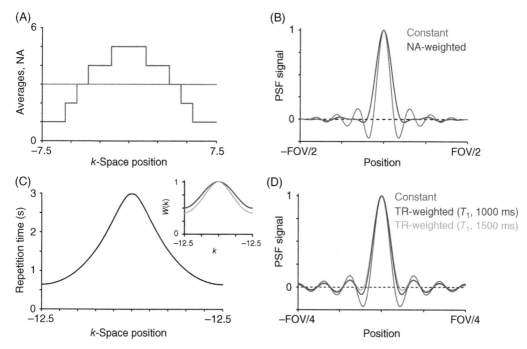

Figure 7.5 **Acquisition-based *k*-space apodization.** (A) *k*-space sampling with constant signal averaging (NA = 3, blue line) and *k*-space sampling with variable signal averaging to approximate a Gaussian *k*-space weighting function for 16 phase-encoding increments (red line). (B) Point spread functions (PSFs) for uniform and discrete Gaussian *k*-space weighting. (C) Repetition time variation to achieve *k*-space weighting according to a modified Hamming function, $W(k) = 0.25 + 0.75\cos(\pi k/k_{max})$ with $-k_{max} \le k \le +k_{max}$, for $T_1 = 1000$ ms. The inset shows the *k*-space weighting function achieved for $T_1 = 1000$ ms (red line) and $T_1 = 1500$ ms (green line). (D) PSFs for *k*-space sampling without apodization (blue line) and TR-weighted apodization.

calculation of the repetition time TR as a function of the *k*-space position. Figure 7.5C shows the *k*-position dependence of TR using the Hamming apodization function (Eq. (7.13)). In order to modulate the transverse magnetization according to the Hamming apodization function, the TR must be varied from 3000 ms in the middle to below 1000 ms at the edges of *k*-space. In the 1D example, the TR-weighted sampling leads to a 35% reduction in measurement time relative to constant-TR sampling (TR = 3000 ms). For 2D and 3D acquisitions the reduction in measurement time will be even greater since the number of low *k*-space coordinates (with a long TR) becomes smaller relative to the number of high *k*-space coordinates (with a short TR). In the original manuscript of Kuhn et al. [19], a 55% reduction in measurement time was achieved for a 2D MRSI acquisition.

A potential concern of TR-weighted *k*-space sampling is that the PSF becomes dependent on the T_1 relaxation time constant. Figure 7.5D compares the PSF of the apodization shown in Figure 7.5C for $T_1 = 1000$ and 1500 ms. There are only minor differences, which typically do not impact signal quantification. Another consideration is the increased RF power deposition at short repetition times, which may limit the applicability for human brain and/or high magnetic fields.

7.5.1.3 Zoom MRSI

The temporal resolution of an MRSI experiment is directly proportional to the number of phase-encoding increments, which for a given spatial resolution, is directly proportional to the FOV.

The FOV in turn is determined by the boundaries of the object under investigation. Therefore, if the FOV can be reduced by restricting the observed signal to a smaller region, the number of phase-encoding increments and hence measurement time can be reduced while maintaining the same nominal voxel size. Numerous methods are available to remove unwanted signal from the object under investigation in order to reduce the FOV. Typically outer volume suppression (OVS) or single-shot localization methods like PRESS or STEAM are used for this purpose. As these techniques are often also used for lipid suppression, they are discussed in detail in Section 7.6.

7.5.2 Methods Based on Fast MRI

So far the reduction in measurement times has been achieved through reduced or modified k-space sampling. However, the limits are fairly quickly encountered since there simply needs to be a minimum amount of data to adequately characterize the object. Rather than aiming for reduced k-space sampling, many of the fast MRI sequences focus on more efficient k-space sampling. This leads in the most extreme case to single-scan echo-planar or spiral imaging in which the entire k-space is sampled in a single acquisition. The principles of fast MRI are often equally applicable to MRSI, which has resulted in MRSI methods based on EPI [20–22], spiral [23–25], concentric circles [26–28], and other [29–32] sequences and trajectories.

7.5.2.1 Echo-planar Spectroscopic Imaging (EPSI)

Echo-planar spectroscopic imaging, or EPSI, and most other ultrafast MRSI methods differ from conventional MRSI methods by the presence of magnetic field gradients during signal acquisition. However, throughout the book it is emphasized that MR spectroscopic acquisitions rely on chemical shift evolution in a *homogeneous* magnetic field. The presence of magnetic field gradients during signal acquisition may therefore appear problematic. Fortunately, the frequencies associated with spatial and spectral information evolve over different time scales during an EPSI sequence, thereby allowing their separation. Figure 7.6A shows a spectroscopic acquisition in the absence of magnetic field gradients. For ^{1}H MRS the chemical shift range for non-exchangeable protons spans about 5 ppm or 640 Hz at 3 T. A spectral width of 640 Hz corresponds to a dwell-time $\Delta t_{\text{spectral}}$ of 1.56 ms. In other words, the spectroscopic information content is completely determined when the time-domain signal is sampled once every 1.56 ms. EPSI relies on the fact that magnetic field gradients can encode spatial information during the spectral dwell-time $\Delta t_{\text{spectral}}$ (Figure 7.6B). The magnetic field gradients do not affect the chemical shift evolution and as long as gradient-induced phases are zero at the point of spectroscopic acquisition, the resulting ^{1}H MR spectrum in Figure 7.6B is indistinguishable from that acquired without magnetic field gradients (Figure 7.6A). Besides demonstrating that the presence of magnetic field gradients does not have to compromise spectroscopic information *per se*, the EPSI sequence in Figure 7.6B has no added value to the regular spectroscopic acquisition shown in Figure 7.6A. The speed advantage of EPSI is realized when multiple points are sampled and acquired during the plateau period of each gradient when echo formation occurs (Figure 7.6C) in analogy with frequency-encoding in MRI. Acquisition of 32 points during a 1 ms gradient plateau period requires a dwell-time $\Delta t_{\text{spatial}}$ of 31.25 μs or an acquisition bandwidth of 32 kHz. The 32 points acquired over 1 ms in the presence of the magnetic field gradient define the spatial MRSI voxels distributed over the FOV in the frequency-encoding direction. One spatially-encoded echo does not contain any spectral information. However, when echoes are repeatedly formed through the EPSI gradient switching, the echoes are encoded with chemical shift information as shown in Figure 7.6B. The EPSI echo train has thus acquired all spectral and spatial points for 1D dimension in a *single* acquisition, thereby

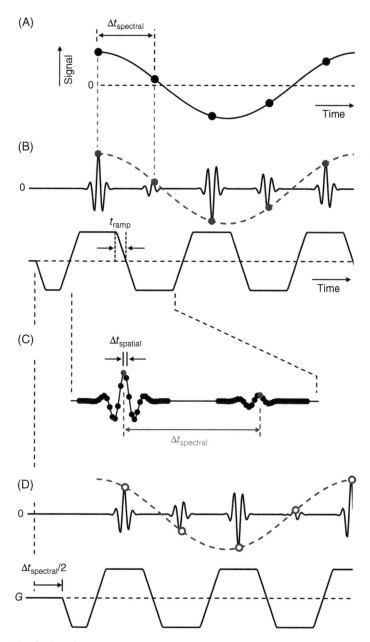

Figure 7.6 Principle of echo-planar spectroscopic imaging. (A) A spectroscopic signal acquired in the absence of magnetic field gradients is sampled with a dwell-time $\Delta t_{spectral}$. (B) Echo-planar spectroscopic imaging signal acquired in the presence of an oscillating, trapezoidal magnetic field gradient leads to repeated echo formation, whereby the echo separation provides the spectroscopic dwell-time $\Delta t_{spectral}$. (C) The individual echoes are sampled at a much higher sampling rate with dwell-time $\Delta t_{spatial}$ in order to acquire all spatial points during the gradient plateau. EPSI with gradient ramp sampling can increase the data sampling duty cycle, thereby reducing $\Delta t_{spectral}$ and increasing the spectral bandwidth. (D) The spectral bandwidth can be increased by acquiring a second EPSI dataset in which data acquisition and sampling are shifted by a delay $\Delta t_{spectral}/2$. Interleaving the EPSI data from (B, C) and (D) increases the spectral bandwidth twofold.

providing an acceleration factor compared to conventional MRSI equal to the number of phase-encoding steps in one direction. From Figure 7.6B and C it follows that the acquisition time for one echo plus the gradient switching time must fit within one spectral dwell-time $\Delta t_{spectral}$ in order to not affect the spectral bandwidth. Depending on the spatial resolution, magnetic field strength, and gradient performance this requirement may not always be satisfied. In that case the spectral dwell-time must be lengthened, leading to a reduced spectral bandwidth and to potential aliasing of resonances. When the spectral bandwidth is insufficient to provide high-quality spectroscopic information, it can be increased by interleaving data from N separate EPSI scans in which the start of spectral acquisition has been delayed by an amount $(n/N)\Delta t_{spectral}$ $(n = 0 \ldots N-1)$ (Figure 7.6D). Interleaving reduces the EPSI temporal resolution N-fold, but also increases the spectral width N-fold, thereby making EPSI feasible even at high magnetic fields.

The EPSI method shown in Figure 7.6C only acquires data during the plateau periods of the magnetic field gradients. Data acquisition during the gradient ramps will shorten the dead time, thereby increasing the acquisition efficiency and the spectral bandwidth. However, ramp sampling will lead to non-Cartesian k-space sampling, which will require regridding before standard processing tools, such as FT, can be used. Furthermore, ramp sampling can lead to severe chemical shift displacement during low-gradient ramp amplitudes.

The data acquired with EPSI represent a 1D series of gradient echoes modulated by chemical shift information (Figure 7.6). The data can be reformatted into a 2D matrix with one temporal and one spatial k-space axis that is equivalent to data acquired with conventional phase-encoded MRSI. However, direct FT of the 2D EPSI data would lead to severe ghosting artifacts in the spatial domain due to differences between even and odd echoes. This effect is similar to that observed in EPI where eddy currents, magnetic field inhomogeneity, and an imbalance in gradient switching leads to ghosting (see Figure 4.9). Several options for the reconstruction of EPSI data exist [21, 22, 33], including the separate processing of data from even and odd echoes or using fly-back EPSI. The first option requires the acquisition of twice as many echoes at twice the sampling rate (referred to as "oversampling") in order to maintain the spectral bandwidth, whereas the second option is characterized by lower sensitivity due to the absence of data acquisition during the fly-back gradient.

Even though EPSI can provide greatly reduced measurement times, a point of concern with EPSI or any high-speed MRSI sequence is sensitivity. This is because the signal is typically acquired using large (20–200 kHz) receiver bandwidths. Ignoring intrinsic parameters as relaxation and experimental imperfections such as residual gradient effects, it can be shown by simple arguments that the S/N per unit of time is the same for conventional and high-speed MRSI sequences. Suppose that a conventional 1D MRSI dataset, acquired with 32-phase encoding increments (one average) and a spectral bandwidth of 4000 Hz, has a signal-to-noise ratio of $(S/N)_{conv}$. The same MRSI dataset can be acquired with a high-speed EPSI sequence using a 32×4 kHz = 128 kHz receiver bandwidth in order to accommodate the 32 spatial points within the spectral dwell time (Figure 7.6). Since noise is proportional to the root of the bandwidth, this results in $(S/N)_{fast} = (1/\sqrt{32})(S/N)_{conv}$. In other words, while an EPSI dataset can be acquired much faster, the corresponding S/N is much lower. In the time needed to acquire the conventional MRSI dataset, one can repeat the fast EPSI sequence 32 times, such that $(S/N)_{fast} = (S/N)_{conv}$ in the same measurement time. This result may be influenced by specific pulse sequence characteristics, including relaxation, RF and/or gradient imperfections, and magnetic field inhomogeneity. However, Pohmann et al. [31] have performed a rigorous comparison between a wide range of MRSI methods, including EPSI and other multi-echo approaches, and they concluded that conventional MRSI is and remains the gold standard for sensitivity per unit time. It is important to realize that while high-speed MRSI sequences do not improve the attainable *S/N*, they do provide a flexible trade-off between sensitivity and measurement time. Conventional MRSI does not have this degree of freedom, since the number of phase-encoding increments determines the minimum achievable measurement time. When sufficient sensitivity can be obtained in a shorter measurement time, fast MRSI methods are ideal to shorten the scan time, obtain whole-head 3D coverage, track and correct magnetic field drift and patient motion [25], or be combined with 2D NMR methods [34].

7.5.2.2 Spiral MRSI

Similar to ultrafast MRI sequences, the echo-planar *k*-space trajectory (Figure 7.7A and B) is only one of many possible choices to traverse *k*-space. Spiral MRSI (Figure 7.7C and D) has been shown as a viable alternative to EPSI. Whereas EPSI samples a 2D (k_x, k_v) plane after which the second spatial dimension is obtained with standard phase encoding along k_y, spiral MRSI is, in principle, capable of acquiring the entire 3D (k_x, k_y, k_v) volume in a single experiment. However, hardware and sensitivity limitations normally prevent the acquisition of all data in a single scan. Spiral MRSI can be interleaved in the (k_x, k_y) dimension to increase the FOV or in the k_v dimension (Figure 7.7C and D) to increase the spectral width or both [33, 35]. Data collected with spiral MRSI cannot be reconstructed directly with FT as the sampling

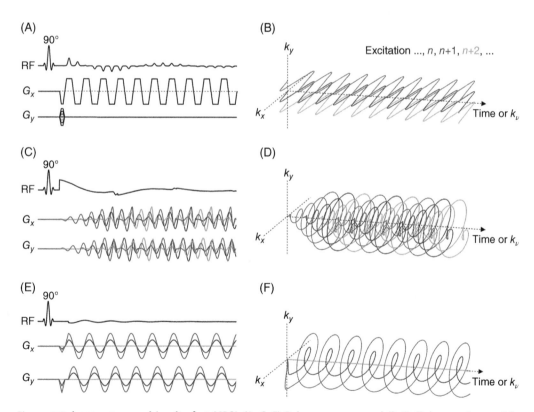

Figure 7.7 *k*-space traversal in ultrafast MRSI. (A, C, E) Pulse sequences and (B, D, F) *k*-space traversal for ultrafast MRSI based on (A, B) echo-planar, (C, D) spiral, and (E, F) concentric ring trajectories. (A, B) For EPSI the second spatial dimension is typically encoded with standard 1D phase encoding along the k_y direction in subsequent experiments. (C, D) Spiral MRSI can, in principle, sample the entire (k_x, k_y, k_v) space in a single scan. However, hardware limitations may require interleaving in the (k_x, k_y) plane or along the k_v dimension (shown). (E, F) Sampling along concentric rings samples the (k_x, k_y, k_v) space in multiple excitations whereby the gradient amplitudes and thus the radius of the concentric *k*-space ring increases.

points do not fall on a rectilinear grid. Similar to spiral MRI, the data first needs to be regridded onto a Cartesian grid before the standard FT can be applied. Among many other possible *k*-space trajectories, acquiring MRSI data along concentric rings [26–28] appears to be robust and efficient with insensitivity towards system imperfections (Figure 7.7E and F). Figure 7.8 shows an example of a complete 3D EPSI dataset of human brain acquired in only 20 min.

7.5.2.3 Parallel MRSI

Phased array receivers are now routine in MRI to provide improved sensitivity and the possibility of image acceleration. For MRSI the sensitivity increase is important as detection of low-concentration metabolites is often sensitivity-limited. To attain the highest possible

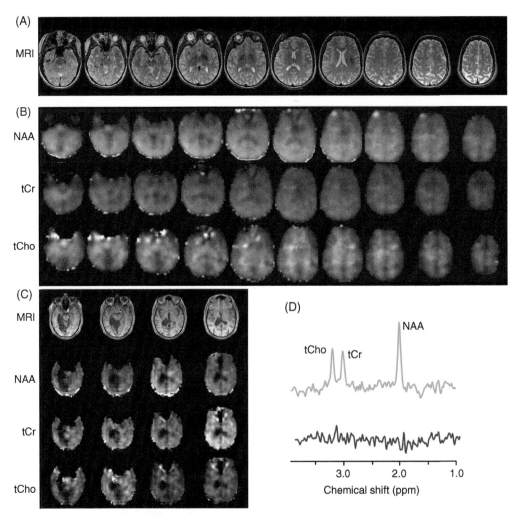

Figure 7.8 **EPSI on human brain *in vivo*.** (A) MRI and (B) MRSI data of a normal subject at 3T acquired with multi-slice 2D MRSI based on echo-planar data acquisition. Data were acquired with a phased-array receiver at a nominal spatial resolution of 4.4×4.4×7.0 mm (=0.14 ml) in 20 min (TE = 70 ms). (C) MRI and echo-planar MRSI data acquired on a stroke patient. (D) Representative ^{1}H NMR spectra from the positions indicated in (C). The abundant [1]H-containing metabolites NAA, tCr, and tCho can be detected with excellent sensitivity. *Source:* Data from Andrew A. Maudsley.

sensitivity, it is important that the contributions of the different receiver coils are properly combined ([36], see also Chapter 10). The most straightforward method is to use the first point of the FID or the integral of (part of) the NMR spectrum to phase-correct and amplitude-scale the MRSI signal for each coil at each spatial location before adding the contributions of the different coil elements. On water-suppressed MRSI data this approach may not always be viable due to the presence of randomly phased, unsuppressed water and lipid signals and/or low spectral S/N. The availability of a non-water-suppressed MRSI dataset from which the phases and weights can be determined is a very robust method to optimally combine the various receiver elements, but does require additional measurement time. The acquisition of sensitivity maps is a general method for the optimal combination of phased array data, but requires the availability of a volume coil with homogeneous sensitivity.

Rather than increasing the sensitivity, the phased array coils may also be used for image acceleration as discussed in Chapter 4. SENSE and GRAPPA MRI are standard MRI acceleration methods used to improve the temporal resolution of MRI several-fold [37, 38]. The use of parallel imaging methods is equally applicable to MRSI acquisitions [36, 39, 40]. By omitting, for example, every other phase-encoding step in MRSI the FOV will be halved, leading to signal aliasing. By combining the MRSI data with the sensitivity profiles of the phased array coils, the ambiguity caused by k-space undersampling can be resolved. Similar to MRI, the data reconstruction quality for parallel MRSI is determined by the ability of the receiver coils to supplement the missing gradient encoding, as expressed by the coil geometry or g-factor in Eq. (4.19) (see also Figure 4.12). In addition, the S/N of the reconstructed image is reduced by the square root of the acceleration factor R. An important consideration for SENSE-MRSI is that the sensitivity maps cover all locations that have a significant PSF contribution in order to avoid incomplete "unfolding" of strong lipid signals near the edge of the sensitivity maps [36]. Parallel MRSI can be combined with ultrafast MRSI methods, such as EPSI to provide unprecedented temporal resolution. However, it should be realized that shorter measurement times always come with a reduced sensitivity.

7.5.3 Methods Based on Prior Knowledge

One of the many strengths of MRSI is that no assumptions are made about the object under investigation. However, this often results in the acquisition of voxels containing no or unwanted signals, for example, from outside the object. In addition, many voxels can arise from a homogenous compartment, like the ventricles, for which it would have been more advantageous to obtain a single high-sensitivity spectrum.

In many cases, some prior knowledge of the object, like spatial position, orientation, and compartments, is available which can be used to increase the efficiency of k-space sampling. The SLIM [41], SLOOP [42], and SLAM [43] techniques use anatomical prior knowledge to limit k-space sampling, thereby greatly increasing the temporal resolution. Theoretically, N compartments of arbitrary shape can be accurately described by N k-space samples, potentially leading to greatly increased temporal resolution.

Suppose that an object is divided into N ROIs and that the MRSI data are acquired with M different phase encoding increments (where $M \geq N$). The amount of signal $P_m(t)$ acquired during phase-encoding step m can then be expressed as

$$P_m(t) = \sum_{n=1}^{N} c_{mn} S_n(t), \quad \text{with } c_{mn} = \int_{\text{compartment } n} e^{ik_m r} d^3 r \qquad (7.15)$$

where $S_n(t)$ represents the signal from compartment n and c_{mn} is a complex-valued weighting factor expressing the acquired phase of all positions within compartment n during phase-encoding increment m, characterized by k-space parameter k_m. The integral in Eq. (7.15) is over all spatial position r in the 3D compartment n. Equation (7.15) can be written in matrix form as $P_M(t) = CS_N(t)$, where $P_M(t)$ designates a $(M \times 1)$ vector of M acquired phase-encoded signals and $S_N(t)$ a $(N \times 1)$ vector of the N signals from each of the compartments in the absence of phase encoding. C is a complex $M \times N$ matrix with the elements c_{mn} given in Eq. (7.15). The desired compartmentalized spectra $S_N(t)$ can then be obtained through matrix inversion of the encoding matrix C according to $S_N(t) = C^{-1}P_M(t)$. For a medium-sized encoding matrix C, the inverse can be calculated through singular value decomposition (SVD). Temporal Fourier transformation of the reconstructed $S_n(t)$ finally provides all compartmentalized spectra.

Figure 7.9 shows an example of the incorporation of prior knowledge to increase the temporal resolution of MRSI data acquisition using the SLIM algorithm. Figure 7.9A shows an anatomical image of a four-compartment phantom containing 50 mM creatine, glycine, and NAA in the smaller tubes and 500 mM acetate in the larger tube. In order to obtain compartmentalized NMR spectra from the four compartments with negligible amounts of signal

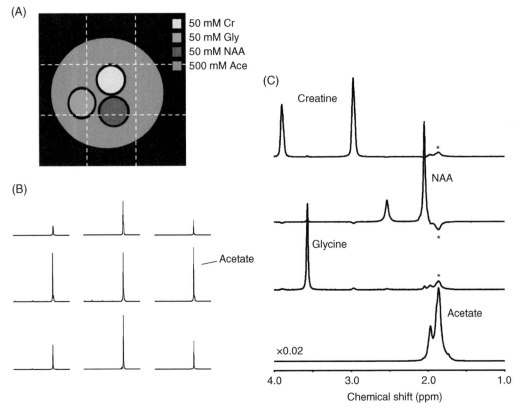

Figure 7.9 Spatial localization using prior knowledge. (A) Four-compartment phantom, containing different metabolites in each compartment. (B) 3×3 ^{1}H MRSI dataset, reconstructed by 2D FFT processing. The low spatial resolution in combination with extensive PSF-related signal bleeding leads to the observation that every pixel is heavily dominated by the concentrated acetate resonance. (C) By utilizing prior knowledge on the spatial positions of the compartments as shown in (A), the SLIM algorithm can reconstruct compartmentalized ^{1}H NMR spectra from a 3×3 k-space with greatly reduced contamination (*) from other compartments.

leakage from other compartments, a roughly 32×32 MRSI dataset would be required. However, under the assumption of homogenous compartments, the SLIM algorithm would only require four phase-encoded FIDs (e.g. a 2×2 matrix). Figure 7.9C shows the compartmentalized NMR spectra calculated from an overdetermined 3×3 MRSI dataset, while Figure 7.9B shows the 3×3 FFT-based spectroscopic image. Clearly by imposing anatomical prior knowledge the SLIM algorithm allows the calculation of accurate compartmentalized NMR spectra with only a small number of phase-encoding increments. The small amount of acetate contamination must be attributed to B_0 and/or B_1 magnetic field inhomogeneity between the compartments. The inter-compartment signal contamination can be further reduced with extended k-space sampling, use of higher-order gradient fields [44], and the incorporation of additional prior knowledge. Other techniques have been described that incorporate various levels of prior knowledge to improve the sensitivity, temporal, or spatial resolution of MRSI. Most noticeable is the spectroscopic imaging by exploiting spatiospectral correlation (SPICE) method [45, 46] that combines low-resolution, high-sensitivity MRSI with high-resolution, low-sensitivity MRSI to generate images that have the best features of both. However, it should always be realized that any method that relies on the incorporation of prior knowledge becomes dependent on the accuracy of that knowledge. In the case of SLIM and its related techniques, incorrect prior knowledge will lead to signal leakage/contamination between compartments (Figure 7.9).

7.6 Lipid Suppression

The PSF as described in Section 7.4 leads to a less than ideal voxel definition which gives rise to intervoxel contamination or bleeding. Fortunately, the contribution of all tissue types to a given voxel can be quantitatively calculated by using the PSF in combination with segmented anatomical MRI data (see Section 7.7). This allows the measurement of metabolism in pure tissue types despite significant voxel contamination [47, 48].

However, when the small metabolite signals are detected in the presence of large signals, like extracranial lipids during MRSI of the brain, the intervoxel contamination can dominate the appearance of the spectra. Figure 7.10 shows ^{1}H MR spectra extracted from a 2D MRSI dataset acquired from human brain. Even though the extracranial lipids are residing outside the brain, the finite intensity of the PSF at positions located well inside the brain is sufficient to cause significant contamination from intense lipid resonances, thereby complicating the quantification of the underlying metabolite resonances. In most cases the solution to this problem is found by removing the lipid signals prior to or during acquisition as will be described for methods that rely on differences in T_1 relaxation (Section 7.6.1), inner volume selection (IVS) and volume prelocalization (Section 7.6.2), and OVS (Section 7.6.3). In addition to these commonly employed methods, a wide range of lipid suppression methods have been reported. These include methods using frequency-selective excitation [49], increased and modified k-space coverage [50, 51], RF-based gradients [52, 53], dedicated crusher coils [54], and higher-order magnetic field gradients [55].

7.6.1 Relaxation-based Methods

Methods based on the differences in relaxation between lipids and metabolites are among the oldest techniques to achieve lipid suppression. The methods are typically straightforward to implement and can give reasonable lipid suppression. However, there are two related methodological considerations that need to be taken into account, namely signal loss of the metabolites of interest and the attainable degree of lipid suppression.

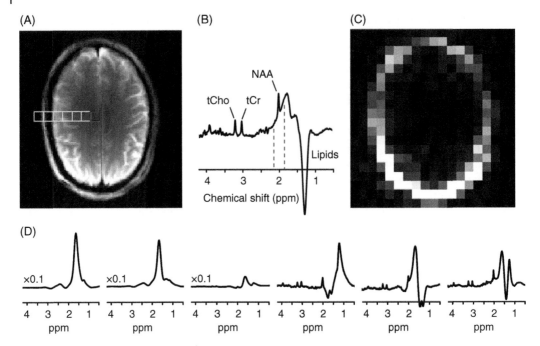

Figure 7.10 Lipid contamination in ^{1}H MRSI of the human brain. (A) MRI showing the location of seven MRSI voxel locations in a 21×21 MRSI dataset (TR/TE = 2000/50 ms) acquired without lipid suppression. (B) ^{1}H MR spectrum extracted from the middle of the brain (red voxel in A) showing large amounts of lipids. (C) Metabolic image of NAA obtained by integrating each MRSI voxel between 1.85 and 2.15 ppm (red lines in B). The metabolic image is dominated by lipid signals residing in the skull regions. (D) ^{1}H MR spectra from the yellow voxel locations in (A) revealing the large lipid signal originating from the skull. Note that the left three spectra are scaled down by a factor of 10.

Transverse T_2 relaxation is typically not used for lipid suppression, since the difference between metabolite and lipid T_2 relaxation is too small. At the echo-time required to achieve sufficient lipid suppression, the metabolites would also have greatly decayed due to T_2 relaxation. This would lead to a low-metabolite SNR, as well as a heavily T_2-weighted spectrum which makes signal quantification more complicated. Even though T_2 relaxation alone is not sufficient for adequate lipid suppression, the decrease in lipid intensity at longer echo-times can definitely aid in achieving high-quality MR spectra with secondary lipid suppression methods.

The difference in longitudinal T_1 relaxation between metabolites and lipids is typically larger, with the metabolite and lipid T_1s being well above 1000 ms and between 200 and 400 ms at most magnetic fields, respectively [56, 57]. This provides an opportunity for selective lipid suppression through the use of an inversion recovery (IR) sequence. By choosing the inversion delay such that the longitudinal lipid magnetization is zero, the lipids are effectively not excited. The metabolites are still along the $-z$ axis due to the much longer T_1 relaxation time constants and will, upon excitation be detected as negative resonances. Figure 7.11C shows a signal recovery curve for a typical situation in which the inversion delay is optimized to null compounds with a T_1 of 300 ms (e.g. inversion delay = $T_1\ln(2)$ = 208 ms) and TR = 2000 ms. It follows that while lipids are suppressed, the metabolites (T_1 = 1000–1600 ms) are observed at circa 50% of the thermal equilibrium magnetization. Note that the T_1 weightings introduced by the short TR and the inversion pulse partially cancel each other, leading to similar metabolite intensities which in turn simplify metabolite quantification. Increasing the TR to $5T_{1max}$, as

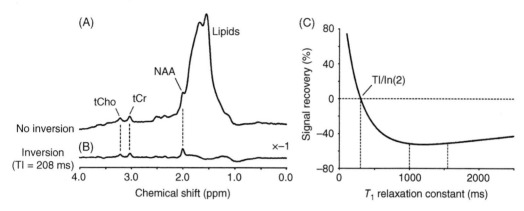

Figure 7.11 **Inversion-recovery-based lipid suppression.** (A) ^{1}H NMR spectrum (TR/TE = 2000/50 ms) acquired without lipid suppression from a single slice through the human brain. (B) The lipid resonances can be significantly reduced through the use of an inversion recovery method in which the inversion time is adjusted to minimize signal with a T_1 of 300 ms. Metabolites with a longer T_1 can be detected at circa 50% of their maximum intensity. (C) Signal recovery curve as a function of T_1 relaxation for an inversion recovery sequence with TR/TI = 2000/208 ms. The T_1 relaxation time constant for which signal nulling is achieved is given by TI/ln(2).

often recommended to minimize T_1 effects in metabolite quantification, would actually result in a strong T_1-dependent metabolite signal recovery. An obvious drawback of a single IR method is that perfect suppression can only be achieved for a single T_1 relaxation constant. It has been shown that extracranial lipids exhibit a range of T_1s [57] thus making the suppression incomplete. Improved suppression can be obtained with a double IR method in which two inversion pulses and two delays can be optimized to achieve suppression over a wider range of T_1 relaxation constants. However, the improved lipid suppression comes at the price of reduced metabolite signal recovery. Figure 7.11B shows a typical result of single IR-based lipid suppression on a single slice in the human brain. While the lipid suppression is not perfect, it is sufficient to reduce the intervoxel bleeding in MRSI to negligible levels. At higher magnetic fields the RF field can be adjusted or "shimmed" in such a manner as to have maximum RF amplitude in the skull region with minimal RF field amplitude in the brain. Under these conditions the T_1-based lipid suppression only affects the skull region, such that the metabolite signals can be detected at full intensity [52, 53].

7.6.2 Inner Volume Selection and Volume Prelocalization

The most commonly employed methods for lipid suppression, namely OVS and inner volume selection (IVS) or volume prelocalization, utilize the difference in spatial origins of lipids and metabolites. Volume prelocalization methods use single-shot localization techniques like STEAM, PRESS, or (semi)-LASER as described in Chapter 6 to select a relatively large rectangular volume inside the brain (Figure 7.12). Since the extracranial lipids are not excited/refocused they do not contribute to the detected signal. All considerations discussed for single-volume localization are also valid for volume prelocalization in MRSI. However, two effects become especially pronounced during volume prelocalization, namely imperfect slice profiles and the chemical shift displacement.

A well-behaved slice profile is characterized by uniform in-slice signal, no signal outside the slice, and a relatively narrow transition zone (ideally <10% of the total slice). However, over a wide slice as used in MRSI volume prelocalization, even a 10% transition zone can cover one or

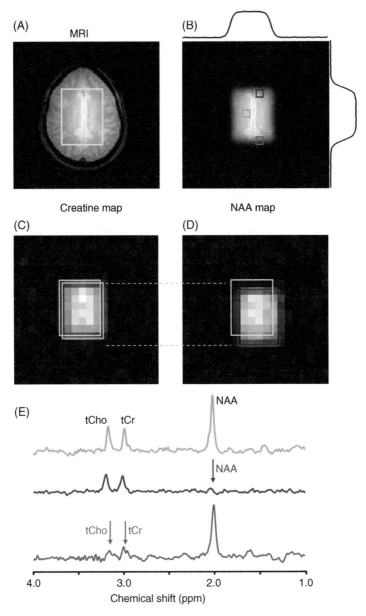

Figure 7.12 MRSI and volume prelocalization. (A) MRI showing the planned volume for prelocalization. (B) MRI showing the actual volume selected by PRESS localization. (C, D) Metabolic maps for (C) total creatine and (D) *N*-acetyl aspartate (NAA) obtained by spectral integration of the respective metabolite resonances in each MRSI voxel. The orange (in (C)) and red (in (D)) boxes indicate the signal boundary of the creatine and NAA metabolic maps and are clearly displaced relative to each other and to the yellow box indicating the planned (water) signal boundary. (E) ^{1}H MR spectra extracted from the positions shown in (B) indicate that voxels at the pre-localized volume edge have anomalous metabolic profiles due to the chemical shift displacement artifact.

more MRSI voxels thereby leading to artificially reduced metabolite levels. One should therefore always use RF pulses with the narrowest transition zones (see also Chapter 5). However, since slice transition zones are unavoidable it is imperative that the slice profiles are measured or calculated and used to correct the MRSI data.

The chemical shift displacement has been extensively discussed in Chapter 6. Since the slices during MRSI volume prelocalization are much wider than in regular single-volume localization, the absolute displacement is much larger and can span one or several spectroscopic volumes. Unlike in single-volume NMR spectroscopy, the chemical shift displacement during volume prelocalization becomes spatially resolved by the phase-encoding gradients, which leads to distorted metabolite ratios even in a spatially homogeneous sample. Figure 7.12 demonstrates the consequences of chemical shift displacement in MRSI with volume prelocalization. The MR image of the pre-localized volume (Figure 7.12B) and the metabolic images of tCr (Figure 7.12C) and NAA (Figure 7.12D) indicate a successful exclusion of lipids, coming at the cost of significant slice transition zones (Figure 7.12B). A closer inspection of the metabolic images (Figure 7.12C and D) reveals that the creatine and NAA images are displaced with respect to the water-based volume definition (yellow box, Figure 7.12A). Furthermore, the creatine (orange box) and NAA (red box) metabolic imaging are displaced relative to each other. When inspecting ^{1}H MR spectra from selected MRSI voxels (Figure 7.12B), one may come to the conclusion that the spectra from the blue and especially the red voxel represent abnormal tissue due to the lack of creatine and NAA. Of course, the distorted metabolic profiles are due to the chemical shift displacement artifact, making all ^{1}H MR spectra from voxels near the edge of the pre-selected volume suspect. As was mentioned in Chapter 6, the only way to minimize the chemical shift displacement artifact is to increase the RF bandwidth. For MRSI volume prelocalization it is therefore always recommended using RF pulses with the largest bandwidth and smallest transition width, giving the limitations imposed by RF power deposition and magnetic field gradient amplitudes. The use of FOCI or GOIA pulses in volume prelocalization is highly recommended due to their greatly improved slice profile and increased bandwidth [49, 58]. However, chemical shift displacement and imperfect slice profile artifacts are *always* present and should be quantitatively assessed.

While volume prelocalization gives excellent lipid suppression and is easy to use and implement, it also has a number of disadvantages. Firstly, using the basic single-volume localization methods, the pre-localized volume must be rectangular in shape. The human (or animal) brain is not rectangular in any slice orientation (see Figure 7.12A), such that volume prelocalization necessarily destroys signal from cerebral tissues on the edge of the brain. This can be problematic when abnormalities are present at the brain periphery, such as encountered in stroke or tumors. Furthermore, whole-brain multi-slice acquisitions are complicated by through-slice interference of the volume prelocalization method. This is especially troublesome when the in-plane volume size and position must change to encompass the changing brain size and position across multiple slices. As a result, volume prelocalization is not commonly used for multi-slice 2D MRSI. However, it is a popular and convenient choice for phase-encoded 3D MRSI [49].

7.6.3 Outer Volume Suppression (OVS)

OVS works essentially opposite to the strategy employed by volume prelocalization. Rather than avoiding the spatial selection of lipids, OVS excites narrow slices centered on the brain in lipid-rich regions (Figure 7.13A). Following slice-selective excitation, the transverse magnetization is dephased ("spoiled") by a magnetic field crusher gradient (Figure 7.14A). Following the OVS modules, the brain water and metabolite signals can be excited without lipid contamination. OVS is a general method to selectively remove unwanted signal. On most clinical MR systems, a minimum of eight OVS slices can be arbitrarily placed around the brain, thereby approximating the roughly elliptical shape of the brain as shown in Figure 7.13A. OVS is readily

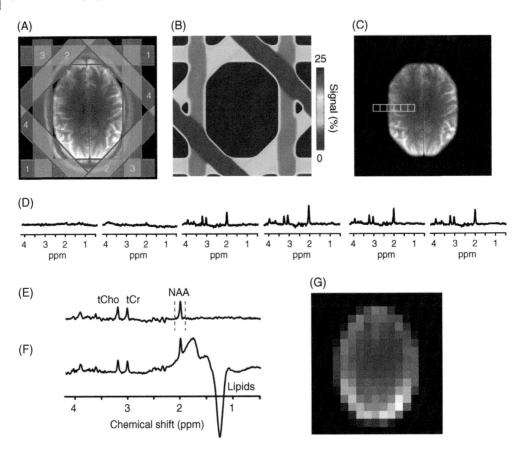

Figure 7.13 **Outer volume suppression on human brain** *in vivo.* (A) MRI showing the eight slices typically used for OVS in an axial plane, together with their excitation order. Parallel slices have an equal excitation order as they are often excited (nearly) simultaneously. (B) Signal intensity for one OVS module (90° excitation for each slice, 10 ms between slices) in the absence of RF magnetic field inhomogeneity for $T_1 = 300$ ms. The 40 ms delay between OVS excitation 1 and global excitation leads to a 12.5% signal recovery in several locations. The shorter 10 ms delay for OVS excitation 4 gives a 3% recovery, or, in other words, a robust 30^+-fold suppression. (C) MRI obtained with three OVS modules with optimized nutation angles and delays. (D) ^{1}H MR spectra extracted from the yellow voxels shown in (C). The OVS-based lipid suppression is robust enough to minimize lipid signal leakage to voxels within the brain. (E, F) ^{1}H MR spectra from the center of the brain extracted from MRSI (E) with and (F) without OVS. (G) NAA metabolic image obtained by integrating each MRSI voxel between 1.85 and 2.15 ppm (red lines in E). The metabolic map provides the spatial NAA distribution without contamination from lipids (compare with Figure 7.10C without OVS). Note that the NAA map has not been corrected for the coil sensitivity profile.

used in multi-slice acquisitions, since the through-slice angle of the OVS slices can be adjusted to minimize slice-to-slice interference [59]. Furthermore, since OVS selects relatively narrow slices, artifacts originating from imperfect slice profiles and chemical shift displacements are typically negligible.

However, OVS is not immune to experimental imperfections and the performance of OVS can be severely compromised by T_1 relaxation and RF magnetic field inhomogeneity. Figure 7.13B shows the performance of OVS in the presence of T_1 relaxation. Since the OVS modules for nonparallel slices are separated in time, the signal recovery due to T_1 relaxation depends on the slicing order and timing. Even in the absence of RF magnetic field inhomogeneity (Figure 7.13B) the signal suppression in some OVS areas is less than 10-fold due to T_1

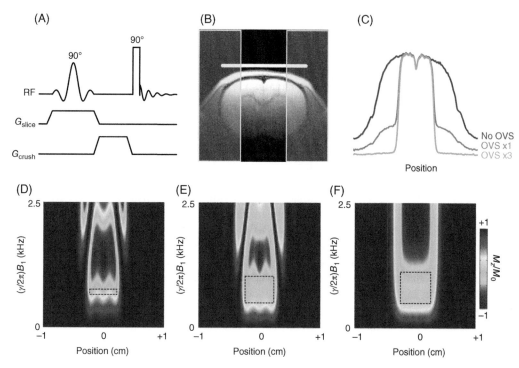

Figure 7.14 **Outer volume suppression and RF inhomogeneity.** (A) General OVS pulse sequence element. For parallel slices the RF pulses can be executed during the same slice-selection gradient. (B) OVS slices perpendicular to the surface coil plane are characterized by severe RF inhomogeneity, (C) leading to poor OVS performance for a single OVS module. The suppression is greatly improved by repeating the OVS module three times. (D–F) Residual magnetization following (D) one and (E) three repeated OVS modules consisting of a 2.5 ms five-lobe sinc pulse or (F) three repeated OVS modules consisting of a 3.33 ms hyperbolic secant adiabatic RF pulse with a 3 kHz bandwidth. Only the adiabatic RF pulse has a well-behaved slice profile over a wide range of RF amplitudes. The dotted boxes in (D–F) indicate the B_1 range over which reasonable OVS performance is expected.

relaxation in between the OVS pulses. Simply repeating identical 90° OVS modules in the absence of RF inhomogeneity will not improve the performance. However, OVS performance can be improved significantly by repeating the modules with varying nutation angles and inter-pulse timings (Figure 7.13C), similar to the strategy used for water suppression (Figures 6.23 and 6.25). Lipid suppression factors between 20- and 50-fold are achievable with careful optimization of the OVS modules [60, 61]. The suppression of lipid resonances does not need to be perfect, as long as intervoxel bleeding of lipid signals due to the PSF is reduced to negligible levels (Figure 7.13D–G).

For many OVS applications the use of a conventional frequency-selective RF pulse is acceptable. However, in the presence of strong RF inhomogeneity (Figure 7.14), the choice of RF pulse becomes important. A single frequency-selective RF pulse has little tolerance for RF inhomogeneity and only provides acceptable suppression for a nutation angle close to 90° (Figure 7.14D). The insensitivity to RF inhomogeneity can be improved by repeating the OVS modules three times (Figure 7.14E), but comes at the expense of a degraded slice profile. An adiabatic hyperbolic secant pulse can achieve a similar level of immunity to RF amplitude, while maintaining its slice profile (Figure 7.14F). Extending the adiabatic OVS pulse train with more RF pulses leads to the BISTRO method [62] for high-quality signal suppression independent of the RF amplitude.

7.7   MR Spectroscopic Image Processing and Display

Multidimensional spectroscopic imaging datasets normally contain a vast amount of spatial and spectral information. This information can only be optimally evaluated when the data are processed correctly and displayed in a proper manner. The processing and display largely depend on the objective of the study and the nature of the acquired data. Here, a number of important considerations concerning the processing and display of MRSI datasets are discussed.

MRSI data acquired with phased array receivers require summation of the separate receiver channels. In order to maximize the sensitivity, the summation needs to be performed in a phase-sensitive manner in which each receiver is weighted by the spectral S/N of the MRSI data at a given spatial location. Access to a non-water-suppressed MRSI dataset on the same subject can greatly simplify the summation by providing high-sensitivity, single-frequency phase correction, and amplitude weighting. In addition, the water MRSI can be used for removal of eddy currents through a post-acquisition B_0 correction (see Chapter 9). More information on phased array data summation can be found in Chapter 10.

Prior to spatial Fourier transformation the MRSI data can undergo apodization as described in Section 7.4. When no apodization is employed, the nominal voxel volume is simply given by the FOV divided by the number of phase-encoding increments. After apodization the real voxel size will be increased and can be obtained by (numerical) integration of the PSF (e.g. see Figure 7.3). To allow comparison between MRSI datasets, it is informative to calculate and report the real voxel volumes.

The time domain signals can be zero-filled to improve the final appearance of the processed MRSI dataset (either in spectral or image form). It should be noted that zero-filling does not affect the PSF, such that the real resolution of the final image remains unchanged. Zero-filling merely allows the calculation of NMR spectra at intermediate spatial locations, but does not enhance the information content. Following the spatial Fourier transformation, an n dimensional MRSI dataset (n = 1, 2, or 3) of spatially resolved FIDs (or echoes) is obtained. These time domain signals can undergo the normal processing, like apodization and zero-filling, as described in Chapter 1. The final spectral Fourier transformation results in the spatially resolved NMR spectra.

Even though a 2D MRSI dataset typically holds information on hundreds of voxels, the spatial resolution is most often insufficient to ignore partial volume effects. This problem can be partially reduced by increasing the number of phase-encoding increments at the expense of an increased experimental measurement time. As an alternative the acquired MRSI dataset may be recalculated at intermediate spatial positions, such that the positions of the voxels coincide more closely with anatomical features. This can be achieved by the so-called shift theorem. Fourier transformation of a 1D MRSI dataset $f(k, t)$ gives the spatially resolved FIDs or echoes $F(r, t)$, where r represents spatial position along any Cartesian direction or any vectorial combination thereof. The spatially resolved FIDs can also be calculated at intermediate positions $r + \Delta r$, by multiplying the original MRSI data, prior to Fourier transformation, by $e^{-i\Delta rk}$, i.e.

$$F\left(r+\Delta r,t\right)=\text{FT}\left[f\left(k,t\right)e^{-i\Delta rk}\right] \tag{7.16}$$

The shift theorem of Eq. (7.16) is readily extended to three spatial dimensions. In fact, one of the advantages of acquiring a complete 4D (i.e. 3D spatial and 1D spectral encoded) MRSI dataset is that the metabolic images can be recalculated in all dimensions to match anatomical and/or pathological features.

Spectroscopic imaging datasets, exhibiting features of both MRI and MRS, can be displayed in several different ways which either focus on the spectroscopic or the spatially resolved

(A)

(B)

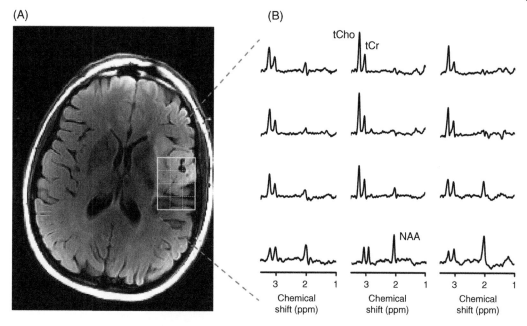

Figure 7.15 **MRSI display with a spectroscopy emphasis.** (A) FLAIR image of a patient bearing a brain tumor showing a 3×4 grid of MRSI voxels as part of a 21×21 MRSI dataset (1 ml volumes, TR/TE = 2000/144 ms). (B) ^{1}H MR spectra from the grid locations shown in (A) provide immediate visual inspection of spectral quality and *S/N*. The characteristic brain tumor metabolic profile of increased choline and decreased NAA can be recognized in the top 3×3 grid, whereby the lower row of three spectra is more indicative of normal brain tissue. *Source:* Data from Henk De Feyter and Robert Fulbright.

information. One of the most straightforward methods of display is to extract and display a number of important spectra from the dataset (Figure 7.15). This allows a rigorous evaluation of the spectral quality and allows the detection of small spectral changes. However, this method of display largely ignores the spatial information of the data and, if one is only interested in these particular voxels, it should be considered if the acquisition of a limited number of single voxel localized spectra (possibly by multi-voxel localization, see Section 7.8) is more time-efficient. For publication of MRSI-based results, one should always show a number of MR spectra from individual MRSI voxels so that the overall spectral quality (line widths, *S/N*, water and lipid suppression) can be evaluated.

Alternatively, one can display spectra from all voxels in a 2D (or 3D) matrix, possibly overlaying an anatomical image. This allows the evaluation of (differences between) spectra in all voxels. However, this type of representation may be too complex in a clinical setting where it is desirable to evaluate spectral changes in voxels at a first glance. For this purpose, a so-called metabolic map or image may be generated of one or more metabolite resonances. This can be accomplished by integrating or fitting the resonance of interest in each voxel and mapping the signal intensity over the entire MRSI dataset. Since a metabolic image obscures the underlying MRSI data quality, it is important to report quality assurance maps in conjunction with the metabolic image (Figure 7.16). Figure 7.16A–C shows three MR spectra extracted from a 2D MRSI at the spatial positions shown in the MR image of Figure 7.16D. The MR spectrum in Figure 7.16A displays the hallmarks of brain tumors, namely increased Cho and decreased NAA. While clearly metabolically abnormal, the spectral quality is excellent with narrow, well-separated resonances, good *S/N*, and the absence of strong water or lipid resonances. The spectrum

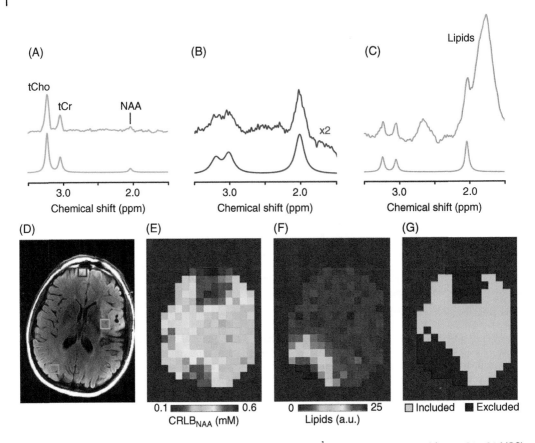

Figure 7.16 **Spectral fitting and quality assurance in MRSI.** (A–C) ¹H MR spectra extracted from a 21 × 21 MRSI dataset (1 ml, TR/TE = 2000/144 ms) at the locations indicated in (D) a FLAIR image. ¹H MR spectra shown are the measured data (top) and the best spectral fit of total choline (tCho), total creatine (tCr), and NAA signals (bottom) in addition to a polynomial baseline. (E) Cramer–Rao lower bound (CRLB) map for the NAA resonance has a strong correlation with the magnetic field inhomogeneity or B_0 map (not shown). (F) Lipid map revealing significant lipid contamination in part of the brain, likely due to an incorrectly placed OVS slice. (G) Binary inclusion/exclusion map based on the NAA CRLB and lipid maps.

provides a high-quality spectral fit and the results are deemed reliable. Note that all considerations for spectral fitting of ¹H MR spectra, such as spectral basis sets, prior knowledge, line shape distortions, and baseline model automatically become part of the considerations underlying metabolic imaging. Spectral fitting is discussed in detail in Chapter 9. The MR spectrum from the frontal cortex (Figure 7.16B) displays broadened resonances due to decreased magnetic field homogeneity. While the spectral fit appears adequate, the reliability of the metabolite levels must be deemed lower than those in Figure 7.16A. The reliability due to S/N, resonance line widths, and spectral overlap can be quantitatively expressed in terms of Cramer–Rao lower bounds (CRLBs) as will be discussed in Chapter 9. Figure 7.16E shows a NAA CRLB map showing that the error or uncertainty on the NAA levels in the frontal cortex is much higher than in more central brain regions. The CRLB map has, in general, a strong correlation with a magnetic field B_0 map (not shown). Finally, the spectrum in Figure 7.16C is compromised by excessive lipid resonances, as confirmed by a lipid metabolic image (Figure 7.16F). While the CRLBs may still be low due to the high S/N and narrow lines, the reliability of the resonance areas must be questioned in the presence of large, unwanted signals.

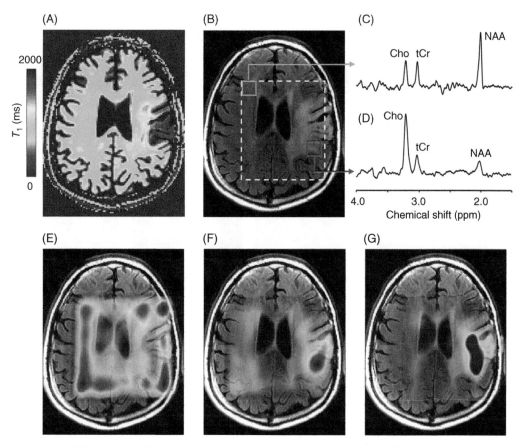

Figure 7.17 **MRSI display with an imaging emphasis.** (A) Quantitative T_1 map and (B) FLAIR image of a patient bearing a brain tumor. 2D MRSI data (TR/TE = 2000/144 ms, 16×16 matrix over 160×160 mm) acquired with semi-LASER volume prelocalization (120×80×10 mm) demonstrate high-quality spectra from (C) normal-appearing brain tissue and (D) tumor tissue. (E, F) Metabolic maps for (E) NAA and (F) total choline and (G) a choline-to-NAA ratio map. *Source:* Data from Henk De Feyter and Robert Fulbright.

During the generation of a metabolic image, spectra from all MRSI voxels (above a minimum intensity threshold) will undergo spectral fitting. However, due to a combination of low S/N, B_0 and B_1 magnetic field inhomogeneity, incomplete water and lipid suppression, and the presence of artifacts, the reliability of a number of spectra will be unacceptable. A data inclusion/exclusion map is the most basic, binary method to visualize the data quality (Figure 7.16G), whereas more gradual, color-coded quality assurance maps represent a more subtle visualization.

Figure 7.17 shows the presentation of MRSI data with an emphasis on metabolic images. The NAA (Figure 7.17E), Cho (Figure 7.17F), and especially Cho-to-NAA ratio (Figure 7.17G) maps clearly reveal an abnormal metabolic profile that correspond to the brain tumor locations visible on the T_1 map (Figure 7.17A) and FLAIR image (Figure 7.17B). Individual ^{1}H NMR spectra (Figure 7.17C) from key spatial locations confirm the metabolic profiles and allow the visualization of spectral quality. Ideally, the metabolic images as shown in Figure 7.17E–G are accompanied by quality assurance or error maps, similar to those shown in Figure 7.16.

A final step in the acquisition, processing, and display of MRSI data is the calculation of absolute metabolite concentrations (in mM or µmol g^{-1} of tissue). Chapter 9 provides a detailed

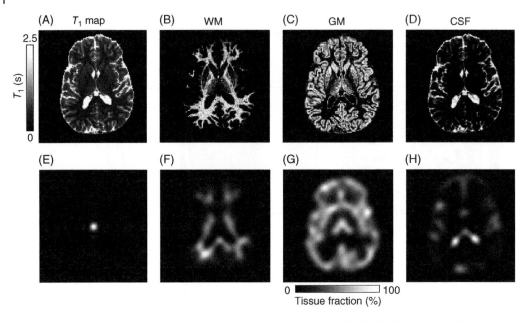

Figure 7.18 Image segmentation for partial volume correction in MRSI. (A) Quantitative T_1 map from human brain, together with segmented images for (B) white matter, (C) gray matter, and (D) cerebrospinal fluid. In order to account for the limited spatial resolution of the MRSI dataset, the (E) point spread function can be convolved with the high-resolution segmented images to produce segmented images at the same resolution as the MRSI dataset (F–H). *Source:* Data from Graeme F. Mason.

summary of the steps necessary for metabolite quantification. Here, some specific considerations for MRSI will be highlighted. Despite the fact that MRSI data can be spatially shifted to match MRSI voxels with tissue boundaries, partial volume effects are the rule rather than the exception. It is therefore crucial to acquire MRSI data in conjunction with high-resolution images that allow tissue segmentation. While there are many available methods for tissue segmentation, the requirements for brain MRSI studies are that the MRI method should be fast relative to the MRSI acquisition time and that it allows for segmentation of gray matter, white matter, and cerebrospinal fluid. Figure 7.18 shows an example of brain segmentation based on quantitative T_1 mapping (see Chapter 4). Since MRSI voxels are not rectangular, it is important to convolve the segmentation maps with the MRSI PSF (Figure 7.18E) to give the real tissue composition in the corresponding MRSI voxels (Figure 7.18F–H).

7.8 Multivolume Localization

In situations where it is required to acquire signal from a small number of voxels, spectroscopic imaging does not provide acceptable localization due to the unfavorable PSF associated with limited k-space sampling. The localization can be improved significantly by performing successive single voxel localization experiments at the cost of a reduction in signal-to-noise per unit time.

Multivolume localization differs from MRSI in that it is based on single-volume methods (i.e. slice selection by RF pulse/gradient combinations) extended for the acquisition of multiple volumes. As such, it will have the localization accuracy of single-volume methods, but without the cost of reduced sensitivity as signal from all multiple volumes is acquired during each repetition time. Multivolume or voxel localization techniques are naturally divided into methods that acquire signal from several voxels simultaneously [63–67] or sequentially [68–70]. When the signal is acquired simultaneously, the voxels are separated post-acquisition by the Hadamard localization principle.

7.8.1 Hadamard Localization

Simultaneous multi-voxel localization is most straightforward when the spatial encoding is performed according to an $N \times N$ Hadamard matrix, where N denotes the order of the square matrix. With longitudinal Hadamard encoding, the magnetization from different spatial locations is stored along the z axis as $\pm M_z$ through selective inversion in subsequent experiments. With transverse Hadamard encoding the magnetization is encoded as $\pm M_y$ by inverting the phase of the selective excitation pulse or shifting the phase of the refocusing pulse by 90° in subsequent acquisitions.

The selective inversion profiles for multiple slices can be generated either by several individual RF pulses or by one RF pulse exhibiting a multifrequency inversion profile. These so-called multifrequency RF pulses simultaneously generate multiple selective excitation or inversion bands and are discussed in Chapter 5. The use of multifrequency RF pulses is limited to low-order Hadamard encoding, since the peak RF amplitude scales linearly with the number of slices (see also Chapter 5). Figure 7.19A shows the four experiments necessary to encode the magnetization according to a 4×4 Hadamard matrix. Slices at different spatial positions are encoded as positive or negative by selectively inverting magnetization prior to excitation. Following acquisition and data storage, the individual volumes can be retrieved through multiplication with a Hadamard matrix. Note that longitudinal fourth-order Hadamard encoding can only resolve three of the four slices. The signal from the fourth slice position is never inverted and can

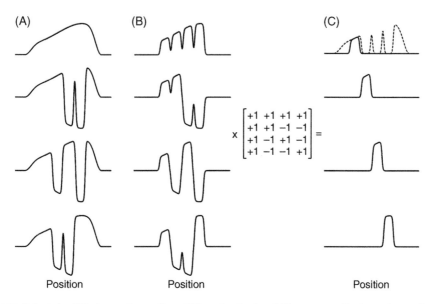

Figure 7.19 **Principle of Hadamard encoding.** (A) Longitudinal and (B) transverse Hadamard encoding of four spatial slices. Multiplication of the four separately acquired signals with a Hadamard matrix allows (C) the selection of the individual slices. Note that longitudinal Hadamard encoding only allows the detection of three separate slices, whereas the fourth slice is convolved with all other signals from outside the four slices (dotted line).

therefore not be distinguished from the signal from outside the four slice positions (Figure 7.19C, dotted trace). Besides this characteristic problem, longitudinal Hadamard encoding is hampered by the same limitations as ISIS localization, i.e. the receiver gain cannot be optimized for the localized volume and manual localized shimming is difficult since multiple acquisitions are required to define the total localized volume. Just as with ISIS localization, longitudinal Hadamard encoding schemes can easily be executed with adiabatic RF pulses, an essential point of consideration when surface coils are used for pulse transmission.

Alternatively, Hadamard encoding can be performed in the transverse plane. Discrimination of transverse magnetization is either achieved by changing the phase of the selective excitation pulse (Figure 7.19B) or by shifting the phase of the selective refocusing pulse by 90°. Note that since signal from outside the four slices is never excited (or refocused), transverse Hadamard encoding can resolve all four slices uniquely.

Hadamard encoding by itself is not frequently used as a spatial localization technique, but the combination of a low-order Hadamard encoding scheme and 2D MRSI is a straightforward method to achieve 3D coverage. When a 2D MRSI experiment requires some signal averaging, one can use the additional scans to spatially encode the object under investigation in the third spatial dimension by Hadamard localization, thereby obtaining multiple 2D MRSI datasets with the same sensitivity without increasing the experimental measurement time.

More recently, the spatial localization in simultaneous double-volume MR spectroscopy was resolved by employing the principle of sensitivity encoding as commonly used for accelerated MRI. By using the receiver sensitivity profiles, the MR spectra from two coplanar volumes could be uniquely resolved with minimal contamination [71].

7.8.2 Sequential Multivolume Localization

Sequential multivolume localization is a direct extension from multi-slice imaging, in which slices from other spatial positions can be excited and acquired during the recovery time of one slice (see Chapter 4). Regular single-volume localization methods like STEAM and PRESS are thus used to acquire signal from multiple volumes with a repetition time that encompasses all volumes (Figure 7.20). Two potential complications arise from intervoxel saturation effects and intervoxel signal contamination due to incomplete dephasing of unwanted signals.

In multi-slice MRI it is crucial that all RF pulses are slice-selective as to avoid slice-to-slice interference and signal saturation. The same principle applies to multivolume localization with the additional complexity that the slice-to-slice interference much be considered in three dimensions. With multiple volume chosen along a Cartesian axis (Figure 7.20A), the slice-selection gradients should select double-oblique slices as to avoid signal saturation [68, 70].

In Chapter 6, the dephasing of unwanted coherences in single-volume localization was discussed. The effect of coherences generated by scan n on the signal of scan $(n + 1)$ was ignored because the repetition time TR was long relative to T_1 and especially T_2 and T_2^* relaxation. However, while the overall repetition time and thus T_1-weighting remains constant, the time between the excitation and acquisition of subsequent volumes during a multivolume sequence becomes smaller as the number of volumes increases. It becomes therefore more likely that coherences from volume n survive and contribute to the signal from volume $(n + 1)$. Transverse coherences are readily destroyed by magnetic field crusher gradients between the acquisition of subsequent volumes, for example, immediately following acquisition. Longitudinal coherences which can potentially generate stimulated echoes can be dephased by having different crusher gradients during the TE (and TM periods) of the different volumes.

One unique aspect of sequential over simultaneous multivolume localization is that sequential acquisition allows a greatly improved magnetic field homogeneity when combined with

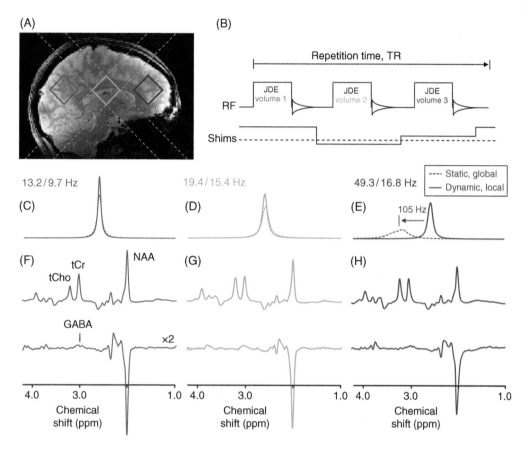

Figure 7.20 **Sequential multivolume localization with dynamic shimming and spectral editing.** (A) Sagittal MRI with three double-oblique volumes (21 ml) placed along the midline. The slice profiles of the central volume (green lines) indicate the absence of mutual slice overlap and signal saturation between the three volumes. (B) Multivolume *J*-difference editing pulse sequence (TR/TE = 3000/68 ms) for GABA detection with dynamic shimming of first- and second-order shims. (C–E) ^{1}H MR spectra of water from the three voxels acquired in the presence of a static shim optimized over the area of the three volumes combined (dotted line) or acquired with dynamically changing shims that are optimal for a given volume at the time it is measured (solid line). Numbers in the upper left corners represent the water line width acquired with static or dynamic shimming. Especially, the frontal volume benefits greatly from dynamic shimming with a narrower line width and a smaller frequency shift. (F–H) Water-suppressed ^{1}H MR spectra (top) and GABA-edited ^{1}H MR spectra (bottom) from the three volumes indicate high spectral quality and GABA detection for all locations. *Source:* Data from Laura I. Sacolick.

dynamic shimming (see Chapter 10). Simultaneously acquired data only allows a static shim setting that has been optimized across the *total* localized volume. However, *in vivo* magnetic field inhomogeneity is characterized by many local, higher-order magnetic field gradients introduced by nearby air-brain tissue boundaries, for example, around the nasal cavity. A single, static shim setting of commonly available second-order shims is generally incapable of compensating the magnetic field inhomogeneity across the entire sample. However, when the data is acquired sequentially, it is possible to establish optimal shim settings over the much smaller and therefore more manageable localized volumes. Dynamically updating the pre-determined shims as the sequence progresses (Figure 7.20B) ensures optimal magnetic field homogeneity for each volume within the limitations imposed by the low-order shim system [68–70].

Exercises

7.1 **A** A PRESS localization sequence (TR = 20 000 ms, TE = 5 ms) is used for volume prelo-calization of a $8 \times 6 \times 1$ cm cubic volume with 5 ms sinc excitation and refocusing pulses ($R = 6.0$). 2D MRSI data are acquired as a 24×24 matrix over a FOV of 24 cm at 3.0 T. For a sample containing 10 mM NAA and 2 mM Cho, calculate the NAA methyl-to-choline methyl MR signal intensity ratio for MRSI voxels in the middle and all four corners of the pre-selected volume. Assume perfect slice and MRSI voxel profiles and an on-resonance condition for the choline methyl resonance.

B Repeat the calculation for a magnetic field strength of 7.05 T.

C Calculate the required RF pulse lengths and the corresponding increase in RF power at 3 T to ensure that the NAA methyl-to-choline methyl MR signal intensity ratio does not change by more than 10% over the selected volume.

7.2 Consider a two-slice OVS method (nonintersecting slices, one RF pulse per slice) with 10 ms between the pulses and 20 ms following the last pulse.

A In the absence of B_1 magnetic field inhomogeneity, calculate the nutation angles for both pulses to achieve optimal lipid suppression. Assume $T_{1,\text{lipids}} = 300$ ms and TR = 3000 ms. Assume perfect 90° nonselective excitation following OVS.

B Recalculate the nutation angles when the OVS modules are aimed at removing inhomogeneously broadened water in the nasal cavity. Assume $T_{1,\text{water}} = 1600$ ms and TR = 3000 ms.

7.3 In a 2D MRSI scan of the human head, extracranial lipid suppression is achieved with a single IR method with TR = 2000 ms and TI = 208 ms.

A Calculate the relative signal recoveries for compounds with $T_1 = 1000$, 1300, and 1600 ms. Assume perfect 90° nonselective excitation following IR lipid suppression.

B In order to avoid saturation effects due to the relative short TR, the spectroscopist decides to repeat the experiment with TR = 6000 ms. Calculate the relative signal recoveries for compounds with $T_1 = 1000$, 1300, and 1600 ms.

C Based on the calculations performed under A/B give a recommendation of the sequence timing parameters for quantitative MRSI studies using IR for lipid suppression.

7.4 A single-slice 2D MRSI dataset (16×16) with a 10 ppm spectral bandwidth can be acquired with a signal-to-noise ratio of 10 for NAA in 8.53 min at 1.5 T (one average).

A Calculate the signal-to-noise ratio for NAA when the MRSI dataset is acquired with single-shot PEPSI (one average). Assume infinitely short-gradient ramp times.

B In order to adequately sample a dynamic time course, the NAA resonance must be sampled with a signal-to-noise ratio of 10 and a time resolution of 7 min. Propose at least three solutions to achieve the required sensitivity and time resolution.

7.5 Derive an analytical expression for the (TR, k) curve shown in Figure 7.7C.

7.6 Describe an experiment to verify the direction of the "half-pixel shift."

7.7 Consider two-point samples placed at (x_1, y_1) and (x_2, y_2). Phase-encoded NMR spectra P are acquired with the phase-encoding gradient in the x-direction, sampling k-space lines k_1 and k_2.

 A Verify that the compartmentalized spectra S_1 and S_2 from the point samples can be uniquely calculated from only two phase-encoded NMR spectra P_1 and P_2.

 B Summarize the exceptions to the answer given under A.

7.8 **A** Calculate the volume sizes for a 2D MRSI experiment (16×16) over a FOV = 19.2 cm using the k-space weighting shown in Figure 7.7C for species with $T_1 = 2000$ ms and $T_1 = 1000$ ms. Note, this problem is readily analyzed numerically, for example, in Matlab.

 B Calculate the scan time reduction for a 2D MRSI experiment (32×32) using a Hamming weighting function and $TR(k = 0) = 1500$ ms, $T_1 = 1500$ ms relative to a non-weighted experiment.

 C Repeat the calculation for when the TR variation is based on a Gaussian weighting function (i.e. Eq. (7.12)).

 D Repeat the calculations under B and C for a 3D acquisition ($16 \times 16 \times 16$).

 E Derive an expression for the weighting function $W(k)$ when the weighting is achieved by adjusting the nutation angle rather than the repetition time TR.

7.9 In a sequential multivolume experiment a spectroscopist wants to acquire three $2 \times 2 \times 2$ cm = 8 ml volumes.

 A Under the assumption of non-oblique slice-selection gradients calculate the nearest volume positions that can be used without generating mutual T_1 saturation when the first volume is placed in the magnetic isocenter.

 B The signal-to-noise ratios of the NAA resonance in the three volumes after circa 10 min of acquisition (TR = 5000 ms, NEX = 128) are 25, 20, and 30, respectively. In the case that multivolume acquisition is not available, calculate the S/N ratios for the NAA resonance that would be expected from single-volume NMR spectra acquired in separate scans in the same amount of total acquisition time (i.e. all three single volumes acquired in circa 10 min).

 C For a NAA T_1 relaxation time constant of 1500 ms, calculate the S/N ratios for a multivolume acquisition (TR = 5000 ms, NEX = 128) when the three volume are placed within each other's excitation slices, such that mutual T_1 saturation cannot be ignored.

7.10 Consider a NMR spectrum acquired from a single volume of volume V_{SV} with repetition time TR and number of averages $N_{A,SV}$.

 A Derive an expression of the S/N ratio per unit of time for the single-volume acquisition.

 Further consider a 3D MRSI experiment for which N_i ($i = 1, 2$, or 3) phase-encoding increments are acquired over a field-of-view FOV_i ($i = 1, 2$, or 3). Given a repetition time TR and $N_{A,MRSI}$ averages per phase-encoding increment.

 B Derive an expression of the S/N ratio per unit of time for the MRSI acquisition.

 C When the single-volume and nominal MRSI volume sizes are identical, discuss how the S/N ratios per unit of time compare.

 D Discuss the advantage and disadvantages of single-volume MRS and MRSI.

7.11 MRSI data acquisition can be significantly reduced through the use of multi-spin-echo methods during which multiple echoes are acquired following a single excitation.

A Discuss the temporal aspects (i.e. scan time reduction) when all echoes are sampled at the same k-space coordinate.

B Discuss the temporal aspects (i.e. scan time reduction) when subsequent echoes are sampled at different k-space coordinates.

C For both methods (A and B) discuss the effects that can arise for scalar-coupled spin-systems.

References

1 Hoult, D.I. (1980). Rotating frame zeugmatography. *Philos. Trans. R. Soc. Lond. B Biol. Sci.* 289 (1037): 543–547.

2 Hoult, D.I., Chen, C.N., and Hedges, L.K. (1987). Spatial localization by rotating frame techniques. *Ann. N. Y. Acad. Sci.* 508: 366–375.

3 Styles, P. (1992). Rotating frame spectroscopy and spectroscopic imaging. In: *NMR Basic Principles and Progress* (ed. P. Diehl, E. Fluck, H. Gunther, et al.), 45–66. Berlin: Springer-Verlag.

4 Hendrich, K., Liu, H., Merkle, H. et al. (1992). B_1 voxel shifting of phase-modulated spectroscopic localization techniques. *J. Magn. Reson.* 97: 486–497.

5 Brown, T.R., Kincaid, B.M., and Ugurbil, K. (1982). NMR chemical shift imaging in three dimensions. *Proc. Natl. Acad. Sci. U. S. A.* 79: 3523–3526.

6 Maudsley, A.A., Hilal, S.K., Perman, W.H., and Simon, H.E. (1983). Spatially resolved high resolution spectroscopy by "four dimensional" NMR. *J. Magn. Reson.* 51: 147–152.

7 Haselgrove, J.C., Subramanian, V.H., Leigh, J.S. Jr. et al. (1983). *In vivo* one-dimensional imaging of phosphorus metabolites by phosphorus-31 nuclear magnetic resonance. *Science* 220: 1170–1173.

8 Pykett, I.L. and Rosen, B.R. (1983). Nuclear magnetic resonance: *in vivo* proton chemical shift imaging. Work in progress. *Radiology* 149: 197–201.

9 Bracewell, R.M. (1965). *The Fourier Transform and Its Applications*. New York: McGraw-Hill.

10 Brooker, H.R., Mareci, T.H., and Mao, J.T. (1987). Selective Fourier transform localization. *Magn. Reson. Med.* 5: 417–433.

11 Wang, Z., Bolinger, L., Subramanian, V.H., and Leigh, J.S. (1991). Errors of Fourier chemical shift imaging and their corrections. *J. Magn. Reson.* 92: 64–72.

12 Mareci, T.H. and Brooker, H.R. (1991). Essential considerations for spectral localization using indirect gradient encoding of spatial information. *J. Magn. Reson.* 92: 229–246.

13 Moonen, C.T.W., Sobering, G., van Zijl, P.C.M. et al. (1992). Proton spectroscopic imaging of the human brain. *J. Magn. Reson.* 98: 556–575.

14 Koch, T., Brix, G., and Lorenz, W.J. (1994). Theoretical description, measurement and correction of localization errors in ^{31}P chemical shift imaging. *J. Magn. Reson. B* 104: 199–211.

15 Maudsley, A.A. (1986). Sensitivity in Fourier imaging. *J. Magn. Reson.* 68: 363–366.

16 Ernst, R.R., Bodenhausen, G., and Wokaun, A. (1987). *Principles of Nuclear Magnetic Resonance in One and Two Dimensions*. Oxford: Clarendon Press.

17 Maudsley, A.A., Matson, G.B., Hugg, J.W., and Weiner, M.W. (1994). Reduced phase encoding in spectroscopic imaging. *Magn. Reson. Med.* 31: 645–651.

18 Hugg, J.W., Maudsley, A.A., Weiner, M.W., and Matson, G.B. (1996). Comparison of k-space sampling schemes for multidimensional MR spectroscopic imaging. *Magn. Reson. Med.* 36: 469–473.

19 Kuhn, B., Dreher, W., Norris, D.G., and Leibfritz, D. (1996). Fast proton spectroscopic imaging employing k-space weighting achieved by variable repetition times. *Magn. Reson. Med.* 35: 457–464.

20 Mansfield, P. (1984). Spatial mapping of the chemical shift in NMR. *Magn. Reson. Med.* 1: 370–386.

21 Posse, S., DeCarli, C., and Le Bihan, D. (1994). Three-dimensional echo-planar MR spectroscopic imaging at short echo times in the human brain. *Radiology* 192: 733–738.

22 Posse, S., Tedeschi, G., Risinger, R. et al. (1995). High speed ^{1}H spectroscopic imaging in human brain by echo planar spatial-spectral encoding. *Magn. Reson. Med.* 33: 34–40.

23 Adalsteinsson, E., Irarrazabal, P., Topp, S. et al. (1998). Volumetric spectroscopic imaging with spiral-based k-space trajectories. *Magn. Reson. Med.* 39: 889–898.

24 Andronesi, O.C., Gagoski, B.A., and Sorensen, A.G. (2012). Neurologic 3D MR spectroscopic imaging with low-power adiabatic pulses and fast spiral acquisition. *Radiology* 262: 647–661.

25 Bogner, W., Gagoski, B., Hess, A.T. et al. (2014). 3D GABA imaging with real-time motion correction, shim update and reacquisition of adiabatic spiral MRSI. *Neuroimage* 103: 290–302.

26 Furuyama, J.K., Wilson, N.E., and Thomas, M.A. (2012). Spectroscopic imaging using concentrically circular echo-planar trajectories *in vivo*. *Magn. Reson. Med.* 67: 1515–1522.

27. Jiang, W., Lustig, M., and Larson, P.E. (2016). Concentric rings *K*-space trajectory for hyperpolarized ^{13}C MR spectroscopic imaging. *Magn. Reson. Med.* 75: 19–31.

28 Emir, U.E., Burns, B., Chiew, M. et al. (2017). Non-water-suppressed short-echo-time magnetic resonance spectroscopic imaging using a concentric ring k-space trajectory. *NMR Biomed.* 30.

29 Dreher, W. and Leibfritz, D. (2002). Fast proton spectroscopic imaging with high signal-to-noise ratio: spectroscopic RARE. *Magn. Reson. Med.* 47: 523–528.

30 Althaus, M., Dreher, W., Geppert, C., and Leibfritz, D. (2006). Fast 3D echo planar SSFP-based ^{1}H spectroscopic imaging: demonstration on the rat brain *in vivo*. *Magn. Reson. Imaging* 24: 549–555.

31 Pohmann, M., von Kienlin, M., and Haase, A. (1997). Theoretical evaluation and comparison of fast chemical shift imaging methods. *J. Magn. Reson.* 129: 145–160.

32 Schirda, C.V., Zhao, T., Andronesi, O.C. et al. (2016). *In vivo* brain rosette spectroscopic imaging (RSI) with LASER excitation, constant gradient strength readout, and automated LCModel quantification for all voxels. *Magn. Reson. Med.* 76: 380–390.

33 Posse, S. and Hu, X. (2015). High-speed spatial-spectral encoding with PEPSI and spiral MRSI. *eMagRes* 4: 587–600.

34 Adalsteinsson, E. and Spielman, D.M. (1999). Spatially resolved two-dimensional spectroscopy. *Magn. Reson. Med.* 41: 8–12.

35 Adalsteinsson, E., Irarrazabal, P., Spielman, D.M., and Macovski, A. (1995). Three-dimensional spectroscopic imaging with time-varying gradients. *Magn. Reson. Med.* 33: 461–466.

36 Schar, M., Strasser, B., and Dydak, U. (2016). CSI and SENSE CSI. *eMagRes* 5: 1291–1306.

37 Pruessmann, K.P., Weiger, M., Scheidegger, M.B., and Boesiger, P. (1999). SENSE: sensitivity encoding for fast MRI. *Magn. Reson. Med.* 42: 952–962.

38 Griswold, M.A., Jakob, P.M., Heidemann, R.M. et al. (2002). Generalized autocalibrating partially parallel acquisitions (GRAPPA). *Magn. Reson. Med.* 47: 1202–1210.

39. Dydak, U., Weiger, M., Pruessmann, K.P. et al. (2001). Sensitivity-encoded spectroscopic imaging. *Magn. Reson. Med.* 46: 713–722.

40 Banerjee, S., Ozturk-Isik, E., Nelson, S.J., and Majumdar, S. (2006). Fast magnetic resonance spectroscopic imaging at 3 Tesla using autocalibrating parallel technique. *Conference proceedings: Annual International Conference of the IEEE Engineering in Medicine and Biology Society IEEE Engineering in Medicine and Biology Society Annual Conference*, New York City, vol. 1, pp. 1866–1869.

41 Hu, X., Levin, D.N., Lauterbur, P.C., and Spraggins, T. (1988). SLIM: spectral localization by imaging. *Magn. Reson. Med.* 8: 314–322.

42 von Kienlin, M. and Mejia, R. (1991). Spectral localization with optimal pointspread function. *J. Magn. Reson.* 94: 268–287.

43 Zhang, Y., Gabr, R.E., Schar, M. et al. (2012). Magnetic resonance spectroscopy with linear algebraic modeling (SLAM) for higher speed and sensitivity. *J. Magn. Reson.* 218: 66–76.

44 Pohmann, R., Rommel, E., and von Kienlin, M. (1999). Beyond k-space: spectral localization using higher order gradients. *J. Magn. Reson.* 141: 197–206.

45 Lam, F. and Liang, Z.P. (2014). A subspace approach to high-resolution spectroscopic imaging. *Magn. Reson. Med.* 71: 1349–1357.

46 Lam, F., Ma, C., Clifford, B. et al. (2016). High-resolution ^{1}H-MRSI of the brain using SPICE: data acquisition and image reconstruction. *Magn. Reson. Med.* 76: 1059–1070.

47 Hetherington, H.P., Mason, G.F., Pan, J.W. et al. (1994). Evaluation of cerebral gray and white matter metabolite differences by spectroscopic imaging at 4.1T. *Magn. Reson. Med.* 32: 565–571.

48 Hetherington, H.P., Pan, J.W., Mason, G.F. et al. (1996). Quantitative ^{1}H spectroscopic imaging of human brain at 4.1 T using image segmentation. *Magn. Reson. Med.* 36: 21–29.

49 Esmaeili, M., Bathen, T.F., Rosen, B.R., and Andronesi, O.C. (2017). Three-dimensional MR spectroscopic imaging using adiabatic spin echo and hypergeometric dual-band suppression for metabolic mapping over the entire brain. *Magn. Reson. Med.* 77: 490–497.

50 Hu, X., Patel, M., and Ugurbil, K. (1994). A new strategy for spectroscopic imaging. *J. Magn. Reson. B* 103: 30–38.

51 Haupt, C.I., Schuff, N., Weiner, M.W., and Maudsley, A.A. (1996). Removal of lipid artifacts in ^{1}H spectroscopic imaging by data extrapolation. *Magn. Reson. Med.* 35: 678–687.

52 Hetherington, H.P., Avdievich, N.I., Kuznetsov, A.M., and Pan, J.W. (2010). RF shimming for spectroscopic localization in the human brain at 7 T. *Magn. Reson. Med.* 63: 9–19.

53 Boer, V.O., Klomp, D.W., Juchem, C. et al. (2011). Multi-slice MRSI of the human brain at 7 Tesla using dynamic B_0 and B_1 shimming. *Proc. Int. Soc. Magn. Reson. Med.* 19: 142.

54 Boer, V.O., van de Lindt, T., Luijten, P.R., and Klomp, D.W. (2015). Lipid suppression for brain MRI and MRSI by means of a dedicated crusher coil. *Magn. Reson. Med.* 73: 2062–2068.

55 de Graaf, R.A., Brown, P.B., De Feyter, H.M. et al. (2018). Elliptical localization with pulsed second-order fields (ECLIPSE) for robust lipid suppression in proton MRSI. *NMR Biomed.* 31: e3949.

56 Hetherington, H.P., Pan, J.W., Mason, G.F. et al. (1994). 2D ^{1}H spectroscopic imaging of the human brain at 4.1 T. *Magn. Reson. Med.* 32: 530–534.

57 Ebel, A., Govindaraju, V., and Maudsley, A.A. (2003). Comparison of inversion recovery preparation schemes for lipid suppression in ^{1}H MRSI of human brain. *Magn. Reson. Med.* 49: 903–908.

58 Sacolick, L.I., Rothman, D.L., and de Graaf, R.A. (2007). Adiabatic refocusing pulses for volume selection in magnetic resonance spectroscopic imaging. *Magn. Reson. Med.* 57: 548–553.

59 Duyn, J.H., Gillen, J., Sobering, G. et al. (1993). Multisection proton MR spectroscopic imaging of the brain. *Radiology* 188: 277–282.

60 Henning, A., Schar, M., Schulte, R.F. et al. (2008). SELOVS: brain MRSI localization based on highly selective T_1- and B_1- insensitive outer-volume suppression at 3T. *Magn. Reson. Med.* 59: 40–51.

61 Henning, A., Fuchs, A., Murdoch, J.B., and Boesiger, P. (2009). Slice-selective FID acquisition, localized by outer volume suppression (FIDLOVS) for ^{1}H-MRSI of the human brain at 7 T with minimal signal loss. *NMR Biomed.* 22: 683–696.

62 Luo, Y., de Graaf, R.A., DelaBarre, L. et al. (2001). BISTRO: an outer-volume suppression method that tolerates RF field inhomogeneity. *Magn. Reson. Med.* 45: 1095–1102.

63 Bolinger, L. and Leigh, J.S. (1988). Hadamard spectroscopic imaging (HSI) for multi-volume localization. *J. Magn. Reson.* 80: 162–167.

64 Souza, S.P., Szumowski, J., Dumoulin, C.L. et al. (1988). SIMA: simultaneous multislice acquisition of MR images by Hadamard-encoded excitation. *J. Comput. Assist. Tomogr.* 12: 1026–1030.

65 Goelman, G., Subramanian, V.H., and Leigh, J.S. (1990). Transverse Hadamard spectroscopic imaging technique. *J. Magn. Reson.* 89: 437–454.

66 Goelman, G. and Leigh, J.S. (1991). B_1-insensitive Hadamard spectroscopic imaging technique. *J. Magn. Reson.* 91: 93–101.

67 Goelman, G. (1994). Fast Hadamard spectroscopic imaging techniques. *J. Magn. Reson. B* 104: 212–218.

68 Ernst, T. and Hennig, J. (1991). Double-volume ^{1}H spectroscopy with interleaved acquisitions using tilted gradients. *Magn. Reson. Med.* 20: 27–35.

69 Theberge, J., Menon, R.S., Williamson, P.C., and Drost, D.J. (2005). Implementation issues of multivoxel STEAM-localized ^{1}H spectroscopy. *Magn. Reson. Med.* 53 (3): 713–718.

70 Koch, K.M., Sacolick, L.I., Nixon, T.W. et al. (2007). Dynamically shimmed multivoxel ^{1}H magnetic resonance spectroscopy and multislice magnetic resonance spectroscopic imaging of the human brain. *Magn. Reson. Med.* 57: 587–591.

71 Oeltzschner, G., Puts, N.A., Chan, K.L. et al. (2017). Dual-volume excitation and parallel reconstruction for *J*-difference-edited MR spectroscopy. *Magn. Reson. Med.* 77: 16–22.

8

Spectral Editing and 2D NMR

8.1 Introduction

Proton NMR spectra from mammalian tissues *in vivo* hold information on a considerable number of metabolites. Even though the *in vivo* NMR detection limit of ~100 μM greatly simplifies the spectral appearance, a typical short TE ^{1}H NMR spectrum from rat or human brain *in vivo* still contains resonances from more than 15 different metabolites [1, 2]. The abundance of resonances in combination with a small proton chemical shift range leads to significant spectral overlap, thereby complicating unambiguous peak assignment and quantification. Prominent examples can be found for lactate, which overlaps with signals from lipids and macromolecules, the inhibitory neurotransmitter γ-amino butyric acid (GABA), which is obscured by total creatine and other resonances, and glutamate and glutamine, which overlap with each other at lower magnetic fields. The collective group of techniques that achieves separation of overlapping resonances is often referred to as spectral editing and this chapter describes the principles of homo- and heteronuclear spectral editing together with practical *in vivo* applications.

In the most general sense, spectral editing may include any technique that can simplify an NMR spectrum in order to limit the detection to specific metabolites. According to this general definition, spectral editing can include water suppression, spatial localization, echo-time variation, use of shift reagents, and selective excitation. Because many of these techniques are discussed in separate chapters, a more specific definition will therefore be employed here. Spectral editing is defined as to include any technique that utilizes the scalar coupling between spins to discriminate scalar-coupled from uncoupled spins. This definition includes 2D NMR, which can be seen as the ultimate spectral editing method, making no assumptions about the spin system under investigation.

8.2 Quantitative Descriptions of NMR

The evolution of scalar-coupled spins and their response to RF pulses and magnetic field gradients must be quantitatively described in order to understand the various aspects of spectral editing. Nuclear spin is an intrinsic quantum property that cannot be described by classical physics. However, the *evolution* of nuclear spins and of macroscopic magnetization originating from a large collection of nuclear spins can be described by classical physics as was done by Felix Bloch in his celebrated equations. The phenomenon of scalar coupling in which the resonance frequency from one spin splits under the influence of the orientation of another spin is also a quantum property that cannot be understood using classical arguments.

In Vivo *NMR Spectroscopy: Principles and Techniques*, Third Edition. Robin A. de Graaf.
© 2019 John Wiley & Sons Ltd. Published 2019 by John Wiley & Sons Ltd.

Furthermore, whereas the Bloch equations described the evolution of non-coupled spins, the evolution of scalar-coupled spins cannot be described by classical equations.

8.2.1 Density Matrix Formalism

A proper description of scalar-coupled spins must be based on quantum mechanics, whereby the density matrix formalism [3–6] is universally applicable. The density matrix formalism is not directly concerned with magnetization, but rather deals with the energy states of the spin system under investigation. A density matrix calculation typically starts with the creation of a $2^N \times 2^N$ thermal equilibrium density matrix (Figure 8.1), where N presents the number of different spins in a spin system. For a two-spin-system with four energy levels the 4×4 thermal equilibrium density matrix contains the energy-level populations on the matrix diagonal (Figure 8.1B). The evolution of the density matrix under the influence of RF pulses and delays is governed by the Liouville–von Neumann equation, similar to how the Bloch equations govern the evolution of magnetization related to non-coupled spins. The effects of RF pulses, delays, and magnetic field gradients are described by operators or Hamiltonians H, which can be seen as a generalization of rotation matrices for larger spin systems. The operators act on the initial density matrix $\sigma(0)$ according to the Liouville–von Neumann equation, $e^{-iHt}\sigma(0)e^{+iHt}$, thereby transforming it into a new density matrix $\sigma(t)$. RF pulses transform the initial diagonal density

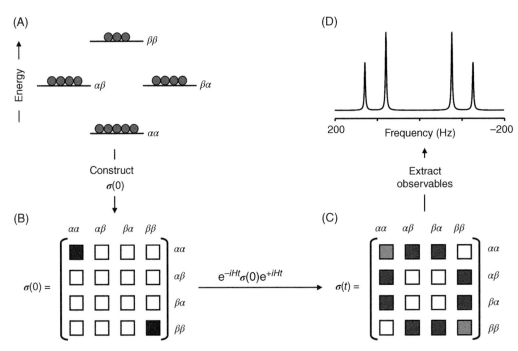

Figure 8.1 **Quantum-mechanical or spectroscopic description of NMR using density matrices.** (A) The energy levels and associated nuclear spin populations for a two-spin-system at thermal equilibrium provide a complete description of (B) the initial density matrix. Note that only the diagonal elements, corresponding to populations, are nonzero (filled squares). (C) Evolution of the density matrix under the influence of chemical shifts and scalar couplings during delays and RF pulses is described by the Liouville–von Neumann equation. Note that following at least two RF pulses all matrix elements can be nonzero, whereby the off-diagonal elements correspond to the various coherences allowed in a two-spin-system. (D) During signal acquisition the observable signal can be extracted from the density matrix and following Fourier transformation displayed in the spectral domain.

matrix into a matrix with off-diagonal elements that correspond to the various coherences (Figure 8.1C). The MR coherences present at the time of signal acquisition can be extracted from the final density matrix and displayed as an NMR spectrum following FT (Figure 8.1D). A detailed treatment of density matrix calculations is well outside the scope of this book and the reader is referred to excellent textbooks and publications covering this subject [3–6].

8.2.2 Classical Vector Model

The majority of NMR text books are devoted to the density matrix formalism because it is the most general method that can describe the evolution of any spin system during any MR pulse sequence and experiment. Unfortunately, as density matrices are concerned with energy states and energy-level populations, they do not provide an intuitive insight into the NMR experiment during the calculation. In Chapter 1, it was demonstrated that a classical description of NMR with many spins distributed over a spin orientation sphere provides a physically intuitive description of NMR. While the classical vector model cannot describe the evolution of scalar-coupled spins, it can form the basis of a modified vector model (the correlated vector model, see Section 8.2.3) that can provide some insights into the evolution of weakly-coupled spins without losing the connection to a physical picture altogether. Figure 8.2A and B shows the small amounts of biased spins that give rise to the thermal equilibrium magnetization of spin I (red vector) and spin S (blue vector), respectively. The biased spins that create magnetization I are, in general, coupled to randomly distributed S spins and vice versa. In other words, the collection of molecules that give rise to macroscopic magnetization of I spins are generally not the same molecules that provide the macroscopic magnetization of S spins. Figure 8.2C shows the situation after the I spins are excited by a 90° RF pulse. Scalar coupling is typically described as a phenomenon where the I spin Larmor frequency depends on the orientation of the scalar-coupled S spin. If scalar coupling would be a classical phenomenon, I spins would acquire a range of Larmor frequencies due to the randomly oriented S spins (Figure 8.2C). This would lead to rapid macroscopic signal dephasing (similar to T_2^* relaxation), which is a feature that is *not* observed in NMR experiments. This mismatch between

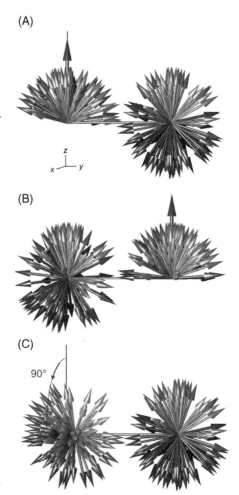

Figure 8.2 **Semiclassical description of NMR using a correlated vector model.** (A, B) Spin orientation spheres for spins I and S in a weakly-coupled IS spin system at thermal equilibrium. Only the small excess of biased spins that contribute to the net macroscopic magnetization are shown. Note that spins that give rise to I magnetization are scalar coupled to randomly oriented S spins and vice versa. In addition, molecules that give rise to macroscopic I magnetization are generally not the same molecules that provide macroscopic S magnetization. (C) Selective excitation of I spins simply results in the rotation of the entire I spin orientation sphere in analogy to non-coupled spins as discussed in Chapter 1.

prediction and experiment is one indication that scalar coupling is a quantum mechanical effect that requires a modified vector model.

8.2.3 Correlated Vector Model

The correlated vector model is similar to the classical vector model described in Section 8.2.2 and Chapter 1 with the addition that when scalar coupling is involved a distribution of randomly oriented spins acts as if half of the spins are in the α or parallel spin state, while the other half are in the antiparallel or β spin state. For a two-spin-system IS this leads to the conversion of Figure 8.2C to Figure 8.3A. The randomly oriented S spins *appear* to the scalar-coupled I spins to be in one of the two pure spin states, α or β. As a result, the transverse I magnetization does not experience a range of magnetic fields from the S spins, but rather resonates at a higher frequency with S spins in the β state or resonates at a lower frequency with S spins in the α state. In other words, I and S spins have evolved from a non-correlated state (Figure 8.2) into a correlated state (Figure 8.3B). In the correlated state, I spins that are biased along the $-y$ axis are coupled to S spins that are biased along the $+z$ axis. Simultaneously, I spins that are biased along the $+y$ axis are coupled to S spins that are biased along the $-z$ axis. In the product operator formalism discussed in the next section, this correlated spin state would be referred to as $-2I_yS_z$, which is also known as a state of antiphase coherence. In the product operator formalism only the red and blue vectors are shown, implying that all spins are perfectly aligned along the $\pm y$ or $\pm z$ axis. However, the correlated vector model of Figure 8.3C shows that the correlated vectors are merely biased in that direction, with very few spins being perfectly aligned with any of the Cartesian axes. Once the spin system is in a correlated state, such as the antiphase coherence state shown in Figure 8.3B, the S spins can be selectively excited to generate a spin state in which correlation exists between the transverse components of the magnetization. In the case of Figure 8.3D following a 90°($-y$) pulse I spins biased along the $+y$ axis are correlated with S spins biased along the $-x$ axis. In the product operator formalism this correlated state would be indicated as $-2I_yS_x$ and is indicative of multiple-quantum-coherences (MQCs). One of the hallmarks of MQCs is that they are NMR invisible or unobservable. Figure 8.3C provides an explanation for this observation. As both spins are correlated in the transverse plane, I and S spins are no longer coupled to a distribution of randomly oriented spins such that scalar coupling evolution is no longer active. In other words, the correlated spin state of Figure 8.3C remains in a correlated spin state in the absence of RF pulses. The correlated I and S spins continue to rotate at their respective Larmor frequencies due to the ever-present main magnetic field. However, when the receiver is opened to acquire MR signal, 50% of I (and S) spins induce a positive signal, while the other 50% of I (and S) spins along the opposite axis induce a negative signal. The sum of all I (and S) spins therefore do not induce any net observable signal. This is different from the correlated state representing antiphase coherence (Figure 8.3B), which is initially also NMR unobservable. However, the antiphase coherences evolve back into an observable, non-correlated state of in-phase coherences under the influence of scalar coupling.

The correlated vector model provides a physically intuitive description of scalar-coupled spins based on the classical vector model modified with an *ad hoc* extension on the quantum mechanics of scalar coupling. Various incarnations of the correlated vector model have been described [6–9] and it is hoped that a semiclassical description can help remove some of the "mystery" involved with evolution of scalar-coupled spins and MQCs. For most every day use, the spin orientation spheres are not an absolute requirement and focus can be given to the macroscopic magnetization vectors (red and blue in Figure 8.3) in which case the correlated vector model provides a similar picture as the product operator formalism.

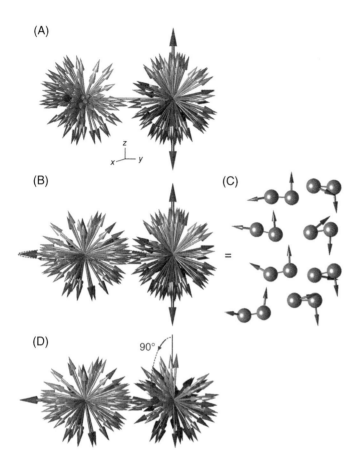

Figure 8.3 Semiclassical description of correlated states in NMR using a correlated vector model. (A) In a classical description, the scalar coupling between *I* and *S* spins would lead to signal dephasing of *I* spins due to the random distribution of *S* spins. Since this prediction is not in agreement with experimental evidence, it can be concluded that scalar coupling is a nonclassical phenomenon. The correlated vector model can still provide a pictorial description of scalar coupling by accepting that when scalar coupling is present, a random spin-orientation sphere should be considered a two-level quantum system where 50% of the spins are in the parallel (α) or antiparallel (β) orientation (blue *S* spin vectors). (B) The α and β *S* spin vectors lead to different resonance frequencies of the scalar coupled *I* spins, which in turn leads to the generation of *I* spin antiphase coherence following an evolution period of $1/(2J)$. (C) The red and blue vectors in (B) provide a convenient summary of the underlying spin orientation spheres and are essentially identical to the picture used in the product operator formalism. However, it should be realized that very few spins lie exactly along the Cartesian axes. The antiphase coherence state in (B) represents a correlated state; when an *S* spin is biased to the positive (or negative) hemisphere of the spin-orientation sphere in the *z* direction, the scalar coupled *I* spin will be biased to the negative (or positive) hemisphere in the *y* direction. (D) Application of a 90° RF pulse along the −*y* axis converts the state of antiphase coherence in (B) into a state of multiple-quantum coherence, whereby the transverse components of the *I* and *S* spins are correlated.

8.2.4 Product Operator Formalism

In most textbooks the product operator formalism [10–12] is not presented as an extension of the correlated vector model, but rather as a simplification of the density matrix formalism for weakly-coupled spin systems. For weakly-coupled spins, the density matrix can be expanded into a linear combination of orthogonal matrices (referred to as product operators), each of which represents an orthogonal component of the magnetization. Various complete orthogonal basis matrix sets can be used, of which product operators and shift (or lowering and raising) operators are most commonly employed. Figure 8.4 shows how specific elements in a density matrix correspond to unique coherences in a two-spin-system. The diagonal elements of a density matrix are characterized by the I_z and S_z product operators and represent longitudinal magnetization for spins I and S (Figure 8.4A). Figure 8.4B shows the I_x product operator, corresponding to magnetization along the x axis. The corresponding I_y product operator can be found as the imaginary component of the same matrix elements. Figure 8.4C–F shows two-spin product operators corresponding to correlated spin states. Note that the graphical representations of the various coherences are closely related to the macroscopic vectors shown in the correlated vector model (Figure 8.3). For a two-spin-system IS there are 16 product operators given by I_x, I_y, I_z, S_x, S_y, S_z, $2I_xS_x$, $2I_xS_y$, $2I_xS_z$, $2I_yS_x$, $2I_yS_y$, $2I_yS_z$, $2I_zS_x$, $2I_zS_y$, and $2I_zS_z$. The unit or identity operator completes the set. Each product operator corresponds to a particular physical state according to

I_z, S_z	Polarization (longitudinal magnetization) of spins I and S.
I_x, I_y, S_x, S_y	In-phase x and y coherence (transverse magnetization) of spins I and S.
$2I_xS_z$, $2I_yS_z$,	
$2I_zS_x$, $2I_zS_y$	Antiphase coherence of one spin with respect to the other (Figure 8.4C).
$2I_xS_x$, $2I_xS_y$	
$2I_yS_x$, $2I_yS_y$	Two-spin or multi-quantum coherence (Figure 8.4E and F).
$2I_zS_z$	Longitudinal two-spin or scalar order (Figure 8.4D).

The evolution of product operators under the effect of chemical shifts and RF pulses is similar to that of regular magnetization and can be calculated through repeated application of rotation matrices applied to each spin sequentially. The evolution of product operators under the effects of scalar coupling involves the interconversion of uncorrelated in-phase coherences with correlated antiphase coherences. The transformation of product operators under the effect of chemical shifts, scalar coupling, and RF pulses are summarized in Appendix A.4. The product operator formalism provides the standard description of NMR for weakly-coupled spins and will be used throughout the remainder of this chapter.

8.3 Scalar Evolution

Spectral editing methods are based on the fact that non-coupled and scalar-coupled spins respond differently to (selective) RF pulses, delays, and magnetic field gradients. Figure 8.5 shows a regular Hahn spin-echo sequence executed with nonselective RF pulses, representing one of the most important elements of any spectral editing method. During the first delay TE/2, a collection of non-coupled I spins will precess in the transverse plane at their Larmor frequency ν_I, while simultaneously losing phase coherence due to magnetic field B_0 inhomogeneity (Figure 8.5B and C). The 180° refocusing pulse inverts the acquired phases, such that after an identical delay TE/2 the phases have been reduced to zero, leading to maximum phase coherence at the top of the spin-echo (Figure 8.5D and E). A weakly-coupled two-spin-system IS will also precess under the effects of chemical shift and magnetic field inhomogeneity. Half of the I spins coupled to S spins in the α spin-orientation precess at a Larmor frequency $\nu_I - J/2$ (blue vectors in Figure 8.5G), whereas the other half are coupled to S spins in the β spin orientation and therefore precess at Larmor frequency $\nu_I + J/2$ (red vectors in Figure 8.5G).

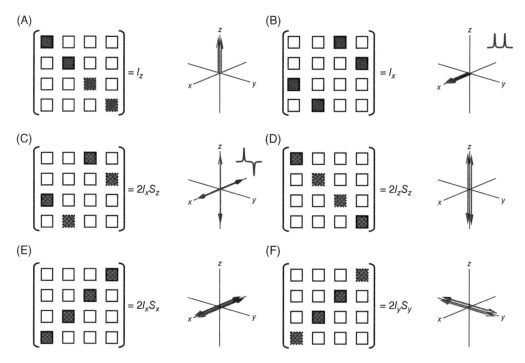

Figure 8.4 **Quantum-mechanical or spectroscopic description of NMR using the product operator formalism.** For a weakly-coupled two-spin-system the general density matrix shown in Figure 8.1 can be decomposed into a linear sum of 16 matrices in which each matrix corresponds to a particular spin coherence or population. (A) The diagonal elements correspond to spin populations which give rise to longitudinal magnetization. Different combinations of diagonal elements give rise to spin I and S longitudinal magnetization, denoted by product operators I_z and S_z. (B) Off-diagonal density matrix elements give rise to single-quantum coherences, or transverse magnetization, denoted by product operators I_x and S_x. The orthogonal component I_y and S_y can be extracted as the imaginary component of the same density matrix elements that characterize I_x and S_x. (C) Off-diagonal elements that give rise to antiphase coherences $2I_xS_z$ containing real (solid) and imaginary (dotted) components. The mixed blue–red color boxes represent a contribution of I and S spin states. (D) Diagonal elements giving rise to longitudinal scalar order, denoted by product operator $2I_zS_z$. (E, F) Various combinations of anti-diagonal elements give rise to multi-quantum-coherences. Note the close correspondence of the graphical representation of product operators and the macroscopic components arising from the correlated vector model in Figures 8.2 and 8.3.

Similar to the situation for non-coupled spins (Figure 8.5B–E), the nonselective 180° pulse inverts the phases acquired during the first TE/2 period. For scalar-coupled I spins the 180° pulse has an additional effect in that it *also* inverts the populations of the scalar-coupled S spins. This means that I spins with a Larmor frequency $\nu_1 - J/2$ (coupled to S spins in the α state) before the 180° pulse will be precessing at a Larmor frequency of $\nu_1 + J/2$ (and coupled to S spins in the inverted β state) following the 180° pulse. Similarly, I spins at Larmor frequency $\nu_1 + J/2$ before the 180° pulse are precessing at $\nu_1 - J/2$ after the 180° pulse. In other words, the phase acquired due to the frequency difference related to scalar coupling is not inverted by the 180° pulse. Phase related to scalar coupling will continue to accumulate during the entire echo-time TE, leading to a final phase difference $\Delta\phi$ between the two I spin resonances equal to $2\pi J$TE. The phases related to the central Larmor frequency ν_1 and magnetic field inhomogeneity are inverted by the 180° pulse and are therefore refocused at the top of the spin-echo. Using the product operator formalism (Appendix A4) it can be shown that the evolution of

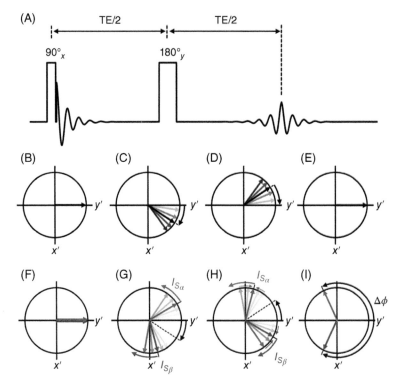

Figure 8.5 **Signal evolution during a spin-echo sequence.** (A) Spin-echo pulse sequence with echo-time TE. (B) Following excitation of non-scalar-coupled spins, (C) signal in various parts of the sample acquire phase proportional to the chemical shift and local magnetic field strength. (D) The refocusing pulse rotates the $x'y'$ plane by 180°, thereby effectively inverting the acquired phases. (E) Following an identical delay TE/2, the total acquired phase for all spins equals zero, leading to echo formation. (F) Excitation of I spins in a scalar-coupled two-spin system IS. The blue and red vectors correspond to I spins coupled to S spins in the α and β states, respectively. (G) During the first TE/2 delay, spins acquire phase due to chemical shift, local magnetic field offsets, and scalar coupling. Note that I spins coupled to S spins in the β state (red) have a higher Larmor frequency ($\nu_I + J/2$) and thus acquire more phase than I spins coupled to S spins in the α state (blue). (H) The refocusing pulse rotates the $x'y'$ plane by 180° around the y' axis, thereby again inverting the acquired phases. This will lead to (I) refocusing of phase evolution due to chemical shifts and magnetic field offsets at the top of the echo, similar to (B–E). The *nonselective* 180° pulse also inverts the spin populations of the S spins, so that I spins coupled to S spins in the α/β state before refocusing are coupled to S spins in the β/α state after refocusing. In other words, the lower frequency I spins before the 180° pulse become the higher frequency I spins after the 180° pulse and vice versa. (I) As a consequence, the phase accumulation due to scalar coupling is not refocused and accumulates to $\Delta\phi = 2\pi J$TE at the time of echo formation.

scalar-coupled spins during a spin-echo sequence can be described as a mixture between in-phase (I_y) and antiphase ($2I_xS_z$) coherences according to

$$\sigma(TE) = I_y \cos(\pi JTE) - 2I_xS_z \sin(\pi JTE) \tag{8.1}$$

Figure 8.6 displays the relation between the coherences at the top of the spin-echo and the appearance of the NMR spectrum for a number of echo-times TE and spin-systems IS_n ($n = 1$, 2, or 3). For a two-spin-system IS, as well as a four-spin-system IS_3, it follows that the signal can be inverted relative to non-coupled spins when TE = $1/J$ (Figure 8.6). For a three-spin-system

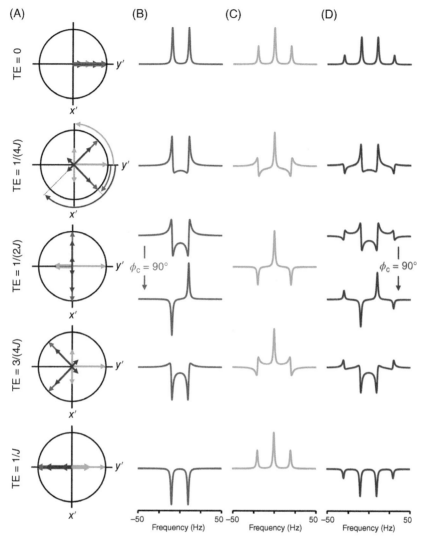

Figure 8.6 Scalar coupling evolution during a spin-echo sequence. (A) Vector diagrams and (B–D) simulated NMR spectra for the *I* spin in (B) a two-spin *IS* system, (C) a three-spin *IS*₂ system, and (D) a four-spin *IS*₃ system. For signals within a multiplet the accumulated phase is directly proportional to the echo-time TE and the frequency relative to the central multiplet Larmor frequency. At TE = 1/(2*J*) the *I* spin signal for *IS* and *IS*₃ spin systems is 90° out-of-phase relative to the *y′* detection axis. Following a 90° phase correction, the antiphase character of both multiplets is clearly visible. Note that the central peak for an *IS*₂ spin system does not modulate with TE.

*IS*₂ it is not possible to invert the entire signal relative to non-coupled spins. However, an echo-time of 1/(2*J*) provides inversion of the outer two resonances accounting for 50% of the signal (Figure 8.6). The most popular form of spectral editing, namely *J*-difference editing, relies on a nonselective spin-echo method in which the signal is inverted at a specific echo-time. For two- and four-spin-systems an echo-time of 1/(2*J*) provide a state of antiphase coherence that is useful for polarization transfer and multiple-quantum methods.

8.4   *J*-Difference Editing

8.4.1   Principle

J-difference editing is arguably the most popular spectral editing method for applications on the human and animal brain *in vivo*. *J*-difference editing is relatively easy to implement, can be combined with standard localization methods, provides good editing performance, and also provides the signal of non-edited metabolites. The nonselective spin-echo sequence shown in Figure 8.5 is one of the requirements for *J*-difference editing. A frequency-selective sequence (Figure 8.7) is needed to complete the spectral editing method. For a two-spin-system *IS* the signal evolution up to the 180° refocusing pulse is identical to that during a nonselective spin-echo (Figure 8.5). The 180° RF pulse in Figure 8.7 is frequency-selective in that it only refocuses the *I* spin without affecting the *S* spin. The normal properties of a 180° pulse, like refocusing of chemical shifts and magnetic field inhomogeneity, are therefore observed for the *I* spin. Since the frequency-selective pulse does not invert the spin populations of the *S* spin, the change in *I* spin frequency (from $\nu_I \pm J/2$ before the 180° pulse to $\nu_I \mp J/2$ after the 180° pulse) does not occur, leading to the refocusing of all phases including those related to scalar coupling. As a result, the TE-dependent scalar coupling evolution as shown in Figure 8.6 (and described by Eq. (8.1) for an *IS* spin-system) does not occur, making the signal of *I* spins positive and in-phase for all echo-times. The different responses of scalar-coupled spins to nonselective and selective spin-echo sequences form the fundamental basis for *J*-difference editing. Figure 8.8A shows a popular implementation of *J*-difference editing based on the Mescher–Garwood (MEGA) water suppression method [13, 14] as used to selectively detect an unknown two-spin-system *IS* (Figure 8.8B). Spin *S* is overwhelmed by the large water signal and is likely to be eliminated by water suppression. As a result, spin *I* will be the detection target for spectral editing. In the absence of the two frequency-selective 180° pulses, the sequence reduces to a nonselective

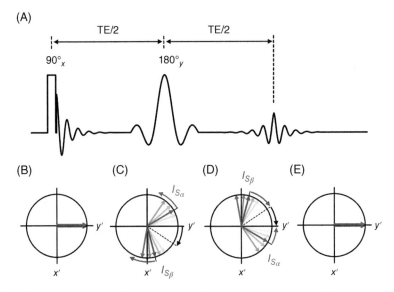

Figure 8.7 **Signal evolution during a frequency-selective spin-echo sequence.** (A) Spin-echo pulse sequence with echo-time TE in which the 180° pulse selectively refocuses the *I* spins without affecting the *S* spins. (B, C) Prior to the selective 180° pulse the situation is identical to Figure 8.5F and G. (D) The 180° pulse provides regular refocusing of chemical shifts and magnetic field offsets for the *I* spins, similar to Figure 8.5. However, since the scalar-coupled *S* spins are not affected, the lower frequency *I* spins (I_{S_α}) before the 180° pulse remain precessing at the lower frequency after the 180° pulse. As a result, (E) phase accumulation due to scalar coupling evolution is refocused during a frequency-selective spin-echo sequence.

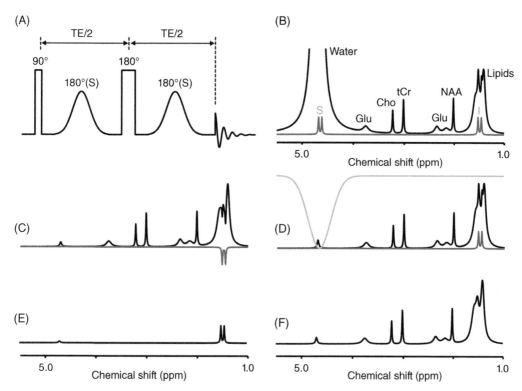

Figure 8.8 **Principle of J-difference spectral editing.** (A) Nonlocalized spin-echo pulse sequence with identical, frequency-selective 180° pulses in the two TE/2 periods. (B) Hypothetical ^{1}H NMR spectrum showing the *IS* spin-system of interest overlapping with intense signals from water and lipids. (C) Water-suppressed ^{1}H NMR spectrum acquired at TE = 1/*J* without the presence of frequency-selective refocusing pulses. Scalar coupling evolution during TE = 1/*J* leads to 180° phase accumulation of the *I* doublet signal relative to signals from uncoupled spins (NAA, tCr, tCho). However, the *I* spin is still overlapping with intense lipid signals, thereby preventing its unambiguous detection. (D) Water-suppressed ^{1}H NMR spectrum acquired at TE = 1/*J* in the presence of frequency-selective refocusing pulses applied at the *S* spin Larmor frequency. The frequency-selective pulses refocus the phase accrual due to scalar coupling (see Figure 8.7), resulting in a positive *I* spin doublet. (E) In calculating the difference spectrum (D) – (C) the signals not directly or indirectly affected by the selective refocusing are subtracted out, thereby revealing the *I* spin doublet. (F) Adding the two sub-spectra (C) + (D) cancels out the *I* spin signal and providing a spectrum of all other resonances.

spin-echo, similar to that shown in Figure 8.5. When the echo-time TE is set to 1/*J* the *I*-spin signal will appear inverted relative to non-coupled spins (Figure 8.8C). Unfortunately, the *I*-spin signal is still overlapping with lipids. In order to separately detect the *I*-spin signal, a frequency-selective spin-echo is needed in which the 180° pulses that are selective for spin *S* are turned on. The selective refocusing inhibits scalar coupling evolution, such that the *I*-spin signal appears positive at the same echo-time of 1/*J* (Figure 8.8D). Since the nonselective and selective experiments are performed at the same echo-time (and hence with the same T_2-weighting) without perturbation of the other signals, it can be expected that the difference spectrum subtracts all resonances except for the signal of the editing target (Figure 8.8E). Since the two experiments are stored separately, they can also be added together, thereby providing the non-edited signals (Figure 8.8F).

8.4.2 Practical Considerations

The *J*-difference editing sequence as shown in Figure 8.8A is not suitable for *in vivo* measurements due to the lack of spatial localization. Figure 8.9 shows two practical implementations based on PRESS and (semi)-LASER spatial localization methods. There are many possible

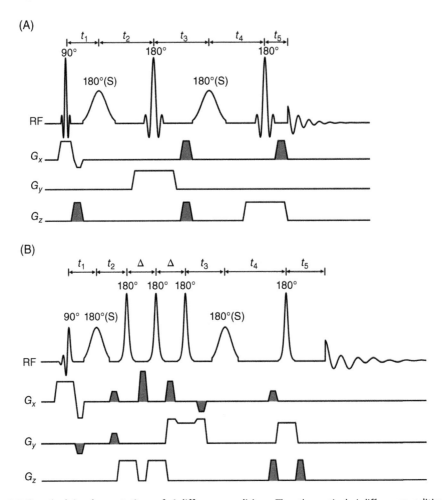

Figure 8.9 Practical implementation of J-difference editing. The theoretical J-difference editing pulse sequence of Figure 8.8A can be extended with spatial localization based on (A) PRESS or (B) semi-LASER. PRESS-based J-difference editing is commonly known as MEGA–PRESS [14]. The delays t_1–t_5 need to be chosen such that they satisfy criteria for spin-echo formation and editing during the entire echo-time as explained in the text. The magnetic field crusher gradients (filled gray) represent the required minimum. Additional crusher gradient may be needed to dephase specific coherence pathways. Both sequences are typically preceded by frequency-selective water suppression technique, such as CHESS, WET, or VAPOR.

combinations of the delays t_1–t_5 that lead to proper echo formation at the echo-time TE. However, only a few specific combinations will simultaneously give optimal editing performance. A general approach for determining the optimal delays t_1–t_5 formulates three conditions that govern proper sequence operation. For the MEGA–PRESS sequence (Figure 8.9A) the spin-echo refocusing condition demands that the phase accumulation due to magnetic field inhomogeneity and chemical shifts is zero, which in turn demands that $2\pi\Delta\nu(t_1 + t_2 - t_3 - t_4 + t_5) = 0$ or $t_1 + t_2 - t_3 - t_4 + t_5 = 0$, where $\Delta\nu$ is a frequency offset due to magnetic field inhomogeneity and/or chemical shift. The sign of the accumulated phase inverts during each 180° refocusing pulse. Secondly, to achieve complete inhibition of scalar coupling evolution over the total echo-time TE and thereby maximize editing efficiency, the following condition must be satisfied, $t_1 - t_2 - t_3 + t_4 + t_5 = 0$. Note that for this condition the accumulated

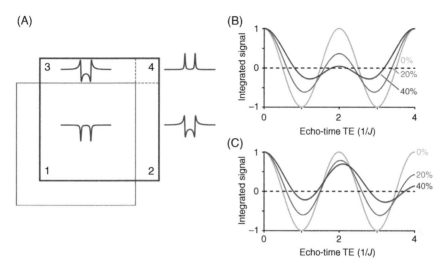

Figure 8.10 **Chemical shift displacement for scalar-coupled spins.** (A) PRESS-localized volume of spin I (red line) and spin S (blue line) in a two-spin IS system. The dimensions shown correspond to the slices selected by the two 180° pulses. The spatial displacement of the two volumes generates four spatial areas that are differently affected by the 180° refocusing pulses. In area 1, both spins experience both 180° pulses, leading to normal scalar coupling evolution which for TE = 2TE1 = 2TE2 = 1/J leads to a negative, in-phase doublet signal for the I spins. In area 4 *only* the I spin experiences both 180° pulses. In effect, the I spins in area 4 experience selective refocusing, leading to inhibition of scalar coupling and a positive in-phase doublet signal. In areas 2 and 3, the I spins experience only one of the two 180° pulses, leading to inhibition of scalar coupling evolution during half the echo-time and hence an antiphase doublet signal. As the detected I spin signal represents the sum of all four areas, the observed I spin signal intensity will be a complicated function of chemical shift displacement and sequence timing. (B, C) Integrated signal intensity of the I spins in an I_nS (n = 1, 2, or 3) spin system as a function of the echo-time for PRESS sequence timings given by (B) TE1 = TE2 and (C) TE2 = 4TE1. The curves represent voxel displacements of 0% (ideal, green), 20% (blue), and 40% (red) of the nominal volume.

phases are only inverted by the frequency-selective RF pulses. Thirdly, the sum of all delays must equal the total echo-time TE, $t_1 + t_2 + t_3 + t_4 + t_5 = $ TE. When setting t_1 to its minimum, constant value and (arbitrarily) assuming $t_3 = t_4$, the aforementioned three conditions provide $t_2 = t_3 = t_4 = $ TE/4 and $t_5 = $ TE/4 − t_1. While the various pulse lengths and magnetic field gradients pose minimum durations on all five delays, the three outlined conditions can be used to determine the delays for optimal signal detection. For the MEGA-(semi)LASER sequence a similar strategy can be used to set the five delays, whereby the delay between the first through third adiabatic full passage (AFP) pulses, Δ, is typically set to a fixed minimum value (see Exercises).

The choice of MEGA–PRESS over MEGA–LASER can be based on a shorter minimum echo-time and a lower amount of RF power deposition. However, at higher magnetic fields the use of LASER or semi-LASER-based sequences is preferable due to the higher bandwidth and hence lower chemical shift displacement achievable by adiabatic RF pulses. While a smaller chemical shift displacement is desirable for any MR spectroscopy study, it is especially important for spectral editing. Figure 8.10A shows two of the spatial dimensions selected by the 180° refocusing pulses in a PRESS localization sequence. For spins with different Larmor frequencies, the limited bandwidth of the 180° RF pulses leads to a spatial displacement of their respective voxel locations, similar to that shown in Figure 6.4. When the spins are also scalar-coupled, additional effects can be observed that complicate the scalar coupling evolution. The voxel of spin I (red in Figure 8.10A) in a scalar-coupled IS spin-system can be divided into four distinct regions, depending on whether the refocusing pulses affect the S spins. Spin I

experiences both 180° refocusing pulses in all four regions. In region 1, the S spins also experience both 180° refocusing pulses, leading to normal, uninhibited scalar evolution for both spins. When the echo-time TE is set to $1/J$, the signal from region 1 would appear inverted. In regions 2 and 3, the S spin only experiences one of the two 180° pulses. In a symmetric PRESS sequence (TE1 = TE2 = TE/2), scalar evolution is not inhibited during half of the echo-time such that signal from regions 2 and 3 would appear in antiphase as expected for scalar evolution over $1/(2J)$. In region 4, the S spins do not experience any 180° RF pulses, such that the sequence appears to contain two frequency-selective 180° pulses for I spins. Signal from region 4 therefore does not exhibit any scalar evolution and will appear in-phase and positive. The total acquired signal for I spins is the sum of the four regions, weighted by the relative volumes of each region. The ultimate consequence for spectral editing is that I spins cannot be fully inverted, even though the echo-time is equal to $1/J$, thus leading to an underestimation of I spin concentration. The effect shown in Figure 8.10A depends on many parameters, including the echo-time TE (Figure 8.10B) and PRESS echo-time spacing (Figure 8.10C).

Incomplete evolution of scalar coupling due to the chemical shift displacement can be greatly minimized in LASER and semi-LASER localization sequences, where AFP RF pulses are employed that can achieve a higher bandwidth at similar RF peak amplitudes. The effects shown in Figure 8.10 can effectively be eliminated when the AFP pulses are replaced by FOCI or GOIA gradient-modulated RF pulses that can achieve a further 10-fold increase in bandwidth without increasing the RF amplitude (see Chapter 6 for details on FOCI and GOIA pulses).

The J-difference editing sequences shown in Figure 8.9 use two frequency-selective 180° refocusing pulses placed around the slice-selective refocusing pulses to achieve spectral editing. In most implementations the editing pulses are always "on," but with different frequencies in the two scans. In the example of Figure 8.8, the editing pulses in one of the two experiments must be placed at the S spin frequency in order to inhibit scalar coupling evolution. The frequency offset of the editing pulses in the second experiment can, in principle, be placed at any offset that (i) does not inhibit scalar coupling evolution and (ii) does not unwillingly perturb resonances that are overlapping with the editing target, spin I. However, there are situations where the correct placement of the editing pulses in the second experiment is critical for the editing performance. For example, when the S spin in Figure 8.8 is resonating at 4.0 ppm (instead of 4.7 ppm as shown) the two editing pulses applied at the S spin frequency would, in addition to the inhibition of scalar coupling evolution, also provide some water suppression. When in the second "control" experiment the spectral editing pulses would be placed far off-resonance (e.g. 10 ppm), the water suppression would not be the same as in the first experiment, leading to a potentially large water residual in the difference experiment. A better offset choice for the editing pulses in the control experiment would be at a position symmetrical around the water resonance (i.e. 5.4 ppm) such that the water suppression is identical in both scans. In other situations the choice of editing pulse offset can reduce the amount of coediting of unwanted signals and will be discussed for GABA editing [15]. In certain fortuitous situations, the control experiment can be used to achieve editing of a second compound as has been shown for simultaneous editing of ascorbic acid and glutathione [16]. This principle has been extended to parallel detection of multiple compounds with Hadamard-based editing (HERMES) [17–19]. The bandwidth of the editing pulses should, in general, be as narrow as possible in order to minimize the perturbation of nearby resonances. Gaussian RF pulses are the standard choice for spectral editing as they have among the narrowest bandwidth for a given pulse length (see Table 5.1). In reality the required echo-time of $1/(2J)$ for IS_2 spin-systems or $1/J$ for IS and IS_3 spin-systems limits the maximum pulse length of the two editing pulses to circa 20 ms, giving a minimum bandwidth on the order of 50–100 Hz.

8.4.3 GABA, 2HG, and Lactate

In the previous sections the theoretical principles of *J*-difference editing have been outlined. However, a successful implementation of *J*-difference editing requires attention to details ranging from the obvious (e.g. the need for spatial localization) to the not so obvious (e.g. frequency drift correction). One of the most popular applications of *in vivo J*-difference editing is the detection of cerebral GABA [14, 20]. GABA is the major inhibitory neurotransmitter in the mammalian central nervous system and alterations in GABA concentration have been found in epilepsy, depression, and schizophrenia. However, the detection of GABA by *in vivo* NMR spectroscopy is not straightforward, as all three resonances of GABA are overlapping with other resonances, making direct observation impossible. Here, the selective detection of GABA by *J*-difference editing will be used to illustrate the requirements of successful spectral editing.

Figure 8.11 shows the three GABA resonances in relation to other commonly observed resonances. The GABA-H2 methylene protons at 2.28 ppm partially overlap with glutamate-H4 protons at 2.34 ppm as well as with broad macromolecular resonances at 2.29 ppm (see also Figure 8.13). The broad GABA-H3 multiplet at 1.89 ppm partially overlaps with macromolecular resonances at 2.05 and 1.72 ppm, as well as with the broad base of the large NAA methyl resonance at 2.01 ppm. The GABA-H4 methylene protons at 3.01 ppm overlap completely with a combined resonance from creatine and phosphocreatine. The first decision that needs to be made is on which resonance to focus in the spectral editing process. Detection of GABA-H3 is not desirable due to the multiple coupling partners and the reduced intensity of the multiplet. Detection of GABA-H2 is not desirable because of the close proximity of glutamate-H3 to GABA-H3. Selective refocusing of GABA-H3 (which is the coupling partner of GABA-H2) will certainly affect glutamtate-H3, leading to partial coediting of glutamate-H4 and hence partial overlap of glutamate-H4 and GABA-H2. To a first approximation GABA-H4 only overlaps with singlet resonances and is therefore typically the choice for spectral editing methods. Since in this case GABA-H3 is also the coupling partner, spectral editing of GABA-H4 typically leads to near-complete GABA-H2 recovery as well (with a small amount of coedited glutamate-H4).

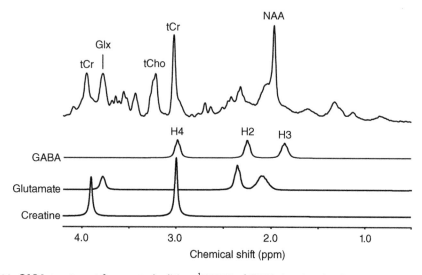

Figure 8.11 GABA as a target for spectral editing. ^{1}H NMR of GABA showing the three resonances in relation to other proton resonances, including those from creatine and glutamate.

Following the selection of the detection target (GABA-H4) and the editing target (GABA-H3) the pulse sequence should be selected. The most popular choices are MEGA–PRESS and MEGA–semiLASER as shown in Figure 8.9. The next decision to be made concerns water suppression. Even though all uncoupled resonances, including water, are eliminated in the subtraction process, *J*-difference editing still requires excellent water suppression for two reasons. Firstly, while the edited resonance (GABA in this case) is the prime focus, the other resonances that can be observed in the two sub-spectra are often also important. Total creatine (creatine + phosphocreatine) is frequently used as an internal concentration reference, while glutamate and glutamine provide complementary information on neurotransmitter metabolism. Reliable detection of the non-edited metabolites requires excellent water suppression, since the large water resonance baseline and vibration-induced sidebands (see Figure 6.26) can obscure the small metabolite resonances. Secondly, small perturbations in the water resonance by motion, magnetic field drift, or other system instabilities can result in a large water subtraction artifact in the GABA-edited spectrum.

CHESS, WET, and VAPOR water suppression (see Chapter 6) are all valid and popular choices, since they selectively perturb the water signal while leaving the metabolites unperturbed prior to excitation. The MEGA–PRESS and MEGA–semiLASER sequences (Figure 8.9) also allow the use of MEGA water suppression. When the water and editing target frequencies are well separated relative to the editing pulse bandwidth, the spectral editing pulses can be replaced with double-banded frequency pulses. One of the frequency bands is placed on the water frequency, providing water suppression, whereby the other frequency band is on the editing target, leading to inhibition of scalar coupling. In the control experiment, a second double-band RF pulse is needed in which the second frequency band is not placed on the editing target, thus allowing regular scalar coupling evolution. The use of multifrequency RF pulses is not recommended when the water and editing target frequencies are in close proximity due to the compromised frequency profile of nearby frequency bands (see Figure 5.13).

Figure 8.12 shows a typical example of *J*-difference editing of GABA in the human brain. The observation of a well-isolated, high-sensitivity signal at 3.00 ppm obtained with good water suppression and a flat baseline is a prerequisite for reliable GABA detection, but there is no guarantee that the detected signal is only GABA. There is always the possibility that the observed signal is contaminated by coedited signals and incompletely subtracted overlapping signals. For each spectral editing experiment it should be determined which metabolites can potentially be coedited and contaminate the edited signal. For GABA-H4 detection the prime candidates for coediting are glutamate, homocarnosine, and macromolecular resonances. From Figure 8.12 it is clear that glutamate coediting does occur, as evidenced by the glutamate-H2 resonance (~3.74 ppm) in the edited spectrum. However, since glutamate-H2 does not overlap with GABA-H4 this form of coediting is of no concern. Homocarnosine is a dipeptide of histidine and GABA and occurs in the human brain at a concentration of circa 0.5 mM. Since the spin-system of the GABA moiety of homocarnosine is essentially identical to that of GABA, the edited signal observed *in vivo* is typically the sum of GABA and homocarnosine (which is sometimes referred to as "total GABA"). This form of complete coediting can be a problem when GABA and homocarnosine need to be separated. Fortunately, in this particular case, homocarnosine can be detected separately through the histidine–imidazole proton resonances at 7.1 and 8.1 ppm observable in short echo-time ^{1}H NMR spectra [21].

The contamination of GABA by signals originating from macromolecules is more complicated and depends on the magnetic field strength, the editing pulse shape and duration, and the specific editing strategy employed. The macromolecular resonances of importance are M4 at 1.72 ppm and M7 at 3.00 ppm with a scalar coupling constant $J_{M4\text{-}M7} = 7.3$ Hz (see Figure 8.13).

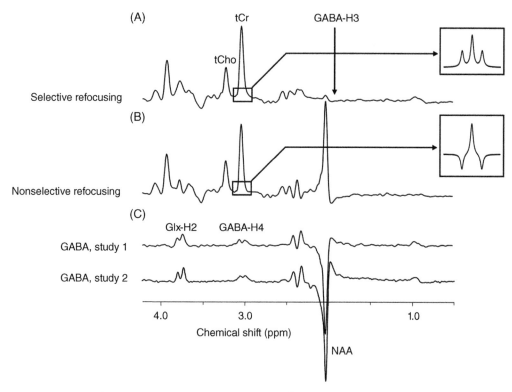

Figure 8.12 **GABA editing on human brain *in vivo*.** Localized ^{1}H NMR spectra (4.0 T, 12 ml, TR/TE = 3000/68 ms) acquired (A) with and (B) without selective refocusing of the GABA-H3 resonance. (C) Difference spectra obtained from the same subject on separate occasions, showing the reproducibility of GABA-H4 detection.

For GABA detection the optimal echo-time of the *J*-difference editing sequence is $1/(2J)$, which equals 68 ms. With two spatially-selective 180° pulses in MEGA–PRESS and a number of magnetic field crusher gradients within the total echo-time, a realistic maximum duration of the two editing pulses is circa 20 ms. A 20 ms Gaussian 180° pulse (truncated at 10%) will have a full bandwidth at half maximum of 75 Hz. Given that the frequency separation between GABA-H3 and M4 is only 0.17 ppm (=12, 24, and 57 Hz at 1.5, 3.0, and 7.0 T, respectively), the editing pulses will certainly affect M4 at low magnetic fields, leading to partial coediting of M7 underlying the GABA-H4 resonance. Different strategies to account for or minimize the contribution of macromolecules have been proposed [14, 15]. Firstly, the nonselective experiment can be modified to apply frequency-selective 180° pulses at a frequency mirrored around the M4 resonance (i.e. at 1.72 – 0.17 = 1.55 ppm). The M4 will be affected in both experiments by the symmetric editing pulses, but because the effect is identical in both experiments the M7 resonance will be subtracted from the edited spectrum [15]. While this method works well at medium-to-high magnetic field strengths, it may not perform satisfactorily at lower magnetic fields due to asymmetry in the M4 resonance and the proximity of GABA-H3 to M4. The macromolecular contribution can also be minimized by utilizing the large T_1 difference between metabolites and macromolecules [22]. Applying a nonselective inversion pulse prior to excitation will "null" the longitudinal magnetization of macromolecules when the recovery delay is adjusted to $T_{1,M7}\ln(2)$. The metabolites, with a much longer T_1 relaxation constant, have only

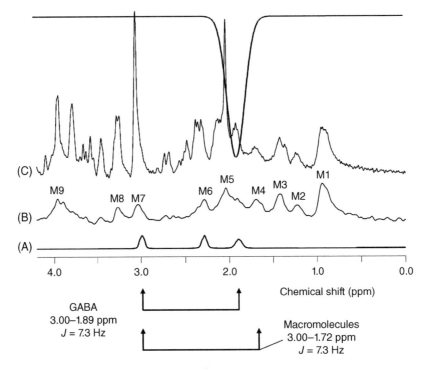

Figure 8.13 **GABA and macromolecules.** (A) *In vitro* ^{1}H NMR spectrum of GABA showing the three resonances relative to (B) macromolecular and (C) other proton NMR resonances acquired from rat brain *in vivo* at 9.4T. The macromolecular NMR spectrum in (B) was obtained with a double-inversion-recovery (TR/TI1/TI2 = 3250/2100/630 ms), which leads to a selective "nulling" of the metabolite signals. The close proximity of the GABA-H3 and macromolecular M4 resonances can lead to coediting of the macromolecular M7 resonance, depending on the bandwidth (in ppm) of the selective refocusing pulses.

partially recovered and can be excited and detected without macromolecule contamination. Unfortunately, both the symmetric editing option and T_1-based signal nulling have not found wide acceptance in the *in vivo* MRS community, likely because they further reduce an already small signal. The common approach that is now followed is to refer to the combined GABA + MM signal as GABA$^+$. While this is certainly the simplest option, it should always be realized that the GABA signal is only a (potentially small) fraction of the detected signal and that changes in GABA$^+$ should be interpreted with caution.

In addition to the standard conditions for successful MRS *in vivo*, like high magnetic field homogeneity, good water suppression and spatial localization, and post-acquisition removal of B_0 eddy currents, a consideration that is particularly important for *in vivo* spectral editing is related to system and subject stability. All MR systems have a basic magnetic field drift by which the magnetic field slowly decreases over time, typically less than 10 Hz per hour. In addition, ultra-long B_0 eddy currents and heating of passive shims can also lead to temporal magnetic field variations [23]. Besides system-related variations, subject motion can introduce temporal alterations of the magnetic field. The main requirement for successful *J*-difference editing is spectral stability between the two acquisitions. Any variation can potentially lead to incomplete subtraction of overlapping resonances, resulting in an overestimation of the edited resonance intensity. For most applications, significant signal averaging (e.g. *N* averages) is required to achieve sufficient signal-to-noise ratio (SNR) of the edited resonances. Rather than

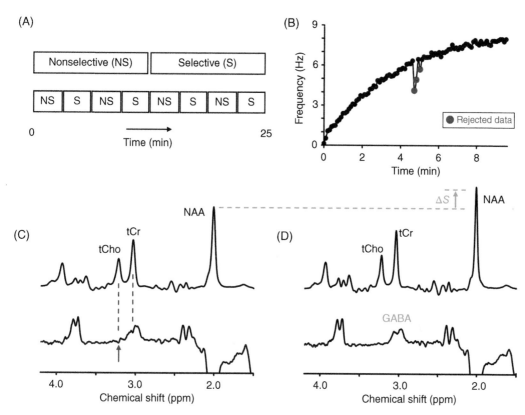

Figure 8.14 **Frequency alignment and data rejection in edited MRS.** (A) The effects of frequency drifts, patient motion, and system instability during an extended edited MRS study can corrupt the subtraction process when acquired and stored in two blocks (top). The effects can be tracked and minimized post-acquisition by acquiring and storing data in smaller blocks (bottom). (B) Frequency variations during a GABA-edited MRS scan on human brain. The gradual frequency variation is largely due to magnet drift, whereby the abrupt variations are due to subject motion (sneezing). (C) Without post-acquisition frequency alignment or data rejection, the GABA-edited spectrum (bottom) is compromised by subtraction artifacts as can be recognized by the residual choline signal. (D) Data rejection (red dots in (B)) and frequency alignment of the individually stored spectra lead to narrower resonance lines and improved editing performance.

acquiring two spectra with $N/2$ averages each, it is preferable to acquire N spectra of 1 average each (Figure 8.14A). This allows frequency alignment of the spectra before summation, thereby significantly reducing the effects of magnetic field drifts on the final edited spectrum. In the realistic example shown in Figure 8.14, ignoring magnetic field drifts resulted in a circa 50% overestimation of the edited GABA resonance (Fig. 8.14C). An additional benefit of post-acquisition frequency alignment is the narrower resonance lines, providing improved S/N and spectral resolution (indicated for NAA in Figure 8.14C and D). In addition to frequency drift correction, the separate storage of each average also allows data quality assurance on subject compliance. Figure 8.14B shows the frequency drift of the creatine signal during a 10 min GABA editing scan. While the majority of data displays a relatively smooth frequency drift, a number of averages are characterized by large frequency variations caused by sneezing of the subject. The same effect can also be observed as an increase in the creatine line width (not shown). When the voxel position is relatively unaffected by the perturbation (sneezing in this case) one can elect to simply eliminate the perturbed averages after which the data can be further processed. However, if the voxel placement is significantly affected, the validity of the spectroscopic results should be questioned. It is advised to collect rapid anatomical MR images prior to and following an MRS study to aid in the evaluation of voxel placement changes.

While spectral editing can be applied to any scalar coupled spin-system the most popular editing target *in vivo* are, besides GABA [24], ascorbic acid [25], glutathione [26], 2-hydroxyglutarate (2HG) [27, 28], and lactate [29–32]. While the basic principles are similar for all compounds, each spin-system has its own unique considerations that will be briefly highlighted for 2HG and lactate.

2HG is an oncometabolite that is unique to brain tumors exhibiting a mutation in isocitrate dehydrogenase (IDH). The biological specificity and significance, as well as its high concentration have made it a popular target for *in vivo* MRS detection. 2HG is structurally similar to glutamate, with a comparable ^{1}H NMR spectral profile. Whereas the 2HG-H4 protons have been the preferred target for non-edited detection ([27], see also Section 8.6), the 2HG-H2 proton offers the best detection target for spectral editing due to its well-separated chemical shift, simpler scalar coupling pattern, and lack of overlap with more intense signals [27, 28]. The 2HG-H3 protons around 1.9 ppm are the scalar coupling partners to the 2HG-H2 proton, thereby forming the target for the editing pulses. Since 2HG forms a complicated, strongly-coupled spin system, the echo-time for optimal 2HG-H2 signal detection needs to be determined through density matrix simulations or experimentally on a phantom containing 2HG. Figure 8.15 shows 2HG detection in an IDH-mutated brain glioma, together with a control measurement in contralateral normal brain tissue. The main challenge of 2HG-H2 detection around 4.0 ppm is its close proximity to the large water resonance. Spurious echo formation of incompletely dephased water can quickly overwhelm 2HG-H2 detection, which is especially true on patients with postsurgical clips that can generate strong, local magnetic field gradients (see also Figure 6.15).

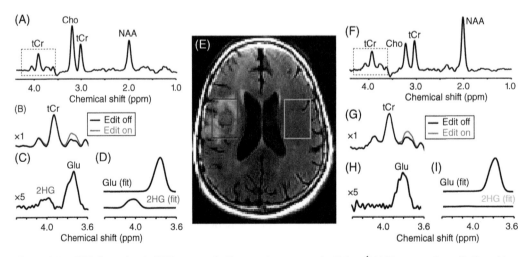

Figure 8.15 **2HG detection in IDH-mutated glioma using spectral editing.** ^{1}H MR spectra from (A–D, red box in E) tumor region and (F–I, green box in E) contralateral control region in a patient with an isocitrate dehydrogenase (IDH)-mutated glioma. (E) T_2-weighted FLAIR (TR/TI/TE = 8000/2100/90 ms) provides tumor-specific image contrast. (A, F) Regular ^{1}H MR spectra (TE = 82 ms) showing the increased choline-to-NAA ratio relative to normal brain. (B, G) Two ^{1}H MR sub-spectra with the editing pulse applied to the 2HG-H3 protons at ~1.9 ppm (gray line) and at a control position (6.1 ppm, black line). (C, H) Subtraction of the two sub-spectra in (B, G) gives the 2HG-edited spectrum. The coedited glutamate-H2 resonance at 3.75 ppm is well separated and does not interfere with 2HG detection in the tumor. (D, I) A spectral fit of (C, H) further confirms the specificity of 2HG for IDH-mutated tissue.

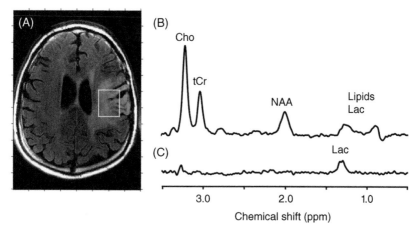

Figure 8.16 **Lactate detection in human brain glioma using spectral editing.** (A) T_2-weighted FLAIR (TR/TI/ TE = 8000/2100/90 ms) image and (B, C) ^{1}H MRS from the localized volume indicated in (A). (B) Semi-LASER ^{1}H NMR spectrum (TE = 144 ms) displays the high choline-to-NAA ratio typical for brain tumors in addition to lactate and lipid resonances around 1.3 ppm. (C) Lactate-edited difference spectrum reveals a clean lactate signal at 1.31 ppm without contamination of overlapping lipids.

Lactate or lactic acid is the classical target for spectral editing due to the weak scalar coupling even at low magnetic fields. As an IS_3 spin system the optimal echo-time for spectral editing equals $1/J$ or 144 ms. Lactate is primarily overlapping with signals from threonine and lipids, whereby lipid contamination is the typical reason for spectral editing. Whereas normal brain tissue does not contain NMR-visible lipid resonances, a number of pathologies (tumors, stroke) and different tissues (muscle, liver) do display pronounced lipid signals. All lipid methylene signals around 1.3 ppm that overlap with lactate have scalar coupling. Fortunately, the scalar coupling partners typically reside in the 1.5–2.5 ppm chemical shift range, such that a frequency-selective editing pulse on lactate-H2 at 4.1 ppm should lead to minimal coediting of lipids. Figure 8.16 shows an example of lactate editing on a patient with a brain tumor. It should be noted that even though *J*-difference editing removes overlapping lipid signals, the editing process is not immune to the appearance of sporadic lipid signals due to subject motion and/ or inadequate spatial localization. For subjects/samples that are not suitable for *J*-difference editing, the use of multiple-quantum-based editing may provide a solution.

8.5   Multiple Quantum Coherence Editing

J-difference editing as detailed in the previous section is a valuable and popular editing method that performs well over a wide range of conditions. However, the requirement for two separate acquisitions is not always desirable and as a result a wide range of spectral editing methods have been developed, including polarization transfer [33–36], longitudinal scalar-order-based editing [37–39], Hartmann–Hahn transfer [40, 41] and MQCs based editing [29–32]. While each editing method has its own merits, the present discussion will be limited to MQCs-based spectral editing since these methods achieve single-scan editing with excellent water suppression.

Figure 8.17A shows a basic pulse sequence to create multiple-quantum coherences (MQCs). The first part of the sequence is a spin-echo with an echo-time of $1/(2J)$. For non-coupled spins the magnetization ends up along the $+y$ axis (Figure 8.17B, top) with full refocusing of magnetic field

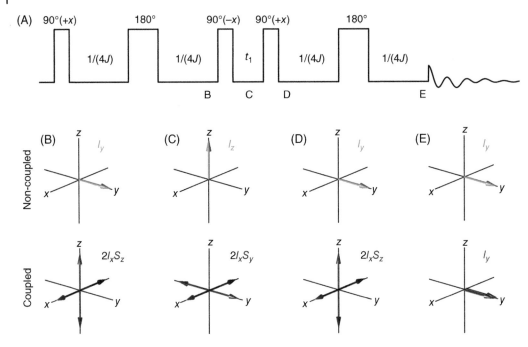

Figure 8.17 **Generation of multiple-quantum coherences.** (A) One possible pulse sequence, among many, to generate multiple-quantum coherences (MQCs) together with product operator states for (middle row) non-coupled and (bottom row) a two-spin, weakly-coupled spin system. (B) Following excitation and a delay of $1/(2J)$ a weakly-coupled spin-system is in a state of antiphase coherence, whereas non-coupled spins are in-phase along the y axis. (C) A second 90° pulse along the −x axis rotates non-coupled spins to the +z axis, whereas scalar-coupled spins are converted to MQCs. (D) A third 90° pulse converts longitudinal magnetization and MQCs backed to transverse magnetization and antiphase coherences, respectively. (E) Following a delay $1/(2J)$, the transverse magnetization for both non-coupled and scalar-coupled spins can be detected along the y axis. Scalar-coupled spins can be detected separately from non-coupled spins by utilizing the difference sensitivity of MQCs present during the t_1 period towards RF phase cycling and/or magnetic field gradients.

inhomogeneity and chemical shifts. For a scalar-coupled *IS* spin system the nonselective spin-echo sequence allows scalar coupling evolution according to Eq. (8.1), thereby ending up in a state of antiphase coherence $2I_xS_z$ (Figure 8.17B, bottom). With a 90° excitation pulse transmitted along the −x axis the transverse magnetization of non-coupled spins, like water, is rotated back to the +z axis (Figure 8.17C, top). The antiphase coherence state $2I_xS_z$ for scalar-coupled spins is transformed into $2I_xS_y$, a correlated spin state representing a combination of zero and double-quantum coherences, collectively known as MQCs. As explained in Section 8.2.3 and Figure 8.3D, MQCs do not generate an NMR observable signal since they are locked in equal and opposing directions that do not evolve under the influence of scalar coupling. However, as will be discussed later, MQCs are present in the transverse plane and can be manipulated by RF pulses, phase cycling, and magnetic field gradients. Application of a third 90° pulse converts the MQCs back into antiphase coherences (Figure 8.17D, bottom), while simultaneously exciting the longitudinal magnetization of non-coupled spins back along the +y axis (Figure 8.17D, top). Following an echo-time delay $1/(2J)$, the antiphase coherences evolve into in-phase coherences (Figure 8.17E, bottom) that can be detected during the signal acquisition period. Note that the pulse sequence in Figure 8.17A did not achieve spectral editing as signal from non-coupled and scalar-coupled spins are equally detected during the acquisition period. The pulse sequence only achieved the generation of MQCs (and subsequent conversion back to regular magnetization, or single-quantum coherences [SQCs]) during a period in which the magnetization from non-coupled spins resided along the z axis. It is this period that is the key to discriminating non-coupled from scalar-coupled spins through the use of phase cycling and/or magnetic field gradients. In order to understand the effect of magnetic field gradients on MQCs, it is convenient to switch from product operators to shift (or lowering and raising) operators.

Transverse magnetization (e.g. M_x) or in general single-quantum-coherence (SQC, e.g. I_x) that is oscillating along a given axis can be seen as the vector sum of anticlockwise and clockwise rotating components (I^+ and I^-). These rotating components can be defined as

$$I^+ = I_x + iI_y \text{ and } I^- = I_x - iI_y. \tag{8.2}$$

which is equivalent to

$$I_x = \frac{1}{2}\left(I^+ + I^-\right) \text{ and } I_y = -\frac{1}{2}i\left(I^+ - I^-\right) \tag{8.3}$$

This is in essence a more formal description of concepts that were already introduced in Chapter 1 regarding quadrature detection and excitation (Figures 1.4 and 1.7). Equation (8.3) states that if signal is detected along a single axis, both anticlockwise (I^+) and clockwise (I^-) components contribute to the signal such that the frequency sign is ambiguous. When quadrature detection is performed along two orthogonal axes (e.g. +x and +y), the detected signal contains only one of the two rotating components (e.g. I^+) such that the frequency sign is determined. While the choice of detected component (I^+ or I^-) is not consequential in any significant manner, most publications describe the detection of the I^- component. The same arguments used for quadrature detection can also be used to explain quadrature excitation in which B_1 magnetic fields are applied along two orthogonal axes, providing a more efficient excitation compared to single axis excitation. Whereas the transverse coherences along the x and y axes are presented by product operators I_x and I_y, the anticlockwise and clockwise rotating coherences are presented by the shift operators I^+ and I^-. The shift operators are also known as ladder or lowering and raising operators. The main advantage of shift operators is that they provide an immediate overview of the coherence order present at any point in time. The shift operator I^+ represents coherence order +1, whereas the shift operator I^- represents coherence order −1. Both operators represent single-quantum-coherence, whereby the sign of the coherence order indicates the direction of precession. In the quantum mechanical description of NMR the lowering and raising operators connect the α and β spin states and thereby represent a spectroscopic transition. In the classical description shown in Figure 8.2 the raising and lowering operators correlate with the detectable magnetization or coherences. The value of shift operators becomes more obvious when the definitions of Eq. (8.3) are applied to the MQC state $2I_xS_y$ encountered in Figure 8.17C. The single product operator can be described as the sum of four shift operators according to

$$2I_xS_y = -\frac{1}{2}i\left(I^+S^+ - I^+S^- + I^-S^+ - I^-S^-\right) \tag{8.4}$$

Equation (8.4) reveals that the term $2I_xS_y$ contains four different types of coherences corresponding to zero-quantum (I^+S^- and I^-S^+) and double-quantum (I^+S^+ and I^-S^-) coherences. Zero-quantum-coherences (ZQCs) have a coherence order of zero, whereas the double-quantum-coherences have a coherence order of +2 for I^+S^+ and a coherence order of −2 for I^-S^-. Figure 8.18 shows a graphical depiction of the decomposition of $2I_xS_y$ (Figure 8.18C)

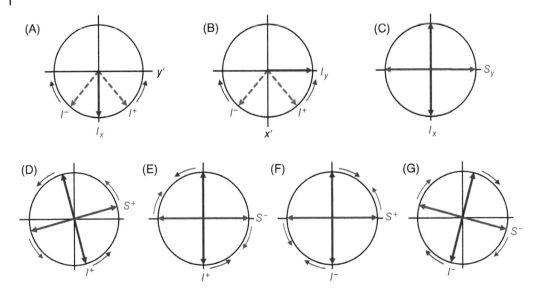

Figure 8.18 Decomposition of product operators into shift or raising/lowering operators. (A, B) Product operators along the *x* and *y* axes, (A) I_x and (B) I_y, can be presented as a sum of anticlockwise rotating I^+ and clockwise rotating I^- shift operators, also known as raising and lowering operators, according to Eqs. (8.2) and (8.3). (C) MQC state $2I_xS_y$ represented as a product operator and as (D–G) a linear sum of four shift operators, (D) I^+S^+, (E) I^+S^-, (F) I^-S^+, and (G) I^-S^-. In the double-quantum I^+S^+ and I^-S^- spin states, the *I* and *S* spins rotate in the same direction such that they are dephased twice as fast under the influence of magnetic field inhomogeneity and gradients as compared to SQCs I^+ or I^-. In homonuclear spin systems the zero-quantum I^+S^- and I^-S^+ spin states are immune to magnetic field gradients.

into the sum of I^+S^+ (Figure 8.18D), I^+S^- (Figure 8.18E), I^-S^+ (Figure 8.18F), and I^-S^- (Figure 8.18G) according to Eq. (8.4). The precession of the different coherences under the effects of chemical shifts, frequency offsets, and magnetic field gradients depends on the coherence order. Coherences of order *p* evolve in the transverse plane according to

$$I^p = I^p \cdot e^{-ip\phi} \tag{8.5}$$

where the phase ϕ acquired over time *t* associated with chemical shifts and frequency offsets $\Delta\nu$ (in Hz) is given by $\phi = 2\pi\Delta\nu t$. The position-dependent phase $\phi(r)$ associated with a rectangular gradient pulse of amplitude *G* (in $Hz\,cm^{-1}$) and duration *t* is given by $\phi(r) = 2\pi rGt$. As a result, DQCs evolve under the *sum* of the chemical shifts, whereas ZQCs evolve under the *difference* of the chemical shifts. More importantly for spectral editing is the fact that for homonuclear spin systems, magnetic field gradients dephase DQCs twice as fast a SQCs. Magnetic field gradients do not dephase homonuclear ZQCs at all. This difference in sensitivity towards magnetic field gradients forms the basis of single-scan spectral editing of MQCs. Note that for heteronuclear applications, the phase in Eq. (8.5) needs to be scaled with the gyromagnetic ratio, in which case ZQCs will be dephased by magnetic field gradients, albeit differently than for heteronuclear DQCs and SQCs. More details of heteronuclear spectral editing using magnetic field gradients can be found in Section 8.7.

A convenient method to keep track of the coherences present during a pulse sequence is provided by coherence transfer pathway (CTP) diagrams (Figure 8.19). The CTP diagram for a two-spin-system is composed of five horizontal lines, representing the possible coherence orders. Prior to any pulse sequence, the coherence order *p* is zero as the magnetization resides along the longitudinal axis. Immediately following excitation, both SQCs ($p = \pm1$) are present

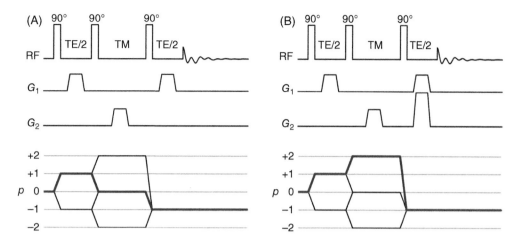

Figure 8.19 **Coherence transfer pathways.** A pulse sequence composed of three RF pulses can generate eight FIDs and echoes (see Chapter 6). A specific echo can be selected through the use of RF phase cycling and/or magnetic field gradients. The pathway of coherences present during the pulse sequence that lead to the final detection signal (Γ by convention) can be visualized in a coherence transfer pathway (CTP) diagram, whereby each horizontal line corresponds to a particular coherence order. (A) Selection of coherence pathway (0, +1, 0, −1) (red line) as achieved with the indicated magnetic field gradients used during STEAM. The coherence transfer pathway (0, −1, 0, −1) could also be selected, albeit without refocusing of magnetic field inhomogeneity or chemical shifts. (B) Selection of coherence pathway (0, +1, +2, −1) (red line) as achieved during MQC-based spectral editing.

and will be dephased by chemical shifts, magnetic field inhomogeneity, and a magnetic field gradient pulse. Application of a second 90° pulse converts the SQCs to ZQCs and DQCs during the TM period. DQCs are dephased further by the magnetic field gradient present during TM, whereas the ZQCs are immune to the effects of that magnetic field gradient. A final 90° pulse converts the coherences into SQCs. Note that both Γ^{+} and Γ coherences are generated by the final 90° pulse. However, by convention a CTP diagram only shows the Γ coherence during the acquisition period, as that represent the detectable signal. It is clear from the CTP diagram in Figure 8.19 that only RF pulses can change the coherence order p. Chemical shift evolution, magnetic field gradients, and even scalar coupling evolution change the phase of the coherence, but not the coherence order. Whereas a CTP diagram shows all *possible* pathways, it is up to the NMR spectroscopist to select the *desired* pathway through a combination of phase cycling and magnetic field gradients.

For a stimulated echo-pulse sequence, such as used in STEAM localization (see Chapter 6), the desired CTP as achieved with the indicated TE and TM magnetic field gradients is shown in red in Figure 8.19A. The unbalanced TM magnetic field gradient ensures that only ZQCs (and longitudinal magnetization) are not dephased at the time of signal acquisition. The CTP (0, +1, 0, −1) for the coherences during (TR, TE/2, TM, TE/2) is the preferable pathway for STEAM and is selected by two equal gradients in both TE/2 period. The CTP (0, −1, 0, −1) can be selected with two equal gradients with opposite sign in each TE/2 period. However, this CTP is not desirable as the dephasing due to chemical shifts and magnetic field inhomogeneity is not refocused. The same three-pulse sequence can also be used to select DQCs during the TM period (Figure 8.19B), as is done during MQC-based spectral editing. The red CTP in Figure 8.19B can be selected by matching the TM magnetic field gradient by a magnetic field gradient in the final TE/2 period with twice the amplitude. The ZQCs and longitudinal magnetization (from uncoupled spins) present during TM are not dephased by the TM magnetic field gradient. However, after conversion to SQCs by the third 90° pulse, the unbalanced final TE/2 magnetic field gradient dephases the signals associated with ZQCs and longitudinal magnetization during TM. In other words, the RF pulse and gradient combination in Figure 8.19B represents a selective filter to pass DQCs and stop ZQCs and SQCs. An excellent textbook on product operators, shift operators, and CTPs is provided by Keeler [42]. The transformations of product and shift operators under the effects of RF pulses, magnetic field gradients, and scalar coupling are summarized in Appendix A.4.

Figure 8.20A shows a practical, single-scan *in vivo* MQC-based spectral editing sequence for the detection of lactate. The selMQC sequence [32] is in essence a homonuclear version of the popular heteronuclear multiple quantum coherence (HMQC) method routinely used in high-resolution, liquid-state NMR [43]. The sequence represents a frequency-selective spin-echo for the signal of interest (spin I or the methyl protons of lactate at 1.31 ppm). Unfortunately, lipid methylene protons also resonate at circa 1.3 ppm (see Chapter 2) and will obscure the lactate signal. Discrimination between lactate and lipids is achieved through the second and third 90° pulses that are frequency selective for the lactate methine proton at 4.10 ppm. During the t_1 period the lactate spin system is present as MQCs. For the particular frequency offsets chosen, lipids do not generate MQCs and are present as SQCs during the t_1 period. The CTP for lactate (Figure 8.20B, red line) can be selected in a single scan with the gradient combination shown in Figure 8.20A. For lipids and non-coupled spins the gradient combination represents an unbalanced, nonzero gradient, leading to dephasing of those signals. Figure 8.20E and F shows a practical example of gradient-enhanced spectral editing of lactate *in vivo*.

MQC-based spectral editing sequences are not limited to ZQC and DQCs only, but can be constructed to selectively observe higher-order coherences. For instance, Wilman and Allen [44] have used triple-quantum-coherence filtering to selectively observe GABA (an $A_2MM'XX'$ spin system) in rat brain extracts. However, in general, it can be stated that higher-coherence-order-based spectral editing is prone to increased signal loss, since it is almost impossible to refocus all coherence pathways simultaneously.

Multiple-quantum-based spectral editing can, besides the methods shown in Figure 8.17, be achieved with a wide variety of pulse sequences. Each method will have considerations concerning spectral selectivity and coediting, signal recovery and sensitivity to experimental imperfections. However, all MQC-based sequences are based on the same principle in that scalar-coupled spins can form MQCs, which are dephased by magnetic field gradients. After conversion to SQCs, the magnetization is refocused by magnetic field gradients of appropriate amplitude. Uncoupled spins cannot form MQCs, such that the combination of magnetic field gradients will lead to dephasing of transverse magnetization.

While MQC-based spectral editing methods are typically very robust and give excellent suppression of unwanted resonances, they have their own drawbacks, typically in terms of signal quantification. Since all uncoupled resonances are dephased in the editing process, they cannot be used as an internal reference unless a separate, non-edited experiment is performed. Furthermore, frequency drift correction as outlined in Figure 8.14 may be difficult as the single-scan SNR of the edited signal is typically insufficient. Fortunately, the performance of MQC-based spectral editing is not sensitive to small frequency drifts.

8.6 Spectral Editing Alternatives

Due to a variety of reasons it may not always be possible or desirable to perform spectral editing. For instance, spectral editing requires a modified MR pulse sequence that may not be available, particularly on clinical MR scanners. For strongly-coupled spins systems, such as

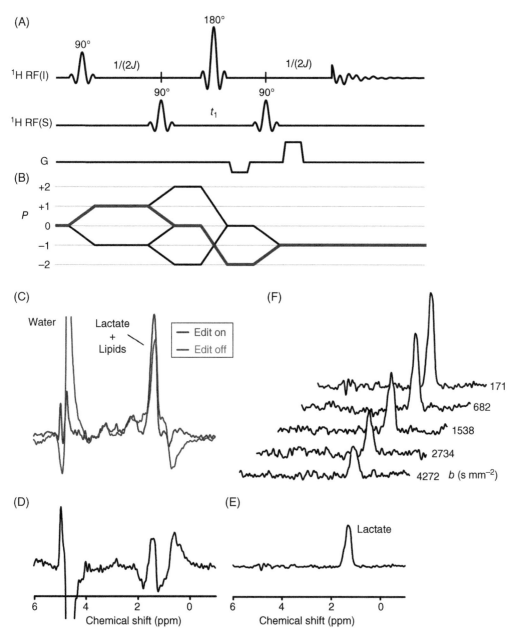

Figure 8.20 **Spectral editing of lactate in a RG2 glioma *in vivo*.** (A) SelMQC spectral editing pulse sequence and (B) coherence transfer pathway (CTP) diagram. The red CTP is selected with a magnetic field gradient combination of −1 and +2. (C) Localized ^{1}H MR spectra (TR/TE = 3000/144 ms, 512 µl) acquired with (red) and without (blue) selective refocusing of the lactate-H2 resonance at 4.10 ppm. (D) *J*-difference editing spectra of lactate as calculated from the two spectra in (C). Lactate is visible, but the spectrum is heavily contaminated with lipids, likely caused by motion of the tumor by breathing. (E) Gradient-enhanced MQC-based edited ^{1}H MR spectrum of lactate. Note the excellent water suppression and clean selection of lactate. (F) The unambiguous detection of lactate allows one to perform more sophisticated experiments, like lactate diffusion measurements. Diffusion sensitization was achieved during an additional (selective) spin echo following the MQC spectral editing sequence. Diffusion sensitizing gradients generated *b*-values of 171, 682, 1538, 2734, and 4272 s mm^{-2} ($\delta = 5$ ms, $\Delta = 35$ ms), respectively, resulting in a calculated ADC of 0.24×10^{-3} mm^2 s^{-1}.

glutamate, glutamine, or 2HG, the scalar evolution is complex and could lead to low editing efficiency. In these circumstances it may be beneficial to use standard MR sequences like PRESS, STEAM, or LASER executed with a specific echo-time that allows optimal detection of the compound of interest. The optimal sequence parameters can be obtained experimentally on phantoms containing the pure compounds or through density matrix simulations as shown in Figure 8.21. Figure 8.21A shows the integrated signal intensities for glutamine (Gln), glutamate (Glu), and 2HG as a function of the echo-times TE1 and TE2 in PRESS. Figure 8.21B shows the Cramer–Rao lower bounds (CRLBs, see Chapter 9) for the same compounds. It follows that the strong scalar coupling in all three compounds leads to complex signal intensity modulations as a function of TE1 and TE2, but also as a function of the RF pulse shape and bandwidth. The curves shown in Figure 8.21A and B can be used to separate more optimal (Figure 8.21C) and less optimal (Figure 8.21D) echo-times for 2HG detection. The simulations shown in Figure 8.21A and B have to be repeated when changing magnetic field strength, pulse sequence, or RF pulses. In addition to echo-time variation, the STEAM sequence provides an additional mechanism for signal modulation through variation of the mixing time TM (Figure 8.22). Figure 8.22A shows a signal intensity plot for lactate as a function of TE and TM. Since lactate is a weakly-coupled spin-system, the plot in Figure 8.22A was obtained with an analytical expression that can be derived with the product operator formalism (see Exercises). The signal intensity modulation as a function of TE is primarily determined by regular scalar evolution combined with polarization transfer between the methine and methyl spins in lactate. The TM dependence shows a high-frequency modulation due to homonuclear ZQCs during TM that are not dephased by magnetic field gradients (see also Figure 8.22B). Homonuclear ZQCs do evolve under the effects of the chemical shift difference between the spins involved, leading to a modulation of several hundred Hertz at most clinical magnetic fields (Figure 8.22B). Experimentally, the ZQC-based modulation can be used to achieve selective detection of lactate by acquiring MR spectra at two slightly different TMs, e.g. 12.0 and 12.5 ms in Figure 8.22C. Note that the glutamate signals show very little TM-dependent signal modulation due to the nonoptimal echo-time for glutamate. Modulation curves as shown in Figure 8.22A can be obtained for all scalar-coupled spin-systems, but require experimental acquisitions or density matrix simulations for strongly-coupled spins. Besides the two examples shown in Figures 8.21 and 8.22, many other methods have been described to achieve modification or simplification of the spectral appearance [27, 35, 45–48].

8.7 Heteronuclear Spectral Editing

Homonuclear spectral editing is aimed at spectral simplification and the unambiguous detection of specific compounds by utilizing scalar evolution based on the homonuclear scalar coupling between protons. Heteronuclear spectral editing utilizes scalar evolution involving the heteronuclear scalar coupling between protons and a non-proton nucleus, typically carbon-13, nitrogen-15, or phosphorus-31. Heteronuclear spectral editing can be aimed at spectral simplification or the detection of specific compounds, but frequently it is used to enhance the NMR sensitivity of the low-sensitivity non-proton nucleus.

8.7.1 Proton-observed, Carbon-edited (POCE) MRS

Heteronuclear spectral editing methods are similar to the homonuclear techniques described in the previous sections, with a noticeable difference that a nonselective RF pulse applied to one type of nucleus is automatically selective for the other nucleus. This is because the bandwidth

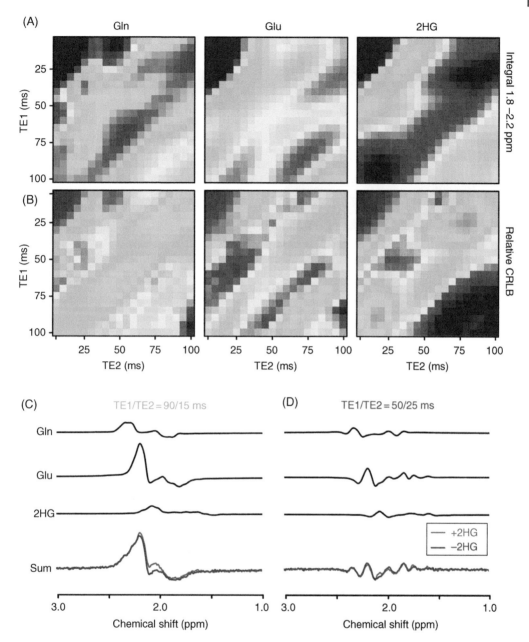

Figure 8.21 **Echo-time optimization for the detection of 2-hydroxyglutarate (2HG).** (A) Integrated signal intensities (1.8–2.2 ppm) for glutamine (Gln), glutamate (Glu), and 2HG simulated as a function of PRESS echo-time TE1 and TE2 at 3 T with an 8 Hz line width. (B) Relative Cramer–Rao lower bounds (CRLBs) for Gln, Glu, and 2HG as calculated from the simulated signals underlying (A). A low CRLB is indicative of a high-intensity signal with low spectral overlap. CRLBs increase with decreasing signal intensity or increasing amount of spectral overlap. See Chapter 9 for details on CRLBs. (C) Optimal and (D) suboptimal TE1/TE2 echo-time combinations for 2HG detection. The summed spectra are calculated with (blue) and without (red) 2HG in the presence of spectral noise.

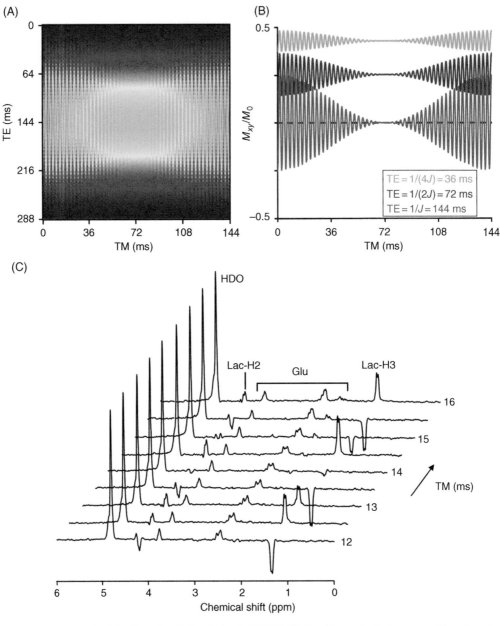

(A)

(B)

(C)

Figure 8.22 **Spectral editing based on TM variation in STEAM.** (A) Signal intensity for lactate at 3T as a function of TE and TM. (B) Signal intensity (M_{xy}/M_0) traces as extracted from (A) for TE = 1/(4J) (green), 1/(2J) (red), and 1/J (blue). The rapid signal modulations are caused by ZQCs during TM that precess at the chemical shift difference between lactate-H2 and H3 protons. (C) Experimental spectra of lactate (1.31 and 4.10 ppm) and glutamate (2.1–2.4 ppm and 3.75 ppm) in D_2O acquired at 4.7T with a stimulated echo sequence with TE = 1/J = 144 ms. Especially, the lactate resonances show a strong TM dependence in agreement with the theoretical predictions (A, B).

of an RF pulse is typically orders of magnitude smaller than the frequency separation between protons and, for example, carbon-13 nuclei. Therefore, a regular ^{1}H spin-echo sequence (Figure 8.23A) *without* ^{13}C pulses is selective for a heteronuclear ^{1}H-^{13}C spin system, thereby leading to selective refocusing of the proton spins without affecting the carbon-13 nuclei. As a result, heteronuclear scalar coupling is refocused for all echo-times and the proton resonances from ^{13}C-labeled metabolites appear in-phase with the rest of the ^{1}H NMR spectrum (Figure 8.23B, top trace). Note that this is an exact opposite to the situation for a homonuclear spin-system which will show TE-dependent scalar evolution during a nonselective spin-echo sequence (compare Figure 8.6). In the heteronuclear case, a nonselective experiment can be created when a ^{13}C 180° inversion RF pulse is applied in conjunction with the ^{1}H 180° refocusing RF pulse (Figure 8.23A). In this case, the phase of the proton magnetization is reset, while the ^{13}C spin populations are inverted, leading to evolution of the scalar coupling according to Eq. (8.1). When the echo-time TE is chosen as $1/J$, where J represents the heteronuclear scalar coupling, the proton resonances from ^{13}C-labeled metabolites appear inverted relative to resonances from protons attached to carbon-12 nuclei (Figure 8.23B, middle trace). Therefore,

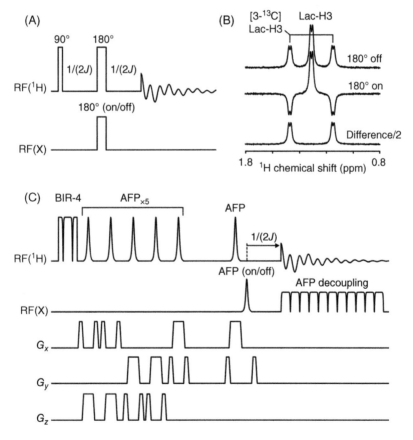

Figure 8.23 Heteronuclear *J*-difference editing. (A) Basic heteronuclear *J*-difference editing, or proton-observed, carbon-edited (POCE) sequence used for indirect ^{13}C NMR detection. (B) ^{1}H MR spectra obtained from a phantom containing lactate and [3-^{13}C]-lactate in the absence (top trace) and presence (middle trace) of a ^{13}C 180° inversion pulse (TE = 1/J). Since only signal phase of [3-^{13}C]-lactate is affected by the ^{13}C 180° pulse, the difference spectrum (bottom trace) represent signal from [3-^{13}C]-lactate only. (C) 3D localized POCE sequence based on the adiabatic LASER method extended with adiabatic editing and decoupling.

subtraction of the two spectra will result in the selective observation of protons attached to carbon-13 nuclei (Figure 8.23B, bottom trace). Note that since the heteronuclear scalar coupling constant is on the order of 125–140 Hz, the echo-time of proton-observed, carbon-edited (POCE) or heteronuclear *J*-difference editing sequence is only 7–8 ms. In analogy to homonuclear spectral editing sequences, the POCE sequence shown in Figure 8.23A requires spatial localization and water suppression to allow meaningful spectral editing *in vivo*. Figure 8.23C shows a practical POCE or ^{1}H-[^{13}C] NMR sequence based on the LASER localization method. The carbon-13 180° is shifted relative to the proton pulse as shown, in order to avoid executing the carbon-13 180° inversion pulse during a magnetic field gradient. The large ^{13}C chemical shift range would lead to an excessive chemical shift displacement and thus incomplete evolution of scalar coupling. Besides spatial localization and water suppression, ^{1}H-[^{13}C] NMR sequences are almost always executed with heteronuclear broadband decoupling during acquisition in order to increase the sensitivity and decrease the spectral complexity. Details on heteronuclear decoupling can be found in Section 8.8.

Figure 8.24 shows a typical POCE or ^{1}H-[^{13}C]-NMR spectrum acquired from rat brain *in vivo*, circa 90 min following the onset of [1,6-^{13}C$_2$]-glucose infusion. The total proton spectrum, acquired in the absence of a ^{13}C 180° inversion pulse, allows the detection and quantification of cerebral metabolites, like glutamate and glutamine. The POCE difference spectrum exhibits large resonances from [4-^{13}C]-glutamate and [4-^{13}C]-glutamine, in close analogy to the spectra acquired with direct ^{13}C NMR (Figure 3.38). Noticeable differences are that the proton spectrum has a greatly increased sensitivity, albeit at a decrease in spectral

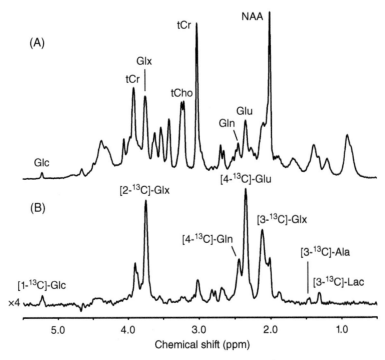

Figure 8.24 **Proton-observed, carbon-edited (POCE) MRS on rat brain *in vivo*.** (A) ^{1}H and (B) ^{1}H-[^{13}C] edited NMR spectra acquired from rat brain (9.4 T, 180 μl volume, TR/TE = 4000/8.5 ms, 512 averages). Data acquisition began circa 90 min after the start of an intravenous [1,6-^{13}C$_2$]-glucose infusion. Adiabatic decoupling based on AFPST4 pulses (*T* = 1.2 ms, *R* = 10) was applied during the entire acquisition period (102 ms) with $(\gamma/2\pi)B_{2max}$ = 1800 Hz. The ^{13}C labeling patterns in the ^{1}H-[^{13}C]-NMR spectrum are similar to those obtained during direct ^{13}C NMR detection (e.g. Figure 3.38).

resolution. However, at higher magnetic fields ($\geq 7\,T$), the spectral resolution is sufficient to separate and quantify resonances from the most important molecules, like glutamate, glutamine, GABA, and lactate. Further note that both the ^1H and ^1H-[^{13}C] NMR spectra hold resonances from glucose and [1-^{13}C]-glucose, respectively, thereby allowing the direct detection of cerebral glucose turnover and thus bypassing possible assumptions underlying the blood-to-brain transport.

The *J*-difference editing method underlying POCE can utilize any combination of homonuclear and heteronuclear scalar couplings, of which proton-observed, phosphorus-edited (POPE) MRS has been described [49].

Besides the *J*-difference method shown in Figures 8.8 and 8.9, essentially all homonuclear spectral editing methods have a heteronuclear counterpart including those based on multiple-quantum coherences [43, 50, 51], some of which have been successfully applied *in vivo* [52, 53]. Heteronuclear methods based on polarization transfer could also be classified as spectral editing techniques. However, since the prime purpose of polarization transfer techniques is sensitivity enhancement over direct heteronuclear NMR detection, they will be discussed separately in the next section.

8.7.2 Polarization Transfer – INEPT and DEPT

Polarization transfer methods are normally not classified as spectral editing methods in the *in vivo* NMR community. Instead, these methods are used to enhance the sensitivity of the insensitive (low-abundance) nucleus in a heteronuclear NMR experiment [54–59]. The simplest heteronuclear polarization transfer sequence is shown in Figure 8.25A and is referred to as insensitive nuclei enhanced by polarization transfer (INEPT) [54]. For the specific example of protons attached to carbon-13 nuclei, the thermal equilibrium magnetization is composed of proton and carbon-13 magnetization, whereby the proton magnetization is (γ_H/γ_C) larger due to the larger energy-level difference of protons (Figure 8.25C). Following excitation on the proton channel (Figure 8.25D), the proton transverse magnetization evolves under the effects of chemical shifts and scalar coupling. The proton 180° pulse ensures refocusing of phase evolution related to chemical shifts whereas the simultaneous carbon-13 180° pulse leads to continued scalar coupling evolution. At an echo-time TE of $1/(2J)$ the proton magnetization has evolved into a state of antiphase coherence (Figure 8.25E). As was shown in Figure 8.3, antiphase coherence is a correlated spin state in which the protons spins along a certain axis (e.g. $+y$) are correlated with carbon-13 spins along the longitudinal axis (e.g. $+z$). The correlation causes the number of carbon-13 spins that are biased along a certain direction to be equal to the number of biased proton spins. In other words, the number of carbon-13 spins that are biased along the $\pm z$ axis is (γ_H/γ_C) larger than the number of carbon-13 spins that are not in a correlated state (Figure 8.25E, bottom). Here, it is important to remember that the molecules containing the correlated carbon-13 spins are, in general, not the same molecules that give rise to the non-correlated, thermal equilibrium carbon-13 magnetization. In Figure 8.25E the enhanced, correlated carbon-13 magnetization is still along the z axis, thereby making it unobservable. The application of two simultaneous ^1H and ^{13}C 90° pulses (Figure 8.25F) leads to polarization transfer in which the correlated proton and carbon-13 spins change their orientation by 90°. The carbon-13 magnetization is now in the transverse plane where it can be detected as an antiphase doublet with (γ_H/γ_C) enhanced sensitivity. Note that the non-correlated, thermal equilibrium carbon-13 magnetization is also excited by the carbon-13 90° pulse. This non-enhanced contribution is generally undesirable and can be removed with a two-step phase cycle on the second proton 90° pulse and the receiver (Figure 8.25A). Comparison of direct ^{13}C pulse-acquire spectra (Figure 8.25B, top) with polarization transfer-enhanced ^{13}C spectra (Figure 8.25B, bottom) demonstrates the improved signal intensities for *IS*, *I₂S*, and *I₃S* spin

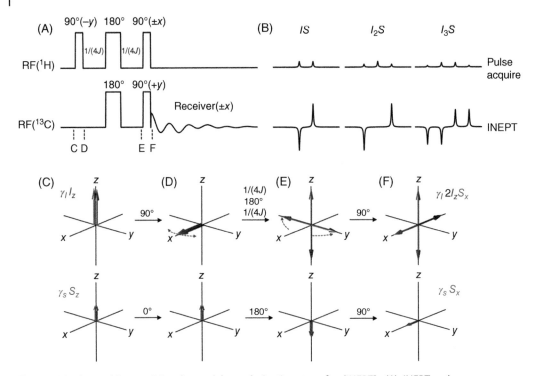

Figure 8.25 Insensitive nuclei enhanced by polarization transfer (INEPT). (A) INEPT pulse sequence as implemented for ^{13}C NMR detection. (B) ^{13}C pulse-acquire and INEPT NMR spectra for IS, I_2S, and I_3S spin systems. (C–F) Principle of heteronuclear polarization transfer for an IS spin system, where I and S represent a proton and a carbon-13 nucleus, respectively. (C) At thermal equilibrium both spins form macroscopic magnetization parallel to the main magnetic field, whereby the proton magnetization is $(\gamma_H/\gamma_C) \sim 4$ times larger than the carbon-13 magnetization. (D) The INEPT sequence starts by exciting the I spins, while leaving the S spins unperturbed. (E) Following a total echo-time of $1/(2J)$ the I spins have formed a correlated, antiphase coherence spin state with the scalar-coupled S spins. Note that the (non-correlated) longitudinal magnetization of the S spins (bottom trace) generally resides on different molecules and can thus exist in parallel with the S spin magnetization that is correlated with the I spins (top trace). (F) Polarization transfer step in which the correlated I spin coherences are converted to correlated S spin coherences. The result is an enhancement of (γ_I/γ_S) in the S spin intensity compared to a pulse-acquire experiment. In addition to enabling polarization transfer, the S excitation pulse also excites the non-correlated, non-enhanced S spin magnetization. In general, this contribution is undesirable and can be removed by phase cycling one of the 90° pulses on the I channel in concert with the receiver.

systems. The basic INEPT sequence as shown in Figure 8.25A is not commonly used for *in vivo* NMR as it leaves the spins in a state of antiphase coherence. This excludes the use of proton decoupling during ^{13}C acquisition, a technique commonly employed to further enhance the signal intensity of ^{13}C resonances (see Section 8.8).

INEPT is most commonly implemented as a refocused INEPT [57] sequence (Figure 8.26A) in which the basic INEPT sequence is extended with a nonselective spin-echo of echo-time t_2. When $t_2 = 1/(2J)$ the antiphase doublet signal of a two-spin-system IS (Figure 8.26B) is completely transformed into an in-phase doublet signal. However, many biological relevant chemical groups form I_2S spin systems (e.g. CH_2 groups in glutamate, glutamine, and GABA) for which $t_2 = 1/(2J)$ is a suboptimal delay. The optimal t_2 delay for I_2S spin systems is $1/(4J)$, leading to refocusing of the outer two resonance lines (Figure 8.26B, red line) representing 50%

of the total triplet signal. A delay t_2 of $1/(4J)$ provides only partial refocusing of scalar evolution in IS and I_3S spin systems, giving ~70% and ~35% signal recovery, respectively. The signals following a refocused INEPT sequence can now be acquired in the presence of heteronuclear decoupling to achieve spectral simplification and enhanced SNR (Figure 8.26B, blue lines).

An alternative to refocused INEPT is provided by the distortionless enhancement by polarization transfer (DEPT, [56]) method (Figure 8.26C). DEPT relies on polarization transfer following a period containing MQCs, whereby the nutation angle β of the final proton pulse determines the polarization transfer efficiency for different spin systems. For IS, I_2S, and I_3S the optimal nutation angle β is equal to 90°, 45°, and ~36°. As eluded to by the acronym, DEPT achieves polarization transfer by which the resulting spectra are in-phase (Figure 8.26D, red lines), without phase distortions as seen for refocused INEPT (Figure 8.26B, red lines). While this represents an advantage for applications without broadband decoupling [60], the resulting spectra in the presence of broadband decoupling are identical to those obtained with refocused INEPT. An additional advantage of DEPT is the smaller number of RF pulses, in general, and the smaller number of 180° pulses, in particular. Refocused INEPT and DEPT have both been implemented for dynamic ^{13}C MR studies on humans and animals [58, 59].

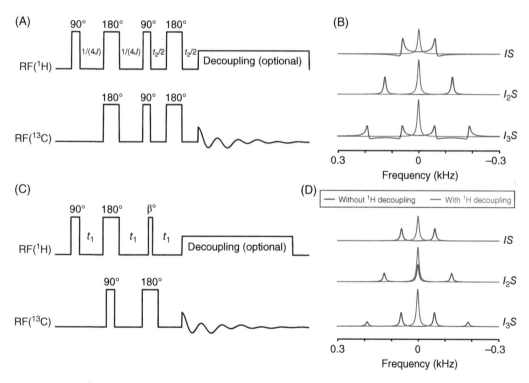

Figure 8.26 Refocused INEPT and distortionless enhancement of polarization transfer (DEPT). (A) Pulse sequence for refocused INEPT with timings set such as to achieve maximum in-phase signal at the time of acquisition. (B) Refocused INEPT NMR spectra for IS, I_2S, and I_3S spin systems without and with broadband decoupling. The signal intensities represent ~75, 100, and 75% of the INEPT signals for IS, I_2S, and I_3S spin systems (Figure 8.25B). (C) Pulse sequence for DEPT and (D) NMR spectra for IS, I_2S, and I_3S spin systems. The high in-phase character of refocused INEPT and DEPT spectra allows broadband decoupling for spectral simplification and signal enhancement. Refocused INEPT and DEPT provide the same signal intensities in the presence of decoupling. In most cases, spatial localization is provided by ISIS preceding the polarization transfer method.

8.8   Broadband Decoupling

The purpose of decoupling is to remove the effects of heteronuclear scalar coupling from NMR spectra for two reasons, namely (i) spectral simplification and (ii) increased sensitivity. Comparing the MR spectra from Figures 8.24 and 8.27 illustrates the utility of broadband decoupling in ^{1}H-[^{13}C] NMR spectroscopy. In the absence of decoupling (Figure 8.27), the edited ^{1}H-[^{13}C] difference spectrum is dominated by doublet resonances as a result of splitting by the single-bond heteronuclear scalar coupling. Heteronuclear scalar coupling over more than one chemical bond (see Table 2.6) leads to broadening of resonances *in vivo*. Applying broadband decoupling effectively merges the doublet resonances into a single resonance line, thereby increasing the sensitivity and simplifying the spectrum (Figure 8.24). Note that in the absence of decoupling, the total spectrum is now a sum of doublet resonances originating from the ^{13}C-labeled metabolites in addition to the regular resonances from the unlabeled metabolites. While the different chemical shifts of proton resonances attached to ^{12}C and ^{13}C nuclei have been used to observe ^{13}C label turnover without the use of a second decoupling channel [47, 61], it is, in general, an undesirable complication that can be removed through broadband decoupling.

The basic principle of decoupling is similar to that already discussed for homonuclear *J*-difference editing, namely that selective refocusing of one spin in a multi-spin-system leads to an inhibition of scalar coupling involving that spin. Figure 8.28A and B shows the theoretical

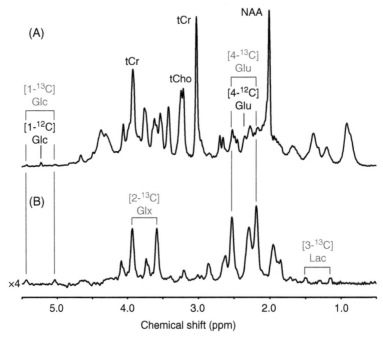

Figure 8.27 **Inverse ^{1}H-[^{13}C] or POCE MRS without decoupling.** (A) ^{1}H and (B) ^{1}H-[^{13}C] edited NMR spectra acquired from rat brain (9.4 T, 180 μl volume, TR/TE = 4000/8.5 ms, 512 averages) *without* broadband decoupling during acquisition. Data acquisition began circa 90 min after the start of an intravenous [1,6-^{13}C$_2$]-glucose infusion. Note that the fractional enrichment of αH1-glucose can be directly obtained from the unedited ^{1}H NMR spectrum. However, the presence of splittings due to heteronuclear scalar couplings generally leads to more complicated spectral patterns and a reduction of the obtainable *S/N* ratio. The effect of broadband decoupling can be evaluated by comparing Figure 8.24.

implementation of this principle for heteronuclear NMR, in this case ^{1}H detection with ^{13}C decoupling, i.e. ^{1}H-[^{13}C]-NMR. On the proton channel no RF pulses are applied, such that the signal can be acquired while evolving under the effects of chemical shift, magnetic field inhomogeneity, and homonuclear as well as heteronuclear scalar coupling. When a short, intense ^{13}C 180° inversion pulse is applied midway between the first and second acquisition points, the effects of heteronuclear scalar coupling will be completely refocused at the time of the second acquisition point. This principle can be continued by applying short, intense ^{13}C 180° inversion pulses in between all following acquisition points, such that even though heteronuclear scalar coupling evolution occurs during the dwell time $\Delta\tau$, it appears constant ("frozen") at the time of data acquisition. As a result, the Fourier transformation of a FID with constant heteronuclear scalar coupling evolution at all data points will result in a spectrum devoid of resonance splitting due to heteronuclear coupling. While theoretically sound, the approach depicted in Figure 8.28B is experimentally not feasible because of RF power restrictions, both in terms of unrealistic RF peak power requirements and excessive RF power deposition. Fortunately, the typical heteronuclear scalar coupling evolution ($^1J_{CH} = 120–170\,Hz$) is much slower than the data acquisition sampling rate, such that scalar coupling evolution during one dwell-time is very small. For example, for a spectral bandwidth of 5000 Hz and $^1J_{CH} = 140\,Hz$, the in-phase component of the transverse magnetization has decreased by <0.5% over one dwell-time $\Delta\tau$ and <3.5% over three dwell-times. Therefore, the ^{13}C 180° inversion pulses can be stretched out over several dwell-times (Figure 8.28C) without detrimental effects on the decoupling performance for on-resonance spins. When the inter-pulse delay becomes zero, the sequence is referred to as continuous wave decoupling [62]. While continuous wave decoupling is experimentally feasible

Figure 8.28 Principle of heteronuclear decoupling during ^{1}H-[^{13}C]-NMR. (A) A continuous ^{1}H time domain signal is sampled at discrete points separated by the dwell-time $\Delta\tau$. (B) The application of short 180° pulse on the ^{13}C channel in the middle of each dwell-time would lead to complete refocusing of heteronuclear scalar coupling evolution at each data acquisition point and theoretically to perfect decoupling. (C) RF power restrictions necessitate lengthening of the 180° ^{13}C RF pulses over several dwell times, which would lead in the extreme case to continuous wave decoupling. (D) In order to improve the off-resonance performance, the regular 180° pulses are typically substituted with pulse combinations, composite or adiabatic RF pulses (denoted R, where the overbar represents a 180° phase inversion) and placed inside decoupling super cycles.

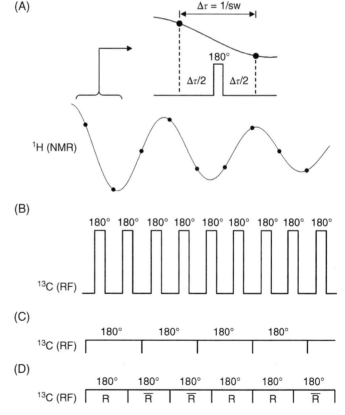

in terms of RF power requirements and deposition, it has one serious drawback in that the decoupling performance is only adequate when the ^{13}C inversion pulses are applied close to on-resonance. For selected applications where only a single resonance requires decoupling (e.g. ^{13}C glycogen detection) continuous wave decoupling is a feasible option. However, for most applications in which multiple metabolites at different chemical shifts require simultaneous decoupling, this approach will fail. Continuous wave decoupling does therefore not belong to the class of broadband decoupling techniques that will be discussed next.

Broadband decoupling methods utilize the same principle as continuous wave decoupling, i.e. the decoupling pulse can stretch over several dwell-times, with the crucial difference that the regular "hard" 180° inversion pulses have been replaced with composite or adiabatic RF pulses that achieve a net rotation angle of 180° over a wider bandwidth. In Chapter 5, composite RF pulses were described as 180° pulses with a built-in compensation towards parameters like RF inhomogeneity and frequency offsets. Figure 5.14 summarizes the performance of the two composite 180° pulses most commonly used for broadband decoupling, namely MLEV $90°_x180°_y90°_x$ [63–65] and WALTZ $90°_x180°_{-x}270°_x$ [62, 66, 67]. It follows that while a regular "hard" 180° pulse only achieves inversion close to on-resonance, MLEV and WALTZ achieve near-complete inversion over a bandwidth slightly over $2B_{1max}$, where B_{1max} is the applied RF amplitude. However, the performance of these pulses is surprisingly poor when executed as a continuous train for broadband decoupling (Figure 8.29A). The origin of the strong oscillations in Figure 8.29A can be found in the imperfect inversion profile of the basic WALTZ element. Despite the greatly improved inversion profile of a WALTZ pulse, the inversion is not perfect at all frequencies (e.g. at frequency offset $\Delta\nu = 0.325B_1$, WALTZ achieves a net rotation of only 155°). These minor imperfections propagate through a decoupling pulse train as shown in Figure 8.29A, giving rise to the oscillatory frequency profile. Since minor imperfections are unavoidable, specific pulse phase schemes have been designed to minimize or compensate error propagation throughout a longer pulse train. These so-called decoupling super cycles

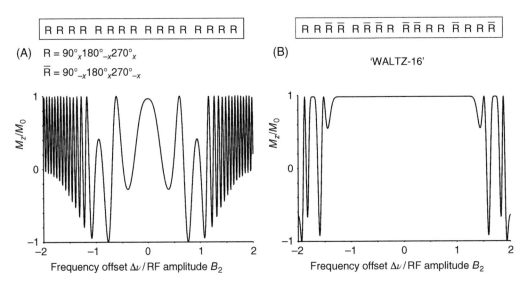

Figure 8.29 Decoupling super cycles. (A) Without an inter-pulse phase variation, small errors in an individual pulse *R* will propagate through a 16-pulse train leading to poor off-resonance performance. (B) With a WALTZ-16 inter-pulse phase variation method (also called a decoupling super cycle), small errors in individual pulses *R* do not propagate, leading to a greatly enhanced off-resonance performance. Note that since *R* represents an inversion pulse, an even number of pulses *R* should ideally lead to a net rotation of 0°. This result is closely approximated in (B) for |frequency offset $\Delta\nu$| < ~|RF amplitude B_2|.

are essential to broadband decoupling. The specific 16-step decoupling super cycle that is used in combination with WALTZ pulses gives rise to the so-called WALTZ-16 decoupling scheme [66, 67], which gives greatly enhanced performance (Figure 8.29B) over the decoupling scheme in the absence of a super cycle (Figure 8.29A). Shorter, but less effective cycles are used in WALTZ-4 and WALTZ-8 decoupling [62].

Up to this point, the fact that decoupling pulses are allowed to run over several dwell-times has been ignored, given the fact that the scalar coupling evolution is a relatively slow process compared to the data acquisition rate. However, RF power deposition restrictions limit the available RF amplitude, typically leading to a lengthening of the RF pulse duration. For example, the RF amplitude for decoupling in human studies is typically limited to 500 Hz, making an individual WALTZ element 3.0 ms in duration. During this relatively long pulse, the coherences will evolve under the effects of heteronuclear scalar coupling giving rise to small signal oscillations in the detected NMR signal, which in turn lead to so-called decoupling or cycling sidebands [68] in the NMR spectrum (Figure 8.30). The decoupling sidebands have two undesirable effects on the appearance of the NMR spectrum. Firstly, the multiple decoupling sidebands increase the effective noise level, leading to a reduced SNR. Secondly, the intensity

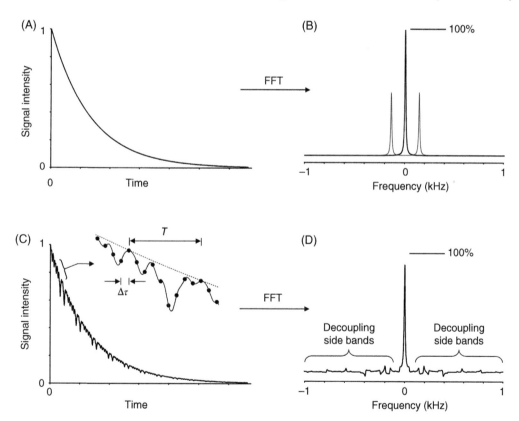

Figure 8.30 **Decoupling sidebands.** (A) In the case of ideal decoupling (e.g. as in Figure 8.28B), heteronuclear scalar coupling evolution would be refocused at each data acquisition point, leading to (B) a single resonance line following Fourier transformation, with a peak height that is exactly twice that of the undecoupled resonances (for a two-spin-system). (C) In the case of realistic decoupling with RF pulses of length *T* that span several dwell times $\Delta\tau$, the effects of heteronuclear scalar coupling evolution are not refocused until the end of the pulse, such that part of the scalar coupling evolution is captured by the data acquisition points covering the length of the RF pulse. (D) These small modulations give rise to so-called decoupling or modulation side bands following Fourier transformation. Furthermore, the peak height of the main decoupled resonance is also reduced.

of the main resonance is redistributed across multiple smaller decoupling sidebands, leading to a reduction in the main decoupled signal. It is therefore desirable to minimize the decoupling sidebands and maximize the main decoupled resonance. Besides these two parameters, other often conflicting considerations include the minimum decoupling bandwidth, the allowable RF power deposition, the maximum available RF peak power, and the sensitivity towards experimental imperfections like RF inhomogeneity. These parameters can be optimized and/or balanced with four basic variables, namely the pulse length (relative to the heteronuclear scalar coupling J_{HX}), amplitude, and shape of the individual decoupling elements, as well as the total decoupling scheme and super cycle. A rigorous comparison and evaluation of decoupling methods most commonly used for *in vivo* NMR has been described [69]. Most noticeably, for RF amplitudes suitable for human applications (i.e. $B_{2max} = 500\,Hz$), the classic MLEV-16 and WALTZ-16 provide the best performance with an effective decoupling bandwidth of circa $1.7B_{2max}$ for a 90% decoupling efficiency (i.e. the main decoupled resonance is at least 90% of a perfectly decoupled resonance). Furthermore, at higher RF amplitudes suitable for animal applications, many other decoupling methods become available [70–74]. However, without reservation it can be stated that for $B_{2rms} > 500\,Hz$, it is always possible to design a decoupling method based on adiabatic RF pulses [69] that outperforms all other methods.

Broadband decoupling is highly recommended for animal studies during which RF power deposition is less critical. However, as the magnetic field strength increases, the use of broadband decoupling in human studies becomes increasingly more difficult. As a result, several studies have explored the use of non-decoupled ^{13}C MRS [60, 61, 75]. As an alternative, ^{13}C MRS has been performed on the non-protonated carbon positions with low-power decoupling [76–79]. The high RF power used to decouple the single-bond ^1H-^{13}C scalar coupling was primarily dictated by the requirement of a short RF pulse length compared to $1/J$. Since the non-protonated carbons only exhibit ^1H-^{13}C scalar coupling over two or three chemical bonds, the scalar coupling constants are on the order of 3–6 Hz (see Table 2.6). The small scalar coupling constants lead to significant line broadening, but are readily removed by low-power stochastic or noise decoupling.

8.9 Sensitivity

The thermal equilibrium magnetization M_0 is proportional to the magnetic field B_0 and the square of the gyromagnetic ratio γ of the nucleus under investigation (see also Eq. (1.4)). One γ term originates from the intrinsic magnetic moment, while another γ term (as well as B_0) entered the equation due to the energy-level difference between the two spin states. The induced electromotive force (emf) is, through the principle of reciprocity, given by $\omega_0 M_0$, making the final detected signal S proportional to $\gamma^3 B_0^2$. The sensitivity of an MR study is almost always determined by the induced signal relative to the induced noise voltage, in other words the SNR. It is well known that the noise voltage arising from conducting samples increases linearly with frequency [80]. In the case that all the noise originates from the sample, as is the case for proton NMR at high magnetic fields, the SNR is proportional to $\gamma^2 B_0$. However, for lower frequency nuclei such as ^{13}C, ^{17}O, or ^{31}P the sample noise may not be dominant, such that the SNR field dependence becomes nonlinear according to B_0^β, where β is a constant between 1 and 2 [81, 82]. When the relative sensitivity is expressed as SNR per (time)$^{1/2}$ the longitudinal T_1 relaxation and line widths (or T_2^*) must be taken into account, leading to

$$SNR \propto \gamma^2 B_0^\beta \sqrt{Q_l} \sqrt{\frac{T_2^*}{T_1}}$$

(8.6)

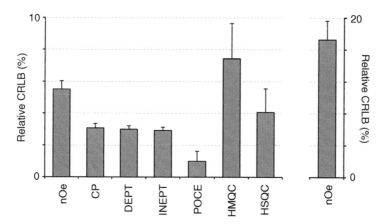

Figure 8.31 **Sensitivity of direct ^{13}C and indirect ^{1}H-[^{13}C] MR detection methods.** Sensitivity is expressed in CRLBs averaged over glutamate, glutamine, GABA, and lactate simulated at 7T in the presence of broadband decoupling. Lower CRLBs are indicative of higher SNR and lower spectral overlap. The direct ^{13}C methods with nOe are calculated for protonated (left) and non-protonated (right) carbon positions. The HMQC and HSQC methods are executed as 2D acquisitions. CP, cross polarization.

where Q_l represents the loaded Q-value of the RF coil. Equation (8.6) has formed the basis for the field dependence of ^{17}O [81] and ^{31}P [82] sensitivity and found β values of 2 and 1.74, respectively. For ^{1}H MRI and MRS studies the magnetic field dependence is close to linear ($\beta \sim 1$, [83]). For direct ^{13}C and indirect ^{1}H-[^{13}C] studies, Eq. (8.6) needs to be extended with pulse sequence-dependent parameters (e.g. nOe, echo delays) and (partial) signal overlap [84]. Figure 8.31 shows the detection sensitivity for ^{13}C-labeled compounds (glutamate, glutamine, GABA, and lactate) for a range of ^{13}C detection methods at 7 T with broadband decoupling. In general, indirect ^{1}H-[^{13}C] or POCE detection provides the highest sensitivity, provided that spectral overlap is minimized through optimal magnetic field homogeneity. The polarization transfer methods (CP, DEPT, and INEPT) achieve a sensitivity that is on average three times lower than POCE, whereas the nOe and 2D MR methods (HMQC, HSQC) provide the lowest sensitivity. The widely used mantra that ^{1}H MR detection provides 16 ($[\gamma_H/\gamma_C]^2$) or even 64 ($[\gamma_H/\gamma_C]^3$) times more SNR than ^{13}C MR detection is based on theoretical grounds that do not consider the dominant sources of noise or sequence details. Under realistic *in vivo* conditions, indirect ^{1}H-[^{13}C] MR detection only provides a several-fold improvement in sensitivity over direct ^{13}C MR detection [84].

8.10   Two-dimensional NMR Spectroscopy

The previous sections described the principles of spectral editing to differentiate scalar-coupled from non-coupled spin-systems by using pulse sequence elements such as frequency-selective RF pulses and magnetic field gradients. While spectral editing methods often lead to the unambiguous detection of one specific compound, much of the information on other compounds is lost in the process. A more general method to manipulate NMR data can be achieved by separating the various characteristics of resonances along orthogonal axes, giving rise to multidimensional NMR. Although high-resolution liquid-state NMR spectroscopy can generate 3D or 4D spectra, the discussion given here will be limited to two dimensions as most commonly encountered for *in vivo* NMR. This section will only review the basic principles of 2D NMR spectroscopy, for more in-depth information the reader is referred to a wide range of texts on multidimensional NMR spectroscopy [4, 6, 85–87].

The first two-dimensional NMR experiment was suggested by J. Jeener in 1971 during a summer school on NMR given in Yugoslavia [88]. This experiment, which is now known as homonuclear correlation spectroscopy or COSY, was designed to obtain information about homonuclear coupling connectivities for the identification of scalar coupled spin systems. In 1976, Ernst and coworkers gave a complete theoretical description of COSY, together with important generalizations about 2D NMR, which opened the way to develop multidimensional NMR into the important technique it is today [89]. Besides homonuclear COSY, many other 2D NMR techniques have been developed and have found routine use in organic chemistry, biochemistry, and structural biology. 2D NMR has not found widespread applications *in vivo*, due to a variety of reasons including longer measurement times, reduced T_2 relaxation times, and reduced sensitivity. In addition, the spectral content for many tissues and conditions is well known such that information on specific metabolites can be obtained through spectral editing and/or spectral fitting. However, 2D NMR has potential for specific *in vivo* applications, where identification and quantification of unknown compounds is required or where 2D NMR can provide unique information not available in regular 1D NMR spectra.

8.10.1 Correlation Spectroscopy (COSY)

The general idea of 2D experiments is to generate a second frequency axis by introducing an evolution delay (and at least one additional pulse) into a pulse sequence, during which the transverse magnetization precesses at a different frequency than during signal acquisition. This principle is highly reminiscent of 2D MRI, where the introduction of a phase-encoding gradient leads to a position-dependent frequency that is encoded as phase in the acquired signal.

The COSY experiment (Figure 8.32) can be seen as an extension of a simple 90°-acquisition experiment, in which an additional delay t_1 and a second 90° pulse are placed between the initial excitation pulse and signal acquisition during t_2 [89]. Following excitation, spins precess in the transverse plane and acquire a phase proportional to their Larmor frequency and the delay t_1. After the second 90° pulse, the phase modulation during t_1 is detected as an amplitude modulation in the acquired signal (Figure 8.32B). The higher Larmor frequency of the red spins leads to an amplitude modulation with an equally high frequency. Fourier transformation of the spectra in Figure 8.32B with respect to the delay t_1 results in a 2D NMR spectrum shown in Figure 8.32C. The spectral dimension that was directly acquired during t_2 shows two resonances at their respective Larmor frequencies. The indirectly sampled spectral dimension that manifests itself as an amplitude modulation in the detected signal contains four resonances at the positive and negative Larmor frequencies. In terms of CTPs (Section 8.5) both I^+ and I^- coherences are present during t_1 and since the COSY sequence in Figure 8.32A does not discriminate between them, they are both present during signal acquisition thereby leading to signals at positive and negative frequencies. Alternatively stated, since signal during the t_1 evolution period is only sampled along a single axis, the detected signal (Figure 8.32B) exhibits a cosine amplitude modulation that cannot discriminate between positive and negative frequencies. Frequency discrimination along the indirect dimension can be achieved by acquiring a second dataset in which the second 90° pulse is applied along an orthogonal axis (Figure 8.32D), such that the amplitude of the detected signal becomes sine-modulated (Figure 8.32E). Combining the cosine- and sine-modulated datasets allows discrimination of the frequency during the t_1 evolution period (Figure 8.32F).

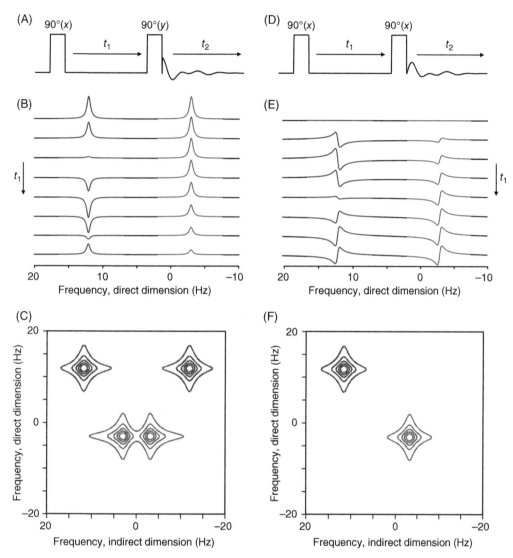

Figure 8.32 Correlated spectroscopy (COSY) for non-coupled spins. (A) Basic COSY pulse sequence providing (B) cosine-modulated signals with increasing evolution time t_1. (C) Fourier transformation with respect to t_1 leads to a 2D spectrum with frequencies directly sampled during the acquisition period t_2 along the vertical axis. Frequencies indirectly sampled during t_1 are along the horizontal axis. The lack of quadrature detection during the t_1 period leads to an undetermined frequency sign along the indirect dimension. (D) Additional COSY scan in which the second pulse is 90° rotated relative to (A) leads to (E) sine-modulated signals with increasing evolution time t_1. (F) Combining the cosine- and sine-modulated datasets provides a 2D MR spectrum with unambiguous frequencies along both dimensions. The lack of scalar coupling for non-coupled spins leads to identical frequencies during t_1 and t_2, thus giving only diagonal peaks. Signals in (C) and (F) are displayed in absolute value.

As an alternative to acquisition of two datasets with different RF phases, quadrature detection in the indirect dimension can also be achieved with magnetic field gradients in a single scan, at the cost of a $\sqrt{2}$ reduction in SNR compared to the RF-cycled 2D NMR experiment. For non-coupled spins the 2D NMR spectrum only contains diagonal signals, such that the 2D experiment does not provide any additional information compared to a conventional 1D

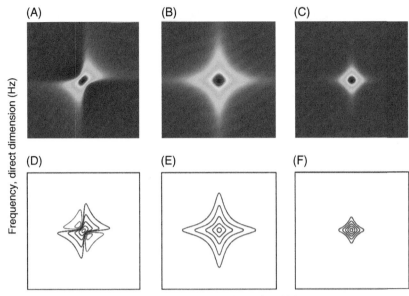

Figure 8.33 **Lorentzian line shapes encountered in 2D NMR spectra.** (A, D) Phase-twisted line shape containing both absorption and dispersion components. This type of line shape can never be phased to a pure absorption or dispersion line shape. (B, E) Magnitude or absolute-valued representation of a 2D resonance line. (C, F) Phase-sensitive, absorption line shape. Note the increased "tails" of the magnitude representation when compared to a pure absorption resonance line. Positive and negative contour lines in (D–F) are red and blue, respectively.

spectrum. For scalar-coupled spins the 2D NMR spectrum will hold off-diagonal or cross signals, indicating a correlation between the two diagonal signals.

Before proceeding with the discussion of scalar-coupled spins, it is instructive to investigate the 2D line shapes in more detail. The 2D Fourier transformation of sine or cosine amplitude-modulated signals will lead to a phase-twisted 2D line shape (Figure 8.33A and D) that contains a mixture of absorptive and dispersive line shapes (see Exercises). The phase-twisted 2D line shape is undesirable since it has positive and negative intensities in addition to long dispersive "tails" that extend over several line widths. Since the phase-twisted 2D line shape cannot be improved by phasing, the spectra are generally displayed with absolute-valued 2D lines (Figure 8.33B and E). While the absolute-valued line shape removes negative intensities, it is still broadened by the dispersive components. The phase-twisted line shape can be transformed into a pure absorption line shape by combining cosine and sine amplitude-modulated datasets. Using the sine-modulated signal as the complementary imaginary component to the cosine-modulated signal achieves quadrature detection, but still leads to a phase-twisted line shape. Quadrature detection and absorptive line shapes are obtained when the datasets are combined according to the processing proposed by States et al. [90]. This so-called hypercomplex method is the standard method to acquire phase-sensitive 2D NMR spectra (Figure 8.33C and F, see Exercises). Other options to obtain phase-sensitive 2D NMR spectra are time-proportional phase incrementation (TPPI, [91]) and magnetic field gradient coherence selection [51, 92–94].

The COSY method works similar for scalar-coupled spins except that spins evolve under the effects of scalar coupling in addition to chemical shifts and magnetic field inhomogeneity. For a weakly-coupled two-spin-system IS, the I_x and S_x coherences present following excitation will evolve into I_x, I_y, $2I_xS_z$, and $2I_yS_z$ coherences for the I spins and S_x, S_y, $2I_zS_x$, and $2I_zS_y$ coherences for the S spins at the end of the t_1 evolution time. Similar to the situation for non-coupled spins, the I_x and S_x coherences are not affected by the second 90°(x) or mixing pulse and remain in the transverse plane. These coherences will give rise to diagonal signals in the 2D COSY spectrum, since the frequencies of the I and S spins are identical during the t_1 and t_2 periods. The I_y and S_y coherences are rotated along the z axis and do not need to be considered. The antiphase coherences $2I_xS_z$ and $2I_zS_x$ are transformed into MQCs, $2I_xS_y$ and $2I_yS_x$, by the mixing 90°(x) pulse. Since MQCs are not directly observable, they also do not have to be considered for the COSY sequence. However, these coherences can be converted back into observable magnetization by a third 90° pulse, as, for example, used in double-quantum-filtered (DQF) COSY (see Exercises). Finally, the antiphase coherences $2I_yS_z$ and $2I_zS_y$ will undergo polarization transfer into the antiphase coherences $2I_zS_y$ and $2I_yS_z$ (Figure 8.34A), respectively. These coherences will lead to off-diagonal or cross-peaks since they evolved at the I or S frequency during the t_1 period, whereas they evolve at the correlated S or I frequency during the t_2 period. Figure 8.34B and C shows the spectra as a function of the t_1 evolution delay for the in-phase coherences I_x and S_x (Figure 8.34B) and for the antiphase coherences $2I_zS_z$ and $2I_zS_y$ present during the t_2 evolution period (Figure 8.34C). Note that in reality the acquired signal represents the sum of Figure 8.34B and C, such that the in-phase and antiphase coherences are not readily separated visually. It follows that the antiphase coherences of the high-frequency I spins (Figure 8.34C, red) have a low-frequency amplitude modulation due to the fact that they originated from low-frequency S spins before the mixing pulse. Similarly the low-frequency antiphase coherences of the S spins (Figure 8.34C, blue) have an amplitude modulation proportional to the higher I spin frequency. With the understanding that a second experiment is required to obtain phase-sensitive, quadrature detection (see Figure 8.32), the spectra in Figure 8.34B and C can be Fourier transformed relative to the t_1 evolution time to yield the 2D COSY spectrum shown in Figure 8.34D and E. For a weakly-coupled IS spin-system the COSY spectrum consists of doublet signals in both spectral domains, whereby the diagonal multiplet is 90° out of phase with the cross peak. Further note that the integrated signal from a cross-multiplet is zero due to the negative and positive resonances within the multiplet. Figure 8.34D shows the 2D COSY spectrum as a color-coded image, whereas Figure 8.34E shows it as a contour plot. Whereas both representations have specific advantages and disadvantages, the contour plot representation is most commonly used in high-resolution NMR.

Additional resonances can appear in a COSY spectrum when magnetization is detected during the acquisition period t_2, which has not been frequency-labeled in the evolution period t_1. This can occur when magnetization recovers during the t_1 period due to T_1 relaxation. The 90° mixing pulse converts the longitudinal magnetization into detectable signal, which will appear at an indirect frequency of zero. These so-called axial peaks can be suppressed by a two-step phase cycle (e.g. +x, −x) on the 90° preparation pulse in concert with a two-step receiver phase cycle (+x, −x). The magnetization recovered by T_1 relaxation does not experience the phase cycle on the preparation 90° pulse and is therefore canceled out by the two-step receiver phase cycle. The two experiments needed for phase-sensitive, quadrature detection together with a two-step phase cycle to eliminate axial peaks leads to a minimum of four averages per t_1 increment.

The multiplets in a 2D COSY spectrum are generally much more complicated than shown for a weakly-coupled IS spin-system (Figure 8.34). For example, the cross-peak between spins A and M in a four-spin-system AMX_2 consists of 36 individual resonances with positive or negative intensities. This has two serious consequences for COSY studies applied *in vivo*. Firstly, as the signal from a single spin is distributed over many smaller resonances, the overall SNR decreases. As *in vivo* MRS is typically sensitivity limited, this often leads to the inability to

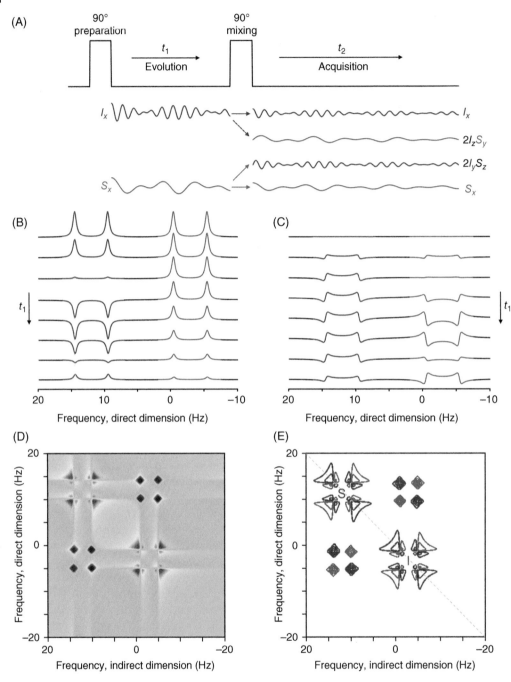

Figure 8.34 **Correlated spectroscopy (COSY) for scalar-coupled spins.** (A) Spin *I* (red) and spin *S* (blue) in a scalar-coupled two-spin-system *IS* precess at their own Larmor frequency during the t_1 evolution period. (B) The in-phase coherences of both spins that are not affected by the mixing pulse retain the same frequencies during the t_2 acquisition period. (C) The antiphase coherences that undergo polarization transfer during the mixing pulse have different frequencies during the t_2 acquisition period. The signals in (B) and (C) lead to diagonal and cross-signals in a (D, E) 2D COSY spectrum, respectively. Note that the phase-sensitive (D) image and (E) contour representations require the acquisition of two separate signals to obtain quadrature detection along the indirect dimension (see also Figure 8.32).

detect lower concentration metabolites in 2D MR spectra *in vivo*. Secondly, the increased line widths encountered *in vivo* due to short(er) T_2 relaxation times and decreased magnetic field homogeneity will lead to signal cancelation of the various positive and negative signals within a 2D MR multiplet. This will lead to a further decrease in SNR and will also have consequences for signal quantification in 2D NMR, whereby the absolute-valued multiplet integral will become dependent upon the magnetic field inhomogeneity. This is in contrast to 1D MRS where the signal integral remains constant with decreasing magnetic field homogeneity. These two effects, in addition to the longer minimum scan time required to acquire a 2D MR spectrum, are the main reasons for the lack of widespread applications of 2D NMR *in vivo*. Nevertheless, 2D NMR has potential for specific applications *in vivo* that require the confirmation and assignment of unknown resonances or provide information that cannot be obtained from 1D MRS.

Figure 8.35 shows simulated 2D COSY spectra from metabolites commonly detected in the human and animal brain *in vivo*. The lactate and threonine methyl resonances overlap perfectly as 1.31 ppm. 2D COSY (Figure 8.35A) provides an opportunity to separate the two compounds based on the small chemical shift difference of their scalar coupling partner (at 4.10 ppm for lactate and 4.25 ppm for threonine). Note that COSY provides cross-peaks for all spins with a nearest neighbor scalar coupling. For beta-hydroxybutyrate (BHB) this results in cross-peaks between BHB-H2 and BHB-H3 protons, as well as between BHB-H3 and BHB-H4 protons. No cross-peak is observed between BHB-H2 and BHB-H4 due to the lack of significant scalar

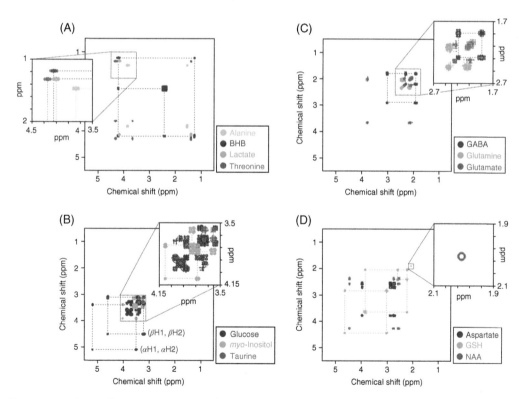

Figure 8.35 **2D correlation spectroscopy of common cerebral metabolites.** All simulations are performed at 7 T with a 1 Hz line width. COSY spectra are shown for (A) alanine, BHB, lactate, and threonine, (B) glucose, *myo*-inositol, and taurine, (C) GABA, glutamate, and glutamine, and (D) aspartate, glutathione (GSH), and *N*-acetyl aspartate (NAA). All spectra are displayed in absolute value.

coupling over four chemical bonds in a nonaromatic molecule. Figure 8.35B shows a COSY spectrum for glucose, *myo*-inositol, and taurine. Despite the large number of cross-peaks, the 2D spectrum does not significantly reduce spectral overlap compared to a conventional 1D MR spectrum. The COSY spectrum for glutamate, glutamine, and GABA (Figure 8.35C) demonstrates the complicated and extended multiplets for compounds with extensive scalar coupling networks, such as glutamate and glutamine. The 2D COSY spectrum for aspartate, GSH, and NAA (Figure 8.35D) shows several well-separated, relatively simple cross-peaks that could be suitable targets for *in vivo* detection. Note that the NAA methyl resonance (Figure 8.35D, inset) is not scalar coupled, thereby only forming a singlet diagonal signal.

The COSY method started the field of 2D NMR and it is still a valuable, commonly used technique. However, a wide range of modifications to the basic COSY have been proposed in order to improve the spectral appearance. DQF COSY [95, 96] is a valuable addition in which the MQCs present during the COSY t_2 period are converted to observable magnetization by a third 90° pulse. DQF-COSY allows the coherences to be filtered based on their coherence order during the MQC evolution period between the second and third 90° pulses. Since non-coupled spins do not form MQCs they are filtered out (i.e. eliminated) in DQF-COSY. In addition, the diagonal peaks in DQF-COSY spectrum are in-phase with cross-peaks, thereby greatly improving the ability to detect small cross-peaks close to the diagonal. The filtering process can be based on RF phase cycling or magnetic field gradient coherence selection similar to MQC-based spectral editing. With magnetic field gradients, coherence selection is achieved in a single scan leading to robust water suppression. Robust water suppression is critical for high-quality 2D NMR. As described earlier, a 2D NMR spectrum is acquired as multiple experiments in which an evolution delay t_1 is linearly incremented. System instabilities or inconsistent water suppression give rise to inter-scan signal and phase variations that lead, upon Fourier transformation, to signal smearing or "t_1-noise" along the indirect dimension. The presence of a significant residual water signal can lead to dominant t_1 noise that can obliterate small signals. Coherence selection with magnetic field gradients, such as in DQF-COSY, typically achieves excellent water suppression and therefore low t_1 noise.

TOCSY or total correlation spectroscopy [41, 97] is another powerful form of 2D COSY. In TOCSY the 90° mixing pulse is replaced with a much longer mixing pulse train, the effect of which is a mixing of coherences throughout the entire scalar-coupled spin-system, rather than just for nearest neighbors. For example, whereas the 2D COSY spectrum of BHB (Figure 8.35A) does not show a cross-peak between BHB-H2 and BHB-H4 protons, the TOCSY experiments "samples" the entire BHB scalar coupling network, thereby producing a cross-peak between BHB-H2 and BHB-H4. In addition, the TOCSY experiment can be setup in such a way as to produce in-phase multiplets, thereby eliminating signal cancelation within a multiplet.

8.10.2 *J*-resolved Spectroscopy (JRES)

Homo- and heteronuclear 2D *J*-resolved NMR is aimed at separating the scalar coupling and chemical shift information along orthogonal axes [98–100]. Compared with COSY and other COSY methods, *J*-resolved or spin-echo NMR methods are conceptually simpler, as are the corresponding pulse sequences. *J*-resolved spectroscopy relies on the fact that a spin-echo pulse sequence (Figure 8.36A) refocuses chemical shifts during the echo-time TE, whereas scalar coupling evolves continuously. The coherences at the top of the echo are given by Eq. (8.1) as described for *J*-difference editing. Whereas *J*-difference editing uses a specific echo-time, typically $1/(2J)$ or $1/J$, *J*-resolved spectroscopy samples many echo-times that are linearly incremented by an amount ΔTE. This results in TE-dependent spectra as shown in Figure 8.36B,

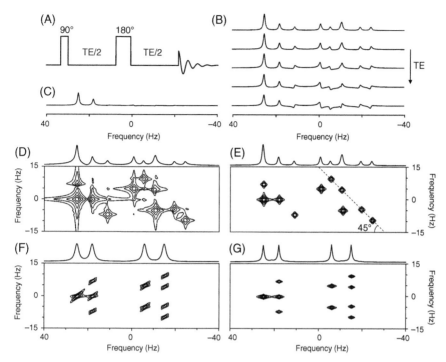

Figure 8.36 **Principle of 2D *J*-resolved spectroscopy.** (A) Spin-echo pulse sequence commonly used for 2D *J*-resolved spectroscopy. (B) NMR spectra acquired with an increasing echo-time TE for singlet and weakly-coupled doublet, triplet, and doublet-of-doublets resonances. (C) TE-averaged NMR spectrum. (D–G) 2D *J*-resolved spectra obtained (D) without and (E) with halfsine time-domain apodization and (F) followed by a 45° spectral tilt and (G) spectral symmetrization. All spectra are processed and displayed in absolute value. The traces above the spectra represent the projections or summations onto the chemical shift dimension.

which after a Fourier transformation with respect to the echo-time TE, provide a 2D *J*-resolved spectrum (Figure 8.36D). In a 2D *J*-resolved spectrum the horizontal axis indicates the chemical shift dimension, similar to a 1D MR spectrum, whereas the vertical axis indicates the scalar coupling dimension. For homonuclear *J*-resolved MRS the bandwidth along the scalar coupling dimension needs to cover the full extent of all multiplets. Bandwidths of circa 50 Hz are typical, leading to an echo-time increment ΔTE of 20 ms. In order to sample a significant fraction of the T_2 relaxation that is active during TE, *J*-resolved MRS requires a relatively small number of TE increments, typically between 16 and 32. The spectral quality of a "raw" 2D *J*-resolved spectrum appears visually low, due to the need to display the data in absolute value. Similar to COSY and other 2D NMR methods (Figure 8.33), 2D *J*-resolved MRS is characterized by phase-twisted line shapes. However, due to the lack of a mixing pulse, 2D *J*-resolved MRS does not provide the opportunity to acquire a second dataset along an orthogonal axis. As a result, 2D *J*-resolved MR spectrum is always displayed in absolute value. The tails of the broad resonances can be suppressed with appropriate time-domain apodization to better reveal the underlying structure of a 2D *J*-resolved spectrum (Figure 8.36E). Multiplets are present along lines making a 45° angle with the chemical shift axis. After application of a 45° spectral tilt, the multiplet structures appear parallel to the vertical axis, whereas the chemical shifts appear along the orthogonal horizontal axis (Figure 8.36F). A projection onto the chemical shift axis will then produce a "homonuclear decoupled" or "pure shift" NMR spectrum. A final processing step that can be employed relates to spectral symmetrization with respect to the horizontal line of zero frequency (Figure 8.36G). While the processing steps shown in Figure 8.36E–G are commonly employed in high-resolution, metabolomics NMR studies on body fluids [101] for visual improvement of the spectral appearance, extreme caution should be exercised when the *J*-resolved NMR data are used for metabolite quantification. The time-domain apodization step (Figure 8.36E) essentially shifts the signal maximum from time point ($t_2 = 0$, TE = 0) to ($t_2 = t_{2max}/2$, TE = $TE_{max}/2$) in order to approximate a time-domain echo signal. Since the Fourier transformation of a symmetrical echo lacks a dispersive component, the *J*-resolved spectrum has sharper, absorption-like line shapes. However, signal at the new echo top has decayed significantly due to T_2^* and T_2 relaxation during the delays $t_{2max}/2$ and $TE_{max}/2$, thus leading to distorted signal intensities. Whereas the 45° spectral tilting (Figure 8.36F) has relatively little effect on signal quantification, the spectral symmetrization (Figure 8.36G) can be detrimental for the detection of small resonances [101]. For quantitative studies it is recommended to use the "raw" *J*-resolved MR data (Figure 8.36D). This is also advocated by ProFit [102, 103], a 2D spectral fitting program tailored for *in vivo* 2D *J*-resolved spectra.

The TE-dependent spectra shown in Figure 8.36B can also be processed by simply averaging, resulting in a so-called "TE-averaged" MR spectrum (Figure 8.36C). This spectrum corresponds to the zero frequency midline in the *J*-resolved spectrum of Figure 8.36D. Therefore, it only contains resonances from singlets and from multiplets that have a zero-frequency contribution, such as triplets and quintets. TE-averaged MRS has been used to achieve a more reliable detection of glutamate [45], by averaging out other unwanted resonances.

Due to its simplicity and high sensitivity, 2D *J*-resolved NMR is one of the more commonly applied methods *in vivo* [102–107]. Figure 8.37 shows simulated 2D *J*-resolved spectra from metabolites commonly detected in the human and animal brain *in vivo*. The weakly-coupled spin systems shown in Figures 8.36 and 8.37A (alanine, BHB, lactate, and threonine) all provide high-quality "homonuclear decoupled" MR spectra. However, for other spin systems (Figure 8.37B–D) additional resonances can be seen that are not expected nor observed under the weak coupling approximation. For example, in the weak coupling limit taurine can be seen as an A_2X_2 spin system, leading to two triplets at 3.27 and 3.44 ppm (Figure 8.37C, inset). However, at a magnetic field strength of 7 T, taurine is more accurately described as a strongly-coupled A_2B_2 spin system, leading to additional resonances at an intermediate chemical shift (asterisk (*) in Figure 8.37C). Strong coupling effects complicate the appearance of the "homonuclear decoupled" or "pure shift" MR spectrum. Figure 8.38 shows an *in vivo* example of 3D localized, 2D *J*-resolved MRS on a F98 glioma in rat.

8.10.3 *In vivo* 2D NMR Methods

The family of multidimensional MR methods has grown from the initial COSY method to include hundreds of different sequences to explore correlations between spins based on scalar couplings, chemical exchange, and relaxation. The majority of those methods are not suitable for direct *in vivo* application due to time and/or sensitivity constraints, limits on RF power deposition, or the absence of appropriate spin systems. Here, a small number of 2D MR methods will be discussed that, while not commonly employed, do have potential in specific *in vivo* applications.

Chemical exchange between phosphocreatine and ATP or between bound and mobile water can be studied with magnetization transfer experiments in which the signal modulations of one exchange partner are observed following perturbation of the other exchange partner by selective RF pulses (see Chapter 3 for more detail). A more general approach for studying (chemical) exchange reactions is provided by 2D exchange spectroscopy [108–110]. Unlike the previously discussed COSY techniques, 2D exchange NMR does *not* require that the spins under

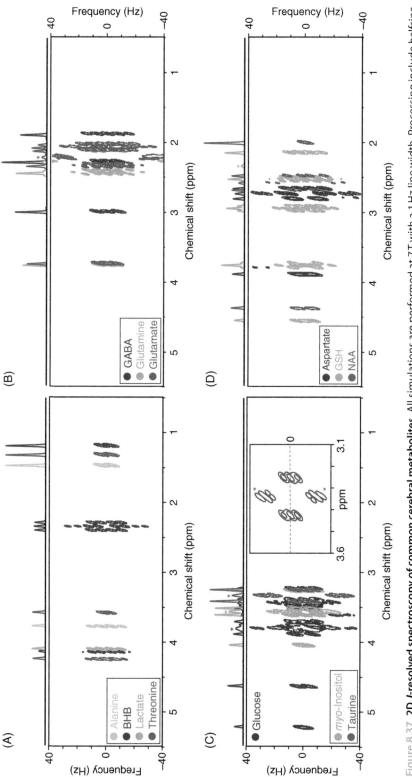

Figure 8.37 **2D J-resolved spectroscopy of common cerebral metabolites.** All simulations are performed at 7 T with a 1 Hz line width. Processing include halfsine apodization and a 45° spectral tilt. 2D J-resolved spectra for (A) alanine, BHB, lactate, and threonine, (B) GABA, glutamine, glutamate, and glutamine, (C) glucose, *myo*-inositol, and taurine, and (D) aspartate, glutathione (GSH), and N-acetyl aspartate (NAA). Projection resonances originating from strong coupling are labeled with an asterisk (*).

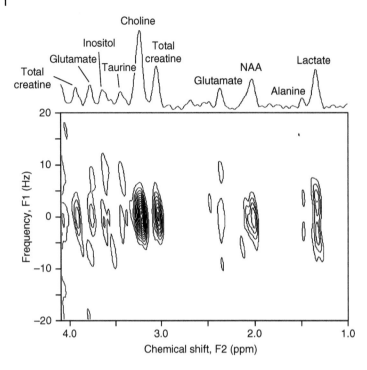

Choline

Inositol Total
creatine
Glutamate | Taurine
Total   NAA   Lactate
creatine   Glutamate
Alanine

Figure 8.38 **2D *J*-resolved spectroscopy from a F98 glioma in rat brain *in vivo*.** Spatial localization (64 µl) was achieved with PRESS in which TE2 was incremented in 16 steps of 17 ms. The spectrum was tilted by 45° after which the projection (summation) of an effectively ¹H-decoupled NMR spectrum onto the F2 axis could be calculated. Although the spectral resolution is reduced due to absolute-valued processing and short *in vivo* T_2 relaxation times, many resonances can be clearly distinguished, like alanine from lactate. *Source:* Data from W. Dreher.

investigation are scalar-coupled. They merely have to be coupled through chemical exchange and/or cross-relaxation. For this reason the principles of 2D exchange NMR experiments are readily explained by a classical vector model. Figure 8.39A shows the 2D exchange NMR (or nuclear Overhauser effect spectroscopy, NOESY) sequence. Prior to excitation, two spin types, *I* and *S*, are at thermal equilibrium. Following excitation, the spins precess at their respective Larmor frequencies, ν_I and ν_S, during the evolution period t_1. The second 90° pulse rotates the magnetization along the longitudinal axis. Note that this is exactly the component that was not used in a COSY experiment. When the *I* and *S* spins are in chemical exchange, some *I* spins turn into *S* spins and vice versa. The spins can also transfer the magnetization through cross-relaxation in which case the spins do not physically exchange. The *I* and *S* spins that do not undergo exchange have the same frequency during the t_1 and t_2 periods, leading to diagonal peaks. The *I* and *S* spins that have exchanged during the mixing period t_m have different frequencies during the t_1 and t_2 periods, leading to cross-peaks. The intensity of the diagonal and cross-peaks is determined by the t_m mixing period, the T_1 relaxation times of the *I* and *S* spins, and the exchange rate between the spins. The diagonal peaks decrease with increasing t_m as a result of signal loss due to exchange and T_1 relaxation ($I_{diagonal}$ curve). The cross-peaks start at zero intensity and see an initial increase with increasing t_m as spins are converted from one spin type (e.g. *I*) into another (e.g. *S*). The cross-peaks' intensity continues to increase with increasing t_m delay until the decay due to T_1 relaxation becomes dominant (I_{cross} curve).

Figure 8.39B shows the application of 2D exchange NMR to study the chemical exchange between phosphocreatine and ATP in rat skeletal muscle *in vivo* as catalyzed by creatine kinase. On the diagonal of the 2D NMR spectrum the resonances of ATP, PCr, and inorganic phosphate P_i can be recognized, as they would also appear in a regular 1D ³¹P NMR spectrum. Off the diagonal, at the intersection between the frequencies for PCr and γ-ATP, two additional resonances can be observed which are due to the chemical exchange of a phosphate group between PCr and γ-ATP. Given the relaxation rates for PCr and ATP, as well as the exchange

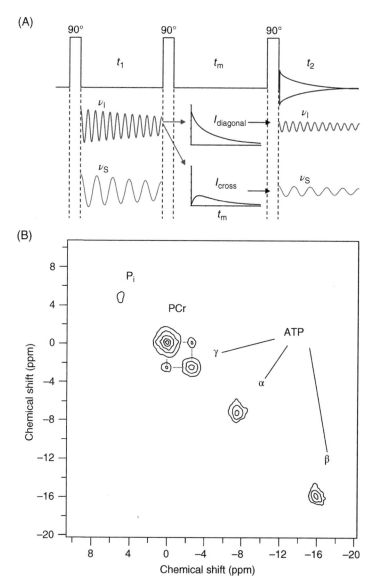

Figure 8.39 2D exchange spectroscopy. (A) Transfer of frequency-labeled longitudinal magnetization in 2D exchange spectroscopy for a two-sided system consisting of spins *I* and *S*. Following excitation, the spins are frequency-encoded during the evolution period t_1 according to their Larmor frequencies, ν_I or ν_S. During the mixing period t_m, the amplitudes of the coherences which give rise to diagonal peaks, $I_{diagonal}$, decay bi-exponentially due to exchange and longitudinal relaxation. The amplitudes of the coherences which give rise to cross-peaks, I_{cross}, first increase due to exchange before decaying because of spin-lattice relaxation. During the acquisition period t_2, spins that have undergone chemical exchange will be acquired at a frequency different from that during the t_1 period, thereby giving rise to a cross-peak. (B) 2D exchange spectrum obtained from rat muscle *in vivo*. The mixing time t_m was 600 ms, while the 2D data matrix was constructed from 64 t_1 increments and 128 complex points during acquisition. Each increment involved 16 averages. The data are presented in absolute-value mode. Because of chemical exchange between PCr and γ-ATP two cross-peaks are visible.

rate, the cross-peak signal intensity reaches a maximum at $t_m = 0.5–1.0\,s$ for the PCr/ATP system in skeletal muscle *in vivo*.

Although 2D NOESY experiments are used to study chemical exchange processes, the main application is in the study of through-space dipolar coupling which can reveal the spatial positions of spins relative to each other. This experiment is crucial in high-resolution NMR studies of macromolecules in which one wants to solve the spatial structure of the compound under investigation. However, since *in vivo* NMR spectroscopy mainly detects resonances from small metabolites (with a known spatial structure), this application has limited value.

The POCE or ^1H-[^{13}C] NMR sequence was discussed as the prime example of heteronuclear spectral editing (Section 8.7). However, in reality, all homonuclear spectral editing sequences and many heteronuclear 2D NMR sequences can be converted and used to achieve heteronuclear spectral editing. The heteronuclear multiple-quantum coherence (HMQC) [43] and heteronuclear single-quantum coherence (HSQC) [50] methods are workhorses in high-resolution NMR to establish correlation between protons and carbon-13 nuclei (or any other

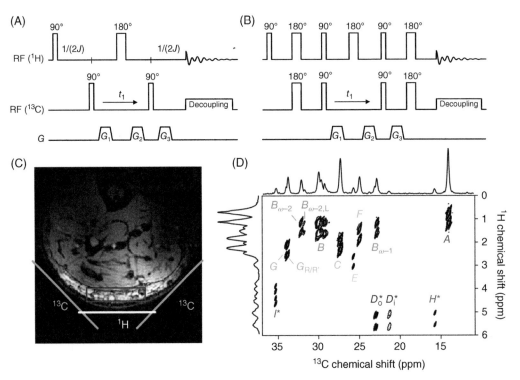

Figure 8.40 **Heteronuclear correlation spectroscopy.** (A) Heteronuclear multiple-quantum coherence (HMQC) and (B) heteronuclear single-quantum coherence (HSQC) pulse sequences using magnetic field gradients G_1–G_3 for coherence selection. The carbon-13 chemical shift is detected indirectly during the evolution period t_1, providing correlation with the directly acquired proton chemical shifts. Coherence selection can be achieved with a number of magnetic field gradient combinations, each with a specific performance regarding signal recovery, dephasing efficiency, and sensitivity towards motion and diffusion (see Exercises). (C) MR image of the human leg at 7T together with ^1H surface coil and ^{13}C quadrature coil geometries. (D) 2D HSQC spectrum acquired from the red volume indicated in (C). Data are acquired as 192 increments over a 2 kHz bandwidth in the indirect dimension in the absence of ^{13}C decoupling during the acquisition period. Note that all resonances are perfectly decoupled along the indirect dimension due to the proton 180° pulse during the evolution period t_1. Assignment of triglyceride resonances follows the nomenclature used for 1D ^1H and ^{13}C NMR spectra of triglycerides shown in Figures 2.35 and 2.40.

8.10 B_1-insensitive spectral editing pulse (BISEP [118]) is a single-scan spectral editing method for heteronuclear spin-system. While originally executed with adiabatic RF pulses, the sequence can also be described with regular pulses as: $90°_{+x}$ (^{1}H) $- t - 180°_{-x}$ (^{1}H/^{13}C) $- t - 90°_{+x}$ (^{1}H).

  A Derive an expression for the coherences of uncoupled and heteronuclear scalar coupled spin-systems at the end of a BISEP sequence.

  B Propose an extension to the sequence that (i) leads to in-phase detection of heteronuclear scalar-coupled spin-systems and (ii) gives an additional suppression of uncoupled spins.

8.11 At the end of a [1-^{13}C]-glucose infusion study, the ^{13}C-labeled metabolic products are detected in the human brain at 3 T with ^{1}H-observed, ^{13}C-edited (POCE) MRS. Broadband decoupling based on WALTZ-16 ($B_1^+ = 450$ Hz) is applied during the 102 ms acquisition period. The ^{13}C carrier frequency is set to Glu-C4 at 34.2 ppm. Strong signals from [4-^{13}C]-glutamate and [4-^{13}C]-glutamine are observed, as expected. However, the signals from [2-^{13}C]-glutamate is very low, whereas the signal from [1-^{13}C]-glucose is completely absent. Provide a likely explanation for the observed effects.

8.12 HMQC and HSQC sequences can achieve coherence selection based on phase cycling and based on magnetic field gradients. For the HMQC and HSQC sequences shown in Figure 8.40 derive a least three gradient combinations that select the CTP for a heteronuclear two-spin-system AX while simultaneously destroying signal from non-coupled spins.

8.13 For the LASER-based JDE sequence shown in Figure 8.9B, derive the three conditions required for maximum editing efficiency during the entire echo-time TE.

8.14 During a 2D NMR acquisition, two datasets are acquired in order to achieve phase-sensitive quadrature detection along the indirect dimension. The two datasets S_1 and S_2 are cosine- and sine-modulated according to:

$$S_1(t_1, t_2) = M_0 \cos(\omega_1 t_1) e^{i\omega_2 t_2} e^{-t_1/T_2} e^{-t_2/T_2}$$

$$S_2(t_1, t_2) = M_0 \sin(\omega_1 t_1) e^{i\omega_2 t_2} e^{-t_1/T_2} e^{-t_2/T_2}$$

  A Show that addition of the signals ($S_1 + iS_2$) provides frequency discrimination along the indirect dimension. Also show that the resonance has a mixture of absorption and dispersion components, characteristic of a "phase-twisted" line.

  B During the "States hypercomplex" method the imaginary part of $S_1(t_1, \omega_2)$ is replaced with the real part of $S_2(t_1, \omega_2)$, whereby both datasets were processed by a 1D FFT with respect to t_2. Show that the hypercomplex signal provides a pure absorption line in the 2D NMR spectrum.

8.15 Double-quantum-filtered (DQF) COSY is an extension of the basic COSY method with an additional 90° pulse surrounded by two magnetic field gradients G_1 and G_2. Derive an expression for the detectable signal of a two-spin-system AX when $G_2 = 2G_1$.

References

1 Pfeuffer, J., Tkac, I., Provencher, S.W., and Gruetter, R. (1999). Toward an *in vivo* neurochemical profile: quantification of 18 metabolites in short-echo-time ^{1}H NMR spectra of the rat brain. *J. Magn. Reson.* 141: 104–120.

2 Govindaraju, V., Young, K., and Maudsley, A.A. (2000). Proton NMR chemical shifts and coupling constants for brain metabolites. *NMR Biomed.* 13: 129–153.

3 Fano, U. (1957). Description of states in quantum mechanics by density matrix and operator techniques. *Rev. Mod. Phys.* 29: 74–93.

4 Ernst, R.R., Bodenhausen, G., and Wokaun, A. (1987). *Principles of Nuclear Magnetic Resonance in One and Two Dimensions*. Oxford: Clarendon Press.

5 Slichter, C.P. (1990). *Principles of Magnetic Resonance*. Berlin: Springer-Verlag.

6 Levitt, M.H. (2005). *Spin Dynamics. Basics of Nuclear Magnetic Resonance*. New York: Wiley.

7 Lowry, D.F. (1994). Correlated vector model of multiple-spin systems. *Concepts Magn. Reson.* 6: 25–39.

8 Freeman, R. (1998). A physical picture of multiple-quantum-coherence. *Concepts Magn. Reson.* 10: 63–84.

9 Stait-Gardner, T. and Price, W.S. (2009). A physical interpretation of product operator terms. *Concepts Magn. Reson.* 34A: 322–356.

10 Sorensen, O.W., Eich, G.W., Levitt, M.H. et al. (1983). Product operator formalism for the description of NMR pulse experiments. *Prog. Nucl. Magn. Reson. Spectrosc.* 16: 163–192.

11 Packer, K.J. and Wright, K.M. (1983). The use of single-spin operator basis-sets in NMR spectroscopy of scalar-coupled spin systems. *Mol. Phys.* 50: 797–813.

12 van de Ven, F.J.M. and Hilbers, C.W. (1983). A simple formalism for the description of multi-pulse experiments. Applications to a weakly-coupled two-spin (I = 1/2) system. *J. Magn. Reson.* 54: 512–520.

13 Mescher, M., Tannus, A., O'Neil Johnson, M., and Garwood, M. (1996). Solvent suppression using selective echo dephasing. *J. Magn. Reson. A* 123: 226–229.

14 Mescher, M., Merkle, H., Kirsch, J. et al. (1998). Simultaneous *in vivo* spectral editing and water suppression. *NMR Biomed.* 11: 266–272.

15 Henry, P.G., Dautry, C., Hantraye, P., and Bloch, G. (2001). Brain GABA editing without macromolecule contamination. *Magn. Reson. Med.* 45: 517–520.

16 Terpstra, M., Marjanska, M., Henry, P.G. et al. (2006). Detection of an antioxidant profile in the human brain *in vivo* via double editing with MEGA-PRESS. *Magn. Reson. Med.* 56: 1192–1199.

17 Chan, K.L., Puts, N.A., Schar, M. et al. (2016). HERMES: Hadamard encoding and reconstruction of MEGA-edited spectroscopy. *Magn. Reson. Med.* 76: 11–19.

18 Saleh, M.G., Oeltzschner, G., Chan, K.L. et al. (2016). Simultaneous edited MRS of GABA and glutathione. *Neuroimage* 142: 576–582.

19 Chan, K.L., Saleh, M.G., Oeltzschner, G. et al. (2017). Simultaneous measurement of aspartate, NAA, and NAAG using HERMES spectral editing at 3 Tesla. *Neuroimage* 155: 587–593.

20 Rothman, D.L., Petroff, O.A., Behar, K.L., and Mattson, R.H. (1993). Localized ^{1}H NMR measurements of gamma-aminobutyric acid in human brain *in vivo*. *Proc. Natl. Acad. Sci. U. S. A.* 90: 5662–5666.

21 Rothman, D.L., Behar, K.L., Prichard, J.W., and Petroff, O.A. (1997). Homocarnosine and the measurement of neuronal pH in patients with epilepsy. *Magn. Reson. Med.* 38 (6): 924–929.

22 Behar, K.L., Rothman, D.L., Spencer, D.D., and Petroff, O.A. (1994). Analysis of macromolecule resonances in ^{1}H NMR spectra of human brain. *Magn. Reson. Med.* 32: 294–302.

23 Henry, P.G., van de Moortele, P.F., Giacomini, E. et al. (1999). Field-frequency locked *in vivo* proton MRS on a whole-body spectrometer. *Magn. Reson. Med.* 42: 636–642.

24 Mikkelsen, M., Barker, P.B., Bhattacharyya, P.K. et al. (2017). Big GABA: edited MR spectroscopy at 24 research sites. *Neuroimage* 159: 32–45.

25 Terpstra, M. and Gruetter, R. (2004). ^{1}H NMR detection of vitamin C in human brain *in vivo*. *Magn. Reson. Med.* 51 (2): 225–229.

26 Terpstra, M., Henry, P.G., and Gruetter, R. (2003). Measurement of reduced glutathione (GSH) in human brain using LCModel analysis of difference-edited spectra. *Magn. Reson. Med.* 50 (1): 19–23.

27 Choi, C., Ganji, S.K., DeBerardinis, R.J. et al. (2012). 2-Hydroxyglutarate detection by magnetic resonance spectroscopy in IDH-mutated patients with gliomas. *Nat. Med.* 18: 624–629.

28 Andronesi, O.C., Kim, G.S., Gerstner, E. et al. (2012). Detection of 2-hydroxyglutarate in IDH-mutated glioma patients by *in vivo* spectral-editing and 2D correlation magnetic resonance spectroscopy. *Sci. Transl. Med.* 4 (116): 116ra114.

29 Sotak, C.H. and Freeman, D. (1988). A method for volume-localized lactate editing using zero-quantum coherence created in a stimulated-echo pulse sequence. *J. Magn. Reson.* 77: 382–388.

30 Sotak, C.H., Freeman, D., and Hurd, R.E. (1988). The unequivocal determination of *in vivo* lactic acid using two-dimensional double-quantum coherence transfer spectroscopy. *J. Magn. Reson.* 78: 355–361.

31 Trimble, L.A., Shen, J.F., Wilman, A.H., and Allen, P.S. (1990). Lactate editing by means of selective-pulse filtering of both zero-and double-quantum coherence signals. *J. Magn. Reson.* 86: 191–198.

32 He, Q., Shungu, D.C., van Zijl, P.C. et al. (1995). Single-scan *in vivo* lactate editing with complete lipid and water suppression by selective multiple-quantum-coherence transfer (Sel-MQC) with application to tumors. *J. Magn. Reson. B* 106: 203–211.

33 Dumoulin, C.L. and Williams, E.A. (1986). Suppression of uncoupled spins by single-quantum homonuclear polarization transfer. *J. Magn. Reson.* 66: 86–92.

34 von Kienlin, M., Albrand, J.P., Authier, B. et al. (1987). Spectral editing *in vivo* by homonuclear polarization transfer. *J. Magn. Reson.* 75: 371–377.

35 Pan, J.W., Mason, G.F., Pohost, G.M., and Hetherington, H.P. (1996). Spectroscopic imaging of human brain glutamate by water-suppressed J-refocused coherence transfer at 4.1 T. *Magn. Reson. Med.* 36: 7–12.

36 Shen, J., Yang, J., Choi, I.Y. et al. (2004). A new strategy for *in vivo* spectral editing. Application to GABA editing using selective homonuclear polarization transfer spectroscopy. *J. Magn. Reson.* 170: 290–298.

37 Reddy, R., Subramanian, V.H., Clark, B.J., and Leigh, J.S. (1991). Longitudinal spin-order-based pulse sequence for lactate editing. *Magn. Reson. Med.* 19: 477–482.

38 Reddy, R., Subramanian, V.H., Clark, B.J., and Leigh, J.S. (1993). *In vivo* lactate editing in the presence of inhomogeneous B_1 fields. *J. Magn. Reson. B* 102: 20–25.

39 de Graaf, R.A. and Rothman, D.L. (2001). Detection of gamma-aminobutyric acid (GABA) by longitudinal scalar order difference editing. *J. Magn. Reson.* 152: 124–131.

40 Choi, I.Y., Lee, S.P., and Shen, J. (2005). Selective homonuclear Hartmann-Hahn transfer method for *in vivo* spectral editing in the human brain. *Magn. Reson. Med.* 53: 503–510.

41 Marjanska, M., Henry, P.G., Bolan, P.J. et al. (2005). Uncovering hidden *in vivo* resonances using editing based on localized TOCSY. *Magn. Reson. Med.* 53: 783–789.

42 Keeler, J. (2005). *Understanding NMR Spectroscopy*. New York: Wiley.

43 Muller, L. (1979). Sensitivity enhanced detection of weak nuclei using heteronuclear multiple quantum coherence. *J. Am. Chem. Soc.* 101: 4481–4484.

44 Wilman, A.H. and Allen, P.S. (1993). *In vivo* NMR detection strategies for γ-aminobutyric acid utilizing proton spectroscopy and coherence pathway filtering with gradients. *J. Magn. Reson. B* 101: 165–171.

45 Hurd, R., Sailasuta, N., Srinivasan, R. et al. (2004). Measurement of brain glutamate using TE-averaged PRESS at 3T. *Magn. Reson. Med.* 51: 435–440.

46 Choi, C., Dimitrov, I.E., Douglas, D. et al. (2010). Improvement of resolution for brain coupled metabolites by optimized ^{1}H MRS at 7T. *NMR Biomed.* 23: 1044–1052.

47 An, L., Li, S., Murdoch, J.B. et al. (2015). Detection of glutamate, glutamine, and glutathione by radiofrequency suppression and echo time optimization at 7 tesla. *Magn. Reson. Med.* 73: 451–458.

48 Toncelli, A., Noeske, R., Cosottini, M. et al. (2015). STEAM-MiTiS: an MR spectroscopy method for the detection of scalar-coupled metabolites and its application to glutamate at 7 T. *Magn. Reson. Med.* 74: 1515–1522.

49 Wijnen, J.P., Klomp, D.W., Nabuurs, C.I. et al. (2016). Proton observed phosphorus editing (POPE) for *in vivo* detection of phospholipid metabolites. *NMR Biomed.* 29: 1222–1230.

50 Bodenhausen, G. and Reuben, D.J. (1980). Natural abundance nitrogen-15 NMR by enhanced heteronuclear spectroscopy. *Chem. Phys. Lett.* 69: 185–189.

51 Ruiz-Cabello, J., Vuister, G.W., Moonen, C.T.W. et al. (1992). Gradient-enhanced heteronuclear correlation spectroscopy. Theory and experimental aspects. *J. Magn. Reson.* 100: 282–302.

52 van Zijl, P.C., Chesnick, A.S., DesPres, D. et al. (1993). *In vivo* proton spectroscopy and spectroscopic imaging of [1-^{13}C]-glucose and its metabolic products. *Magn. Reson. Med.* 30: 544–551.

53 Kanamori, K., Ross, B.D., and Tropp, J. (1995). Selective, *in vivo* observation of [5-^{15}N] glutamine amide protons in rat brain by ^{1}H-^{15}N heteronuclear multiple-quantum-coherence transfer NMR. *J. Magn. Reson. B* 107: 107–115.

54 Morris, G.A. and Freeman, R. (1979). Enhancement of nuclear magnetic resonance signals by polarization transfer. *J. Am. Chem. Soc.* 101: 760–762.

55 Burum, D.P. and Ernst, R.R. (1980). Net polarization transfer via a J-ordered state for signal enhancement of low-sensitivity nuclei. *J. Magn. Reson.* 39: 163–168.

56 Doddrell, D.M., Pegg, D.T., and Bendall, M.R. (1982). Distortionless enhancement of NMR signals by polarization transfer. *J. Magn. Reson.* 48: 323–327.

57 Sorensen, O.W. and Ernst, R.R. (1983). Elimination of spectral distortion in polarization transfer experiments. Improvements and comparison of techniques. *J. Magn. Reson.* 51: 477–489.

58 Gruetter, R., Adriany, G., Merkle, H., and Andersen, P.M. (1996). Broadband decoupled, ^{1}H-localized ^{13}C MRS of the human brain at 4 Tesla. *Magn. Reson. Med.* 36: 659–664.

59 Shen, J., Petersen, K.F., Behar, K.L. et al. (1999). Determination of the rate of the glutamate/glutamine cycle in the human brain by *in vivo* ^{13}C NMR. *Proc. Natl. Acad. Sci. U. S. A.* 96: 8235–8240.

60 Deelchand, D.K., Ugurbil, K., and Henry, P.G. (2006). Investigating brain metabolism at high fields using localized ^{13}C NMR spectroscopy without ^{1}H decoupling. *Magn. Reson. Med.* 55: 279–286.

61 Boumezbeur, F., Besret, L., Valette, J. et al. (2004). NMR measurement of brain oxidative metabolism in monkeys using ^{13}C-labeled glucose without a ^{13}C radiofrequency channel. *Magn. Reson. Med.* 52: 33–40.

62 Shaka, A.J. and Keeler, J. (1987). Broadband spin decoupling in isotropic liquids. *Prog. Nucl. Magn. Reson. Spectrosc.* 19: 47–129.

63 Levitt, M.H. and Freeman, R. (1981). Composite pulse decoupling. *J. Magn. Reson.* 43: 502–507.

64 Levitt, M.H. (1982). Symmetrical composite pulse sequences for NMR population inversion. I. Compensation of radiofrequency field inhomogeneity. *J. Magn. Reson.* 48: 234–264.

65 Levitt, M.H. (1982). Symmetrical composite pulse sequences for NMR population inversion. II. Compensation for resonance offset. *J. Magn. Reson.* 50: 95–110.

66 Shaka, A.J., Keeler, J., and Freeman, R. (1983). Evaluation of a new broadband decoupling sequence: WALTZ-16. *J. Magn. Reson.* 53: 313–340.

67 Shaka, A.J., Keeler, J., Frenkiel, T., and Freeman, R. (1983). An improved sequence for broadband decoupling: WALTZ-16. *J. Magn. Reson.* 52: 335–338.

68 Shaka, A.J., Barker, P.B., Bauer, C.J., and Freeman, R. (1986). Cycling sidebands in broadband decoupling. *J. Magn. Reson.* 67: 396–401.

69 de Graaf, R.A. (2005). Theoretical and experimental evaluation of broadband decoupling techniques for *in vivo* NMR spectroscopy. *Magn. Reson. Med.* 53: 1297–1306.

70 Fujiwara, T. and Nagayama, K. (1988). Composite inversion pulses with frequency switching and their application to broadband decoupling. *J. Magn. Reson.* 77: 53–63.

71 Fujiwara, T., Anai, T., Kurihara, N., and Nagayama, K. (1993). Frequency-switched composite pulses for decoupling carbon-13 spins over ultrabroad bandwidths. *J. Magn. Reson. A* 104: 103–105.

72 Bendall, M.R. (1995). Broadband and narrowband spin decoupling using adiabatic spin flips. *J. Magn. Reson. A* 112: 126–129.

73 Kupce, E. and Freeman, R. (1995). Adiabatic pulse for wideband inversion and broadband decoupling. *J. Magn. Reson. A* 115: 273–276.

74 Starcuk, Z., Bartusek, K., and Starcuk, Z. (1994). Heteronuclear broadband spin-flip decoupling with adiabatic pulses. *J. Magn. Reson. A* 107: 24–31.

75 De Feyter, H.M., Herzog, R.I., Steensma, B.R. et al. (2018). Selective proton-observed, carbon-edited (selPOCE) MRS method for measurement of glutamate and glutamine ^{13}C-labeling in the human frontal cortex. *Magn. Reson. Med.* 80: 11–20.

76 Li, S., Yang, J., and Shen, J. (2007). Novel strategy for cerebral ^{13}C MRS using very low RF power for proton decoupling. *Magn. Reson. Med.* 57: 265–271.

77 Sailasuta, N., Robertson, L.W., Harris, K.C. et al. (2008). Clinical NOE ^{13}C MRS for neuropsychiatric disorders of the frontal lobe. *J. Magn. Reson.* 195: 219–225.

78 Li, S., Zhang, Y., Wang, S. et al. (2009). *In vivo* ^{13}C magnetic resonance spectroscopy of human brain on a clinical 3 T scanner using [2-^{13}C]-glucose infusion and low-power stochastic decoupling. *Magn. Reson. Med.* 62: 565–573.

79 Li, S., An, L., Yu, S. et al. (2016). ^{13}C MRS of human brain at 7 Tesla using [2-^{13}C]glucose infusion and low power broadband stochastic proton decoupling. *Magn. Reson. Med.* 75: 954–961.

80 Hoult, D.I. and Richards, R.E. (1976). The signal-to-noise ratio of the nuclear magnetic resonance experiment. *J. Magn. Reson.* 24: 71–85.

81 Zhu, X., Merkle, H., Kwag, J. et al. (2001). ^{17}O relaxation time and NMR sensitivity of cerebral water and their field dependence. *Magn. Reson. Med.* 45: 543–549.

82 Lu, M., Chen, W., and Zhu, X.H. (2014). Field dependence study of *in vivo* brain ^{31}P MRS up to 16.4 T. *NMR Biomed.* 27: 1135–1141.

83 Vaughan, J.T., Garwood, M., Collins, C.M. et al. (2001). 7T vs. 4T: RF power, homogeneity, and signal-to-noise comparison in head images. *Magn. Reson. Med.* 46: 24–30.

84 Chen, H., De Feyter, H.M., Brown, P.B. et al. (2017). Comparison of direct ^{13}C and indirect ^{1}H-[^{13}C] MR detection methods for the study of dynamic metabolic turnover in the human brain. *J. Magn. Reson.* 283: 33–44.

85 Bax, A. (1982). *Two-dimensional NMR in Liquids*. Dordrecht: Delft University Press.

86 Cavangh, J., Fairbrother, W.J., Palmer, A.G., and Skelton, N.J. (1996). *Protein NMR Spectroscopy. Principles and Practice*. San Diego: Academic Press.

87 van de Ven, F.J.M. (1995). *Multidimensional NMR in Liquids*. New York: Wiley.

88 Jeener, J. (1971). *Ampere Summer School*. Yugoslavia: Basko Polje.

89 Aue, W.P., Bartholdi, E., and Ernst, R.R. (1976). Two-dimensional spectroscopy. Application to nuclear magnetic resonance. *J. Chem. Phys.* 64: 2229–2246.

90 States, D.J., Haberkorn, R.A., and Ruben, D.J. (1982). A two-dimensional nuclear Overhauser experiment with pure absorption phase in four quadrants. *J. Magn. Reson.* 48: 286–292.

91 Marion, D. and Wuthrich, K. (1983). Application of phase sensitive two-dimensional correlated spectroscopy (COSY) for measurements of ^{1}H-^{1}H spin-spin coupling constants in proteins. *Biochem. Biophys. Res. Commun.* 113: 967–974.

92 Bax, A., de Jong, P.G., Mehlkopf, A.F., and Smidt, J. (1980). Separation of the different orders of NMR multiple-quantum transitions by the use of pulsed field gradients. *Chem. Phys. Lett.* 69: 567–570.

93 Barker, P. and Freeman, R. (1985). Pulsed field gradients in NMR. An alternative to phase cycling. *J. Magn. Reson.* 64: 334–338.

94 Hurd, R.E. (1990). Gradient-enhanced spectroscopy. *J. Magn. Reson.* 87: 422–428.

95 Piantini, U., Sorensen, O.W., and Ernst, R.R. (1982). Multiple quantum filters for elucidating NMR coupling networks. *J. Am. Chem. Soc.* 104: 6800–6801.

96 Davis, A.L., Laue, E.D., Keeler, J. et al. (1991). Absorption-mode two-dimensional NMR spectra recorded using pulsed field gradients. *J. Magn. Reson.* 94: 637–644.

97 Braunschweiler, L. and Ernst, R.R. (1983). Coherence transfer by isotropic mixing: application to proton correlation spectroscopy. *J. Magn. Reson.* 53: 521–528.

98 Aue, W.P., Karhan, J., and Ernst, R.R. (1976). Homonuclear broadband decoupling in two-dimensional J-resolved NMR spectroscopy. *J. Chem. Phys.* 64: 4226–4227.

99 Bodenhausen, G., Freeman, R., and Turner, D.L. (1976). Two-dimensional J spectroscopy: proton-decoupled carbon-13 NMR. *J. Chem. Phys.* 65: 839–840.

100 Muller, L., Kumar, A., and Ernst, R.R. (1977). Two-dimensional carbon-13 spin-echo spectroscopy. *J. Magn. Reson.* 25: 383–390.

101 Parsons, H.M., Ludwig, C., and Viant, M.R. (2009). Line-shape analysis of J-resolved NMR spectra: application to metabolomics and quantification of intensity errors from signal processing and high signal congestion. *Magn. Reson. Chem.* 47: S86–S95.

102 Schulte, R.F. and Boesiger, P. (2006). ProFit: two-dimensional prior-knowledge fitting of J-resolved spectra. *NMR Biomed.* 19: 255–263.

103 Fuchs, A., Boesiger, P., Schulte, R.F., and Henning, A. (2014). ProFit revisited. *Magn. Reson. Med.* 71: 458–468.

104 Ryner, L.N., Sorenson, J.A., and Thomas, M.A. (1995). 3D localized 2D NMR spectroscopy on an MRI scanner. *J. Magn. Reson. B* 107: 126–137.

105 Ryner, L.N., Sorenson, J.A., and Thomas, M.A. (1995). Localized 2D J-resolved ^{1}H MR spectroscopy: strong coupling effects *in vitro* and *in vivo*. *Magn. Reson. Imaging* 13: 853–869.

106 Dreher, W. and Leibfritz, D. (1995). On the use of two-dimensional-J NMR measurements for *in vivo* proton MRS: measurement of homonuclear decoupled spectra without the need for short echo times. *Magn. Reson. Med.* 34: 331–337.

107 Thomas, M.A., Ryner, L.N., Mehta, M.P. et al. (1996). Localized 2D J-resolved ^{1}H MR spectroscopy of human brain tumors *in vivo*. *J. Magn. Reson. Imaging* 6: 453–459.

108 Jeener, J., Meier, B.H., Bachman, P., and Ernst, R.R. (1979). Investigation of exchange processes by two-dimensional NMR spectroscopy. *J. Chem. Phys.* 71: 4546–4553.

109 Macura, S. and Ernst, R.R. (1980). Elucidation of cross relaxation in liquids by two-dimensional NMR spectroscopy. *Mol. Phys.* 41: 95–117.

110 Balaban, R.S., Kantor, H.L., and Ferretti, J.A. (1983). *In vivo* flux between phosphocreatine and adenosine triphosphate determined by two-dimensional phosphorus NMR. *J. Biol. Chem.* 258: 12787–12789.

111 Watanabe, H., Ishihara, Y., Okamoto, K. et al. (2000). 3D localized ^{1}H-^{13}C heteronuclear single-quantum coherence correlation spectroscopy *in vivo*. *Magn. Reson. Med.* 43: 200–210.

112 Watanabe, H., Umeda, M., Ishihara, Y. et al. (2000). Human brain glucose metabolism mapping using multislice 2D ^{1}H-^{13}C correlation HSQC spectroscopy. *Magn. Reson. Med.* 43: 525–533.

113 de Graaf, R.A., De Feyter, H.M., and Rothman, D.L. (2015). High-sensitivity, broadband-decoupled ^{13}C MR spectroscopy in humans at 7T using two-dimensional heteronuclear single-quantum coherence. *Magn. Reson. Med.* 74: 903–914.

114 de Graaf, R.A., Rothman, D.L., and Behar, K.L. (2007). High resolution NMR spectroscopy of rat brain *in vivo* through indirect zero-quantum-coherence detection. *J. Magn. Reson.* 187: 320–326.

115 de Graaf, R.A., Klomp, D.W.J., Luijten, P.R., and Boer, V.O. (2014). Intramolecular zero-quantum-coherence 2D NMR spectroscopy of lipids in the human breast at 7 T. *Magn. Reson. Med.* 71: 451–457.

116 He, Q., Richter, W., Vathyam, S., and Warren, W.S. (1993). Intermolecular multiple-quantum coherences and cross correlations in solution nuclear magnetic resonance. *J. Chem. Phys.* 98: 6779–6800.

117 Faber, C., Pracht, E., and Haase, A. (2003). Resolution enhancement in *in vivo* NMR spectroscopy: detection of intermolecular zero-quantum coherences. *J. Magn. Reson.* 161: 265–274.

118 Garwood, M. and Merkle, H. (1991). Heteronuclear spectral editing with adiabatic pulses. *J. Magn. Reson.* 94: 180–185.

9

Spectral Quantification

9.1 Introduction

Spectra obtained by NMR spectroscopy can, in principle, be used to derive absolute concentrations, expressed in $mmol\,l^{-1}$ or $\mu mol\,g^{-1}$ of tissue, in animal and human tissues *in vivo*. This originates from the fact that the thermal equilibrium magnetization M_0 is directly proportional to the number of spins N (see Eq. (1.4)), which is proportional to the molar concentration. However, in an NMR experiment the thermal equilibrium magnetization M_0 is not detected directly, but rather an induced current proportional to the transverse magnetization is observed. The signal from a metabolite M induced in a receiver coil following a particular NMR sequence is given by

$$S_M = NA \times RG \times v_0 \times [M] \times V \times f_{sequence} \times f_{coil} \tag{9.1}$$

where NA and RG equal the number of averages and receiver gain setting, v_0 is the Larmor frequency, $[M]$ is the molar concentration, V is the volume size, and $f_{sequence}$ and f_{coil} are functions describing the signal modulations due to the NMR pulse sequence and RF coil, respectively. $f_{sequence}$ will depend on the repetition time TR, the echo-time TE, the number and type of RF pulses, as well as the T_1 and T_2 relaxation times. f_{coil} contains factors related to the geometry and quality of the RF coil, like the quality factor Q and the filling factor (see also Chapter 10). Since several of the coil-related factors in the f_{coil} function are not readily available, the direct calculation of the metabolite concentration $[M]$ from the detected signal S_M is not possible. In practice, all quantification methods utilize a calibration or reference compound of known concentration $[R]$ to which the metabolite signals are referenced such that the metabolite concentration can be calculated according to

$$[M] = [R] \frac{S_M}{S_R} C_{MR} \tag{9.2}$$

where S_R is the detected signal from the reference compound and C_{MR} is a correction factor accounting for differences in relaxation times T_1 and T_2, diffusion, gyromagnetic ratio, magnetic susceptibility, spatial position relative to the coil, and many other differences between the reference compound and the metabolites. Since the calculation of a reliable correction factor C_{MR} can be very time-consuming, many reports on *in vivo* MRS, especially in the clinical field, have used metabolite ratios as an alternative. Since the correction factor C_{MR} is identical for both metabolites, a ratio will not be sensitive to several of the unknown parameters in Eq. (9.1). However, unlike absolute concentrations, metabolite ratios cannot give

In Vivo NMR Spectroscopy: Principles and Techniques, Third Edition. Robin A. de Graaf.
© 2019 John Wiley & Sons Ltd. Published 2019 by John Wiley & Sons Ltd.

unambiguous information about metabolic changes, as encountered in many disorders and pathologies studied with *in vivo* NMR. For example, a choline-to-creatine ratio change from 1.0 to 2.0 could mean that (i) the choline concentration increased, (ii) the creatine concentration decreased, (iii) both the choline and creatine concentration changed, or (iv) relaxation parameters for either or both metabolites changed. In order to reach conclusive statements about metabolic changes it is therefore crucial that absolute concentrations are obtained. Furthermore, in order to obtain quantitative fluxes through metabolic pathways as detected from ^{13}C-label turnover studies (see Chapter 3), knowledge of absolute metabolite concentrations is imperative.

Figure 9.1 shows a flowchart for the absolute quantification of metabolites, which can generally be decomposed into four separate steps. During the data acquisition step, the longitudinal magnetization in the volume of interest is converted to transverse magnetization, which generates a free induction decay (FID) in one or multiple receiver coils. While data-acquisition strategies are discussed throughout the book, Section 9.2 summarizes the most critical considerations for quantification purposes. Before the NMR resonance areas can be quantified, the acquired FID signal is often preprocessed (Section 9.3) during which several improvements can be made, typically related to the resonance line-shapes and baseline. In order to reduce the processing time, as well as eliminate ambiguous quantification results, data exclusion is an essential step of data preprocessing. Section 9.3 summarizes several criteria that can be used for automated data exclusion. Acceptable time- or frequency-domain data are subsequently quantified to yield relative metabolite resonance areas. The data-processing step, as well as the data-calibration step, often needs additional data acquired in a separate experiment. Section 9.4 will give an overview of the most commonly used data-processing algorithms. The final step concerns the calculation of the C_{MR} correction term in Eq. (9.2) for the metabolite and reference signals. This includes corrections for T_1, T_2, B_1, and related parameters and will be summarized in Section 9.5. Finally the metabolite concentration can be calculated according to Eq. (9.2), possibly followed by a partial volume correction to account for compartmental metabolism. While Figure 9.1 shows the flowchart for single-volume MRS data, a similar flowchart can be constructed for MRSI data.

9.2 Data Acquisition

The first and arguably most important step in obtaining reliable metabolite concentrations concerns proper acquisition of the raw time-domain signal. The quality of the acquired signal in terms of sensitivity, resolution, artifacts, and general information content directly determines the accuracy and reliability by which metabolite levels can be estimated. While many of these considerations are discussed throughout the other chapters, a short summary of important features concerning optimal data acquisition will be given.

9.2.1 Magnetic Field Homogeneity

The most important parameter determining the spectral quality is the magnetic field homogeneity across the region-of-interest. Inhomogeneity in the external magnetic field leads to a spread in resonance frequencies, which will directly lead to a broadening of spectral lines and thus a decrease in spectral resolution. In order to separate creatine from phosphocreatine the homogeneity needs to be better than 0.01 ppm, which translates to 3.0 Hz at 7.0 T. Less stringent requirements are encountered for separate detection of glutamate and glutamine ($\Delta \sim 0.1$ ppm)

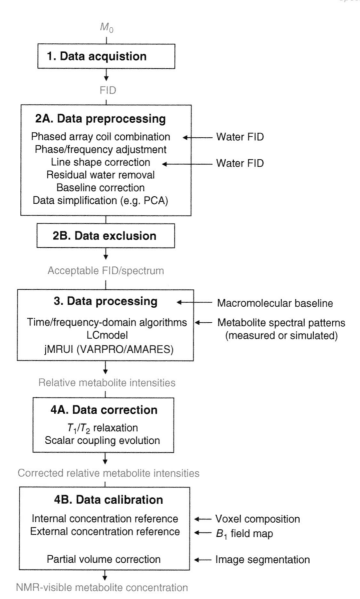

Figure 9.1 Flow chart for processing and quantification of NMR signals. Following data acquisition of a time-domain signal (e.g. FID), the signal undergoes a preprocessing and exclusion step to obtain signals that are acceptable for further processing. In the data-processing step, the relative amplitudes of the metabolite signals are obtained, which are then corrected for a number of effects, including relaxation. The corrected relative metabolite intensities are then compared to a reference signal to arrive at a metabolite concentration as detected by NMR. The necessity of the input signals on the right depends on the exact nature of each step and is discussed in the text.

and an even larger spread of frequencies can be allowed for the separation of choline from creatine ($\Delta \sim 0.2\,\text{ppm}$). However, regardless of the particular resonances, the uncertainty in the amplitude estimation of a broader line is always higher than that for a narrower line (see Figure 9.12). Improving the magnetic field homogeneity and hence the spectral resolution is therefore crucial for the most reliable parameter estimation.

The principle of magnetic field homogeneity optimization through shimming is discussed in Chapter 10. In general, it can be stated that optimal magnetic field homogeneity across small single volumes can be adequately optimized with first- and second-order spherical harmonic shims. When acquiring data from 2D slices or 3D slabs as is common in MR spectroscopic imaging, the magnetic field homogeneity can be further improved by third-order spherical harmonic shims. With the current state of spherical harmonic shimming technology, perfect magnetic field homogeneity across the human or animal brain cannot be expected, as the encountered inhomogeneity cannot be approximated with low-order spherical harmonic shims. Chapter 10 will discuss alternative strategies to achieve improved whole-brain magnetic field homogeneity.

9.2.2 Spatial Localization

The quality of spatial localization is important for several reasons. Firstly, accurate spatial localization eliminates unwanted signals from outside the volume-of-interest, for example, signal from extracranial lipids, as well as signals from areas with an inhomogeneous magnetic field. Secondly, accurate spatial localization allows the determination of the exact voxel content (e.g. fraction of gray matter, white matter, and cerebrospinal fluid [CSF]) and allows a reproducible positioning of the voxel. Both considerations are crucial for longitudinal or population-based studies. Chapter 6 describes the considerations involved with spatial localization, including the chemical shift displacement artifact, effects of imperfect RF pulses, and insufficient magnetic field gradient dephasing.

9.2.3 Water Suppression

The quality of water suppression is often directly related to the magnetic field homogeneity, as well as the quality of spatial localization. Water suppression techniques are discussed in Chapter 6 and can typically achieve suppression factors of >1000. It should be noted that while many papers on water suppression strive for perfect water suppression, this is not required in reality. Water suppression should achieve a reduction in the water resonances to a degree that (i) water does not limit the dynamic range of the receiver, (ii) water does not significantly affect the spectral baseline, and (iii) vibration-induced sidebands of the water are negligible. More important than the absolute degree of suppression is the shape of the water resonance. A relatively large, but approximately Lorentzian-shaped water resonance is readily removed by post-acquisition methods (see Section 9.3.4). However, an incompletely dephased water resonance, spread out over a wide frequency range will lead to a significant baseline distortion, which is not readily removed by post-acquisition methods.

9.2.4 Sensitivity

The sensitivity of the FID or spectrum directly determines the accuracy by which parameters can be estimated from the data. Given a certain noise level, parameters like amplitude and frequency can only be estimated to a certain accuracy, independent of the quantification method employed. Therefore, when the minimum signal-to-noise ratio to estimate a certain parameter with a given accuracy cannot be achieved, only a change in data acquisition can improve the data. SNR can be improved through prolonged signal averaging, using larger volumes or improved RF receivers or achieve an improved magnetic field homogeneity.

While the *in vivo* NMR community currently does not have a consensus on guidelines for acceptable NMR spectra, a number of criteria for rejection of unacceptable spectra can readily be formulated. Unless specific processing methods are followed (e.g. post-acquisition lipid removal from MRSI data) MRS or MRSI spectra (from brain) should not be considered for further automated data processing when:

1) The residual water resonance is not well defined and dominates the baseline underlying the metabolite resonances.
2) The spectrum is characterized by large signals from extracranial lipids.
3) The spectrum contains unexplained signals, like spikes, ghosts, or other artifacts.
4) The NMR resonances are highly asymmetric, even following post-acquisition eddy current correction.
5) The metabolite resonances are wider than a certain ppm range. The exact number will strongly depend on the application. For example, for short-echo-time proton MRS of neurotransmitter metabolism, the line widths should be smaller than 0.03 ppm in order to reliably separate glutamate from glutamine at high magnetic fields. However, for long-echo-time proton MRS, the line width can be as wide as ~0.1 ppm before the total choline and total creatine resonances overlap.

While acceptable MR spectra also need to have a certain level of SNR, a general threshold on SNR is difficult as it depends on the metabolites of interest. The Cramer–Rao lower bounds (CRLBs, see Section 9.4.4) provide an objective measure for the SNR, spectral resolution, and overlap for each metabolite that can function as a measure for data rejection. MR spectra containing one or more of the mentioned features should undergo a rigorous manual inspection by an expert NMR spectroscopist to confirm data rejection.

9.3 Data Preprocessing

Following data acquisition, the next step in the metabolite quantification process concerns the estimation of resonance areas, as these are ultimately related to the metabolite concentration. However, before spectral quantification algorithms are applied to the data, as will be discussed in Section 9.4, a data preprocessing step is often desired in which several aspects of the acquired data can be improved. Data preprocessing can include combining data from phased-array coils, phasing and frequency alignment of multiple spectra, eddy current and line-shape corrections, removal of the residual water resonance, baseline correction, and the exclusion of additional unacceptable data.

9.3.1 Phased-array Coil Combination

Phased-array receivers are fully integrated into clinical MR, whereby 16–32 independent receive channels are standard. As phased-array receivers are often composed of surface coils, it can be expected that the sensitivity and phase of each receiver at a given (spectroscopic) voxel location is different. In order to avoid phase cancelation, the signal of each receiver should be phase-corrected before signal summation. However, a simple (non-weighted) summation does, in general, not provide the highest SNR. A receiver that is far from the (spectroscopic) voxel location contributes primarily noise and should be weighted less than a nearby receiver that contributes primarily signal. Roemer et al. [1] have described several methods in which signal from the various receivers can be weighted and combined. For proton MR spectroscopic applications, it is often convenient to estimate the required phase corrections and amplitude weights from non-water-suppressed MRS data. See Chapter 10 for more details on phased-array coil combination.

9.3.2 Phasing and Frequency Alignment

Depending on the application, phasing, and frequency alignment of NMR spectra can range from a cosmetic adjustment to a crucial procedure. During extensive data-acquisition periods, individual storing and phasing NMR spectra prior to summation can become important in the presence of motion. As discussed in Chapter 3, for diffusion measurements, macroscopic motion can lead to zero-order phase changes across the spectrum. While phase changes do not have to lead to signal loss *per se*, the addition of spectra with different phases during extensive signal averaging will lead to signal loss due to phase cancelation. Phase correction prior to summation can eliminate this signal loss.

Phase correction can also help with the convergence of spectral fitting algorithms. While most algorithms can accommodate zero- and first-order phase variations, the spectral fitting is typically faster and more reliable when phase and frequency variations can be (largely) eliminated in a preprocessing step. Phase correction can be implemented as a manual operation, whereby the user visually maximizes the integral over the real component of the spectrum. The large amount of data associated with MRSI or dynamic measurements necessitates the use of automated phasing and frequency alignment algorithms.

9.3.3 Line-shape Correction

Many spectral fitting algorithms model or approximate the measured MR spectrum as the sum of theoretical (Lorentzian, Gaussian, Voigt) line-shapes. Unfortunately, resonance lines obtained *in vivo* are typically not well-behaved due to residual eddy-currents, magnetic field inhomogeneity, and multi-exponential relaxation. While some algorithms can impose an arbitrary line-shape common to all metabolites [2], the line-shape distortions can also be removed or reduced in a preprocessing step. Residual eddy-currents lead to a time-varying magnetic field during signal acquisition, which distorts the NMR resonances as shown in Figure 9.2A. While preemphasis and B_0 compensation offer a hardware-based solution to reducing the time-varying magnetic fields (see Chapter 10), residual eddy-currents are typically unavoidable. A simple and convenient post-acquisition correction method [3–5] utilizes the fact that the time-varying magnetic fields are the same for all resonances. Therefore, when an unsuppressed on-resonance water signal is acquired, any temporal frequency variations in the water time-domain signal can be attributed to residual eddy-currents (Figure 9.2B). Applying the opposite phase modulation to the metabolite spectrum therefore automatically cancels phase modulations due to residual eddy-currents (Figure 9.2C). Since the water and metabolite spectra are acquired with the same pulse sequence, the eddy current correction automatically performs a zero-order phase correction, which provides a convenient tool for the automated phasing of large MRSI datasets.

Line-shape distortions due to magnetic field inhomogeneity can also be removed by dividing the metabolite FID by the water signal envelope [6], as technique referred to as Quantification Improvement by Converting Line shapes to the Lorentzian Type (QUALITY). Alternatively, information about local magnetic field homogeneity can be obtained from high-resolution magnetic field maps [7].

9.3.4 Removal of Residual Water

With the use of first- and second-order shims, the magnetic field homogeneity over a small single voxel in the human brain can, in general, be close to perfect. As a result, the large water resonance can typically be suppressed to well below the metabolite levels. However, in the

Figure 9.2 **Principle of post-acquisition correction of temporal magnetic field variations.** (A) ^{1}H NMR spectrum acquired from rat olfactory bulb at 9.4 T. The presence of residual magnetic field variations originating from eddy currents manifest themselves as frequency shifts and oscillations which lead to non-Lorentzian lines. (B) Phase evolution of the water signal obtained with the same sequence as used to acquire the signal in (A). The inset shows the water spectrum before (black) and after removal (gray) of the temporal phase. (C) Phase-corrected metabolite ^{1}H NMR spectrum.

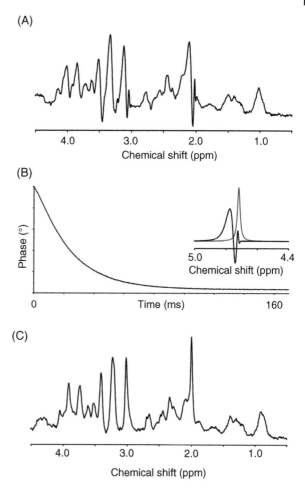

case of magnetic resonance spectroscopic imaging (MRSI) or for single-volume MRS in organs outside the brain, the magnetic field homogeneity is typically not perfect, leading to a significant residual water resonance (Figure 9.3A). A large unsuppressed water resonance often leads to an elevated and varying baseline underneath the metabolite resonances that can lead to increased uncertainty in the estimation of metabolite levels. In the case where the water resonance is relatively well-behaved, it can be reliably removed with post-acquisition methods. A residual water resonance that has many components with randomly varying phase and center frequencies in the metabolite chemical shift range (e.g. Figure 6.15G) cannot be readily removed post-acquisition. In this case the water suppression should be improved during the experiment. One of the more reliable techniques that readily allows automation relies on a singular value decomposition (SVD) of the FID signal [8–12]. The largest singular values correspond to the largest resonances in the MR spectrum. When, following a SVD, the singular values with frequencies corresponding to the water spectral region (indicated in Figure 9.3A) are selected, the corresponding reconstructed NMR spectra largely resembles the water resonance (Figure 9.3B). Subtraction of the reconstructed water from the original spectrum yields a water-suppressed metabolite spectrum as shown in Figure 9.3C. As SVD requires little to no user input, it is often referred to as a "black-box" method. And while

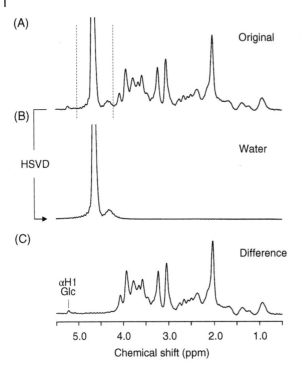

(A)

Original

(B)

HSVD

Water

(C)

Difference

αH1
Glc

5.0   4.0   3.0   2.0   1.0

Chemical shift (ppm)

Figure 9.3 Post-acquisition water removal through the use of a HSVD algorithm. (A) Short-echo-time ^1H NMR spectrum acquired from human brain at 4.0T, displaying a relatively small amount of residual water signal. (B) Theoretical water spectrum calculated from the singular values generated by a HSVD of spectrum (A) of which the frequencies fall within the indicated boundaries. (C) Difference between (A) and (B).

"black-box" methods do typically not allow the enforcement of prior knowledge information, they do lend themselves perfectly for automated signal removal [8, 9], especially when executed as the fast Hankel–Lanczos (HLSVD) variant [11, 13].

9.3.5 Baseline Correction

NMR spectra from tissues *in vivo* often display sharp resonances from metabolites superimposed on a "baseline" of broader resonances. While the baseline is often approximated and removed by polynomial or cubic-spline-based fitting, it is important to realize that most baselines have physical or physiological origins. Targeting the baseline origins, either through measurement or sequence design, often leads to a more robust data acquisition and quantification method. In unlocalized NMR spectroscopy (i.e. pulse-acquire methods), a significant part of the baseline can often be attributed to solids (e.g. plastics) in the RF coil assembly [14, 15]. In unlocalized ^{31}P NMR, the solids in the human and animal skull lead to a very broad and intense baseline, especially when acquired with surface coil reception. In these cases, the baseline can be significantly reduced by full 3D localization or delayed acquisition. In ^1H NMR spectra, the metabolites are superimposed on a baseline of macromolecular compounds, as discussed in Chapter 2. Delayed acquisition (e.g. TE > 80 ms) removes the macromolecules due to their shorter T_2 relaxation times (~30 ms), at the expense of loss of information of many scalar-coupled resonances. Macromolecular resonances can also be reduced by utilizing the difference in T_1 relaxation between metabolites and macromolecules [16]. However, the macromolecules hold valuable information in their own right, having demonstrated altered spectral patterns in stroke [17] and tumors [18]. Instead of suppressing the macromolecules through T_1 differences, they can also be enhanced by the same mechanism (see Figure 2.26).

Figure 9.4 **Spectral integration and baseline correction.** (A) Without additional preprocessing, direct integration of an unlocalized ^{31}P NMR spectra from piglet brain will result in an overestimation of the metabolite resonances since the integral signal is dominated by a broad, underlying resonance. (B) Baseline correction based on cubic-spline interpolation results in an acceptable baseline, making integration of resonances feasible. Note that partially overlapping resonances cannot be accurately quantified by integration (e.g. PME and P_i or α-ATP and NAD$^+$/NADH).

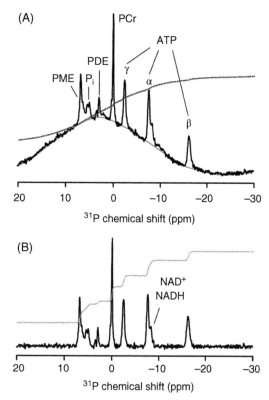

Once the macromolecular resonances are accounted for, the actual baseline in short-echo-time proton NMR spectra is insignificant.

However, in some cases, like unlocalized ^{31}P NMR spectroscopy of the brain (Figure 9.4A) a large, undesired baseline is unavoidable. In these cases, the baseline can be removed in a preprocessing step by approximating and subtracting the baseline with a polynomial fit (Figure 9.4B).

9.4 Data Quantification

The third step in the estimation of metabolite levels from *in vivo* NMR spectra concerns the calculation of the relative resonance areas, as these are directly proportional to the number of spins and hence the concentration. The quantification of NMR resonances covers a wide range of methods including integration, non-iterative "black-box" methods [8, 19], and iterative, user-dependent model fitting algorithms imposing various amounts of prior knowledge [2, 19–21]. Before these methods are discussed in more detail, a brief summary of the spectral parameters in the frequency and time-domain will be given.

9.4.1 Time- and Frequency-domain Parameters

A decision encountered early on in the quantification of NMR resonances is whether the analysis should be performed in the time or frequency domain. However, with the Fourier transformation being a linear and reversible operation, it should be realized that the two domains are equivalent and that all parameters can be equally well estimated in either domain [22]. Nevertheless, experimental imperfections, proper visualization, and computational considerations may favor one domain over the other. For example, removal of distorted data points at the beginning of the time-domain signal does not significantly affect time-domain analysis as the removed data points can be recovered through data extrapolation. In the frequency domain, however, missing time-domain data points can lead to complicated modulations throughout the entire spectrum. Furthermore, time-domain calculations are computationally less demanding (i.e. multiplication) than equivalent calculations in the frequency domain (i.e. convolution). Regardless of the analysis domain, the visualization of the data and quantification results is always performed in the frequency domain.

Figure 9.5 and Table 9.1 summarize the relations between the time- and frequency parameters for Lorentzian and Gaussian line-shapes. The imaginary part of the Gaussian frequency

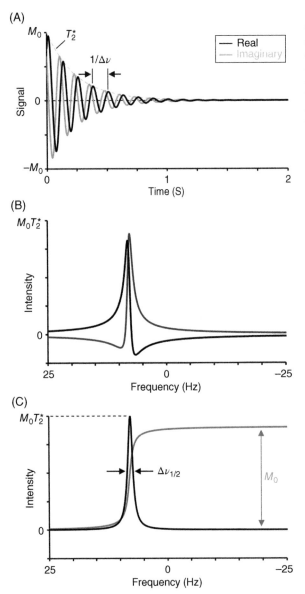

Figure 9.5 **Relations between time- and frequency-domain parameters.** (A) An exponentially damped sinusoidal signal is characterized by four parameters, namely the amplitude M_0, the frequency ν (or frequency difference $\Delta\nu$ relative to the RF frequency), the relaxation time constant T_2^*, and the phase ϕ. M_0 and ϕ can both be obtained from the first data point. The T_2^* relaxation time constant is equal to the time in which the signal decays to $1/e$ or 36.8% of the initial value, while the frequency difference is equal to the reciprocal of the signal maxima. (B) A nonzero phase leads to a mixture of absorptive and dispersive components in the NMR spectrum, thereby obscuring the other three parameters. (C) Following a phase correction, the resonance frequency equals the frequency where the real NMR resonance is maximum. The relaxation time constant T_2^* is proportional to the reciprocal of the line width (i.e. $T_2^* = 1/\pi\Delta\nu_{1/2}$) at half maximum and M_0 is equal to the resonance area, which can be obtained by integration. Note that the resonance height is equal to $M_0 T_2^*$. Table 9.1 provides the quantitative relationships between time- and frequency-domain parameters for Lorentzian and Gaussian lineshapes.

Table 9.1 Relationships between time and frequency-domain parameters for Lorentzian and Gaussian spectral line shapes.

Time domain (FID)	Frequency domain (spectrum)
$M_0 e^{i\Delta\omega t} e^{i\phi_0} e^{-t/T_2}$ [a]	$\dfrac{M_0 T_2}{1+\Delta\omega^2 T_2^2}\cos\phi + i\dfrac{M_0\Delta\omega T_2^2}{1+\Delta\omega^2 T_2^2}\sin\phi$
$M_0 e^{i\Delta\omega t} e^{i\phi_0} e^{-t^2/T_2^2}$ [b]	$M_0 T_2\sqrt{\dfrac{\pi}{4}}\,e^{-(\Delta\omega T_2/2)^2}\cos\phi$
Amplitude M_0 (first point)	Total integrated area of $A(\omega)$ [c]
Frequency ω $(=2\pi\nu)$	Frequency ω $(=2\pi\nu)$
Phase ϕ	Phase ϕ
T_2 relaxation time $(=1/R_2)$	Line width (i.e. FWHM) [d]

[a]Time-domain signal which gives rise to a Lorentzian spectral line shape.
[b]Time-domain signal which gives rise to a Gaussian spectral line shape. Only the real component in the frequency-domain is shown.
[c]Peak heights for Lorentzian and Gaussian lines equal $M_0 T_2$ and $M_0 T_2\sqrt{\pi/4}$.
[d]See Exercises for the exact relationship.

domain function is a rather complicated function, but can be calculated through a Hilbert transformation of the real component. It follows that all four relevant parameters have equivalent representations in both domains. The time-domain function of a Voigt line-shape is simply given by the multiplication of Lorentzian and Gaussian functions, which is equivalent to the frequency-domain convolution of Lorentzian and Gaussian line-shapes.

For relatively simple NMR spectra, containing only one or a few non-overlapping signals, the resonance areas can be obtained by numerical integration (e.g. see Figure 9.4B). The only required user interaction consists of indicating the frequency boundaries of the spectral region over which numerical integration will be performed. However, even for single resonances this procedure can lead to significant underestimation of the resonance area due to truncation of the long "tails" for Lorentzian lines. Furthermore, the procedure relies heavily on proper baseline correction, making it is unreliable for most *in vivo* NMR applications. For spectra with multiple overlapping resonances, integration does not provide a reliable estimate of the individual resonance areas and more sophisticated methods are required. The majority of spectral fitting algorithms can roughly be divided into non-iterative "black-box" or iterative methods. The primary difference is that iterative methods typically allow the incorporation of constraints and prior knowledge on the fitted parameters, whereas non-iterative methods do not. Figure 9.6 demonstrates the importance of imposing relevant prior knowledge. Following extensive ^{13}C labeling during an intravenous infusion of $[1\text{-}^{13}C]$-glucose, the $[2\text{-}^{13}C]$-glutamate and $[2\text{-}^{13}C]$-glutamine resonances around 55 ppm are split into a number of isotopomers, giving rise to the multiplet pattern shown in Figure 9.6A. Fitting the spectrum with a non-iterative, "black-box" method (SVD in this case) gives a mathematically adequate fit, as can be judged from the small residuals. However, the fit is biophysically useless as the fitted resonances vary widely in phase and width, i.e. rather than corresponding to a single compound, the fitted resonances hold information on multiple compounds. Iterative methods allow the inclusion of prior knowledge, which in the case of Figure 9.6 may include equal phase for all resonances, scalar couplings, and multiplet patterns. When the fit is performed with an iterative method imposing prior knowledge (VARPRO in this case), the fit is mathematically adequate (Figure 9.6B) as well as biophysically relevant. Each of the fitted signals in Figure 9.6B directly corresponds to an isotopomer resonance (Figure 9.6C), thereby allowing isotopomer analysis (see also Chapter 3).

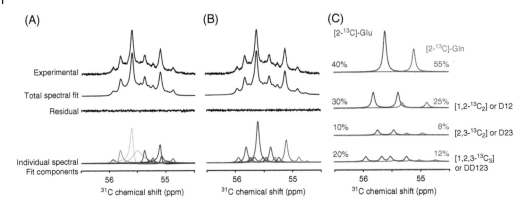

Figure 9.6 Black box fitting and prior knowledge. (A, B) Simulated ^{13}C NMR spectrum of a mixture of single and double ^{13}C-labeled glutamate and glutamine. Time-domain curve fitting was either performed using (A) SVD without prior knowledge or (B) VARPRO using extensive prior knowledge on chemical shifts and scalar couplings. No large differences can be observed from the residuals, indicating a good mathematical fit in both cases. However, when the individual resonances are inspected, it can be concluded that SVD produces a physically and biologically meaningless fit. (C) The extensive prior knowledge on ^{13}C–^{13}C splitting patterns as imposed by the VARPRO algorithm produces a biophysical meaningful fit that allows quantitative isotopomer analysis.

In summary, the inclusion of prior knowledge is the primary means to impose biophysically relevant information in a spectral fitting algorithm. Furthermore, as will be discussed in Section 9.4.4, the inclusion of prior knowledge is the only method of improving the accuracy of parameter estimation for a given noise level.

9.4.2 Prior Knowledge

The incorporation of biophysical prior knowledge in a spectral fitting algorithm is of crucial importance to improve the performance (Figure 9.6). However, there are a number of considerations involved with obtaining and incorporating prior knowledge. First and foremost it should be realized that the incorporation of incorrect prior knowledge will generally lead to systematic bias. Secondly, prior knowledge can be imposed as "hard" or "soft" constraints. Here, the decisions involved with prior knowledge will be demonstrated for adenosine triphosphate (ATP) acquired with ^{31}P NMR. Figure 9.7 shows the theoretical ATP ^{31}P NMR spectrum. The spectrum is dominated by scalar coupling between the three phosphate groups, leading to doublet resonances for γ- and α-ATP and a triplet resonance for β-ATP (or more accurately a doublet of doublets). In an unconstrained fit without prior knowledge a total of 28 parameters are involved, namely 7 amplitudes, 7 frequencies, 7 phases, and 7 line widths (or T_2^* relaxation time constants). To a first approximation the following prior knowledge can be obtained:

1) The overall, integrated intensities of the α-, β-, and γ-ATP resonances are identical, since they all originate from one phosphorus nucleus. In terms of model parameters this translates to

$$M_{0,1} + M_{0,2} = M_{0,3} + M_{0,4} = M_{0,5} + M_{0,6} + M_{0,7} = 2M_0 \tag{9.3}$$

2) The intensity ratios between the individual resonance lines within a multiplet are coupled to each other. In a doublet, the individual resonance lines should be of equal intensity, while

a triplet exhibits a 1 : 2 : 1 ratio between the resonance lines. This leads to the following prior knowledge in terms of model parameters:

$$M_{0,1} = M_{0,2} = M_{0,3} = M_{0,4} = M_{0,6} = M_0 \quad \text{and} \quad 2M_{0,5} = 2M_{0,7} = M_0 \qquad (9.4)$$

3) The transverse T_2 relaxation rates between the α-, β-, and γ-ATP resonances differ slightly. However, since the observed line widths *in vivo* are largely dominated by magnetic field inhomogeneity it is reasonable to assume identical line width or T_2^* relaxation time constants for all resonances, i.e.

$$R_{2,1}^* = R_{2,2}^* = R_{2,3}^* = R_{2,4}^* = R_{2,5}^* = R_{2,6}^* = R_{2,7}^* = R_2^* \qquad (9.5)$$

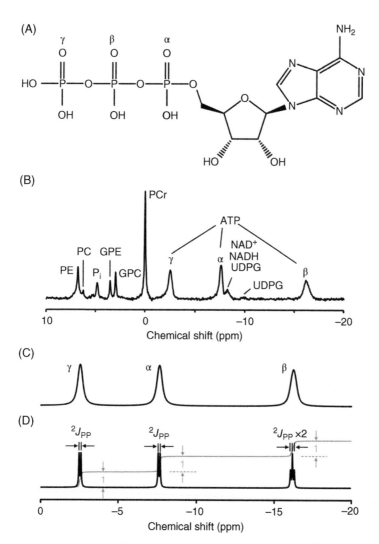

Figure 9.7 **Prior knowledge for ATP in ^{31}P NMR.** (A) Chemical structure of ATP. (B) ^{31}P NMR spectrum of human brain at 7 T prominently displays the three phosphate groups from ATP. (C, D) Theoretical, simulated ^{31}P NMR spectra of ATP under conditions of (C) low and (D) high magnetic field homogeneity and spectral resolution. The scalar couplings affecting the β-ATP resonance are the combination of the scalar couplings affecting the α- and γ-ATP resonances. Under ideal conditions (e.g. uniform excitation) the integrals of all three resonances are identical.

4) The spin–spin or scalar coupling is constant irrespective of the magnetic field strength or magnetic field homogeneity. Therefore, the frequencies of the resonance lines within individual multiplets are linked according to

$$|v_1 - v_2| = |v_3 - v_4| = |v_5 - v_6| = |v_6 - v_7| = \Delta v \tag{9.6}$$

where $\Delta v = {}^2J_{PP}$.

5) When the experiment is adequately performed it can be argued that the phases of all resonances should be equal, leading to

$$\phi_1 = \phi_2 = \phi_3 = \phi_4 = \phi_5 = \phi_6 = \phi_7 = \phi \tag{9.7}$$

Therefore, with the incorporation of prior knowledge the number of free-fitting parameters is reduced from 28 to 6 (one amplitude, one relaxation rate, three frequencies, and one phase).

However, it must be stressed that each stated point of prior knowledge is accompanied by one or more assumptions, which must be rigorously validated for a given application. For the prior knowledge involved with ATP these assumptions include:

1) The $1:1:1$ ratio for the ATP multiplet resonances only holds when the FID is acquired with $TR \gg T_1$ and TE = 0. If this is not the case, differences in T_1 and T_2 relaxation constants may influence the ratio. Furthermore, in a spin-echo experiment the observed ratio will be influenced by scalar coupling evolution, which differs between the three multiplets. In addition, the frequency profile of the excitation and refocusing pulses may distort the signal amplitudes in a frequency-dependent manner. With the RF transmit frequency centered on phosphocreatine at 0 ppm, the β-ATP resonance is typically only partially excited, leading to a reduced signal intensity. When the sequence is acquired in the presence of proton saturation, the ratio may be different as the nuclear Overhauser enhancement is not the same for all three resonances *per se*.

2) Second-order quantum mechanical effects (i.e. "strong coupling") will disturb the resonance ratio within a multiplet. This effect can be significant on low-field commercial MR systems, but can be accounted for through density matrix simulations.

3) The assumption of equal transverse relaxation rates does not hold when magnetic field inhomogeneity is not the dominating factor determining the observed line width. This is because the T_2 relaxation for β-ATP differs from that of α- and γ-ATP.

4) As indicated earlier, the β-ATP resonance is not a true triplet, but is more accurately described by a doublet of doublets. This is because $J_{\alpha\beta} \neq J_{\beta\gamma}$. Furthermore, $J_{\alpha\beta}$ and $J_{\beta\gamma}$ may vary during an experiment due to physiological changes (pH, temperature, magnesium levels). It is therefore typically better to treat $J_{\alpha\beta}$ and $J_{\beta\gamma}$ as independent model parameters. Of course, the β-ATP multiplet is still described by a linear combination of $J_{\alpha\beta}$ and $J_{\beta\gamma}$.

The constraints given by Eqs. (9.3)–(9.7) can be referred to as "hard" or "absolute" constraints. While hard constraints can make the largest improvement to a given fitting algorithm, the assumptions underlying the constraint are rarely true under all conditions. Often a given parameter is approximately known, e.g. $J_{\alpha\beta} \sim 16$ Hz and a reasonable range can be estimated, e.g. $15 \leq J_{\alpha\beta} \leq 17$ Hz. Therefore, instead of using hard constraints ($J_{\alpha\beta} = 16$ Hz), most fitting algorithms work with so-called "soft" constraints in which a parameter is allowed to vary over a restricted range (e.g. $15 \leq J_{\alpha\beta} \leq 17$ Hz). Soft constraints can still greatly improve the performance of a fitting algorithm, while at the same time allowing for variations in parameters due to, for example, physiology.

9.4.3   Spectral Fitting Algorithms

Since the early *in vivo* MRS experiments over 30 years ago, the processing of *in vivo* NMR spectra has advanced from simple integration and line fitting to advanced spectral fitting algorithms incorporating prior knowledge. Popular fitting algorithms include the jMRUI package [23, 24], which includes the VARPRO [20] and AMARES [21] programs, LCmodel [2, 25], and others [26]. Rather than discussing all available fitting algorithms, the principles are discussed using methods that are based on modeling linear combinations of metabolite spectra [2, 25].

NMR signal reception is a linear operation in that the induced signal is directly proportional to the number of spins in the sample. In that light, a given NMR spectra can be seen as a linear combination of the NMR signals from the pure, isolated compounds. This forms the basis for so-called linear combination (LC) modeling algorithms [25] as popularized by Provencher through the LCmodel program [2]. With LC modeling algorithms the measured NMR spectrum (typically ^{1}H, but extensions have been made to ^{13}C [27]) is approximated as a linear combination of a "basis set" of metabolite NMR spectra. There are in general two methods of obtaining a basis set, namely by measurement or through simulation. In the measurement approach, high-concentration solutions (e.g. 100 mM) of each individual metabolite in water are measured with the exact same sequence as used for the *in vivo* NMR measurement. To approximate *in vivo* conditions, the measurements should be made at physiological temperature (310 K) and pH. Furthermore, the signal-to-noise ratio of the resulting metabolite spectra should be high. The main advantage of the method is that all aspects of the pulse sequence, like eddy currents and spatial profiles, are identical between the *in vivo* and *in vitro* measurements. In addition, the details of the pulse sequence do not need to be known, which is especially important on less accessible, clinical vendor platforms. Obvious disadvantages of the measurement approach are that (i) the measurement of 15–20 metabolite solutions is time-consuming and potentially expensive and (ii) needs to be repeated for each pulse sequence. Furthermore, obtaining a uniform, raised temperature across a large phantom is nontrivial.

As an alternative to *in vitro* measurement strategy, the metabolite basis set can be simulated with the aid of density matrix calculations. As discussed in detail in Chapter 8, density matrices provide the most general approach for the theoretical description of NMR. A limited number of density matrix platforms are available [28] that are primarily geared towards non-localized NMR methods executed with ideal RF pulses. Figure 9.8A shows a non-localized, double-spin-echo sequence reminiscent of a localized PRESS sequence (Figure 9.8D). For short echo-times the simulated MR spectra obtained with an ideal, non-localized sequence (Figure 9.8B) are comparable to those simulated with the 3D localized PRESS sequence (Figure 9.8E). However, for longer echo-times the non-localized (Figure 9.8C) and 3D localized (Figure 9.8F) MR spectra strongly deviate. When the non-localized basis set (Figure 9.8C) would be used to fit experimental MR spectra obtained with a 3D localized PRESS sequence, large errors in the estimated metabolite levels and an overall poor spectral fit can be expected. The differences between the basis sets shown in Figure 9.8C and F can be explained by chemical shift displacements in 3D localized MRS and to a lesser degree by nonideal slice selection profiles. Since the ideal pulse sequence does not account for chemical shift displacement, the lactate methyl resonance appears in anti-phase at an echo-time of 80 ms (Figure 9.8C). However, since the chemical shift displacement for lactate for the PRESS sequence used in this example is circa 70% per spatial dimension, lactate scalar evolution is largely inhibited (see also Figure 8.10) leading to a mostly in-phase lactate resonance (Figure 9.8F).

Besides including all essential details of the pulse sequence, such as delays, RF pulse shapes, and magnetic field gradients, the success of the density matrix simulations to generate an appropriate basis set is largely dependent on knowledge of the appropriate chemical shifts and

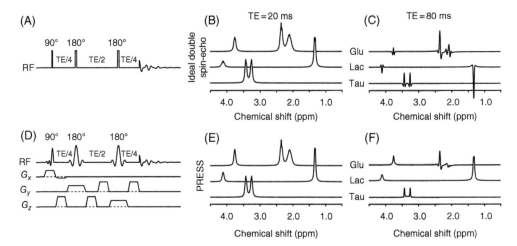

Figure 9.8 Spectral simulation of basis sets for localized MR spectroscopy. (A) Ideal, non-localized double-spin-echo sequence and (B, C) the corresponding, simulated 7T ^{1}H NMR spectra for glutamate (Glu), lactate (Lac), and taurine (Tau) with an echo-time TE of (B) 20 ms and (C) 80 ms, respectively. (D) 3D localized PRESS sequence and (E, F) the corresponding, simulated ^{1}H NMR spectra. At short TE the non-localized and localized spectra are almost identical. However, at longer TE the localized ^{1}H NMR spectra deviate substantially from the ideal, non-localized ^{1}H NMR spectra due to a combination of chemical shift displacement, nonideal slice profiles, and polarization transfer.

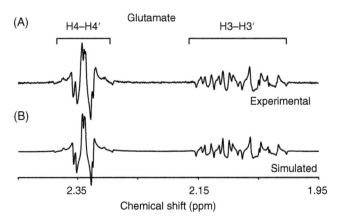

Figure 9.9 Experimental and simulated ^{1}H NMR spectra of 3D localized glutamate. (A) Experimentally measured and (B) simulated ^{1}H NMR spectra of glutamate at 11.74T acquired with a 3D LASER sequence (TE = 45 ms). The close agreement between experiment and simulation can only be achieved when all details of the sequence, like shaped RF pulses and slice selection gradients, are included in the simulation.

scalar coupling constants for each compound. Several publications [29, 30] as well as Table 2.1 provide good estimates for the majority of compounds detected *in vivo*. When faced with a new compound, prior knowledge can be obtained from high-resolution NMR spectra obtained from the pure compound, acquired at the appropriate pH, temperature, and ionic composition. When simulation of complete 3D pulse sequences is combined with reliable prior knowledge on chemical shifts and scalar couplings, the resulting basis set can be highly accurate as demonstrated in Figure 9.9 for glutamate. The calculation of a basis set for 15–20 metabolites using a complete 3D pulse sequence can take several hours. Fortunately, the simulations can be performed on an off-line computer without using valuable MR system time. Furthermore, the simulations are always noiseless, artifact-free (e.g. absence of water), and are readily modified

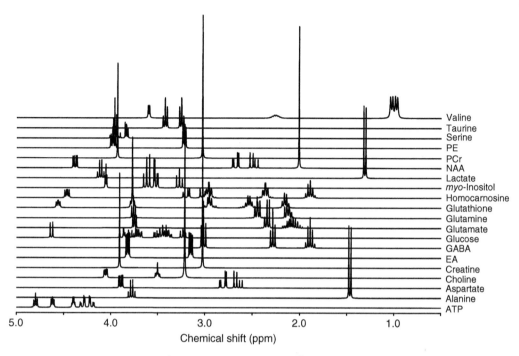

Figure 9.10 **Spectral basis set for ¹H MRS of brain.** Simulated ¹H NMR spectra of 20 metabolites that form a typical basis set for spectral quantification of *in vivo* short TE ¹H NMR spectra of animal or human brain. The spectra are simulated using the density matrix formalism for a 3D LASER sequence (TE = 10 ms) at 7.0 T. The basis set is often extended with an experimentally determined macromolecular baseline.

to accommodate different pulse sequences. Obvious disadvantages can be traced back to the use of incomplete or incorrect prior knowledge. However, the rather tedious and time-consuming preparation and measurement of *in vitro* samples (which can also contain incorrect prior knowledge) has fueled the switch to the flexible density matrix simulation approach.

Figure 9.10 shows a typical metabolite basis set for spectral fitting of short TE ¹H NMR spectra of human or animal brain. Depending on the application, the basis set may have to be extended with additional metabolites, like beta-hydroxybutyrate (BHB) which can be observed during periods of fasting. Most important is that the basis set is complete, i.e. all metabolites that are present in the *in vivo* NMR spectrum should be included in the basis set. When a metabolite is omitted from the basis set it will typically lead to a systematic bias in the estimated parameters and, in particular, the amplitudes of the other metabolites [31]. This is because metabolite amplitudes are typically correlated depending on the amount of spectral overlap. Section 9.4.4 discusses the construction of parameter correlation matrices to evaluate the mutual dependence of metabolites. The inclusion of basis set metabolites that are not present in the *in vivo* MR spectrum does typically not lead to error or bias, because the amplitude of the nonexistent parameter can simply be set to zero. However, the inclusion of nonexistent metabolites does lead to longer calculation times and a slower convergence.

LC modeling algorithms essentially adjust the amplitudes, frequencies, line widths, and phases of the metabolite basis set to match the *in vivo* MR spectrum as close as possible. In most cases, the algorithm imposes soft constraints on the fitting parameters in order to achieve faster convergence. Short-echo-time ¹H NMR spectra are typically characterized by a significant macromolecular baseline in addition to the linear combination of metabolite signals (Figure 9.11A). In order to achieve a meaningful spectral fitting result, the macromolecular resonances have to be taken into account. In the original LCmodel paper by

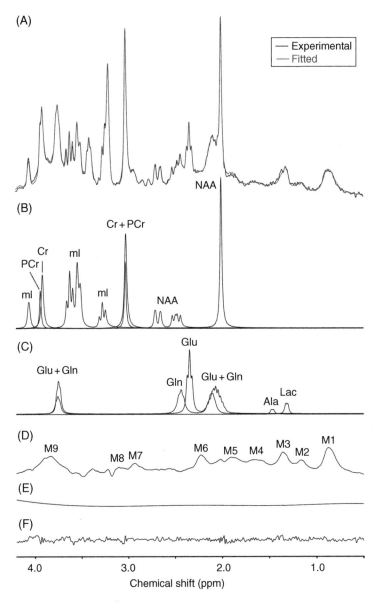

Figure 9.11 **Spectral quantification of ^{1}H NMR spectra with a LC modeling algorithm.** (A) Experimental ^{1}H NMR spectrum acquired from rat brain *in vivo* at 9.4 T (100 μl, STEAM localization, TR/TE/TM = 4000/8/25 ms) overlaid with the fitted NMR spectrum calculated by the LC modeling algorithm. (B, C) NMR spectra of individual components as extracted from the LC modeling fit shown in (A) for (B) *myo*-inositol (ml), phosphocreatine (PCr), creatine (Cr), and N-acetyl aspartate (NAA) and (C) glutamate (Glu), glutamine (Gln), alanine (Ala), and lactate (Lac). (D) Experimentally measured macromolecule baseline (double inversion recovery with TR/TI1/TI2 = 6000/1950/550 ms) as included in the basis set of the LC modeling algorithm. (E) Additional cubic-spline spectral baseline and (F) residual difference between the calculated and measured ^{1}H NMR spectra.

Provencher [2], the macromolecular resonances were approximated with a cubic-spline function. However, this introduces a large number of additional fitting parameters that are not based on biophysical parameters. A more robust method is to measure the macromolecular resonances by utilizing differences in T_1 relaxation (Figure 9.11D). When the macromolecular resonances are taken into account it can be seen that the residual background or baseline signal is very small and can be readily modeled with a low-order function (e.g. second-order polynomial). The difference between the fitted and measured NMR spectra (Figure 9.11A) can be a good indication for any problems during the spectral fit. Exclusion of metabolites typically leads to a coherent difference at well-defined frequencies, whereas the dispersive component throughout the difference can indicate frequency differences between the fitted and measured spectra. As was demonstrated in Figure 9.6, a small residual is no guarantee for an accurate and meaningful spectral fit. Only the opposite is true in that a large residual indicates a poor spectral fit. In order to assess the accuracy of the spectral fit it is important to find a measure for the error on the estimated parameters, as will be discussed next.

9.4.4   Error Estimation

Error estimation is an essential part of metabolite quantification, since in the absence of error estimates one cannot have confidence in estimated parameters. For a given noise level the lowest possible estimator-independent errors are given by the so-called Cramer-Rao lower bounds (CRLBs) [32–34]. For a proper evaluation of CRLBs the experimental data must be modeled with an exactly known model function. For real data, the exact model functions are not known by definition, and as such the calculated CRLBs are approximate. This is especially problematic for data with a large nonparametric component, such as macromolecular resonances in short-TE ^{1}H NMR spectra. The CRLBs are independent of the spectral fitting algorithm so that they apply equally well to the time and frequency domains. Figure 9.12 shows several aspects of error estimation on time and frequency domain signals. The four time domain signals shown in Figure 9.12A have an equal signal S to noise N ratio of 15. The decreasing T_2^* relaxation time constants manifest themselves as increasing frequency width at half maximum (FWHM) in the frequency domain (Figure 9.12B). The increasing FWHMs in turn lead to decreasing signal height S to noise N ratios, even though the signal integral to noise ratio remains constant. As a result, the error estimates on the signal integral increase, scaling proportional to $(\text{FWHM})^{\frac{1}{2}}$. This shows that a

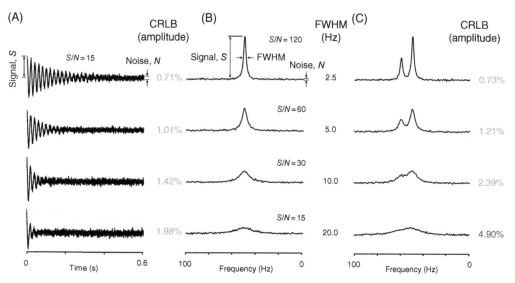

Figure 9.12 Time- and frequency-domain signals, spectral overlap, and Cramer–Rao lower bounds (CRLBs). (A) Time-domain signals of identical signal-to-noise ratio (*S/N*), but decreasing magnetic field homogeneity (from top to bottom). (B) The corresponding frequency-domain signals show decreasing *S/N* due to the decreasing peak height. For a single resonance, the CRLB is proportional to the time-domain *S/N* and FWHM$^{\frac{1}{2}}$. (C) When the signal in (B) is partially overlapping with a second (smaller) signal, the CRLB becomes also dependent on a spectral overlap term which causes a dramatic increase in CRLB in the presence of significant spectral overlap.

high magnetic field homogeneity that provides longer T_2^* relaxation times and narrower FWHMs is synonymous with lower errors on the signal intensity. In the presence of multiple resonances, the errors on the signal amplitudes become dependent on the amount of spectral overlap (Figure 9.12C). With minimal spectral overlap the error estimate for the amplitude of the lower-frequency resonance is essentially identical to that obtained in the absence of other signals (Figure 9.12B, top two spectra). With increasing FWHMs (Figure 9.12B, bottom two spectra) the amount of spectral overlap increases leading to a dramatic increase in the error on the amplitude. In the case of Figure 9.12C, the error increases proportional to $f(\nu, T_2^*) \times (\text{FWHM})^{1/2}$, where $f(\nu, T_2^*)$ is a function describing the spectral appearance of resonances, including chemical shifts, scalar couplings, and line widths. The CRLBs provide an objective error estimate for a given noise level and spectral overlap. Evaluation of the CRLB requires inversion of the Fisher information matrix $\boldsymbol{F}$ whose size is equal to the number of real-valued parameters to be estimated. The Fisher matrix can be calculated as

$$F = \frac{1}{\sigma^2}\left(\boldsymbol{P}^T \boldsymbol{D}^H \boldsymbol{D} \boldsymbol{P}\right) \tag{9.8}$$

where σ is the standard deviation of the measurement noise and T and H denote transposition and Hermitian conjugation, respectively. The matrix $\boldsymbol{D}$ hold the derivatives of the model function x_i with respect to the parameters p_j, whereas the matrix $\boldsymbol{P}$ is known as the prior knowledge matrix as it holds the derivatives of one parameters p_m with respect to another parameter p_n, i.e.

$$D_{ij} = \left(\frac{\partial x_i}{\partial p_j}\right) \quad \text{and} \quad P_{mn} = \left(\frac{\partial p_m}{\partial p_n}\right) \tag{9.9}$$

For a given model function (e.g. exponentially damped sinusoids), Eqs. (9.9) are known and the Fisher matrix of Eq. (9.8) is readily calculated. The CRLBs are obtained from the covariance matrix, which in turn is obtained from inverting the Fisher matrix according to $\text{CRLB}_{Pi} = (F_{ii}^{-1})^{1/2}$. As a rule of thumb it is generally accepted that metabolite concentrations with $\text{CRLB} < 10\%$ are measured with sufficient precision. Metabolites determined with $\text{CRLB} < 20\text{–}30\%$ should be considered with caution, whereas those obtained with $\text{CRLB} > 20\text{–}30\%$ are not reliable. However, relative CRLBs should be used with caution [35]. Suppose that glutamate and GABA are quantified from a ^{1}H MR spectrum with CRLB error estimations of 4 and 40%, respectively. One might be quick to conclude that the GABA estimate is unreliable and reject the GABA data. However, when the absolute concentrations of glutamate and GABA are quantified as 10 and 1 mM, respectively, the absolute CRLB is 0.4 mM in both cases. In other words, the uncertainty for glutamate and GABA is identical and is simply caused by an insufficient SNR for the low-concentration GABA. In this case there is no fundamental problem with GABA detection and the CRLB can be reduced by improving the SNR through increased signal averaging and/or increasing the volume.

Metabolites with very similar chemical structures and thus similar NMR spectra, like creatine and phosphocreatine or glutamate and glutamine, typically have high CRLBs for the individual compounds. The CRLB for the combined pool (e.g. tCr = PCr + Cr) is often dramatically lower as the sum has a unique spectral pattern. For this reason, it is often advisable to report combined metabolic pools like total choline, total creatine, total NAA (NAA + NAAG), and total lactate (lactate + threonine), especially at lower magnetic fields where the separation between individual metabolite NMR spectra can be very small.

The correlation between parameters in a NMR spectrum can be quantitatively described by the so-called parameter correlation matrix that describes the correlation ρ_{mn} between parameters p_m and p_n according to

$$\rho_{mn} = \frac{F_{mn}^{-1}}{\sqrt{F_{mm}^{-1} F_{nn}^{-1}}} \tag{9.10}$$

Figure 9.13A gives a typical correlation matrix for the amplitude of metabolites detected by short TE ^{1}H MRS on human brain under conditions of optimal magnetic field homogeneity. Most resonances display a low correlation due to the absence of significant spectral overlap. A number of metabolites like creatine and phosphocreatine display a strong correlation. This indicates that the estimated amplitudes for creatine and phosphocreatine can vary wildly between datasets, whereby a lower creatine level is compensated by a higher phosphocreatine level and vice versa (hence the negative correlation). In other words, a strong correlation indicates that the two metabolites cannot be reliably quantified individually. The sum (e.g. creatine + phosphocreatine) is often more reliable, with greatly reduced correlation. Figure 9.13B shows the correlation matrix in the presence of suboptimal magnetic field

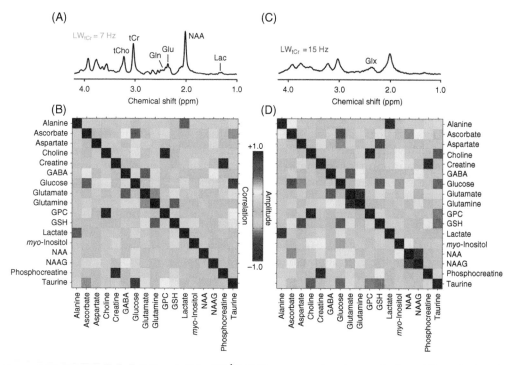

Figure 9.13 **Amplitude correlation matrix.** (A) ^{1}H NMR spectrum from human brain at 4T acquired under conditions of high magnetic field homogeneity (line width (LW) of total creatine (tCr) equals 7 Hz). (B) The amplitude correlation matrix shows small correlations between most metabolites with a few noticeable exceptions. Creatine (Cr) and phosphocreatine (PCr) are strongly correlated, such that each individual compound cannot be reliably quantified. However, their sum (Cr + PCr = tCr) is highly reliable. An identical conclusion can be reached for a number of cerebral metabolites Cho, GPC, and their sum tCho. The situation for other pairs is more complex. NAA and NAAG as well as Glu and Gln have only a mild correlation under conditions of high magnetic field homogeneity. (C, D) When the magnetic field homogeneity decreases to LW$_{tCr}$ = 15 Hz, both NAA/NAAG and Glu/Gln become strongly correlated. The reliable detection and quantification of Glu and Gln therefore becomes dependent on the magnetic field homogeneity, whereas the separate detection of, for example, tCr and tCho is less sensitive to the experimental conditions.

homogeneity. Overall, the correlations have increased due to increasing spectral overlap, again underlining the importance of attaining high magnetic field homogeneity.

It should be realized that the macromolecular baseline can normally not be described by an analytical model function, and as such cannot be included in the determination of the Fisher matrix. However, via introduction of so-called nuisance parameters, Ratiney et al. [34, 36] were able to show that the macromolecular baseline has strong correlations with the majority of metabolites. When the baseline is completely unknown, it will lead to bias in the estimated parameters. Inclusion of a predetermined macromolecular baseline can significantly reduce the metabolite CRLBs.

9.5 Data Calibration

The fourth and final step in metabolite quantification is the conversion of relative resonance areas to absolute concentrations expressed in $mmol\,l^{-1}$ ("molar" concentration) or $\mu mol\,g^{-1}$ of tissue ("molal" concentration). The molar and molal concentration can be converted if the brain water density is known [37]. There are two general approaches to calibrate or reference the relative numbers to absolute concentrations [37–45]. The first group utilizes a so-called external concentration reference, in which a compound of known concentration is positioned outside the object under investigation but within the sensitive volume of the coil. The other group uses an internal concentration reference, which can be a stable metabolite naturally occurring in the tissue (endogenous) or an appropriate compound, which can be introduced into the organ being studied (exogenous). Exogenous concentration references are not frequently used for safety reasons and the lack of appropriate (non-toxic, stable, selective for one organ) compounds. Figure 9.14 schematically shows the use of concentration references for the quantification of brain metabolites, but the methods are equally applicable for a wide range of organs. Figure 9.14A shows the strategy of internal concentration referencing in which a spectrum (localized or non-localized) is acquired from the region of interest. When the resonance of the internal concentration reference compound is present in this spectrum (e.g. NAA or total creatine for ^1H MRS or ATP for ^{31}P MRS), the concentration of the other metabolites can be directly calculated by comparison of resonance areas (after correction for factors like relaxation, as will be discussed next). If this is not the case, a second spectrum containing the resonance of the calibration compound (e.g. unsuppressed water) from the same VOI has to be acquired. Figure 9.14B and C shows two possible calibration strategies utilizing an external concentration reference. In Figure 9.14B the external concentration reference compound is placed in the coil together with the object under investigation. Acquisition of two spectra as shown in Figure 9.14B makes quantification possible. The strategy of Figure 9.14C differs from B in that the two spectra are acquired from the exact same spatial position. The specific advantages, disadvantages, and correction factors of the different calibration strategies will be discussed in the next sections. First, the general factors affecting the acquired signals during a calibration procedure will be described.

The resonance area of a metabolite resonance is, in principle, proportional to the concentration, which makes the application of a reference compound with known concentration a convenient method of quantification. However, any difference between the metabolite and the reference compound needs to be taken into account in order to obtain reliable concentration values. Here, the factors which could differ between the two compounds will be summarized.

Figure 9.14 **Calibration strategies for the quantification of cerebral metabolite concentrations.** (A) Internal concentration reference, (B) external concentration reference, and (C) phantom-replacement concentration reference. The left figures indicate the acquisition of the metabolite spectra, while the right figures indicate the acquisition of the reference compound spectra. In (C, right column) a small saline-filled bottle is inserted in (or retracted from) the coil, in order to equalize the *in vivo* and *in vitro* coil loads. Alternatively, correction of coil loading can be obtained through the principle of reciprocity, as explained in the text.

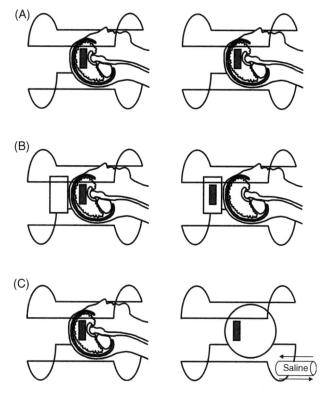

9.5.1 Partial Saturation

When the repetition time TR of a pulse sequence is shorter than 4–5 times the longitudinal relaxation time T_1, the magnetization cannot completely recover before the following excitation, leading to reduction of the steady-state longitudinal magnetization, which is given by

$$M_z(TR) = M_0 \frac{1 - e^{-TR/T_1}}{1 - \cos\theta \cdot e^{-TR/T_1}} \tag{9.11}$$

where θ is the nutation angle. With different T_1 relaxation times between metabolites and reference compound, the acquired signal intensity of each resonance must be corrected for partial saturation. This can simply be achieved by using Eq. (9.11) as a correction factor, but it requires knowledge of the T_1 relaxation time. Note that Eq. (9.11) only holds for a simple pulse-acquisition experiment. For more complicated experiments involving spin-echo delays, Eq. (9.11) needs to be modified to account for the additional RF pulses, especially for short T_1 relaxation times and long echo-times TE. With surface coils correcting for partial saturation is further complicated since the nutation angle θ, and consequently the saturation factor, becomes dependent on the position relative to the coil. This problem can be alleviated by executing the entire pulse sequence with adiabatic RF pulses. The correction for partial saturation can be omitted completely if the experiments are performed with $TR > 5T_{1max}$ (T_{1max} being the longest T_1 relaxation time present), such that $M_z \sim 1$ for all resonances. Even though this increases the experimental duration, the use of long repetition times is strongly recommended for single-volume spectroscopy, since it eliminates systematic errors caused by application of an empirically determined T_1 saturation factor. For MRSI applications the use of a long repetition time is unrealistic, such that a partial T_1 saturation correction is required.

9.5.2 Nuclear Overhauser Effects

In homo- or heteronuclear double resonance spectroscopy experiments, the signal intensities depend on the nuclear Overhauser enhancement as described in Chapter 3. By determining, for a given pulse sequence, the nuclear Overhauser enhancement factor for each resonance, this effect can be corrected. However, for the sake of simplicity (and accuracy) it is recommended to design pulse sequences in such a manner that nuclear Overhauser effects are completely eliminated.

9.5.3 Transverse Relaxation

Any experiment utilizing spin-echo delays induces signal losses due to T_2 relaxation. Many publications have minimized this effect by using short echo times (TE < 20 ms). However, for compounds such as ATP or macromolecules even the shortest TE can lead to significant signal reduction. On the other hand, many groups have used long echo times (TE > 100 ms) in order to reduce baseline oscillations, simplify the appearance of spectra, and improve water suppression. In all cases a proper signal intensity correction can only be made if the transverse relaxation time T_2 is known for each resonance.

9.5.4 Diffusion

Pulse sequences utilizing strong magnetic field crusher gradients to achieve accurate spatial localization or experiments with long echo times are almost always affected by diffusion. This effect can be pronounced when large molecules like ATP (with a low diffusion constant $D \sim 0.2 \times 10^{-3}$ mm^2 s^{-1}) are being calibrated against a low molecular weight compound like water ($D \sim 0.7 \times 10^{-3}$ mm^2 s^{-1}). In analogy to T_2 relaxation, this effect can be minimized by using shorter echo times (and minimal magnetic field crusher gradients), but in many cases the effect can only be corrected for when the apparent diffusion constant is quantitatively known.

It should be realized that these four parameters and especially relaxation and diffusion, may change over time due to development of pathology or changes in temperature. For instance, in stroke the apparent diffusion coefficient of water decreases almost immediately after the onset of ischemia. In the more chronic phase of the ischemic lesion, the T_2 relaxation time of water significantly increases.

9.5.5 Scalar Coupling

The signal amplitude of scalar-coupled resonances is a complicated function of echo-time, scalar coupling constant, chemical shift (for strongly-coupled spins), nutation angle, and other pulse sequence elements, such as the TM period in STEAM. Several approaches can be followed to eliminate or compensate this effect. The use of short echo times (TE < 20 ms) reduces the amplitude- and phase modulation for some coupled metabolites (e.g. see Figure 9.8). However, even a very short echo time can lead to significant modulation when the scalar coupling constant is large (e.g. inositol, glucose). Another approach which can be followed is to choose TE as a multiple of $1/J$, such that the metabolite of interest is completely refocused. Unfortunately, unless the J coupling constants are similar, this approach only works for one metabolite (or one resonance of a metabolite). The amplitude- and phase distortions caused by J-modulation can also be determined experimentally in model solutions or calculated using the density matrix formalism. When the scalar coupling modulation is part of a spectral basis set, the effects are automatically accounted for during the spectral fitting algorithm [2, 21, 26].

9.5.6 Localization

Since the absolute concentration is directly proportional to the localized volume, it is clear that the spatial localization needs to be accurate and identical for metabolites and reference compound. Especially for heteronuclear internal calibration (e.g. calibration of ^{31}P metabolites with the water signal) that is a challenging task since the two nuclei often use different localization methods (e.g. STEAM for ^{1}H and ISIS for ^{31}P).

Even if the localized volumes were of identical shape, differences in resonance frequency between metabolites will lead to a chemical shift displacement artifact. When measurements are made in heterogeneous tissues, like cerebral gray and white matter, the voxel composition between the reference compound and the metabolite of interest may differ, leading to quantification errors.

9.5.7 Frequency-dependent Amplitude- and Phase Distortions

This effect is most pronounced when using binomial pulses for water suppression in ^{1}H MRS. The spectrum is amplitude modulated according to a sinusoidal profile, whereby the higher-order binomial pulses also exhibit nonlinear phase distortions (Figure 6.21). By performing simulations based on the Bloch equations these effects can be completely compensated. Other areas where amplitude distortions can play a role are in ^{31}P and ^{13}C MRS, where the effective chemical shift range is much larger than for ^{1}H MRS. When the RF amplitude is small with respect to this chemical shift dispersion, the nutation angle becomes frequency dependent. This can result in substantial errors when, for instance, the β-ATP resonance (which is normally on the edge of the *in vivo* ^{31}P chemical shift range) is used for quantification. The use of smaller nutation angles will reduce these effects.

The application of adiabatic half passage pulses is very popular in (non-localized) ^{31}P MRS (and to a lesser degree ^{13}C MRS). But also for adiabatic pulses, the RF amplitude should be large enough to excite the entire chemical shift range uniformly. When the (off-resonance) adiabatic condition is not satisfied, substantial errors will arise if these effects are not taken into account. Also in this case can the Bloch equations can be used to simulate the effects of RF pulses and achieve complete compensation.

9.5.8 NMR Visibility

The line width of resonances are inversely proportional to the T_2^* relaxation time, which is related to the rotational mobility of the metabolite. Metabolites with low mobility (e.g. bound to macromolecular structures) give rise to very short T_2 relaxation times and hence broad resonances, which can be unobservable in conventional NMR spectra. This will in turn lead to an underestimation of the concentration. Furthermore, if a water suppression technique like presaturation is used, the narrow resonance line arising from the mobile component of the metabolite under investigation may decrease due to magnetization transfer effects (see Chapter 3) leading to a further underestimation of the true concentration.

Considering the above-described factors, a general applicable formula can be constructed for the concentration of a metabolite.

$$S_{\mathrm{m}} = S_{\mathrm{mm}} C_{T_1,\mathrm{m}} C_{T_2,\mathrm{m}} C_{\mathrm{nOe,m}} C_{\mathrm{ADC,m}} C_{\mathrm{J,m}} C_{\mathrm{loc,m}} C_{\mathrm{RF,m}} \tag{9.12}$$

$$S_{\mathrm{r}} = S_{\mathrm{rm}} C_{T_1,\mathrm{r}} C_{T_2,\mathrm{r}} C_{\mathrm{nOe,r}} C_{\mathrm{ADC,r}} C_{\mathrm{J,r}} C_{\mathrm{loc,r}} C_{\mathrm{RF,r}} \tag{9.13}$$

$$[m] = \left(\frac{S_m}{S_r}\right)[r]C_nC_{av}C_{gain} \qquad (9.14)$$

where

S_{mm} = measured metabolite signal.

S_m = corrected metabolite signal.

S_{rm} = measured reference signal.

S_r = corrected reference signal.

C_{T_1} = correction factor for partial saturation due to incomplete signal recovery due to finite T_1 relaxation (Eq. (9.11)).

C_{T_2} = correction factor for T_2 relaxation (Eq. (1.5)).

C_{nOe} = correction factor for nuclear Overhauser effects (Eq. (3.7)).

C_{ADC} = correction factor for diffusion (Eqs. (3.25)–(3.31)).

C_J = correction factor for amplitude- and phase modulations due to scalar-coupling evolution. This factor can include the effects of frequency-selective RF pulses on scalar-coupled spins [46].

C_{loc} = correction factor for deviations from the ideal localization profile.

C_{RF} = correction factor for amplitude- and phase distortions due to specific RF pulse combinations like binomial RF pulses.

$[m]$ = concentration of the metabolite under investigation.

$[r]$ = concentration of the reference compound.

C_n = correction for the number of equivalent nuclei for each resonance.

C_{av} = correction for the number of averages.

C_{gain} = correction for receiver gain setting.

The factor of partial NMR invisibility is difficult to correct for and can only be established by comparing the *in vivo* MRS data with invasive *in vitro* results like chemical analysis of biopsies or high-resolution NMR of tissue extracts. When some factors do not affect the measured signal, the corresponding correction factor equals one. Each individual calibration technique as shown in Figure 9.14 requires, in addition to the general applicable factors in Eqs. (9.12)–(9.14), specific correction factors which will be described next.

9.5.9 Internal Concentration Reference

The strategy of using an internal concentration reference is straightforward. The (corrected) resonance areas in the acquired spectrum are compared with that of a stable endogenous reference compound. For ^{1}H MRS, water, total creatine, and NAA have been proposed as endogenous concentration references. However, it should always be kept in mind that the concentration of endogenous concentration references can change during the development of pathologies. NAA shows a substantial decrease in a wide range of neurodegenerative pathologies ranging from ischemia to Alzheimer's disease. Total creatine and water are relatively stable metabolites, although changes in their rotational mobility may change their NMR visibility. Furthermore, the concentration is often a function of spatial position. Especially important is the discrimination between water from tissue and CSF, which may differ by 30–40% in water content. This discrimination can be achieved on the basis of a double-exponential decay of the water [41] (Figure 9.15B). Alternatively, the compartments could be retrieved from high-resolution MR images with appropriate contrast (e.g. Figure 9.15A). Despite the potential difficulties, water has been used by several MR groups with some degree of success, judging from the favorable comparison with concentrations obtained with other techniques. For ^{31}P MRS, ATP and water have been used as endogenous concentration markers. ATP is only a

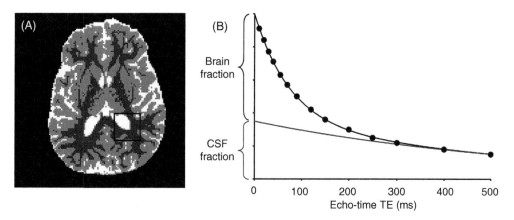

Figure 9.15 **Correction of cerebrospinal fluid (CSF) water partial volume.** CSF water can be detected separately from brain water based on differences in T_1, T_2 relaxation, diffusion, or magnetization transfer. (A) Using quantitative T_1 mapping, the CSF contribution to a spectroscopic volume (black box) can be calculated directly from tissue segmented MR images (see also Figure 3.15). (B) The CSF contribution to a spectroscopic volume can also be obtained from a multi-exponential analysis of (spectroscopic) T_2 relaxation data whereby brain and CSF water correspond to the short and long T_2 components, respectively.

suitable reference compound in those applications where the system under investigation is only mildly challenged. With severe pathologies like ischemia, anaerobic glycolysis rapidly consumes the available PCr pool (through the creatine kinase equilibrium) after which the ATP resonances start to decline.

In applications where the peak ratios are used, the total phosphate pool, defined as

$$[P_{tot}] = [PME] + [P_i] + [PDE] + [PCr] + 3[NTP] + 2[NAD + NADH] \qquad (9.15)$$

can be used. The total phosphate pool should in principle be constant (assuming that none of the metabolites becomes NMR invisible and none of the metabolites are transported out of the cells), such that changes in peak ratio (e.g. $[PCr]/[P_{tot}]$) can be attributed to a single metabolite (i.e. $[PCr]$).

Internal water referencing is typically used for ^{1}H MRS, although it can also be used for non-proton nuclei. For internal water referencing, Eq. (9.14) should be modified and extended to

$$[m] = \left(\frac{S_m}{S_{water}}\right)[water] C_n C_{av} C_{wc} C_{HX} \qquad (9.16)$$

where

S_{water} = corrected water signal.

[water] = water concentration (55.14 mmol l^{-1} at 310 K).

C_{wc} = correction for the water content in the VOI. C_{wc} equals ~0.82 for gray matter, ~0.73 for white matter, >0.98 for CSF, and ~0.78 for skeletal muscle.

C_{HX} = correction factor for the relative sensitivities between the proton and X channels (X = ^{31}P or ^{13}C).

The mentioned factors C_{wc} for different tissue types are from biochemical measurements. It should be realized that this factor will be reduced if NMR-invisible pools are present. The factor C_{HP} (i.e. X = ^{31}P) can be assessed by performing a phantom experiment with a phosphorus metabolite of known concentration in water. C_{HP} is then calculated as the ratio of the ^{1}H signal per mM of protons over the ^{31}P signal per mM of phosphorus. Since C_{HP} depends on a number of factors, including coil load, it is advisable to mimic the *in vivo* conditions as close as possible. For the homonuclear calibration strategy C_{HH} equals one.

9.5.10 External Concentration Reference

One of the methods that utilizes an external concentration reference is shown in Figure 9.14B. After the collection of the *in vivo* spectrum, a reference spectrum from a calibration sample is obtained. To minimize the effects of B_1 magnetic field inhomogeneity, the two voxels are chosen symmetrically about the center of the coil. Alternatively, the B_1 magnetic field inhomogeneity can be accounted for by measuring the B_1 distribution of the particular RF coil used. The metabolite concentration can be calculated according to

$$[m] = \left(\frac{S_m}{S_r}\right)[r]C_nC_{av} \tag{9.17}$$

where
S_r = corrected reference signal.
$[r]$ = concentration of external reference.

Equation (9.17) immediately reveals the main distinction with the method of internal water calibration, in that no assumption needs to be made for the internal water concentration (which may vary with pathology, age, and voxel composition). Furthermore, for heteronuclear experiments no calibration factor for the relative ^{1}H and ^{31}P (or ^{13}C) sensitivities is required.

The method of external concentration referencing is relatively simple since the experimental setup or the position of the patient need not be changed. However, the performance of external concentration referencing is dependent on the B_1 magnetic field distribution. Especially at higher magnetic fields (>3 T) the B_1 magnetic fields between subject and phantom are likely different. Even though the transmit B_1^+ magnetic field can be mapped with B_1 mapping methods (Chapter 4), variations in the receive B_1^- magnetic fields are difficult to account for. Therefore, external concentration referencing is not recommended for ^{1}H MRS above 3 T.

9.5.11 Phantom Replacement Concentration Reference

Another method of quantification is aimed at simulating (human) tissue as close as possible with a (spherical) phantom of known composition. Because almost all systematic errors like B_1 magnetic field inhomogeneity (at low magnetic field) and localization affect the tissue and phantom in identical ways, this calibration method is in principle very robust. The method is complicated by differences in coil loading between the tissue and the phantom. Two methods are available to compensate for differences in coil loading, i.e. load adjustment and load correction through the principle of reciprocity. For load adjustment, the electrical conductivity of the solution in the phantom is slightly lower than that of human tissue ($\sigma \sim 0.64\,S\,m^{-1}$) such that it allows for fine adjustment of the coil load with a second (smaller) phantom containing, for example, saline. During the procedure of load adjustment, the matching capacitance of the RF coil is left unchanged at the end of the *in vivo* experiment. After removal of the patient and accurate positioning of the phantom, the matching is optimized by slowly inserting the saline bottle (i.e. increasing the coil load). When the *in vitro* matching equals the previous *in vivo* matching, the *in vivo* and *in vitro* coil loads are identical. The concentration can then simply be calculated with Eq. (9.17).

The method of load correction involves the addition of an external capillary which is measured non-selectively (non-localized) during the *in vivo* and during the *in vitro* experiments. Most conveniently a compound is used which falls outside the spectral region of interest (e.g. tetramethylsilane [TMS] for ^{1}H or ^{13}C MRS and phenylphosphonicacid [PPA] or hexamethylphosphorustriamide [HMPT] for ^{31}P MRS). The correction factor for difference in coil load is then calculated from the capillary signals obtained from the *in vivo* and *in vitro* experiments according to

$$C_{\text{load}} = \frac{S_{\text{in vivo}}}{S_{\text{in vitro}}} \tag{9.18}$$

after which the concentration can be calculated as

$$[m] = \left(\frac{S_{\text{m}}}{S_{\text{ref}}} \right) [\text{ref}] C_n C_{\text{av}} C_{\text{load}} \tag{9.19}$$

Alternatively, a load correction term can be obtained by determining the power/voltage to obtain a 90° nutation angle *in vivo* and *in vitro*. Through the principle of reciprocity the difference in B_1 magnetic field strength is then directly proportional to the difference in acquired signal strength. Similar to the external concentration referencing method, the simulated phantom replacement method is complicated by B_1 magnetic field variations between tissue and phantom, such that it is not recommended for ^{1}H MRS for magnetic fields above 3 T.

While there is no single calibration strategy that is optimal under all experimental conditions, the method of internal concentration referencing using the water signal appears to be the most convenient method with a reasonable accuracy. Whereas the water concentration can be assumed under many conditions, including mild pathologies, there are methods based on proton density MRI and T_1 and T_2 relaxation-based segmentation that can be used to fine-tune the water concentration in more severe pathologies [47, 48].

Exercises

9.1 A proton NMR spectrum is acquired from a 12 ml volume positioned in the human occipital lobe. A surface coil is used for RF pulse transmission as well as signal reception. The spectrum is acquired with the double spin-echo localization method PRESS with TR = 2000 ms, TE = 100 ms, and number of averages = 64. The integrals of NAA (CH$_3$), creatine (CH$_3$), and choline (N(CH$_3$)$_3$) are 100, 70, and 60, respectively. To obtain the absolute concentrations, water will be used as an internal concentration reference. The water spectrum is acquired with a single average from the same volume with the same sequence and has an integral of 5000.

The relaxation parameters are given by:

NAA: $T_1 = 2000$ ms, $T_2 = 150$ ms
Creatine: $T_1 = 2500$ ms, $T_2 = 200$ ms
Choline: $T_1 = 2250$ ms, $T_2 = 150$ ms
Water: $T_1 = 1650$ ms, $T_2 = 75$ ms

From quantitative T_1 images it can be deduced that the voxel contains 40% CSF, 50% gray matter, and 10% white matter. The water content of CSF, GM, and WM is 100, 87, and 83% of pure water, respectively. The water content of pure water is 55.6 mol l^{-1}.

    A Calculate the metabolite integrals corrected for T_1 and T_2 relaxation losses.
    B Calculate the water concentration in the selected volume.
    C Calculate the average metabolite concentrations inside the volume.
    D Calculate the average metabolite concentrations inside the brain.

9.2   Consider a CHESS sequence with six 10 ms Gaussian excitation pulses ($R = 2.7$).
    A Assuming a Gaussian-shaped excitation profile for individual RF pulses, calculate the signal loss for αH1-glucose at 4.7 and 9.4 T.
    B Under the same assumptions, calculate the relative signal intensities for αH1-glucose and the two (non-decoupled) ^{13}C–^{1}H satellites of [1-^{13}C]-α-H1-glucose for a 50% fractional enrichment at 4.7 and 9.4 T.
    C Calculate the experimental fractional enrichment (^{13}C/(^{12}C + ^{13}C)) when the RF-induced distortions are not taken into account.

9.3   Consider the pulse-acquire ^{1}H NMR spectrum of aspartic acid.
    A Determine the number of free parameters when the spectrum is fitted unconstrained with a sum of single Lorentzian lines.
    B Determine the number of free parameters when the spectrum is fitted with full prior knowledge of the NMR characteristics and chemical structure of aspartic acid. Discuss the underlying assumptions.
    C Repeat the calculation under (B) when the ^{13}C fraction of aspartate in rat brain *in vivo* is fitted after 40 min of intravenous infusion of [1-^{13}C]-glucose.

9.4   The presence of IMCL and EMCL lipids in rat skeletal muscle necessitates the use of spectral editing to unambiguously detect lactic acid. Furthermore, it is known that lactate displays bi-exponential T_2 relaxation with 70 and 30% of the lactate having T_2s of 50 and 150 ms, respectively. The lactate T_1 equals 1500 ms, whereas water T_1 and T_2 relaxation time constants are 1200 and 30 ms, respectively. RF magnetic field inhomogeneity reduces the lactate editing efficiency from 100 to 82%.
    A During a spectral editing experiment (TR = 2500 ms, TE = 144 ms), lactate is observed with an intensity of 45 in 128 scans. When water is observed with a relative intensity of 620 in 16 scans, calculate the absolute lactate concentration.
    B Determine the echo time that gives the optimal detection of lactate and calculate the signal gain with respect to TE = 144 ms. Assume single-exponential T_2 relaxation with T_2 = 100 ms.

9.5   During a ^{1}H MRSI study of lactate detection in cerebral tumors, the spectroscopist suspects that the tumor water content is different than for regular gray matter. Describe an imaging experiment by which the tumor water content can be estimated.

9.6   Name at least three methods for the measurement of the T_2 of glutamate and three methods for the T_2 of lactate at 11.74 T. Explain why increasing the echo-time of a PRESS sequence is likely to give an incorrect estimate of T_2.

9.7   **A** In the presence of a ±10° transmitter phase instability, calculate the maximum integrated signal loss for a FID averaged over 128 transients.
    B In the presence of a gradual (linear) +8 Hz frequency drift over 1 hour, calculate the maximum line width at half maximum for a FID averaged over 1024 transients (assume T_2^* = 100 ms and TR = 2500 ms).
    C Describe a method to minimize the signal loss and line broadening calculated under A and B.

9.8 Consider a jump-return sequence echo-sequence (TR – JR90 – TE/2 – JR180 – TE/2) where the intrapulse pulse delays are adjusted to give maximal excitation and refocusing 500 Hz off-resonance.

A Using the NAA, tCho, and tCr relaxation parameters given under Exercise 9.1, calculate the steady-state metabolite signals when TR = 1500 ms and TE = 50 ms. Ignore T_1 relaxation during the echo-time and assume a magnetic field of 4.7 T.

B Due to residual B_0 eddy currents, the first pulse of the JR sequence experiences a +20° additional phase rotation relative to the second pulse. Recalculate the metabolite signals for this situation.

C Continuing with condition B, the unsuppressed water signal is acquired by placing both JR pulses +500 Hz off-resonance. The water intensity equals 4000. The acquired NAA, tCho, and tCr intensities are 100, 50, and 75. Calculate the absolute metabolite concentration under the assumption of a 40 M water concentration.

9.9 A MRS study on GABA detection with spectral editing in the human brain finds a CRLB for GABA of 40% following 10 min of signal averaging.

A How much should be signal averaging be extended to reduce the CRLB for GABA to 20%?

B Doubling the volume size is accompanied by a doubling of the spectral line width. Predict the GABA CRLB for 10 min of signal averaging.

References

1 Roemer, P.B., Edelstein, W.A., Hayes, C.E. et al. (1990). The NMR phased array. *Magn. Reson. Med.* 16: 192–225.

2 Provencher, S.W. (1993). Estimation of metabolite concentrations from localized *in vivo* proton NMR spectra. *Magn. Reson. Med.* 30: 672–679.

3 Ordidge, R.J. and Cresshull, I.D. (1986). The correction of transient B_0 field shifts following the application of pulsed gradients by phase correction in the time domain. *J. Magn. Reson.* 69: 151–155.

4 Jehenson, P. and Syrota, A. (1989). Correction of distortions due to the pulsed magnetic field gradient-induced shift in B_0 field by postprocessing. *Magn. Reson. Med.* 12: 253–256.

5 Klose, U. (1990). *In vivo* proton spectroscopy in presence of eddy currents. *Magn. Reson. Med.* 14: 26–30.

6 de Graaf, A.A., van Dijk, J.E., and Bovee, W.M. (1990). QUALITY: quantification improvement by converting lineshapes to the Lorentzian type. *Magn. Reson. Med.* 13: 343–357.

7 Webb, P., Spielman, D., and Macovski, A. (1992). Inhomogeneity correction for *in vivo* spectroscopy by high-resolution water referencing. *Magn. Reson. Med.* 23: 1–11.

8 Pijnappel, W.W.F., van den Boogaart, A., de Beer, R., and van Ormondt, D. (1992). SVD-based quantification of magnetic resonance signals. *J. Magn. Reson.* 97: 122–134.

9 van den Boogaart, A., van Ormondt, D., Pijnappel, W.W.F. et al. (1994). *Mathematics and Signal Processing III*. (ed. J.G. McWhirter). Oxford: Clarendon Press.

10 Vanhamme, L., Fierro, R.D., Van Huffel, S., and de Beer, R. (1998). Fast removal of residual water in proton spectra. *J. Magn. Reson.* 132: 197–203.

11 Cabanes, E., Confort-Gouny, S., Le Fur, Y. et al. (2001). Optimization of residual water signal removal by HLSVD on simulated short echo time proton MR spectra of the human brain. *J. Magn. Reson.* 150: 116–125.

12 de Beer, R., van Ormondt, D., Pijnappel, W.W.F., and van der Veen, J.W.C. (1988). Quantitative analysis of magnetic resonance signals in the time domain. *Isr. J. Chem.* 28: 249–261.

13 de Beer, R. and van Ormondt, D. (1992). Analysis of NMR data using time domain fitting procedures. In: *NMR Basic Principles and Progress*, vol. 26 (ed. P. Diehl, E. Fluck, H. Gunther, et al.), 201–248. Berlin: Springer-Verlag.

14 Babcock, E.E., Vaughan, J.T., Lesan, B., and Nunnally, R.L. (1990). Multinuclear NMR investigations of probe construction materials at 4.7 T. *Magn. Reson. Med.* 13: 498–503.

15 Marjanska, M., Waks, M., Snyder, C.J., and Vaughan, J.T. (2008). Multinuclear NMR investigation of probe construction materials at 9.4T. *Magn. Reson. Med.* 59: 936–938.

16 Behar, K.L., Rothman, D.L., Spencer, D.D., and Petroff, O.A. (1994). Analysis of macromolecule resonances in ^{1}H NMR spectra of human brain. *Magn. Reson. Med.* 32: 294–302.

17 Hwang, J.H., Graham, G.D., Behar, K.L. et al. (1996). Short echo time proton magnetic resonance spectroscopic imaging of macromolecule and metabolite signal intensities in the human brain. *Magn. Reson. Med.* 35: 633–639.

18 Opstad, K.S., Murphy, M.M., Wilkins, P.R. et al. (2004). Differentiation of metastases from high-grade gliomas using short echo time ^{1}H spectroscopy. *J. Magn. Reson. Imaging* 20: 187–192.

19 van den Boogaart, A., Howe, F.A., Rodrigues, L.M. et al. (1995). *In vivo* ^{31}P MRS: absolute concentrations, signal-to-noise and prior knowledge. *NMR Biomed.* 8: 87–93.

20 van der Veen, J.W., de Beer, R., Luyten, P.R., and van Ormondt, D. (1988). Accurate quantification of *in vivo* ^{31}P NMR signals using the variable projection method and prior knowledge. *Magn. Reson. Med.* 6: 92–98.

21 Vanhamme, L., van den Boogaart, A., and Van Huffel, S. (1997). Improved method for accurate and efficient quantification of MRS data with use of prior knowledge. *J. Magn. Reson.* 129: 35–43.

22 Abildgaard, F., Gesmar, H., and Led, J.J. (1988). Quantitative analysis of complicated nonideal Fourier transform NMR spectra. *J. Magn. Reson.* 79: 78–89.

23 Naressi, A., Couturier, C., Castang, I. et al. (2001). Java-based graphical user interface for MRUI, a software package for quantitation of *in vivo*/medical magnetic resonance spectroscopy signals. *Comput. Biol. Med.* 31: 269–286.

24 Naressi, A., Couturier, C., Devos, J.M. et al. (2001). Java-based graphical user interface for the MRUI quantitation package. *MAGMA* 12: 141–152.

25 de Graaf, A.A. and Bovee, W.M. (1990). Improved quantification of *in vivo* ^{1}H NMR spectra by optimization of signal acquisition and processing and by incorporation of prior knowledge into the spectral fitting. *Magn. Reson. Med.* 15: 305–319.

26. Slotboom, J., Boesch, C., and Kreis, R. (1998). Versatile frequency domain fitting using time domain models and prior knowledge. *Magn. Reson. Med.* 39: 899–911.

27 Henry, P.G., Oz, G., Provencher, S., and Gruetter, R. (2003). Toward dynamic isotopomer analysis in the rat brain *in vivo*: automatic quantitation of ^{13}C NMR spectra using LCModel. *NMR Biomed.* 16: 400–412.

28 Smith, S.A., Levante, T.O., Meier, B.H., and Ernst, R.R. (1994). Computer simulations in magnetic resonance. An object oriented programming approach. *J. Magn. Reson. A* 106: 75–105.

29 Govindaraju, V., Young, K., and Maudsley, A.A. (2000). Proton NMR chemical shifts and coupling constants for brain metabolites. *NMR Biomed.* 13: 129–153.

30 Govind, V. (2016). ^{1}H-NMR chemical shifts and coupling constants for brain metabolites. *eMagRes* 5: 1347–1362.

31 Hofmann, L., Slotboom, J., Jung, B. et al. (2002). Quantitative ^{1}H-magnetic resonance spectroscopy of human brain: influence of composition and parameterization of the basis set in linear combination model-fitting. *Magn. Reson. Med.* 48: 440–453.

32 Cavassila, S., Deval, S., Huegen, C. et al. (2000). Cramer-Rao bound expressions for parametric estimation of overlapping peaks: influence of prior knowledge. *J. Magn. Reson.* 143: 311–320.

33 Cavassila, S., Deval, S., Huegen, C. et al. (2001). Cramer-Rao bounds: an evaluation tool for quantitation. *NMR Biomed.* 14: 278–283.

34 Ratiney, H., Coenradie, Y., Cavassila, S. et al. (2004). Time-domain quantitation of ^{1}H short echo-time signals: background accommodation. *MAGMA* 16: 284–296.

35 Kreis, R. (2016). The trouble with quality filtering based on relative Cramer-Rao lower bounds. *Magn. Reson. Med.* 75: 15–18.

36 Ratiney, H., Sdika, M., Coenradie, Y. et al. (2005). Time-domain semi-parametric estimation based on a metabolite basis set. *NMR Biomed.* 18: 1–13.

37 Kreis, R., Ernst, T., and Ross, B.D. (1993). Absolute quantitation of water and metabolites in the human brain: II. Metabolite concentrations. *J. Magn. Reson. B* 102: 9–19.

38 Tofts, P.S. and Wray, S. (1988). A critical assessment of methods of measuring metabolite concentrations by NMR spectroscopy. *NMR Biomed.* 1: 1–10.

39 Roth, K., Hubesch, B., Meyerhoff, D.J. et al. (1989). Noninvasive quantitation of phosphorus metabolites in human tissue by NMR spectroscopy. *J. Magn. Reson.* 81: 299–311.

40 Buchli, R. and Boesiger, P. (1993). Comparison of methods for the determination of absolute metabolite concentrations in human muscles by ^{31}P MRS. *Magn. Reson. Med.* 30: 552–558.

41 Kreis, R., Ernst, T., and Ross, B.D. (1993). Absolute quantification of water and metabolites in the human brain: I. Compartments and water. *J. Magn. Reson. B* 102: 1–8.

42 Buchli, R., Martin, E., and Boesiger, P. (1994). Comparison of calibration strategies for the in vivo determination of absolute metabolite concentrations in the human brain by ^{31}P MRS. *NMR Biomed.* 7: 225–230.

43 Danielsen, E.R., Michaelis, T., and Ross, B.D. (1995). Three methods of calibration in quantitative proton MR spectroscopy. *J. Magn. Reson. B* 106: 287–291.

44 Hajek, M. (1995). Quantitative NMR spectroscopy. Comments on methodology of *in vivo* MR spectroscopy in medicine. *Q. Magn. Reson. Biol. Med.* 2: 165–193.

45 Kreis, R. (1997). Quantitative localized ^{1}H MR spectroscopy for clinical use. *Prog. Nucl. Magn. Reson. Spectrosc.* 31: 155–195.

46 Slotboom, J., Mehlkopf, A.F., and Bovee, W.M.M.J. (1994). The effects of frequency-selective RF pulses of J-coupled spin-1/2 systems. *J. Magn. Reson. A* 108: 38–50.

47 Neeb, H., Zilles, K., and Shah, N.J. (2006). A new method for fast quantitative mapping of absolute water content *in vivo*. *Neuroimage* 31: 1156–1168.

48 Neeb, H., Ermer, V., Stocker, T., and Shah, N.J. (2008). Fast quantitative mapping of absolute water content with full brain coverage. *Neuroimage* 42: 1094–1109.

10

Hardware

10.1 Introduction

The objectives of this chapter are to describe the instrumentation involved in *in vivo* NMR. The complete NMR system can roughly be divided in four categories, namely (i) a magnet having a bore size that is large enough to accommodate entire living subjects, including humans. For convenience, the bore of animal and human systems is normally horizontal. Besides the superconducting coil to produce the main magnet field, a set of superconducting shim coils are supplied to adjust the homogeneity of the "raw" magnet (i.e. without the presence of a sample). (ii) A gradient coil system to create time-dependent magnetic field gradients for spatial encoding as used in MRI, localized spectroscopy, and many other NMR experiments. Room temperature shim coils to adjust the magnetic field homogeneity on a subject-specific basis are often an integral part of the gradient coil assembly. (iii) A radiofrequency (RF) transmitter/receiver (transceiver) system for generating the RF magnetic field B_1 and for detecting the NMR signal. (iv) A computer system for managing the magnet, shims, gradient, and transceiver components of the entire NMR system. Furthermore, a computer is necessary for processing and storing the raw NMR signal (FID or echo) and processing and displaying the final NMR signal (spectrum or image).

A complete overview of the hardware involved in an NMR experiment would represent an enormous amount of work and could easily fill a number of books. Furthermore, an extensive discussion of, for example, high-field superconducting magnet design would be inappropriate and outside the scope of a book on NMR techniques and principles. Therefore, the choice was made to focus on the general aspects of a complete MR system, like the magnet, magnet field gradients, and RF coils. Furthermore, some specific components of the NMR system, which require attention during experiments, like tuning and matching of RF coils, optimizing the magnetic field homogeneity, and the effects of time-varying magnetic fields (eddy currents), will be discussed. More detailed description of MR-related hardware can be found in Refs. [1–3].

10.2 Magnets

The main magnetic field is the most essential component of a complete NMR system. The magnetic field strength determines the intrinsic NMR sensitivity, while the magnet design (i.e. bore size, orientation) largely determines its applications [1, 2, 4–7].

There are essentially three types of magnet designs that are suitable for NMR, that is, the magnet should have a relatively intense (0.1–20 T), homogeneous, and stable field. The oldest

In Vivo *NMR Spectroscopy: Principles and Techniques*, Third Edition. Robin A. de Graaf.
© 2019 John Wiley & Sons Ltd. Published 2019 by John Wiley & Sons Ltd.

magnet design is a permanent magnet, which is constructed of ferromagnetic materials such as iron, nickel, cobalt, and alloys thereof. Once the polycrystalline, ferromagnetic material is aligned in a suitable external magnetic field, the material will be permanently magnetized. However, the generation of a suitable homogeneous magnetic field for human applications requires large amounts of ferromagnetic material, easily in excess of 10 tons. Although this problem can be somewhat reduced by using rare earth alloys, it nevertheless limits the magnetic field strength to ~0.2 T. An advantage of low-field permanent magnetic fields is that they can be constructed in a variety of configurations, some of them with an easy patient access and an open structure, which eliminates an occasional claustrophobic reaction induced by the high-field MR systems.

A resistive magnet is an electromagnet in which the magnetic field is generated by the passage of current through a wire that is a good electrical conductor (but with a finite electrical resistance). Typically, resistive whole-body magnets are constructed in a four-coil Helmholtz configuration with the outer coils smaller than the inner as to approximate a spherical geometry. At normal temperatures, the finite resistance of the coils for the passage of electrical current places high-power requirements on the system to produce sufficient electrical current for the generation of the desired magnetic field. The required power for a 0.15 T resistive magnet is circa 50 kW. The power to magnetize a resistive magnet is dissipated as heat in the coils, which must be removed by passing cooled water along the coils. Although power and cooling requirements are easily met for low-field strengths, they limit the magnetic field strength since the power increases with the square of the magnetic field strength. The stability (expressed in $ppm\,h^{-1}$) of resistive magnets is not nearly as good as superconducting magnets, which will be described next. Just as permanent magnets, resistive magnets can be designed in a number of open, patient-friendly configurations.

Almost all modern MRI systems are based on a superconducting magnet design. Superconductivity is a phenomenon occurring in certain materials at low temperatures, characterized by zero electrical resistance and the exclusion of internal magnetic fields (the Meissner effect). Figure 10.1A shows a graphical depiction of the resistance of a superconductor and a copper conductor as a function of temperature. At a material-specific critical

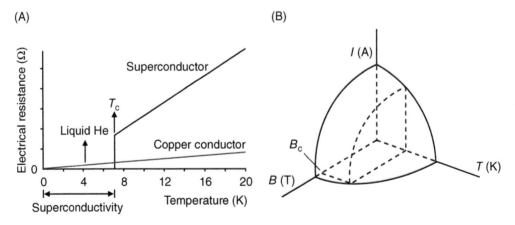

Figure 10.1 **Superconductivity.** (A) Electrical resistance as a function of absolute temperature for a typical superconductor (black line) and a copper conductor (gray line). While the copper conductor retains a finite electrical resistance at all temperatures, the resistance for a superconductor drops to zero below a critical temperature T_c. (B) The critical temperature for a given material decreases with increasing magnetic field strength B and increasing current density I, such that superconductivity can only be maintained behind the 3D surface.

temperature T_c, the finite resistance of the superconductor becomes zero. The resistance of a copper conductor also decreases with decreasing temperature, but never reaches zero, thereby preventing superconductivity. As long as the temperature of the superconductor remains below the critical temperature, current can continue to flow without heat dissipation. This in turn can maintain a constant magnetic field indefinitely. However, the critical temperature of a material is lowered when placed in a strong external magnetic field. Figure 10.1B shows a superconductor boundary surface for the variables temperature T, magnetic field B, and current I. Only behind the surface does the material maintain superconductive properties. This means that at a given temperature (e.g. 4.2 K, liquid helium at atmospheric pressure) and current, there is a maximum magnetic field strength that can be achieved before the material becomes resistive. The magnetic field can only be further increased by lowering the temperature or using a different material.

Common materials for superconducting wire are niobium-titanium (NbTi) and niobium-tin (Nb$_3$Sn) alloys. NbTi filaments are normally placed within a copper matrix (Figure 10.2A), which aids in electrically and mechanically stabilizing the superconductor. Friction or mechanical motion makes superconductors susceptible to local heating, which could potentially lead to a local loss of superconductivity. Local heat could bring the temperature of the superconducting wire above the critical temperature, giving it finite electrical resistance. Passing of current through resistive wire leads to Joule or Ohmic heating, which can expand the area of electrical resistance to adjacent wires. Once this process has been initiated, it may propagate throughout the entire magnet. The heat will also be transferred to the helium bath, leading to a rapid boil-off of the helium in a matter of minutes. The total and abrupt discharge of a magnet and the associated boil-off of the helium bath are referred to as a quench. In order to ensure that the boiled-off helium gas does not replace the air inside the magnet room, magnets are always equipped with a quench manifold, which leads the helium gas to a safe outdoor location. The superconducting properties of NbTi stretch to magnetic fields up to 10 T at liquid helium (4.2 K) temperatures. NbTi-based magnets can be stretched to 12 T by cooling the helium bath to 2.2 K through active pumping. For higher magnetic fields, the Nb$_3$Sn alloy offers more favorable properties. However, because Nb$_3$Sn is brittle rather than ductile, it must be formed at its final shape and position. In an unreacted state (Figure 10.2B) an Nb–Sn wire is flexible enough to be pulled into thin wires to be used for winding of the main magnet coils. The superconductive properties of the pure Nb in the unreactive state are poor. The superconductive Nb$_3$Sn alloy is formed by exposing the wire to temperatures in excess of 700 °C during which the tin in the bronze matrix diffuses to the niobium where it forms superconducting Nb$_3$Sn

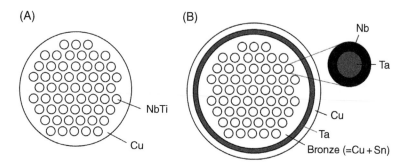

Figure 10.2 **Superconducting magnet wire.** Cross-sectional views of (A) niobium-titanium (NbTi) and (B) niobium-tin (Nb$_3$Sn) superconducting wires. The Nb$_3$Sn wire is shown in an unreacted state. During the high-temperature bronze diffusion process, the tin diffuses to the niobium and forms the superconducting Nb$_3$Sn crystals. The copper acts as a thermal and structural stabilizer. See text for more details.

crystals. This process is referred to as the bronze diffusion technique. The tantalum acts as a barrier to prevent tin from diffusing into the copper layer, which would greatly reduce its stabilizing effect.

With the rapidly increasing global helium prices, magnet manufacturers have started to use alternatives to helium-cooled superconductors. Magnets build from high-temperature superconducting (HTS) wire or tape provide a promising direction, in which magnets can be operated at liquid nitrogen temperatures (77 K) or higher. While current HTS wires have their own specific challenges that prevent persistent magnets build from HTS wire only, it seems likely that HTS-based magnets will play an increasingly important role in NMR. A completely different solution to the helium problem is the construction and use of cryogen-free magnets. By thermally attaching cryocoolers directly to the magnet coils, it is possible to cool the superconducting wire to below the critical temperature without using any helium or nitrogen cryogens [2].

Following transportation and installation, the magnet needs to be energized to its final magnetic field. Typically, the first step is to cool the magnet coils with liquid nitrogen, followed by liquid helium, a process that may take several days. Next, a power supply is connected to the magnet through the use of a superconducting switch. A small portion of the superconducting wire inside the switch is made resistive by an adjacent heater. Since the magnet has no resistance, all current from the power supply runs through the magnet coil. Once the desired current has been established, the heater is turned off and the wire of the switch becomes superconductive, thereby closing the current loop of the main magnet coil. After the driving current has slowly been reduced to zero, the power supply can be disconnected, leaving the magnet in a persistent mode.

The homogeneity of the static magnetic field is, besides the strength, the most important parameter characterizing an NMR magnet. It can be shown that a magnet designed as an infinitely long solenoid will produce a perfectly homogeneous magnetic field. Short truncated solenoidal coils with correction coils at either end, or optimized coil constellations or wire patterns are the practical implementation of this theoretical solution. Figure 10.3A and B shows the magnetic field generated by a commonly employed six-ring magnet. The magnetic field homogeneity is optimized over a diameter spherical volume (DSV), which is typically on the order of 40–50% of the magnet bore diameter. From Figure 10.3A and B it follows that a carefully designed magnet produced without manufacturing errors can generate a magnetic field with a homogeneity of better than 1 ppm over its DSV. However, minor manufacturing errors

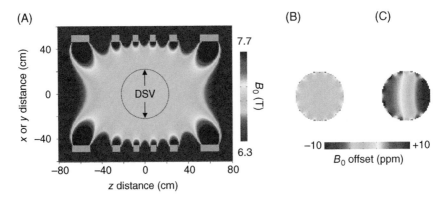

Figure 10.3 **Magnet homogeneity.** (A) Magnetic field contours of a 7 T magnetic field (Ø 100 cm) generated by a six-coil design. (B) Across the diameter spherical volume (DSV) of 45 cm the magnetic field homogeneity is better than 1 ppm. (C) When the current in the outermost right coil is 99.99% of its nominal value, the magnetic field homogeneity deteriorates to more than 10 ppm.

in the placement of miles of superconducting wire for a typical 7 T magnet can have a devastating effect on the magnetic field homogeneity. An error of 0.01% in the current density of one of the outer rings will lead to a magnetic field inhomogeneity of >10 ppm over the DSV (Figure 10.3C). Similarly, large (metallic) objects placed close to the magnet (structural beams, laboratory equipment) or even inside the magnet (gradient and shim coils, patient bed) can greatly perturb the magnetic field homogeneity. The magnetic field inhomogeneity of the magnet without a sample or subject is, in principle, constant and is typically optimized once during MR system installation using superconductive shim coils. As most MR users will not directly be involved with magnet shimming this aspect will only be mentioned in passing when relevant. Magnetic field inhomogeneity generated by the sample is something that all MR users are affected by and will be the primary focus of Section 10.3.

Other considerations involved with the magnet include temporal stability of the magnetic field, the spatial extend of the magnetic field outside the magnet, and the boil-off of the cryogenics. The temporal stability of the magnetic field, also referred to as "magnet drift," is an important parameter for longer experiments. Modern superconducting magnets typically have a drift of less than $-0.1\,ppm\,h^{-1}$. However, larger apparent magnet drifts can, for example, be observed when passive shims are heating up due to insufficient thermal isolation from the gradients. When an experiment requires signal averaging, significant broadening of spectral resonances can result, even when the magnet drift is as low as 0.1 ppm. It is therefore essential to either correct for magnet drifts by magnetic field locks [8], similar to those used in high-resolution NMR, or by post-acquisition frequency alignment as detailed in Chapter 8.

The spatial extend of the magnetic field outside the magnet, also known as the "fringe field," is important for magnet placement and study-related logistics, like placement of physiology monitoring equipment and computers. The fringe field is often specified as the distance where the magnetic field has dropped to 5 Gauss or 0.5 mT. A conventional, non-shielded 1.5 T magnet can have a 5 Gauss fringe field stretching out for 12 and 10 m in the axial and radial directions, respectively. This poses serious limitations for placement of the MRI system and placement of peripheral equipment. As a result, many modern magnets are of a self-shielded design in which additional shield coils on the outside of the main magnetic coils (see Figure 10.4) are constructed to minimize the magnet fringe field without sacrificing too much of the desired internal magnetic field. The 5 Gauss fringe field of a shielded 1.5 T magnet is reduced to 4 and 3 m in the axial and radial directions, respectively. Since magnet shielding invariably leads to a reduction in the desired magnet field, as a rule of thumb magnetic shielding is not available at the highest available magnetic fields. Other methods of reducing the fringe field include surrounding the magnet directly with steel or placing the magnet in a room with steel walls. Magnetic materials like steel have a smaller resistance to magnetic flux than air and as a result magnetic flux tends to focus inside the steel, thereby reducing the fringe field. This method remains effective as long as the steel does not become magnetically saturated. Figure 10.4 shows an actively-shielded superconducting magnet. The liquid helium bath holds the main magnet superconducting coils, as well as the superconducting shim coils. In addition, active shielding superconducting coils are visible that are specifically designed to minimize the fringe field.

A final consideration with superconducting magnets concerns the boil-off of the cryogenics and in particular the liquid helium. Due to a number of reasons, most importantly heat leaks of the helium Dewar to the surrounding environment, there will be a constant boil-off of helium. Since the magnet must remain in a superconductive state this necessitates regular filling of the magnet with cryogenics. To relieve the constant need for expensive cryogens (i.e. helium), modern magnets are equipped with cold heads, which extract excess heat from the magnet through the Joule–Thomson effect. While cold heads can reduce the helium boil-off several

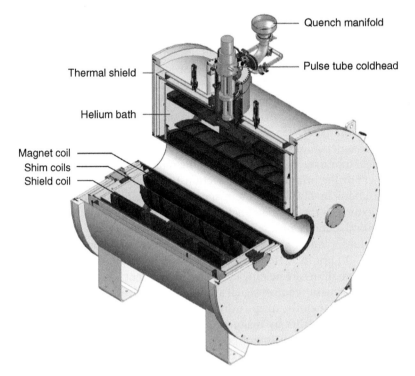

Figure 10.4 **3D cross-sectional view of a preclinical 9.4 T magnet.** Besides the primary magnetic and secondary shield coils, the helium bath also contains the superconducting shim coils. Thermal insulation is provided by vacuum and a thermal shield. The pulse tube cold head virtually eliminates net helium boil-off. *Source:* Data from Simon Pittard.

orders of magnitude, there will be a finite boil-off, which necessitates replenishment of liquid helium (typically only once a year). Many manufacturers now also offer "zero-boil-off" magnets, which supposedly have no net helium boil-off.

10.3 Magnetic Field Homogeneity

10.3.1 Origins of Magnetic Field Inhomogeneity

In a perfectly homogeneous magnetic field B_0, the total magnetic field B_{total} inside a continuous and homogeneous material is given by

$$B_{\text{total}} = B_0 + \mu_0 M \quad \text{with} \quad M = \left(\frac{\chi}{\mu_0} \right) B_0 \tag{10.1}$$

where B_0 is the magnetic field in a vacuum and M represents the magnetization induced inside the material. The amount of magnetization that a material can acquire is proportional to the magnetic susceptibility χ, a dimensionless parameter that describes how the magnetic permeability μ of the material deviates from the vacuum permeability μ_0 according to $\chi = (\mu/\mu_0) - 1$. From an NMR point of view, materials can be classified based on the sign and magnitude of the magnetic susceptibility. Materials with a negative susceptibility are referred to as diamagnetic

Table 10.1 Magnetic susceptibility of materials encountered in MRI.

Material	Magnetic susceptibility (ppm)
Water	−9.05
Air	+0.36
Lipids	−8.0
Copper	−9.63
Aluminum	+20.7
Bismuth	−164
Zirconium	+109
Chromium	+320
Titanium	+182
Cu^{2+} solution[a]	$-9.05 + 0.021[Cu^{2+}]$
Mn^{2+} solution[a]	$-9.05 + 0.180[Mn^{2+}]$

[a] Copper and manganese concentrations in $mmol\,l^{-1}$.

materials and decrease the magnetic field inside the material. As water is diamagnetic ($\chi = -9.05 \times 10^{-6} = -9.05$ ppm) most tissues have a negative magnetic susceptibility between −7 and −11 ppm ([9], Table 10.1). Materials with a positive susceptibility are referred to as paramagnetic materials and increase the magnetic field inside the material. The most commonly encountered paramagnetic material is air ($\chi = +0.36 \times 10^{-6} = 0.36$ ppm), whereas other paramagnetic materials like titanium and chromium may be encountered as part of metallic prostheses. Both diamagnetic and paramagnetic materials do not retain their magnetic properties when the external magnetic field is removed. Ferromagnetic materials have a very large, positive magnetic susceptibility and retain some of their magnetic properties even when the external magnetic field is removed. Although used for passive shimming the bare magnet [10], ferromagnetic materials are generally incompatible with MRI applications.

The presence of a homogeneous material will, according the Eq. (10.1), slightly alter the magnitude of the magnetic field but will not lead to magnetic field inhomogeneity *per se*. Magnetic field inhomogeneity is created at boundaries between components of different magnetic susceptibilities, for example, between air and tissue. The difference in induced magnetization $\Delta M = (\Delta\chi/\mu_0)B_0$ between two susceptibilities ($\Delta\chi = \chi_1 - \chi_2$) of volume ΔV leads to a spatial magnetic field distribution that is given by the z-component of a magnetic dipole field [9]:

$$\Delta B_z(r) = \Delta\chi \frac{B_0}{4\pi} \frac{3z^2 - r^2}{r^5} \Delta V \tag{10.2}$$

with $r^2 = x^2 + y^2 + z^2$. Figure 10.5 shows the magnetic field lines and amplitude of a magnetic dipole. Note that an identical magnetic dipole field can be generated by electrical current in a closed loop or by intrinsic magnetic dipoles such as nuclear spins or bar magnets. The magnetic field generated by an arbitrary magnetic susceptibility distribution $\Delta\chi(r)$ can be calculated by integration of $\Delta B_z(r)$ over all positions r. Figure 10.6A–F shows the susceptibility-induced

(A)

(B)

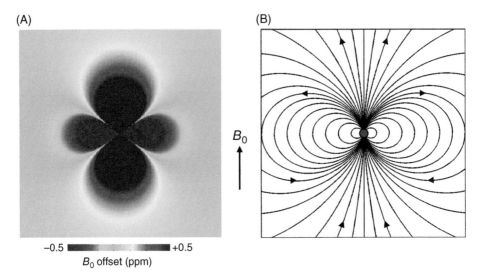

−0.5 ▄▄▄ +0.5

B_0 offset (ppm)

Figure 10.5 **Magnetic dipole fields.** (A) Magnetic dipole field distribution as calculated with Eq. (10.2) over a 20×20 mm field-of-view, $\Delta\chi = 20$ ppm, and $\Delta V = 25$ μl. (B) Magnetic field lines corresponding to the magnetic dipole field distribution in (A).

magnetic field perturbations of geometrically simple objects such as paramagnetic (Figure 10.6A and B) and diamagnetic (Figure 10.6C and D) ellipsoids. It follows that paramagnetic and diamagnetic ellipsoids increase and decrease the magnetic field strength within the material, respectively. Whereas the magnetic field in both cases is perfectly homogeneous within the ellipsoid, the magnetic field outside the object is highly inhomogeneous. From Figure 10.6A and C it can be seen that the magnetic field inhomogeneity around the ellipsoids is caused by the demand for continuous, non-crossing magnetic field lines. As the magnetic field within the sample and very far away from the sample is homogeneous, but non-equal, the only way to enforce continuity of magnetic field lines is through magnetic field inhomogeneity around the sample as shown in Figure 10.6A and C.

Besides the amplitude, the spatial distribution of the magnetic susceptibility (i.e. the geometry of the object) also has a strong effect on the observed magnetic field inhomogeneity. Figure 10.6E shows the magnetic field distribution of a paramagnetic sphere. While the general features outside the sphere are similar to those displayed in Figure 10.6B for an ellipsoid, the magnetic field inside the sphere is identical to the applied magnetic field despite the paramagnetic nature of the sphere. This demonstrates that the spatial geometry as well as the magnetic susceptibility of the object are equally important in determining the final magnetic field perturbation. Especially for phantom design, small modifications like changing a flat bottom tube to a round bottom tube can have large consequences for the final magnetic field homogeneity. Figure 10.6F shows the magnetic field distribution created by a small, air-filled sphere inside a large, water-filled sphere. It follows that the smaller, paramagnetic sphere creates large magnetic field disturbances inside the larger, diamagnetic sphere. This simplified scenario is encountered in MR applications of the human brain. Small, air-filled cavities from the nasal passages and auditory tract surround the brain and cause strong, localized magnetic field distortions.

For simple objects like spheres and tubes, Eq. (10.2) can be expanded to provide an analytical expression [11, 12]. For more complicated objects, Eq. (10.2) can, in principle, be integrated numerically over all spatial positions. However, this approach becomes too time-consuming

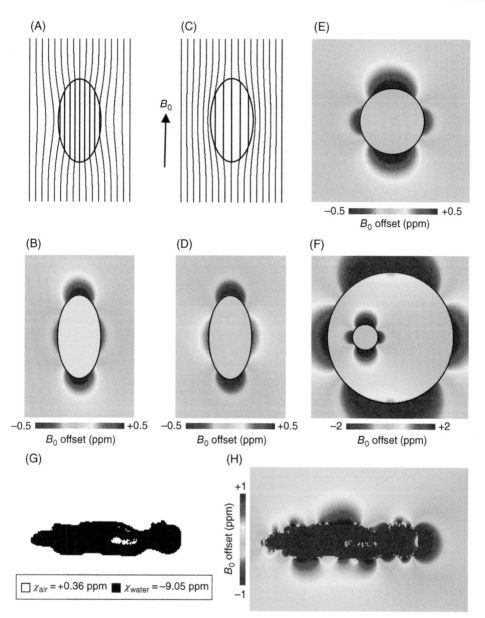

Figure 10.6 **Magnetic field lines and strengths of various magnetic susceptibility distributions.** (A, C) Qualitative magnetic field lines and (B, D) quantitative induced magnetic field amplitudes of (A, B) paramagnetic ($\chi = +1.0$ ppm) and (C, D) diamagnetic ($\chi = -1.0$ ppm) ellipsoids in a vacuum. (E) Induced magnetic field of a paramagnetic sphere ($\chi = +1.0$ ppm) in a vacuum. (F) Induced magnetic field of an air-filled sphere ($\chi = +0.36$ ppm) inside a larger, water-filled sphere ($\chi = -9.05$ ppm). (G) Simplified magnetic susceptibility distribution of the human body solely composed of water and air. (H) Magnetic field induced by the magnetic susceptibility distribution shown in (G). In addition to the many local magnetic field inhomogeneities, the magnetic field inside the human body is overall reduced due to the diamagnetic nature of water. The main magnetic field B_0 is in the horizontal direction in (H).

for larger magnetic susceptibility distributions. Marques [13] and Salomir [14] have independently described a fast method for the numerical evaluation of Eq. (10.2) based on efficient fast Fourier transforms given by

$$\Delta B_z(r) = B_0 \mathrm{FT}^{-1}\left[\left(\frac{1}{3}-\frac{k_z^2}{k^2}\right)\mathrm{FT}\left(\Delta\chi(r)\right)\right]$$

(10.3)

where FT and FT^{-1} represent forward and inverse Fourier transformations, k is the coordinate in reciprocal k-space, and $k^2 = k_x^2 + k_y^2 + k_z^2$. Using Eq. (10.3) allows calculation of the magnetic field $\Delta B_z(r)$ from an arbitrary magnetic susceptibility distribution $\Delta\chi(r)$ in seconds for a 128^3 matrix. Equation (10.3) has been extensively evaluated [13–15] and is a valuable tool in the prediction and characterization of magnetic field inhomogeneity. It has also found an important application in quantitative susceptibility mapping [16]. Figure 10.6G shows a simplified magnetic susceptibility distribution map of the human body composed only of air and water, together with the calculated magnetic field B_0 distribution (Figure 10.6H) as calculated using Eq. (10.3). Besides the general decrease in magnetic field strength due to the diamagnetic nature of water, the magnetic field is characterized by strong, localized magnetic field inhomogeneity originating from the many magnetic susceptibility transitions visible in Figure 10.6G. The magnetic field inhomogeneity that the MR user ultimately encounters is the sum of the perturbations created by magnet imperfections and environment (Figure 10.3) and the human body (Figure 10.6G and H). Magnetic field effects arising from motion, blood oxygenation, and respiration also contribute to the final magnetic field inhomogeneity. However, these dynamic effects are outside the scope of this chapter and will only be mentioned in passing.

10.3.2 Effects of Magnetic Field Inhomogeneity

The general effects of magnetic field inhomogeneity can be summarized as loss in sensitivity and resolution. However, the exact nature of these losses manifest themselves differently depending on the application and examples will be given here for MRI and MRS of the human brain. Figure 10.7A shows an anatomical gradient-echo MRI of a sagittal slice through the human head. One of the most significant air–tissue interfaces is found between the nasal sinus

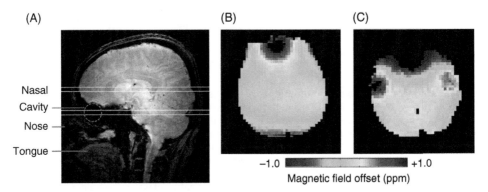

Figure 10.7 **Magnetic field inhomogeneity in the human brain.** (A) Anatomical gradient-echo MRI of a human head in the sagittal plane. The nasal cavity can be seen as an air-filled sphere positioned below a water-filled brain compartment, similar to that shown in Figure 10.6F. (B, C) Magnetic field B_0 maps acquired from the two axial slices indicated in (A) following whole-brain, second-order shimming. Significant magnetic field inhomogeneity can be observed (B) above the sinus cavity and (C) around the auditory tracts.

cavity and the lower frontal cortex. The magnetic susceptibility difference between air and tissue leads to a perturbation of the magnetic field as can be seen from the quantitative magnetic field map shown in Figure 10.7B. Other areas of decreased magnetic field homogeneity due to susceptibility differences can be found in the temporal cortices and around the auditory tracts (Figure 10.7C), while smaller effects can be observed around ventricles, close to the skull and between gray and white matter. Note that since the magnetic field inhomogeneity scales linearly with the magnetic field strength B_0, according to Eq. (10.2), the inhomogeneity encountered in the human head increases from ±60 Hz at 1.5 T to ±280 Hz at 7.0 T. It can therefore be expected that artifacts and signal losses associated with magnetic field inhomogeneity will be more severe at higher magnetic fields.

Figure 10.8 shows the effects of magnetic field inhomogeneity in MR spectroscopy. For single-volume MRS, high-quality, high spectral resolution ^{1}H MR spectra can be obtained from many areas of the brain with a homogeneous magnetic field (Figure 10.8E acquired from green volume in Figure 10.8A). However, especially in the frontal brain areas the magnetic field homogeneity is low, resulting in lower sensitivity ^{1}H NMR spectra with poor spectral resolution (Figure 10.8F acquired from red volume in Figure 10.8A). For MR spectroscopic imaging, the magnetic field B_0 map (Figure 10.8B) and the derived local gradient map (Figure 10.8C) can be used to predict MRSI data quality and guide-automated data inclusion and exclusion (Figure 10.8D).

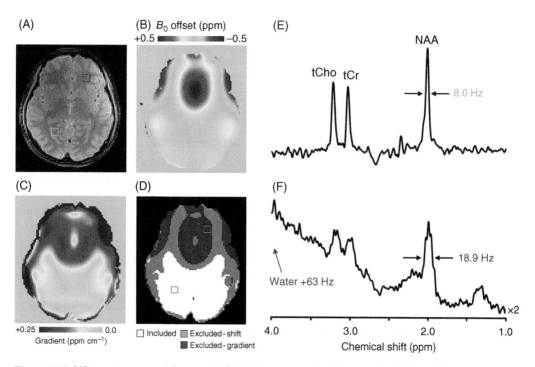

Figure 10.8 MR spectroscopy and magnetic field inhomogeneity. (A) Anatomical MRI and (B) quantitative B_0 map indicating two MRS volumes in areas of low (red) and high (green) magnetic field homogeneity. (C) Magnetic field gradient map $G = ((dB_0/dx)^2 + (dB_0/dy)^2 + (dB_0/dz)^2)^{1/2}$ calculated from the B_0 map in (B). (D) Data inclusion/exclusion map based on a combination of local B_0 shifts, local magnetic field gradients, and possibly other parameters, such as lipid or water suppression. (E, F) MR spectra obtained from areas indicated in (A) of (E) high and (F) low magnetic field homogeneity. In addition to the broader lines, spectrum (F) displays an overall shift of +63 Hz, leading to poor water suppression.

(A) Spin-echo   (B) B_0 map   (C) Gradient-echo   (D) EPI

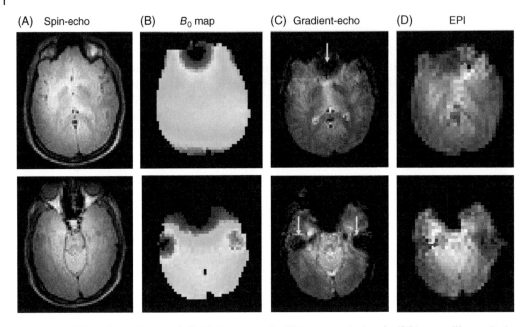

Figure 10.9 **MR imaging and magnetic field inhomogeneity.** (A) Anatomical spin-echo (SE) images, (B) quantitative B_0 maps, and (C) gradient-echo (GE) images (TE = 20 ms) acquired from the human brain. Whereas the SE images are largely unaffected, the GE images display signal loss in areas of low magnetic field homogeneity (yellow arrows). (D) Simulated echo-planar images (EPI, bandwidth = 100 kHz, 64×64 matrix) display geometric image distortion (red and blue arrows) proportional to the magnetic field offset in (B). Experimental EPI images would also exhibit through-slice signal loss as shown in (C) in addition to the image distortion.

Whereas the local magnetic field homogeneity is strongly correlated with the MRS data quality, image artifacts encountered in MRI range from negligible to extreme depending on the MRI pulse sequence. MR images based on spin-echo signals are largely immune to magnetic field inhomogeneity (Figure 10.9A) providing signal in areas of poor magnetic field homogeneity (Figure 10.9B). This is because a spin-echo sequence refocuses the effects of magnetic field inhomogeneity at the time of echo formation. Gradient-echo-based MRI methods are characterized by signal loss in areas of low magnetic field homogeneity (Figure 10.9C). A pixel that has significant phase dispersion along any spatial dimension will show signal loss. In many gradient-echo MRI methods, the through-plane slice selection is thicker than the in-plane voxel dimensions such that the through-plane signal loss often dominates. As the phase dispersion increases linearly with increasing echo-time, signal loss can become dominant in applications such as BOLD fMRI. One straightforward method to decrease signal loss is to increase the spatial resolution, such that the phase dispersal across a pixel decreases. For ultra-fast MR sequences, such as echo-planar imaging (EPI), magnetic field inhomogeneity leads to image distortion in addition to signal loss. Figure 10.9D shows simulated, single-shot echo-planar images acquired with a typical acquisition bandwidth of 100 kHz. The prolonged EPI image acquisition along the phase-encoding dimension is characterized by severe geometric distortions (i.e. pixel shifts), as discussed in detail in Chapter 4. For applications that require accurate knowledge of spatial positions, like diffusion tensor imaging (DTI) and functional imaging (fMRI), EPI may not be the best choice. However, the superior temporal resolution of EPI makes it an ideal candidate for DTI and fMRI. The conflicting features of EPI have led to the development of a wide range of alternative fast MRI methods, each with restrictions towards temporal resolution, RF power deposition, and sensitivity towards magnetic field inhomogeneity.

10.3.3 Principles of Spherical Harmonic Shimming

In the previous section it has been demonstrated that magnetic field inhomogeneity has two primary origins, namely (i) imperfections in the main magnetic field due to manufacturing errors and environmental disturbances and (ii) magnetic susceptibility boundaries of the sample or subject. The early papers on shimming were primarily concerned with the optimization of the magnet's homogeneity [17]. The magnetic field inside an NMR magnet can be described mathematically by Laplace's equation $\nabla^2 B_z = 0$, which states that for a current-free region of interest (ROI), magnetic field lines cannot form closed loops. In other words, Laplace's equation is valid when the magnetic field inhomogeneity inside the ROI originates from perturbations outside the ROI. For a bare NMR magnet without a sample or subject this assumption is perfectly valid and Laplace's equation describes the magnetic field to high accuracy. The solution to Laplace's equation for a spherical region around the magnet isocenter is given by an infinite series of spherical harmonic (SH) functions, $Y_{n,m}(\theta, \phi)$, according to

$$B(r,\theta,\phi) = \sum_{n=0}^{\infty} \sum_{m=-n}^{+n} C_{n,m} r^n Y_{n,m}(\theta,\phi) = \sum_{n=0}^{\infty} \sum_{m=-n}^{+n} C_{n,m} r^n P_{n,m}(\theta)\cos(m\phi - \phi_m) \tag{10.4}$$

where θ and ϕ are the polar and azimuthal angles, respectively, and r is the radius. $P_{n,m}(\theta)$ are polynomial functions known as Legendre polynomials for $m = 0$ and associated Legendre polynomials for $m \neq 0$. $C_{n,m}$ and $\phi_{n,m}$ are constants, whereby $\phi_m = 0$ for $m \geq 0$ and $\phi_m = \pi/2$ for $m < 0$. n and m are referred to as the order and degree of the polynomials, respectively. When $m = 0$, the variation of the magnetic field B with ϕ disappears and the field has complete cylindrical symmetry. The remaining Legendre polynomials are then also referred to as zonal harmonics giving rise to a zonal magnetic field. For $m = 0$ and $\theta = 0$ (i.e. along the z axis) Eq. (10.4) reduces to a simple polynomial sum of r, i.e. $B(r) = C_{0,0} + C_{1,0}r + C_{2,0}r^2 + C_{3,0}r^3 \cdots$. When $m \neq 0$ the variation of the magnetic field B becomes oscillatory around the z axis with a frequency $m\phi$ and initial phase ϕ_m. The associated Legendre polynomials are referred to as tesseral harmonics giving rise to a tesseral magnetic field. Note that along the z axis ($\theta = 0$) the contribution of the tesseral harmonics to the total magnetic field is zero. Table 10.2 provides the polynomials for $n \leq 5$ and Figure 10.10 shows a graphical depiction for $n \leq 3$.

In many circumstances, such as magnetic field plotting of the bare magnet or shim coil design via analytical evaluation of SH expansions, it is advantageous to express the magnetic field and the SH fields in spherical coordinates as in Eq. (10.4). However, since modern MRI-based magnetic field B_0 mapping is almost always based on a Cartesian grid, it is often convenient to express the fields in terms of Cartesian coordinates. Using the relationships $x = r\sin\theta\cos\phi$, $y = r\sin\theta\sin\phi$, $z = r\cos\theta$, and $r^2 = x^2 + y^2 + z^2$ the magnetic field can be expressed in Cartesian coordinates according to

$$B(x,y,z) = B_0 + \sum_{n=1}^{\infty} \sum_{m=-n}^{+n} C_{n,m} F_{n,m}(x,y,z) \tag{10.5}$$

The functions $F_{n,m}(x,y,z)$ are given in Table 10.2 up to $n = 5$. It should be noted that the common name of most spherical harmonics terms is an incomplete representation of the actual function in Cartesian coordinates. For example, a "Z3" function ($n = 3$, $m = 0$) is not limited to the z axis but also has contributions in the x–y plane. Other terms, like "ZC2" ($n = 3$, $m = 2$), represent a mix between spherical and Cartesian coordinates and nomenclature.

The magnetic field inside an MR magnet can be quantitatively described by a SH expansion given by Eq. (10.4) or Eq. (10.5). The overall strategy for improving the magnetic field

Table 10.2 Spherical and Cartesian representation of low-order ($n \leq 5$) spherical harmonics functions.

Order n	Degree m	$P(\theta)$[a]	$F(x,y,z)$[b]	Common name
0	0	1	1	Z0
1	0	$\cos\theta$	z	Z
1	+1/−1	$\sin\theta$	x/y	X/Y
2	0	$\tfrac{1}{2}(3\cos^2\theta-1)$	$z^2-\tfrac{1}{2}R^2$	Z2
2	+1/−1	$3\sin\theta\cos\theta$	$3zx/3zy$	ZX/ZY
2	+2/−2	$3\sin^2\theta$	$3(x^2-y^2)/6xy$	X2Y2 (or C2)/XY (or S2)
3	0	$\tfrac{1}{2}(5\cos^3\theta-3\cos\theta)$	$z^3-\tfrac{3}{2}zR^2$	Z3
3	+1/−1	$\tfrac{3}{2}\sin\theta(5\cos^2\theta-1)$	$6x(z^2-\tfrac{1}{4}R^2)/6y(z^2-\tfrac{1}{4}R^2)$	Z2X/Z2Y
3	+2/−2	$15\sin^2\theta\cos\theta$	$15z(x^2-y^2)/30zxy$	ZX2Y2 (or ZC2)/ZXY (or ZS2)
3	+3/−3	$15\sin^3\theta$	$15x(x^2-3y^2)/15y(3x^2-y^2)$	X3 (or C3)/Y3 (or S3)
4	0	$\tfrac{1}{8}(35\cos^4\theta-30\cos^2\theta+3)$	$z^4-3z^2R^2+\tfrac{3}{8}R^4$	Z4
4	+1/−1	$\tfrac{5}{2}\sin\theta(7\cos^3\theta-3\cos\theta)$	$10zx(z^2-\tfrac{3}{4}R^2)/10zy(z^2-\tfrac{3}{4}R^2)$	Z3X/Z3Y
4	+2/−2	$\tfrac{15}{2}\sin^2\theta(7\cos^2\theta-1)$	$45(x^2y^2)(z^2-\tfrac{1}{6}R^2)/90xy(z^2-\tfrac{1}{6}R^2)$	Z2C2/Z2S2
4	+3/−3	$105\sin^3\theta\cos\theta$	$105zx(x^2-3y^2)/105zy(3x^2-y^2)$	ZC3/ZS3
4	+4/−4	$105\sin^4\theta$	$105(x^2-y^2)^2-420x^2y^2/420xy(x^2-y^2)$	X4/Y4
5	0	$\tfrac{1}{8}(63\cos^5\theta-70\cos^3\theta+15\cos\theta)$	$z^5-5z^3R^2+\tfrac{15}{8}zR^4$	Z5
5	+1/−1	$\tfrac{15}{8}\sin\theta(21\cos^4\theta-14\cos^2\theta+1)$	$15z^2x(z^2-\tfrac{3}{2}R^2)+\tfrac{15}{8}xR^4/15z^2y(z^2-\tfrac{3}{2}R^2)+\tfrac{15}{8}yR^4$	Z4X/Z4Y
5	+2/−2	$\tfrac{105}{2}\sin^2\theta(3\cos^3\theta-\cos\theta)$	$105z(x^2-y^2)(z^2-\tfrac{1}{2}R^2)/210zxy(z^2-\tfrac{1}{2}R^2)$	Z3C2/Z3S2
5	+3/−3	$\tfrac{105}{2}\sin^3\theta(9\cos^2\theta-1)$	$420x(x^2-3y^2)(z^2-\tfrac{1}{8}R^2)/420y(3x^2-y^2)(z^2-\tfrac{1}{8}R^2)$	Z2C3/Z2S3
5	+4/−4	$945\sin^4\theta\cos\theta$	$945z(x^2-y^2)^2-3780zx^2y^2/3780zxy(x^2-y^2)$	ZC4/ZS4
5	+5/−5	$945\sin^5\theta$	$945(x^5-5x^3y^2(2x^2-y^2))/945(y^5+5x^2y(x^2-2y^2))$	X5/Y5

[a] Note that only $P(\theta)$ is given. The complete spherical harmonics function requires multiplication with $rn\cdot\cos(m(\phi-\phi_m))$ according to Eq. (4.2).
[b] $R^2=x^2+y^2$.

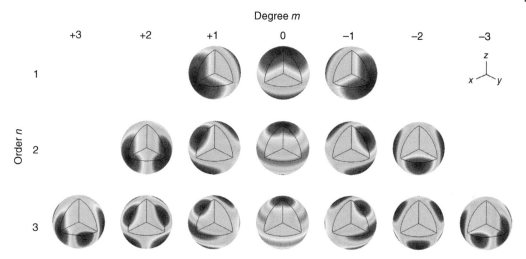

Figure 10.10 Spherical harmonic magnetic fields. Graphical representation of all spherical harmonic fields up to the third-order n and degree m ($m = -n, -n+1, \ldots, +n$) on the surface of a unit sphere. In order to evaluate the spherical harmonic functions internal to the spherical surface the positive octant has been removed. The trivial spherical harmonic function for $n = 0$, i.e. a pure offset, is not displayed.

homogeneity relies on the application of additional magnetic fields that cancel out the magnetic field inhomogeneity terms expressed in Eqs. (10.4) and (10.5), leaving in an ideal scenario only the homogeneous B_0 magnetic field term. The adjustment of the magnetic field homogeneity with small additional magnetic fields is referred to as shimming. In general, there are only two ways of modifying the magnetic field inside an MR magnet, namely by (i) current-carrying wire or by (ii) additional magnets. In the latter case the magnets do not have to be permanently magnetized, but could also come in the form of diamagnetic, paramagnetic, or ferromagnetic materials magnetized by the main magnetic field. Shimming with magnetic pieces of steel is typically used to improve the homogeneity of the bare magnet [10]. Since the adjustment is only performed once during magnet installation, it is often referred to as passive shimming. While passive shimming has been used to adjust the magnetic field homogeneity on human and animal brain *in vivo* [18–22], it is not commonly used and will not be discussed further. The standard method to generate additional magnetic fields with spatial distributions according to specific spherical harmonic functions relies on electrical currents in well-defined current loops. Since the current and thereby the magnetic fields can be adjusted on every sample, this method for optimizing the magnetic field homogeneity is referred to as active shimming.

The magnetic field generated by a circular, current-carrying loop of radius R, n number of turns, and carrying current I placed in the magnet isocenter with the coil axis parallel to the z (or B_0) axis can be obtained through integration of the Biot–Savart law yielding

$$B_x = \frac{xzC}{r_{xy}^2 \left(R^2 + r^2 + 2Rr_{xy}\right)^{1/2}} \left[\frac{R^2 + r^2}{R^2 + r^2 - 2Rr_{xy}} E\left(\kappa^2\right) - K\left(\kappa^2\right) \right]$$

$$B_z = \frac{C}{\left(R^2 + r^2 + 2Rr_{xy}\right)^{1/2}} \left[\frac{R^2 - r^2}{R^2 + r^2 - 2Rr_{xy}} E\left(\kappa^2\right) + K\left(\kappa^2\right) \right] \tag{10.6}$$

Degree *m*

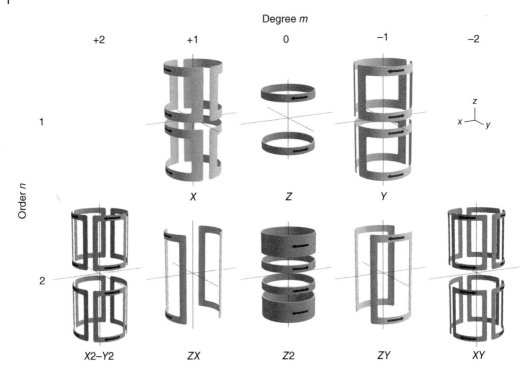

Figure 10.11 Spherical harmonic shim coils. First- and second-order SH shim coils designed based on current loops and arcs as described by Romeo and Hoult [17]. Note that wire parallel to the main magnetic field B_0 (and the z axis) does not contribute to the generated SH fields, but are essential in connecting various arcs. Black arrows indicate the current directions. *Source:* Data from Dan Green and Simon Pittard.

where $r^2 = x^2 + y^2 + z^2$, $r_{xy}^2 = x^2 + y^2$, and $C = \mu_0 nI/2\pi$. $K(\kappa^2)$ and $E(\kappa^2)$ are the complete elliptical integrals of the first and second kind, respectively, with $\kappa^2 = 4Rr_{xy}/(R^2 + r^2 + 2Rr_{xy})$. The magnetic field along the y direction can be found from B_x, according to $B_y = (y/x)B_x$. Whereas NMR is only concerned with static magnetic fields along the z direction, i.e. B_z, B_x, and B_y are nonetheless important since B_x, B_y, and B_z are converted into each other upon rotation of the coil. For example, rotating the coil about the x axis over an angle α leads to the transformation $B_y = B_y\cos(\alpha) + B_z\sin(\alpha)$ and $B_z = B_z\cos(\alpha) - B_y\sin(\alpha)$. Every rotation of the coil can be described by standard 3×3 rotation matrices. Early work on shim field design and creation analyzed the magnetic field produced by steel pieces, current loops, and arcs in terms of analytical SH expansions [17, 23]. The pieces or currents are then spatially arranged in such a way that undesired SH contributions are canceled. The analytical evaluation can also be performed numerically and practical solutions can be obtained in seconds with modern-day computer optimization routines. Figure 10.11 shows shim coil designs for the first- and second-order SH functions largely based on analytical evaluation of SH expansions. Note that the tesseral SH functions require the use of arcs placed with a periodicity mϕ. Modern shim coil designs are based on sophisticated optimization algorithms that can incorporate additional criteria with respect to coil inductances (and hence current rise times), power requirements, improved DSV, and nonsymmetrical designs. However, many optimized designs still have a strong resemblance to the earlier, basic shim coils shown in Figure 10.11.

10.3.4 Practical Spherical Harmonic Shimming

The final element in the process of active shimming is the availability of a method to quantitatively map the magnetic field homogeneity within the sample. MRI-based B_0 mapping is the standard tool to achieve a 3D characterization of the magnetic field homogeneity in, for example, the human brain (Figure 10.12A and B). The minimum spatial resolution of the B_0 map is largely dictated by the ROI over which the magnetic field homogeneity should be optimized and the level of magnetic field inhomogeneity encountered. For smaller ROIs the spatial resolution of the B_0 map should be high enough to provide a significant number of MRI pixels across the ROI. For example, when optimizing the third-order SH shims consisting of 16 SH terms over a 1 ml volume, the number of pixels should be at least $4 \times 4 \times 4 = 64$ in order to achieve a robust fit on an overdetermined system. For a typical $25 \times 25 \times 25$ cm field-of-view, this translates into a B_0 map with a $100 \times 100 \times 100$ data matrix. When the experimental duration of the B_0 map acquisition becomes too long, it can be switched to a multi-slice 2D acquisition over a $100 \times 100 \times 4$ data matrix. When shimming the entire human brain, the spatial resolution of the B_0 map can be drastically reduced, provided that the lower resolution does not lead to excessive signal loss in areas of low magnetic field inhomogeneity. A spatial resolution of $64 \times 64 \times 64$ over $24 \times 24 \times 24$ cm can be acquired in under 2 min and provides a reasonable default setting that is adequate for most shimming challenges on the human head.

For single-volume MRS, the FASTMAP [24, 25] method and its variants [26–28] provide a fast and often a more robust alternative to 3D B_0 mapping. The FASTMAP method relies on the fact that specific SH terms can be determined quantitatively from 1D projections along a thin column. For example, the magnetic field in the magnet isocenter along a thin column in

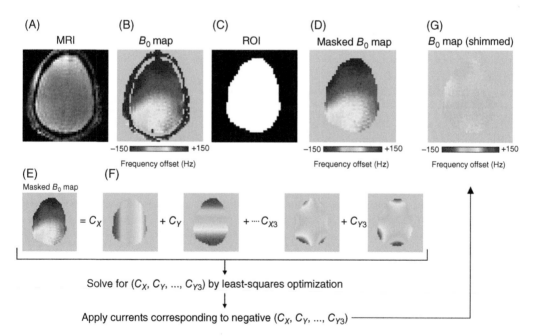

Figure 10.12 Practical *in vivo* spherical harmonic shimming. (A) Anatomical MR image and (B) magnetic field B_0 map of the human head. (C) Region-of-interest (ROI) over which the magnetic field homogeneity is to be optimized. (D, E) Masked B_0 map calculated as the product of (B) and (C). (F) The masked B_0 map is modeled as a linear sum of the calibrated magnetic field produced by each shim coil. (G) Application of the corresponding shim currents and reacquisition of a B_0 map shows the improved magnetic field homogeneity over the ROI.

the z direction can be described as a polynomial according to $B(z) = B_0 + C_{z1}z + C_{z2}z^2 + C_{z3}z^3 + \cdots$. Unfortunately, several SH shims (Z, ZX, and ZY) appear as a linear z magnetic field along a narrow column in the z direction. In other words, the Z, ZX, and ZY shim fields are degenerate (or identical) along a 1D projection in the z direction. In order to remove the shim degeneracy and allow the determination of all eight zero-to-second-order SH terms, FASTMAP acquires six projections along different directions. The advantage of FASTMAP is that (i) the measurement is very fast as only 6×2 projections (for $t = 0$ and $t > 0$) need to be acquired and that (ii) a large number of data points can be acquired along the 1D column. While FASTMAP only allows the determination of up to second-order SH terms, this is in general sufficient for single-volume MRS. FASTMAP has also been extended to shim over 2D slices [26].

Following the acquisition of a robust B_0 map of sufficient spatial resolution, arguably the most important step is the selection of a proper ROI (Figure 10.12C). In the case of whole-brain shimming, the skull area needs to be excluded from the ROI in order to avoid artifactual B_0 offsets due to lipids in the skull region (blue pixels in Figure 10.12B, see also Chapter 4). For single-volume MRS the ROI is typically limited to the immediate volume dimensions. Depending on the specific area of the brain, the ROI dimension can extend beyond the volume dimensions in order to include more B_0 map pixels for improved determination of the SH coefficients. However, the ROI should never increase to the extent that the magnetic field inhomogeneity in regions outside the volume dimensions becomes detrimental to the shimming performance. The allowable extension of the ROI beyond the volume dimensions is heavily dependent on the brain area and is largely a matter of experience. As a general rule-of-thumb, the ROI can typically be extended beyond the VOI by 25% per dimension, provided that no artifactual pixels are included. The selected ROI (Figure 10.12C) can be combined with the measured or "raw" B_0 map (Figure 10.12B) to provide a "masked" B_0 map (Figure 10.12D) from which the optimal SH shims will be determined. On most clinical and preclinical MRI systems, the SH shim coils have been calibrated during the installation of the gradient/shim coils. The calibration procedure entails a detailed spatial mapping of the magnetic field generated by unit current in each SH coil. In an ideal world, a "Z2" shim coil will only produce a Z2 shim field. However, imperfections during coil construction, mounting, or installation can lead to small contributions from other SH terms. The calibration quantitatively measures the total magnetic field produced by each SH coil, which automatically includes the minor imperfections in addition to the desired SH shim term. The calibration step is typically only performed once during system installation, making it largely transparent to most MR users.

The next step in active SH shimming is to model (or approximate) the experimental, masked B_0 map (Figure 10.12E) as a linear sum of the available, calibrated SH magnetic fields (Figure 10.12F). The modeling is typically based on a linear least-squares optimization routine that enforces the limits on the maximum available current in each SH shim coil. Once the best mathematical fit has been obtained, the required currents can be calculated from calibrated shim maps. Application of the (negative) currents to the actual SH shim coils within the bore of the magnet then results in improved magnetic field homogeneity (Figure 10.12G). The performance of the SH shimming can be checked by acquiring an additional B_0 map following the application of the correction currents. When the measured B_0 map after shimming does not resemble the predicted B_0 map (Figure 10.12G), possible explanations can be found in (i) subject movement, (ii) incorrectly working shim amplifiers, or (iii) incorrect calibration maps.

In clinical (and preclinical) MR platforms, the steps outlined above are part of an automated workflow whereby the only user interaction involves the VOI placement. However, given the importance of high magnetic field homogeneity (e.g. see Figures 10.8 and 10.9) and the many potential complications that lead to a suboptimal homogeneity, it is advisable that every MR users is aware of the fundamental steps involved in SH shimming.

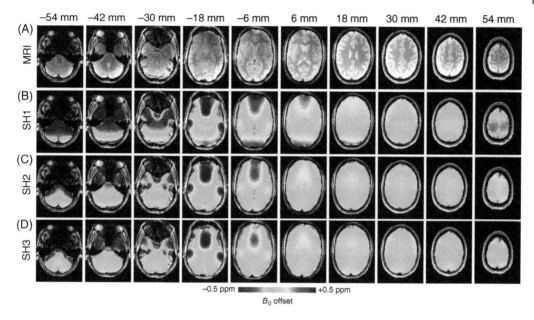

| −54 mm | −42 mm | −30 mm | −18 mm | −6 mm | 6 mm | 18 mm | 30 mm | 42 mm | 54 mm |

(A) MRI

(B) SH1

(C) SH2

(D) SH3

−0.5 ppm ▬▬▬▬▬ +0.5 ppm

B_0 offset

Figure 10.13 **Static spherical harmonic shimming of the human brain** *in vivo.* (A) MR images and (B–D) B_0 maps selected from a multi-slice, whole brain dataset acquired at 4T (66 slices of 2 mm, isotropic 2×2 mm in-plane resolution). Residual B_0 maps were calculated following removal of spherical harmonic terms corresponding to (B) $n \leq 1$, (C) $n \leq 2$, and (D) $n \leq 3$.

Figure 10.13 summarizes the performance of first-, second-, and third-order SH shimming on the human brain. The SH shim terms are optimized over the entire human brain and applied in a static manner, representing the default mode of SH shimming. Following compensation of the three first-order SH terms (Figure 10.13B), the magnetic field homogeneity across the human brain is poor. Compensation of the eight zero-, first-, and second-order SH terms (Figure 10.13C) and the 16 zero-through-third-order SH terms (Figure 10.13D) improves the global magnetic field homogeneity, with excellent homogeneity in the top half of the brain. However, significant magnetic field inhomogeneity remains in the bottom half of the brain due to the close proximity to magnetic susceptibility boundaries from the nasal and auditory tracts. As the magnetic field inhomogeneity in the lower slices is highly localized, it requires higher SH terms beyond those available on standard clinical MR systems. The inability of SH shimming to provide high magnetic field homogeneity has sparked the development of alternative shimming strategies that will be discussed next.

10.3.5 Alternative Shimming Strategies

The poor magnetic field homogeneity across the human brain following compensating of all zero-to-second-order SH shims is primarily caused by the localized character of the residual inhomogeneity. Magnetic susceptibility boundaries between air and tissue in the nasal passages cause local magnetic field homogeneity in the frontal cortex and temporal lobes. The auditory tracts lead to additional inhomogeneity in the temporal lobes and other parts of the lower brain. SH analysis of the magnetic field distribution in the human brain shows that higher-order contributions dominate the residual magnetic field inhomogeneity shown in Figure 10.13D. The most straightforward method to improve the magnetic field homogeneity is therefore the extension of SH shim set with higher-order ($n > 3$) SH shim coils. This strategy has been pursued [29] and has demonstrated greatly improved homogeneity with the inclusion of fourth- and fifth-order SH coils.

A practical observation from single-volume MRS is that the magnetic field homogeneity over a small volume can be much better than that over a large volume like the entire human brain. In other words, the magnetic field inhomogeneity over a small volume can be approximated by a SH expansion (Eq. (10.5)) of a much lower-order than required for the entire brain. This explains the fact that the magnetic field homogeneity across small MRS volumes can be close to optimal with only second-order SH coils. This observation forms the basis of dynamic shimming [30–35], whereby signal from a large object (e.g. human brain) is acquired in multiple, smaller sections (e.g. 2D slices). The magnetic field homogeneity of the smaller ROI (e.g. 2D slice) can be improved with lower-order SH terms, whereby an optimal slice-specific shim setting is dynamically updated when the slice is excited and signal is acquired. In this manner, each 2D slice has its own optimized SH shim settings, guaranteeing optimal magnetic field homogeneity across the entire human brain. Figure 10.14 shows the implementation of dynamic shimming

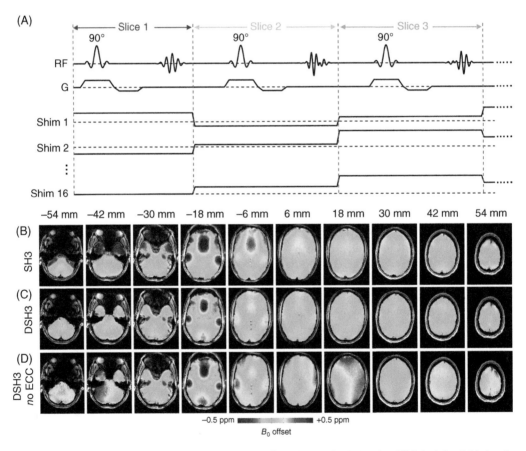

Figure 10.14 **Dynamic spherical harmonic shimming of the human brain *in vivo*.** (A) Principle of third-order dynamic shimming for a gradient-echo MRI sequence. Optimal shim fields present during the excitation and acquisition of a spatial slice are dynamically updated to shim values that are optimal for the next slice. (B) Residual B_0 map as calculated following removal of static, global third-order SH terms and (C) following slice-by-slice dynamic shimming of all third-order SH shims. (D) In the absence of eddy current compensation (ECC) through shim preemphasis, the residual B_0 maps are dominated by residual low-order SH eddy current fields, thereby largely negating the advantages of dynamic shimming.

for a multi-slice, gradient-echo MRI sequence. Before excitation of a given slice the slice-optimized shim coil currents are updated, thereby providing optimal magnetic field homogeneity during excitation, the echo-time TE, and acquisition. Before excitation of the next slice, the optimized shim coil currents for that slice are updated. Figure 10.14C compares the performance of dynamic, third-order SH shimming to static, third-order SH shimming (Figure 10.14B). Clearly the magnetic field homogeneity across all slices has improved, with very significant improvements in the lower two slices.

A time-varying magnetic field will generate a time-varying current and a related magnetic field in a nearby conductor that will oppose the original magnetic field changes. These so-called eddy currents are well known from rapidly switching currents in gradient coils used for MR imaging (see Section 10.4). However, these effects will also be present when switching currents in shim coils as done during dynamic shimming. When these effects are ignored, the resulting magnetic field homogeneity will be dominated by low-order magnetic fields generated by the shim-induced eddy currents (Figure 10.14D). The eddy currents can be reduced by shim coil preemphasis, similar to that performed for magnetic field gradients (Section 10.4). However, for the 16 zero-to-third-order SH coils, the preemphasis circuitry may have to span the entire 16×16 SH matrix for complete eddy current compensation [35], including shim cross-terms (i.e. a pulsed Z2 shim generating first-order or other second-order SH eddy currents).

A completely different approach to shimming the human brain is provided by multi-coil (MC) shimming [36–42]. MC shimming achieves improved magnetic field homogeneity by combining the magnetic fields generated by simple (circular) direct-current loops (Figure 10.15A). A small circular coil generates a magnetic field that is localized to the immediate vicinity of the coil. The intrinsic localization of the magnetic field provides an inherent match to the localized magnetic field inhomogeneity encountered in the human brain. In a proof-of-principle demonstration it was shown that the magnetic field inhomogeneity in the frontal cortex can be eliminated with only six small DC coils [36]. In order to provide whole brain coverage, a general-purpose MC setup has to contain a larger matrix of DC coils, each with independent control of the current (Figure 10.15B). Application of a static, whole-brain current setting in a 48-element MC setup provides excellent magnetic field homogeneity across the human brain (Figure 10.15D). Using the MC setup dynamically on a slice-by-slice basis improves the homogeneity even further (Figure 10.15E). It should be noted that MC shim-induced eddy currents are not significant due to the large distance between the MC setup and the magnet cryostat. MC shimming has been successfully used on mouse brain [37, 38], rat brain [41], human brain [36, 39, 42], human spine [43], and at ultra-high magnetic fields [44]. With appropriate modifications the DC shim currents can run on the same physical wires as a RF receive array [45–49].

It should be realized that the alternative shimming strategies presented here have been largely applied in a nonclinical, research setting. The default shimming on clinical MR scanners is still largely limited to first- or second-order SH shimming (Figure 10.13B and C). Significant magnetic field inhomogeneity will therefore be present across large parts of the human brain, such that the artifacts discussed for MRS (Figure 10.8) and MRI (Figure 10.9) are the rule and not the exception.

10.4 Magnetic Field Gradients

Magnetic field gradients are linearly varying magnetic fields that are applied in addition to the main magnetic field B_0 to achieve spatial encoding. The principle of magnetic field gradients is discussed in detail in Chapter 4. Slice selection, frequency, and phase encoding in MRI are all

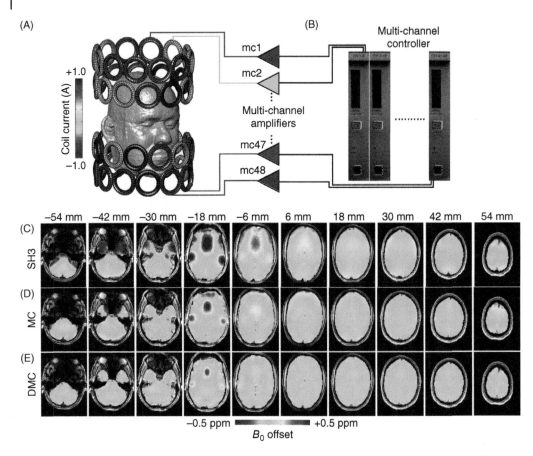

Figure 10.15 **Multi-coil shimming of the human brain** *in vivo.* (A) Multi-coil (MC) setup consisting of four rings of 12 direct current (DC) coils in relation to a human head. Each DC coil is independently controlled by (B) a dedicated waveform controller and low current (±1 A) amplifier. Courtesy of Terry Nixon. (C–E) Residual B_0 maps following (C) removal of global, static third-order SH terms, (D) removal of global, static MC shim terms, and (E) removal of dynamic, slice-by-slice MC shim terms. The absence of nearby cold-conducting structure obviates the need for MC shim preemphasis.

performed with the aid of magnetic field gradients, as are single volume localization, MR spectroscopic imaging, and many types of water suppression. Magnetic field gradients are further used to select specific coherence transfer pathways in spectral editing and two-dimensional NMR, thereby often providing a single-scan alternative to phase cycling. For these reasons almost all clinical and research MR systems are equipped with a set of three orthogonal X, Y, and Z magnetic field gradients.

The basic gradient designs are still based on the coils that were initially developed as first-order spherical harmonic shim coils for the homogenization of magnetic fields [17, 23]. A linear magnetic field gradient in the z-direction, collinear with the main magnetic field is generated with a so-called Maxwell pair (Figure 10.11). Essentially, it consists of two parallel coils perpendicular to the main magnetic field direction in which the DC currents flow in opposite directions. As a consequence, the direction of the generated magnetic field is different for the two coils leading to complete cancelation of the z-component of the magnetic field at the center between the coils. This point corresponds to the gradient (and normally also magnet and shim coil) isocenter. On one side of the gradient isocenter the magnetic field of one coil dominates, making the magnetic field direction, for instance, positive and the amplitude dependent on the distance from the isocenter. On the other side of the gradient isocenter, the reverse is true, i.e. the magnetic field direction is negative. In practice, more than two coils are used to improve the linearity of the gradient magnetic field. The design for x and y magnetic field gradients is more complicated than for z gradients. Most commonly used x and y magnetic field gradients are based on the Golay gradient coil design [23] and are shown in Figure 10.11. The principle is essentially the same as for the Maxwell pair in that opposite currents in the four different saddle-shaped coils produce magnetic fields, which partially cancel each other in order to generate a linear magnetic field gradient. The linearity of the magnetic gradient fields is improved by using more loops. In reality, gradient coils are not made from wire, but are designed as a continuous pattern of copper sheets as optimized by computer simulations [50]. Figure 10.16A shows a schematic of the integrated design of magnetic field gradient coils and water cooling, whereas Figure 10.16B shows a realistic example of a Y gradient coil.

Magnetic field gradients are characterized by several parameters. The amplitude, expressed in $G\,cm^{-1}$ or $mT\,m^{-1}$, is probably the most important, since it largely determines the minimum voxel sizes and acquisition time in MRI and voxel sizes and dephasing capabilities in MRS. On most clinical systems, the gradient amplitude is limited to around $50\,mT\,m^{-1}$, while on animal research systems, gradients can be as high as $1\,T\,m^{-1}$. At a fundamental level the amplitude of a magnetic field or magnetic field gradient is governed by the Biot–Savart law for electromagnetism, which states that for a simple current loop the generated magnetic field is inversely proportional to the square of the radius of the loop and linearly proportional to the current. Therefore, for the same current, gradient coils designed for human applications necessarily generate smaller magnetic fields than gradient coils designed for animals. To counteract the reduced efficiency due to the increased coil radius, human MR systems often are equipped with powerful high-voltage and high-current amplifiers. However, increasing the current density is limited by heat removal considerations. As can be seen in Figure 10.16B, magnetic gradient coils are typically made of copper sheets. Since copper has a finite resistance, the increased current density will lead to increased power deposition and hence heat generation. The excess heat is removed by an extensive water cooling network built in between the coil windings (Figure 10.16A). However, the heat-removal capacity of the cooling water is limited. When the heat dissipation exceeds the heat removal, the temperature of the gradient coil will rise, which in the extreme case may damage the gradient coil. While the gradient amplitude is the parameter of interest for the user, it is the gradient coil efficiency that ultimately determines the performance of the gradient set. Gradient coil efficiency η is defined as the gradient strength G produced by a current I, i.e. $\eta = G/I$. The efficiency η should be as large as possible, but is inherently linked to two fundamental parameters in gradient coil design, namely the inductance L and the resistance R.

The spatial linearity (i.e. homogeneity) is another crucial characteristic of magnetic field gradients. If a magnetic field gradient varies nonlinearly across the field-of-view, the spatial information is not encoded linearly by the imaging gradients and consequently the image will appear distorted. With gradient systems presently available, nonlinearities have a negligible effect on image resolution across the DSV of the gradient coils. Distortions may be visible when the field-of-view approximates the dimensions of the gradient coil as in the case of whole-body scanning. In these extreme cases, distortions may be corrected by experimental or theoretical methods. Note that it is especially important that the magnetic field gradient amplitude does not tend towards zero within the sensitive volume of the RF coil, as this can lead to image artifacts that cannot be corrected by post-processing. Especially head gradient coils are sensitive to this phenomenon, during which signal from the chest region can fold back on top of signal from the head.

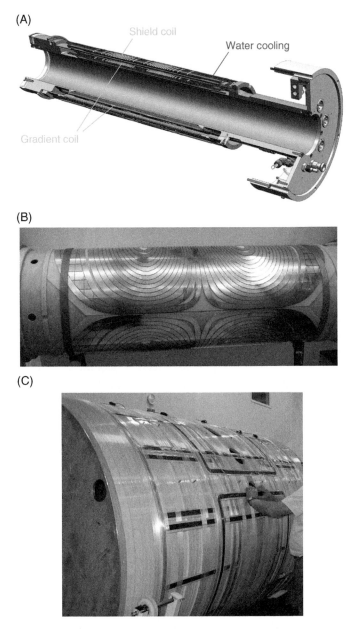

Figure 10.16 **Magnetic field gradient and shim coils.** (A) Cross-sectional 3D drawing of an actively-shielded, water-cooled magnetic field gradient set. (B) Photograph of a *Y* magnetic field gradient. (C) Photograph of higher-order shim coils being placed on a human-sized MR magnet. *Source:* Data from Dan Green and Simon Pittard.

Another important characteristic of magnetic field gradient coils is their rise time, i.e. the time it takes to switch a gradient from zero to full amplitude. Ideally, the gradient switching time is as short as possible, since this minimizes echo-times, improves the performance of fast MRI methods, like EPI, and allows for more efficient diffusion-weighting as required in DTI. However, for human applications, in particular, there are physiological limitations on the rate at which the magnetic field can change (i.e. dB/dt). A time-varying magnetic field generates an electric field in any sample placed within the magnet, including human and animal subjects. At sufficiently high amplitudes, the electric field, which is proportional to dB/dt, can stimulate peripheral nerves and muscles (i.e. twitching). At higher levels, painful nerve stimulation has been observed. At extremely high levels, cardiac stimulation or even ventricular fibrillation are of concern. However, these extreme cases are not encountered in human MRI systems due to the previously mentioned hardware restrictions. These observations and potential concerns have led to safety standards for magnetic field gradient switching rates. A commonly used safety standard was developed by the International Electrotechnical Commission (IEC) and differentiates between two operating modes. In the normal mode, the upper dB/dt limit is specified as $20\,\mathrm{T\,s^{-1}}$ for gradient ramp times, t_{ramp}, exceeding 120 μs. For ramp times between 12 and 120 μs, the dB/dt limit is given by $(2400/t_{\mathrm{ramp}})\,\mathrm{T\,s^{-1}}$. In the normal mode, patient discomfort is minimal and no peripheral nerve stimulation is expected. In the first controlled mode, the dB/dt limit is raised to $(60\,000/t_{\mathrm{ramp}})\,\mathrm{T\,s^{-1}}$. In this mode, the operator should maintain contact with the subject at all times, as patient discomfort and peripheral nerve stimulation may be observed. The dB/dt limit in the first operating mode is primarily designed to prevent cardiac stimulation.

Besides restrictions dictated by physiological considerations, the minimum gradient ramp time is also limited by the hardware, i.e. gradient coil and amplifier. The ramp time, t_{ramp}, of a gradient coil is governed by the voltage V and current I supplied by the gradient amplifier, as well as the inductance L and efficiency η of the coil according to

$$t_{\mathrm{ramp}} = \frac{IL}{V} = \frac{GL}{\eta V} \tag{10.7}$$

Equation (10.7) indicates that short ramp times can be achieved with low inductance gradient coils in combination with high-voltage amplifiers. The inductance of a gradient coil can be lowered by using a small number of turns, n, in the gradient coil since L is proportional to n^2. However, the gradient coil efficiency η is proportional to n, such that a small number of turns compromises the maximum achievable gradient strength. A useful parameter to characterize the gradient coil performance is the ratio η^2/L, which is independent of n and indicates the efficiency that can be achieved from a given inductance. The power, $P_{\mathrm{amplifier}}$, that a gradient amplifier must generate in order to produce a gradient strength G within a ramp time t_{ramp} is given by

$$P_{\mathrm{amplifier}} = \frac{G^2}{t_{\mathrm{ramp}}} \frac{L}{\eta^2} \tag{10.8}$$

This indicates the importance of maximizing the ratio η^2/L in order to reduce the power requirements. It can be shown that the ratio η^2/L depends on the gradient coil radius r according to $1/r^5$, indicating the importance of minimizing the gradient size to achieve optimal performance. Besides the inductance L, the resistance R of the gradient coil is also an important design parameter as it is directly related to the amount of power dissipated in the gradient coil (i.e. $P = I^2 R$). The resistance can also be lowered by reducing the number of turns, n, at the cost of reduced coil efficiency. Resistance also depends on the average conductor cross-section in the coil (i.e. the wire thickness), which should be made as large as possible. The interplay of efficiency η, inductance L, resistance R, and gradient coil uniformity dominates gradient coil design and necessarily always leads to a compromise.

10.4.1   Eddy Currents

Magnetic field gradients are most commonly used as short pulses during which the current in the gradient coil is quickly maximized, maintained for a short delay, and quickly reduced to zero amplitude (Figure 10.17A). Ideally, the generated magnetic field gradient would follow the current instantaneously, resulting in a desired trapezoidal magnetic field gradient pulse. However, rapid switching of the gradients induces so-called eddy currents in all nearby conducting structures, such as the cryostat, heat shields, magnet, and shim coils. These eddy currents are a manifestation of Faraday's law of induction, which states that a time-varying magnetic field will induce a current (and consequently a second time-varying magnetic field) in a nearby conductor that opposes the effect of the time-varying magnetic field (Figure 10.17B). High-gradient amplitudes and/or fast gradient switching will produce the largest eddy currents. As a result of the additional, time-varying magnetic fields, the desired magnetic field gradient is heavily distorted with decreased amplitude at the beginning and an additional decaying field following the actual pulse (Figure 10.17C). Unless measures are taken to eliminate or compensate for these effects, they will severely degrade image quality and may preclude MRS experiments altogether (Figure 10.18), due to the much higher magnetic field homogeneity requirements of MRS.

The induced time-varying magnetic fields are to first approximation composed of two main components, a magnetic field gradient $G(t)$ opposite to the applied gradient and a shift in the main magnetic field $\Delta B_0(t)$. Additional higher-order effects can be present, but are usually negligible [51]. The main magnetic field shift $\Delta B_0(t)$ is identical for all positions, while the induced gradient varies linearly with position. Besides $G(t)$ and $\Delta B_0(t)$ the eddy currents may also hold so-called cross terms, e.g. induction of gradients in directions perpendicular to that of the applied gradient. A theoretical analysis of eddy currents in terms of LCR circuits

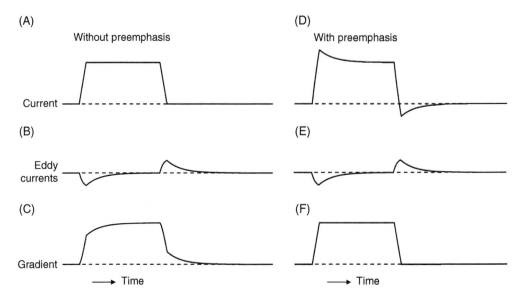

Figure 10.17 **Principle of preemphasis.** (A) A time-varying current, serving as input to a gradient coil, will generate eddy currents in surrounding conducting structures that oppose the current input. These eddy currents will in turn generate (B) time-varying B_0 and gradient magnetic fields. (C) The total magnetic field gradient seen by the spins is a superposition of the desired gradient and the undesired, eddy-current-related gradients. (D) During preemphasis the current waveform is distorted, such that the sum of (E) the eddy-current-related gradient and the gradient generated by (D) leads to the desired gradient waveform (F).

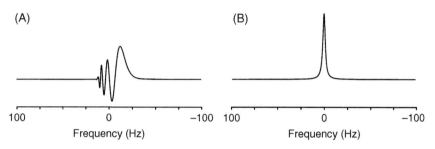

Figure 10.18 **Effects of eddy currents on MR spectra.** 3D localized ^{1}H NMR spectrum of water (A) without and (B) with eddy-current compensation by preemphasis and B_0 magnetic field compensation. The eddy currents originate from the magnetic field gradients required for 3D spatial localization (see Chapter 6).

[52] reveals that (i) eddy currents decay exponentially, whereby (ii) the decay constant depends on the resistance and inductance of the structure in which the eddy current in induced. The long-lasting eddy currents typically originate from cold structures within the magnet cryostat. In general, the time-varying eddy currents can by described as a (limited) sum of decaying exponentials with time constants up to 1000 ms. Several methods to (partially) compensate eddy currents are available of which preemphasis and active screening (or shielding) are the most robust and universally applicable.

10.4.2 Preemphasis

The time-varying eddy currents can be significantly reduced by a process referred to a preemphasis [52–54], which involves cancelation of the exponentially decaying gradient fields by overdriving the applied current (Figure 10.17D). Most commercial MR systems are equipped with several preemphasis circuits (usually up to three per gradient direction) in which a high-pass RC filter generates an exponentially decaying signal with adjustable amplitude and time constant. Summation of this additional signal to the input current step function gives a current input, which will produce the desired rectangular magnetic field gradient (Figure 10.17F).

The actual time-varying magnetic field gradients need to be quantitatively measured if a significant reduction of eddy currents is to be achieved. As mentioned earlier, eddy currents are to a first approximation comprised of time-varying magnetic field gradients $G(r, t)$ and magnetic field shifts $\Delta B_0(t)$ that need to be compensated separately, leading to 12 or more parameters (i.e. 6 amplitudes and 6 time constants) per channel to be optimized. In order to distinguish between $G(r, t)$ and $\Delta B_0(t)$, the total magnetic field $B(r, t)$ must be measured in, at least, two spatial positions. Several methods have been reported for the measurement of eddy currents, either using NMR signals from separate samples, using NMR signals along a spatially-encoded column [35], with small pickup coils [54], or with field cameras [51, 55–58].

A convenient hybrid method to demonstrate the principles of eddy current measurements is shown in Figure 10.19. A water-filled tube is placed in the magnet (and gradient) isocenter, parallel to the direction in which the eddy currents need to be measured. The tube should be narrow in the other two dimensions in order to minimize signal loss due to cross-term eddy currents (e.g. eddy currents generated in the y direction by a gradient pulsed in the x direction). A length-to-diameter ratio of ≥ 10 is recommended. Reference and test spatial profiles are obtained in the absence and presence of the test gradient, respectively, after a delay τ following the end of the test gradient (Figure 10.19A). During the echo-time TE, phase accumulation occurs due to eddy currents generated by the imaging and test gradients, magnetic field inhomogeneities, and frequency offsets. However, since the accumulated phase due to the imaging gradients, magnetic field inhomogeneities, and frequency offsets is identical between the

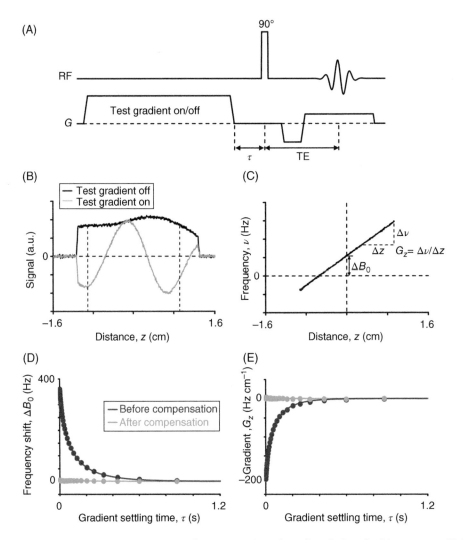

Figure 10.19 Quantitative measurement and compensation of gradient-induced eddy currents. (A) Pulse sequence generating (B) a spatial readout profile along the direction of a long, narrow cylindrical phantom in the absence or presence of a test gradient. (C) The phase difference between the profiles in (B) is, to first approximation, linear, whereby the intercept is indicative of a residual magnetic field offset generated by the test gradient. The slope is indicative of a residual gradient from the test gradient. Repeating the measurements for a range of gradient settling delays τ provides a complete description of the dynamic eddy currents in terms of (D) magnetic field offset and (E) magnetic field gradient (red lines). Following B_0 compensation and gradient preemphasis, the eddy currents can be reduced to negligible levels (D, E green lines).

reference and test spatial profiles, the phase difference will hold information about the eddy currents.

The final step is the adjustment of the preemphasis unit to give the desired amplitudes and time constants. However, the real-time constants being generated by the unit may be different from the entered time constants, depending on the tolerances and specifications of the individual R and C components used to generate the exponentials. Furthermore, the amplitudes are typically completely unknown as they are entered as a percentage of a maximum (often arbitrarily set) value. While the amplitude and time constant adjustments can be performed iteratively, a more reliable and consistent method is to calibrate the preemphasis unit before compensation. Suppose that gradient coil generates an eddy current gradient of amplitude $20\,\mathrm{Hz\,cm^{-1}}$ with a 100 ms time constant. The unit can be quickly calibrated by performing two measurements with preemphasis settings (G1, $\mathrm{TC_{G1}}$) = (0%, 100 ms) and (5%, 100 ms). While both measurements are affected by the system eddy currents, the difference will reveal a single exponential as generated by the preemphasis unit. Fitting of the difference data with a single exponential will give the calibrated amplitude $\mathrm{G1_{cal}}$ and time constant $\mathrm{TC_{G1cal}}$ (e.g. $-10\,\mathrm{Hz\,cm^{-1}}$/% and 110 ms). Therefore, a time constant of 90.91 ms and amplitude of +2% are required to completely cancel the measured eddy current gradient. When the unit has an inherent offset (e.g. $\mathrm{TC_{G1}}$ = 0 ms gives $\mathrm{TC_{G1cal}}$ > 0 ms) the calibration can be extended with a third measurement (G1, $\mathrm{TC_{G1}}$) = (5%, 200 ms) to characterize the intercept. The outlined calibration can be repeated for other time constants, for B_0 compensation, and for other channels. While labor intensive, the outlined calibration is completely quantitative and can set the preemphasis and B_0 compensation non-iteratively.

Similar to magnetic field homogeneity, the requirements for eddy current compensation differ widely for MRI and MRS applications. Typically, DTI has the most stringent requirements for MRI applications. Multiple images with strong diffusion weighting ($b > 0\,\mathrm{s\,mm^{-2}}$) in different directions are compared to a base image ($b = 0\,\mathrm{s\,mm^{-2}}$). Since DTI data are typically acquired with EPI methods, the effects of eddy currents lead to image distortion and signal loss and subsequently errors in the estimation of the diffusion tensor. For MRS applications, the requirements for eddy current compensation are dictated by the spectral resolution. As a general rule of thumb the effects of eddy currents should be minimized to the point that the spectral lines are not significantly broadened or distorted. For a line broadening of less than or equal to 25% of the line width at half maximum, $\Delta v_{\frac{1}{2}}$, this quantitatively translates into $|\Delta B_0| \le (\Delta v_{\frac{1}{2}}/4)$ and $|G(x)| \le \Delta v_{\frac{1}{2}}/(4\Delta x)$, where Δx represents the voxel dimension in the direction of the eddy current-induced gradient field [59]. For ^{1}H NMR experiments at 11.74 T, the optimal spectral resolution corresponds to metabolite line widths of 10–14 Hz in a volume of $3 \times 3 \times 3\,\mathrm{mm} = 27\,\mu l$ [60]. Therefore, the eddy current induced B_0 and gradient fields need to be compensated to better than circa 3 Hz and $10\,\mathrm{Hz\,cm^{-1}}$, respectively. While residual eddy currents will always lead to line broadening and/or signal loss, residual B_0 variations can be taken out during a post-acquisition correction step, as detailed in Chapter 9. While the method detailed in Figure 10.19 provides quantitative and accurate characterization of temporal magnetic field variations, the method is limited to one spatial dimension per measurement. A multichannel magnetic field monitor or magnetic field camera ([55, 56, 61], Figure 10.20) is a generalization of the aforementioned concepts. By measuring the magnetic field variations with multiple probes in parallel (Figure 10.20A) one can derive the magnetic field along all spatial dimensions simultaneously. The key to magnetic field cameras is in the construction of the individual probes (Figure 10.20B). The probe and sample should be small enough such that the dephasing caused by MRI magnetic field gradients is minimal across the sample. Furthermore, to achieve a long lifetime of the NMR signal, the RF probe, sample, and surrounding materials should be susceptibility matched. During a typical measurement the signal from a given probe is measured in the absence and presence of a magnetic field gradient sequence (Figure 10.12C). The phase difference is directly proportional to the k-space trajectory (Figure 10.12D) and can quickly and quantitatively reveal any difference with the ideal k-space trajectory. A four-channel magnetic field camera with a tetrahedral distribution can simultaneously determine magnetic field gradients along the three Cartesian axes, as well as the magnetic field offset. A 16-channel field camera can decompose the magnetic field in terms of a third-order SH expansion. Magnetic field cameras have been used to characterize shim and gradient eddy currents [57, 58], measure and correct k-space trajectories [51], and monitor the magnetic field for subject motion [62, 63].

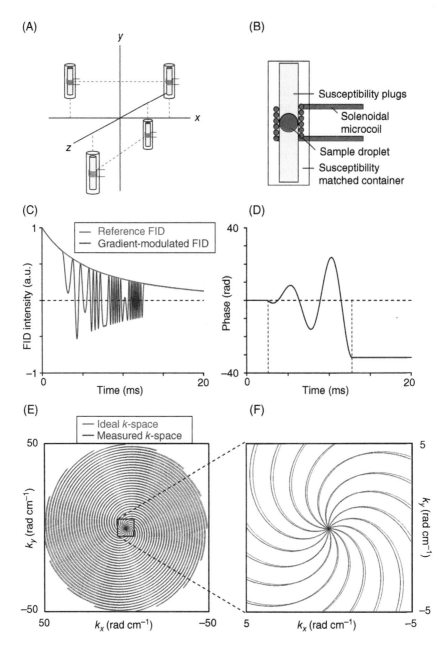

Figure 10.20 **Principles and applications of magnetic field cameras.** (A) Tetrahedral arrangement of a four-probe magnetic field camera. (B) Each probe is constructed as to maximize the signal lifetime (i.e. T_2^* relaxation time constant). Macroscopic magnetic field inhomogeneity is minimized through miniaturization (sample droplet Ø 1 mm), whereas microscopic magnetic field inhomogeneity is minimized through careful magnetic susceptibility matching of the various materials in close proximity to the sample droplet. (C) Typical data from a single probe shows a long lifetime in the absence of magnetic field gradients (blue) and phase modulation in the presence of magnetic field gradients (red). (D) The phase difference between the curves in (C) provides the k-space trajectory of the spins. (E, F) Magnetic field cameras are ideally suited to characterize the exact k-space trajectories, including deviations due to eddy currents and delays. A four-probe camera (A) allows the determination of all three linear gradients and the magnetic field offset. A 16-probe camera is required for characterizing imperfections up to the third SH order.

10.4.3   Active Shielding

The problem of eddy currents can also be reduced through a completely different approach. Instead of compensating the effects of eddy currents (for example, by preemphasis), eddy currents could be minimized by so-called active shielding [64–67]. As described in the beginning of this paragraph, eddy currents are a manifestation of Faraday's law of induction, i.e. time-varying magnetic field gradients induce a current and consequently a second opposing magnetic field in surrounding conductors like the magnet cryostat. If one could reduce (or even eliminate) the extraneous magnetic fields outside the active volume of the magnetic field gradient coil system, the problem of eddy currents would be eliminated. This is achieved by active shielding or screening. Although a detailed description of active screening is beyond the scope of this book, the principle of active screening will briefly be discussed and is shown in Figure 10.21 for a single loop of current-carrying wire. Without any screening, the magnetic field lines extend far beyond the dimensions of the loop. However, by strategically positioning a discrete wire array, which carries an opposite current, the magnetic field lines outside the active volume will be drastically reduced. For the simple example of a single current loop, Figure 10.21C shows the magnetic fields in the absence and presence of magnetic shielding. It follows that active shielding can reduce the magnetic field outside the coil by more than two orders of magnitude for positions >60 cm (coil radius = 25 cm). Note that the homogeneous volume is typically much smaller than the gradient coil dimensions. While active shielding is crucial for a wide range of experiments, in particular EPI and NMR spectroscopy, the methodology comes at the price of a reduced magnetic field inside the coil. This reduction is typically 30–50% of the amplitude of the unshielded coil geometry. However, for pulse sequences with high gradient demands like EPI, this reduction is often acceptable when the substantial gain in gradient performance is taken into account. The relatively simple principle is readily extended to three dimensions and more complex wire positions as encountered for Maxwell and Golay-type magnetic field gradient coils. Figure 10.21D shows a realistic coil design for an active shielded Y magnetic field gradient. In practice, active shielding is always combined with preemphasis, since active shielding is never perfect. The finite length of gradient coils compromises the shielding performance, as do fabrication errors.

10.5   Radiofrequency (RF) Coils

RF coils and associated hardware represent the third and final component of an MR system. The two primary functions of RF coils are (i) to deliver short RF pulses to manipulate and excite the magnetization into the transverse plane after which (ii) the same (or another receive-only) RF coil can be used to detect the precessing magnetization through electromagnetic induction. Both functions are performed optimally when the RF amplifier delivers *all* the available RF power to the coil during transmission, while *all* of the MR signal induced in the coil during reception is transferred to the preamplifier and associated receive chain. An RF coil that can be made resonant with the spin Larmor frequency and matched to the impedance of the MR system can achieve both functions simultaneously. While a detailed discussion on RF coil design is outside the scope of this book, the next section explains the need for tuning and matching of RF coils, the effect of sample loads, and the evaluation of RF coil performance.

10.5.1   Electrical Circuit Analysis

Figure 10.22A shows one of the simplest RF coils used in NMR, namely a circular surface coil made from copper wire and connected to a coaxial cable. One end of the coil is connected to the inner conductor of the cable, whereas the other end is connected to the outer conductor.

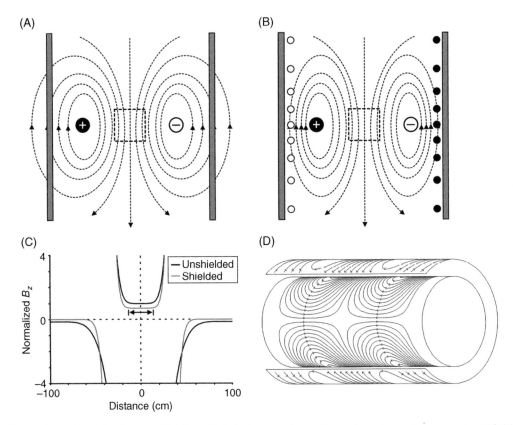

Figure 10.21 Principle of active shielding. (A) A single loop of current-carrying wire produces a magnetic field that can stretch far outside the magnet bore (vertical gray bars), thereby causing eddy currents in conducting structures like the cryostat. (B) By strategically placing a mesh of current-carrying wires on the outside of the single loop, the magnetic field outside the magnet can be greatly reduced. (C) Typical magnetic field profiles for unshielded (black lines) and shielded (gray lines) magnetic field gradients. While the magnetic field gradient amplitude of a shielded design is typically lower inside the magnet than that of an unshielded variant, the greatly reduced magnetic fields and thus eddy currents outside the magnet often greatly outweigh that drawback. (D) Schematic drawing of an actively shielded Y magnetic field gradient. *Source:* Data from Simon Pittard.

The two conductors within the coaxial cable are separated by a dielectric insulator. The dielectric constant of the inner insulator and cross sectional geometry of the coaxial cable determines the impedance of the cable, which for most MR applications is set to 50 Ω. A fixed capacitor is placed in parallel with the coil for reasons that will be explained shortly. The surface coil can be represented schematically by the electrical circuit diagram as shown in Figure 10.22B. The RF coil, like many other electrical circuits, can be described by three basic elements, namely resistors, inductors, and capacitors. The resistance (R, in Ohm Ω) is a measure of how much an object opposes the passage of electrons or electrical current. With the exception of superconducting wire, all materials have resistance or resistivity. The resistivity ρ (in $\Omega \cdot m$) is an intrinsic property of a material, independent of the shape, size, or length. The object-specific resistance can be calculated from the resistivity according to $R = \rho(l/A)$, where l and A represent the length (in m) and cross-sectional area (in m^2) through which the current is flowing. For a circular coil of round copper wire with loop diameter D_{loop} and wire diameter d_{wire}, the resistance can be calculated according to $R = \rho_{\text{copper}}(D_{\text{loop}}/\delta d_{\text{wire}})$, whereby δ is the skin depth given by [68]

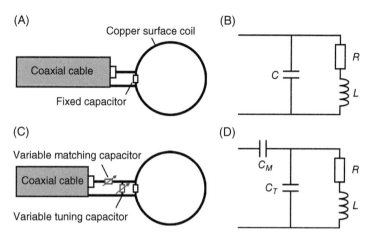

Figure 10.22 Basic elements of RF coil circuits. (A) A fixed capacitor single loop surface coil connected to a coaxial cable can be represented by (B) a capacitor C placed in parallel with an inductor L and resistance R. (C) A variable capacitor coil that allows sample-specific tuning and matching can be represented by the LCR circuit shown in (D).

$$\delta = \sqrt{\frac{2\rho_{copper}}{\omega\mu_0}} \qquad (10.9)$$

where $\mu_0 = 4\pi \times 10^{-7}$ H m^{-1} is the magnetic permeability of free space. The skin depth expresses the fact that alternating current (AC) ($\omega > 0$) has the tendency to run close to the surface (or skin) of a conductor. The skin effect arises from eddy currents induced by the changing magnetic field resulting from the AC. For a copper coil ($\rho_{copper} = 2.2 \times 10^{-8}$ Ωm) at 3 T (128 MHz), the skin depth is 6.6 μm, showing that current runs only in a very thin region close to the surface of the coil wire. For a 90 mm diameter coil made from 4 mm diameter copper wire, the resistance at 3 T is equal to 0.075 Ω. During RF transmission the system shown in Figure 10.22 is driven by a time-varying voltage (or driving electromotive force, emf, ε) with angular frequency ω according to $\varepsilon(t) = \varepsilon_{max}\sin(\omega t)$. The current I in the RF coil is then given by $I(t) = I_{max}\sin(\omega t + \phi)$, where ϕ denotes a phase difference between the driving voltage and current. The various elements of the RF coil (resistors, inductors, and capacitors) affect the maximum achievable current I_{max} and phase difference ϕ and it is their interplay that can be used to achieve optimal performance in terms of maximum power delivery and sensitivity. A quantitative analysis starts with Kirchhoff's rule, which states that the applied emf ε equals the sum of voltages over the resistor, inductor, and capacitor. The voltage ΔV_R over a resistor is in phase with the current and can be calculated with Ohm's law according to

$$\Delta V_R = IR = I_{max}R\sin(\omega t + \phi) \qquad (10.10)$$

When a current I flows through a conductor, it generates a magnetic field around that conductor. If the current is changing, the magnetic flux through the circuit will change and by Faraday's law of induction leads to the induction of an emf (electromotive force or voltage). Lenz's law states that the direction of the induced voltage V will be such as to oppose the change in current that created it. Self-inductance and inductance L is the ratio between the induced voltage $V(t)$ and the rate of change of the current ($dI(t)/dt$) according to $V(t) = L \cdot dI(t)/dt$.

An inductor therefore opposes *change* in current flow. All conductors have some inductance, whereby the inductance of a circuit depends on the geometry of the current path and on the magnetic permeability of nearby materials. For a circular loop of wire, the inductance L (in Henry, H) is given by [69]

$$L = \mu_0 \left(\frac{D}{2}\right)\left[\ln\left(\frac{8D}{d}\right) - 2\right] \tag{10.11}$$

For a 90 mm diameter coil made from 4 mm diameter copper wire, the inductance is equal to 180 nH. The voltage ΔV_L over an inductor is a quarter of a wavelength ahead (i.e. +90°) relative to the current and is given by

$$\Delta V_L = \omega L I_{max} \cos(\omega t + \phi) = X_L I_{max} \cos(\omega t + \phi) \tag{10.12}$$

Inductive reactance X_L is the opposition of an inductor to AC and is defined similarly to resistance as the peak voltage over the peak current according to $X_L = \omega L$. The 90° phase difference relative to the resistance is a direct consequence of the fact that energy is stored rather than dissipated in an inductor.

The final element in the electrical circuit diagram of an elementary RF coil (Figure 10.22B) is a capacitor. While some capacitance exists between any electrical conductors in close proximity to one another, a capacitor is specifically designed to add an exact amount of capacitance to a circuit. Most capacitors contain two electrical conductors or plates separated by a small distance containing a nonconducting, dielectric medium (glass, ceramics). Capacitors are widely used as part of electrical circuits to block direct current while simultaneously allowing AC to run in the circuit. A capacitor stores energy in the form of an electric field by collecting positive charge on one plate and negative charge on the other plate, whereby the dielectric medium determines the charge capacity. When driven by a time-varying voltage the capacitor is continuously charged and discharged. The voltage over a capacitor is a quarter of a wavelength behind (i.e. −90°) relative to the current and is given by

$$\Delta V_C = -\frac{I_{max}}{\omega C}\cos(\omega t + \phi) = -X_C I_{max} \cos(\omega t + \phi) \tag{10.13}$$

where X_C is the capacitive reactance, a measure for the resistance against changes in voltage. Similar to inductors, the −90° phase difference implies that energy is stored and not dissipated. In fact, the only electrical element that dissipates energy (normally as heat) is a resistor. C is the capacitance (in Farad) and is proportional to the conductor or plate area within the capacitor and inversely proportional to the distance between the plates. Typical capacitance values for RF coils in MRI are in the picoFarad (pF) range.

Upon application of Kirchhoff's rule in a phasor diagram, the maximum current I_{max} and phase angle ϕ can be calculated according to

$$\tan\phi = \frac{X_L - X_C}{R} \quad \text{and} \quad I_{max} = \frac{\varepsilon_{max}}{|Z|} \tag{10.14}$$

with $Z = R + i(X_L - X_C)$ and $|Z| = (R^2 + (X_L - X_C)^2)^{1/2}$. Z is known as the impedance and indicates the total opposition of the circuit to current flow. The impedance of an AC circuit can be regarded as the analog of resistance for a DC circuit. Note that resistance R and reactance X are both in units of Ohms (Ω).

The current within the circuit depends on the frequency ω of the applied emf. The current is maximized when the impedance Z is minimized, which corresponds to

$$\left(X_L - X_C\right) = 0 \quad \text{or} \quad \omega = \omega_0 = \frac{1}{\sqrt{LC}} \tag{10.15}$$

In other words, the current is maximized when the frequency of the applied emf ω equals the natural frequency ω_0 of the circuit. In this case a state of resonance is achieved. When a series RF circuit is brought into resonance, the impedance is minimum and equal to R, the current is maximum and equal to ε_{max}/R, and the voltage and current are in phase ($\phi = 0$). A given RF coil can be brought into resonance for a given frequency by adjusting either L or C. Because the inductance L can essentially only be altered by changing the coil geometry, in practice, the capacitance C is varied in a process referred to as "tuning."

However, bringing the LCR circuit in a state of resonance is not sufficient, since the RF coil needs to be connected to the preamplifier and the rest of the receiver system. Standard coaxial cables with 50 Ω impedance are typically employed to connect the coil to the preamplifier. In general, the impedance of the RF coil (and the sample) differs from 50 Ω, resulting in an inefficient power transfer from the amplifier (and coaxial cable) to the RF probe [70]. This would result in a decrease of the attainable current and consequently the B_1 field strength. Furthermore, if the same cable was used to transport the induced NMR signal from the RF probe to a preamplifier (with a 50 Ω input impedance) again, the signal transfer efficiency would be decreased, resulting in a decreased S/N. In order to avoid significant signal losses, the impedance of the RF coil must therefore be matched to 50 Ω. The principle of impedance matching can be understood by considering the response of a LCR circuit to an impulse. When a capacitor is charged (by a battery) and then connected to an inductor, an AC is generated, which is most intense at the resonant frequency ω_0. The capacitor is continuously charged and discharged as the electric energy is transferred between capacitor and inductor. Because some energy will be dissipated as heat in the resistor, the current (and voltage) will decrease as a function of time. A practical measure for the resistance of a coil is the quality factor Q, which is the ratio of reactance over resistance and equals $\omega L/R$ and $1/\omega CR$ for pure inductors and capacitors, respectively. Q is the time constant describing the exponential decay of voltage and current following an impulse. The Fourier transformation with respect to time will give information on the impedance as a function of frequency (Figure 10.23). Figure 10.23 reconfirms that impedance is a complex parameter, holding a resistance and a reactance component. On-resonance, i.e. at the Larmor frequency, the reactance is zero and the impedance is a pure resistance. However, this resistance is typically not 50 Ω, such that the matching condition is not satisfied. At a lower frequency ω_1 the resistance is equal to 50 Ω, but comes with a nonzero inductive reactance. At a higher frequency ω_2 the resistance is again equal to 50 Ω, but now comes with a nonzero capacitive reactance. The so-called matching capacitor (Figure 10.22) is used to cancel out the imaginary reactance component, leaving a pure 50 Ω resistance. Besides capacitive impedance matching, the 50 Ω matching condition can also be obtained through inductive impedance matching and using transmission lines [2]. Following impedance matching, the resonance condition at frequency ω_0 can be reestablished by adjusting the tuning capacitor (Figure 10.22).

Since the resonance frequency and complex impedance of a RF coil are influenced by the load, tuning and matching must be performed for each load individually through the use of variable tuning and matching capacitors. Figure 10.24 shows a commonly employed method for coil tuning and matching. The setup requires a variable frequency generator with frequency sweep capabilities, a RF bridge that directs a small RF signal to the probe, and an oscilloscope to monitor the output signal. These elements are fully combined and integrated in modern

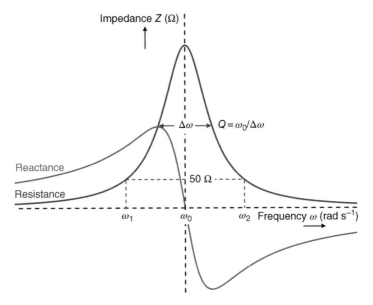

Figure 10.23 Impedance of an LCR circuit as a function of frequency. The complex impedance is composed of a resistance and a reactance component. On-resonance, the reactance is zero and the impedance is maximum. At a certain off-resonance frequency ω_1 or ω_2, the circuit can be considered a 50 Ω resistance plus a certain amount of reactance. The reactance may be canceled with capacitive or inductive matching, respectively.

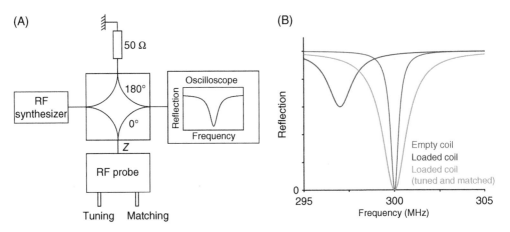

Figure 10.24 RF coil tuning and matching. (A) Tuning and matching setup using a hybrid junction. The RF synthesizer feeds the RF probe with a small signal. When the probe is properly matched, signals reflected by the probe and a 50 Ω resistance will have a 180° phase difference, leading to signal cancelation that can be detected and visualized on an oscilloscope. When the probe is also correctly tuned, the signal cancelation occurs at the desired resonance frequency. (B) An empty RF coil (i.e. without a sample) that is correctly tuned and matched is typically characterized by a narrow, selective tuning curve corresponding to a high Q value (blue). Introduction of a magnetically lossy sample (i.e. a biological sample) results in an impedance mismatch, Q reduction, and a shift in the resonance frequency (red). The mismatch in tuning and matching can be removed through adjustment of the variable tuning and matching capacitors (green). However, as the sample adds to the overall resistance of the combined coil/sample system, the reduction in the coil quality or Q value remains.

network analyzers. When the impedance of the RF coil equals the impedance of the 50 Ω load, no signal is reflected to the oscilloscope and the curve is minimized. Therefore, by changing the matching capacitor, the oscilloscope shows a minimum when the impedances are matched. Tuning is then performed by adjusting the minimum to the Larmor frequency. The main advantage of this particular setup for tuning and matching is that the coil impedance is known as a function of frequency rather than at the Larmor frequency only, as would be the case for simpler methods. This makes measurement of other parameters, like the coil quality factor, possible.

10.5.2 RF Coil Performance

Upon loading of the RF probe with a subject or sample, the subject and the coil will exchange energy. There will be inductive (i.e. magnetic) and capacitive (i.e. electric) interactions between the subject and the RF probe. The inductive interaction consists of an inductive coupling between current loops in the conducting sample and the RF probe. The subject–probe interaction is responsible for an efficient exchange of energy during RF pulse transmission and signal reception. This desirable interaction can easily be verified with a test sample. For an unloaded (correctly tuned and matched) RF coil, the quality factor Q is

$$Q_{unloaded} = \frac{\omega L}{R_c} \tag{10.16}$$

where R_c represents the resistance of the RF coil. On a network analyzer the Q value can be measured as Larmor frequency ω_0 divided by the bandwidth $\Delta\omega$, as shown in Figure 10.23. Maximization of the unloaded Q value is desirable. Introducing a magnetically lossy test sample (i.e. the sample conductivity $\sigma \neq 0$), like a physiological salt solution, reduces the Q value to

$$Q_{loaded} = \frac{\omega L}{\left(R_c + R_m\right)} \tag{10.17}$$

where R_m represents the effective resistance to magnetically induced currents within the sample/subject (see Figure 10.24B). A large drop in quality factor upon the introduction of a magnetically lossy sample indicates large magnetically induced currents in the sample (when dielectric losses can be excluded as will be explained shortly hereafter) and a high B_1 magnetic field within the sample. A concentrated B_1 is, through the principle of reciprocity (see Section 10.5.4), indicative of a sensitive coil making a large drop in Q value, one of the design criteria for a well-constructed RF probe. As the coil sensitivity S is proportional to $(Q_{loaded})^{1/2}$ [45], Eqs. (10.16) and (10.17) can be combined to give

$$\frac{S}{S_0} = \sqrt{1 - \frac{Q_{loaded}}{Q_{unloaded}}} \tag{10.18}$$

where S is the experimental coil sensitivity and S_0 represents the maximal sensitivity for an ideal RF probe (i.e. $R_c \rightarrow 0$). Equation (10.18) provides a good basis to evaluate the sensitivity of a given RF coil. For instance, when Q drops by a factor of 5 upon introduction of the sample, the achieved S/N is to within 10% of the maximum S/N for a probe with a given, unloaded Q value. Loading a coil with a magnetically lossy sample reduces, besides the quality factor Q, also the frequency (Figure 10.24B) because the coupling between the coil and the load is equivalent to a supplementary capacitance added to the coil capacitances. However, the tuning can easily be restored to the Larmor frequency by readjusting the tuning capacitance (Figure 10.24B).

Besides inductive coupling, capacitive (i.e. electric) coupling (which explains the reduction in frequency upon coil loading) occurs between sample and probe because electric fields produced by the coil cause a current flow within the sample. Since electric fields do not contribute to signal excitation it is generally desirable to minimize them as much as possible. The presence of significant electrical fields can easily be verified by introducing a dielectrically lossy test sample (e.g. a bottle of distilled water or a plastic) into the probe. The resulting Q value is identical to Eq. (10.17) except that R_m should be replaced with R_e, the effective resistance to dielectric coil/load interactions, because dielectrically lossy samples only interact with the electric fields in the probe. A large Q drop is indicative of strong electrical fields, significant energy dissipation, and a less efficient RF probe.

Therefore, a large Q drop upon introduction of a magnetically lossy sample only demonstrates the performance of a well-designed probe when there is no significant Q drop upon introduction of an exclusively dielectrical lossy sample. Electrical fields within the sensitive volume of the coil can be minimized, for instance, by proper (e.g. symmetrical) distribution of capacitors. Additional reductions in electrical fields may be obtained by distributing several capacitors along the coil wiring (instead of just using one).

10.5.3 Spatial Field Properties

An understanding of the electrical properties of an RF coil, as expressed in terms of an LCR circuit (Figure 10.22), is important to ensure maximum RF power delivery and efficient B_1 magnetic field generation during transmission and maximum signal transfer and optimal S/N during reception. An equally important aspect of RF coils relates to the spatial distribution of the magnetic and electric fields in relation to the sample. Before proceeding with a summary of the spatial field properties of various classes of RF coils, a basic description regarding electromagnetic waves is given.

First and foremost, it is emphasized that the NMR phenomenon is based on *magnetic* fields. During excitation a magnetic field is used to rotate the spins into the transverse plane, whereas during signal reception the rotating magnetization induces a voltage though magnetic induction. The magnetic fields rotate at a frequency equal or very close to the spin Larmor frequency, which for protons on most commercial MR systems (10^7–10^9 Hz) falls within the radiofrequency or RF range. This fact is continuously expressed when working with RF pulses, during which a short pulse is applied to the sample in order to achieve spin excitation. The RF part refers to the fact that the pulse is composed of a magnetic field rotating at a frequency in the RF range.

In the spectroscopic or quantum-mechanical description of NMR (Figure 1.13B), the energy-level difference between antiparallel or β spins and parallel or α spins is treated like any other two-level quantum system, such as that found in infrared (IR) spectroscopy (Figure 1.13A). With IR spectroscopy, electromagnetic waves in the IR frequency range are used to induce transitions between the various energy levels. Since the quantum-mechanical description of NMR uses the same two-level quantum representation (Figure 1.13B), it is only a small step to extend the analogy and describe the NMR phenomenon with RF electromagnetic waves transmitted by RF antennas, similar to those used by commercial radio stations. This description of NMR is not in agreement with practical NMR studies and leads to confusion, as described in detail by Hoult [71, 72].

In reality, RF pulses are transmitted by RF coils that are specifically designed to minimize the electrical field component. This is in stark contrast to RF antennas (Figure 10.25A) that are designed to transmit electromagnetic waves. Strong magnetic and electric fields are present in close proximity to the RF antenna that decrease inversely as $1/r^3$ and $1/r^2$ with distance r,

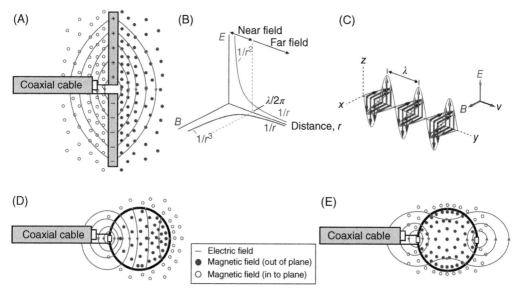

Figure 10.25 Magnetic and electric fields associated with RF antennas and RF coils. (A) A dipole antenna is designed to create electromagnetic waves or radiation in the far-field region of space. (B) In the near-field region the magnetic and electric fields decrease as the inverse cube and square of distance, respectively. (C) In the far-field region the magnetic and electric fields are orthogonal to each other and in a fixed amplitude and phase relation, whereby both fields decrease inversely with distance. (D) As the NMR phenomenon is based on magnetic fields, RF coils commonly used in NMR are designed as to minimize the electric field component, while simultaneously maximizing the magnetic field component. (E) A symmetrical distribution of capacitors provides a surface RF coil with a zero electric field in the center of the loop.

respectively (Figure 10.25B). At some distance away from the antenna, typically one or several wavelengths, both field components decrease inversely as $1/r$ with distance. At this point, when the magnetic and electric fields are orthogonal and in a fixed ratio, the components are referred to as an electromagnetic wave (Figure 10.25C). The far majority of MR studies performed today do *not* use electromagnetic waves present in the far-field region. Instead, the RF coils used for NMR are typically placed immediately adjacent to the sample and therefore operate in the near-field region of space. While the near-field components generated by an RF antenna as shown in Figure 10.25A could be used for NMR, the RF antenna is typically modified as to minimize the electric field component. Since NMR is based on the interaction between spins and magnetic fields, the electrical field component is not relevant. However, the electrical fields do lead to heating in conductive samples, such that minimization of the electrical fields is one of the prime design criteria of RF coils used in NMR. Bending the dipole RF antenna in Figure 10.25A into a circular loop (Figure 10.25D) reduces the electric field lines, while simultaneously enhancing the magnetic field line density. The circular RF coil can be improved further by breaking the loop into several pieces (Figure 10.25E) connected with appropriate capacitors to maintain the resonance condition. The even distribution of capacitance around the loop leads to a reduction in local electric field density and a more uniform magnetic field distribution. The electrical fields associated with the charge distribution on the RF coil are referred to as conservative electrical fields. During RF pulse transmission the time-varying, oscillating magnetic field induces a time-varying electric field in a conducting medium as a consequence of Maxwell's equations. These induced or nonconservative electrical fields are unavoidable and will lead, in combination with the conservative electrical fields, to sample or tissue heating (see Section 10.5.6).

The wealth of known RF probe designs, all of which utilize a cylindrical symmetry, can roughly be divided into two categories, according to the magnetic field orientation with respect to the main cylindrical axis. In this section the spatial field properties of different RF probe designs will be discussed. Spatial effects originating from the dielectric and conductive properties of the sample are ignored (i.e. the RF coils are empty), as these are the subject of Section 10.5.4.

10.5.3.1 Longitudinal Magnetic Fields

A perfectly homogeneous magnetic field can be generated by running a current tangentially on the surface of an infinitely long cylinder. One way to approximate this ideal current distribution is by winding a wire equidistantly on a cylinder (Figure 10.26D). Within the cylinder, parallel with the cylinder axis a very homogeneous magnetic field is generated (Figure 10.26E). This so-called solenoidal RF probe is, besides a completely spherical probe, the best probe design in terms of sensitivity and magnetic field homogeneity. Despite the outstanding electromagnetic properties of a solenoid, it is not frequently used for *in vivo* NMR studies because of the geometry of the coil relative to the main magnetic field B_0. Superconductive magnets themselves are normally built as solenoids (also because of the excellent magnetic field homogeneity), thereby offering a cylindrical bore with an axial main magnetic field. The magnetic field created by the RF coil should be perpendicular to the main magnetic field (e.g. Figure 1.4), which results

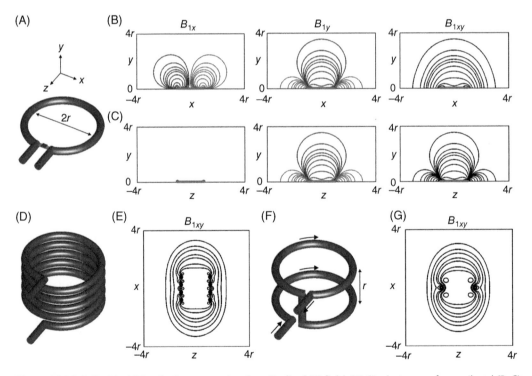

Figure 10.26 Cylindrical RF coils that generate a longitudinal RF field. (A) Single-turn surface coil and (B, C) the magnetic field components B_{1x}, B_{1y} and $B_{1xy} = (B_{1x}^2 + B_{1y}^2)^{1/2}$ in (B) the *xy* and (C) the *yz* planes perpendicular to the plane of the surface coil. (D) Five-turn solenoidal coil and (E) the B_{1xy} magnetic field distribution. (F) Two-turn Helmholtz coil and (G) the B_{1xy} magnetic field distribution. Maximum B_{1xy} magnetic field homogeneity is achieved when the two loops in (F) are separated by the coil radius *r*. The capacitors in (D) and (F) are not shown.

in a geometry conflict for solenoidal RF probes. Since the main cylindrical axis must be perpendicular to the main magnetic field B_0, the sample cannot easily be positioned in the RF probe. In addition, the large axial magnet bore space is not used efficiently.

This problem can be partially alleviated by reducing the number of windings to two as shown in Figure 10.26F. The two parallel, in series connected windings, known as a Helmholtz coil [73], allow a much better access to the interior of the RF coil. When the distance between the two coil planes equals the coil radius, maximum field homogeneity is obtained. Even though the Helmholtz coil provides lower sensitivity and homogeneity than a solenoidal coil of the same dimensions, it is more frequently used for *in vivo* NMR applications since the sensitive volume is better accessible.

When a solenoidal coil is reduced to a single loop, the RF coil is known as a surface coil ([74–79], Figure 10.26A). As a surface coil can be placed immediately adjacent to the sample, it provides high RF transmit efficiency and receive sensitivity. In the static regime (i.e. direct currents), the magnetic field distribution of a surface coil can be obtained through integration of the Biot–Savart law and is given by Eq. (10.6). Figure 10.26B and C shows the magnetic field distribution of a surface coil. It is clear that the transmit efficiency and sensitivity advantages of a surface coil are offset against a very inhomogeneous magnetic field distribution and limited volume coverage of the sample (Figure 10.26B and C). The problem of inhomogeneous transmission is often alleviated through the use of a separate, homogeneous transmit coil, whereby the surface coil is used in a receive-only mode. Alternatively, RF pulse transmission can rely on adiabatic RF pulses that achieve a constant nutation angle despite large variations in RF amplitude (see Chapter 5). The limited volume coverage of a single surface coil is often overcome by using a phased array, consisting of an array of independent surface coil receivers that can cover a large volume with the same sensitivity as a single surface coil.

10.5.3.2 Transverse Magnetic Fields

A homogeneous magnetic field perpendicular to the axis of an infinitely long cylinder can be generated by running a (co)-sinusoidal current distribution along the surface of the cylinder (Figure 10.27A and B). While this theoretical prediction cannot be achieved in reality, many RF coil designs are aimed at achieving an approximation of the ideal current distribution. The Alderman–Grant or slotted tube resonator [80–82] represent an early approximation of the ideal current distribution (Figure 10.27C). Unlike the longitudinal magnetic field coils shown in Figure 10.26, the RF magnetic field B_1 of a slotted tube resonator is perpendicular to the main magnetic field B_0, thereby providing full and easy access to the sensitive volume of the RF coil. A saddle coil (Figure 10.27D) provides another approximation of the ideal current density and is still used today in high-resolution NMR probes. In both the Alderman–Grant and saddle resonators the parallel conductors provide the homogeneous RF field, whereby the arcs are largely there for support and to ensure continuity. Extending the number of parallel conductors will improve the approximation of the ideal current density, thereby naturally leading to the so-called birdcage resonator [83–85]. In a birdcage resonator (Figure 10.27E and F), a large number of equidistant conductors or "rungs" (typically 16–32) are connected to two end rings. The birdcage resonator can be made to work at lower frequencies in a "low-pass geometry" (Figure 10.27E) or at higher frequencies in a "high-pass geometry" (Figure 10.27F). The birdcage coil, as well as many other volume coils, is typically surrounded by an RF shield to minimize external effects on the RF coil performance. The RF shield is typically very thin or segmented in order to avoid eddy currents. The birdcage coil is the workhorse for clinical MR at 1.5 and 3.0 T, both as a transceiver and as a transmitter in combination with phased-array receivers. At higher magnetic fields, birdcage coils are typically replaced by Transverse Electromagnetic Mode (TEM) resonators [86, 87].

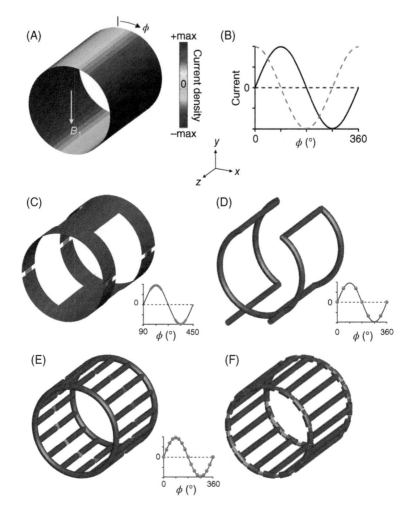

Figure 10.27 **Cylindrical RF coils that generate a transverse RF field.** (A, B) Current density on an infinitely long cylinder as a function of azimuthal angle ϕ to generate a homogeneous RF field orthogonal to the cylinder length (solid line, B_1 in up/down direction, dotted line, B_1 in left/right direction). (C–F) Cylindrical RF coils that achieve an approximation of the ideal current density distribution in (A, B). (C) Alderman–Grant or slotted tube resonator achieves a very good approximation over small ranges of the azimuthal angle ϕ. (D) Saddle coil. (E) Low-pass and (F) high-pass birdcage coil.

10.5.4 Principle of Reciprocity

The MR signal or emf induced in a RF probe can be calculated by the principle of reciprocity [88–90], which states that the emf ξ induced by a magnetic moment **M** from a voxel dV at a spatial position r is determined by the magnetic field B_1 at that position when unit current flows through the coil. Mathematically this can be represented by

$$d\xi(dV) = -\frac{d(B_1 \cdot M)dV}{dt} \tag{10.19}$$

Therefore, knowledge of the B_1 magnetic field during transmission will give direct information about the induced MR signal in that same RF coil during reception. Following substitution of the three orthogonal component of B_1 and M, the induced MR signal is given by (see Exercises)

$$d\xi = M_0\omega B_1 \sin(\gamma B_1 T)e^{+i\omega t}dV \tag{10.20}$$

where T represents the length of a square excitation RF pulse. Equation (10.20) shows the characteristic ωB_1 product common to the principle of reciprocity [90] and confirms that the MR signal observed during signal reception is directly proportional to the B_1 magnetic field during transmission.

An important assumption underlying Eq. (10.20) is that NMR can be treated as a near-field phenomenon (i.e. the wavelength is much longer than the sample dimensions). At the time of the original papers on the principle of reciprocity as applied to NMR [88, 89], the magnetic field strengths were relatively low making this a perfectly valid approximation. However, as magnetic fields strengths for *in vivo* applications steadily increase, the corresponding wavelengths become comparable or even smaller than the sample size. Under these conditions the wave behavior inside the sample must be considered, making it necessary to utilize the complete equations for the principle of reciprocity. As clearly pointed out by Hoult [90], despite the additional complications due to sample-induced wave effects, the principle of reciprocity remains valid under all realistic conditions. The origin of sample-induced wave effects will be discussed following a brief summary of electromagnetic wave propagation.

10.5.4.1 Electromagnetic Wave Propagation
A complete and concise treatment of electromagnetic wave propagation in materials is well outside the scope of this book and can be found in standard text books on electromagnetism [11]. This section merely summarizes some of the most important findings in order to aid the reader in forming a basic understanding into the origins of wave behavior and its effects on high-field MR images of the human brain.

The magnitude, orientation, and spatial distribution of magnetic and electric fields produced inside a sample by an RF coil are completely described by Maxwell's equations. The equations state that as a time-varying electric or E-field propagates through the sample it induces a time-varying magnetic or B-field. The time-varying magnetic B-field in turn gives rise to a time-varying electric E-field, underlining the observation from Figure 10.25C that the magnetic and electrical fields are intrinsically coupled. Whereas the B and E-fields are in a fixed amplitude and phase relationship in the far-field regime, the relationship in the near-field regime relevant for NMR is complex and dependent on the electrical properties (conductivity and permittivity) of the sample. Flowing in the sample are both conduction currents associated with the sample conductivity and displacement currents associated with the dielectric permittivity constant. Note that the displacement current is not a real current (i.e. movement of charge), but rather refers to the fact that a changing electric field generates a changing magnetic field. These currents have a 90° phase difference and also produce their own fields, thereby modifying the magnetic field inside the sample. When the RF coil applies a sinusoidal time-dependent magnetic field B to the sample (i.e. $B = B_{max}\cdot e^{-i\omega t}$), the magnetic field generated within the sample is given by $B = B_{max}\cdot e^{-(ikr+i\omega t)}$ with

$$k = \sqrt{\omega^2\mu\varepsilon - i\omega\mu\sigma} = k_R + ik_I \tag{10.21}$$

where μ represents the permeability of free space ($4\pi \times 10^{-7}\,\mathrm{H\,m^{-1}}$), $\varepsilon = \varepsilon_0\varepsilon_r$ represents dielectric permittivity, and ε_0 and ε_r are the dielectric permittivity of free space and the relative dielectric permittivity, respectively. σ is the conductivity and is inversely related to the resistivity ρ

($\rho = 1/\sigma$). The imaginary part of k, k_I, is often referred to as an attenuation factor. The reciprocal of k_I determines the distance over which the amplitude of the magnetic field is attenuated by a factor ($1/e$) and is commonly known as the skin depth of the medium. The real part of k is inversely proportional to the wavelength λ in the medium ($\lambda = 2\pi/k_R$) and gives rise to a change in phase with position in the medium.

Figure 10.28A shows the effects of conductivity and permittivity on the appearance of MR images at 64 MHz (protons at 1.5 T) and 298 MHz (protons at 7.0 T). Transmit and receive magnetic fields B_1^+ and B_1^- are calculated for an infinitely long cylinder with uniform electrical properties for a homogeneous, sinusoidally varying RF magnetic field. The low permittivity of mineral oil ($\varepsilon_r = 2.5$) and virtual absence of conductivity ($\sigma \sim 0\,\mathrm{S\,m^{-1}}$) leads to a 69.2 cm wavelength at 298 MHz, resulting in homogeneous magnetic fields both during transmission and reception. RF frequencies in distilled water ($\varepsilon_r = 81$, $\sigma = 0.01\,\mathrm{S\,m^{-1}}$) have an 11.2 cm wavelength at 298 MHz. The short wavelength compared to the sample size (12 cm diameter) results in standing waves, also known as "dielectric resonances," that lead to local minima and maxima in the magnetic field distribution. This effect is similar to the constructive and destructive interference of waves caused by stones thrown in a water pond. Adding salt to the water sample in Figure 10.28A results in a greatly increased conductivity ($\sigma = 1.96\,\mathrm{S\,m^{-1}}$ for 150 mM NaCl solution), which together with a similar permittivity ($\varepsilon_r = 78$) leads to a 9.6 cm wavelength at 298 MHz. Despite the shorter wavelength the constructive interference in the center of the sample is minimal, because the conductivity has significantly attenuated the RF magnetic field. A "brain-equivalent" sample ($\varepsilon_r = 53.0$, $\sigma = 0.69\,\mathrm{S\,m^{-1}}$) leads to a 13.0 cm wavelength at 298 MHz, leading to a mixture of constructive interference and attenuation. The "brain-equivalent" sample at 1.5 T ($\varepsilon_r = 94.7$, $\sigma = 0.48\,\mathrm{S\,m^{-1}}$) leads to a 40.9 cm wavelength and relatively homogenous transmit and receive magnetic fields. The results in Figure 10.28A indicate that at high magnetic fields, (i) it is not possible to obtain homogeneous magnetic fields in commonly encountered samples and (ii) the transmit B_1^+ and receive B_1^- magnetic fields diverge due to phase effects caused by sample conductivity. The effects simulated in Figure 10.28A are readily observed in high-field MR studies on the human brain (Figure 10.28B and C). While perfectly homogeneous on a mineral oil phantom, the RF transmission B_1^+ magnetic field in the human brain at 4 T (170 MHz for protons) shows clear constructive and destructive interference in the middle and edge of the brain (Figure 10.28B). A coronal MR image of the human head at 7 T acquired with a TEM coil shows signal voids in the cerebellum due to destructive interference (Figure 10.28C).

The principle of reciprocity can be extended to include sample-induced phase and attenuation effects [90], such that the detected MR signal given by Eq. (10.20) becomes

$$d\xi = M_0 \omega B_1^{-*} \sin\left(\gamma B_1^+ T\right) e^{+i\omega t} dV \qquad (10.22)$$

where B_1^+ and B_1^- are the magnetic fields effective during transmission and reception, respectively. * denotes complex conjugation. At low magnetic field strengths, and hence low frequencies, the rotating magnetic fields are real and equal, i.e. $B_1^+ = B_1^- = B_1$, such that Eq. (10.22) reduces to Eq. (10.20). When the sample is conductive, B_1^+ and $(B_1^-)^*$ are not identical. The experimental measurement of B_1^+ is relatively straightforward and several MRI-based mapping methods are described in Chapter 4. Once B_1^+ is known, B_1^- can, in principle, be calculated using Eq. (10.22). While this is straightforward on a homogeneous phantom, the *in vivo* B_1^- calculation is complicated by T_1, T_2, and proton density contrast between tissues and by non-RF-related phase effects. The distinction between B_1^+ and B_1^- becomes important during accelerated parallel imaging (Chapter 4), B_1 shimming, and signal quantification using external concentration references (Chapter 9).

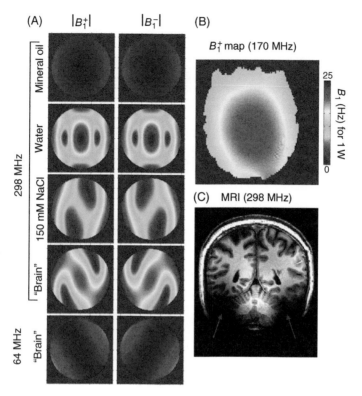

Figure 10.28 Standing wave effects at high magnetic fields. (A) Simulated images for the transmit B_1^+ and receive B_1^- magnetic fields for different materials at 64 MHz (1.5 T) and 298 MHz (7 T). The RF wavelength for the cylindrical mineral oil ($\varepsilon_r = 2.5$, $\sigma \sim 0\,S\,m^{-1}$), water ($\varepsilon_r = 81$, $\sigma = 0.01\,S\,m^{-1}$), and 150 mM NaCl ($\varepsilon_r = 78$, $\sigma = 1.96\,S\,m^{-1}$) phantoms are 69.2, 11.2, and 9.6 cm, respectively. The short wavelength in water compared to the 12 cm diameter phantom leads to large standing wave effects and a central bright spot. The RF wavelength for a brain-like phantom ($\varepsilon_r = 94.7$, $\sigma = 0.48\,S\,m^{-1}$ at 1.5 T and $\varepsilon_r = 53.0$, $\sigma = 0.69\,S\,m^{-1}$ at 7 T) is 40.9 and 13.0 cm at 1.5 and 7 T. (B) Experimentally measured transmit B_1^+ map on the human brain at 4 T using a TEM coil illustrates the focusing of the B_1^+ magnetic field in the center of the brain. (C) MP-RAGE image of human brain at 7 T acquired with a TEM coil shows signal dropout in the lower parts of the brain, including the cerebellum (red arrows).

10.5.5   Parallel Transmission

Magnetic field inhomogeneity in the B_1^+ RF transmit field is a significant problem at higher magnetic fields, especially for larger body regions such as the abdomen. The standard method to improve the B_1^+ magnetic field homogeneity is RF shimming through parallel transmission. This method requires a multichannel transmit coil in which each channel is connected to an independent RF amplifier with full amplitude and phase control (Figure 10.29A). Simultaneous transmission on all channels with equal amplitude and phases (incremented by 45° on each channel) in order to simulate a conventional volume coil results in poor magnetic field homogeneity (Figure 10.29C), as expected. RF shimming entails the acquisition of a quantitative B_1^+ map for each individual transmit channel (Figure 10.29B), together with the acquisition of RF transmit phase maps relative to channel 1 (Figure 10.29D). Using a least-squares minimization routine, similar to that used for B_0 shimming, the eight B_1^+ amplitudes and seven relative phases can be optimized to provide a homogeneous B_1^+ map across an entire slice (Figure 10.29E). When the transmit magnetic field homogeneity only needs to be optimized over a small region, for example, for single-volume MRS, the eight B_1^+ amplitude and seven relative phases can be

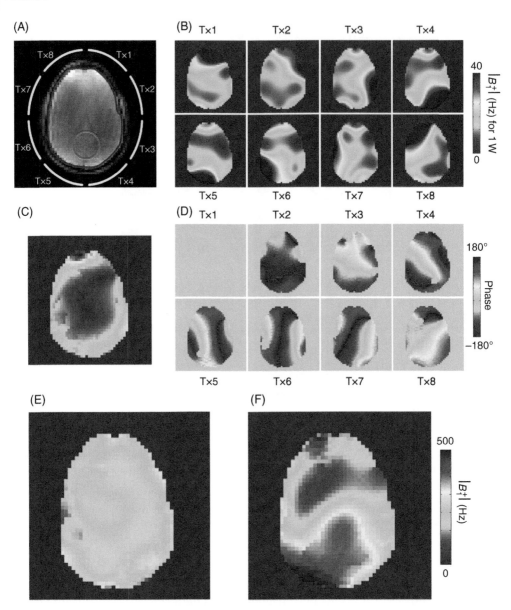

Figure 10.29 Principle of RF shimming through parallel transmission. (A) Eight-element transmit coil (yellow) together with brain (red) and localized (blue) ROIs. (B) Quantitative B_1^+ transmit maps for each of the eight channels shown in (A). (C) Adding the B_1^+ maps in (B) without amplitude scaling or phase adjustment leads to a highly inhomogeneous B_1^+ map across the brain. (D) Quantitative phase maps of each transmit channel relative to channel 1. (E, F) B_1^+ map following amplitude scaling and phase adjustment of each RF channel optimized to achieve a homogeneous B_1^+ map across (E) the brain ROI (200 Hz target B_1^+ amplitude) and the localized ROI (400 Hz target B_1^+ amplitude). The total summed RF power in (E) and (F) are approximately equal.

optimized to provide a higher maximum B_1^+ amplitude within the ROI. While RF shimming has been successfully used for MRS [91, 92], MRSI [93, 94], and MRI [95, 96] applications at high magnetic field, it is generally an expensive and time-consuming option.

An alternative to multichannel RF shimming is provided through passive modification of the RF transmit field by high permittivity materials [97–99] such as calcium titanate. The high dielectric materials modify the RF transmit and receive magnetic fields, such that strategic placement of high permittivity blocks or bags can improve the local RF magnetic field homogeneity and/or amplitude.

10.5.6 RF Power and Specific Absorption Rate (SAR)

In earlier sections it was emphasized that the NMR phenomenon is based on magnetic fields and that the primary function of an RF coil is to generate magnetic fields. However, the presence of electric fields is unavoidable. Conservative electric fields associated with the charge distribution of the RF coil can be reduced, but not completely eliminated. Nonconservative or induced electric fields are intrinsically linked to the magnetic fields necessary for NMR. The electric fields will lead to heating in conductive samples, like humans and animals. The amount of power absorbed by the sample is directly proportional to the electric field according to

$$P = \frac{1}{2} \int_v \sigma |E|^2 \, dv \tag{10.23}$$

where σ is the conductivity and v the sample volume. The electric field term in Eq. (10.23) represents the three orthogonal components E_x, E_y, and E_z and includes both conservative and nonconservative electric fields. For homogenous objects like spheres and cylinders, the electric field distribution can be calculated analytically, similar to the calculation of B_1^+ and B_1^- magnetic fields shown in Figure 10.28A. However, for biological samples like the human brain, the calculation of the RF power deposition and electric fields requires a numerical calculation.

Among several available methods [2], the finite difference time domain (FDTD) method is a popular choice in the MRI community. As an initial step, a model of the human head is gridded into various tissue types, each with specific conductivity and permittivity values. Standard human heads from the virtually family [100] provide a realistic and consistent option. The RF coil is also digitized, after which the various B and E fields are calculated based on Maxwell's equations. Figure 10.30 shows typical results produced by FDTD calculations on the human head at 3 and 7 T for a birdcage resonator. Similar to the analytical results of Figure 10.28, the B_1^+ and B_1^- homogeneities are lower at 7 T (Figure 10.30B and C) than at 3 T (Figure 10.30G and H). The B_1^+ efficiency at 7 T (in terms of the amount of B_1^+ produced per Watt of input power) is lower than at 3 T, likely due to radiated losses and limited penetration depth. The total E fields at 3 and 7 T are shown in Figure 10.30D and I, respectively. While the magnetic B_1^+ field can be experimentally measured, it shows little resemblance to the corresponding E fields, thereby further underlining the need for numerical simulations. Since E fields are closely related to the absorbed power (Eq. (10.23)) and since RF power is readily monitored during MRI, the most commonly used criteria to regulate RF heating is the specific absorption rate (SAR). SAR can be defined as the time- and volume-averaged power absorbed in the whole body, the whole head, or per 10 gram of tissue (SAR_{10g}). The first two SAR measured can be estimated from the power delivered by the RF amplifier (corrected for losses and reflected power) and the total tissue volume or mass "seen" by the RF coil. The SAR_{10g} measure is a local estimate and requires numerical simulations as shown in Figure 10.30E and J. Figure 10.30

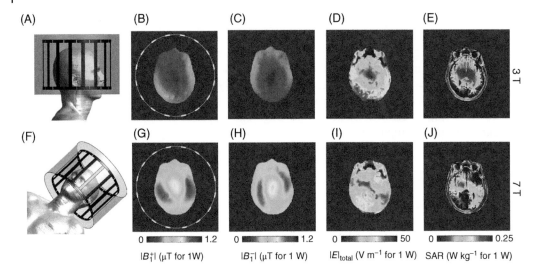

0 ▬▬ 1.2   0 ▬▬ 1.2   0 ▬▬ 50   0 ▬▬ 0.25

$|B_1^+|$ (μT for 1W) $|B_1^-|$ (μT for 1 W) $|E|_{total}$ (V m^{-1} for 1 W)   SAR (W kg^{-1} for 1 W)

Figure 10.30 **Finite difference time domain (FDTD) simulations of a birdcage coil at 3 and 7 T.** (A, F) 3D model of the head of human model Duke [100] relative to a 16-rung, quadrature high-pass birdcage coil with 32 capacitors of 31.5 pF (3 T) and 4.95 pF (7 T) placed along the end rings. Simulations were performed in Sim4Life (Zurich Medtech, Zurich, Switzerland) with a 1.5 mm minimum voxel size, leading to 22.18 million cells describing the 3D model. All simulations were performed for a 1 W input power. (B, G) Quadrature mode, clockwise rotating magnetic field B_1^+ and (C, H) anti-quadrature, counter-clockwise rotating magnetic field B_1^- at (B, C) 3 T and (G, H) 7 T. (D, I) Total electric field, including conservative and nonconservative contributions, at (D) 3 T and (I) 7 T. (E, J) Specific absorption rate (SAR) at (E) 3 T and (J) 7 T. *Source:* Data from Bart Steensma and Dennis Klomp.

shows that global SAR estimates should be used with great caution as the local SAR$_{10g}$ values can vary several fold. The Food and Drug Administration (FDA) has set SAR limits on RF heating as 4 W kg^{-1} for whole body averaged over 15 min, 3 W kg^{-1} over the head for 10 min, and 8 W kg^{-1} in any 1 g of tissue. SAR limits in the extremities are more lenient and are set at 12 W kg^{-1} averaged over 15 min.

Even though RF heating is governed by guidelines on SAR limits, the absorbed power has only an indirect correspondence to the real parameter of interest, namely local tissue temperature. As real-time monitoring of temperature *in vivo* is very difficult, numerical models are used for the estimation of local temperature (Figure 10.31). The Pennes bioheat equation [101] relates the change in local temperature to the local tissue heat capacity, metabolic heat generation, local perfusion and, in the case of NMR, also the RF power deposition. Thermal simulations (Figure 10.31C and F) show that tissue heating at 7 T is larger than at 3 T due to the lower B_1^+ efficiency at higher magnetic fields. The relevance of local temperatures has led the IEC to issue limits on core body temperature and maximum local temperatures [102].

10.5.7 Specialized RF Coils

In the previous sections the general design and operating criteria of a variety of RF coils have been described. In this section, a number of more specialized RF coils are described that have utility for *in vivo* NMR spectroscopy. These include RF coils in which RF transmission and reception are achieved by separate coils, RF coils for heteronuclear applications, phased-array coils, and cooled (superconducting) RF coils.

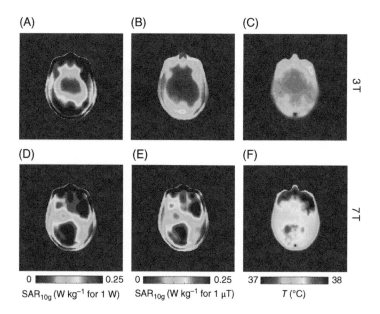

(A)　　　　　　(B)　　　　　　(C)

3T

(D)　　　　　　(E)　　　　　　(F)

7T

0 ▰▰▰▰ 0.25　　0 ▰▰▰▰ 0.25　　37 ▰▰▰▰ 38

SAR$_{10g}$ (W kg^{-1} for 1 W)　　SAR$_{10g}$ (W kg^{-1} for 1 µT)　　T (°C)

Figure 10.31 **RF power deposition and temperature in the human brain at 3 and 7 T.** (A, D) Specific absorption rate (SAR) averaged over 10 g tissue (SAR$_{10g}$) as calculated from the SAR map in Figure 10.30E and J at (A) 3 T and (D) 7 T for a 1 W input power. (B, E) SAR$_{10g}$ maps resulting from a magnetic field B_1^+ of 1 µT in the center of the brain at (B) 3 T and (E) 7 T, showing the increased power requirements at 7 T to achieve an identical B_1^+. (C, F) Temperature maps at (C) 3 T and (F) 7 T following a 180 s heating period with a continuous power level to achieve a 10 µT B_1^+ magnetic field in the center of the brain. Note that the duty cycle in realistic MR studies is far below 100%. Thermal simulations were performed in Sim4Life (Sim4Life [Zurich Medtech, Zurich, Switzerland]) using the Pennes bioheat equation. *Source:* Data from Bart Steensma and Dennis Klomp.

10.5.7.1　Combined Transmit and Receive RF Coils

The many different RF coil designs can partially be explained by the fact that each coil is often a compromise between its desirable and undesirable features. There is, in general, not a single coil optimal for all applications. A volume coil (e.g. a birdcage coil) provides a homogeneous transmit B_1^+ magnetic field, such that spin excitation is uniform across the sample. However, the large size and often poor filling factor of volume coils compromises their sensitivity. Surface coils on the other hand are very sensitive due to their high filling factor and their optimal size relative to the object under investigation. A drawback of surface coils is that the generated B_1 field is extremely inhomogeneous, which will lead to signal loss when conventional amplitude-modulated RF pulses are transmitted. In other words, a volume coil is desirable for RF transmission, but undesirable for signal reception. A surface coil, to the contrary, is desirable for signal reception, but performs poorly during RF transmission.

A more optimal RF coil can be constructed when transmission and reception can be achieved by volume and surface coils, respectively. The construction of a volume coil transmission–surface coil reception combination is more complicated as any interaction between the two RF coils will lead to artifacts, signal loss, or inefficient power transfer. As a first step to minimize coil interactions, the coils should be positioned orthogonal relative to each other. Secondly, the mutual coupling between the coils is minimized by so-called active detuning. During RF transmission, the surface coil is actively detuned (for example, by temporarily changing the capacitance of the surface coils circuitry through the use of a PIN-diode), so that it becomes very inefficient to interact with RF fields in the Larmor frequency range. During signal acquisition, the surface coil is quickly returned to a state of resonance, so that signal can be efficiently received. Depending on the spatial resolution, the sensitivity obtained with this combined coil setup as compared to a single transceiver surface coil in combination with adiabatic RF pulses could be slightly reduced. This is because the B_1 flux lines between transmit and receive coils are spatially dependent [77], which will lead to some phase cancelation. For MRI this effect is minimal, as the phase variation across an MRI pixel will be negligible. For localized NMR spectroscopy, the effect can be minimized by limiting the voxel to the sensitivity volume of the surface coil where the variation in B_1 flux lines is minimal. In analogy to signal reception with regular surface coils, the penetration depth of the dual coil setup is limited to about one coil radius (see Figure 10.26).

10.5.7.2 Phased-Array Coils

The primary advantage of surface coil reception is the high sensitivity that can be obtained immediately adjacent to the coil. This has made surface coils the primary choice of applications that are inherently sensitivity limited, like MR spectroscopy. Besides the fact that the B_1 magnetic field is highly inhomogeneous, the main disadvantage of surface coils is their limited spatial coverage. Hyde [103] proposed using multiple, noninteracting surface coils and receivers to independently and simultaneously obtain signals from multiple coils, achieving improved signal-to-noise ratio (SNR) from an expanded spatial region with no increase in time. Roemer [104] implemented a system using four receiver coils and four receiver channels, which was referred to as an NMR-phased array and is schematically shown in Figure 10.32. In order to minimize noise correlation, each coil in a phased-array assembly is connected to its own preamplifier and receiver. Under the assumption that the noise is uncorrelated, the combined signal $S_c(x,y,z)$ can be calculated from the signals from the N individual coils, $S_n(x,y,z)$, multiplied by a weighting coefficient $w_n(x,y,z)$ that is proportional to the SNR for coil n according to

$$S_c\left(x,y,z\right)=\frac{\sum_{n=1}^{N}w_n\left(x,y,z\right)S_n\left(x,y,z\right)}{\sqrt{\sum_{n=1}^{N}\left[w_n\left(x,y,z\right)\right]^2}}\qquad(10.24)$$

Prior to the weighted addition given by Eq. (10.24), the spectra must be phase-corrected in order to avoid phase cancelation. Phased-array receiver coils are often used in combination with a homogeneous volume transmission coil and can achieve significant improvements in the obtainable SNR, especially in areas close to the surface coils. Current implementations of phased-array coils use 16–32 coils, while extensions to 64–128 coils and have been described [105, 106]. Rather than using the phased-array assembly to increase the SNR of the experiment, the multiple coils can also be used to increase the temporal resolution of the experiment through parallel data acquisition strategies, like GRAPPA [107], SENSE [108], and SMASH [109]. The principles of parallel data acquisition are discussed in Chapter 4.

10.5.7.3 ^{1}H-[^{13}C] and ^{13}C-[^{1}H] RF Coils

NMR spectroscopy is a unique tool to measure important metabolic fluxes like the TCA cycle flux noninvasively in human and animal brain *in vivo*. Chapter 3 provided a detailed description of the experimental considerations and analysis of flux measurements through the infusion of ^{13}C-labeled substrates and the *in vivo* detection of the ^{13}C-labeled metabolites. Chapter 8 provided a description of the NMR pulse sequences necessary to perform these state-of-the-art experiments. For almost all experiments, be it through indirect ^{1}H-[^{13}C] or direct ^{13}C NMR detection, a double RF coil setup is required. Figure 10.33 shows the most commonly used ^{13}C-[^{1}H] (Figure 10.33A) and ^{1}H-[^{13}C] RF coils (Figure 10.33B and C) for human and animal applications *in vivo*. Since direct or polarization transfer-enhanced ^{13}C NMR spectroscopy is

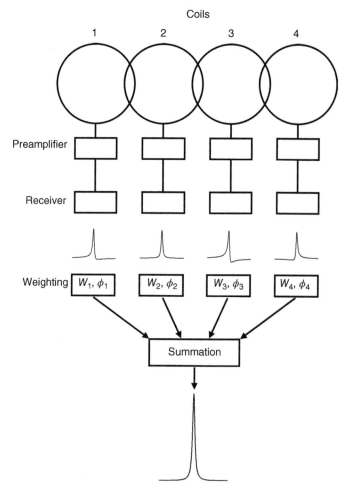

Figure 10.32 Four-element, phased-array receiver coil. Partial overlap between the coils and preamplifier decoupling strategies are used to minimize coil-to-coil interactions in a noise correlation matrix. A wide range of factors, including cable lengths and sample position relative to an individual surface coil, can lead to signal amplitude and phase differences between the phased-array channels. When used to increase SNR, a weighted, phase-corrected summation must be performed to ensure optimal sensitivity.

hampered by low sensitivity, the ^{13}C reception coil is typically a relatively small surface coil (circa 80–100 and 12–15 mm for human occipital cortex and rat brain applications, respectively). The ^{1}H RF coil, required for decoupling and other spin manipulations, was traditionally a larger surface coil coplanar with the ^{13}C RF coil. However, this geometric arrangement leads to significant coil interactions as well as unacceptable RF power deposition during proton decoupling. A more favorable, and currently the most commonly used design was described by Adriany and Gruetter [110] in which the ^{1}H RF coil is split into two surface coil driven in quadrature (Figure 10.33A). The quadrature requirement is met by a hybrid coupler in which the incoming RF power is split evenly over the two ^{1}H coils, while simultaneously offsetting the RF phase by 90° for one ^{1}H coil. During reception the hybrid splitter combines the two out-of-phase NMR signals, such that one in-phase signal arrives at the preamplifier. Besides the RF power advantage of a quadrature design, the two ^{1}H RF coils also cover the volume excited by the ^{13}C RF coil more homogeneously. Furthermore, the B_1 and B_2 magnetic field flux

(A)

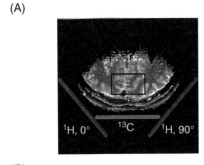

¹H, 0°   ¹³C   ¹H, 90°

(B)

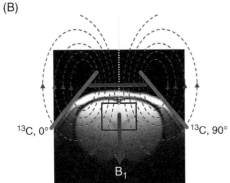

¹³C, 0°   ¹³C, 90°

B₁

(C)

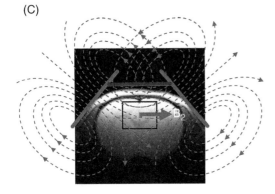

B₂

Figure 10.33 **Direct ^{13}C-[^{1}H] and indirect ^{1}H-[^{13}C] MR coil setups.** (A) Typical coil setup direct ^{13}C MRS on human brain with the option for ^{1}H decoupling or polarization transfer methods. The two ^{1}H surface coils are 90° out of phase in order to achieve quadrature transmission. (B, C) RF coil setup for ^{13}C-decoupled, ^{1}H NMR (i.e. POCE) on rat brain. (B) Magnetic field flux lines of the ^{1}H surface coil point approximately downwards in the volume-of-interest, whereas (C) the magnetic field flux lines of the quadrature ^{13}C coils are roughly horizontal, thereby achieving an inherent, geometric decoupling of the ^{1}H and ^{13}C coils.

lines of the two coils are approximately perpendicular over a large fraction of the useable volume of the ^{13}C RF coil, leading to an intrinsic decoupling of the mutual coil interactions (Figure 10.33B and C). The ^{1}H and ^{13}C RF coils in the ^{13}C-[^{1}H] design of Figure 10.33A can simply be reversed to give a ^{1}H-[^{13}C] RF coil (Figure 10.33B and C).

Despite the reduced interaction as a result of the geometric configuration of the two coils, the interference between the two coils is typically strong enough to cause a significant increase in the noise level when broadband decoupling is attempted. This additional noise on the observe channel typically originates from noise on the decoupling channel. For almost all heteronuclear experiments involving broadband decoupling, it is therefore crucial to apply filters on the decoupling channel to filter out the frequencies that lead to noise on the observe channel. A typical setup for ^{1}H-[^{13}C] NMR is shown in Figure 10.34. At least one, but often two filters are required on the decoupling ^{13}C channel to remove spurious signal at the ^{1}H frequency. Even though band-pass filters can be used, low-pass filters typically have a better performance in

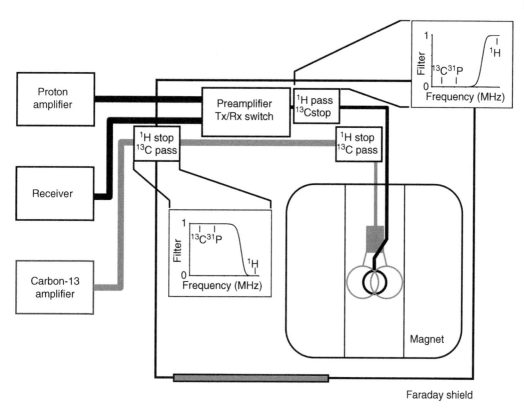

Figure 10.34 RF coil, filter, and amplifier setup for a ¹H-[¹³C] NMR experiment. In order to minimize noise interference on the ¹H receive channel during RF transmission on the ¹³C channel (i.e. ¹³C decoupling during ¹H NMR reception), a combination of low-pass or ¹³C-pass/¹H-stop and high-pass or ¹³C-stop/¹H-pass filters are required as shown. See text for more details.

terms of reduced signal loss at the desired frequency (this so-called insertion loss should be <0.5 dB for a good filter). A good low-pass filter achieves at least 60 dB attenuation at the ¹H NMR frequency. While somewhat counterintuitive, a high-pass filter on the observe channel prior to the preamplifier often greatly decreases the noise injection during ¹³C decoupling. This is due to nonlinear elements in the preamplifier/TR-switch that can multiply spurious signals at the ¹³C NMR frequency to the observed ¹H NMR frequency. While the exact filter combination and placement is often site-specific, the combination of filters should lead to negligible degradation of the SNR during decoupling.

10.5.7.4 Cooled and Superconducting RF Coils

The available SNR is the dominant limiting factor in the majority of MR experiments, preventing the acquisition of higher-resolution images or the detection of metabolites from smaller volumes. Many strategies have evolved with the aim of increasing the detected signal in order to increase the SNR, including the use of higher magnetic fields, improved RF coil designs, and hyperpolarized materials. However, being a ratio, the noise plays an equally important role. As the size of the sample decreases, the thermal noise of the receiver coil represents a greater proportion of the total system noise. In the extreme case, the RF coil rather than the sample dictates the noise at the preamplifier. In an effort to limit receiver coil noise for small samples, as well as low magnetic field applications, the use of cooled [111–113] or superconducting [114–116] materials in the construction of low-noise RF coils have been explored. The overall aim is to reduce the noise contributions of the RF coil (i.e. R_c in Eq. (10.17)) and preamplifier such that the sample noise once again becomes the dominant contributor of the SNR. Cooling the RF coil and preamplifier with liquid nitrogen can aid an increase in SNR of 2 to 3. However, it should be realized that these improvements are only realized when the sample does not significantly load the RF coil at room temperature (i.e. $Q_{loaded}/Q_{unloaded} \sim 1$). This typically limits the applications of cooled and superconducting RF coils to small surface coils or low-frequency MR systems [112, 117].

10.6   Complete MR System

10.6.1   RF Transmission

Besides the commonly encountered components such as gradients, RF coils, and magnet, a complete MR system holds many more essential components related to RF pulse transmission, signal reception, and overall system integration. In Chapter 5, the characteristics of RF pulses were described, while Section 10.5 detailed the RF coils utilized to transfer the RF pulse power to a sample. Figure 10.35 shows a simplified flow chart for the steps involved in converting the desired shape into a high-powered RF pulse, i.e. the RF transmission chain. The synthesizer module generates a continuous frequency of constant amplitude. The exact frequency is under computer control and can thus be adjusted to coincide with the Larmor resonance frequency to create an on-resonance condition. Next, only the time period of the continuous frequency that corresponds to the length and position of a RF pulse is selected, after which the uniform pulse is modulated with the desired shape. Note that the pulse shape only determines the modulation of the base frequency, the actual pulse is still transmitted in the RF (MHz) range. The final step is an overall amplitude scaling, which allows fine adjustment of the nutation angle. The scaled, shaped RF pulse is then multiplied to the kW range by a RF amplifier after which it

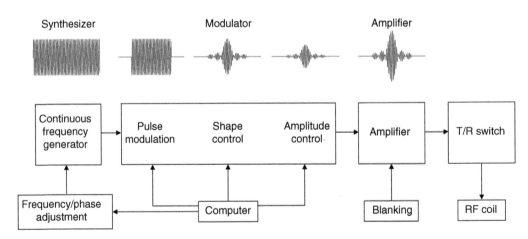

Figure 10.35 RF transmission subsystem. A continuous frequency generator generates a signal of which the frequency and phase are under computer control. During an NMR pulse sequence, a RF pulse is selected from the continuous signal (pulse modulation), shaped (e.g. sinc), and amplitude-adjusted, after which the signal is amplified to the kW range by a RF amplifier. Since many experiments use the same coil for RF pulse transmission and signal reception, a T/R switch is required to protect the sensitive receiver subsystem from the high-powered RF pulse that traverses along the same coaxial cable to the RF coil.

can be transmitted to the sample via the RF coil. Note that, even in an idle mode, RF amplifiers generate noise in the Larmor frequency range, which can quickly overwhelm the small NMR signal. It is therefore imperative that the RF amplifier is only on during the RF pulses and blanked (i.e. blocked) during the rest of the NMR sequence. In case of significant RF amplifier nonlinearities, RF pulses may be pre-distorted in order to compensate the distortions introduced by the RF amplifier [118].

10.6.2 Signal Reception

Following excitation the precessing magnetization can be detected as an induced voltage in a receiver coil, which can be the same coil as used for RF pulse transmission. The main task of the receiver chain is to amplify the small NMR signal with minimal degradation in the SNR, determine the absolute frequency by quadrature detection, and digitize the analog signal for further computer processing. Figure 10.36 shows a simplified flow chart of the steps involved with NMR signal reception. The NMR signal, detected as an induced voltage in the receiver coil, is typically very weak in the μV range. The first, and arguably the most important step is to amplify the NMR signal to a more manageable range (mV) where it is less sensitive to sources of interference or additional noise. This amplification is achieved by a low-noise preamplifier. The low-noise characteristic cannot be overstressed, as any noise introduced at this level directly degrades the SNR of the measurement. A measure of the spectrometer preamplifier to amplify the NMR response without the introduction of additional noise is given by the noise figure F. Good preamplifiers have a noise figure $F < 1 \, dB$, corresponding to a S/N degradation of <12%. A practical method of determining the noise figure is through a so-called hot/cold resistor measurement during which the noise level

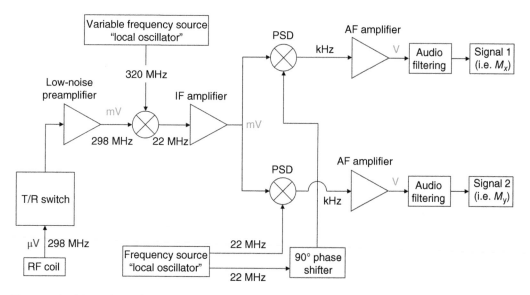

Figure 10.36 RF receiver subsystem. Upon entering the receiving chain following the T/R switch, the small MR signal (typically in the μV range) is amplified to the mV by a low-noise preamplifier. This early step is essential to minimize SNR degradation by noise contributions from other elements in the receive chain. The radiofrequency signal is then down-sampled to a constant intermediate frequency, where it is split for phase-sensitive quadrature detection. Following further down-sampling to the audiofrequency range (kHz), the signal can be digitized by an analog-to-digital converter. See text for more details.

produced by a 50 Ω resistor at cold and hot temperatures is measured. The noise figure can then be calculated according to

$$F = -1.279 - 10\log_{10}\left(1 - \frac{N_{77K}^2}{N_{293K}^2}\right)$$
(10.25)

where N_{77K} and N_{293K} are the noise levels at 77 K (liquid nitrogen) and 293 K (room temperature). Following pre-amplification the signal amplitude is in the mV range. Note that the frequency of the NMR signal following pre-amplification is still in the RF range (MHz). This poses several challenges for further processing of the signal. Firstly, analog-to-digital converters (ADCs) typically used in MRI are not fast enough to directly digitize the signal in the RF range. Secondly, unwanted frequencies generated by the electronics at the Larmor frequency can lead to additional noise and artifacts. For these reasons the NMR frequency is often down-sampled to an intermediate frequency (IF) (e.g. 22 MHz) that does not coincide with the Larmor frequency of any of the commonly studied nuclei. An additional advantage of working at an IF is that all NMR frequencies, from protons to carbon-13 and oxygen-17, can be down-sampled to the same IF such that any subsequent processing can be performed in the same manner. Down-sampling involves mixing of the NMR frequency ω_0 with a reference frequency ω_{ref} according to

$$\cos(\omega_0 t)\xrightarrow{\text{mixers}}\cos(\omega_0 t)\cos(\omega_{ref} t)$$
$$= \frac{1}{2}\cos\left[(\omega_0 + \omega_{ref})t\right] + \frac{1}{2}\cos\left[(\omega_0 - \omega_{ref})t\right]$$
(10.26)

Note that the NMR signal at this stage only has a single component, which was arbitrarily chosen as $\cos(\omega_0 t)$. The mixing of two signals with frequencies ω_0 and ω_{ref} leads to signals oscillating at the sum and difference of ω_0 and ω_{ref}. While actual mixers lead to additional frequencies or harmonics, Eq. (10.26) holds well to a first approximation [119]. The high-frequency component ($\omega_0 + \omega_{ref}$) can be removed by applying a low-pass filter to the data, such the final down-sampled or demodulated signal is given by

$$\xrightarrow{\text{low-pass filter}} \frac{1}{2}\cos\left[(\omega_0 - \omega_{ref})t\right] = \frac{1}{2}\cos\left[\omega_{IF}t\right]$$
(10.27)

where the difference frequency is now an IF. Unfortunately, detecting only one component of the rotating magnetization (i.e. the cosine term) does not allow the discrimination between positive and negative frequencies. As described in Chapter 1, both cosine and sine components of a harmonic signal must be recorded in order to determine the sign of the frequency. Sampling a signal such that both components are recorded is known as quadrature detection.

10.6.3 Quadrature Detection

Quadrature detection during acquisition is accomplished by splitting the IF signal originating from a single coil into two separate channels. The two IF signals are compared to cosine- and sine-modulated reference signals according to

$$\cos(\omega_{IF}t)\xrightarrow{\text{mixers}}\cos(\omega_{IF}t)\cos(\omega_{ref}t)$$
$$= \frac{1}{2}\cos\left[(\omega_{IF} + \omega_{ref})t\right] + \frac{1}{2}\cos\left[(\omega_{IF} - \omega_{ref})t\right]$$
(10.28)

and

$$\cos(\omega_{IF}t)\xrightarrow{\text{mixers}}\cos(\omega_{IF}t)\sin(\omega_{ref}t)$$
$$=\frac{1}{2}\sin\left[(\omega_{IF}+\omega_{ref})t\right]-\frac{1}{2}\sin\left[(\omega_{IF}-\omega_{ref})t\right] \tag{10.29}$$

Following further amplification and low-pass filtering, the two quadrature signals in the audio frequency (AF) range are obtained which then uniquely define the absolute frequency, i.e.

$$\frac{1}{2}\cos\left[(\omega_{IF}-\omega_{ref})t\right]+\frac{i}{2}\sin\left[(\omega_{IF}-\omega_{ref})t\right]=\frac{1}{2}\exp\left[i(\omega_{IF}-\omega_{ref})t\right]=\frac{1}{2}\exp\left[i\omega_{AF}t\right] \tag{10.30}$$

The AF signal undergoes a band pass filtering step to remove all signals, and in particular noise, outside the spectral bandwidth which would otherwise fold back into the spectrum and degrade the attainable SNR. The amplitude of the analog signals before analog-to-digital conversion is in the order of several Volts and essentially insensitive to minor system noise sources. Note that with modern AD converters, the signal can be directly digitized at the IF, thereby eliminating the need for an AF stage.

10.6.4   Dynamic Range

The gain of an NMR receiver determines how well the dynamic range of a given ADC is utilized. An ADC is supplied with a limited range, e.g. a 16 bit ADC has a dynamic range of $2^{16}=65\,536$ components (i.e. the amplitude of the FID can be described by a maximum of 2^{16} different numbers (bits)). The optimal (maximum) dynamic range is obtained when the highest intensity data point of the FID corresponds to one of the highest bits (Figure 10.37A and B). If the gain is set too low (Figure 10.37C and D), only a small part of the available dynamic ADC range is used and the FID is described by only a few numbers (bits). Consequently, the accuracy of the digital representation of the FID is not optimal due to digitization errors. The Fourier transformation of the coarsely digitized FID signal leads to a greatly reduced SNR in the spectrum. Furthermore, FID components of low-intensity resonances may not be digitized at all, leading to a loss of information. On the other hand, when the gain is set too high (Figure 10.37E and F) the ADC is said to be overloaded and the situation becomes even worse. With an ADC overload the strongest signal in the FID is larger than the highest number in the ADC and consequently the first few points of the FID are unreliable. The spectrum obtained after Fourier transformation of the FID shows strong fluctuations in all resonance lines and in the baseline, making quantification of resonances unreliable. The setting of the gain is therefore crucial for a proper digital representation of the analog FID after analog-to-digital conversion. A guideline for gain setting is that the highest signal should fill the ADC for 50–70%. In this way, most resonances are digitized accurately, while maintaining some room for increase in signal intensities (e.g. due to less optimal water suppression) to prevent an ADC overload.

The Nyquist condition (see Chapter 1) is a minimum requirement, which also imposes the minimum hardware (i.e. memory) requirements. On many NMR spectrometers, the analog signal may be digitized at sampling rates much higher than the Nyquist frequency. This is referred to as oversampling and is typically performed in order to improve the dynamic range and sensitivity. NMR experiments on aqueous solutions typically encounter a large range of signals; small metabolite signals in the presence of an overwhelming water resonance. The receiver gain is typically adjusted to digitize the largest signal over the full range of ADC bits. However, with typical 12- or 16-bit ADCs the metabolite signals may be incompletely described

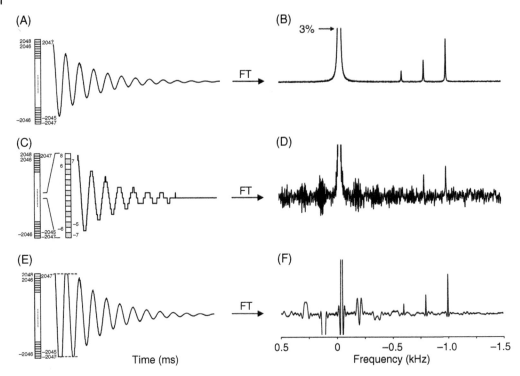

(A)

(B) 3% →

FT →

(C)

(D)

FT →

(E)

(F)

FT →

Time (ms)

0.5   0   −0.5   −1.0   −1.5
Frequency (kHz)

Figure 10.37 **Dynamic range and analog-to-digital conversion (ADC).** (A) A 12-bit ADC can digitize an analog signal over 4096 numbers, ranging from −2.047 to +2.048. Signal detection is optimal when the full ADC range is utilized, resulting in (B) a high-quality MR spectrum with good SNR and Lorentzian lines. (C) When the MR receiver gain is set too low, only a few bits of the total ADC range are used to digitize the analog MR signal. (D) The edges in the digitized signal lead to digitization noise in the resulting MR spectrum. (E) When the analog signal exceeds the ADC range, the signal will be truncated at the highest ADC value. (F) Upon Fourier transformation the spectrum is characterized by distorted lines and an apparent lower SNR.

by the lowest few bits, leading to loss of information and degradation of SNR. While water suppression can minimize this problem for many applications, digital oversampling is a more general approach to increase the effective ADC size. It has been shown [120] that for an oversampling factor of $n = (F_{oversampling}/F_{Nyquist})$ the gain in the ADC dynamic range is $^2\log(n)$. For example, for an oversampling factor of 8 (i.e. digitize 8 times faster than the minimum Nyquist frequency) the gain in ADC dynamic range is 3 bits.

10.6.5 Gradient and Shim Systems

The characteristics of magnetic field gradients and the corresponding coils were discussed in detail in Section 10.4, whereas shimming-related considerations were the topic of Section 10.3. Figure 10.38 gives a schematic overview of the interfacing between gradient- and shim controllers and the currents ultimately produced in the gradient- and shim coils.

The gradient controller determines the gradient timings, shapes, and relative amplitudes. Due to the generation of eddy currents, the gradient input waveforms are modified in a preemphasis step to counteract the eddy currents generated within the magnet. The preemphasis unit is normally under computer control and can generate several exponential functions with variable amplitude and time constants. In addition to preemphasis, the gradient waveforms also serve as input to a B_0 compensation unit, which generates exponential modulations of the

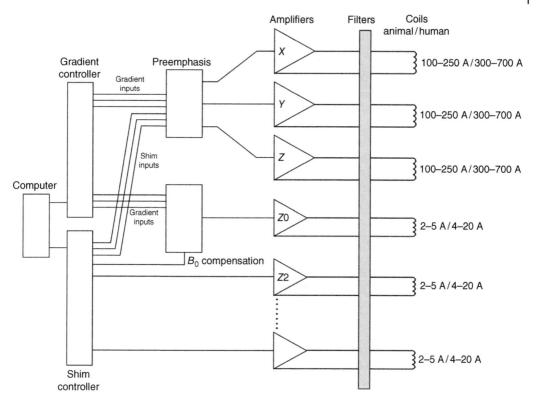

Figure 10.38 Gradient and shim subsystems. Most MRI systems are equipped with three linear *X, Y, Z* gradient coils, and 12 or more higher-order shim coils as well as a constant B_0 coil. Each coil has its own independent amplifier, whereby the linear gradient coil amplifiers are significantly stronger than those for the other coils. Eddy currents generated by pulsed gradient fields require preemphasis and B_0 field compensation. Note that shims normally do not require preemphasis and will as such be connected directly to the *X, Y,* and *Z* gradient amplifiers. However, for dynamic shimming applications, the linear as well as higher-order shims generate constant (B_0) and linear eddy current field that can be corrected by preemphasis.

B_0 magnetic field to counteract the B_0 terms generated by eddy currents. Note that many clinical MR systems lack a dedicated B_0 coil for eddy current compensation. In that case the B_0 eddy currents can be compensated in software by modulating the RF frequency. The preemphasized gradient waveforms are then amplified and sent to the corresponding gradient coil.

Unlike the gradient system, the shim system is much simpler as the shim controller directly operates on the shim coils. Note, however, that when shims are being used in a pulsed manner, as for example with dynamic shim updating ([30, 31, 33, 35], see also Figure 10.14), the shim system has to be extended with shim preemphasis, B_0 compensation, and possibly shim-to-gradient cross-term preemphasis [33, 35].

Exercises

10.1 Consider the following shim coil characteristics for a 100% shim change:

$\Delta Z^0 = 5000\,\text{Hz}$,
$\Delta X, \Delta Y$, and $\Delta Z = 1000\,\text{Hz}\,\text{cm}^{-1}$ and associated $\Delta Z^0 = 20\,\text{Hz}$,

$\Delta Z^2 = 500\,\text{Hz}\,\text{cm}^{-2}$ and associated $\Delta Z^0 = 200\,\text{Hz}$,
$\Delta ZX = 400\,\text{Hz}\,\text{cm}^{-2}$ and associated $\Delta Z^0 = 40\,\text{Hz}$, and
$\Delta Z^3 = 80\,\text{Hz}\,\text{cm}^{-3}$ and associated $\Delta Z^0 = 150\,\text{Hz}$.

A In a given spatial volume the required shim fields to counteract the B_0 magnetic field inhomogeneity are $\Delta Z = 250\,\text{Hz}\,\text{cm}^{-1}$, $\Delta Z^2 = -450\,\text{Hz}\,\text{cm}^{-2}$, and $\Delta Z^3 = 60\,\text{Hz}\,\text{cm}^{-3}$. Calculate the required change in Z^0 shim (in %) to maintain the water frequency on-resonance.

B Repeat the calculation in the presence of a Z^3-to-Z coupling, i.e. a change in Z^3 of $80\,\text{Hz}\,\text{cm}^{-3}$ leads to a $-40\,\text{Hz}\,\text{cm}^{-1}$ change in Z.

C For a third-order B_0 magnetic field inhomogeneity of $30\,\text{Hz}\,\text{cm}^{-3}$ in a 3D voxel located in the magnetic isocenter, calculate the required shim changes (in %). Repeat the calculation for a volume located at $(X, Y, Z) = (1, -2, +3)$ cm.

10.2 Calculate the nondegenerate first- through third-order shims for infinitesimally thin slices in

A the coronal plane (slice selection in the y direction).

B the sagittal plane (slice selection in the x direction).

10.3 For a circular RF coil placed in the xz plane, the magnetic field along the y axis, B1y, through the center of the coil is proportional to $IR^2/(R^2 + y^2)^{3/2}$, where I is the current running in the coil and R is the coil radius.

A For a 14 mm diameter surface coil calculate the distance along the coil axis at which the generated B_1 field has dropped to 33% of the B_1 field at the coil center.

B For a given current I, the 14 mm diameter coil generates a B_1 field at the coil center of 0.58 Gauss. Calculate the required coil diameter to produce the same B_1 field 10 mm away from the coil center, assuming an equal current.

10.4 **A** Under the assumption that eddy current-related magnetic fields decay exponentially, derive an expression for eddy currents following a magnetic field gradient of finite length T. Assume instantaneous gradient switching.

B A NMR spectroscopist wants to investigate if gradient preemphasis and B_0 magnetic field compensation can be eliminated by balancing each magnetic field gradient with an equal gradient of opposite amplitude. Derive an expression for this experiment (i.e. a gradient of amplitude $+G$ for duration T followed by a gradient of amplitude $-G$ for duration T).

C For a gradient duration of 2 ms, calculate the range of eddy current time constants for which the proposed method gives at least a 10-fold suppression of the temporal B_0 and gradient magnetic fields. Discuss the applicability of the proposed method.

D Following the optimization of preemphasis in all three directions, as well as the B_0 compensation, the NMR spectroscopist is ready to perform high-quality MRS studies. However, while the magnetic field homogeneity achieved with room temperature shimming is excellent, the water spectrum obtained with a 3D LASER sequence is still very broad with spectral distortions pointing towards residual eddy currents. Give a likely explanation for this observation and propose a possible solution.

10.5 **A** In a vacuum the wavelength at 300 MHz is 100 cm. Calculate the wavelength in the human brain ($\mu = 4\pi \times 10^{-7}\,\text{N}\,\text{A}^{-2}$, $\varepsilon_r = 50$, $\sigma = 0.66\,\text{S}\,\text{m}^{-1}$) at 300 MHz.

B Calculate the skin depth in human brain at 300 MHz.

10.6 Careless mounting of a second-order shim coil assembly inside a superconducting magnet has resulted in a 15° rotation in the x–y plane and a 20 mm displacement in the z direction relative to the magnet isocenter and the first-order gradient coils.

   A Derive expressions for the spherical harmonic fields generated by the five second-order shim coils in the reference frame of the magnet.

   B Magnetic field mapping reveals a magnetic field distribution given by $+20Z^2 + 33XZ + 10Y$. Calculate all first- and second-order shim corrections necessary to achieve a homogenous magnetic field.

10.7 **A** A volume coil tested on a phantom of distilled water arrives from the factory. In order to perform experiments on human brain, discuss the need for readjusting the tuning and matching capacitors.

   B Attaching the coil, loaded with a human head, to a network analyzer reveals that the coil is well matched and that the circuit resonance frequency appears at 123.4 MHz. For experiments at 3.0 T, discuss the need for readjustment of the tuning capacitor.

   C A bench measurement of the coil reveals an unloaded Q value of 300. Upon loading the coil with a solid plastic human-sized phantom the loaded Q value drops to 290. Discuss the performance of this coil in light of these results.

   D Further experiments reveal a loaded Q value of 120 when a real human head is inserted. Discuss the performance of this coil in light of these additional results.

   E Another, home-made coil shows an unloaded to loaded Q drop from 500 to 200. Discuss the preferred choice of RF coil.

10.8 A given gradient coil has an inductance of 0.4 mH and is connected to a gradient power supply capable of delivering 350 V and 150 A. The efficiency is $0.3 \, \text{mT} \, \text{m}^{-1} \, \text{A}^{-1}$.

   A Calculate the gradient ramp time (Hint, perform the calculation in S.I. units).

   B Calculate the gradient strength.

   C When the gradient power amplifier is replaced with one capable of delivering 600 V and 200 A, calculate the new gradient ramp times and strengths.

   D Discuss the advantage of the new amplifier for EPI.

10.9 Open MR magnets can be designed in a variety of patient-friendly configurations. By changing the direction of B_0 relative to the head, open MR magnets could potentially reduce susceptibility-induced magnetic field inhomogeneity.

   A Consider the nasal sinus cavity a 3 cm diameter sphere located 5 cm below the frontal cortex (as measured from the center of the sphere). Given the susceptibilities of air and water as +0.37 and –9.05 ppm, calculate the magnetic field offset for a regular patient orientation (i.e. B_0 is parallel to the foot–head direction) at 7.05 T. Use the magnetic dipole field given by Eq. (10.2) for all calculations.

   B Calculate the magnetic field offset when B_0 is changed to the ear–ear orientation.

   C Discuss the considerations involved with this approach to improve magnetic field homogeneity.

   D It has been suggested that susceptibility-induced magnetic field inhomogeneity can be temporarily reduced by breathing nitrogen gas. Given that the magnetic susceptibility of nitrogen gas is –0.0063 ppm, calculate the improvement in magnetic field homogeneity in the regular patient orientation when breathing nitrogen instead of oxygen.

     A considerable amount of research has been dedicated to the study of breathing-related magnetic field changes in the human brain. Assume that the human lungs can be approximated as 19 cm diameter sphere filled with 18% O_2/82% N_2 during exhaling and a 23 cm diameter sphere filled with 22% O_2/78% N_2 during inhaling.

E Giving the magnetic susceptibility of oxygen gas of +1.8 ppm, calculate the change in magnetic field between inhaling and exhaling in a slice through the human brain, 30 cm above the center of the "lung sphere."

F During a gradient-echo MRI sequence with an echo-time TE of 25 ms, calculate the maximum phase difference between subsequent echoes.

G Discuss the importance of respiration-induced magnetic field perturbations in the human brain for *in vivo* NMR measurements.

10.10 During phase-sensitive quadrature detection one channel produces a signal with a relative amplitude and phase of 100 and 20°, whereas the other channel produces a signal with a relative amplitude and phase of 104 and 110°.

A Calculate the relative size of the "ghost" signal that appears at the negative frequency of the main resonance due to the amplitude/gain imbalance.

B Calculate the relative size and phase of the "ghost" signal when the phase of the second channel was 115° instead of 110°.

C Discuss the spectral effects when both channels have a constant receiver offset, i.e. constant amplitude signal is generated even when no NMR signal is acquired.

References

1 Chen, C.-N. and Hoult, D.I. (1989). *Biomedical Magnetic Resonance Technology*. Bristol: Adam Hilger.

2 Webb, A.G. (2016). *Magnetic Resonance Technology. Hardware and System Component Design*. Cambridge: Royal Society of Chemistry.

3 Vaughan, J.T. and Griffiths, J.R. (2012). *RF Coils for MRI*. Chichester: Wiley.

4 Hoult, D.I. and Richards, R.E. (1975). Critical factors in the design of sensitive high resolution nuclear magnetic resonance spectrometers. *Proc. Roy. Soc. Lond. A* A344: 311–340.

5 Hanley, P. (1984). Magnets for medical applications of NMR. *Br. Med. Bull.* 40: 125–131.

6 Gordon, R.E. and Timms, W.E. (1984). Magnet systems used in medical NMR. *Comput. Radiol.* 8: 245–261.

7 Laukien, D.D. and Tschopp, W.H. (1994). Superconducting NMR magnet design. *Concepts Magn. Reson.* 6: 255–274.

8 Henry, P.G., van de Moortele, P.F., Giacomini, E. et al. (1999). Field-frequency locked *in vivo* proton MRS on a whole-body spectrometer. *Magn. Reson. Med.* 42: 636–642.

9 Schenck, J.F. (1996). The role of magnetic susceptibility in magnetic resonance imaging: MRI magnetic compatibility of the first and second kinds. *Med. Phys.* 23: 815–850.

10 Hoult, D.I. and Lee, D. (1985). Shimming a superconducting nuclear-magnetic-resonance imaging magnet with steel. *Rev. Sci. Instrum.* 56: 131–135.

11 Jackson, J.D. (1999). *Classical Electrodynamics*. New York: Wiley.

12 Chu, S.C., Xu, Y., Balschi, J.A., and Springer, C.S. (1990). Bulk magnetic susceptibility shifts in NMR studies of compartmentalized samples: use of paramagnetic reagents. *Magn. Reson. Med.* 13: 239–262.

13 Marques, J.P. and Bowtell, R. (2005). Application of a Fourier-based method for rapid calculation of field inhomogeneity due to spatial variation of magnetic susceptibility. *Concepts Magn. Reson.* 25B: 65–78.

14 Salomir, R., de Senneville, B.D., and Moonen, C.T.W. (2003). A fast calculation method for magnetic field inhomogeneity due to an arbitrary distribution of bulk susceptibility. *Concepts Magn. Reson. B* 19: 26–34.

15 Koch, K.M., Papademetris, X., Rothman, D.L., and de Graaf, R.A. (2006). Rapid calculations of susceptibility-induced magnetostatic field perturbations for in vivo magnetic resonance. *Phys. Med. Biol.* 51: 6381–6402.

16 Wang, Y. and Liu, T. (2015). Quantitative susceptibility mapping (QSM): decoding MRI data for a tissue magnetic biomarker. *Magn. Reson. Med.* 73: 82–101.

17 Romeo, F. and Hoult, D.I. (1984). Magnet field profiling: analysis and correcting coil design. *Magn. Reson. Med.* 1: 44–65.

18 Juchem, C., Muller-Bierl, B., Schick, F. et al. (2006). Combined passive and active shimming for *in vivo* MR spectroscopy at high magnetic fields. *J. Magn. Reson.* 183: 278–289.

19 Wilson, J.L., Jenkinson, M., and Jezzard, P. (2002). Optimization of static field homogeneity in human brain using diamagnetic passive shims. *Magn. Reson. Med.* 48: 906–914.

20 Wilson, J.L., Jenkinson, M., and Jezzard, P. (2003). Protocol to determine the optimal intraoral passive shim for minimisation of susceptibility artifact in human inferior frontal cortex. *Neuroimage* 19: 1802–1811.

21 Koch, K.M., Brown, P.B., Rothman, D.L., and de Graaf, R.A. (2006). Sample-specific diamagnetic and paramagnetic passive shimming. *J. Magn. Reson.* 182: 66–74.

22 Yang, S., Kim, H., Ghim, M.O. et al. (2011). Local in vivo shimming using adaptive passive shim positioning. *Magn. Reson. Imaging* 29: 401–407.

23 Golay, M.J.E. (1958). Field Homogenizing coils for nuclear spin resonance instrumentation. *Rev. Sci. Instrum.* 29: 313–315.

24 Gruetter, R. (1993). Automatic, localized *in vivo* adjustment of all first- and second-order shim coils. *Magn. Reson. Med.* 29: 804–811.

25 Gruetter, R. and Boesch, C. (1992). Fast, noniterative shimming of spatially localized signals, in vivo analysis of the magnetic field along axes. *J. Magn. Reson.* 96: 323–334.

26 Shen, J., Rothman, D.L., Hetherington, H.P., and Pan, J.W. (1999). Linear projection method for automatic slice shimming. *Magn. Reson. Med.* 42: 1082–1088.

27 Gruetter, R. and Tkac, I. (2000). Field mapping without reference scan using asymmetric echo-planar techniques. *Magn. Reson. Med.* 43: 319–323.

28 Shen, J., Rycyna, R.E., and Rothman, D.L. (1997). Improvements on an *in vivo* automatic shimming method [FASTERMAP]. *Magn. Reson. Med.* 38: 834–839.

29 Pan, J.W., Lo, K.M., and Hetherington, H.P. (2013). Role of very high order and degree B_0 shimming for spectroscopic imaging of the human brain at 7 Tesla. *Magn. Reson. Med.* 68: 1007–1017.

30 Blamire, A.M., Rothman, D.L., and Nixon, T. (1996). Dynamic shim updating: a new approach towards optimized whole brain shimming. *Magn. Reson. Med.* 36: 159–165.

31 Morrell, G. and Spielman, D. (1997). Dynamic shimming for multi-slice magnetic resonance imaging. *Magn. Reson. Med.* 38: 477–483.

32 de Graaf, R.A., Brown, P.B., McIntyre, S. et al. (2003). Dynamic shim updating (DSU) for multislice signal acquisition. *Magn. Reson. Med.* 49: 409–416.

33 Koch, K.M., McIntyre, S., Nixon, T.W. et al. (2006). Dynamic shim updating on the human brain. *J. Magn. Reson.* 180: 286–296.

34 Koch, K.M., Sacolick, L.I., Nixon, T.W. et al. (2007). Dynamically shimmed multivoxel ^{1}H magnetic resonance spectroscopy and multislice magnetic resonance spectroscopic imaging of the human brain. *Magn. Reson. Med.* 57: 587–591.

35 Juchem, C., Nixon, T.W., Diduch, P. et al. (2010). Dynamic shimming of the human brain at 7 Tesla. *Concepts Magn. Reson. B* 37: 116–128.

36 Juchem, C., Nixon, T.W., McIntyre, S. et al. (2010). Magnetic field homogenization of the human prefrontal cortex with a set of localized electrical coils. *Magn. Reson. Med.* 63: 171–180.

37 Juchem, C., Nixon, T.W., McIntyre, S. et al. (2010). Magnetic field modeling with a set of individual localized coils. *J. Magn. Reson.* 204: 281–289.

38 Juchem, C., Brown, P.B., Nixon, T.W. et al. (2011). Multicoil shimming of the mouse brain. *Magn. Reson. Med.* 66: 893–900.

39 Juchem, C., Nixon, T.W., McIntyre, S. et al. (2011). Dynamic multi-coil shimming of the human brain at 7T. *J. Magn. Reson.* 212: 280–288.

40 Juchem, C., Green, D., and de Graaf, R.A. (2013). Multi-coil magnetic field modeling. *J. Magn. Reson.* 236: 95–104.

41 Juchem, C., Herman, P., Sanganahalli, B.G. et al. (2014). Dynamic multi-coil technique (DYNAMITE) shimming of the rat brain at 11.7 T. *NMR Biomed.* 27: 897–906.

42 Juchem, C., Umesh Rudrapatna, S., Nixon, T.W., and de Graaf, R.A. (2015). Dynamic multi-coil technique (DYNAMITE) shimming for echo-planar imaging of the human brain at 7 Tesla. *Neuroimage* 105: 462–472.

43 Topfer, R., Starewicz, P., Lo, K.M. et al. (2016). A 24-channel shim array for the human spinal cord: design, evaluation, and application. *Magn. Reson. Med.* 76: 1604–1611.

44 Aghaeifar, A., Mirkes, C., Bause, J. et al. (2018). Dynamic B_0 shimming of the human brain at 9.4 T with a 16-channel multi-coil shim setup. *Magn. Reson. Med.* doi: 10.1002/mrm.27110.

45 Stockmann, J.P., Witzel, T., Keil, B. et al. (2016). A 32-channel combined RF and B_0 shim array for 3T brain imaging. *Magn. Reson. Med.* 75: 441–451.

46 Han, H., Song, A.W., and Truong, T.K. (2013). Integrated parallel reception, excitation, and shimming (iPRES). *Magn. Reson. Med.* 70: 241–247.

47 Truong, T.K., Darnell, D., and Song, A.W. (2014). Integrated RF/shim coil array for parallel reception and localized B0 shimming in the human brain. *Neuroimage* 103: 235–240.

48 Darnell, D., Truong, T.K., and Song, A.W. (2017). Integrated parallel reception, excitation, and shimming (iPRES) with multiple shim loops per radio-frequency coil element for improved B_0 shimming. *Magn. Reson. Med.* 77: 2077–2086.

49 Stockmann, J.P. and Wald, L.L. (2018). *In vivo* B_0 field shimming methods for MRI at 7T. *Neuroimage* 168: 71–87.

50 Turner, R. (1993). Gradient coil design: a review of methods. *Magn. Reson. Imaging* 11: 903–920.

51 Wilm, B.J., Barmet, C., Pavan, M., and Pruessmann, K.P. (2011). Higher order reconstruction for MRI in the presence of spatiotemporal field perturbations. *Magn. Reson. Med.* 65: 1690–1701.

52 Jehenson, P., Westphal, M., and Schuff, N. (1990). Analytical method for the compensation of eddy-current effects induced by pulsed magnetic field gradients in NMR systems. *J. Magn. Reson.* 90: 264–278.

53 van Vaals, J.J. and Bergman, A.H. (1990). Optimization of eddy-current compensation. *J. Magn. Reson.* 90: 52–70.

54 Jensen, D.J., Brey, W.W., Delayre, J.L., and Narayana, P.A. (1987). Reduction of pulsed gradient settling time in the superconducting magnet of a magnetic resonance instrument. *Med. Phys.* 14: 859–862.

55 Barmet, C., De Zanche, N., and Pruessmann, K.P. (2008). Spatiotemporal magnetic field monitoring for MR. *Magn. Reson. Med.* 60: 187–197.

56 De Zanche, N., Barmet, C., Nordmeyer-Massner, J.A., and Pruessmann, K.P. (2008). NMR probes for measuring magnetic fields and field dynamics in MR systems. *Magn. Reson. Med.* 60: 176–186.

57 Vannesjo, S.J., Haeberlin, M., Kasper, L. et al. (2013). Gradient system characterization by impulse response measurements with a dynamic field camera. *Magn. Reson. Med.* 69: 583–593.

58 Vannesjo, S.J., Dietrich, B.E., Pavan, M. et al. (2014). Field camera measurements of gradient and shim impulse responses using frequency sweeps. *Magn. Reson. Med.* 72: 570–583.

59 Terpstra, M., Andersen, P.M., and Gruetter, R. (1998). Localized eddy current compensation using quantitative field mapping. *J. Magn. Reson.* 131: 139–143.

60 de Graaf, R.A., Brown, P.B., McIntyre, S. et al. (2006). High magnetic field water and metabolite proton T_1 and T_2 relaxation in rat brain *in vivo*. *Magn. Reson. Med.* 56: 386–394.

61 Barmet, C., De Zanche, N., Wilm, B.J., and Pruessmann, K.P. (2009). A transmit/receive system for magnetic field monitoring of *in vivo* MRI. *Magn. Reson. Med.* 62: 269–276.

62 Duerst, Y., Wilm, B.J., Dietrich, B.E. et al. (2015). Real-time feedback for spatiotemporal field stabilization in MR systems. *Magn. Reson. Med.* 73: 884–893.

63 Duerst, Y., Wilm, B.J., Wyss, M. et al. (2016). Utility of real-time field control in T_2*-Weighted head MRI at 7T. *Magn. Reson. Med.* 76: 430–439.

64 Mansfield, P. and Chapman, B. (1986). Active magnetic screening of coils for static and time-dependent magnetic field generation in NMR imaging. *J. Phys. E* 19: 540–545.

65 Mansfield, P. and Chapman, B. (1986). Active magnetic screening of gradient coils in NMR imaging. *J. Magn. Reson.* 66: 573–576.

66 Mansfield, P. and Chapman, B. (1987). Multishield active magnetic screening of coil structures in NMR. *J. Magn. Reson.* 72: 211–223.

67 Bowtell, R. and Mansfield, P. (1991). Gradient coil design using active magnetic screening. *Magn. Reson. Med.* 17: 15–21.

68 Kumar, A., Edelstein, W.A., and Bottomley, P.A. (2009). Noise figure limits for circular loop MR coils. *Magn. Reson. Med.* 61: 1201–1209.

69 Grover, F.W. (1973). *Inductance Calculations*. Mineola, NY: Dover Publications, Inc.

70 Traficante, D.D. (1989). Impedance: what it is, and why it must be matched. *Concepts Magn. Reson.* 1: 73–92.

71 Hoult, D.I. (1989). The magnetic resonance myth of radio waves. *Concepts Magn. Reson.* 1: 1–5.

72 Hoult, D.I. (2009). The origin and present status of the radio wave controversy in NMR. *Concepts Magn. Reson.* 34: 193–216.

73 Link, J. (1992). The design of resonator probes with homogenous radiofrequency fields. In: *NMR Basic Principles and Progress*, vol. 26 (ed. P. Diehl, E. Fluck, H. Gunther, et al.). Berlin: Springer-Verlag.

74 Ackerman, J.J.H., Grove, T.H., Wong, G.G. et al. (1980). Mapping of metabolites in whole animals by ^{31}P NMR using surface coils. *Nature* 283: 167–170.

75 Evelhoch, J.L., Crowley, M.G., and Ackerman, J.J.H. (1984). Signal-to-noise optimization and observed volume localization with circular surface coils. *J. Magn. Reson.* 56: 110–124.

76 Haase, A., Hanicke, W., and Frahm, J. (1984). The influence of experimental parameters in surface-coil NMR. *J. Magn. Reson.* 56: 401–412.

77 Crowley, M.G., Evelhoch, J.L., and Ackerman, J.J.H. (1985). The surface-coil NMR receiver in the presence of homogeneous B_1 excitation. *J. Magn. Reson.* 64: 20–31.

78 Ackerman, J.J.H. (1990). Surface (local) coils as NMR receivers. *Concepts Magn. Reson.* 2: 33–42.

79 Bosch, C.S. and Ackerman, J.J.H. (1992). Surface coil spectroscopy. In: *NMR Basic Principles and Progress*, vol. 27 (ed. P. Diehl, E. Fluck, H. Gunther, et al.), 3–44. Berlin: Springer-Verlag.

80 Alderman, D.W. and Grant, D.M. (1979). An efficient decoupler coil which reduces heating in conductive samples in superconducting spectrometers. *J. Magn. Reson.* 36: 447–451.

81 Schneider, H.J. and Dullenkopf, P. (1977). Slotted tube resonator: a new NMR probe head at high observing frequencies. *Rev. Sci. Instrum.* 48: 68–73.

82 Leroy-Willig, A., Darrasse, L., Taquin, J., and Sauzade, M. (1985). The slotted cylinder: an efficient probe for NMR imaging. *Magn. Reson. Med.* 2: 20–28.

83 Hayes, C.E., Edelstein, W.A., Schenck, J.F. et al. (1985). An efficient, highly homogeneous radiofrequency coil for whole-body NMR imaging at 1.5 T. *J. Magn. Reson.* 63: 622–628.

84 Tropp, J. (1989). The theory of the bird-cage resonator. *J. Magn. Reson.* 82: 51–62.

85 Jin, J., Shen, G., and Perkins, T. (1994). On the field inhomogeneity of a birdcage coil. *Magn. Reson. Med.* 32: 418–422.

86 Vaughan, J.T., Hetherington, H.P., Otu, J.O. et al. (1994). High frequency volume coils for clinical NMR imaging and spectroscopy. *Magn. Reson. Med.* 32: 206–218.

87 Vaughan, J.T., Adriany, G., Garwood, M. et al. (2002). Detunable transverse electromagnetic (TEM) volume coil for high-field NMR. *Magn. Reson. Med.* 47: 990–1000.

88 Hoult, D.I. and Richards, R.E. (1976). The signal-to-noise ratio of the nuclear magnetic resonance experiment. *J. Magn. Reson.* 24: 71–85.

89 Hoult, D.I. and Lauterbur, P.C. (1979). The sensitivity of the zeugmatographic experiment involving human samples. *J. Magn. Reson.* 34: 425–433.

90 Hoult, D.I. (2000). The principle of reciprocity in signal strength calculations: a mathematical guide. *Concepts Magn. Reson.* 12: 173–187.

91 Emir, U.E., Auerbach, E.J., Van De Moortele, P.F. et al. (2012). Regional neurochemical profiles in the human brain measured by ^{1}H MRS at 7 T using local B_1 shimming. *NMR Biomed.* 25: 152–160.

92 Henning, A., Koning, W., Fuchs, A. et al. (2016). ^{1}H MRS in the human spinal cord at 7 T using a dielectric waveguide transmitter, RF shimming and a high density receive array. *NMR Biomed.* 29: 1231–1239.

93 Hetherington, H.P., Avdievich, N.I., Kuznetsov, A.M., and Pan, J.W. (2010). RF shimming for spectroscopic localization in the human brain at 7 T. *Magn. Reson. Med.* 63: 9–19.

94 Boer, V.O., Klomp, D.W., Juchem, C. et al. (2011). Multi-slice MRSI of the human brain at 7 Tesla using dynamic B_0 and B_1 shimming. *Proc. Int. Soc. Magn. Reson. Med.* 19: 142.

95 Van de Moortele, P.F., Akgun, C., Adriany, G. et al. (2005). B_1 destructive interferences and spatial phase patterns at 7 T with a head transceiver array coil. *Magn. Reson. Med.* 54: 1503–1518.

96 Metzger, G.J., Snyder, C., Akgun, C. et al. (2008). Local B_1^{+} shimming for prostate imaging with transceiver arrays at 7T based on subject-dependent transmit phase measurements. *Magn. Reson. Med.* 59: 396–409.

97 Yang, Q.X., Mao, W., Wang, J. et al. (2006). Manipulation of image intensity distribution at 7.0 T: passive RF shimming and focusing with dielectric materials. *J. Magn. Reson. Imaging* 24: 197–202.

98 Webb, A.G. (2011). Dielectric materials in magnetic resonance. *Concepts Magn. Reson.* 38A: 148–184.

99 Teeuwisse, W.M., Brink, W.M., and Webb, A.G. (2012). Quantitative assessment of the effects of high-permittivity pads in 7 Tesla MRI of the brain. *Magn. Reson. Med.* 67: 1285–1293.

100 Christ, A., Kainz, W., Hahn, E.G. et al. (2010). The Virtual Family-development of surface-based anatomical models of two adults and two children for dosimetric simulations. *Phys. Med. Biol.* 55: N23–N38.

101 Pennes, H.H. (1948). Analysis of tissue and arterial blood temperatures in the resting human forearm. *J. Appl. Physiol.* 1: 93–122.

102 IEC (2010). *Medical Electrical Equipment. Part 2-33: Particular Requirements for the Safety of Magnetic Resonance Equipment for Medical Diagnosis*, IEC 60601-2-33. Geneva: International Electrotechnical Commission.

103 Hyde, J.S., Jesmanowicz, A., Froncisz, W. et al. (1987). Parallel image acquisition for noninteracting local coils. *J. Magn. Reson.* 70: 512–517.

104 Roemer, P.B., Edelstein, W.A., Hayes, C.E. et al. (1990). The NMR phased array. *Magn. Reson. Med.* 16: 192–225.

105 Schmitt, M., Potthast, A., Sosnovik, D.E. et al. (2008). A 128-channel receive-only cardiac coil for highly accelerated cardiac MRI at 3 Tesla. *Magn. Reson. Med.* 59: 1431–1439.

106 Wiggins, G.C., Polimeni, J.R., Potthast, A. et al. (2009). 96-Channel receive-only head coil for 3 Tesla: design optimization and evaluation. *Magn. Reson. Med.* 62: 754–762.

107 Griswold, M.A., Jakob, P.M., Heidemann, R.M. et al. (2002). Generalized autocalibrating partially parallel acquisitions (GRAPPA). *Magn. Reson. Med.* 47: 1202–1210.

108 Pruessmann, K.P., Weiger, M., Scheidegger, M.B., and Boesiger, P. (1999). SENSE: sensitivity encoding for fast MRI. *Magn. Reson. Med.* 42: 952–962.

109 Sodickson, D.K. and Manning, W.J. (1997). Simultaneous acquisition of spatial harmonics (SMASH): fast imaging with radiofrequency coil arrays. *Magn. Reson. Med.* 38: 591–603.

110 Adriany, G. and Gruetter, R. (1997). A half-volume coil for efficient proton decoupling in humans at 4 Tesla. *J. Magn. Reson.* 125: 178–184.

111 Styles, P., Soffe, N.F., Scott, C.A. et al. (1984). A high-resolution NMR probe in which the coil and preamplifier are cooled with liquid helium. *J. Magn. Reson.* 60: 397–404.

112 Jerosch-Herold, M. and Kirschman, R.K. (1989). Potential benefits of a cryogenically cooled NMR probe for room-temperature samples. *J. Magn. Reson.* 85: 141–146.

113 Wright, A.C., Song, H.K., and Wehrli, F.W. (2000). In vivo MR micro imaging with conventional radiofrequency coils cooled to 77 degrees K. *Magn. Reson. Med.* 43: 163–169.

114 Black, R.D., Early, T.A., Roemer, P.B. et al. (1993). A high-temperature superconducting receiver for nuclear magnetic resonance microscopy. *Science* 259: 793–795.

115 Miller, J.R., Hurlston, S.E., Ma, Q.Y. et al. (1999). Performance of a high-temperature superconducting probe for *in vivo* microscopy at 2.0 T. *Magn. Reson. Med.* 41: 72–79.

116 Hurlston, S.E., Brey, W.W., Suddarth, S.A., and Johnson, G.A. (1999). A high-temperature superconducting Helmholtz probe for microscopy at 9.4 T. *Magn. Reson. Med.* 41: 1032–1038.

117 Darrasse, L. and Ginefri, J.C. (2003). Perspectives with cryogenic RF probes in biomedical MRI. *Biochimie* 85: 915–937.

118 Chan, F., Pauly, J., and Macovski, A. (1992). Effects of RF amplifier distortion on selective excitation and their correction by prewarping. *Magn. Reson. Med.* 23: 224–238.

119 Hoult, D.I. (1978). The NMR receiver: a description and analysis of design. *Prog. Nucl. Magn. Reson. Spectrosc.* 12: 41–77.

120 Delsuc, M.A. and Lallemand, J.Y. (1986). Improvement of dynamic range in NMR by oversampling. *J. Magn. Reson.* 69: 504–507.

Appendix A

A.1 Matrix Calculations

A matrix is a collection of numbers in rows and columns, which can, for example, characterize a set of linear equations. For example, the set of linear equations

$$
\begin{aligned}
a_{11}x_1 + a_{12}x_2 + a_{13}x_3 &= b_1 \\
a_{21}x_1 + a_{22}x_2 + a_{23}x_3 &= b_2 \\
a_{31}x_1 + a_{32}x_2 + a_{33}x_3 &= b_3
\end{aligned}
\tag{A.1.1}
$$

is completely characterized by the matrices

$$
\begin{pmatrix} a_{11} & a_{12} & a_{13} \\ a_{21} & a_{22} & a_{23} \\ a_{31} & a_{32} & a_{33} \end{pmatrix},
\begin{pmatrix} x_1 \\ x_2 \\ x_3 \end{pmatrix}, \text{and}
\begin{pmatrix} b_1 \\ b_2 \\ b_3 \end{pmatrix}
\tag{A.1.2}
$$

The elements in a matrix are normally labeled according to their column and row position. A matrix with k rows and n columns is normally referred to as a k by n (or $k \times n$) matrix. When $k = n$ the matrix is referred to as a square matrix of order n. One of the most common manipulations in matrix algebra is the product (multiplication) of two matrices. This is computed as follows:

$$
\begin{pmatrix} a_{11} & a_{12} \\ a_{21} & a_{22} \end{pmatrix} \cdot \begin{pmatrix} b_{11} & b_{12} \\ b_{21} & b_{22} \end{pmatrix} = \begin{pmatrix} a_{11}b_{11} + a_{12}b_{21} & a_{11}b_{12} + a_{12}b_{22} \\ a_{21}b_{11} + a_{22}b_{21} & a_{21}b_{12} + a_{22}b_{22} \end{pmatrix}
\tag{A.1.3}
$$

In general, the product of a $k_1 \times n_1$ and a $k_2 \times n_2$ matrix is a $k_1 \times n_2$ matrix. Note that n_1 should always equal k_2. A diagonal matrix is a square matrix in which all off-diagonal elements (i.e. elements a_{ij} for which $i \neq j$) are zero and at least one of the diagonal elements (i.e. elements a_{ij} for which $i = j$) is nonzero. A diagonal matrix in which all diagonal elements are equal to 1 is called a unity matrix I. For example, a third-order unity matrix is

$$
\mathbf{I} = \begin{pmatrix} 1 & 0 & 0 \\ 0 & 1 & 0 \\ 0 & 0 & 1 \end{pmatrix}
\tag{A.1.4}
$$

In Vivo NMR Spectroscopy: Principles and Techniques, Third Edition. Robin A. de Graaf.
© 2019 John Wiley & Sons Ltd. Published 2019 by John Wiley & Sons Ltd.

The product of a unity matrix $\mathbf{I}$ with another matrix $\mathbf{A}$ does not have any net effect, i.e. $\mathbf{I}\cdot\mathbf{A} = \mathbf{A}$. Multiplication of $\mathbf{A}$ with the inverse matrix $\mathbf{A}^{-1}$ gives a unity matrix, i.e. $\mathbf{A}\cdot\mathbf{A}^{-1} = \mathbf{I}$.

To simplify calculations for particular NMR applications, it is often convenient to write the phenomenon under investigation in matrix form. As an example, consider the rotation of transverse magnetization about the z-axis. The clockwise rotations are completely described by

$$
\begin{aligned}
M_{x,\text{after}} &= M_{x,\text{before}}\cos\theta + M_{y,\text{before}}\sin\theta \\
M_{y,\text{after}} &= M_{y,\text{before}}\cos\theta - M_{x,\text{before}}\sin\theta \\
M_{z,\text{after}} &= M_{z,\text{before}}
\end{aligned}
\tag{A.1.5}
$$

where "before" and "after" refer to the magnetization components before and after the rotation. The set of linear equations of Eq. (A.1.5) can be expressed in matrix form as

$$
\begin{pmatrix} M_x \\ M_y \\ M_z \end{pmatrix} = \begin{pmatrix} \cos\theta & \sin\theta & 0 \\ -\sin\theta & \cos\theta & 0 \\ 0 & 0 & 1 \end{pmatrix} \cdot \begin{pmatrix} M_x \\ M_y \\ M_z \end{pmatrix}, \quad \text{or}
\tag{A.1.6}
$$

$$
\mathbf{M} = \mathbf{R}_z(\theta)\mathbf{M}
\tag{A.1.7}
$$

where $\mathbf{R}_z(\theta)$ is the rotation matrix for a rotation about the z-axis through an angle θ (see also Chapter 5). For some applications, like RF pulse simulation or quantitative description of diffusion, one would like to describe the rotations of $\mathbf{M}$ in a frame $\mathbf{A}'$, which is rotated with respect to the original frame $\mathbf{A}$. This can be achieved if the rotation matrix $\mathbf{R}$ connecting both frames is known according to

$$
\mathbf{A}' = \mathbf{R}^{-1}\cdot\mathbf{A}\cdot\mathbf{R}
\tag{A.1.8}
$$

For other applications one is not interested in the rotations in either frame, but would like to obtain a parameter independent of the frame orientations. One parameter (among many others) that is rotationally invariant is the trace of the matrix and is defined as

$$
\text{Tr}(\mathbf{A}) = \text{Tr}\begin{pmatrix} a_{11} & a_{12} & a_{13} \\ a_{21} & a_{22} & a_{23} \\ a_{31} & a_{32} & a_{33} \end{pmatrix} = a_{11} + a_{22} + a_{33}
\tag{A.1.9}
$$

One can easily verify that the trace of a matrix is rotationally invariant, i.e.

$$
\text{Tr}(\mathbf{A}') = \text{Tr}(\mathbf{R}^{-1}\cdot\mathbf{A}\cdot\mathbf{R}) = \text{Tr}(\mathbf{A})
\tag{A.1.10}
$$

Another feature encountered in matrix algebra that is frequently applied in NMR applications is transposition. The transpose matrix $\mathbf{A}^{\text{T}}$ of a $(k \times n)$ matrix $\mathbf{A}$ is a $(n \times k)$ matrix in which the rows and columns of $\mathbf{A}$ have been interchanged, i.e.

$$
\mathbf{A}^{\text{T}} = \begin{pmatrix} a_{11} & a_{21} \\ a_{12} & a_{22} \\ a_{13} & a_{23} \end{pmatrix}, \quad \text{when } \mathbf{A} = \begin{pmatrix} a_{11} & a_{12} & a_{13} \\ a_{21} & a_{22} & a_{23} \end{pmatrix}
$$

Besides the above-cited characteristics of matrices, a wealth of other properties and rules exist for which the reader is referred to standard textbooks on mathematics.

A.2 Trigonometric Equations

In Fourier transform NMR many calculations and manipulations involve sums and products of sines and cosines and complex combinations thereof. Therefore, some of the most important trigonometric equations will be summarized here. The sums and differences of (co)sines are given by

$$\sin(a+b)=\sin a\cos b+\cos a\sin b \tag{A.2.1}$$

$$\sin(a-b)=\sin a\cos b-\cos a\sin b \tag{A.2.2}$$

$$\cos(a+b)=\cos a\cos b-\sin a\sin b \tag{A.2.3}$$

$$\cos(a-b)=\cos a\cos b+\sin a\sin b \tag{A.2.4}$$

Other trigonometric equations, which are frequently used are

$$\sin^2 a+\cos^2 a=1 \tag{A.2.5}$$

$$\cos^2 a-\sin^2 a=\cos 2a \tag{A.2.6}$$

$$2\sin a\cos a=\sin 2a \tag{A.2.7}$$

$$\sin a=-\sin(-a) \tag{A.2.8}$$

$$\cos a=\cos(-a) \tag{A.2.9}$$

When complex notation is involved, the Euler relations are invaluable and are given by

$$e^{+ia}=\cos a+i\sin a \tag{A.2.10}$$

$$e^{-ia}=\cos a-i\sin a \tag{A.2.11}$$

with $i^2=-1$. Many other useful relations can be derived from Eqs. (A.2.1)–(A.2.11). For a more complex overview of trigonometry the reader is referred to standard textbooks on mathematics.

A.3 Fourier Transformation

The Fourier transformation (FT) is at the heart of modern NMR. A thorough understanding of its characteristics and definitions is essential to describe the actions of many tools and techniques encountered in NMR, like data processing, RF pulses, and magnetic resonance imaging. Here, some of the basic definitions concerning FT are given and how they relate to NMR. For more thorough discussions, the reader is referred to the literature [1].

A.3.1 Introduction

Any periodic function $f(t)$ can be decomposed into an infinite harmonic series according to

$$f(t)=\frac{A_0}{2}+\sum_{n=1}^{\infty}\left[A_n\cos\left(\frac{2\pi nt}{T}\right)+B_n\sin\left(\frac{2\pi nt}{T}\right)\right] \tag{A.3.1}$$

where T is the period and the Fourier components A_n and B_n are given by

$$A_n=\frac{1}{T}\int_{-T/2}^{T/2}f(t)\cos\left(\frac{2\pi nt}{T}\right)dt \tag{A.3.2}$$

$$B_n = \frac{1}{T} \int_{-T/2}^{T/2} f(t) \sin\left(\frac{2\pi n t}{T}\right) dt \tag{A.3.3}$$

This decomposition is called a Fourier series. The frequency components $\omega_n = 2\pi n/T$ and amplitude A_n and B_n can be retrieved from an unknown signal $f(t)$ by performing a FT of signal $f(t)$ according to

$$F(\omega) = \int_{-\infty}^{+\infty} f(t) e^{-i\omega t} dt \tag{A.3.4}$$

The inverse FT is defined as

$$f(t) = \int_{-\infty}^{+\infty} F(\omega) e^{+i\omega t} d\omega \tag{A.3.5}$$

Note that Eqs. (A.3.4) and (A.3.5) often include a normalization factor, depending on the exact definition of Eqs. (A.3.1)–(A.3.3). For simplicity these factors have been omitted here. The interpretation of Eqs. (A.3.4) and (A.3.5) is straightforward. The time-domain function $f(t)$ is a linear combination of orthonormal basis functions (i.e. sines and cosines) $e^{-i\omega t} = \cos \omega t - i \sin \omega t$. Each basis function denotes a circularly polarized oscillation (frequency). Corresponding to each frequency is an amplitude $F(\omega)$, which relates how much the component $e^{-i\omega t}$ contributes to $f(t)$. FT of $f(t)$ is therefore a decomposition of $f(t)$ into its frequency components and their corresponding amplitudes.

A more practical, alternative way of looking at Eqs. (A.3.4) and (A.3.5) is the following. Eq. (A.3.4) expresses the fact that the FID, $f(t)$, is multiplied by a monochromatic reference frequency ω and then integrated over the entire time domain. Only frequency components at or near the reference frequency give a finite integral, while the rest gets quickly out of phase with the reference signal giving a (near) zero integral. The procedure is repeated for different values of the reference frequency until all appropriate frequencies (i.e. the entire spectral bandwidth) have been explored.

A.3.2 Properties

Several useful properties of the FT are summarized below.

A.3.2.1 Linearity
For any integrable functions $f(t)$ and $g(t)$ and for any constant a, the FT is a linear operation

$$FT(f(t) + g(t)) = FT(f(t)) + FT(g(t)) \tag{A.3.6}$$

$$FT(a \cdot f(t)) = a \cdot FT(f(t)) \tag{A.3.7}$$

The same holds true for the integrable functions $F(\omega)$ and $G(\omega)$.

A.3.2.2 Time and Frequency Shifting
The frequency domain signal $F(\omega)$ can be shifted to $F(\omega - \omega_0)$ by applying a phase shift to the time-domain signal $f(t)$ according to

$$FT(f(t) \cdot e^{-i\omega_0 t}) = F(\omega - \omega_0) \tag{A.3.8}$$

A time-domain signal $f(t)$ can be shifted in an analogues manner

$$FT\left(F(\omega)\cdot e^{+i\omega t_0}\right)=f\left(t-t_0\right)$$

(A.3.9)

The time and frequency properties have found widespread applications in NMR. In MRI and MRSI, Eqs. (A.3.8) and (A.3.9) are regularly used to shift the image under investigation. In MRS, time-frequency shifting is often employed with RF pulses and localization. Adding a phase-ramp to a RF pulse shifts the frequency excitation profile without the need for frequency switching.

A.3.2.3 Scaling

When a time-domain function $f(t)$ is scaled to $f(at)$ then its FT is given by

$$FT\left(f(at)\right)=\frac{1}{|a|}F\left(\frac{\omega}{a}\right)$$

(A.3.10)

This phenomenon is also observed in many parts of *in vivo* NMR. For example, reducing the pulse length of a selective pulse by a factor of 2 ($a=1/2$) increases the excitation bandwidth $F(\omega/a)$ by a factor 2 and doubles the required RF amplitude (proportional to $|a|^{-1}$).

A.3.2.4 Convolution

The convolution $h(t)$ of two functions $f(t)$ and $g(t)$ is a broadening of one function by the other. When convolution is used for filtering, one function $g(t)$ is called the weighting function, which is convolved with the original data $f(t)$ to give the filtered data $h(t)$. Mathematically this can be described as

$$h(t)=\int_{-\infty}^{+\infty}f(t)\cdot g(t-\tau)\mathrm{d}\tau=f(t)\times g(t)$$

(A.3.11)

Convolutions obey the commutative, associative, and distributive laws. In combination with the FT, convolution leads to the following results:

$$FT\left(f(t)\cdot g(t)\right)=FT\left(f(t)\right)\times FT\left(g(t)\right)$$

(A.3.12)

$$FT\left(f(t)\times g(t)\right)=FT\left(f(t)\right)\cdot FT\left(g(t)\right)$$

(A.3.13)

Identical equations can be derived for an inverse FT. Equations (A.3.12) and (A.3.13) reveal that a convolution in one domain is a simple multiplication in the other domain. One of the most commonly used convolutions is apodization of the FID signal as described in Chapter 1. To make a relative enhancement of the signal in the beginning of the FID over the noisy end, the original FID can be multiplied by an exponentially decaying function. This results in a convolution of the two functions in the frequency domain, i.e. a broadening of resonances.

A.3.3 Discrete Fourier Transformation

All modern NMR experiments are performed with the aid of computers, which require that the FID signal is sampled (digitized) at discrete intervals. This consequently makes the FT also a discrete process in which the continuous integration of Eq. (A.3.4) is replaced by a discrete summation. The discrete Fourier transformation (DFT) of a discretely sampled time-domain signal $f(t)$ is then given by

$$F(\omega)=\sum_{t=0}^{N-1}f(t).e^{-i(\omega/N)t}$$

(A.3.14)

The computation of Eq. (A.3.14) can be performed in an analogous manner to that described for the continuous FT. For simple 1D spectra this is an acceptable task, but for larger 1D datasets and all 2D and 3D MRI and MRSI datasets, this would lead to unacceptable calculation times. By using symmetry and recursive characteristics of DFT, Cooley and Tukey proposed in 1965 [2] a new algorithm for the special case of N being a power of 2. This fast Fourier transform (FFT) algorithm requires $N \log_2 N$ calculations on a dataset containing N datapoints, whereas the conventional DFT of Eq. (A.3.14) requires N^2 calculations. For $N = 256$ this already leads to a reduction in calculation time of 32, while for a 256×256 2D dataset the efficiency is 1024 higher. Even though modern algorithms give substantial improvements for any N, the matrix size in MRI and MRS are still often chosen as a power of 2.

A.4   Product Operator Formalism

As has been demonstrated throughout the book, the use of classical vectors to describe NMR phenomena is a valuable method to visualize certain NMR experiments. However, experiments that involve scalar coupling such as those encountered in 2D NMR, spectral editing, and polarization transfer cannot be adequately described by the classical vector model.

A variety of quantum mechanical methods exist, which are capable of a quantitative description of all NMR phenomena. In particular, the density matrix formalism is universally applicable. In the density matrix formalism one is not directly concerned with magnetization, but rather with the energy states of the spin system under investigation. Therefore, a density matrix holds all possible energy states of a given spin system. Just as one is interested in the time evolution of magnetization during an NMR experiment, the time evolution of the density matrix needs to be calculated, which is governed by the Liouville–von Neumann equation. It is well outside the scope of the book to give a detailed treatment of density matrix calculations. For further reading, the reader is referred to Refs. [3–10]. Because the density matrix formalism is concerned with energy states, it does not provide any physical insight into the NMR experiment during the calculation. Furthermore, for a spin system of more than a few spins, density matrix calculations become rapidly cumbersome. In certain cases, for instance, in the case of non- or weakly coupled spin systems, it is possible to simplify the density matrix formalism by expanding the density matrix into a linear combination of orthogonal matrices (or product operators) [11–13], each of which represents an orthogonal component of the magnetization (some of which are not directly observable as will be explained below). Orthogonality is defined using the trace relation, i.e. the trace of the product of two orthogonal matrices is zero. Various complete orthogonal basis matrix sets can be used, including single-transition operators, spherical tensor operators, shift (or lowering and raising) operators, and product operators. The last two basis sets are most straightforward to use and will be discussed in more detail.

A.4.1   Cartesian Product Operators

The complete basis set for N spins ½ consists of 2^{2N} product operators. For a two-spin system IS, the 16 product operators are (besides the unity operator) I_x, I_y, I_z, S_x, S_y, S_z, $2I_xS_x$, $2I_xS_y$, $2I_xS_z$, $2I_yS_x$, $2I_yS_y$, $2I_yS_z$, $2I_zS_x$, $2I_zS_y$, and $2I_zS_z$. Each of these product operators corresponds to a particular physical state, according to

I_z	Polarization of spin I (longitudinal magnetization)
I_x, I_y	In-phase x and y coherence of spin I (transverse magnetization)
$2I_xS_z, 2I_yS_z$	x and y coherence of spin I in antiphase with respect to spin S
$2I_xS_x, 2I_xS_y$	
$2I_yS_x, 2I_yS_y$	Two-spin coherence of spins I and S
$2I_zS_z$	Longitudinal two-spin order of spins I and S

Using the product operator formalism [11–17], the rotation of magnetization due to an RF pulse applied along the x axis can be described as

$$I_x \xrightarrow{\theta I_x} I_x \tag{A.4.1}$$

$$I_y \xrightarrow{\theta I_x} I_y \cos\theta + I_z \sin\theta \tag{A.4.2}$$

$$I_z \xrightarrow{\theta I_x} I_z \cos\theta - I_y \sin\theta \tag{A.4.3}$$

where θ is the nutation angle. Note that these transformations are for counter-clockwise rotations. Clockwise rotations can be obtained by using $-\theta$ instead of $+\theta$. In an equivalent manner, the rotation about the y axis can be described by

$$I_x \xrightarrow{\theta I_y} I_x \cos\theta - I_z \sin\theta \tag{A.4.4}$$

$$I_y \xrightarrow{\theta I_y} I_y \tag{A.4.5}$$

$$I_z \xrightarrow{\theta I_y} I_z \cos\theta + I_x \sin\theta \tag{A.4.6}$$

The effects of chemical shift evolution can be described by a rotation about the z axis over an angle ωt:

$$I_x \xrightarrow{\omega t I_z} I_x \cos\omega t + I_y \sin\omega t \tag{A.4.7}$$

$$I_y \xrightarrow{\omega t I_z} I_y \cos\omega t - I_x \sin\omega t \tag{A.4.8}$$

$$I_z \xrightarrow{\omega t I_z} I_z \tag{A.4.9}$$

So far the transformations given by Eqs. (A.4.1)–(A.4.9) are similar to the rotation matrices given by Eqs. (5.4)–(5.7) for regular (non-scalar-coupled) magnetization. One of the most powerful characteristics of the product operator formalism is that it remains intuitive even for scalar-coupled spin-systems. For a weakly-coupled two-spin system IS, the evolution of coherence due to scalar coupling is given by

$$I_x \xrightarrow{\pi J t 2 I_z S_z} I_x \cos\pi J t + 2 I_y S_z \sin\pi J t \tag{A.4.10}$$

$$I_y \xrightarrow{\pi J t 2 I_z S_z} I_y \cos\pi J t - 2 I_x S_z \sin\pi J t \tag{A.4.11}$$

$$I_z \xrightarrow{\pi J t 2 I_z S_z} I_z \tag{A.4.12}$$

where J is the coupling constant between I and S. The evolution of antiphase coherence due to J coupling can easily be derived from Eqs. (A.4.10)–(A.4.12), using the relation $4I_z^2 = 4S_z^2 = 1$, and is given by

$$2I_x S_z \xrightarrow{\pi J t 2 I_z S_z} 2 I_x S_z \cos\pi J t + I_y \sin\pi J t \tag{A.4.13}$$

$$2I_y S_z \xrightarrow{\pi J t 2 I_z S_z} 2 I_y S_z \cos\pi J t - I_x \sin\pi J t \tag{A.4.14}$$

Up to this point, the product operator formalism has (superficially) not been more than an alternative to the classical or correlated vector model. However, consider a scalar-coupled two-spin system IS for which a pure antiphase coherence state $2I_x S_z$ has been created by a spin-echo sequence with $t = TE = 1/(2J)$. When at this point a nonselective $90°_x$ pulse is executed, a spin-state described by $-2I_x S_y$ is created which cannot be described by the classical vector model. The product operator formalism offers a convenient method to deal with these so-called multi-quantum-coherences, as will be described next.

A.4.2 Shift (Lowering and Raising) Operators

Because terms such as $2I_xS_x$, $2I_xS_y$, $2I_yS_x$, and $2I_yS_y$ contain a combination of zero and double-quantum coherences, that behave differently to external perturbations, it is convenient to replace the product operators by raising and lowering operators given by

$$I^+ = I_x + iI_y \tag{A.4.15}$$

$$I^- = I_x - iI_y \tag{A.4.16}$$

$$I_0 = I_z \tag{A.4.17}$$

making

$$I_x = \frac{1}{2}\left(I^+ + I^-\right) \tag{A.4.18}$$

$$I_y = -\frac{i}{2}\left(I^+ - I^-\right) \tag{A.4.19}$$

Using raising and lowering operators one can easily keep track of the coherence order. Single-quantum coherences (i.e. transverse magnetization) are those spin states with only one raising or lowering operator (e.g. I^+, $2I^-S_0$), zero-quantum coherences are spin states in which the net quantum number is zero (i.e. I^+S^-), while for double-quantum coherences the spin quantum number adds to two (e.g. I^+S^+). In this formalism, chemical shift evolution takes a simpler form:

$$I^+ \xrightarrow{\;\omega t I_z\;} I^+ e^{-i\omega t} \tag{A.4.20}$$

$$I^- \xrightarrow{\;\omega t I_z\;} I^- e^{+i\omega t} \tag{A.4.21}$$

$$I_0 \xrightarrow{\;\omega t I_z\;} I_0 \tag{A.4.22}$$

The effect of scalar evolution is described by

$$I^+ \xrightarrow{\;\pi Jt\,2I_zS_z\;} I^+ \cos \pi Jt - 2iI^+ S_0 \sin \pi Jt \tag{A.4.23}$$

$$I^- \xrightarrow{\;\pi Jt\,2I_zS_z\;} I^- \cos \pi Jt + 2iI^- S_0 \sin \pi Jt \tag{A.4.24}$$

$$I_0 \xrightarrow{\;\pi Jt\,2I_zS_z\;} I_0 \tag{A.4.25}$$

and

$$2I^+ S_0 \xrightarrow{\;\pi Jt\,2I_zS_z\;} 2I^+ S_0 \cos \pi Jt - iI^+ \sin \pi Jt \tag{A.4.26}$$

$$2I^- S_0 \xrightarrow{\;\pi Jt\,2I_zS_z\;} 2I^- S_0 \cos \pi Jt + iI^- \sin \pi Jt \tag{A.4.27}$$

The effect of an RF pulse with phase ϕ (where 0 corresponds to the $+x$ axis, $\pi/2$ to the $+y$ axis, and so on) generating a nutation angle θ is governed by

$$I^+ \longrightarrow \frac{I^+}{2}\left(\cos\theta + 1\right) - \frac{I^-}{2}\left(\cos\theta - 1\right)\cdot e^{-2i\phi} + iI_0 \sin\theta \cdot e^{-i\phi} \tag{A.4.28}$$

$$I^- \longrightarrow \frac{I^-}{2}\left(\cos\theta + 1\right) - \frac{I^+}{2}\left(\cos\theta - 1\right)\cdot e^{-2i\phi} - iI_0 \sin\theta \cdot e^{+i\phi} \tag{A.4.29}$$

$$I_0 \longrightarrow \frac{iI^+}{2}\sin\theta \cdot e^{-i\phi} - \frac{iI^-}{2}\sin\theta \cdot e^{+i\phi} + I_0\cos\theta \qquad (A.4.30)$$

Up to this point only single-quantum coherences ($p = \pm1$) have been discussed. However, the product operator formalism is particularly valuable when higher-order coherences are involved. The two-spin operators $2I_xS_x$, $2I_yS_y$, $2I_xS_y$, and $2I_yS_x$ can be written in terms of raising and lowering operators according to

$$2I_xS_x = \frac{1}{2}\left(I^+S^+ + I^+S^- + I^-S^+ + I^-S^-\right) \qquad (A.4.31)$$

$$2I_yS_y = -\frac{1}{2}\left(I^+S^+ - I^+S^- - I^-S^+ + I^-S^-\right) \qquad (A.4.32)$$

$$2I_xS_y = -\frac{i}{2}\left(I^+S^+ - I^+S^- + I^-S^+ - I^-S^-\right) \qquad (A.4.33)$$

$$2I_yS_x = -\frac{i}{2}\left(I^+S^+ + I^+S^- - I^-S^+ - I^-S^-\right) \qquad (A.4.34)$$

Clearly all four 2-spin terms contain both double (I^+S^+ and I^-S^-) and zero (I^+S^- and I^-S^+) quantum coherences. Linear combinations of two-spin terms can reveal pure double and zero-quantum coherences according to

$$\frac{1}{2}\left(I^+S^+ + I^-S^-\right) = \frac{1}{2}\left(2I_xS_x - 2I_yS_y\right) = DQC_x \qquad (A.4.35)$$

$$-\frac{i}{2}\left(I^+S^+ - I^-S^-\right) = \frac{1}{2}\left(2I_xS_y + 2I_yS_x\right) = DQC_y \qquad (A.4.36)$$

$$\frac{1}{2}\left(I^+S^- + I^-S^+\right) = \frac{1}{2}\left(2I_xS_x + 2I_yS_y\right) = ZQC_x \qquad (A.4.37)$$

$$-\frac{i}{2}\left(I^+S^- - I^-S^+\right) = \frac{1}{2}\left(2I_yS_x - 2I_yS_x\right) = ZQC_y \qquad (A.4.38)$$

From Eqs. (A.4.35)–(A.4.38) it can be seen that selective inversion of one spin operator (e.g. I) results in the interconversion of double- and zero-quantum coherences. This is an important aspect in spectral editing using multiple-quantum-coherences. Double-quantum coherences between spins I and S evolve in the transverse plane under the sum of the chemical shifts ($\omega_I + \omega_S$):

$$DQC_x \xrightarrow{\omega_I tI_z + \omega_S tS_z} DQC_x \cos\left[(\omega_I + \omega_S)t\right] + DQC_y \sin\left[(\omega_I + \omega_S)t\right] \qquad (A.4.39)$$

$$DQC_y \xrightarrow{\omega_I tI_z + \omega_S tS_z} DQC_y \cos\left[(\omega_I + \omega_S)t\right] - DQC_x \sin\left[(\omega_I + \omega_S)t\right] \qquad (A.4.40)$$

Using Eqs. (A.4.39) and (A.4.40) it can be shown that double-quantum coherences have twice the sensitivity to magnetic field gradients as compared to regular transverse magnetization (i.e. single-quantum coherences). Zero-quantum coherences evolve under the difference of the chemical shift ($\omega_I - \omega_S$):

$$ZQC_x \xrightarrow{\omega_I tI_z + \omega_S tS_z} ZQC_x \cos\left[(\omega_I - \omega_S)t\right] + ZQC_y \sin\left[(\omega_I - \omega_S)t\right] \qquad (A.4.41)$$

$$ZQC_y \xrightarrow{\omega_I tI_z + \omega_S tS_z} ZQC_y \cos\left[(\omega_I - \omega_S)t\right] - ZQC_y \sin\left[(\omega_I - \omega_S)t\right] \qquad (A.4.42)$$

such that *homonuclear* zero-quantum coherences do not evolve under magnetic field gradients (i.e. *homonuclear* ZQCs are *not* dephased by magnetic field gradients!). A final aspect of multiple-quantum coherence that needs to be taken into account when performing product operator calculations is the evolution of MQCs under coupling with active and passive spins. For example, consider [3-^{13}C]-lactate with heteronuclear scalar coupling between carbon-13 and the methyl protons, as well as homonuclear scalar coupling between the methyl and methine protons. Further consider a heteronuclear multiple-quantum editing sequence that utilizes the heteronuclear MQCs between [3-^{13}C] and [3-^{1}H]. These two nuclei are actively involved in the MQC transition and the evolution of MQCs is not influenced by active spins. The homonuclear proton–proton couplings are not actively involved in the MQC transition and are therefore referred to as passive spins. Passive spins lead to an evolution of MQCs according to

$$\text{ZQC}_x^{IS} \xrightarrow{\pi J_{IL}t 2I_z L_z + \pi J_{IS}t 2L_z S_z} \text{ZQC}_x^{IS} \cos \pi K_{IS} t + 2\text{ZQC}_y^{IS} L_z \sin \pi K_{IS} t \tag{A.4.43}$$

where $K_{IS} = |J_{LS} - J_{IL}|$, which is known as the zero-quantum splitting.

With the rules presented in this section for evolution of coherences under chemical shifts, magnetic field gradients, scalar coupling, and during RF pulses, the theoretical outcome of any NMR pulse sequence can be quantitatively calculated (ignoring relaxation and other physical processes).

References

1 Bracewell, R.M. (1965). *The Fourier Transform and Its Applications*. New York: McGraw-Hill.
2 Cooley, J.W. and Tukey, J.W. (1965). An algorithm for machine calculation of complex Fourier series. *Math. Comput.* 19: 297–301.
3 Ernst, R.R., Bodenhausen, G., and Wokaun, A. (1987). *Principles of Nuclear Magnetic Resonance in One and Two Dimensions*. Oxford: Clarendon Press.
4 Slichter, C.P. (1990). *Principles of Magnetic Resonance*. Berlin: Springer-Verlag.
5 Farrar, T.C. and Harriman, J.E. (1992). *Density Matrix Theory and Its Applications in NMR Spectroscopy*. Madison: Farragut Press.
6 Cavanagh, J., Fairbrother, W.J., Palmer, A.G. III, and Skelton, N.J. (1996). *Protein NMR Spectroscopy. Principles and Practice*. San Diego: Academic Press.
7 Keeler, J. (2005). *Understanding NMR Spectroscopy*. New York: Wiley.
8 Levitt, M.H. (2005). *Spin Dynamics. Basics of Nuclear Magnetic Resonance*. New York: Wiley.
9 Goldman, M. (1991). *Quantum Description of High-Resolution NMR in Liquids*. Oxford: Oxford University Press.
10 Hore, P.J., Jones, J.A., and Wimperis, S. (2000). *NMR: The Toolkit*. Oxford: Oxford University Press.
11 Sorensen, O.W., Eich, G.W., Levitt, M.H. et al. (1983). Product operator formalism for the description of NMR pulse experiments. *Prog. NMR Spectrosc.* 16: 163–192.
12 Packer, K.J. and Wright, K.M. (1983). The use of single-spin operator basis-sets in NMR spectroscopy of scalar-coupled spin systems. *Mol. Phys.* 50: 797–813.
13 van de Ven, F.J.M. and Hilbers, C.W. (1983). A simple formalism for the description of multi-pulse experiments. Applications to a weakly-coupled two-spin (I = 1/2) system. *J. Magn. Reson.* 54: 512–520.
14 Shriver, J. (1992). Product operators and coherence transfer in multi-pulse NMR experiments. *Concepts Magn. Reson.* 4: 1–34.

15 Kingsley, P.B. (1995). Product operators, coherence pathways and phase cycling. Part I: product operators, spin-spin coupling, and coherence pathways. *Concepts Magn. Reson.* 7: 29–48.

16 Kingsley, P.B. (1995). Product operators, coherence pathways and phase cycling. Part II: coherence pathways in multipulse sequences: spin echoes, stimulated echoes and multiple-quantum-coherences. *Concepts Magn. Reson.* 7: 115–136.

17 Kingsley, P.B. (1995). Product operators, coherence pathways and phase cycling. Part III: phase cycling. *Concepts Magn. Reson.* 7: 167–192.

Further Reading

General NMR Spectroscopy

Abragam, A. (1961). *Principles of Nuclear Magnetism*. Oxford: Clarendon Press.

Bax, A. (1984). *Two-dimensional Nuclear Magnetic Resonance in Liquids*. Delft: Delft University Press.

Cavanagh, J., Fairbrother, W.J., Palmer, A.G. III et al. (1996). *Spectroscopy. Principles and Practice*. San Diego: Academic Press.

Chandrakumar, N. and Subramanian, S. (1987). *Modern Techniques in High-resolution NMR*. Berlin: Springer-Verlag.

Ernst, R.R., Bodenhausen, G., and Wokaun, A. (1987). *Principles of Nuclear Magnetic Resonance in One and Two Dimensions*. Oxford: Clarendon Press.

Freeman, R. (1987). *A Handbook of Nuclear Magnetic Resonance*. Harlow: Longman.

Fukushima, E. and Roeder, S.B.W. (1984). *Experimental Pulse NMR: A Nuts and Bolts Approach*. New York: Addison-Wesley.

Grant, D.M. and Harris, R.K. (eds.) (1996). *Encyclopedia of NMR* (8 vols.). New York: Wiley.

Harris, R.K. (1987). *Nuclear Magnetic Resonance Spectroscopy. A Physicochemical View*. Harlow: Longman.

Homans, S.W. (1989). *A Dictionary of Concepts in NMR*. Oxford: Clarendon Press.

Keeler, J. (2005). *Understanding NMR Spectroscopy*. New York: Wiley.

Levitt, M.H. (2005). *Spin Dynamics. Basics of Nuclear Magnetic Resonance*. New York: Wiley.

Munowitz, M. (1988). *Coherence and NMR*. New York: Wiley.

Slichter, C.P. (1990). *Principles of Magnetic Resonance*. Berlin: Springer-Verlag.

Biomedical NMR Spectroscopy and Imaging

Bernstein, M.A., King, K.F., and Zhou, X.J. (2004). *Handbook of MRI Pulse Sequences*. New York: Academic Press.

Cady, E.B. (1990). *Clinical Magnetic Resonance Spectroscopy*. New York: Plenum Publishing.

Callaghan, P.T. (1993). *Principles of Nuclear Magnetic Resonance Microscopy*. Oxford: Clarendon Press.

Chen, C.-N. and Hoult, D.I. (1989). *Biomedical Magnetic Resonance Technology*. New York: IOP Publishing.

de Certaines, J.D., Bovee, W.M.M.J., and Podo, F. (eds.) (1992). *Magnetic Resonance Spectroscopy in Biology and Medicine. Functional and Pathological Tissue Characterization*. Oxford: Pergamon Press.

Diehl, P., Fluck, E., Gunther, H. et al. (eds.) (1992). *In Vivo Magnetic Resonance Spectroscopy, NMR Basic Principles and Progress*, vol. 26–28. Berlin: Springer-Verlag.

Gadian, D.G. (1992). *Nuclear Magnetic Resonance and Its Applications to Living Systems.* Oxford: Oxford University Press.

Haacke, E.M., Brown, R.W., Thompson, M.R., and Venkatesan, R. (1999). *Magnetic Resonance Imaging: Physical Principles and Sequence Design.* New York: Wiley-Liss.

Mansfield, P. and Morris, P.G. (1982). *NMR Imaging in Biomedicine, Advances in Magnetic Resonance,* Suppl. 2 (ed. J.S. Waugh). New York: Academic Press.

Morris, P.G. (1986). *Nuclear Magnetic Resonance Imaging in Medicine and Biology.* Oxford: Clarendon Press.

Stark, D.D. and Bradley, W.G. Jr. (1988). *Magnetic Resonance Imaging.* St. Louis: C. V. Mosby.

Webb, A.G. (2016). *Magnetic Resonance Technology. Hardware and System Component Design.* Cambridge: Royal Society of Chemistry.

Index

In Vivo *NMR Spectroscopy: Principles and Techniques*, Third Edition. Robin A. de Graaf.
© 2019 John Wiley & Sons Ltd. Published 2019 by John Wiley & Sons Ltd.

Printed and bound by CPI Group (UK) Ltd, Croydon, CR0 4YY

20/03/2024

14473700-0001